Seventh International Conference on

AC-DC Power Transmission

28 – 30 November 2001

ORGANISERS

The Conference is being organised by the IEE Power Professional Networks with the support of the

Institute of Electrical & Electronic Engineers – Power Engineering Society

SPONSORS

VENUE

IEE, Savoy Place, London

Author Disclaimer

"While the author and the publisher believe that the information and guidance given in this work are correct, all parties must rely upon their own skill and judgement when making use of it. Neither the author nor the publisher assume any liability to anyone for any loss or damage caused by any error or omission in the work, whether such error or omission is the result of negligence or any other cause. Any and all such liability is disclaimed."

Copyright and Copying

"All Rights Reserved. No part of this publication may be produced, stored in a retrieval system, or transmitted in any form or by any means - electronic, mechanical, photocopying, recording or otherwise - without the prior written permission of the publisher."

Conditions of Acceptance of Advertisements

The Institution reserves the right to refuse to insert any advertisements (even if ordered and paid for) and/or to make alterations necessary to maintain its standards.

It is not guaranteed that any advertisement will be placed in any specified position or on any specified page unless agreement has been entered into and the agreed surcharge paid.

Every effort will be made to avoid errors but no responsibility will be accepted for any mistakes that may arise in the course of publication of any advertisement. These mistakes may include non-insertion, insertions other than those ordered and errors and omissions within the advertisement.

Notice to cancel any advertisement must be received in writing ten days prior to its next scheduled appearance, otherwise a charge will be made.

No responsibility will be accepted for repetitive errors unless the advertiser's correction has been accepted in respect of that error.

No responsibility will be accepted for loss or damage alleged to arise from errors within advertisement copy, non-appearance of an advertisement or delay in forwarding box number replies.

Advertisers are required to ensure that the content of advertisements conforms with all legislation currently in force affecting such matters. They shall further indemnify the publisher in respect of any claims, costs and expenses that may arise from anything contained within the advertisement and published on their behalf by the Institution of Electrical Engineers.

The placing of an order or contract for insertion of an advertisement in any newspaper or journal published by the Institution of Electrical Engineers whether in writing or by verbal or telephone instructions will be deemed an acceptance of each and all the above conditions.

Published by the Institution of Electrical Engineers, London ISBN 0 85296 745 4   ISSN 0537-9989

ORGANISING COMMITTEE

T Adhikari, India
Bjarne Andersen (Chairman), UK
Ralph Barone, Canada
Håkon Borgen, Norway
Sep Boshoff, South Africa
Lennart Carlsson, Sweden
Maurice Dwek, UK
Mircea Eremia, Romania
Maria Garcia, Philippines
Nick Jenkins, UK
Peter Lips, Germany
Martin Luckett, Japan
Joseph Mutale, Zambia
Peeter Muttik, Australia
Alan Rainey, UK
Colin Ray, UK
Tadashi Senda, Japan
Paul Smith, Ireland
Kent Søbrink, Denmark
Vijay Sood, Canada
Marcio Szechtman, Brazil
Alan Wood, New Zealand

Thanks are also given to Clarence Thio of Canada, who sadly died during the planning of this event.

CONTENTS

NEW TECHNOLOGY II

POSTER PAPERS

EFFECT OF DE-REGULATION ON OPERATION OF ELECTRICAL GRID AND NUCLEAR POWER GENERATING STATIONS

N K Trehan
US Nuclear regulatory Commission, USA

SYNOPSES

In the UK, the Electricity Act of 1989 was amended in 2000 to remove the distinction between private and public electric supply companies. The customers in the UK saved £ 700 million in one year under de-regulation. In the US, Congress passed the Energy Policy Act in 1992 that de-regulated the wholesale electric market and mandated electric utilities to open up the transmission system on an equal basis to all who utilize it. During the past decade, the demand for electricity in the United States has increased by 35 percent, whereas, the generation has increased only by 18 percent. With the bulk power transactions and reservations, the transmission system now operates closer to its stability limits. Voltage instability occurs when the balance between the reactive power producing sources and the VAR losses cannot be maintained. The limited construction of new transmission lines has pushed the power industry toward the development of advanced technologies such as flexible ac transmission system (FACTS), static VAR compensators (SVC), static synchronous compensators (STATCOM), unified power controllers (UPFC), and distributed superrconducting magnetic energy storage (D-SMES).

In a nuclear power plant, offsite and onsite electric power systems are required to provide sufficient capacity and capability to assure that the containment integrity be maintained during power operation or in case of a postulated accident. Analyses must verify that the electric grid remains stable in case of loss of the nuclear unit, the largest other unit on the grid or the most critical transmission line and a change it due to deregulation would affect those analyses. Nuclear Regulatory Commission (NRC) staff ensures that the licensees maintain adequate voltages to the Class 1E (safety) buses. If the operator has failed to restore adequate voltages, the Class 1E buses should be automatically separated from the offsite power system and connected to the emergency diesel generators. Most utilities use on-line contingency analysis and the control centers notify them when the offsite power system in their nuclear plant is degraded. The control centers use load conservation procedures to provide the plant a stable offsite power system. The control centers have assured the NRC staff that the nuclear power plants are their preferred customers and that they will be provided with an adequate offsite power system even under reduced grid generation availability. This paper discusses the voltage stability, power transfer limits, reactive reserves, methods of improving voltage stability (such as superconductivity), load conservation procedures and effects of deregulation on nuclear power generating stations.*

Keywords: De-regulation, voltage stability, VAR control, transfer capability, nuclear power plants.

I. INTRODUCTION

In the UK, the Electricity Act of 1989 was amended in 2000 to remove the distinction between private and public electric supply companies. In the US, Congress passed the Energy Policy Act in 1992 that de-regulated the wholesale electric market and mandated electric utilities to open up the transmission system on an equal basis to all who utilize it. Federal Energy regulatory Commission (FERC) issued Order Numbers 888 and 889 in 1996 that mandated electric utilities to open up the transmission system. Under deregulation, the electric power system is divided into separate domains namely generation, transmission, and distribution. The independent power producers can sell electric power directly to the customers using electric utility s transmission and distribution lines. Voltage instability may occur when a system is so inadequate in VARs that a voltage decay over time is unavoidable. Use of extensive non-linear electronic power loads is one of the major causes of voltage instability. The recent example is the August 10, 1996 blackout in the western North American grid where faults at the 500 kV and 230 kV lines resulted in the separation of the North-South Pacific inter-tie and caused blackouts in 11 U.S. states and 2 Canadian provinces. During this event, nuclear power plants did not lose preferred offsite power completely, however, the following plants tripped due to the grid perturbations during the event:

---

* DISCLAIMER -- This paper represents the views of the author and does not represent a Nuclear Regulatory Commission (NRC) position on the subject covered in the paper. This paper is not a substitute for regulations, and compliance with it is not required. Methods and solutions different from that set out in the paper could be applied.

* Diablo Canyon 1 and 2 (1075 MW each), and Palo Verde 1 and 3 (1270 MW each) tripped
* San Onofre 2 and 3 (1100 MW each), Palo Verde 2 (1270 MW), and WNP-2 (1100 MW) did not trip.

Diablo Canyon 1 tripped as a result of bus undervoltage while Diablo Canyon 2 tripped on the trip of reactor coolant pump breaker. Palo Verde 1 and 3 tripped on a high Linear Power Distributed (KW/Ft) signal. San Onofre 1, 2, and 3 registered frequency oscillations from 61 Hz to 58.5 Hz but did not trip. WNP-2 experienced frequency disturbances but remained operating.

II. OPERATION OF GRID AND SYNCHRONOUS GENERATORS

It is essential to operate the electric power system within its design limits if the grid security (stability) is to be maintained. To achieve it, the system must be adequate and secure. The system is adequate if it can supply the consumers load requirements taking into account the scheduled and unscheduled outages of the generators and the transmission lines. The system is secure if the electric power system can withstand sudden disturbances including a single worst contingency.

AC synchronous generators are the primary sources of voltage and frequency controls in the bulk power transmission system. Synchronous generator capability (supplied by the manufacturer) is derived from the generator real and reactive capability curves. The generators should be operated with their excitation system in automatic control voltage mode, governor control with appropriate speed/load characteristics to regulate frequency and maintaining VARs output within their reactive capability. Adjustments to real power (Watts) are made by throttle adjustments (governor settings) and adjustments to reactive power (reactive volt-ampere or VAR) are made by the generator field, voltage regulators and transformer taps. For the speed/load characteristics, the governor "droop" is generally set at 5%. The "droop" governor normally maintains a full load speed that is 5% less than no-load speed. If all the synchronous generators are set in droop-droop operation, the generator prime-mover units will share the load proportionally regardless of their sizes. The "isochronous" governor on the other hand, has zero droop because it maintains a constant speed throughout the load range of equipment. The excitation system that controls the reactive power flow by influencing the flux level, is designed to give close voltage regulation under transient conditions; the voltage regulators are made quick acting with a small time constant exciter. Modern voltage regulators can bring the generator back within its capabilities when the generator is overexcited (volts per hertz or underexcited. If the generator is overexcited, the stator may be damaged due to the increase in core losses. The overexcitation can also be caused be a blown potential transformer fuse, or the malfunction of automatic voltage regulator. Volts per hertz protection is provided to prevent generator damage from overexcitation. Underexcitation is caused by malfunction of the excitation system, or problems with the field winding. The transient stability analysis is generally performed under the severest fault condition. The results should show that the system stability is maintained, and the power angle swing damps immediately. The transient stability analysis should not include the reactive power compensation, though reactive power compensation is available to maintain the transient stability.

III. VOLTAGE STABILITY

The study of voltage stability (longer term stability) is done as a steady state problem using conventional power flow diagrams. Electric power systems that experience heavy loadings on the grid with limited VARs are vulnerable to voltage instability. There is a delicate balance between large reactive power losses and reactive power producing sources. A trip of a generator or transmission line can easily upset this balance. During a system disturbance, the system voltage goes down due to VAR deficiency and the transformer taps mounted on the primary side of the transformer are raised to provide the VAR deficiency. As a result, the voltage on the low voltage side of the transformer is depressed which depresses the stator voltage. The stator voltage is restored by the automatic voltage regulator by increasing the generator excitation. The load tap changers on the secondary side of the distribution station transformers improve the voltage on the low voltage side but in doing so, depress the voltage on the high voltage side of the distribution station transformer that aggravates the voltage instability situation. Therefore, the load tap changers on the distribution transformers should be blocked during declining system voltages as they may add to the instability if the grid is already experiencing instability.

During the past decade, the demand for electricity in the United States has increased by 35 percent, whereas, the generation has increased only by 18 percent. With the bulk power transactions and reservations, the transmission system now operates closer to its stability limits. Voltage instability occurs when the balance between the reactive power producing sources and the losses cannot be maintained. Moreover, the system should be able to withstand the worst credible contingency (loss of the most critical transmission line or the loss of the largest unit) while maintaining the security of the system. The spinning reactive reserve activated during a contingency should be replenished to maintain the minimum reserves so that the system is always ready for the next contingency. In other words, the system should always be one contingency ahead. Most of the utilities are using diagnostic and self-check features. Every 2 seconds, dispatch signals are calculated and sent to control centers and every 2 minutes, the thermal and voltage contingencies are modeled.

IV. POWER TRANSFER

Beside the question of voltage stability, there is a limit to the power that can be transferred across a transmission line. With deregulation, open transmission access has resulted in increased electric power transfers. For the highest power transfer, each additional MW transferred (transfer limits are represented in MW) results in much higher VAR losses.

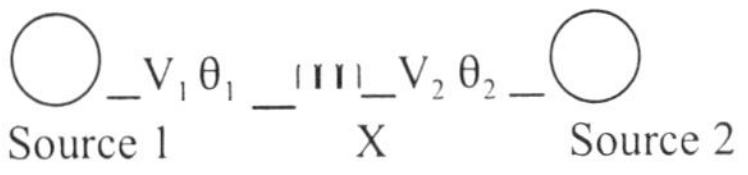

$\bigcirc$ — $V_1\,\theta_1$ — | | | | — $V_2\,\theta_2$ — $\bigcirc$

Source 1   X   Source 2

Figure 1. Power Transfer Equation

Power Transferred $= V_1\,V_2 \times 1/X \times \sin(\theta_1 - \theta_2)$

There are three basic parts in the power transfer equation namely: transmission voltages V1 and V2 at both ends; impedance Z; and phase angle θ between them (Figure 1). The shunt capacitors are used for voltage control; series compensation is used for impedance control; and the phase shifting transformers are used for phase angle control. Voltage transfer limits are pre-defined and are set for the most severe single contingency by off-line studies and simulations. Thermal transfer limits are calculated every day, using transmission analysis for the most severe single contingency. The system can be operated in the thermal limit range (long term or short term) only for a limited time and should be brought back to normal. The capacity in long-term mode can be increased above 15% of normal, but it has to be brought back to normal in 8 hours. The capacity in short-term
mode can be increased above 25% of normal, but it has to be brought back to long-term mode in 15 minutes. The rating of a transmission line can be increased by operating the line at higher temperatures. For conductors of the

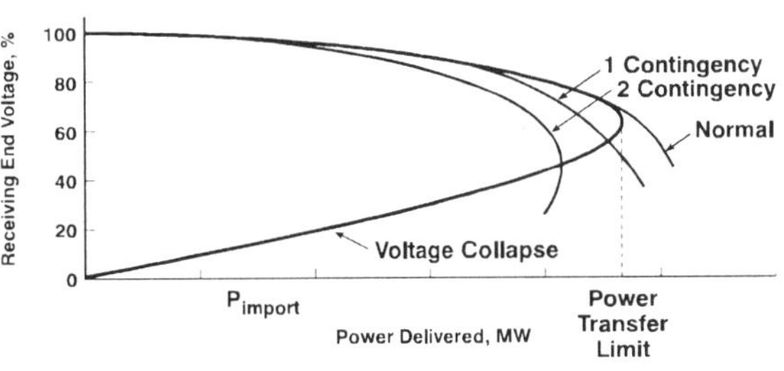

Transmission line capability

Figure 2.

size of 795 kcmil and larger, the load capability increases by about 20% if the conductor's temperature of 75^0 C is raised to the 100^0 C level. Though, the higher temperatures result in greater sag in the transmission lines which can touch the trees if enough clearance is not provided. Two major grid disturbances in the western United States in July and August 1996 are attributed to the transmission lines touching the trees that led to a voltage collapse and major blackouts. The lines can be upgraded by reconductoring and increasing the voltage or raising the transmission line conductors so that they can operate in the thermal range and still provide enough clearance from the ground. The system should not be scheduled to operate in the thermal range as it lessens the life of the conductor due to a loss in its tensile strength.

Complex systems with a large number of voltage sources and loads behave on the P-V characteristic (Figure 2). If the demand is increased at the receiving end when operating at the knee point of the transmission line capability curve, the additional voltage drop in the line reduces the voltage at the receiving end and, thereby, reduces the power delivered. A more complex system analysis utilizes V-Q characteristics (nose curves) for VAR compensation and gives better results because reactive power is related to voltage stability. The V-Q curves are widely used by Western System Coordinating Council utilities in the U.S. A set of V-Q curves should be developed to illustrate cases for the loss of a most critical transmission line or the largest unit (single contingency). Dynamic voltage stability study should also include tap changing transformers, automatic governor control, generator maximum excitation limiters.

V. REACTIVE RESERVES

Operating Reserves (5-7%) consist of Spinning Reserve Service and Supplemental Reserve Service. A Spinning Reserve Service consists of generation synchronized to the power system and responsive to Automatic Generation Control. A Supplemental Reserve Service consists of generation synchronized or capable of being synchronized to the power system that is fully available within ten minutes of the first contingency. If all on-line units can not contribute the required reserve, then the off-line units are brought on-line in 10-minutes to provide additional power.

Due to high inductive reactance of the transmission lines, it is very difficult to push (transport) VAR along the transmission lines over long distances and, therefore, VAR producing devices should be distributed throughout the electric grid at regular intervals with a balance between static and dynamic VAR support. Static VAR support is provided by line charging, series and switched shunt capacitors. Dynamic VAR support is provided during transients by synchronous condensers, synchronous generators, SVCs, and STATCOM. Due to the limited construction of new transmission lines, advanced technologies are being introduced for reactive and voltage control of the electrical grid. Flexible AC Transmission System (FACTS) was introduced to make sure that there is flexibility in energy system so that it can adjust to swings in supply and demand without getting instability. FACTS provides transient voltage support and system regulation to damp system oscillations, enhance system security, and allow increased power transfer capability. FACTS technology uses high speed electronic controllers, advanced control technology, fibre optics, and microcomputers to provide the correct transmission voltage, line impedance and the phase angle between them. Unified power flow controllers can provide active real power as well as reactive power compensation. The Electric Power Research Institute and other vendors are in the process of developing methods to assist in on-line dynamic security assessment tools that could help in assessing the power system in a dynamic situation. Wide Area Monitoring Systems (WAMS) using globe positioning satellites (GPS) transmit signals of disturbances to the control centers throughout the interconnected system.

VI. OTHER METHODS OF IMPROVING VOLTAGE STABILITY

VI a. Superconductivity

Another approach to provide better voltage control is to upgrade the conventional cables with the high temperature superconducting (HTS) cables. Superconductivity refers to the conduction of electricity with almost zero resistance. In 2001, Detroit Edison Co. U.S. installed HTS cables at Frisbie station that will increase the power carrying capacity of cables by a factor of three. The superconductors will be cooled by liquid nitrogen and made of Bismuth, Strontium, Calcium, Copper, and Oxygen (BSSCO). The designs of either warm dielectric or cold dielectric are being developed. In the former design, only the main phase conductors are operated at low temperatures and in the latter case, the entire cable is operated at low temperatures. The latter type can carry about five times the current of conventional cables with two-thirds of the losses.

Superconducting Magnetic Energy storage (SMES) allows the storage of large amounts of electricity in a superconducting electromagnetic coil until required. Complemented with inverters, Distributed SMES (D-SMES) can inject both real and reactive power quickly into the system during a contingency. D-SMES introduced in 1998 consisted of shunt connected, four quadrant inverter with a nominal capacity of 2.8 MVA and a 1-second overload rating of over 5 MVA. A group of these devices can protect a transmission line against a voltage collapse. The world s first commercial D-SMES was installed at substations in Wisconsin in the U.S. The Texas grid is installing four (transportable) D-SMES units (two in 2001 and two in 2002). Superconductivity applied in the manufacturing of superconducting generators, and superconducting motors will increase their ratings.

VI b. Power Uprates

The power uprates (ranging from 2.4% to 20%) in nuclear power generating stations is a cost effective method to increase their MW ratings. The U.S. electric supply can be increased by about 12,000 MW by power uprates. The nuclear steam supply system is generally capable of supporting an increase in the thermal power (MWt) or MWt can be increased by changing the steam generators in the pressurized water reactors. Turbine generator is generally the limiting component. Under power uprate, with the increase in MWe, there is a decrease in the MVAR because the generator MVA is fixed. Shunt capacitors can compensate for the decrease in MVAR. The MWe output of the generator can also be increased by raising the hydrogen pressure (from 60 to 75 psig).

VI c. License Renewal

Of the existing 103 nuclear power plants in the U.S., five units have received approval from the Nuclear Regulatory Commission (NRC) for another 20 years of operation increasing their life span from 40 to 60 years. Three units have formal application under review and 26 other units intend to renew their licenses to 60 years. This will help in improving the grid security.

VI d. Distributed Power

Distributed power (DP) such as microturbines, solar cells, wind turbines, fuel cells, diesel generators augment the power supply system. DP ranges from less than a kilowatt to tens of MW in size. Nuclear Power Plants being cheaper to run are generally base loaded. Gas turbine plants have higher running costs, but are used as peaking units with a fast start capability. Fossil power plants need a minimum of 1 hour to stabilize expansion in the boiler and turbine generator. Synergy has gas-fired peaking units on its nuclear plant site. During a contingency, gas turbine units are automatically started to produce reactive power and pumped storage plants, if operating in the pumping mode are tripped and put in generating mode to produce reactive power. In the pumping mode, pumped storage plants are used to provide an additional system load and in the generating mode, they supply reactive power during peak load demands.

VII. LOAD CONSERVATION PROCEDURES

During a contingency, the operators may not have sufficient time to stabilize the electric system, automatic undervoltage load shedding can be used to provide reactive power and under-frequency load shedding is applied to deal with real power deficiency. During severe generation shortage situations, the electric utilities curtail interruptible loads to the extent required to maintain the 10-minute reserve. This is followed by implementing a voltage reduction of 5% of normal operating voltage. An automatic under-frequency islanding scheme may also remove the burden on the transmission lines. If the individual system was importing power before the separation, the generation would be insufficient to match the demand and as a result, the frequency within that system would drop and under-frequency load shedding can restore the frequency to most of the customers. On the other hand, if the individual system was exporting power before the separation, the frequency within the system would rise and the frequency would be restored by tripping some generating units. If these actions are not enough, they implement brownouts and rotating blackouts. A five percent voltage reduction does not affect electrical equipment as they are manufactured with +/- 10% change in voltage. Manual load shedding should be exercised in extreme cases.

VIII. EFFECT OF DEREGULATION ON NUCLEAR POWER GENERATING STATIONS

In a nuclear power generating station, two offsite power circuits with adequate capacity and capability are required in accordance with General Design Criteria 17 of Appendix A to 10 CFR 50. As a result of de-regulation, the voltages on the offsite power sources may be degraded and may cause degraded voltages at the safety buses. In case of accident, motor-operated valves (MOVs) must quickly operate to isolate the reactor and open the flow path into the reactor core by the safety injection pumps. The electrical contactors may not have enough power (80% or 85% Voltage is required) under degraded voltage conditions to close the contacts, or the contacts could rapidly make and break (i.e., chatter). In either case, the motors would fail to start and the sustained inrush current could cause blowing of the control circuit fuses. Also, the torque developed by a motor at any speed is proportional to the square of the voltage applied to the terminals. Too low a voltage during starting may prevent a motor from reaching its rated speed. Moreover, reduced voltage causes longer starting time for the induction motors. Typical accelerating time for a 850-HP motor at 4.16 kV bus is 3 seconds at 80% of rated voltage vs. 0.5 second at 100% voltage. For larger motors, the starting time could be as high as 3.5 to 4 seconds at 75% of rated voltage. Voltage degradation during pump starting is a safety concern if the condition delays the flow beyond the times used in the accident analysis.

SVC can be used on the secondary (4.16 kV) side of the auxiliary and start-up transformers to ensure sufficient voltages at Class 1E loads and 120 Vac contactors. At one such plant, each SVC was rated at +28/-14 MVAR and included a Thyristor Controlled Reactor bank rated at 21 MVAR, a Thyristor Switched Capacitor bank (TSCB) rated at 21 MVAR, and a Harmonic Filter Capacitor (HFC) bank rated at 7 MVAR. The HFC is always connected to the 4.16 kV bus provides a capacitive 7 MVAR supply boosting the 4.16 kV bus voltage. If the boost raised the 4.16 kV voltage beyond its setpoint, the voltage is lowered by inserting the reactance from the reactor bank. If the voltage dips below the capability of the HFC, the TSCB is inserted to maintain the voltage.

IX. CONCLUSION

In the UK, the Electricity Act of 1989 was amended in 2000 to remove the distinction between private and public electric supply companies. The customers in the UK saved £ 700 million in one year under de-regulation. In the US, Congress passed the Energy Policy Act in 1992 that de-regulated the wholesale electric market and mandated electric utilities to open up the transmission system on an equal basis to all who utilize it. With the bulk power transactions and reservations, the transmission system now operates closer to its stability limits. Voltage instability occurs when the balance between the reactive power producing sources and the VAR losses cannot be maintained. The limited construction of new transmission lines has pushed the power industry toward the development of advanced technologies such as FACTS, SVCs, STATCOM, UPFC, and D-SMES.

In a nuclear power plant, offsite and onsite electric power systems are required to provide sufficient capacity and capability to assure that the containment integrity be maintained during power operation or in case of a postulated accident. Analyses must verify that the electric grid remains stable in case of loss of the nuclear unit, the largest other unit on the grid or the most critical transmission line and a change it due to deregulation would affect those analyses. The NRC staff ensures that the licensees maintain adequate voltages to the Class 1E buses. If the operator has failed to restore adequate voltages, the Class 1E buses should be automatically separated from the offsite power system and connected to the emergency diesel generators. Most utilities use on-line contingency analysis in their control centers that notify them when the offsite power system in their nuclear plant is inadequate. The control centers use load conservation procedures to provide the plant a stable offsite power system. The control centers have assured the NRC staff that the nuclear power plants are their preferred customers and that they will be provided with an adequate offsite power system even under reduced grid generation availability.

X. REFERENCES

1. Narinder K. Trehan, 1997, "Electric Grid Stability and its Impact on Nuclear Power Generating Stations, " <u>Proceedings of the IEEE Nuclear Science Symposium and Medical Imaging Conference</u>

2. Narinder K. Trehan, 2000, "Ancillary Services - Reactive and Voltage Control, " <u>Proceedings of the IEEE Power Engineering Society, Winter Meeting</u>

3. M.A. Lamoureux, 2001, "Evolution of Electric Utility Restructuring in the UK," <u>IEEE Power engg. Review</u>

4. N. Trehan, J.D. Kueck, et al, 1998, " A discussion of Degraded Voltage relaying for Nuclear Generating Stations," IEEE Transactions on Nuclear Science, vol 45

5. P. Kundur, 1994, "Power System Stability and Control," <u>McGraw-Hill</u>

6. R.T. Byerly and E.W. Kimbark, 1999, " Stability of Large Electric Power System," <u>IEEE Press</u>

7. Elizabeth A. Bretz, 2000, "PJM Interconnection: Model of a Smooth Operator," <u>IEEE Spectrum</u>

8. William V. Hassenzahl, 2000, "Applicat. of Superconductivity to Electric Power Systems," <u>IEEE Pwr.E.Review</u>

9. M. Shan Griffith, 1978, "IEEE Transactions on Industry Applications," <u>Vol 1A-14, No.6</u>

10. H. K. Clark, 1990, " New Challenge: Voltage Stability," <u>IEEE Power Engineering Review</u>

11. <u>Conference</u>"NERC Planning Standards," 1997, <u>North American Electric Reliability Council</u>

12. A Position Paper of the Electric-System Reliability Task Force -Secretary of Energy Advisory Board, 1998 "Ancillary Services and Bulk- Power Reliability"

13. A. Edris, 2000, "FACTS Technology Development: An Update," <u>IEEE Power Engineering Review</u>

14. R. D. Rosevear, 2000,"Power Cables in 21st Century Energy Development," <u>IEEE Pwr Engg. Review</u> 15. Mario Rabinowitz, 2000, "Power Systems of the Future," <u>IEEE Power Engineering Review</u>

16. W. Buckles and W.V. Hasenzahl, 2000, "Superconducting Magnetic Energy Storage," <u>IEEE Power Engineering Review</u>

17. Gabriel Ejebe, 1998, "Two New Functions for Assessing Security in Power Systems," <u>Powerlines</u>

18. JW. Pope, 1997, "Reliability in Transition," Southern Services Inc., <u>Transmission & Distribution world</u>

19. Earl Hazan, 1997, "A Word of Caution," <u>Transmission & Distribution World</u>

20. North American Electric Reliability Council, 2001, "2001 Summer Assessment"

21. Department of Energy, 1996, "The Electric Power Outages in the Western United States, July 2-3, 1996," <u>Report to the President</u>

22. "Disturbance Report for the Power Outage that Occurred on the Western Interconnection August 10, 1996, " <u>Western Systems Coordinating Council</u>

23. " Modern Power Station Practice," <u>British Electricity International, 1990, Volume L System Operation</u>

24. Task Force, 91 "Survey of the Voltage Collapse Phenomenon," <u>North American Electric Reliab. Council</u>

25. Criterion 17, "Electric Power Systems," of <u>10CFR50</u>.

26. Starr,"Gen, Trans&Utiliz. of Elect Pwr," <u>Pitman, UK.</u>

INVESTMENT DECISIONS FOR BULK POWER TRANSMISSION SCHEMES IN LIBERALISED MARKETS

W. Fischer, W. Braun

VDE ETG, Germany

Project financing based on private initiative becomes important even in the regulated electricity sector of many countries. Whereas in former times the laws were directed to safe, reliable and macro economic power supply, the aims under deregulated market conditions are micro-economic and competitive.

This report deals with the micro economics of an investment focusing on the three outputs

payback period and payoff time, return on equity before tax and transmission costs.

It shall allow a quick scan of the financial feasibility of a project. A simplified model is taken to prove whether the investment is attractive to investors and lenders, how long it takes to pay the loans back and which minimum transfer charge must be taken from the market in order to run the operator company successfully.

The model is demonstrated by taking a power transmission scheme in HVDC technology. Figures taken from the practise will allow an insight in the "secrets of profitability of a transmission scheme". Varying the model will show when and if it is justified by economic reasons to invest in series capacitors in order to increase the transmission capacity of the AC overhead line.

PROJECT FINANCING BASED ON PRIVATE INITIATIVE

Since the EU-Power guideline came into force, privatisation and liberalisation are guaranteed by law in a deregulated market. The monopoly regarding power generation, transmission, distribution and sales was abolished and the German Ministries of Commerce and trade of the federal states do no longer act as a regulation and control authority, also valuing the economic benefit and approving power tariffs for the private user. Since then, profit orientated investment decisions are to the fore. Consequently, an initiator / developer thinks and acts as follows: After analysing his market and estimating the financial risk he has to choose the adequate type of financing for his project. Among others he can decide for "project financing based on private initiative". This kind of private financing is most convenient for "really deregulated markets" and countries with free market economy, taking into

consideration the respective investment laws as well as the convertibility of the national currency and liberality of monetary transfer (in the case of investment in foreign countries). If these general conditions are guaranteed, a "special purpose company" for one single project can be founded. Such companies can work as private financed Utilities and/or lessors during the construction and operation period. Depending on the individual circumstances, they can be transferred with their capital assets after the payback period.

In any case, it is necessary to take into consideration the following three financial parameters, in order to ensure that the investment will be attractive to investors and lenders and that the "special purpose company" can be run successfully:

- *Payback Period and Payoff Time*, which means in the case of the above described type of financing: The time the new company will have to pay off the debts.

- *Return on Equity:* What will be the rate of return of the investment?

- *Compensation for use:* Minimum rate the market has to turn out in order to run the operator company successfully.

THE HVDC TRANSMISSION SYSTEM MODEL

Model Assumptions

The present model to be taken as an example is based on the following assumptions:

Technical Assumptions:

It is designed for a power of 2000 MW, consists of two Converter Stations and a +/- 500 kV Overhead Transmission Line with 1000 km rights of way. The connected three-phase systems are adapted.

Financial Assumptions:

In order to realise the planned transmission line there will be a capital demand (capital costs C_c) of

AC-DC Power Transmission, 28-30 November 2001
Conference Publication No. 485 © IEE 2001

720 million €, composed of Hardware Costs and "Soft Costs" as follows:

Hardware Costs: 600 Mio. €

Soft Costs (20 % of Hardware Costs):   120 Mio. €

Total: 720 Mio. €

The "Soft Costs" include expenses for project development, bank charges, interest rates during construction, debt reserve as well as consultation for technology, contracts and financing.

Basic Financial Parameters

Based on the model assumptions the investor will realise his project always visualising the three above mentioned financial parameters:

Payback Period and Payoff Time. In order to avoid that the "special purpose company" will go bankrupt a constant amount, called annuity A_{ZR}, has to be earned per annum which is composed of interest rate z and redemption t (in %). The Payback Period n which is independent of the capital demand can be calculated according to the following mathematical equation:

$$n = \frac{\ln (1 + \text{interest rate } z \,/\, \text{redemtion } t)}{\ln (1 + \text{interest rate } z)}$$

ln stands for natural logarithm.

Dependent on the risk estimation of the lenders, interest rates of up to 15 % are charged. For low risk projects you can assume for example a 7 % interest rate as well as a redemption of 4 %. Consequently, the payoff time amounts to 15 years.

Return on Equity. As the expenses for interests and the rate of return are pre-selected in this simplified model the return on equity as a planning figure for investment decisions is given, provided that the transmission system is used as planned.

Consequently, the return on the invested equity capital before tax can be derived from the ratio equity capital (EC) to outside capital (OC) if the interest rate is known:

$$\text{Return on equity in } \% = A_{ZR} + (A_{ZR} - z) * OC / EC$$

Under safe business circumstances at a minimum risk, which means long term delivery and/or pay or take contracts, a capital ratio of 25 % / 75 % (EC / OC) can be assumed. Otherwise, a higher equity capital proportion of up to 50 %, as for example in the case of "pure merchant plants" is more realistic.

As you can see from the above equation a low equity capital proportion might lead to a higher rate of return.

Transmission Costs. The question of interest for every investor is: Will there be a market for my services? In the case of current transmission projects this means: How much will be the transmission costs?

The transmission costs U can be derived from the ratio of annual expenses for interests and redemption A_{ZR} and the outside capital weighted by the redemption t to the amount of energy E sold, multiplied by the Capital Costs C_c.

$$U = \frac{A_{ZR} + t*OC}{E} * C_C \text{ in } € / MWh$$

The amount of energy supplied E itself depends on the load profile, that means on the maximum use of the transmission line as well as on the workload factor and corresponds to the integral of power over time.

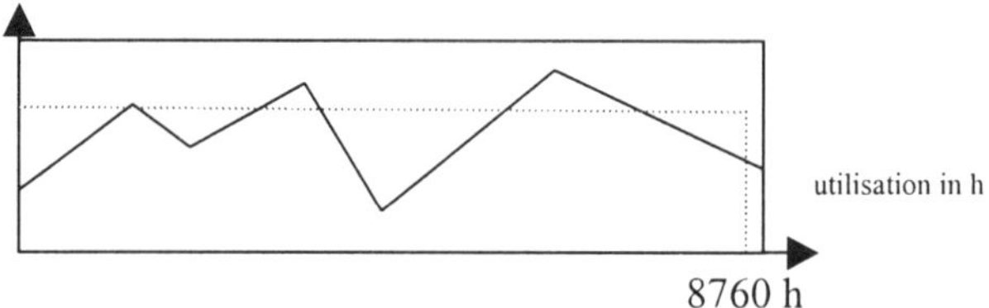

Figure 1: Delivered Energy E = Maximum Use x Workload x Power

The example is based on a constant power of 1333 MW which corresponds to a maximum use of 66 % at 2000 MW. Typical of an HVDC system is the high workload factor of for example 97 %, that means 8500 hours per annum.

Annual expenses A_{ZR} and the outside capital weighted by the redemption t in % therefore amount to 13 % of the original capital demand. With these figures, taken from the practice, transmission costs then amount to approximately

8,3 € / MWh or 0,83 ct / KWh. This would be the minimum amount the running company would have to charge as compensation for transfer.

After the payback period the transmission rate can either be reduced, if the return on equity shall remain constant, or a higher profit than before can be expected at the same transmission rate.

There are other parameters such as costs for maintenance, rights of way, wayleave rights, current losses as well as depreciation, tax laws, mixed financing or inflation which also have an impact on the payoff time, the rate of return and the transmission rate. They are subject to further observation.

MORE POWER THROUGH ADDITIONAL SERIES CAPACITOR

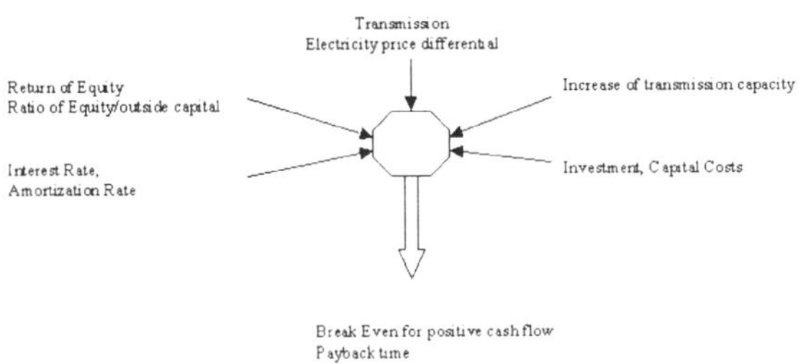

Figure 2: Chapter Power transmission increased by Series Capacitors

The basic considerations of investments in deregulated markets can apply to other investments as well which shall be clarified by taking a series capacitor as an example. The total hardware costs for the series capacitor (325 MVAr) amount to 6 million US $. This corresponds to a specific price of 18,5 $ per KVAr. The "soft costs" are calculated at another 12 % leading to a total capital demand of 6'72' US $. The series capacitor partly compensates the inductivity of the long line which makes it possible to transmit more power .

The line has a length of 200 miles equal to 320 km from Shiprock, New Mexico to Glen Canyon, Arizona. The incremental in power is 100 MW (from 300 MW to 400 MW).

The utility had the option to build a new line requesting investment capital of 70 ' US $ and a time consuming approval procedure for rights of way and environmental study.

The very economic and time saving alternative was to install the series capacitor.

How long did it take to pay the investment back ? Applying the same methodology as shown above and assuming real commercial parameters such as a 7 % interest rate and a price differential of 2 US $ / MWh between the markets at each end of the line you will find the new installation being free of debt after only 6 years.

In the case of pre-selected expenses A_{ZR} (annuity) of 9 % a yearly return on equity of 20 % can be expected, satisfying at the same time the needs of the private investor. After the loan will have been paid off, the owner will have the option of higher earnings or lowering the transmission rates.

Figure 3: Series Capacitor Installation at Kayenta, Az, USA

Literature

1. Schwanfelder W., 1987, "Exportfinanzierung für Großprojekte", Gabler, Wiesbaden

2. Hartmann T., Glausinger W., 1993, "Primärenergietransport versus Stromübertragung über größere Entfernungen", EW, 14

3. RAO EES of Russia; Ministry of Belarus; PPGC of Poland; VEAG, Preussen Electra, 1994, "Feasibility Study on East-West High Power Transmission System"

4. Schmidt Dr., Capellin R., Mosters M., Fischer W., 1994, "HVDC Essentials for Economic Bulk Power Transmission", 2nd Afro Asian Conference

5. Ernst R., 1995, "New Approaches to International Infrastructure and Plant Engineering Business"

6. Fischer W., 1998, "Power Bridge into the New Millennium"

7. Renz K., 1993, "Power Transmission: Leading the Way", Siemens EV Report, 4

8. Prof. Spar R., 1999, "Capital Investment Analysis for Profit and Nonprofit Organisations"

THE APPLICATION OF FACTS CONTROLLERS TO A DEREGULATED SYSTEM

J.W.M.Cheng, F.D.Galiana, D.McGillis

McGill University, Canada

Abstract

This paper is in two parts. The first part deals with the representation of random bilateral transactions in the form of matrices which are, first of all, feasible (Kirchhoff's Law) and of which some are secure (Ohm's Law).The ratio of secure to feasible is the index POST and is a measure of the transmission system adequacy. It is interesting to note that as the adequacy of the generation system increases, the adequacy of the transmission system decreases simply because there is greater scope for these bilateral transactions to be entered into.

The second part of this paper deals with the application of FACTS controllers to improve the security of one or many transactions. It is generally acknowledged that a sufficient number of these controllers can achieve the objective of 100 percent adequacy or close to it. The question is at what cost. It may well turn out that an adequacy of 90 percent is all that can be justified. In any case, it is now evident that the transmission system has not been designed to accommodate unrestricted bilateral trading and that some form of external intervention is required. This paper proposes security indices that could support and justify such intervention.

Introduction

Deregulation of the electricity supply industry involves the isolation of its major components and their acquisition by business entities for the stated purpose of creating a competitive environment. In this pursuit of profit, the generation facilities have been acquired by the major players who have smothered the smaller ones in the process. Likewise, the distribution systems have been privatized so that the final consumer can theoretically choose the supplier with the most attractive rates. Indeed, it can happen that the generation and distribution supplier are one and the same entity.

With respect to transmission, it should be noted that this system is a complicated network and generally requires some form of external intervention, such as the National Grid Company, to make it work. With transmission facilities made available to any proposed supplier, namely open access, there results an obligation on the part of the transmission supplier to guarantee delivery usually in the form of a power contract between producer and consumer.

The emergence of such power transactions suggests a new approach to the understanding of transmission system reliability.

The objective of this paper is to present a methodology based on the concept of feasible and secure transactions which estimates the adequacy of a deregulated network. From this estimate, the remedial measures required to improve this adequacy, such as FACTS controllers, are examined. The idea of adequacy to describe the reliability of a transmission system arises from the nature of the deregulated environment where specific contracts between producer and consumer come face-to-face with one would normally refer to as the Laws of Electricity.

The Concept of the Transaction Matrix

In a given power system one can assume a number of transactions taking place, limited only by the size of the network, where each transaction has an equal probability of occurrence. One can then construct a matrix of such transactions wherein the rows represent the generation outputs and the columns represent the bus loads. Assuming a lossless system for the sake of simplicity, the sum of the generation and loads will be equal but there must be some reserve in the generation capacity otherwise the transaction matrix will be static.

An example of a transaction matrix, representing all possible bilateral transactions, is shown in (1). It is evident that such a transaction matrix is an abstraction in that it is quite independent of the power system topology and its inherent constraints.

$$T \quad = \quad \begin{bmatrix} t_{1,1} & \cdots & t_{1,j} \\ \vdots & t_{i,j} & \vdots \\ t_{n,1} & \cdots & t_{n,n} \end{bmatrix} \qquad (1)$$

where:

$t_{i,j}$=bilateral transaction from generator i to load j

n = number of nodes

AC-DC Power Transmission, 28-30 November 2001
Conference Publication No. 485 © IEE 2001

The Development of the Transaction Matrix

The relation between the power system and the transaction matrix is established by two rules governing its development, namely the Row and Column Rules. The rules are intended to relate the elements of the transaction matrix to the generation and load characteristics of the power system.

The Row Rule states that the sum of the elements in any row must be equal to or less than the generation available at that bus. The Column Rule states that the sum of the elements in any column must equal the load at that bus. This requirement can be achieved by deterministically fixing the last entry or by simply normalizing the whole column.

Because of the dimensions of the problem, a Monte-Carlo approach has been adopted to generate each element of the transaction matrix [1]. Once a transaction matrix has been developed that satisfies the Row and Column Rules, it is declared feasible and is retained for verification. Those matrices which do not satisfy the rules are rejected. In this way, a number of feasible transaction matrices are developed depending on the size of the problem.

The Verification of the Transaction Matrix

There is always the possibility that a feasible transaction matrix may gave rise to a congestion in the network. To verify if such a condition exists, a DC load flow is carried out on every feasible transaction matrix. Those that exhibit no congestion are declared secure. The line limits that characterize each element of the network could be the thermal limit or the stability limit or simply an equipment rating. The result of this exercise is to determine the ratio of the number of secure   transaction matrices to the number of feasible transaction matrices. This ratio then becomes the basis of the probabilistic approach to the assessment of transmission system adequacy.

In a graphical sense, this ratio of secure to feasible transaction matrices can be understood as the ratio of the secure to feasible operating areas of the system from the point of view of power transactions. In all of this work, the security of the transmission system itself is a given, based on its prescribed design criteria

The Analysis of the Transaction Matrix

In a mathematical sense, the ratio of the number of secure to feasible transaction matrices is a probabilistic index whose validity depends on the number of feasible matrices generated. This index is termed POST or the Probability of Secure Transactions [2]. There are two indices of particular interest which have been identified, namely the BIPOST and the SYPOST [3].

BIPOST (Bilateral POST) is the probability of a feasible transaction matrix being secure, given that one individual transaction is fixed. The index BIPOST is, in fact, a measure of the probability of success of a proposed transaction of a given value. The index thus indicates the viability of such a transaction which can be converted into financial terms. This process of determining BIPOST offers the possibility of adjusting the amount of a contract to improve its acceptability.

SYPOST (System POST) is the probability that all bilateral transactions having a value within their feasible range are secure. The index POST is, in fact, a measure of the adequacy of the transmission system in the face of random bilateral transactions. The concept of system adequacy has usually been applied to generation planning so that its application to transmission planning presents a new approach to the understanding of power system performance in the deregulated environment.

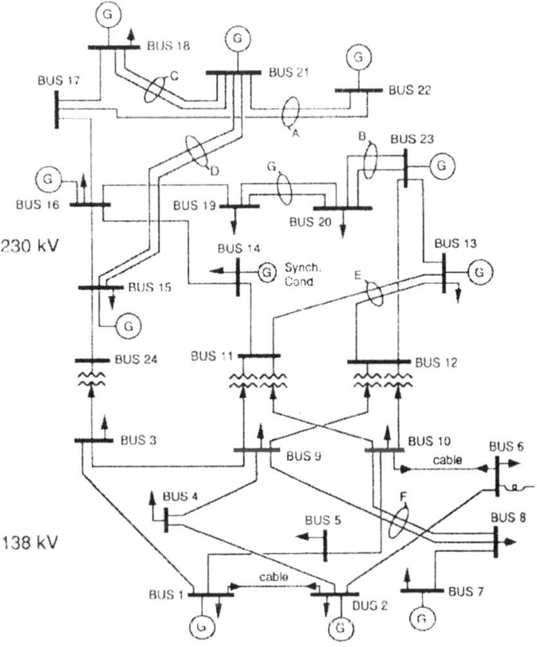

Figure 1: IEEE 24-BUS SYSTEM

Transmission System Adequacy

In the vocabulary of reliability, adequacy is defined as the ability to supply the aggregate load with a prescribed risk of failure. As such, adequacy is a probabilistic concept. Generally, the transmission system is more than adequate in this respect based on its design criteria. It is only when specific loads must be supplied by specific generators that the transmission system may not be capable of accommodating these bilateral transactions. This problem of transmission system adequacy in a deregulated environment has been studied extensively [4,5]. Two of such studies are described here to illustrate how SYPOST varies with system operating conditions.

In the IEEE 24 bus system shown in Figure 1, the

generation reserve is increased by 50 percent while the load remains constant. It can be seen from Table 1 that the value of SYPOST decreases from 100 percent to 43 percent. In graphical terms, it can be explained that the feasible region increases while the secure region does not or at least not at the same rate. Thus the ratio of secure to feasible decreases. With more possible bilateral contracts to accommodate, the transmission system finds it difficult to cope. Some would call this anomaly the irony of deregulation since it is widely held that the greater the generation reserve, the greater the opportunities for bilateral transactions to be entered into.

Table 1:DIFFERENT SYPOST VALUES UNDER DIFFERENT NETWORK CONDITIONS

CASE	GENERATION PROFILE	RESERVE	SYPOST	VIOLATED LINES
1	No change	16.1%	100%	Nil
2	Pg*1.3	35.6%	43%	3
3	Pg*1.5	44.2%	43%	3

In the IEEE 118 bus system, the value of SYPOST was calculated with all feasible bilateral contracts allowed to take place. This is shown in Table 2 along with the values of SYPOST when the number of fixed transactions is increased from zero to 50 percent and 90 percent. At 90 percent pool operation, the value of SYPOST increases from 13.5 percent to 92.3 percent. In carrying out this exercise, only the non-pool transactions are permitted but all transactions are included in the DC load flow. This result illustrates the fact that power systems originally designed to operate with a centralized control structure are now being asked to operate under conditions for which they are not designed. In other words, some form of external control is required to limit bilateral transactions to a level that is reflected in the desired value of SYPOST.

Table 2: SYPOST CHANGES FOR 118-BUS SYSTEM FOR DIFFERENT POOL/BILATERAL MIXTURES

POOL SHARING FACTOR	0%	50%	90%
SYPOST	13.5%	35.5%	92.3%

Returning to the IEEE 24 bus system, Figure 2 shows the lines that are most heavily loaded (in percent of their rating). It can be seen that line 10, the cable from bus 6 to bus 10, is the worst performer which can be expected since cables generally tend to become overloaded. This phenomenon will be the starting point of the investigation into the effect of FACTS controllers. This exercise will be concerned not so much with relieving a specific circuit as with the remedial effect of FACTS controllers in

improving the value of SYPOST or the transmission system adequacy.

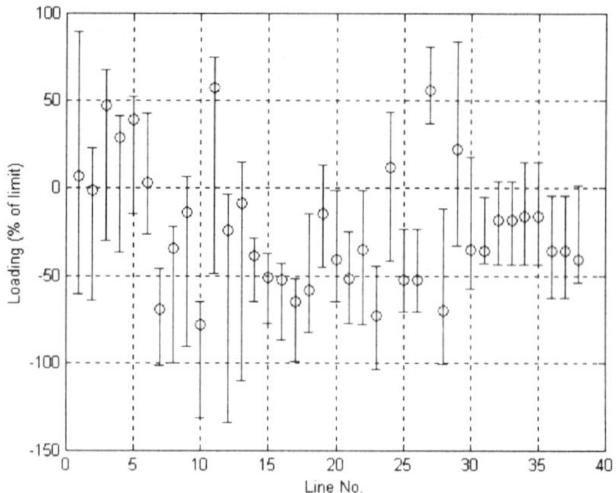

Figure 2: 24-BUS LINE FLOW LOADING DISTRIBUTION UNDER RANDOM TRADING

Simulating FACTS Controllers in System Adequacy Studies

Suppose one wants to study the effects of adding a FACTS controller to an existing line, a simplified model based on the DC load flow can be described as follows. Assuming the system voltage profile can always be maintained at unity (e.g. using static var compensators), a FACTS controller designed to enforce a desirable power flow between its terminals can be modelled by placing a constant generation and load at the respective ends and removing the line connection as shown in Figure 3. The placement of a generation or load at the sending or receiving end depends on the desirable flow direction. For example, in Figure 3, the constant flow will be from bus j to bus i. The amount of constant generation and load will depend on the desirable power flow one wants to attain under different operating conditions, e.g. 80-100% circuit rating.

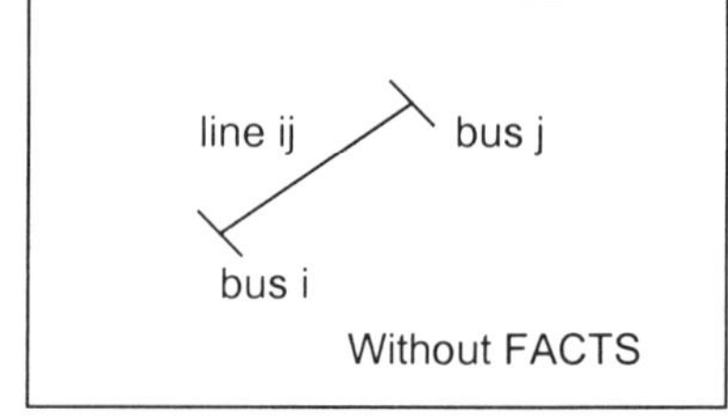

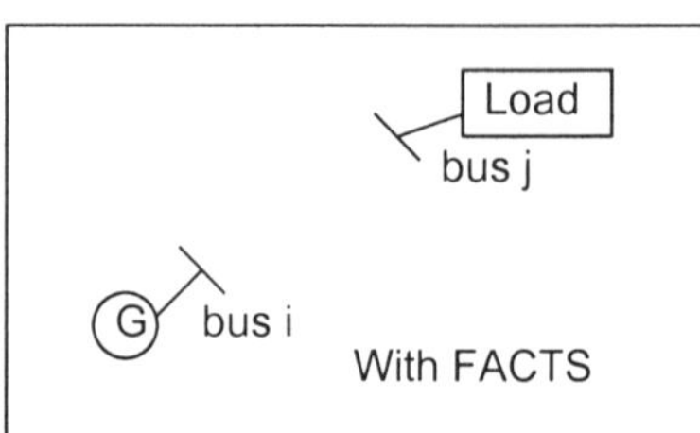

Figure 3: REPRESENTING A FACTS CONTROLLER BY A FIXED GENERATION AND LOAD

With this representation, one can simulate the effects of FACTS controller in a deregulated environment, especially under random trading. Noted that the generation and load added to the FACTS model does not participate in the random transaction matrices generation process because they are considered as fixed. However, their effects on the load flow calculation are included. Two interesting questions are: (1) How would the FACTS controller affect system adequacy in a deregulated environment? (2) What is the most desirable rating of the FACTS controller under deregulation? The following studies are designed to answer these questions by using POST.

Table 3:SYPOST VALUES FOR DIFFERENT SCHEMES OF ADDING A FACTS CONTROLLER IN THE 24-BUS SYSTEM

IMPACT of Single FACT Controller on System Adequacy

Using the modified 24-bus system, a series of numerical

CASE	LINE NO. WITH FACTS CONTROLLER (50% RATING)	FROM BUS	TO BUS	SYPOST (%)	Remarks
1	10	10	6	90.6	improved
2	18	13	11	89.5	improved
3	4	4	2	0	degraded
4	24	16	15	27.8	degraded
5	29	19	16	30.6	degraded
6	10	6	10	0	degraded
7	18	11	13	0	degraded
8	4	2	4	77.5	degraded
9	24	15	16	84.0	marginal
10	29	16	19	65.4	degraded

studies was conducted. Each of these studies basically adds a FACTS controller to each existing line (1-38) one at a time and the desirable power flow is fixed at 50% of the circuit rating. Table 3 shows some selected results indicating the system adequacy as measured by SYPOST (SYPOST for the system without any FACTS controllers is 87%).

It is interesting to note that almost all single FACTS additions result in a negative impact on the system adequacy as measured by SYPOST. Only line number 10 and 18 show a significant improvement since the probability of having a secure transaction matrix under random trading will be higher compared to a system without the FACTS controller. It is more interesting to note that if the direction of the flow control is reversed, the resulting SYPOST is changed drastically (compare cases 1 and 6, 2 and 7 etc.).

Flow Level Control Studies

In Table 4, it is shown how the amount of flow control will affect system adequacy as measured again by SYPOST. In these simulations, a FACTS controller is applied on line 10 and the direction is always from bus 10 to bus 6. The only parameter that is being changed is the amount of the flow control, i.e. constant generation and load.

Table 4: SYPOST VALUES OF DIFFERENT LEVELS OF FLOW CONTROL

CASE	FROM BUS	TO BUS	FLOW CONTROL LEVEL (% OF CIRCUIT)	SYPOST
A	10	6	100%	90.7%
B	10	6	90%	89.8%
C	10	6	50%	90.5%
D	10	6	20%	86.5%
E	6	10	20%	0%

From Table 4, one can observe the following: (1) SYPOST seems to have most significant improvement when the control flow amount is between 20-50% of the circuit rating. (2) Anything beyond that amount does not seem to provide further improvement to the system. (3) SYPOST is reduced to zero as soon as the desirable flow direction is forced to go from bus 6 to bus 10. Further investigation shows that this is because line 5 (bus 2 to bus 6) will always be overloaded to support the FACTS.

FACTS Rating Studies

The methodology introduced here can also be used to study the appropriate rating of a FACTS controller under random trading. Based on a DC load flow model where the line under study is represented by a fixed load and generator of value $P_{i,j}$. The angle between buses i and j can be called $\delta_i - \delta_j$ and these are fixed by the system. Since $P_{i,j}$ is a fixed value, e.g. at 90% of the circuit rating, and $X_{i,j}$ (impedance of the line) is also known, one can thus calculate the angle correction δ^f that is required to be injected by the FACTS device in each simulation case as in (2).

$$\delta^f = (\delta_i - \delta_j) \pm P_{i,j} * X_{i,j} \qquad (2)$$

For example, by placing a FACTS controller on the cable between buses 10 and 6 in the 24-bus system and setting its flow control at 90% rating, the angular changes from the Monte Carlo simulations can be studied. Figure 4 shows the histogram of the angular corrections required by the FACTS controller. For instance, the maximum angle correction is –0.1252 radian which is approximately 7 degrees.

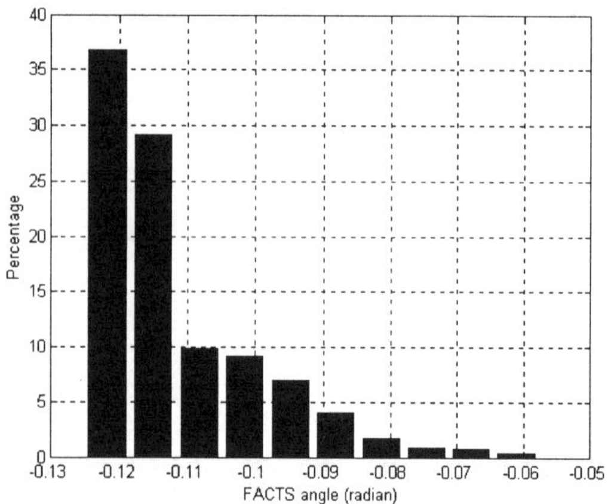

Figure 4: HISTOGRAM OF ANGULAR CORRECTIONS

Conclusions

It is possible to represent bilateral contracts in a matrix form which respects the generation and load constraints of the network. It is then possible to test such feasible matrices for security (no violations) and the result is an index (or ratio) of the secure to feasible operating areas of the network. This index is referred to as POST, the probability of secure transactions.

With this methodology, it is possible to determine a value of this index for a variety of market conditions ranging from all random bilateral transactions to a limited number. This index can be considered a measure of the adequacy of the transmission system to accommodate such transactions.

The application of FACTS Controllers to a deregulated network is an affective way of improving transmission system adequacy while relieving congested circuits in the process. Indeed, this is simply two sides of the same coin but such an effort can have only limited success since the transmission system has not been designed for this purpose.

One must accept the constraints in the power system based on the Laws of Electricity. As a result, one must accept some form of regulation or external intervention to preserve the integrity of the power network. Having said this, there are still positive ways in which some form of regulation can be introduced into the market structure if it is equipped with performance indices that reflect the inherent limitations of the power system.

Reference

[1] Cheng, J., Galiana, F., McGillis, D., *Studies of Bilateral Contracts with respect to Steady-state Security in a Deregulated Environment,* Proceedings of 20th PICA Conference, Columbus, Ohio, 1997, pp.31-36

[2] Cheng, J., McGillis, D., Galiana, F., *Bilateral Transactions Considered as Interconnections in a Deregulated Environment,* Proceedings of IEEE-CCECE 1998, Waterloo, Canada, May 1998.

[3] Cheng, J., McGillis, D., Galiana, F., *Probabilistic Security Analysis of Bilateral Transactions in a Deregulated Environment,* IEEE Transactions on Power Systems, Vol. 14. No.3 August 1999, pp.1153-1159.

[4] Cheng, J., McGillis, D., Galiana, F., *The Impact of Deregulation on System Security,* Proceedings of Seventeenth COPIMERA Conference, San Salvador, October 1999

[5] Cheng, J., McGillis, D., Galiana, F., *Power System Reliability in a Deregulated Environment,* Proceedings of IEEE-CCECE 2000, Halifax, Canada, May 2000.

VALUE OF CONTROLLABLE DEVICES IN A LIBERALIZED ELECTRICITY MARKET

Ch. Schaffner, G. Andersson

Swiss Federal Institute of Technology (ETH), Zurich, Switzerland

Abstract - This paper presents an overview over different areas where Flexible AC Transmission Systems (FACTS) devices can help system operators run the transmission system more efficiently especially under the consideration of liberalized electricity markets. Furthermore, an example network, based on realistic data, is shown where the installation of a TCSC increases the transfer capacity into a region considerably while maintaining security margins.

Keywords - FACTS, liberalized electricity market, economic value, available transfer capability (ATC).

INTRODUCTION

In a liberalized electricity market, the transmission capability of a transmission system represents an economical value to the network company. This company has a natural monopoly combined with the commission to maximize the benefit for its customer while giving a reasonable profit to its owners. Due to physical constraints in the surrounding network, its lines are often only utilized at a fraction of their individual limits. To increase customer benefit one possibility would be to increase the *value* of the transmission lines by increasing the amount of transported energy over these lines. Additionally, there will be a gain in overall market efficiency since more energy trading can take place between competing regions with different price structures. Flexible AC Transmission Systems (FACTS) devices allow the increase of the overall utilization of an electrical power network by controlling the power flow.

Since installations of FACTS devices require huge investments with costs similar to new transmission lines (see Mutale and Strbac (1)) the effect of higher transfer capability only, can not necessarily justify these new installations. It is therefore evident that one has to consider all possible aspects, that add to the value of FACTS devices in a transmission system: static and dynamic stability, increased transfer capacity, increased system reliability, and regained controllability over the power flow for Independent System Operators (ISO) or Transmission System Operators (TSO) in a liberalized electricity market.

In this paper, we will discuss an approach to determine the *economic value* of such controllable device to the network company (ISO or TSO) in a liberalized electricity market. This includes presenting an integral view over the benefits of using flexible systems for transmission.

The paper is divided into three sections: First, we will give a systematic overview over all areas where controllable devices such as FACTS can contribute to the value of an electric transmission system. Second, the aspects of increased transfer capacity will be illustrated in an example based on a "real-world" problem in the field of inter-regional electricity transmission. In the third section the *results* of the analysis of the problem described in the example will be given. Apart from the quantitative simulation results, we will also discuss qualitative issues that may add to the *value* of controllable devices.

The paper presents first results of an ongoing research project in the area of the value of controllable devices in a liberalized electricity market. At the end, we discuss the future work to be carried out on this topic.

THE VALUE OF A CONTROLLABLE DEVICE

Except for static VAR compensators (SVCs) FACTS technology is very young. SVCs are installed at many places all over the world since they offer an economic way to flexibly compensate long lines interconnection distant regions. But even if not many of the other FACTS devices - such as unified power flow controller (UPFC) or controlled series capacitors (CSC) - are installed today there will definitely be an increasing demand for controllable devices in the new electricity transmission systems. Today there are a few installations which help smooth operations mainly in countries with long distance interconnections (e.g. UK, USA, Scandinavia): They help to increase the transfer capacity of the congested links. The overall value of such a device can easily be linked to the gained transfer capacity between the interconnected regions.

In highly meshed grids, e.g. in continental Europe, the added value of a FACTS device is more difficult to determine. Each case has to be evaluated individually (1). To justify the investments, it will not be sufficient to consider only one aspect, such as the increased transfer capacity. To evaluate the value it will be necessary to include different aspects such as increased stability considerations into the investigations. Furthermore, the liberalized electricity market will make it increasingly difficult for ISOs or TSOs to operate their system in an optimal way since they no longer have the means to redispatch or control (e.g. the reactive power output)

AC-DC Power Transmission, 28-30 November 2001
Conference Publication No. 485 © IEE 2001

generation units. FACTS devices can help ISOs to regain control over the power flow in their system.

Not many publications can be found on that integral view on the value of a FACTS device. However, there are several references about the ideal placement of devices with specific objectives:

Singh and David (2), Orfanogianni (3), and Herbig (4) are using different sensitivity indices to show ideal placement options to reduce either real power flow over a particular line or total system power losses, which will decrease loop flows. It is evident that the results differ vastly depending on the objective function chosen.

De Oliveira et al. (5) discuss the special case where FACTS devices are used to optimise a hydrothermal coordination problem. They conclude that overloaded lines are not always the best candidates for installing controllable devices.

These examples show clearly that a methodology to determine the value of a FACTS device is needed. This methodology should help decision makers to get an integral view over the different areas where such a device adds value to an existing system.

In the following we will discuss the five main areas: *Static stability* - retrieve line overloading, *dynamic stability* - improve damping of system oscillations, *transfer capacity* - increase transfer capacity over certain links, *reliability* - reduce risk of loss of load or generation, and *added value for ISO* - provide ancillary services.

Static Stability

Today, electric distribution systems are normally designed based on a (n-1)-security criterion. That means that the system must have enough security margins to operate even if one of the elements, e.g. a transmission line, fails. With congested inter regional links this normally leads to the maximum allowed transfer capacity being considerably below the maximum power flow physically possible.

Lu and Abur (6) propose the use of TCSC to relieve line overloads during contingencies and thus increase the static stability of the whole systems. They show the feasibility with different configurations on a 14 bus network. However, they make also clear that not only network configuration and parameters influence functionality of the controllable devices but also load and generation patterns. Therefore accurate load and generation forecasts will be an important part in the decision to invest in FACTS devices.

An other way to increase the static security is given by Billinton et al. (7): By installing a TCSC or an UPFC at one end of a parallel path, the security of the system can be increased considerably, especially the loss of load probability.

Dynamic Stability

Surprisingly, there are more publications in the area of increased dynamic stability by FACTS devices.

Chen et al. (8) show with a three-machine system that the voltage stability can effectively be improved by installing an UPFC. Especially the shunt branch of the FACTS device contributes to stability even if only local signals are used.

Ghandhari et al. (9) also show that it is possible, with the help of control Lyapunov functions, to use series devices for system damping only by using locally measurable signals. However, under certain circumstances, it is possible that a change in controller parameters can excite inter-area modes (see (4)).

On the other hand, Kohno et al. (10) suggest that UPFC controllers using global information are more effective for power system damping enhancement than those using local information. They are arguing that global information has stronger observability for power system oscillations than local information. It has to be shown with more practical test systems whether local signals are adequate to improve dynamic stability with controllable devices.

FACTS can also be used to improve power quality such as reducing voltage dips, phase shifting etc. In Hara et al. (11) there is a good overview with simulation results where an UPFC is used to reduce voltage dips and harmonics at the node.

Transfer Capacity

In the new world of liberalized electricity markets system operators have no longer direct means - neither in short nor long term - to control the power flow by generator dispatch, since generating companies are free to choose how much energy when and where they want to produce. This implies changes in the geographic generation-load pattern and result in the need to change network topology since certain paths will get congested, and consequently traded transactions cannot always take place. For the system operator there are two fundamentally different solutions: Either reinforce the network by building new transmission lines or add flexible devices to leverage and control power flow.

Building new lines especially in highly populated regions is due to environmental concerns often not feasible.

As Lin et al. (12) and Xiao et al. (13) show it is possible to use FACTS (CSC or UPFC) to improve network performance und thus reduce load or generation curtailments and, at the same time, reduce system losses by minimizing loop flows.

Reliability

Electric power distribution system reliability is defined as the ability to deliver uninterrupted service to customers. Fotuhi-Firuzabad et al. (14) show that it is possible to increase reliability for the consumer considerably by installing TCSCs at the distribution delivery point without increasing short circuit current levels.

The same authors (7) suggest a method to determine the change in loss of load probability and loss of load expectation. They show that the installation of an UPFC in one of two different parallel transmission lines significantly improves the reliability for the network fed by those two lines.

Added Value for ISO/TSO

In today's interconnected electric transmission systems the problem of timely generation is often solved by automatic generation control (AGC): Electro-mechanical machines change their electric power output automatically if the system frequency or tie-line powers deviate from desired values in order to restore scheduled operation. Even if the time constants are quite long (more than ten seconds), this worked quite well until today since loads are mostly frequency and voltage dependent. In addition, tap-changing transformers have a long response time to voltage changes in the feeding network.

But the trend moves into another direction: Due to increased use of power electronics in medium and low voltage systems the loads will be less sensitive to changes in frequency and voltage which will decrease stability margins. In addition, in some countries (e.g. Denmark) considerable amount of energy is produced by renewable power generation (e.g. wind power), which usually does not provide mechanical inertia to help stabilize the system.

Also damping of inter-area oscillations will be considerably decreased in future liberalized systems since transmission paths will be loaded at a higher level.

The network operator (ISO, TSO) has to cope with these emerging problems. FACTS devices will be one opportunity to provide short time active power to stabilize the system (Hingorani (15)).

Concluding Remarks

In the preceding paragraphs, it was shown that FACTS devices could be used in a wide area of applications. Static VAR compensation is already widely used, but other fields will become more important as the liberalization of transmission systems advances further and the TSO or ISO has no longer direct control over generation.

The different types of FACTS devices can be used for different applications. While the static VAR compensator (SVC) can be used for stability improvements, it is not well suited for increasing transfer capacity over a congested link. However, as the name implies, it is normally used for shunt compensation of long transmission lines. To control power flow for increasing transfer

capability the thyristor controlled series capacitor (TCSC) or the unified power flow converter (UPFC) are best suited. The UPFC is of course the most versatile device and can be used for all areas, but it is also the most expensive since it needs a series and a parallel transformer. (See Song and Johns (16))

In table 1, we provide an overview of most used FACTS devices and their typical applications. The decision of what device to install has to be made depending on the system configuration, the actual needs of an ISO or TSO.

TABLE: 1 - Typical use of FACTS devices

FACTS	Static. stability	Dyn. stability	Incr. transfer	VAR comp.
SVC	+	+		+
Statcom	+			+
SSSC	+	+		
TCSC	+	+	+	
UPFC	+	+	+	+

EXAMPLE

Introduction

One aspects of how a TCSC can improve inter-regional transmission capacity will be illustrated by an example based on a "real-world" problem. A simplified representation of this network configuration is illustrated in Figure 1: In region A there is a large amount of nuclear power generation installed leading to low electricity prices. Region C has high prices due to lacking local generation.

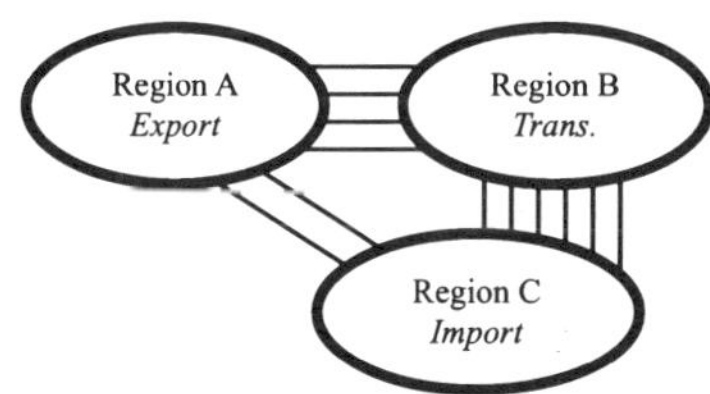

Figure 1: Overview over example system

It is not permissible for the network operators in Region B to load the lines to region C much over 50 % due to security reasons: Since the link between B and C is crucial for secure operation of region C, a (n-2)-security is maintained by the authorities which means that even if two lines fail the scheduled transmission power can still be maintained without overloading any lines.

We will show how the installation of a controllable device in this configuration makes it possible to increase the maximum power imported into region C without decreasing the security limits. This will increase the overall market efficiency since trading is less constrained by unavailable transfer capacity.

The data used for lines and generation and load characteristics is chosen to match a realistic scenario.

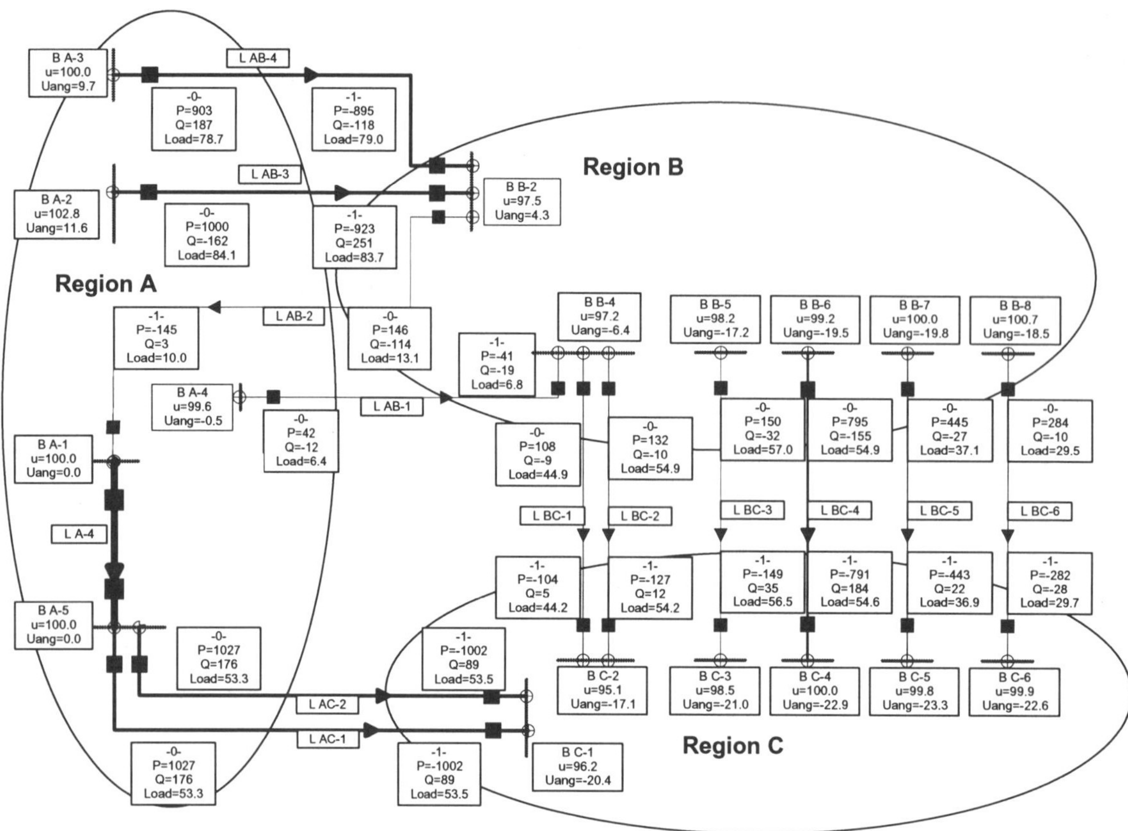

Figure 2: Base Case without FACTS: Lines are loaded only around 50% to maintain (n-2)-security.

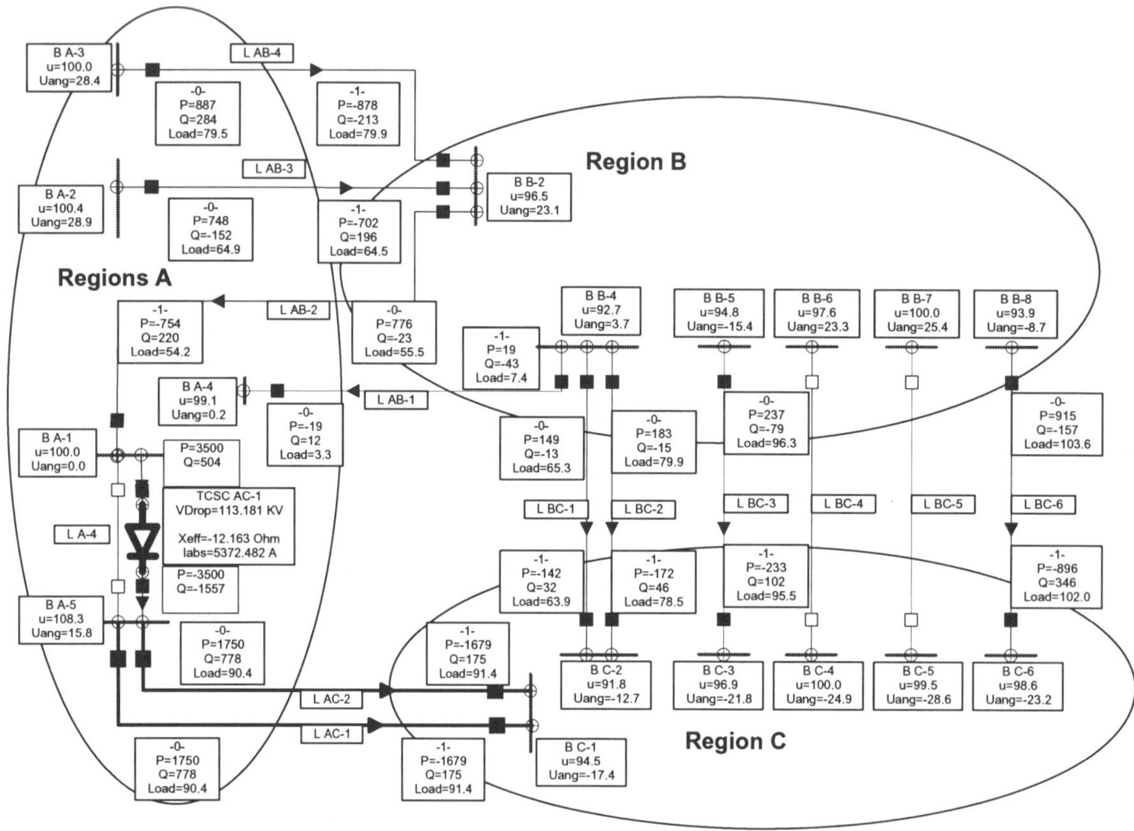

Figure 3: Two lines at fault with TCSC: No overloads above 105% with higher import in Region C vs. base case.

Description of example system

In Figure 2 the base case of the system is represented. All lines are in operation. Region A is connected to region C through two high capacity parallel lines at 380 kV (AC-1 and AC-2). Region B is connected to C through six lines of different capacities where line BC-1 to BC-3 is at 220 kV and lines BC-4 to BC-6 are at 380 kV. In the presented case 3900 MW are imported into region C. This is the maximum amount allowed if the (n-2)-security criterion is to be maintained meaning that if any two of the lines BC-1 to BC-6 fail, there will be no major overloads.

One way to increase the imported power flow to region C would be to reinforce transmission by installing new lines. However, this is impractical due to reasons stated above. It is therefore interesting to see if the installation of a FACTS device can improve the situation.

There are many studies on where the best location would be to install such a system (see above). We used an empirical approach to determine the most suitable position to increase transfer capacity while maintaining (n-2)-security. Since the strongest link represent the two parallel lines connecting region A to C (AC-1 and AC-2) it is evident that placing a FACTS device to the sending bus (A-5) shown in Figure 3 leads to good results.

RESULTS

We used a power system analysis software (BCP (17)) to do the analysis of the static load flow with a TCSC of the example system described above. We compared the base case where a maximum of 3900 MW can be imported into region C to the case, where a TCSC is installed in region A at the feeding bus for lines AC-1 and AC-2. In both cases, we did an analysis of the system running under normal operation and when lines BC-4 and BC-5 between region B and region C fail. The latter is a worst-case scenario since those two lines are feeding the most power from region B to C. The results are represented in Figure 2 (base case), Figure 3 (TCSC with double line failure), and numerically in table 2 for all cases.

In the base case lines are only loaded at a fraction of the physical limits with a maximum of 56.5 % for line BC-3 at a total of 3900 MW imported into region C. TSOs can't transfer above this value since lines would be overloaded if two lines fail. If lines BC-4 and BC-5 fail, line BC-3 is loaded 93 %, which is almost at the physical limit. This shows clearly that it is impossible to import more power into region C without any additional means like line reinforcement or installation of a FACTS device.

If a TCSC is installed at the mentioned position in region A, the maximum possible transfer power can be augmented to 4800 MW, an increase of about 23 %. We can see from the results (last section of table 2 and Figure 3) that even if we import 4800 MW into region C we can keep line loads at a reasonable rate (maximum 102 % at line BC-6) during a double line failure if the TCSC is installed.

TABLE: 2 -   Results

Case	Lines							
Base Case	Import Region C: 3900 MW							
Normal Op.	**AC-1**	**AC-2**	**BC-1**	**BC-2**	**BC-3**	**BC-4**	**BC-5**	**BC-6**
Load [%]	53.5	53.5	44.2	54.2	56.5	54.6	36.9	29.7
Trans.[MW]	1002	1002	104	127	149	791	443	282
Base Case	Import Region C: 3900 MW							
Dbl. line fail.	**AC-1**	**AC-2**	**BC-1**	**BC-2**	**BC-3**	**BC-4**	**BC-5**	**BC-6**
Load [%]	71.2	71.2	69.1	84.9	93.1	0.0	0.0	82.1
Trans.[MW]	1297	1297	153	186	229	0	0	738
TCSC	Import Region C: 4800 MW							
Normal Op.	**AC-1**	**AC-2**	**BC-1**	**BC-2**	**BC-3**	**BC-4**	**BC-5**	**BC-6**
Load [%]	65	65	43.4	53.1	60.5	69.7	47.8	42.8
Trans.[MW]	1214	1214	102	124	158	1008	573	407
TCSC	Import Region C: 4800 MW							
Dbl. line fail..	**AC-1**	**AC-2**	**BC-1**	**BC-2**	**BC-3**	**BC-4**	**BC-5**	**BC-6**
Load [%]	91.4	91.4	63.9	78.5	95.5	0.0	0.0	102.0
Trans.[MW]	1679	1679	142	172	233	0	0	896

This also results in a better overall utilization of the transmission path during normal operation. E.g. the loading of line BC-4, the line transporting the most energy from region B to C, increased from 54.6 % to 69.7 %.

CONCLUSIONS

FACTS devices can help ISOs or TSOs to increase the efficiency of the network in various ways: To improve static and dynamic stability, to increase transfer capacity over congested links, to reduce the risk of loss of load etc. All these fields have to be defined as ancillary services to the network customers such as generation companies or load aggregators. The value of a device can only be determined if an integral view over the gains in all mentioned areas can be achieved.

The results of the simulations show clearly that the installation of just one TCSC can considerably increase the available transfer capacity over a congested link. However, it is important to have an overall view over the network configuration of the inter-connected regions since otherwise sub-optimal solutions will be found. Due to the same reason TSOs of different areas will need to work in a partnership: If a congestion occurs between two regions it is sometimes more efficient to install devices in a third region to achieve good results.

On the economic side it will be necessary that cross border tariffs are well defined in order to determine the value of a new installation. It is not enough to define prices for net transfers since often transfer assets of one region (see region B in example above) are used to transmit energy between two other regions.

A very important question is: What is the value of transactions that cannot take place due to congested inter-connections? The answer to this question gives important input to evaluating the value of FACTS devices. However, it gives also important information when designing a congestion management system that is fair and gives the right signals to the different actors on the market.

FUTURE WORK

This paper has reported the first results from a project aiming at developing a systematic method to evaluate the value in various aspects (including economic) of FACTS devices in an open electricity market. FACTS devices constitute different advantages for different stakeholders, e.g. generators, TSO/ISO, consumer, and it is the aim to include all these into a "global" assessment.

REFERENCES

1. Mutale J and Strbac G, 2000, "Transmission network reinforcement versus FACTS: an economic assessment", IEEE Transactions on Power Systems, 15(3), 961-967

2. Singh S N and David A K, 2000, "Congestion management by optimising FACTS device location", Electric Utility Deregulation and Restructuring, and Power Technologies, 23 - 28

3. Orfanogianni T, 2000, "A flexible software environment for steady-state power flow optimization with series FACTS devices", Power Systems Group, Swiss Federal Institute of Technology (ETH), Zurich, 127

4. Herbig A, 2000, "On Load Flow Control in Electric Power Systems", Department of Electric Power Engineering, Royal Institute of Technology, Stockholm, 154

5. De Oliveira E I, Marangon Lima I W and De Almeida K C, 2000, "Allocation of FACTS devices in hydrothermal systems", Power Systems, IEEE Transactions on, 15(1), 276 - 282

6. Lu Y and Abur A, 2001, "Improving System Static Security via Optimal Placement of Thyristor Controlled Series Capacitors (TCSC)", IEEE Power Engineering Society Winter Meeting

7. Billinton R, Fotuhi-Firuzabad M, Faried S O and Aboreshaid S, 2000, "Impact of unified power flow controllers on power system reliability", Power Systems, IEEE Transactions on, 15(1), 410 - 415

8. Chen H, Wang Y and Zhou R, 2000, "Analysis of voltage stability enhancement via unified power flow controller", Power System Technology, International Conference on, 403-408

9. Ghandhari M, Andersson G and Hiskens I A, 2000, "Control Lyapunov Function for Controllable Series Devices", VII SEPOPE, 11

10. Kohno H, Nakajima T and Yokoyama A, 2000, "Power system damping enhancement using Unified Power Flow Controller (UPFC)", Electrical Engineering in Japan, 133(3), 35-47

11. Hara Y, Masada E, Miyatake M and Shutoh K, 2000, "Application of unified power flow controller for improvement of power quality", IEEE Power Engineering Society Winter Meeting, 2600 -2606

12. Lin W-M, Chen S-J and Su Y-S, 2000, "An application of interior-point based opf for system expansion with FACTS devices in a deregulated environment", Power System Technology, International Conference on, 1407 - 1412

13. Xiao Y, Song Y H and Sun Y Z, 2000, "Application of stochastic programming for available transfer capability enhancement using FACTS devices", Power Engineering Society Summer Meeting, 508 - 515

14. Fotuhi-Firuzabad M, Billinton R and Faried S O, 2000, "Subtransmission system reliability enhancement using a thyristor controlled series capacitor", Power Delivery, IEEE Transactions on, 15(1), 443 - 449

15. Hingorani N G, 2000, "Role of FACTS in a deregulated market", IEEE Power Engineering Society Winter Meeting, 1463-1468

16. Song Y H and Johns A T, 1999, "Flexible ac transmission systems (FACTS)". IEE Power and Energy Series, ed. Johns A T, Ter-Gazarian A and Warne D F, The Institution of Electrical Engineers, London

17. BCP, 2001, "Neplan", ver. 5.0, www.neplan.ch

Dielectric response measurements in time and frequency domain of different stressed low pressure oil-filled cables

T. Kumm   W. Kalkner   P. Jacobsen

Technical University of Berlin / Federal Republic of Germany

ABSTRACT

Power transmission has become commercial relevant to an increasing degree in the recent past. Therefore it is necessary to find out strategies for an assessment of the reliability of cable systems, especially after many years under operating conditions which have changed the ageing state of the insulation owing to electrical, thermal, environmental or mechanical influences. With the aim of gaining a better understanding of these problems from a technical point of view, the dielectric properties of low pressure oil-filled cable cores, which have been equally produced but differently used, are investigated. Results are presented and discussed in this paper and should lead to some new experiences of diagnostic tools for ageing mechanisms in oil-paper insulation.

Keywords: low pressure oil-filled cable, ageing of oil-paper insulation, dielectric diagnosis

1.   INTRODUCTION

Power cable systems are an important part of AC and DC power transmission which must have a high reliability. In the field of power cables oil-paper insulation was developed and used at first, while extruded cables became important after World War II, especially at medium voltage power transmission.

Only with thermally stable cables power transmission at high voltage was possible. In 1924 the first thermally stable cable, a low pressure oil-filled cable, has been in use. Besides pipe type cables (e.g. external gas pressure cables [1]) oil filled cables have a high relevancy in the field of high voltage power cables and transmission, respectively. The service experiences with this cable type (also known as low pressure self-contained fluid-filled cable) are salient, however, the development of the power industry in the last years, a radical change with respect to a deregulation in an open market has taken place, requires new economical strategies and, of course, new respectively intensified precautions to evaluate the condition of cable systems. Moreover, ageing factors and mechanisms as well as their measurability by diagnostic tests are to be found out in order to establish adequate criteria [2].

The assessment of the ageing state of cable systems could be done by using non-destructive or destructive test methods in the laboratory or on-site. The aim of current investigations is to find out the effect of ageing mechanisms on the dielectric response in frequency and/or time domain. The determination of the dielectric response usually does not mean a destruction of the system, and thus it seems to be particularly suitable for on-site diagnosis.

This paper reports on a comparison of the dielectric properties of differently aged low pressure oil-filled cable cores. Some of the equally produced cables have been in service for about 35 years, while a single phase was laid as a reserve cable. This electrically non-stressed cable could be regarded as a reference, based on its dielectric properties those of the electrically stressed should be discussed. Measurements of the dielectric response, like dissipation factor, capacitance, polarisation current and depolarisation current are the main focus of this evaluation. Moreover, some basic physical and chemical investigations were done to study possible changes of the insulation.

2.   EXPERIMENTAL TECHNIQUES AND PROCEDURES

2.1.   Test objects

The investigations were carried out on low pressure oil-filled cable cores manufactured in 1966 that were withdrawn from a cable system of the type NÖKuDEY 1 x 625 rm/v 12 h used at a voltage level of 18/30 kV.

A three-phase current system performed of this cable type has supplied a steel smelting plant having a load of about 70 MW for about 25 years. After the quiescence of the steel production the cable system was stressed with a distinctly fewer load for another 10 years.

Parallel to this service-stressed three-phase system a reserve cable was laid that was not electrically stressed. All cables were connected with the same oil container, whereas the impregnating medium was pure mineral oil.

AC-DC Power Transmission, 28-30 November 2001
Conference Publication No. 485 © IEE 2001

The diameter of the hollow conductor is 33 mm and the insulation thickness of a single core is about 3 mm.

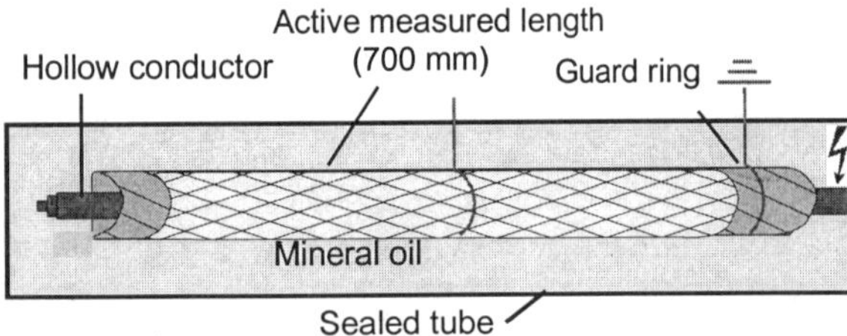

Figure 1: Schematic drawing of experimental assembly (sectioning)

The principle assembly of core samples as used for dielectric measurements is shown schematically in figure 1. Polymeric protective jacket, reinforcement, drained jute bedding and lead sheath were removed, and a guard-ring core sample was prepared as shown in figure 1. This core sample was put in a mineral oil filled sealed tube whose rounded metallic bottom is used as electric connection between cable conductor and test voltage generation.

Table 1 summarizes the investigated cable cores and their properties.

Since oil-paper insulation is degraded significantly when moisture comes into contact with hygroscopic cellulose a determination of the water content gives first indications about possibly proceeded ageing processes inside the insulation. Therefore the method of Karl-Fischer-titration was used. Some oil-paper layers were taken from each withdrawn cable core and their water content was measured. This investigation has shown that due to the relatively small insulation thickness water content of the insulation near the conductor is similar to this of outer layers. Moreover, an axial dependence of moisture content was not found. Therefore, all measured values of water content of each core were averaged and quoted as this value in table 1.

Cable/ condition	Notation	No. of samples	Water content (average)
Reference electr. non-stressed	Cable 0	1	0,59 %
Phase 1 electr. stressed	Cable 1-x	2	0,61 %
Phase 2 electr. stressed	Cable 2-x	2	0,61 %
Phase 3 electr. stressed	Cable 3-x	2	0,81 %

Table 1: Investigated samples and properties

2.2. Measurement techniques

Dissipation factor: The dissipation factor measurement system used consists of a high voltage source (HVS) and a potential-free tan δ/capacitance measurement part, both variable in frequency.

The dissipation factor is obtained extremely precise by calculating the phase difference of the base frequency wave by DFT (discrete fourier transformation) from the current through the test object (C_x) and the current through the reference capacitor (C_{ref}, pressurised gas capacitor) [3, 6]

Polarisation respectively depolarisation current: The system used for time domain measurements makes possible a determination of the polarisation current versus time curve using a self-developed sensor and of the depolarisation current curve using an external high-resistance-input electrometer in dependence on different eligible charging and discharging parameters [6].

2.3. Measurement procedures

Prior to the diagnostic measurements the assembly of each cable core was checked for its operativeness. Primarily, the guard-ring resistance was measured because of its influence to the dielectric response. Only high values (> 0,1 TΩ) were accepted. Furthermore, within the scope of preliminary investigations no partial discharges up to V_0 occurred in the sample set-up due to figure 1 were detected.

The dissipation factor was measured frequency dependent at voltage levels of $0,1 \cdot V_0$ (0,01 Hz to 1 kHz) $0,3 \cdot V_0$, $0,5 \cdot V_0$ and $1 \cdot V_0$ (0,01 Hz to 100 Hz).

Measurements in time domain are based on a DC-charging of the sample dielectric that was realised in consideration of usually "high" responses of the oil-paper dielectric with voltages of 0,2 kV, 0,5 kV, 1 kV and 2 kV. Charging as well as measurement time were chosen to 900 s, while short-circuit time was 1 s.

All dielectric measurements were carried out at normal conditions (room temperature, atmospheric pressure).

3. RESULTS AND DISCUSSION

3.1. Frequency domain measurements

Figure 2 represents voltage and frequency dependent dissipation factor curves of the electrically non-stressed reference cable core (cable 0).

First of all, it is conspicuous that the minimum of detected losses at room temperature is at about 500 Hz, while dissipation factor increases towards lower frequencies. The former is likely attributed to the moisture content of the insulation that has risen to a value of about 0,6 % (see table 1). Though this water content increases dielectric losses, especially at service condition temperatures, with respect to a failure it is considered to be not critical [5].

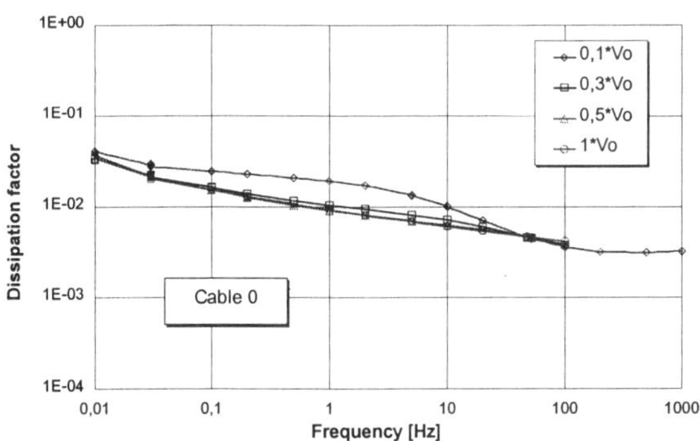

Figure 2: Dissipation factor of non-electrically stressed reference cable

Similar values of water content were measured for the service aged cables 1 and 2, while insulation of cable 3 has absorbed somewhat more water (table 1). Detected moisture contents of both electrically stressed and non-stressed cables indicate that humidity got into insulation, either during handling or oil refilling after a leakage repair due to the use of not completely dried oil.

Furthermore, in figure 2 one can see that dissipation factor curve at $0,1 \cdot V_0$ differs from the other, nearly voltage independent ones. The increase of dissipation factor to smaller electrical field strengths is usually known as Garton- effect [4]. This behaviour is typically for oil-paper dielectric, and, as Garton has been proved, dissipation factor obtains its maximum at a field strength of about 0,3 kV/mm (e.g. $0,1 \cdot V_0$ corresponds to 0,6 kV/mm). The Garton- effect, respectively its characteristic, indicates changes of the dielectric due to ageing processes or high temperatures. In case of cable 0, the authors suppose that not completely dried oil was used after a leakage repair that has flown through all cables and has led to a rise of ion concentration.

Figure 3 makes possible a comparison of dissipation factor curves of investigated core samples at $0,1 \cdot V_0$ and $1 \cdot V_0$. Compared with the reference object, measured dissipation factor values of the electrically stressed cable cores are higher, especially towards lower frequencies.

In case of cable 1 and cable 2, conductivity respectively conduction part of dissipation factor has increased. In contrast to cable 0, this is probably the result of the heating of electrically stressed cables due to the flowing current. This thermal stress deteriorates cellulose molecules and products of separation are created. Furthermore, this 'true' DC-conductivity is not voltage dependent (see cable 1-2).

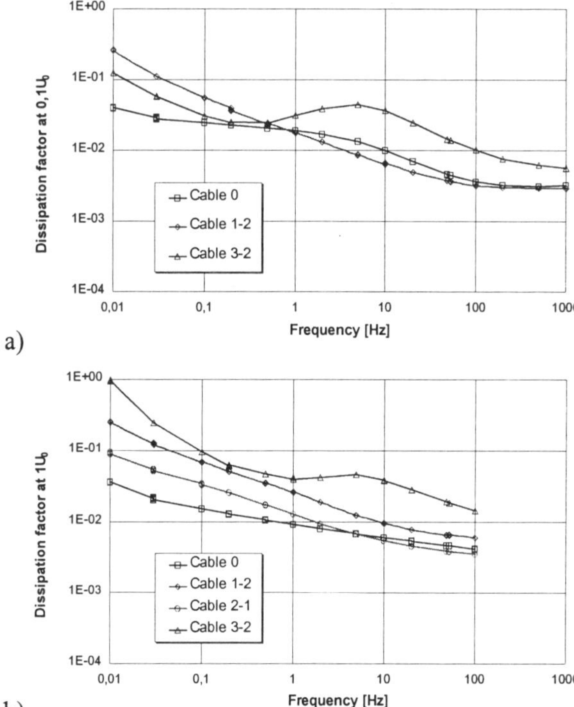

a)

b)

Figure 3: Dissipation factor curve shapes of investigated core samples
a) at $0,1 \cdot V_0$ b) at $1 \cdot V_0$

The dielectric behaviour in frequency domain of the third service aged phase, represented by cable 3-2, is different to that of above described cable samples.
Figure 3 shows a distinct loss peak at a frequency of about 5 Hz at $0,1 \cdot V_0$ as well as V_0. Due to this, the frequency dependence of the real part of the permittivity, here calculated as the relative change of the measured capacitance (1), supports this statement.

$$relC_x = \frac{C_x - C_{100Hz}}{C_{100Hz}} \qquad (1)$$

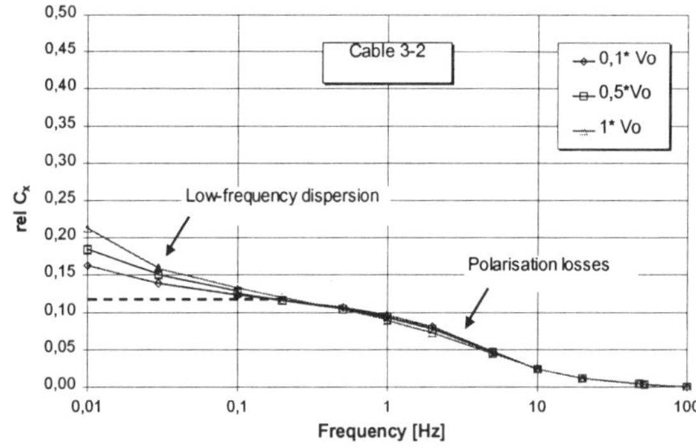

Figure 4: Relative change of capacitance of cable 3-2

As one can see in figure 4, relative capacitance increases towards lower frequencies up to about 0,2 Hz as a result of polarisation processes inside the insulation, and by neglecting low-frequency dispersion effect it would obtain its stationary value (figure 4, dashed curve). The rise of capacitance respectively real part of the permittivity below 0,1 Hz is caused by the low frequency dispersion (LFD-) effect [7], which should not be mistaken for 'true' DC-conductivity, that is mainly responsible for the rise of dissipation factor towards lower frequencies of cable 1-2 and cable 2-1. The LFD- effect is probably caused by electrochemical polarisation processes, more precisely by interacting dipoles and collective motions and characterised by a positive dependence on voltage [7, 8].

3.2.   Time domain measurements

Measurements in time domain were aimed at creating alternative tools for an appraisal of dielectric behaviour compared to conventional destructive test methods as well as dissipation factor measurement which is usually not possible at service frequency on-site. The non-existing of universal and trusty criteria is opposed to the relatively low effort due to a voltage supply of time domain based measurements.

First of all, results of polarisation current measurements of investigated cable cores should be discussed (figure 5). By means of the interpretation of polarisation currents it is obvious that cable 0 has the smallest magnitude at every measured time. The rise of dissipation factor towards lower frequencies of cable 1-2 and cable 2-1 (figure 3) was attributed to a rise of conductivity losses. Polarisation current curves of the same cables confirm this behaviour since they have the highest magnitudes and a fewer time dependency like cable 0 and, due to it, are in the same order as dissipation factor curves.

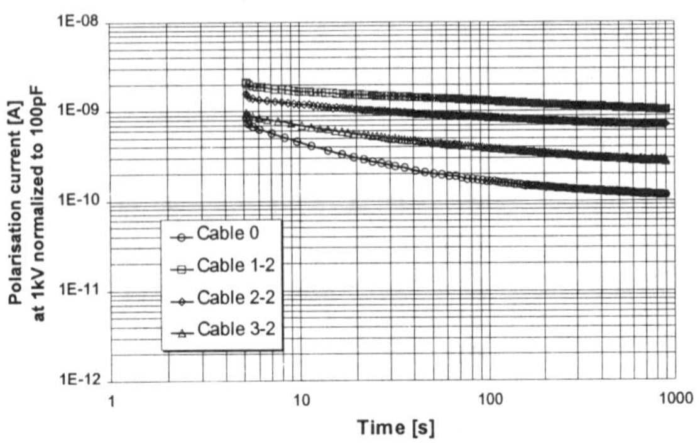

Figure 5: Polarisation current curves of investigated core samples at $U_L = 1\ kV$

However, polarisation current of cable 3-2 is not in accordance with the above statement that the order in dissipation factor curves at low frequencies is like that of polarisation currents, especially at long times where polarisation processes are irrelevant and a charge carrier current dominates.

To explain that different behaviour of cable 3-2 the voltage levels used for time domain measurements

have to be taken into account. Electric field strengths hereby are much smaller than these at service conditions and only comparable with that of the $0,1 \cdot V_0$-dissipation factor measurement. According to the dissipation factor measurement results, polarisation current of cable 3-2 has a positive voltage dependence, too, which is represented in figure 6 where the degree of non-linearity (DoNL) is shown. It is obtained by (2),

$$DoNL = \frac{i_p(t_1, U_L) * 1kV}{i_p(t_1, 1kV) * U_L}$$

(2)

where t_1 is a specific point of time (e.g. 10 s) and U_L the loading voltage. As reference voltage 1 kV was set. The 10 s- as well as the 100 s– polarisation current values of cable 3-2 increase with voltage, while in case of cable 2-2 'true' DC-conductivity does not cause this effect.

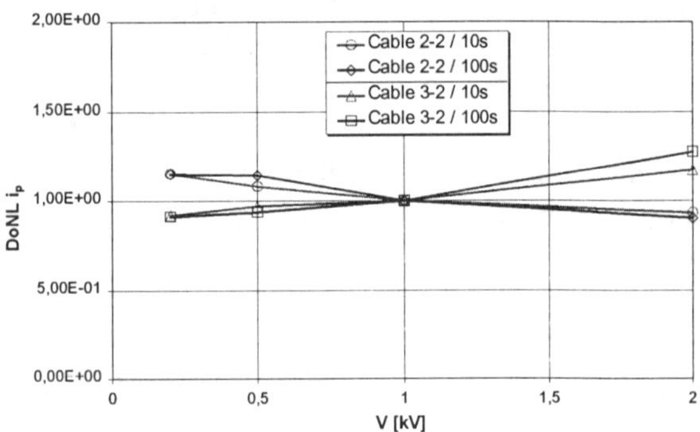

Figure 6: Degree of non-linearity of cable 2-2 and cable 3-2

Figure 7 makes possible a comparison of polarisation vs. depolarisation current of cable 3-2. At very short times both polarisation (i_p) and depolarisation current (i_{dep}) converge to the term $\varepsilon_r \cdot \delta(t)$ due to (3) and (4),

$$i_p(t) = U * C_0 * (\varepsilon * \delta(t) + f(t) + \frac{\sigma}{\varepsilon_0})$$

(3)

$$i_{dep}(t) = U * C_0 * (\varepsilon * \delta(t) + f(t))$$

(4)

where C_0 is the geometrical capacitance, ε_r the relative permittivity, $\delta(t)$ the delta function and f(t) the response function [9].

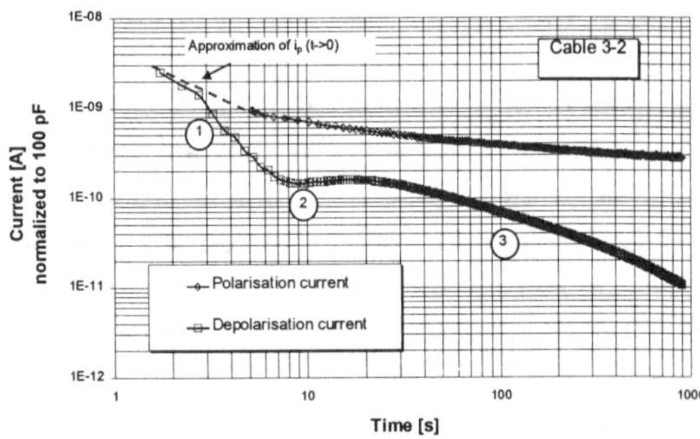

Figure 7: De- and polarisation current of cable 3-2

First section (section 1 in figure 7) of depolarisation current is given by polarisation processes and characterised by two slopes, one at short times and another, 'bigger one' at long times, than (section 2) depolarisation current follows due to Jonscher [7, 8] the carrier-dominated low-frequency dispersion. Section 3 in figure 7 is once again given by a dipolar character of the response but to a lesser extent [7, 8] that is similar to the other investigated core samples (figure 8).

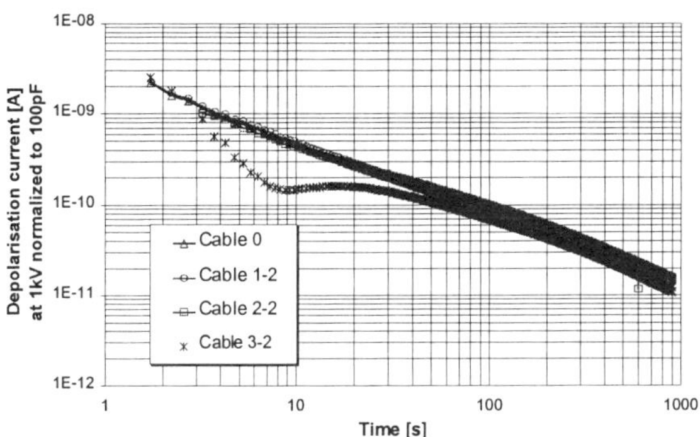

Figure 8: Depolarisation current curves of investigated core samples

Depolarisation current measurements of core samples taken from the electrically stressed cables 1 and 2 and from the electrically non-stressed cable 0 do not show differences (figure 8). Due to an evaluation of the ageing state by means of depolarisation current curves or values at specific times and the assumption that cable 0 can be regarded as an 'unaged' reference no ageing processes inside the insulation could be detected. More concretely, the addition of an electrical stress respectively a current load inducing thermal stress does not create polar impurities that can be generally detected by depolarisation current measurements. To what extent further, for all cables similar environmental influences lead to an change of dielectric properties compared with the condition immediately after manufacturing respectively laying could not be clarified, however a slightly raised water content (table 1) has caused supposedly an increase of polarisation losses.

4. CONCLUSIONS

Based on the assumption that investigated core samples, on the one hand service stressed low pressure oil-filled cables and on the other hand a not electrically stressed one, are in a different ageing condition, dielectric measurements were carried out to get new experiences with respect to an assessment of life expectation by means of dielectric diagnosis. Particularly, a thermal ageing in case of the electrically stressed cables and a moisture ageing in all cables were expected, even though measured water contents are at a low level.

By means of dissipation factor measurements the following conclusions can be drawn:

- Ageing processes in a minor degree lead to an increase of conductivity losses that can be distinctly detect at frequencies below 10 Hz and are independent on voltage.

- Additionally, polarisation processes owing to a deterioration of oil-paper dielectric can occur at a frequency range of about 1 Hz and 100 Hz. However, the latter means that a detection on-site that is usually done at 0,1 Hz is not successful.

- Furthermore, the low-frequency dispersion effect has been observed at frequencies below 0,2 Hz. It differs distinctly from the 'true' DC-conductivity since it is strongly voltage dependent and connected with a change of the relative permittivity. This dependence on voltage is to take into account at on-site measurements, where only limited voltage levels are available.

The time domain measurements carried out at voltage levels distinctly fewer than service ones reveal above described ageing mechanisms too, however it has to be mentioned that at these levels the low-frequency dispersion- effect possibly could not be detected. Conductivity losses are to be seen in polarisation current behaviour that is almost independent on time and voltage, while polarisation peaks cause a depolarisation current curve shape consisting of two slopes [7].

Low frequency dispersion- effect is detectable in both polarisation and depolarisation current behaviour, in case of polarisation current it depends strongly on voltage, while depolarisation current becomes more flat.

5. REFERENCES

[1] Kumm T et al., 2001, "Dielectric behaviour of different impregnated and aged oil-paper insulated HV cables in dependence on temperature", 12th ISH 2001, Bangalore, India

[2] Densley J, 2001, "Ageing mechanisms and diagnostics for power cables – an overview", IEEE Electrical Insulation magazine, 17/1, 14-22

[3] Emanuel H et al., 1996, "A new High Voltage Dielectric Test System for insulation diagnosis and partial discharge measurement", Nord-IS, Bergen, 1996, 283-290

[4] Garton C.G., 1941, "Dielectric loss in thin films of insulation liquids", Journal of the Institution of Electrical Engineers, Part II, 1941, 103-120

[5] Neimanis R et al., 2000, "Determination of moisture content in mass impregnated cable insulation using low frequency dielectric spectroscopy", IEEE PES Summer meeting, Seattle, 2000

[6] Kuschel M, 2000, "Diagnose des Alterungs-zustands von PE/VPE-isolierten Kabeln mittels Verlustfaktormessung bei Niedrigfrequenz (VLF) und Depolarisationsstrom- und Rückkehr-spannungsmessung", Doctoral thesis, Technische Universität Berlin, 2000

[7] Jonscher A.K., 1992, "The universal dielectric response and its physical significance", IEEE transactions on Electrical Insulation, 27/3, 407-423

[8] Jonscher A.K., 1996, „Universal relaxation law", Chelsea Dielectrics Press, London, UK

[9] Ildstad et al., 2000, „Condition assessment of water treed service aged XLPE cables by dielectric response measurements", CIGRE 2000 Group 21 Session papers, 21-201

Advantages of HV Cable Transmission Links
by using Real Time Thermal Rating (RTTR) and Water Monitoring

L. Goehlich   M. Kuschel   F. Donazzi   R. Gaspari
Pirelli Kabel und Systeme   Pirelli Cavi e Sistemi
Berlin, Germany   Milan, Italy

In modern high voltage cable systems temperature overloads or water ingress are often limitation parameters for the lifetime. However, two highly sophisticated monitoring techniques are available to control these parameters. A system based on distributed temperature and load monitoring to enable cables to be more effectively loaded without exceeding thermal limits and a system for monitoring the dry condition of the insulation and alarming the ingress of water into the cable in the event of external sheath damage.

Both systems are already introduced into commercial plants and have shown their excellent performance. The measuring principles, hardware- and software concepts of these systems will presented here as well as different cable designs, sensors and cable laying arrangements. Furthermore, the advantages and yet available experiences of these systems are evaluated and discussed. General aspects of monitoring and individual steps helpful to users for checking if monitoring offers advantages in their application complements this contribution.

1. Introduction

Utilities expect highest reliability and lowest prices of modern cable systems. The general world wide development of cable design has focussed on HV-cables with XLPE-insulation to meet these tough requirements. An obvious but often hidden demand is the need for little or no maintenance in spite of highest utilisation. This aspect becomes more and more important as utilities are asked to improve their efficiency in order to become more competitive.

Technical performance can be achieved either by "material" or by "intelligence" and for optimisation by both methods. Regarding HV-cables it is not possible to maximise the power transmission capability by an increase of cross section and to make a cable inviolably by heavy armouring only by means of "material". This cable would not fulfil any commercial demand. In the contrary cable plants with integrated "intelligence" by means of monitoring are expected to become the most profitable in an over-all consideration.

Utilities all over the world have trusted their responsibility to this principles of cable monitoring which are shown in the following applications.

2. Cable Designs, sensors and cable laying arrangements

Fitting of HV-cables and cable plants with sensors is widely used for many centuries. But just in the recent decade sophisticated sensors and monitoring techniques have been developed to promise reliable applications in practical field operation. Therefore each cable design requires a particular sensor design and arrangement.

One of the most critical parameters to be monitored is the conductor temperature, because exceeding specified values can reduce the life time of the cable. To measure distributed temperatures along the cable route, only

optical fibre sensors are suitable. Cables can be up to 10 km or 30 km long, depending on the type of optical fibre. The fibres can be integrated in the cable or be arranged separately close to the cable. In both cases conductor temperatures can be determined with high accuracy by a sophisticated computerised measuring system, called "RTTR".

The other harmful parameter to cables is water, even in modern times of plastic insulation. Water can harm a cable in the long term and must therefor be excluded from the cable core. This can be done by means of electrical water sensors, which must be arranged inside the cable. Depending of the task – detection of ingress of water in the screen or early detection of risky corrosion of a metal sheath - the sensors are placed between the respective layers.

The following gallery of sensor applications shows the variety of possible individual solutions:

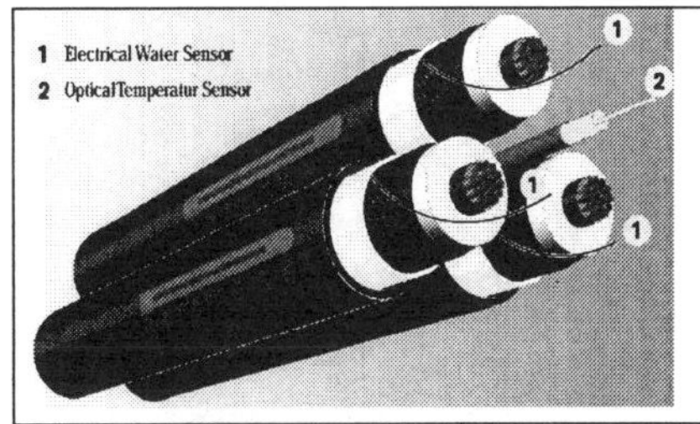

Figure 1: HV-cables with temperature and water sensors in trefoil-laying formation

Figure 1 shows a typical arrangement of temperature and water sensors in a high voltage cable link. While for electrical reasons the trefoil formation laying is used, this also helps for economical application of external temperature sensors in the area of uniform temperature in the centre of the cables. The electrical water sensors are required in the screen area of each individual cable.

AC-DC Power Transmission, 28-30 November 2001
Conference Publication No. 485 © IEE 2001

Figure 2 presents a similar task, but temperature sensors are required integrated in the cable, because a higher cross section and mainly laying of the cable in parallel formation and in ducts.

Figure 2: XLPE-cable with Al/PE laminated sheath and with temperature and water sensors in the screen area

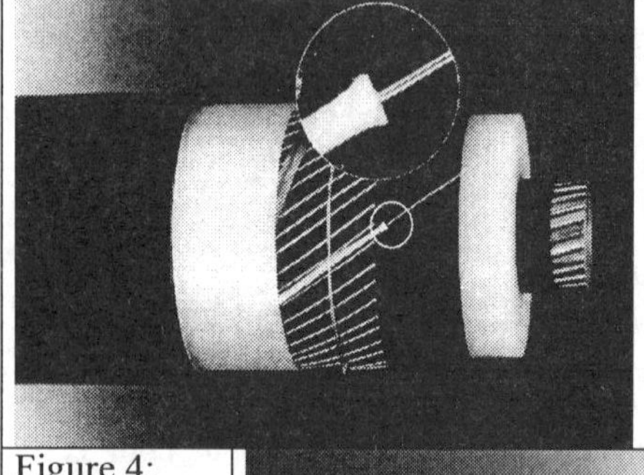

Figure 3: XLPE-cable with Al/PE laminated sheath and water sensor (magnified) in the screen area

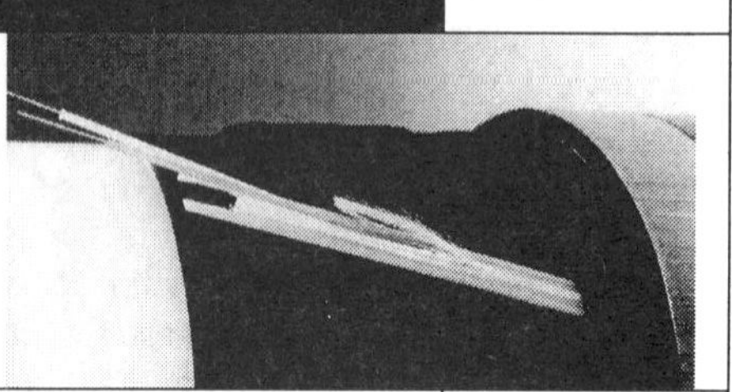

Figure 4: XLPE-cable with metal sheath and temperature sensor in the screen area

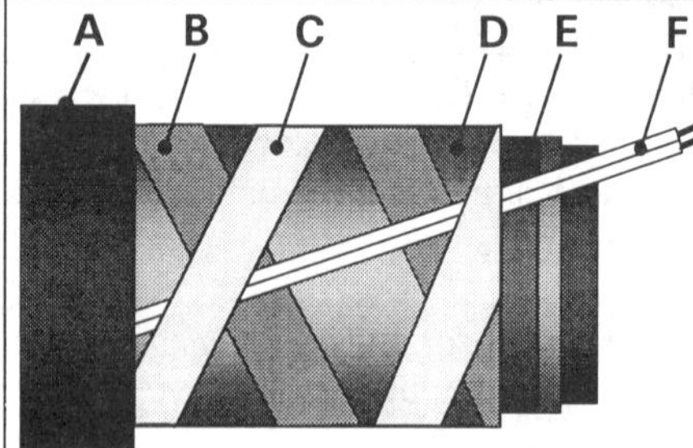

Figure 5: Metal sheathed cable with water sensors (design study)

Figure 6: Heavy armoured sub marine cable with temperature sensor

Figure 3 demonstrates economical application of water sensors in the screen of cables with Al/PE laminated sheaths.

Figure 4 is an example for optimised temperature sensor application: the commonly used fibre in metallic tube is replaced by a special reinforced and impact resistant design which is necessary for heavy metal sheathed cables.

Figure 5 is a design study of a metal sheathed cable for monitoring the integrity of the anti corrosion protection itself with integrated water sensors. (A=outer plastic sheath, B, C=water blocking tapes, D=inner plastic sheath, E=cable core, F=water sensors). This application is advantageous in case of oil filled cables in brackish water areas.

Figure 6 relates to sub marine cables. Even if cables are lateral cooled by water, temperature monitoring is still useful because of temperature dependant insulation properties which dictate a reduced thermal gradient across the insulation. The temperature sensors are mechanically reinforced, water tight and placed in the bedding of the cable.

Water monitoring is usually used for HV and EHV cables because of the need to keep water away from the cable core. In case of MV cables this is not so mandatory, and on the other hand, very hard to achieve in a three core cable design because of water seeping into the inter phase spaces. Water sensors in this areas and a water monitoring system would exactly detect water ingress in case of a cable damage.

3. Measuring principals, hard- and software concepts of Monitoring Systems

3.1 RTTR-Temperature Monitoring

The principle of RTTR continuous temperature and load monitoring is shown in figure 7. The temperature is measured by a commercially available system using distributed optical-fibre sensors. The current is measured by commonly used inductive current transformers which are already installed in most transmissions plants and therefore available for monitoring purposes. Further details can be found in [1], [2] and [3].

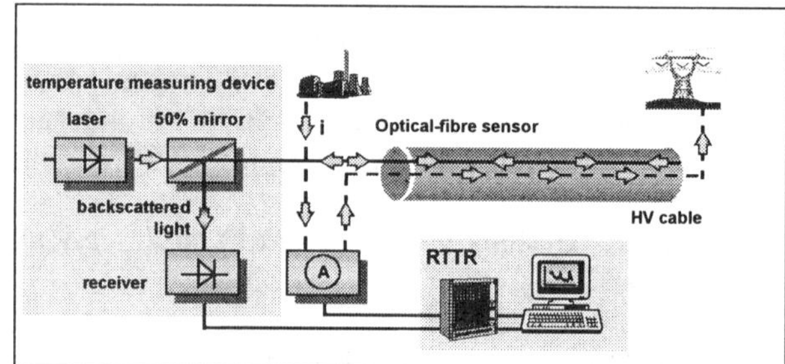

Figure 7: Principle of RTTR Measurement

From the measured quantities as well as predefined cable link parameters (e.g. cable type, installation geometry) the system calculates in real time:
- the conductor temperature,
- the actual thermal transient evolution,
- the heat dissipation (e.g. the thermal resistance of the surrounding soil),

- the steady state permissible current rating and
- the permissible power overloads (short time as well as long term) depending on actual temperatures

These calculations are based on IEC standards [4], [5] and enable the complete management of cable systems with optimum and safe thermal cable loading [6].

3.2 Water Monitoring

In case of water monitoring the measuring principle is shown in Figure 8. The system consists of the HV cable system (C) with distributed water sensors which are integrated in the cables, some link and protection boxes (B and D) and finally the measuring device (A). This device measures the DC insulation resistance of the water sensor as an indication of water penetration and the voltage across the three measuring resistors of one cable link. DC measurement improves the accuracy by filtering off all AC noise. Measurement results are evaluated automatically in order to recognise any faulty phase and calculate the fault location.

Details of scientific research and application of this system are reported on earlier ISH conferences [7], [8] and [9], [10].

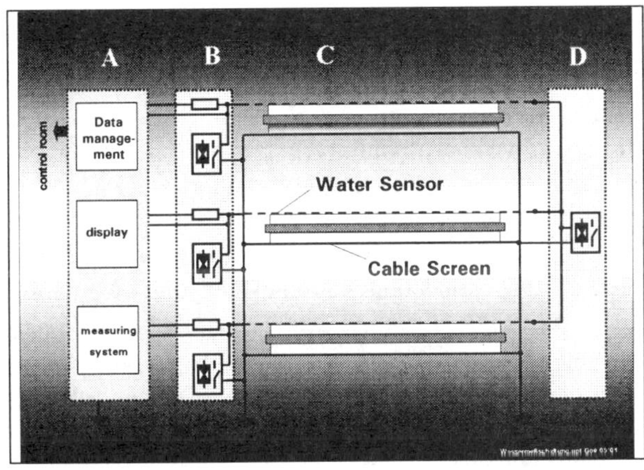

Figure 8: Principle of water monitoring
A=measuring device, B and D=link box, C=HV cable

The measurement results are stored internally by the measuring device and generate alarms on a contact panel. Depending on the need of the utility, the measurement results are provided via a modem to be displayed in the utility control centre. Furthermore these information can be transmitted to the RTTR-system as will be presented later on.

In case of cross-bonded screens water monitoring is as well applicable. In that case, water sensor link boxes are installed additionally to the screen link boxes next to the respective joints. These boxes enable connections of the water sensors in the same sequence as the cable screens in order to avoid differences of induced voltages between these two electrical layers on the cable core. The design of the water sensor link boxes sustain high voltage DC tests of the outer sheaths, which could be

necessary for commissioning or intermediately after possible repair work.

The utmost challenge to every electrical operated measuring device linked to a high voltage line is the sudden presence of high voltage or high current impulses which will destroy not proper protected equipment. In case of monitoring the sensors are exposed to induced and direct high currents in a short circuit condition, which are of the electrical characteristic of a constant current source. For that purpose an over voltage protection of the sensors as well as of the measuring device had to be realised. Commonly used MO surge arresters would be far too big as a market study (Figure 9) has shown. A maximum voltage of 100V was allowed during short circuit only of the sensor of 100A for 1s, which is represents an energy of 10.000Ws and would have required an arrester volume of 1000cm^3 for each phase! The problem was solved by means of a circuit equipped with thyristors, which switch within micro seconds, every time a safe voltage limit is exceeded, and ground the sensors. Thus the values of voltage and current are still the same but the induced energy at the end of the sensors is nearly zero, because in case of voltage there is no current and in case of current there is no voltage. The circuit does not require any external power.

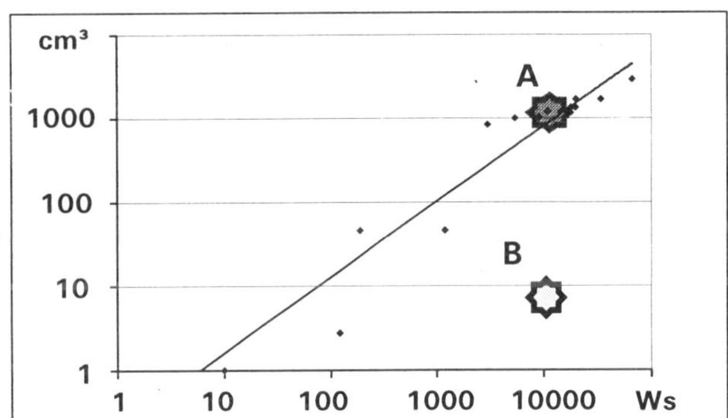

Figure 9: Volume of devices for protection of water sensors versus energy
A= MO arrestors (market study)   B=thyristor circuit

3.3 User friendliness of the Monitoring Systems

System performance includes Temperature and Water Monitoring: Both monitoring systems (RTTR Temperature and Water Monitoring) are designed for automatic operation without any maintenance. Their task is to relief the cable operator instead of requiring attention.

Anyway there are cases where the operator needs to know that the systems are properly working or, in case of failure, it is necessary to provide to get all failure information directly on the computer screen. The system provides the operator with information about fundamental automatic operations: e.g. acquisition from the overall set of sensors, data validation, archiving and processing by the mathematical model, alarms generation to highlight dangerous or out of limit conditions, statistical analysis of historical data, data routing between the different units and users.

Full automatic start-up and cold restart capabilities have been embedded into the systems. In parallel safekeeping procedures and redundancies improve the overall system reliability in order to minimise necessary monitoring of the systems.

Customer related System Mathematical Description: The different elements and laying conditions of the related customer installation are with regard to the RTTR Temperature Monitoring System described inside the system as an assembly of elements: types and geometry of laying, cables and accessories description, materials characteristics and utilised sensors and their location.

All elements can be upgraded and modified easily to take into account retrofits and or indications by the Statistical analysis module.

All the laying conditions, the retrofits and custom updates are stored on board the System Description Data Base to be presented to users whenever requested.

Interfacing with external SCADA: In most installations there is the particular requirement expressed by the client, to be able to interface with its own system.

That capability was developed even for the early installations of the RTTR system and is now available for both RTTR- and Water Monitoring.

The input/output from and toward the customers own system can be of two kind:

- analogue signals give a direct reading of the most relevant evaluated quantities for each one of the monitored links like maximum conductor temperature and permissible loads for selected duration
- digital outputs are instead used to flag/activate alarms directly on the customers system. If the system status or behaviour exceeds predefined limits, in addition the on board diagnostics will monitor the system status and will generate specific alarms in order to avoid the system running with malfunctions.
- Moreover direct interfacing using dedicated interfaces is a viable and adopted option giving direct access to systems and SCADAS in a transparent way.

Customisation: Besides reliable hardware and mathematical treatment of the measured data in real-time and according to IEC standards the most important part of the RTTR is the so-called Man Machine Interface (MMI). The MMI consists of a real-time Graphical User Interface (GUI), an Alarm server and historical data displays.

The GUI is customised for the end-user's particular application and enables mainly changes to be identified in the cables system and displays predictions of system capability

Furthermore, with regard to the RTTR Temperature Monitoring System for each section, supplementary results and information on the laying arrangements of the cable circuits are accessible by clicking on the relevant section of the plant map. A result is shown in Figure 10. The Water Monitoring System does not require this volume of data because of its more damage display characteristic.

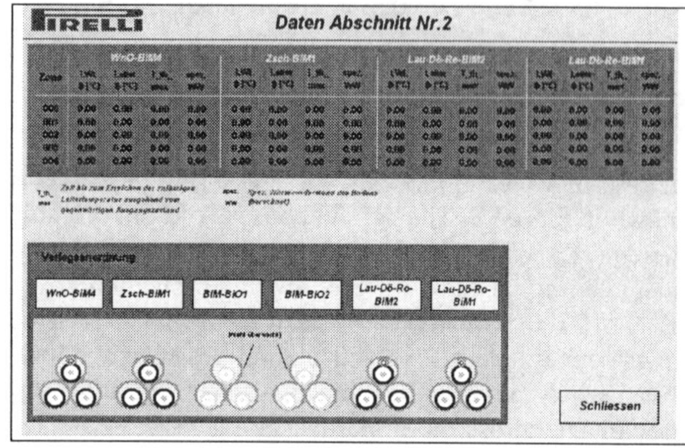

Figure 10: Screen image of a selected section no.2 as an example of the Graphical User Interface

The alarm server of the RTTR System is capable of managing and generating graphical and visual alarms to warn the operators of dangerous or unexpected conditions and of trends nearing customer pre-set thresholds. All alarms statuses are stored on-board into a dedicated alarm historical archive for off-line operation analysis, for procedures control and for contingency recovery. Finally, the historical data displays give the possibility to perform on-line or off-line analyses of stored trend information.

Since the user is able to access the data and all data access are performed in a client-server the user can request a connection through a dial-up remote connection and visualise, analyse and download any data which is possible for both monitoring systems.

It is obvious that the RTTR-temperature monitoring has to handle much more data than the water monitoring system, anyway in case of water monitoring it is comfortable also to get additional information. An advantage in user friendliness is given by the possibility of utilising route maps and synoptics, available for both RTTR and water monitoring, for convenient data visualisation (Figure 11).

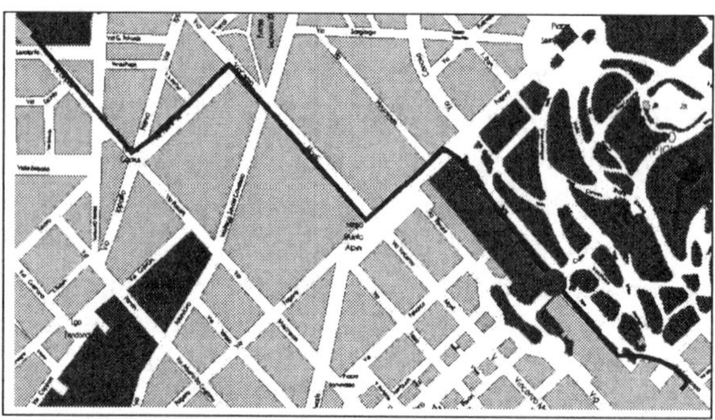

Figure 11: Example of cable route and automatically indicated sheath damage (red dot)

In comparison to temperature monitoring - where during operation of the cable an exceeding of the specified conductor temperature has to be avoided – in case of water monitoring the past behaviour of the cables has no influence on the occurrence of a failure. Thus the capabilities of the system are required very seldom. But in case of a failure the time related change of insulation resistance in the past may be interesting to track back to a possible reason.

4. Advantages and experiences

The technologies for Water Monitoring and RTTR system performance management are now fully developed and are being applied in many commercial system operations.

RTTR has more than 5 years field experience. Five systems have been installed in five different countries.
The cable system reliability has been enhanced through RTTR by monitoring the cable for identification of thermal constraints. Conversely water monitoring will detect external damage in case of water entering the cable and thus provide the possibility for a quick repair.

RTTR will enable the dynamic current rating capability of the system to be continuously available, providing full system loading flexibility against actual conditions.
The system has demonstrated the capability to calculate the desired status with a high accuracy, e.g. the conductor temperature is calculated within 1 K of its value when measured in field trials.

As an example of the gain in day to day rating, in one of the installations the admissible load increase over the safe design limits has been in the order of 60% (Figure 12, case A). By evaluation of the data, the RTTR provided for all sections of the cable link, the observed limit to the rating was not the cable itself but another part of the system. This part is currently up rated to the new limits and the result will monitored again by RTTR.
On a different installation the gain in "allowed" rating was instead of the order of 25% (Figure 12, case B).

A further advantage was experienced in a different installation where the nominal current carrying capacity was too high due to changed thermal environmental condition. In this case the moisture content of the soil changed due to a decrease of the level of ground water caused by temporarily construction measures (Figure 12 case C). The advantage of the use of the real time management of the line is that RTTR was able to prevented an over heating of the cable and thus a reduced risk of costly damages and repairs.

This is an indication also that RTTR will provide a better understanding of overall system loading and real operating conditions, which may be applicable when specifying future cable systems.

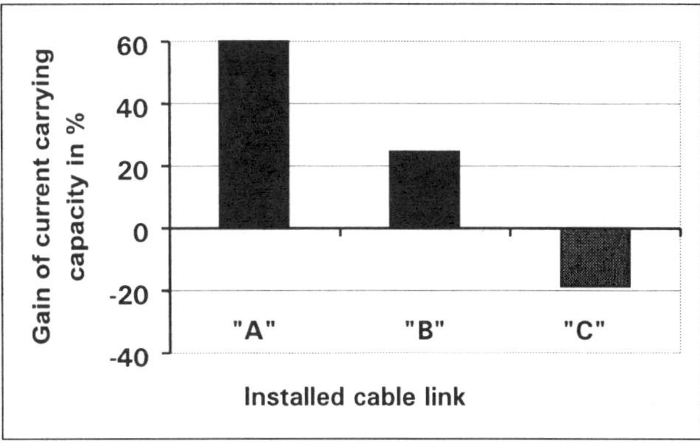

Figure 12: Different RTTR-controlled cable links
Installation "A" and "B" = Gain in power rating
Installation "C" = Avoidance of damage by over heating

In case of water monitoring utilities relay on this system if there is a very important power link to be operated with minimised risks of power cut off by maintenance, repair or short circuit.
The advantages of a water monitoring system are obviously if the cable is laid in a reconstruction area or in near city areas with plenty of cables close together where the risk of a third party cable damage is a likely one. Also cables laid in ground with high ground water level are a subject to consider about water monitoring.

Tests in laboratory and under field operation have shown the reliability of the complete water monitoring system when operated on a simulated three-phase HV cable system and in several practical applications. E.g. it is possible to detect and to locate failures some hours after ingress of water into the cable screen with an accuracy of approx. 1% even under AC noise.

Water monitoring reduces the repair cost and time and increases the mean lifetime and safety of the energy supply.

CONCLUSIONS

This paper describes two monitoring systems developed for high voltage cable systems:
1. the Real Time Thermal Rating (RTTR) Performance Management System to enable a more efficient cable system loading without exceeding thermal limits
2. the Water Monitoring for detection and location the ingress of water into the cable in the event of sheath damage

The following customer advantages can be pointed out:

RTTR Performance Management:
- transfer of the maximum permissible power without thermal overload of the cable link
- determination of thermal bottlenecks (remedy step)

- determination of thermal change of the cable link during the operation
- prediction of the temperature trend in case of power overload
- controlled power overload is possible, e.g. in case of emergency
- gathering of experience, improvement for new circuit design

Water Monitoring:
- minor additional investments
- continuous water detection under operating conditions
- no need of routine HV-tests of cable sheaths
- locating of failure with high accuracy of <1%
- immediate detection – that means less water penetration
- dry cables – that means max. life time

The RTTR and Water Monitoring System are now in service for several years. Several projects where both techniques are applied either separately or combined in the same cable, demonstrate and confirm that the systems are fully functional and are working as expected for the satisfaction of the customers.

Monitoring of HV cables will increase in future with respect to tougher demands for delivery quality of electric power as a result of deregulation of energy markets. Utilities will be able to improve their efficiency using the hereby presented techniques.

Pirelli contact:
Lothar.Goehlich@pirelli.com

References

1. P.R. Orrell: „Distributed Fibre Optic Temperature Sensing", Sensor Review, Vol. 12 No. 2, 1992, page 27-31, MCB University Press

2. F. Donazzi, R. Gaspari: „Method and System for the Management of Power Links", Cigre 1998, Session 21-203

3. R.G. Pragnell, G. Williams: „Real time thermal rating of HV cables", IEE Medpower Conference, 1998

4. IEC 853 „Recommendations for the calculation of cyclic and emergency current rating of cables", 1985

5. IEC 287 „Calculation of the continuous current rating of cables", 1994

6. M. Kuschel, R. Gaspari: „Real Time Thermal Rating (RTTR) – ein Managemant- und Optimierungssystem der Energieübertragung in Hochspannungskabeln", E-Wirtschaft, Heft 26, 2000

7. U. Glaese, E Gockenbach, L. Goehlich,: "Water Sensor as an Integral Part of a Cable Monitoring system", 9th ISH, Graz, Austria, 1995

8. W. Rungseevijitprapa et al.: "Principle and Practical Experiences with a Three-Phase Water Monitoring System for XLPE High-Voltage Cables", 11th ISH, London, UK: 23-27 August 1999

9. L. Goehlich et al.: "First Commercial Application of New Monitoring Systems Against Water Ingress and Overtemperatures in HV-XLPE-Cables", IEE Conference on Dielectric Materials; Measurement and Applications, Edinburgh, UK: 17-21 September 2000

10. L. Goehlich et al.: „Optimierung der Stromübertragung mit Hochspannungskabeln durch Monitoring – Wasser- und RTTR-Temperaturmonitoring in einer ersten Anwendung in Deutschland bei der MEAG in Bitterfeld", Elektrizitätswirtschaft, Heft 26, 2000

Cost-efficient XLPE Cable System Solutions

J. Karlstrand, G. Bergman and H-Å Jönsson

ABB Power Technology Products, Sweden

Keywords: XLPE, cable, OH-line, transmission, reliability, cost, rating

ABSTRACT

No complete, fair and impartial analysis is easy to make between different power transmission solutions. An advantage for one solution may be a disadvantage from one time to another, due to the prevailing situation. However, no other power transmission solution has had such a high rate of improvement as XLPE Cable System Technology during the last decade. Furthermore, improvements have been possible by decreasing the overall costs. Environmental focus, de-regulation of electricity markets and the improvements in technology of XLPE Cable Systems make them more attractive in situations where they were not even optional, in the past.

Due to large thermal time constants for direct buried XLPE Cable Systems, it is possible to optimise the rating to the real need of the network. Lean Cable designs with reduced insulation wall thickness and optimised conductor areas can enhance the attraction of XLPE Cable Systems from technical, economical and ecological points of view. The cost ratio between XLPE cable Systems and OH-line systems is decreasing and can reach 2:1 if using latest technologies. The land value of ROW (Right of Ways) can sometimes finance the costs for undergrounding.

With respect to reliability, the trend is positive for XLPE Cable Systems with pre-moulded accessories. One extra single-phase reserve can improve the reliability. However, double circuits do not improve the reliability significantly, compared to the 1-single phase reserve alternative.

INTRODUCTION

Overhead line systems have proven to be a technical, economical and reliable solution for power transmission and distribution in the past. Still they are therefore the primary choice for power transmission. During the whole 20th century several overhead line networks have been built and a large electrical infrastructure rely on these systems. No other competitive system solution was available when the construction rate was very high in mid of the 20th century. During this time and even later, there was little focus on the environmental aspects when new electrical networks were constructed. However, both deregulation of electrical markets and attention to environmental aspects, have changed the way to plan, purchase and construct next-coming generations of electrical infrastructures.

XLPE HV Systems are today less expensive than 10 years ago. Therefore modern XLPE Cable System Solutions can be competitive with respect to OH-lines, not only technically and environmentally, but also economically. Thanks to improved installation techniques, longer dispatch lengths and less number of joints, the investment cost ratio between OH-line systems and XLPE Cable Systems is about 1:2 to 1:3 for the 132 kV level.

To make a complete comparison between different transmission systems like OH-lines, oil-filled cables, XLPE cables, GIL (Gas Insulated Lines) etc, is a challenging task. The comparison must be performed in terms of a defined system problem rather than isolate facts from each other. Known differences in capacitance and inductance between the different transmission alternatives may play a role for the rating, load flow, short circuit rating, etc. However, an increase or decrease of these factors may be an advantage or disadvantage, depending on the prevailing situation.

In this paper, different XLPE Transmission System Solutions are therfore described in terms of technology, ecology and economy. Based on the growing experience of XLPE Cable Systems and their qualitative imrovements during the years, a condensed reliability study between OH-lines and XLPE cable Systems with pre-moulded accessories is presented. Other aspects are commented with regard to the following question: "Which consequences are present if another power transmission system solution is chosen?"

AC-DC Power Transmission, 28-30 November 2001
Conference Publication No. 485 © IEE 2001

TECHNICAL ASPECTS

During a 30 year time-period a lot of experience of XLPE Cable Systems has been gained. The design criteria in terms of rating margins, conductor temperature, insulation wall thickness etc. were established during a time when this technology was in its cradle. Thanks to improvements in material handling systems, cleanliness of insulation materials, pre-moulded accessories and a higher degree of understanding and experience, a critical review of traditional thinking is necessary.

Insulation wall thickness-lean cable design

During a 30-year period the electrical stress has increased. This was necessary due to moves up in voltage. The 400-500 kV level is now established and comercialised by many manufacturers. The electrical stress level has therefore increased with 3-4 times (from 4-5 to 15-16 kV/mm) when moving in voltage from 100 to 500 kV. At the same time service reliability records and n-factor tests from different manufactures show continuous improvements. However, the insulation wall thickness for HV XLPE Cables (52-275 kV) is now being reduced and type tests and commercial installations show satisfactory results. A stress level ≤ 10 kV/mm for the 132 kV level may soon become established.

The trend to decrease the insulation wall thickness opens up several opportunities that were only suitable for MV XLPE cables in the past [3]. These are;

♦ Longer dispatch lengths. Due to smaller outer diameter the number of reels can be reduced. The freight cost is therefore decreased as well. A 25-30 % increase in length is not unusual when decreasing the outer diameter from say 90 to 80 mm.

♦ Less number of joints. Of course the number of joints are decreasing. Not only material but also installation cost, civil work cost due to excavation etc. Even if pre-moulded joints are very reliable today, the less components the higher reliability of the system.

♦ Easier installation. New and mixed installation methods can be used in order to optimise the installtion both technically, economically and environmentally. Thanks to smaller dimension of cables, tehniques like e.g. plowing can be utilised. A lower weight will also ease the installation. With smaller outer diameter replacement of oil-filled cables installed in ducts or pipes (gas-pressure cables, HPFF-cables, LPOF-cables) is possible.

♦ Less thermal expansion/contraction. Modern XLPE accessories are taking care of the axial and radial thermal movements within the cable. However, extra insulation thickness does not necessarily confer additional reliability on the cable, due to the thermomechanical complications.

♦ Environment. Less material scrap in the end of use phase, less emissions during transportation and installation and less emissions during manufacturing. Somewhat lower magnetic fields, at least in trefoil configuration.

With regards to OH-lines there are solutions for shortening the phase distance. However, since air is the insualting medium there are certain restrictions. The driving force for decreasing the phase distance has been to reduce the magnetic fields from the lines. So far, the tendancy is that compact OH-poles and structures are more costly.

The GIL has a $SF6/N_2$ – mixture as insualtion medium. There are so far, small opportunities to make the pipes smaller and more compact as long as there is no better insualtion alternative. The potential for cost reduction of GILs is probably to find cheaper enclosures, transportation and installation. The trend to enhance the electrical field stress and making a lean XLPE cable system design will continue and the potential is obvious from technical, environmental and economical aspects.

Rating and overloading

The rating of a XLPE cable system is historically determined according to IEC 60287 and consequently 100% continuous loading. What not always is taken into account in the design phase, is the load shape or the real use of the cable. By taking this into account not only a reduced insualtion thickness can be achieved, but also a reduced conductor area. An example of a 132 kV XLPE cable with an old design (1000 mm^2) and a new design (630 mm^2), is given below in Figure 1.

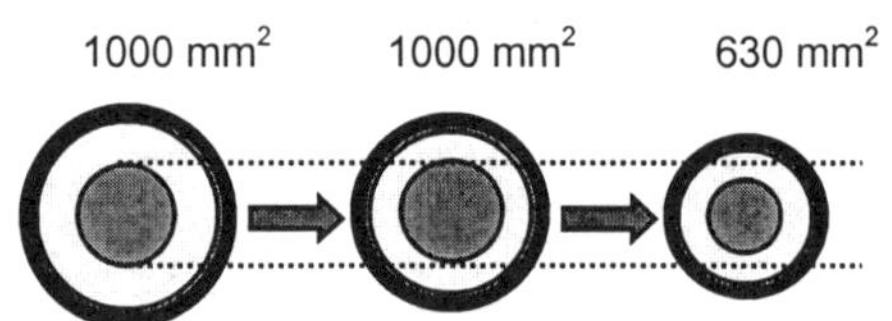

Figure 1. Optimizing the cable design with regards to the real use or load shape decreases the outer diameter.

According to Table 1 the following effects are outlined:

Table 1. Comparison of different sizes of XLPE cables. U=132 kV, copper conductor. Direct buried at 1 m depth, T=15 °C. Thermal resistivity 1 Km/W. Trefoil formation – both-ends bonding.

	1000 mm² Old design	1000 mm² Present design	630 mm² Present design
Insulation thickness [mm]	17.5	13	13
Rating at 70 °C [A]	850	840	720
Rating at 90 °C [A]	980	960	825
Overload during 12 hours starting from 70 °C up to 90 °C [A]	1120	1120	930
Weight [kg/m]	14.5	13.5	9.5
Outer diameter [mm]	89	79	69
Length on K30 reel [m]	850	1250	1550
Number of joints per reel	1.00	0.68	0.55

The current carrying capacity is decreasing margnially if reducing the insulation wall thickness. If also decreasing the conductor area when taking into account a changing load, the current carrying capacity is still almost the same for 630 mm².

Decreasing the conductor area is possible due to a considerably longer thermal time constant of a buried cable system than a corresponding OH-line system. The customary way of designing OH-lines is to select a conductor area so large, that normal loads due to different system configurations can be handled without exceeding the highest permissible temperature. Most of the time an OH-line is therefore operated at 40-60% [1] of its maximum load. The margin up to 100 % is used for the expected or unexpected extra loads, that has to be taken up by the system for various events.

A direct buried XLPE Cable system has a rather stable rating, regardless the temperature in air. The soil temperature is varying only +/- 5 °C from an average during a year in most cases. However, the soil thermal resistivity may be changing and therefore temperature monitoring systems are increasingly used for underground cables. As can be seen in Figure 2, the XLPE cable system has a flatter loading profile than an OH-line. Due to the increasing use of computers and air-conditioning equipment the electricity demand tends to increase and be smoothed out during the year even more. An XLPE Cable system matches such a power demand (load profile) very well. Figure 2 below is adopted to Nordic conditions and may be somewhat different in countries having higher temperatures in the wintertime.

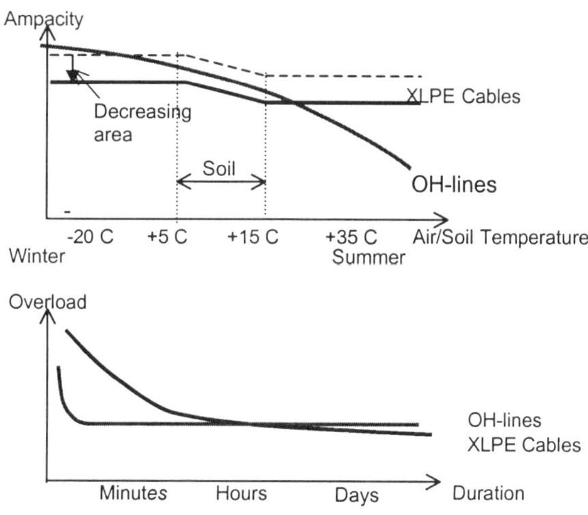

Figure 2. The dashed line symbolizes a higher rating if taking into account the real load shape of the system. The cable rating is not due to the air temperature. Instead the rating remains unaffected during summer or winter.

When taking varying loads into account (dashed line), one also notices the possibility to decrease the conductor area. The cable has normally a high short overloading capacity. The base load can be carried without overheating the cable. The magnitude and the duration of the overload (or a change from base load to a higher load) is however due to the actual conductor temperature before the overload starts.

An XLPE Cable System 132 kV/1000 mm² Cu and a corresponding OH-line system AlMgSi 3x1x593 mm², reflect completely different overload capabilities. To reach 120 % load it takes about 18 hours if originating at 50 % load. To reach 120 % overload for the OH-line takes about 14 minutes, i.e. a factor of 75 longer duration for the XLPE cable System. A surrounding temperature of 15 °C is used.

In Figure 3 a comparison of 100 % rating is made between OH-lines and XLPE Cable Systems. The rating/circuit is decreasing for a XLPE Cable System but is constant for a OH-line system. Anyway the former discussion regarding the overload capability of XLPE cable Systems has to be incorporated in a fair comparison.

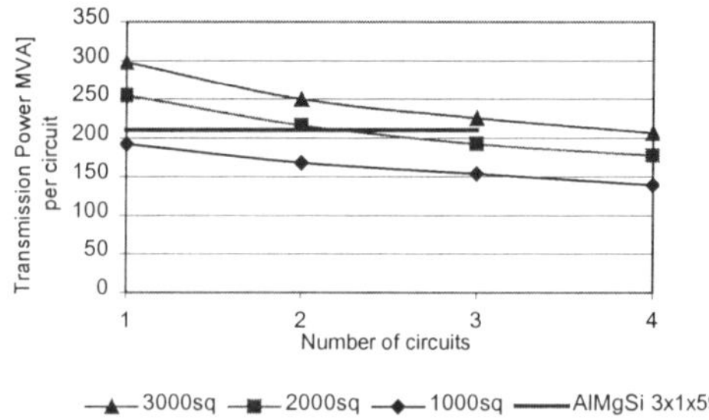

Figure 3. Typical 100% rating of XLPE cable Systems and an OH-line system for 132 kV. Power per circuit [MVA/circuit]

ECOLOGICAL ASPECTS

The manufacturing of transmission systems may be an important factor, from an environmental point of view. However, most concern has been put on the effects regarding installation and the maintenance required. Firstly, the manufacturing of XLPE cables is becoming more environmentally friendly. By going from lead to Al- or Cu-laminated sheaths as well as from PVC to PE-coverings, the trend towards environmentally friendly systems is promising. In fact XLPE Cables, dry type terminations and joints make it possible to build transmission systems, that have insignificant impact on the environment.

The required Right-Of Way (ROW) is smaller for a cable transmission system than a corresponding GIL- or OH-line system. Most GIL-systems are installed in tunnels or laid on the ground. OH-line systems have their visual impact on the environment, which sometimes in combination with the magnetic field debate obstruct permissions for new or upgraded lines. With respect to environmental legislation, the same procedures apply for re-approval of permits as for new approval of permits.

In principle, the following environmental characteristics recognize a modern XLPE Cable system:

- Smaller dimensions means less material use during manufacturing and less scrap after the end of life.

- Low magnetic fields. GIL-systems have even lower magnetic fields. OH-line systems may have a higher magnitude but not necessarily. However, the magnetic field has a very slow decay.

- No maintenance. Dry XLPE Cable Systems do not need much maintenance. This has a positive effect on both cost and environment. Both GIL and OH-line systems need some kind of regular maintenance.

- An XLPE Cable System is free from emissions to air and water. No oil leaks, noise, etc.

- When direct buried, the XLPE Cable system (or cable system in general) does not disturb the natural scene. In other words there is no visual impact.

In summary, the smaller dimensions of XLPE Cables lead to positive environmental effects. The trend to improved insulation characteristics (lean insulation) will continue. It is unclear how the trend will develop for GIL and OH-line systems. No other transmission system has so far such long and good records, concerning direct burial installation. The environmental legislation and consequences due to electricity de-regulation will continue. XLPE Cable Systems with dry type accessories are suited for such a trend.

ECONOMICAL ASPECTS

The total cost of transmission system consist mainly of material cost (cables and accessories), installation, civil work, maintenance and losses. The two latter costs are normally re-calculated according to the present value method. However, it is extremely difficult to exactly calculate the loss costs. It is however, clear that a XLPE Cable system has lower losses than a corresponding OH-line system. However, the loss costs for a GIL-system is considerably lower but since the GILs mostly are rather short the loss cost will not affect the total cost since the investment is very high. The maintenance cost for OH-lines in Sweden lies between 500 to 1000 USD/km and year for transmission voltages.

A more interesting approach is probably to look at the values of land and buildings in existing ROW's. In many cities, like in Stockholm, the land is becoming more valuable. If using the ROW for construction of new buildings, indications show that selling the land of the ROW can finance a whole XLPE Cable Project. If undergrounding takes place, the tax values for buildings will increase and the community may gain more money, if selling the land. The urbanisation is still ongoing and this trend will probably continue. When the permission period is ended, it might be possible to find new interesting solutions for undergrounding.

The total investment cost (cable, accessories, installation and direct burial installation in a trench) of a 132 kV XLPE Cable System compared to an OH-line system is shown in Table 2 below.

Table 2. Investment Costs for 3 different 132 kV XLPE Cable Systems and 1 OH-line AlMgSi 3x1x593 mm². The length of the system is 10 km-1 circuit and direct buried.

	OH-line	1000 mm² Old std	1000 mm² New std	630 mm² New std
Cost kUSD/km	90	230	215	183
Cost Ratio	1	2.56	2.39	1.83

If calculating with 750 USD per km for the maintenance and say 50 USD per km for the cable the capitalized cost during 40 years for 10 km becomes about 13 and 1 kUSD/km, respectively. The losses are as said before very unfair to take into account. However, a thumb role is that the XLPE Cable System has about 10 % lower losses than a corresponding OH-line system.

Instead it is more interesting to estimate toady's and future cost for environmental impacts, cost for application of permits and land values. The cost ratios decrease to 2.24, 2.10 and 1.78 respectively, if taking only maintenance into account. Assuming that 30 m of the ROW can be used for new building constructions, parking places etc. Assume further that only 1/5 of 10 km can be used for new constructions. To balance the whole undergrounding cost, the square meter price would lay between 30-40 USD. That seems to be a reasonable price.

RELIABILITY ASPECTS

The performance of XLPE Cable Systems has been improved a lot during 30 years. Thanks to a cleaner insulation material, improved material handling, pre-molded accessories and a lot of improvements in processing, the trend towards even better system performance will continue. In failure statistics from ABB and utilities in Japan and Sweden, one can see the trend of decreasing failure rates. On the other hand the electrical stress is tripled since the beginning of 70's.

From the failure statistics the total amount of XLPE Cables from ABB and the utilities is almost 17000 km*year 3-phase. From the Japanese utility there are no faults in service to insulation defect. From the Swedish power utility there was one early fault in 1976. Table 3 summarizes some data.

Table 3. Service statistics from ABB, Swedish and Japanese power utilities. The voltage range is 110-170 kV.

	ABB	Utility in Japan	Utility in Sweden
Number of km 3-ph installed	1250	350	157
XLPE Cable in km*yr 3-ph	13500	3000	600
Number of pre-molded joints	2000	645	--
Pre-molded Joints in yr*pcs	12500	≈3200	--
Number of terminations	7600	378	--
Terminations in yr*pcs	67000	≈2300	--

At least 90% of the total amount of XLPE cables and accessories are installed in the voltage range of 110-170 kV.

Pre-molded accessories show very good performance in this study. Only 1 failure in a pre-molded joint from a Japanese utility has been observed. The total time of service experience is about 10-12 years. A number of 4 failures in terminations have occurred on the 132 kV level. Because of a large number of terminations and few faults in pre-moulded joints, the failure rate for accessories show very low values.

Even if the failure rate has decreased the last decade a mean value, which includes the accumulated failures in the past, has been used in calculations. For the cable 0.072 faults/100 km*yr has been used. For premolded joints, 0.00043 faults per 100 pcs/year and for terminations 0.0061 per 100 pcs/yr have been used. Sensitivity analysis has been performed in order to see the influence of changing repair times, failure rates in accessories etc.

With regards to OH-lines most failures are very short (< 1 s), mostly caused by surge strikes. However, permanent faults are mostly caused by bad weather conditions like, ice, snow, wind etc. In other words the influence from the surrounding plays an important role in this sense. The potential for XLPE Cable Systems, which are almost protected from the surrounding wheather conditions, is obvious. The failure rate is decreasing. For OH-line systems failure rates are about the same during the last decades so the possibilities to even more decrease the failure rate is not to be expected [2]. A lot of good development work has already been done.

If connecting a 132 kV XLPE Cable system into an existing OH-line system one may ask whether a single cable reserve or a double circuit improves the reliability significantly or not. From a loading point of view there may of course be a reason for a double circuit, however. In Figure 4 a XLPE cable system is installed in a OH-line network. The cable length is a parameter that may be varying.

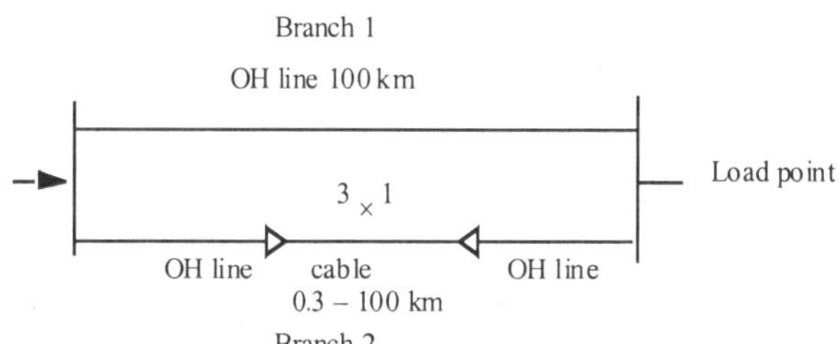

Figure 4. OH-line and 132 kV XLPE Cable System network. The cable line can be a single circuit, single circuit + single phase reserve or double circuit.

Base Case – 1 single circuit

Evaluting the base case with 1 single circuit shows an increased outage duration (t) with cable length but a decreased unavailability (U) and failure rate(f). See Figure 5.

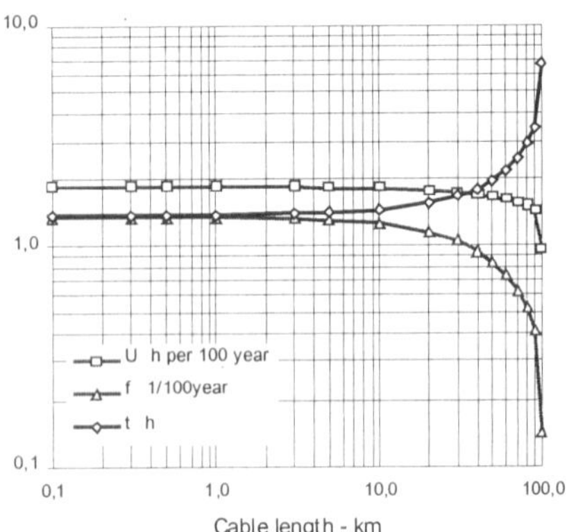

Figure 5. Meshed 132 kV network- base case with 1 single circuit.

The break-point of the curves at the last step 90-100 km depend on the fact that the overlapping events – OH-line failure and OH-line maintenance – disappears when the cable is extended to 100 km and completely replaces the OH-line in the branch containing the XLPE cable.

Single phase reserve and double circuit

In these cases it is assumed that the cable reserve or the extra double circuit are equipped with jumpers and that it takes about 6 hours to re-arrange the system. In other words the repair time for the cable (60 hours) is replaced by a re-arrangement time of 6 hours. In Figure 6 the total unavailability versus the cable length can be seen.

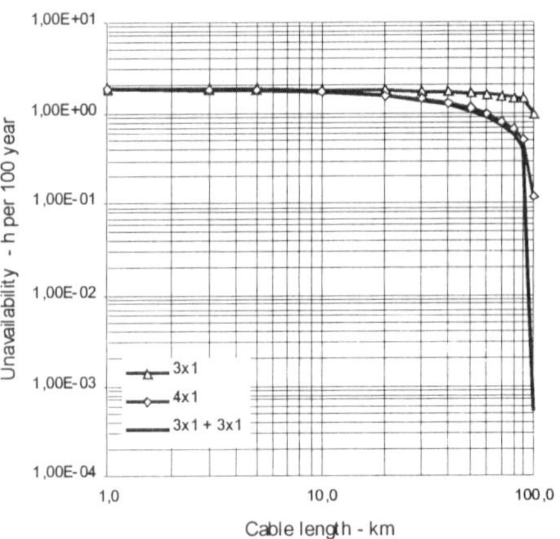

Figure 6. Unavailability for single circuit (3x1), single-circuit with single-phase reserve (4x1) and double circuit (3x1 + 3x1).

For cable lengths below 20 km (of the total distance 100 km) the cable failures have little effect on the total unavailability. With a single phase cable reserve the unavailability is significantly reduced for cable length above 30 km (of 100 km). Two complete parallell 3-phase cables are, however, not feasible to improve reliability.

CONCLUSION

No complete, fair and impartial analysis is easy to make between different power transmission solutions. An advantage for one solution may be a disadvantage from one time to another, due to the prevailing situation. However, no other power transmission solution has had such a high rate of improvement as XLPE Cable System Technology during the last decade. Furthermore, improvements have been possible by decreasing the overall costs. Environmental focus, de-regulation of electricity markets and the improvements in technology of XLPE Cable Systems make them more attractive in situations where they were not even optional, in the past.

REFERENCES

1. Jung M, Peurian S. LE.,Optimising high voltage underground cable links.

2. Fault Statistics, 1986-1998 from Nordel (Nordic co-operation for fault statistics in electrical power system 40 400 kV.

3. Dellby et al, XLPErformance, 2000, ABB Review(4).

GAS-INSULATED LINE (GIL) OF THE 2ND GENERATION

H. Koch

Siemens Aktiengesellschaft, Germany

Keywords:
Gas-insulated transmission line (GIL), laying methods, tunnel-laid, directly buried, gas mixture, applications

INTRODUCTION

With the commissioning of the 500 m long gas-insulated transmission line (GIL) at the Geneva Airport, Siemens is the first manufacturer of the second generation GIL world-wide. This project concluded successfully the recent development works on this future transmission technology. The redesign of GIL brought a reduction of costs by more than 50 %. The main changes were replacing pure SF_6 by a gas mixture of N_2 and SF_6 as insulating medium and by adopting pipeline laying methods into the laying process.

Based on the now more than 30 years of experience in gas-insulated high voltage technology the redesign of the first generation GIL installed in 1974 at the Hydro Power Plant Wehr in Germany was started in 1995 with a first feasibility study together with EDF. In the following for both types (directly buried and tunnel-laid) a prototype set-up was built at IPH, an independent test lab in Berlin. The development and test program have been carried out in co-operation with three major German utilities. The main intentions of the tests were to simulate the electrical and mechanical stresses of a lifetime of more than 50 years and to prove the assembly and laying process under on-site conditions. Both types of GIL, directly buried and tunnel laid, passed the tests successfully without any technical problem and without any time delay. The GIL of the second generation is now certificated. The assembly and laying was carried out under real in-site conditions and are proved laying techniques. The GIL is qualified as a long distance power transmission system laid in a tunnel or directly buried, so that distances of 100 km and more are possible to build.
The main advantages of the GIL compared to other transmission systems, overhead lines and cables are the capability of very high power transmission ratings due to the low resistive losses, very low electromagnetic field, no risk of fire or external harm and no need for reactive power compensation.

AC-DC Power Transmission, 28-30 November 2001
Conference Publication No. 485 © IEE 2001

In this article the main results of the development will be outlined and the realization of the first project in Geneva, Switzerland, will be explained.

CONTENT

1. Experience
2. Technique
3. Laying Methods
4. Safety and Gas Handling
5. Application at Geneva
6. Conclusion
7. Literature

1. EXPERIENCE

Gas-insulated transmission lines (GIL) are world-wide used since more than 25 years in the voltage range of 145 up to 550 kV. Two different types of assembling are in parallel use: the flanged GIL, and the welded GIL. Flanged GIL are using flanges and bolts with sealings to connect the single sections of the transmission line. Welded GIL are using orbital welding to connect these sections. From both systems more than 120 km are installed world-wide. So far no major failure has been reported for more than 25 years of operation.

Typical applications of GIL until today are links within power plants to connect the high voltage transformer with the high voltage switchgear, or with cavern power plants to connect high voltage transformers in the cavern with the overhead line on the outside, or to connect gas-insulated substations (GIS) with overhead lines, and finally to use it as a bus duct within gas-insulated substations. The applications are carried out under all different climate conditions from the low temperature in Canada to the high ambient temperatures in Saudi Arabia or Singapore, or severe conditions in Europe or in South Africa. The GIL transmission system is independent from environmental conditions because the high voltage system is completely sealed inside a metallic enclosure.

The reason why in the past GIL was not used for long transmission distances was the high price per kilometre. To cut down the price by at least 50 % was the goal of the development program of Siemens for the creation of the 2nd generation GIL.

This breakthrough in cost and the successful passed through all the long duration testing of tunnel laid and directly buried GIL makes the now called 2nd generation GIL a very interesting transmission system for high power transmission over long distances, especially if high power ratings are needed.

The 2nd generation GIL has been built for PALEXPO at the Geneva Airport in Switzerland. Since January 2001 the GIL is in operation as part of the overhead line connecting France with Switzerland. In this project it has been successfully shown, that the new laying techniques are suitable to build also very long GIL transmission links of several tenth of kilometres within a suitable time schedule. The second project is now under construction in Thailand as a connection to the 550 kV transmission net, and a third project will start with construction in October 2001 in India and will add more experience to the 2nd generation GIL.

2.　TECHNIQUE

The GIL of the 2nd generation is suitable for high voltage transmission systems up to 550 kV and transmission currents of up to 4000 A and a rated short time current of 63 kA. The layout and design criteria are in accordance with the standard IEC 61640 "Rigid high-voltage, gas-insulated transmission lines for rated voltages of 72.5 kV and above" [1].

One step on cutting the cost for the 2nd generation GIL was to create a modular system with only four different modules to build a transmission line. One basic module is the so-called straight unit which is shown in figure 1. One straight unit has a length of 100 to 150 m, depending on the temperature range in operation. The straight unit will be assembled by GIL segments of 11 to 14 m transport length. The connection of the enclosure pipe (1) and the conductor pipe (2) is done by orbital welding. Each 100 to 150 m the conductor is fixed towards the enclosure by a conical insulator (3). To allow the movement of the conductor according to the temperature rise during operation. At the end of each straight unit a plug and socket sliding contact system (5a, b) will take care of the movement of the conductor. Post type insulators (4) are able to slide inside of the enclosure.

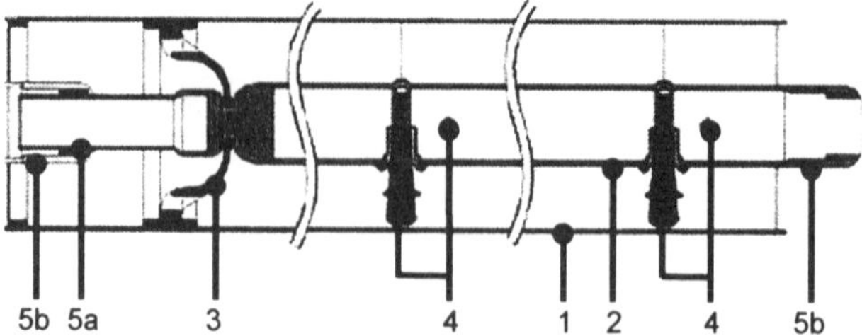

Fig. 1:　Straight unit
1	enclosure	4	support insulator
2	inner conductor	5a	male sliding contact
3	conical insulator	5b	female sliding contact

This straight unit can be bended using the elasticity of the aluminium material with a minimum bending radius of 400 m. This allows to follow the landscape or the route without using solid angle elements. If the straight unit will be laid directly into ground the outer enclosure will be covered with a corrosion protection coating.

If a directional change cannot be met with the elastic bending an angle element which covers angles of 4 to 90° can be added as a second type of basic module. The angle element is connected by orbital welding with the straight unit.

The third component is the compensator of the enclosure. In the tunnel-laid version or in an underground shaft the enclosure of the GIL is not fixed, so it will extend with the thermal heat-up during operation. The thermal extension of the enclosure will be compensated by the compensation unit. If the GIL is directly buried in the soil, the compensation unit is not needed. Because of the weight of the soil and the friction of the surface of the enclosure the GIL is fixed in the ground.

The fourth and last basic module used is the so-called disconnecting unit which will be used any 1 to 1.5 km to separate the GIL in gas compartments. Also the disconnecting unit is used to carry out sectional high voltage commissioning testing.

An assembly of all these elements as a typical setup is shown in figure 2.

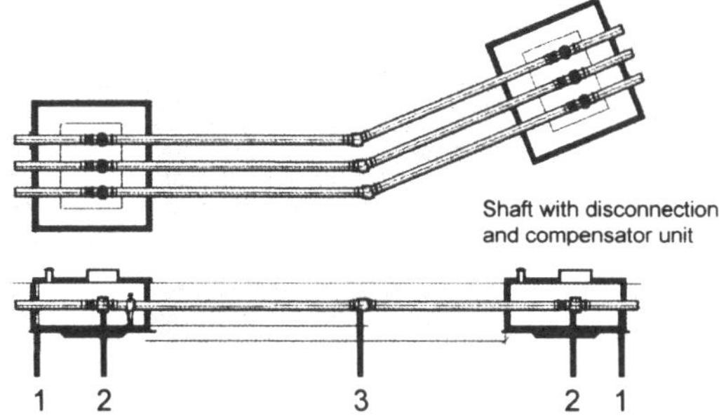

Fig. 2: Directly buried GIL system components

Figure 2 shows a section of a GIL between two shafts (1). The underground shafts are housing the disconnecting and compensator unit (2). The distance between the shafts is between 1000 to 1500 m and represents one single gas compartment. Also in the middle the directly buried angle (3) unit is shown as an example. Each angle unit has also a fix point, where the conductor is fixed towards the enclosure.

3.   LAYING METHODS

The GIL can be laid above ground on structures, in a tunnel or directly buried into the soil like an oil or gas pipeline. The all over costs for the directly buried version of the GIL in most cases is the least expensive version of GIL laying. Usually, for this laying method a certain accessibility for working on site is necessary, so that this directly buried laying will generally be used in open landscape crossing the countryside, similar to overhead lines.

If enough space is available laying the GIL into a tunnel will be the most appropriate method. This tunnel laying method will be used in cities or metropolitan areas to cross a river or to connect islands. A new way of application is now available with the GIL by using traffic tunnels, for example in the mountains, where existing or new build street or railway tunnels can be used to insert the GIL.

Above ground structures also may be used to fix a GIL just above ground or using a bridge to cross a river. Because of the rigid metal enclosure structures are only needed in distances of 20 to 40 m, depending on the mechanical layout of the GIL.

Directly buried GIL

The most economical and fastest method of laying cross-country is the directly buried GIL. Similar to pipeline laying the GIL will be continuously laid as an open trench with the nearby preassembly site to reduce the transportation of GIL units. With the elastic bending of the metallic enclosure the GIL will flexibly adapt to contours of the landscape.

In the soil the GIL is continuously anchored, so that no additional compensation elements are needed. In figure 3 the laying procedure of a GIL is shown.

Fig. 3: Laying procedure of a directly buried GIL

Tunnel laid GIL

Tunnels are usually the shortest connection between two points and therefore reduce the cost of the transmission systems. Modern tunnelling techniques have been developed in the last years with improvements in drilling speed and accuracy. So-called microtunnels of a diameter of about 3 m are a today economical solution in cases when directly buried GILs are not possible.

In figure 4 a view into a tunnel shows two systems of GIL as realized at PALEXPO at Geneva Airport in Switzerland. The tunnel dimensioning in this case was 2.4 m wide and 2.6 m high. In case of a round tunnel a diameter of 3 m will be acceptable.

Fig. 4: Tunnel laid GIL for voltages up to 550 kV

In both laying methods the elastic bending of the GIL is used as it can be seen in figure 3 for the directly buried version and in figure 4 for the tunnel laid GIL. The minimum acceptable bending radius is 400 m.

4. SAFETY AND GAS HANDLING

The GIL is a gas-filled high-voltage system. The gases used, SF_6 and N_2, are inert and non-toxic. The filling pressure of the GIL is with 7 bar relatively low. The metallic enclosure is solidly grounded and delivers because of the wall thickness of the outer enclosure high personal safety. The automized orbital welding process makes sure that the connections of the GIL segments are gas-tight for the life time.

In case of an internal failure which is very unlikely the metallic encapsulation withstands the internal arc so that no influence is given to the surrounding. In arc fault tests in a laboratory it was proven that no burn-through occurs and also the increase of internal pressure during an arc fault is very low, because the gas compartments are very large. Even under the arc fault condition usually no insulating gas will be released into the tunnel or into the soil for directly buried GIL.

For the gas handling of the N_2/SF_6 gas mixture devices are available for emptying, separation, storing and filling of the N_2/SF_6 gas mixtures. In figure 5 the devices for the gas handling are shown.

From the GIL system with a vacuum pump (1) the gas will be pumped out of the system, then separated (2) with a filter into pure SF_6 and a rest of the N_2/SF_6 gas mixture. The rest of the N_2/SF_6 gas mixture will have an SF_6 content of a few percent (1 - 5 %) so that it can be stored (3) under high pressure up to 200 bar in standard steel bottles. Three sets of steel bottles as shown in the photo (3) can take the gas content of a 1 km section for storage. The pure SF_6 will be stored (4) in fluid condition also under high pressure. To fill or refill the GIL system a gas mixing device (5) is available, including a continuous gas monitoring system for temperature, humidity, SF_6 percentage, and gas flow. The gas mixing device is using pure N_2 (6), SF_6 (4) and the stored gas mixture with a few percent of SF_6. The mixing device will adjust the chosen N_2/SF_6 gas percentages used in the GIL, e.g., 80 % N_2.

With these gas handling devices a complete cycle of use and reuse the gas mixture is available. In normal use the SF_6 and N_2 will not be separated completely because the gas mixture will be reused again. A complete separation into pure SF_6 as used e. g. in Gas-Insulated Substations (GIS) can be done at the SF_6 manufacturers fulfilling the IEC 60480 [2] and IEC 61634 [3] requirements.

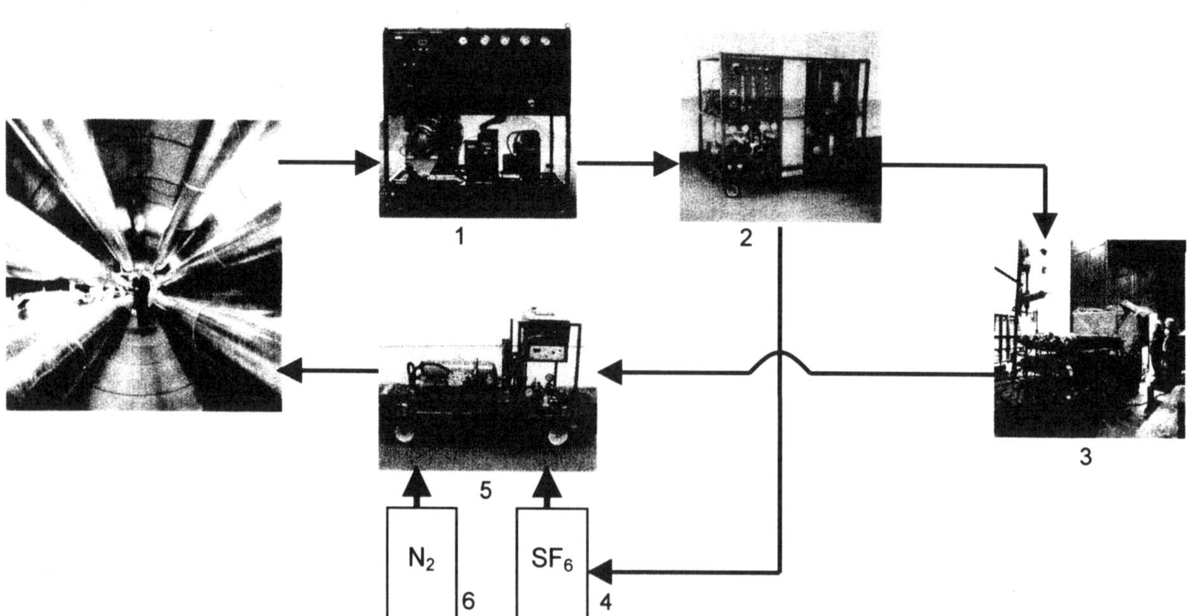

Fig. 5: Gas handling devices

5. APPLICATION AT GENEVA

The first application of the 2nd generation of GIL has been carried out between September and December 2000. Only in 3 months erection time the overhead line was brought underground into a tunnel (see figure 4) and connected to the net again in January 2001. In figure 6 a site view shows the delivery of GIL transport units to the preassembly area.

Fig. 6: Delivery of transport unit to the preassembly area

The preassembly tent has been placed directly under the overhead line and was positioned directly above the shaft connected to the tunnel right under the street (see figure 6). The narrow space between an airport access road on one side and the highway to France on the other side makes it only possible to use the space directly under the overhead line for the site works. The laying procedure had proven to be applicable also for long distance connections. With the site experience the productivity for assembling the GIL sections could be very much increased from 2 connection per shift and day to 4 connections per shift and day. These are very positive experiences for future projects, especially if very long distances for GIL links have to be carried out. The highly automated laying process had proven to deliver a very constant quality over the complete laying time so that the commissioning of the system could be carried out without any failure.

6. CONCLUSION

Environmental requirements by national laws about landscape protection or health of the public will make it necessary to build underground transmission systems in special areas with special requirements. Limiting values for the use of overhead lines could be magnetic field values, or just invisibility of the overhead line, or limited space. The GIL is one technical solution to allow the transmission of electricity under ground at high transmission ratings. The GIL reaches the lowest magnetic field values of today available high voltage and high power transmission systems. The GIL has also the lowest transmitting losses and has a high safety level towards the surrounding because of the metallic enclosure.

Installed in line with the overhead line the GIL can be operated in the same way as the overhead line, for example switching the overhead line or using the autoreclosure function. Because of the low capacitive load a compensation of the phase angle is not needed below 200 to 300 km of GIL system length.

The high safety level of the GIL makes it also possible to use public accessible tunnels like street, highway or railway tunnels to implement the GIL system.

The cost reduction of the GIL compared to the first generation GIL and the increased productivity in laying of GIL directly buried or in a tunnel makes the GIL to a high power rating underground transmission solution for the future.

7. LITERATURE

[1] IEC 61640 "Rigid high-voltage, gas-insulated transmission lines for rated voltages of 72.5 kV and above"

[2] IEC 60480 "Guide to the checking of sulphur hexafluoride (SF6) taken from electrical equipment"

[3] IEC 61634 "High-voltage switchgear and controlgear - Use and handling of sulphur hexafluoride (SF6) in high-voltage switchgear and controlgear"

RECENT ADVANCES IN SPACE CHARGE MEASUREMENT AND EXTRUDED HVDC CABLE DESIGN

P. Mirebeau[1], A. Toureille[3], R. Coelho[1], H. Janah[1], J.Matallana[2], P. Hourquebie[2], S. Agnel[3], D. Sy[1]

[1] : Nexans, France
[2] : CEA/Le Ripault, France
[3] : Université Montpellier 2, LEM, France

INTRODUCTION

The design of high power HVDC cables is a challenge. The field and space charge distributions over the insulation thickness, which somehow control the cable behaviour and its life expectancy have never been measured up to now and the cable designer can only rely on models and calculations.

The calculation of the field and space charge distribution in the permanent state is a classical problem that has been treated in many publications (1 – 8). On the other hand, the transient state has been scarcely considered, see Law (1) , McAllister et al (3). Recently, Aladenize et al (9) extended the pioneering work of Law to the case of a conductivity that depends both of temperature and field through a Poole-Frenkel mechanism.

On the other hand, space charge and field distribution on flat samples have been studied and measured since the 80's using methods such as the Laser Induced Pressure Pulse (LIPP) (10), the Pulsed Electro Acoustic (PEA) (11) and the Thermal Wave (TSM) (12).

In this paper, we relate for the first time space charge measurements made on cables while they are energised and loaded. We used the thermal wave method for this measurement.

EXPERIMENTS

Experiment were performed on a prototype cable. Insulation system was extruded through a triple head using state of the art machinery. Electrodes were made of a proprietary semiconductive formulation and insulation was High Pressure Low density polyethylene.

Space charge measurements were performed in both the isothermal and in the "loaded" stages. Samples used for the experiments were 15m long.

Conductivity measurements were performed to provide input data to calculate the intrinsic space charge in the "loaded" stage

Cable dimensions were : conductor diameter 11,39mm, inner semicon thickness : 0,52mm, Insulation thickness : 5.34mm, outer semicon

thickness : 0,71mm. A copper tape screen and a sheath were screening and protecting the insulation system.

Space charge measurement

Principle of space charge measurements by the Thermal Step Method. The space charge measurements described in this paper have been made by the thermal step method (TSM), which is based on applying a temperature step to an insulating sample and on measuring a capacitive current response. This signal is a function of the charge distribution within the dielectric. As the charge is not evacuated, the evolution of the electrical state of the sample can be surveyed. TSM is easily adaptable to various geometries and therefore applicable both to flat dielectric samples and to insulating components with thickness up to 20 mm (cables, electrical machines windings, circuit boards, capacitor films).

The technique for generating the thermal perturbation in a *cable* consists of using a ring-shaped radiator adjoined to the outer semicon of the cable, in which the temperature step is produced by the arrival of a cold liquid (Fig. 1). This allows to analyse the part of insulator placed below the radiator. By moving the radiator along the cable, a cartography of the space charge both in the radius direction and in the length direction can be achieved.

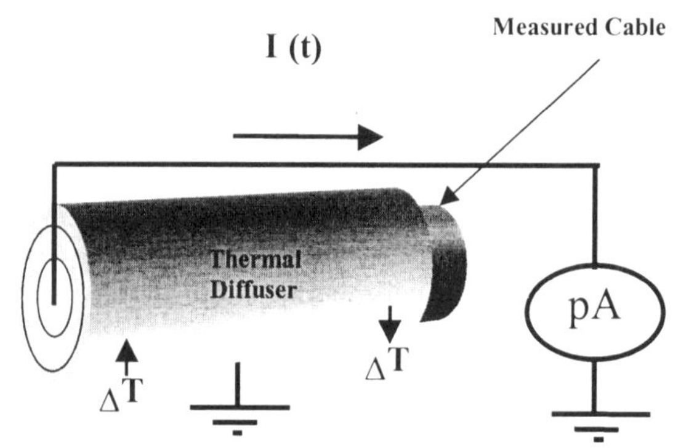

Figure 1: Principle of TSM

The obtained thermal step signal is a function of the electric capacitance of the measured region C, of the electric field repartition within the material in the radius direction $E(r)$, and of the temperature distribution across the insulator $\Delta T(r,t)=T(r,t)-T_0$ (T_0

AC-DC Power Transmission, 28-30 November 2001
Conference Publication No. 485 © IEE 2001

is the temperature of the cable before applying the thermal step):

$$I(t) = -\alpha C \int_{R_i}^{R_e} E(r)\frac{\partial \Delta T(r,t)}{\partial t}\,dr \qquad (1)$$

where α is a constant of material related to the sample's contraction (or expansion) and to the variation of its permittivity with the temperature, and R_e and R_i are the external and internal radii of the insulator. The electric field dependence on the space charge density is given by the Poisson's equation :

$$\rho(r) = \varepsilon\left(E(r)/r + \partial E(r)/\partial r\right). \qquad (2)$$

Space charge measurement under applied DC field For *measurements under DC field*, the set up described in *Figure 1* presents two major disadvantages :
- the current amplifier must not be in contact with the high voltage;
- if a set up with the current amplifier placed between the sample submitted to the high voltage and the ground is used (as for, e.g., conduction current measurements), the conduction and the polarization currents are likely to mask the thermal step current.

A solution to these problems is a "compensation sample", of identical dimensions as the measured specimen and placed oppositely to this latter.

By connecting one side of the compensation sample to the current amplifier and the other to the measured specimen via en electrode, a "double capacitor" is obtained. A voltage can then be applied to the cable core, and the thermal step to the measured sample. The TSM current will be acquired via the "compensation sample". Using two identical specimens offers the considerable advantage of compensating the polarization and conduction currents which can occur under high fields. The measured current is then due solely to the internal field of the measured sample.

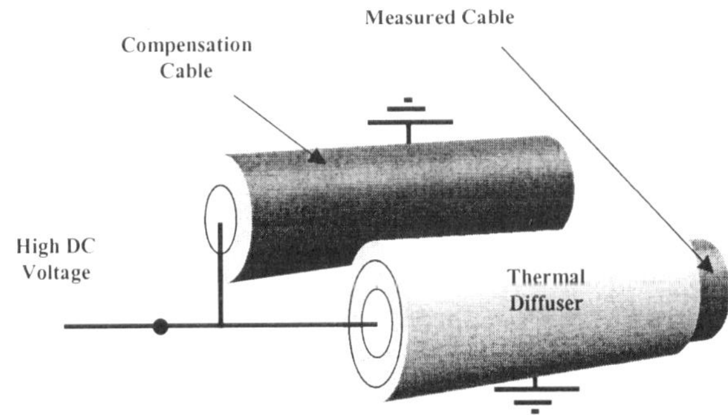

Figure 2.a : "Under field" space charge measurement by TSM on a power cable

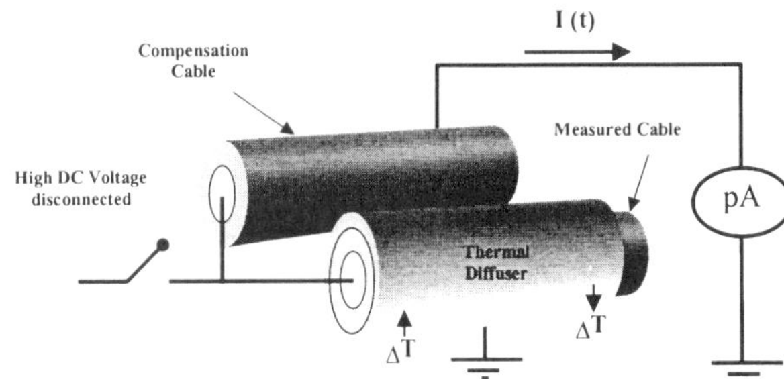

Figure 2.b : "Under field" space charge measurement by TSM on a power cable

The measurement "under applied field" is made in two steps, as shown in *Fig. 2* :
- during the polarization time, the voltage is applied to the middle electrode (cable core) and the current amplifier is short circuited. Thus, the two samples constitute two identical capacitances placed in parallel with respect to the voltage source (*Fig 2.a*).
- during the measurement, in order to prevent the transport of influence charges at electrodes via the HV generator rather than via the current amplifier, the power source has to be disconnected (*Fig 2.b*). The thermal step current is then measured by exciting thermally the studied specimen (in contact with the thermal diffuser), while the current amplifier is connected to the compensation sample. This time, the two samples are in series with respect to the current amplifier, and the short-circuit conditions are fulfilled.

Conductivity measurements as a function of DC field and temperature :

The conductivity measurements were performed as described in Alabenize et al. (13).
An electrode system with a rounded guard ring, minimising edge field enhancement was designed. The electrode system (fig 3) was placed in an oven and the temperature was measured in the vicinity of the sample. DC voltage was applied using a highly stabilised Spellman voltage supply.

Figure 3 : Photo of the electrode system.
Conductivity measurements have to be performed when the steady state is reached. The chosen procedure is described in fig.4. The sample was first conditioned at the highest field and the highest temperature to avoid long transient responses and the measurements were performed keeping the field constant and decreasing the temperature.

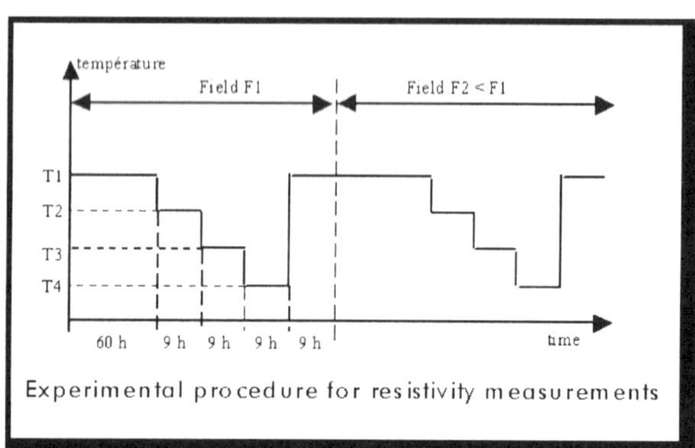

Fig 4 : experimental procedure for resistivity measurements.

RESULTS

Space charge

Space charge measurements were performed in both isothermal and "loaded" cases.

Space charge in the case of an isothermal cable. The cable was submitted to 15kV DC (conductor at the negative polarity) during 4hours. Then the space charge were measured while the sample was kept under voltage.

Space charge in the loaded case. The cable conductor was heated up to 47°C by Joule effect (I=330A) the outer semi conductor temperature was 42°C . The cable was submitted to 15kV DC (conductor at the negative polarity) during 62h. Then the thermal step was applied under voltage

on the outer semi conductive layer by means of cold water (+5°C). Recorded current is plotted fig.5.

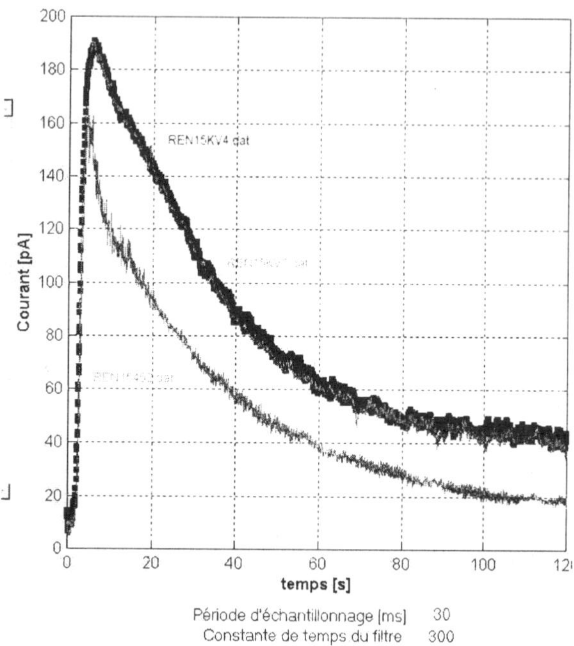

Fig 5 : measured current during the thermal step in loaded condition (red curve).

Conductivity

Conductivity results have been reported by Aladenize and al. (13). Let us note hereunder from this paper the conductivity vs temperature dependence under 60kV/mm and the γ coefficient at 40°C (from $\rho = \rho o\ E^{-\gamma}$) : $\gamma = 1.43$.

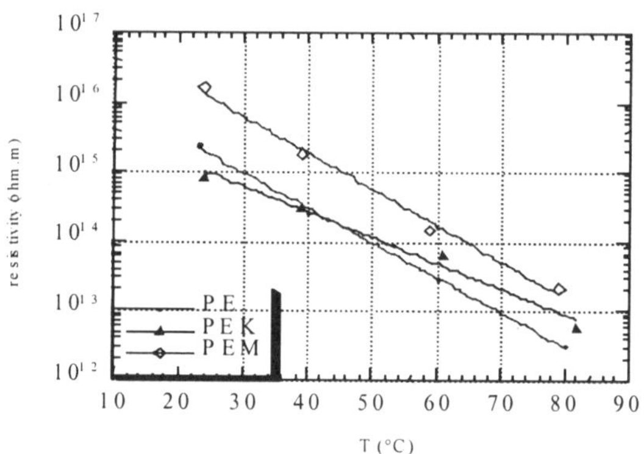

Fig 6 : Resistivity of PE, PEK and PEM under 60 kV/mm

DISCUSSION

The isothermal case

The measured current from the thermal step was deconvoluted and gave the space charge pattern represented fig 7. The integration of the charge along the cable radius lead to the space charge field that was added to the Laplace field to get the total field in the cable. Both fields are represented fig 8.

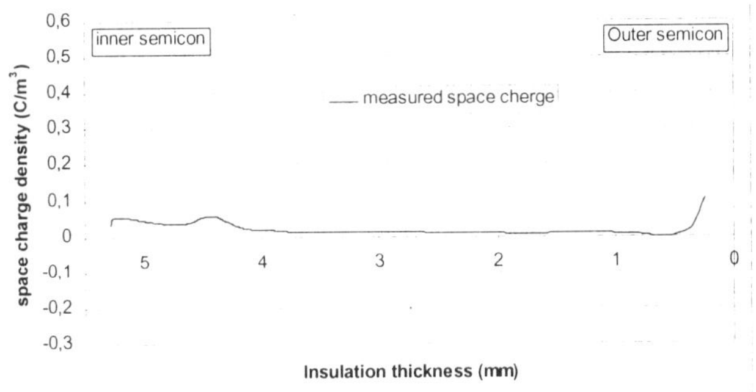

Fig 7 : measured space charge pattern

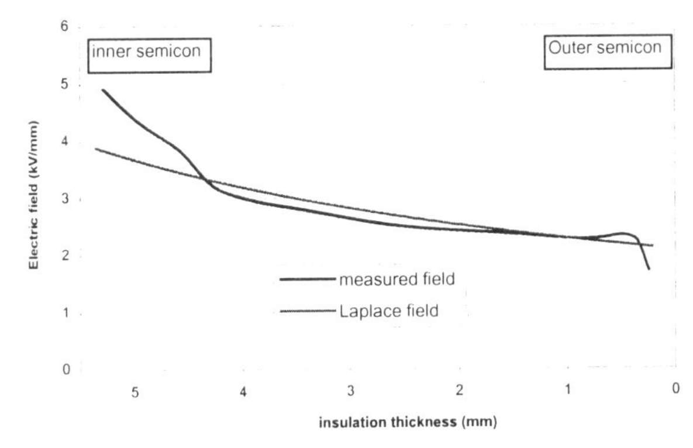

Fig 8 : measured field and Laplace field on the cable (isothermal case)

The measurement shows that there is an important increase of the field on the conductor side, where the Laplace field is already maximum. This increase is due to positive charges that come towards the negative electrode.

The loaded case

When polarising a DC cable with a thermal gradient inside the insulation, the time constant for the setting up of the space charge field due to the thermal gradient must be compared to the experiment time. In our case, referring to the conductivity measurements at 45°C and 3kV/mm, the time constant ε/σ is about 100hours. The temperature gradient has been applied during about one time constant.
The temperature difference across the insulation is only 4°C (5° from the conductor to the outer semicon surface).
As a consequence, the so-called inversion of the field was not reached during this experiment.
Field calculation according to Aladenize and al (9) are represented on fig. 9.
We can see that the general shape of the total field is not much changed as compared to the isothermal case, but the field dependence vs the radius is less pronounced than that of the Laplace field.

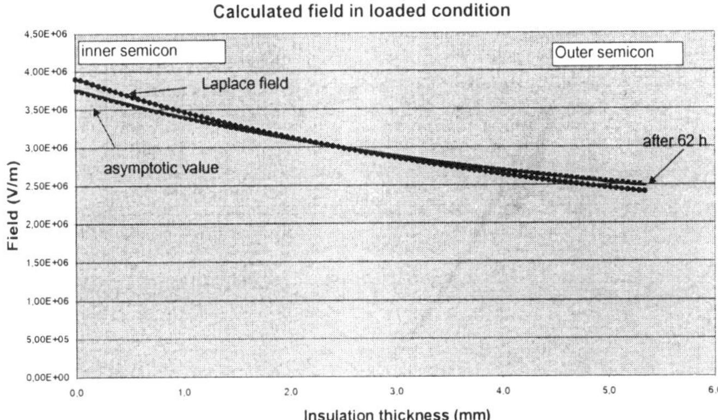

Fig. 9 : calculated field in loaded conditions

Fig 10 gives the experimental total field, i.e. the space charges field added to the Laplace field.

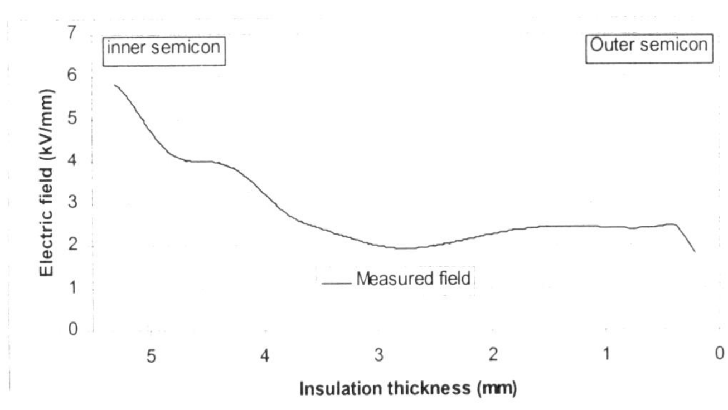

Fig 10 : Measured field on the cable (loaded case)

We do not see on the measured field pattern any flattening of the Laplace field due to the temperature gradient, but on the opposite an increase of the field close to the conductor.
We believe that this fact is due to the increase of the positive charges quantity close to the conductor. This increase is most probably linked to the temperature increase of the conductor. It is predominant over the space charge field due to the thermal gradient.

CONCLUSION

In this paper, we have reported the measurement of space charges in a cable under voltage both in isothermal and "loaded" conditions (i.e. with a temperature gradient in the insulation).
We have shown that the space charge field due to the injection and migration of charge carriers under DC polarisation cannot be neglected against the space charge field due to the thermal gradient.
This type of experiment should be considered when designing a cable or any other device that works under high voltage and direct current.

List of the references

1. Lau H.L. 1970, "Aufbau der Raumladung in einem betriebswarmen Gleichspannungskabel", Archiv für Electrotechnik, 53, 265

2. Beers B.L. and O'dwyer J.J.,1983,"Thermal dielectric breakdown with cylindrical electrodes", Journal of Applied Physics, 54, 4084

3. Mcallister I.W., Critchon G.C. and Pedersen A., 1994, "Charge accumulation in DC cables : a macroscopic approach", IEEE Intern.Symp on Electr. Insul., Pittsburg, 212

4. Wintle H.J., "Conduction processes in polymers", 1983, in Bartnikas (ed.) "Engineering dielectrics" vol IIA, ASTM, 239-354

5. Coelho R., and Goffaux R.,1981, "Dissipation et claquage dans les solides non métalliques", Revue de physique appliquée, 16, 67

6 Oudin J.M. et Fallou M., 1966, "Etude et développement des câbles à courant continu", Rev. Gen. De l'électricité, 75, 257

7 Eoll C.K., 1975, "Theory of stress distribution in insulation of high voltage DC cables, part 1", Trans. Electr. Insul, 10 (1), 27

8 Coelho R. and Mirebeau P., 1995, "On the field distribution in a DC power cable", Trans. Jicable, 570

9 Aladenize B., Coelho R., Guillaumond F., Mirebeau P., 1997, "On the intrinsic space charge in a DC power cable", Journal of Electrostatics, 39, 235 - 251

10 Laurenceau P., Dreyfus G. , Lewiner J., 1977, , J. Phys. Lett., 38, 46

12 Toureille A., Reboul J.P., Merle P., 1991, J. Phys. III, 1, 111 - 123

13 Aladenize B., Coelho R., Assier J-C., Janah H., Mirebeau P., 1999, "Field distribution in HVDC cables : dependence on insulation material". Jicable 99 proceedings.

A COMPARISON OF CONVENTIONAL AND CAPACITOR COMMUTATED CONVERTERS BASED ON STEADY-STATE AND DYNAMIC CONSIDERATIONS

M. Meisingset
Statnett SF, Norway

A.M. Golé
University of Manitoba, Canada

Abstract: The Capacitor Commutated HVDC Converter (CCC) is evaluated using steady-state as well as transient analysis. The steady-state analysis indicates that the CCC device is superior to the conventional converter when operating into very weak ac-networks. This is also confirmed by electromagnetic transient simulation. However, transient simulation shows that the CCC's performance is poorer than its conventional counterpart when recovering from unbalanced disturbances such as single-phase faults.

Keywords: Capacitor Commutated Converter, Long Cable HVDC Transmission, Weak AC Networks.

INTRODUCTION

Conventional HVDC converters appear to have a drawback in that they rely on the ac-network voltage for the turn-off of the thyristor valves. This imposes a serious limitation particularly when the converter is applied in extremely long dc cable transmission or feeds a very weak ac-network. The CCC topology (1,2) includes capacitors in series with the valve side transformer windings as shown in Fig. 1. The voltages on these capacitors aid in the commutation process thus resulting in a more robust converter, which is potentially less dependent on the ac-network strength and more robust against network disturbances. This feature makes the CCC attractive for long cable dc transmission schemes. In a conventional scheme with a long dc cable, a lowering of the inverter voltage can cause a large dc current surge due to the discharge of the capacitor, which increases the likelihood of commutation failure. The CCC, on the other hand, is better able to withstand sudden dc current increases. This technology is, thus, a potential candidate for use in long cable submarine transmission such as the HVDC-connections with cable lengths of about 600 km, which are currently in the planning stage in Northern Europe.

The paper begins with an analytical steady-state study, which is used to examine the operational characteristics for both the CCC and the conventional HVDC converter. An evaluation of the transient performance using PSCAD/EMTDC simulation is then conducted in order to critically examine the performance of the CCC in a typical cable transmission application.

STEADY-STATE PERFORMANCE

An analytical formulation is developed in order to investigate and compare the steady-state behaviour of

the CCC and conventional HVDC converter topologies. This formulation allows the conduct of preliminary investigations regarding applicability and steady-state control characteristics.

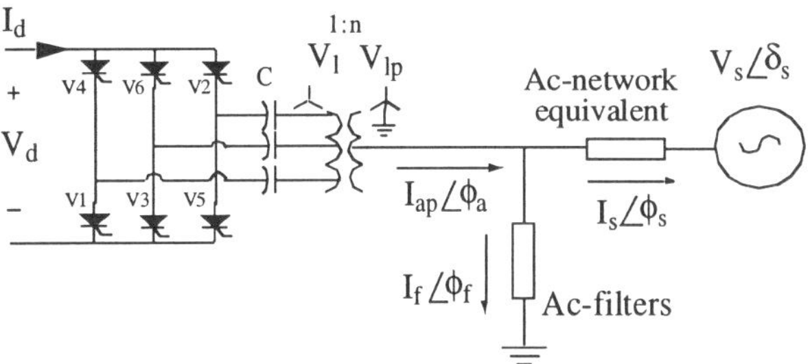

Fig. 1. The CCC inverter connected to an ac-network.

The formulation represents a 6-pulse 800 MW (500 kV, 1.6 kA) inverter connected to an ac-network equivalent, as shown in Fig. 1. It consists of 14 non-linear power flow equations and 18 unknown variables, which are listed in the Appendix. The numerical solution is obtained by first specifying any four variables and then using the Newton-Raphson iteration technique.

The formulation is based on fundamental frequency quantities only. It is valid for the representation of steady-state phenomena in the time-scale after the actions of the HVDC-controls, but prior to the response of voltage controlling devices in the ac-network (such as transformer tap-changers, synchronous machines and SVCs).

As mentioned earlier, the CCC is an HVDC converter topology that shows promise for use in long distance transmission via cables. It is interesting to study the consequences of increased current when the cable capacitance discharges. This is particularly important when the inverter is connected to a weak ac-network. The analytical formulation has the capability of analysing the inverter for various control modes and for different levels of SCR. The short-circuit-ratio (SCR) is a measure of the ac-network strength via-a-vis the transmitted dc-power.

Parametric plots can be generated from several successive steady-state solutions where one of the pre-specified variables is incremented from one solution to the next. Typical studies include the calculation of the Maximum Power Curve (MPC), the stability limits for the ac/dc- system and ac-voltage variations including load-rejection overvoltages as the dc-current varies.

A base case solution is required in order to produce a parametric plot. In this case V_d=500 kV, I_d=2.4 kA,

AC-DC Power Transmission, 28-30 November 2001
Conference Publication No. 485 © IEE 2001

$\gamma=20.4$ degrees and $V_{lp}=300$ kV are chosen as the pre-specified variables. The base case solution represents the rated condition and is indicated in the graphs. In each parametric plot, where a given variable is plotted as a function of I_d, the transformer turns ratio n and the network equivalent voltage V_s are held constant at the values found in the base case solution. Each parametric plot consists of a number of successive solutions where I_d, γ_{app}, n and V_s are the pre-specified variables and where I_d is slightly increased from one solution to the next.

The objective here is to compare the steady-state behaviour of the CCC and the conventional HVDC-converters. The comparison between the two converter types is carried out when they are connected to a relatively weak ac-network (SCR=1.82). The formulation is equally adept at handling the conventional inverter as a degenerate case with the impedance of the CCC series-capacitors set to zero.

The results presented in Fig. 2, show that the dc-voltage regulation for the CCC is far more superior to that of the conventional option as evidenced by the flatness of the voltage profile. The CCC also maintains the inverter ac-voltage closer to rated value in comparison to the conventional inverter. It is evident from the MPC-curve that that the CCC is superior to the conventional converter in terms of dc-power capability. The plots clearly show that the CCC has both a higher Maximum Available Power (MAP) and a larger dc-current stability limit in comparison to the conventional converter. Hence, the margin from the stability limit is larger in the CCC-option, indicating more robust operation. The results also show that the natural tendency for the real extinction angle is to increase when the dc-current becomes larger. This is caused by the larger series-capacitor voltage, which results in a larger real extinction angle, since the apparent extinction angle is kept constant by the controls. This characteristic of the CCC reduces the likelihood for commutation failure, particularly for long cables, where any lowering of the inverter ac-voltage results in a surge of current from the discharge of the cable capacitance.

The results obtained by the analytical study for the CCC-inverter type was verified by electromagnetic transient simulation (indicated by crosshair markers). As may be seen from the graphs, the simulated and the analytical results agree well up to rated conditions ($I_d=1.6$ kA). There is, however, a small deviation at larger currents, but the curves demonstrate a very similar trend. This indicates that the derived steady-state CCC-equations and the iteration algorithm have been correctly implemented in the computer program.

The primary point to be recognised is that the CCC-option, based on steady-state analysis, demonstrates a more robust operation against a sudden increase in current.

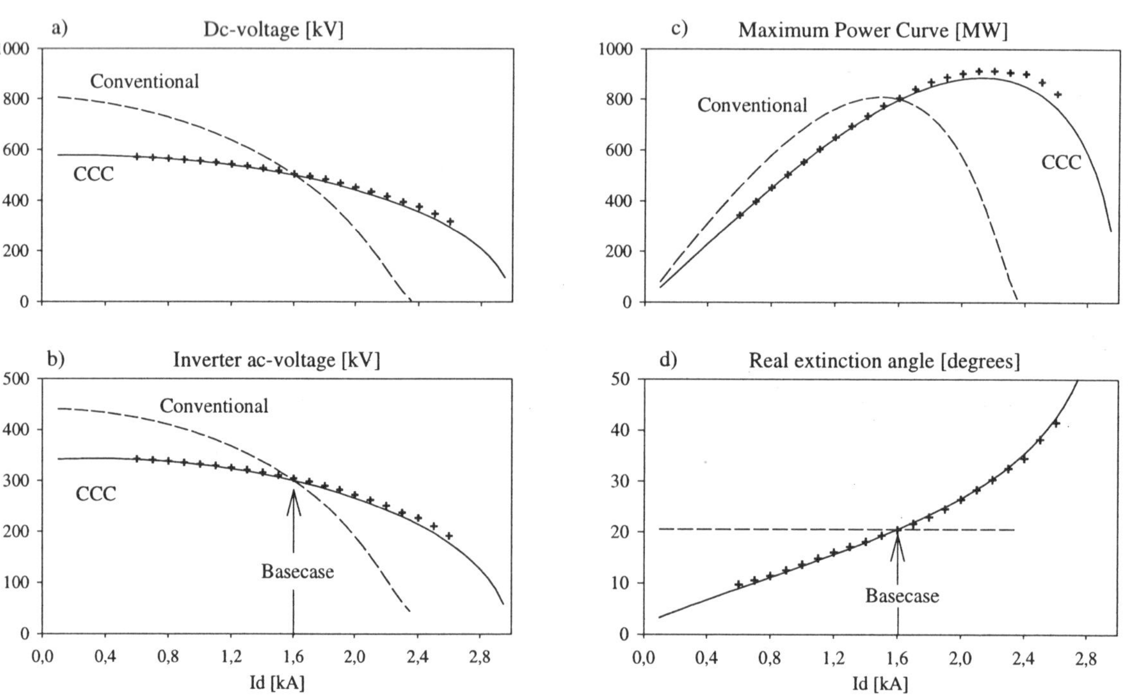

Fig. 2. Steady-state comparison between the CCC and the conventional inverter.

TRANSIENT PERFORMANCE

The analytical formulation is based on steady-state equations at fundamental frequency. In order to make a more thorough comparison between the two converter types, a study using electromagnetic transient simulation is necessary. Such a study will also take into account the transient nature (gains, time-constants, limits, etc.) of the control system when various types of disturbances are applied in the ac-network.

Modelling

The suitability of the CCC-inverter is investigated in a 600 km long cable 12-pulse HVDC-scheme of 1200 MW (500 kV, 2.4 kA) power rating, as shown in the Appendix. The conventional rectifier is connected to a very strong sending end, whereas the inverter is connected to a relatively weak receiving ac-network (SCR=2.05). At the CCC-inverter's ac-bus, there are bandpass filters to eliminate harmonic components of 11th, 13th, 23rd and 25th order, as well as an highpass filter for removal of higher order harmonics. The total reactive power installation in the ac-filters and the shunt capacitors at the conventional inverter bus is 55 % of its rated dc-power. The ac-filters at the CCC-inverter bus have a reactive power installation of only 14.5 % of the rated dc-power, because this converter type consumes significantly less reactive power. The ac-network on the inverter side is represented by an RRL-type equivalent with a SCR equal to 2.05 and a damping angle of 85 degrees at fundamental frequency. The selection of the series-capacitor impedance represents a trade-off between a good power factor and a sufficient margin to commutation failure versus low voltage stress on the thyristor valves. The 144 µF value selected in this design is based on 10 % additional voltage stress on the valves compared to the conventional converter. The over-voltages across the series-capacitors are limited to 3 pu by surge arrestors. Smoothing inductors of 0.4 H are located at both the rectifier and inverter side of the dc-connection. The dc-cable is modelled as simple π-equivalent. The rectifier operates in voltage control and the inverter operates in current control during normal operation in both types of HVDC-schemes. Additional control details are described in (3).

Results

The transient performance of the CCC and the conventional HVDC scheme was investigated when various faults were applied in the inverter ac-network and when setpoint changes were made in the controls.

Inverter load rejection overvoltage. A sudden change in dc-power transmitted, for instance caused by an inverter load rejection, results in an immediate surplus of reactive power. This will temporarily raise the inverter bus voltage particularly if the ac-network is weak. The steady-state overvoltage resulting from the load rejection of the CCC-inverter is recorded to 350.5 kV (i.e. 1.17 pu). This is significantly smaller than the 448.5 kV (i.e. 1.50 pu) overvoltage generated by load rejection of the conventional inverter.

The CCC is able to operate at significantly lower extinction angles than the conventional option. The lower load rejection overvoltage for the CCC is simply due to the smaller amount of reactive power installed in ac-filters and shunt capacitors at its inverter bus.

Robustness against remote ac-fault. The robustness against two types of remote faults is investigated for both HVDC-schemes. A remote fault is, in this context, defined as a fault occurring electrically distant from the inverter bus. The remote fault is simulated by connecting a shunt reactor (due to the inductive impedance of transmission lines) to the inverter bus. The minimum value of fault inductance under which the system remains operating without suffering commutation failure, is identified by trial and error. Any fault with more severity (lower inductance) than this limiting value causes commutation failure. Because a low inductance corresponds to a more severe fault, the lower the limiting inductance, the more robust is the scheme. The results are obtained with a fault clearance time of 200 ms.

First, the inverters' robustness against a single-phase-to-ground remote fault is investigated. This is the most typical type of fault that occurs in overhead lines and is by Thio et al (4) considered more severe than a three-phase-fault in terms of commutation failure. The reason for this is due to the fact that the single-phase faults result, contrary to the balanced three-phase fault, in phase-shifts in the zero-crossings of the commutation voltages. These phase-shifts decrease the commutation margin for some of the thyristor valves and increase it for other valves.

The results do not support this theory since both inverter types demonstrate a larger degree of robustness against single-phase faults in comparison to three-phase faults. The reason could be the presence of the large cable capacitance, since the single-phase fault results in a smaller lowering of the dc-voltage and thus also a smaller discharge current from the cable. The results for the single-phase remote fault show that the conventional inverter is able to withstand a more severe fault (0.75 H) in comparison to the CCC (1.04 H).

The steady-state impact of a three-phase-to-ground remote fault is first investigated theoretically. It is evident that the fault causes the inverter ac-voltage to drop in magnitude and its phase to move in the leading direction. Both these effects bring the inverter closer to commutation failure. They happen immediately and occur at both inverter types, but to a larger extent for the conventional inverter than for the CCC. The theoretical results therefore indicate that the CCC is the most robust option against this type of remote fault.

Let us consider the immediate impact caused by the remote fault prior to any response from the controls. The drop in inverter ac-voltage magnitude clearly increases the overlap angle and hence increases the likelihood of commutation failure. The positive phase shift in the ac-voltage takes place instantaneously, whereas the phase-lock-loop (PLL) based firing scheme needs, due to its time constants, a certain time to lock on to the ac-voltage again. The ac-voltage leads its PLL-signal during this period of time, which means that the firing angle is transiently larger than the ordered firing angle. A larger firing angle at the inverter results in a smaller commutation margin since the valves becomes forward biased sooner before they cease conducting. Both the above-mentioned effects are thus bad from the commutation point of view, but they have opposite impact on the inverter dc-voltage. The drop in magnitude permanently reduces the dc-voltage, whereas the positive phase-shift temporary increases the dc-voltage. It is evident from the results that the dominant effect is due to the drop in magnitude, since the response from the controls is to compensate for the reduction in dc-voltage by increasing the firing angle. The results for the three-phase remote fault show that the CCC inverter is able to withstand a more severe fault (1.16 H) in comparison to the conventional inverter (1.22 H).

The results therefore show that the CCC is more robust against a three-phase remote fault, but less robust against a single-phase fault than the conventional inverter.

Recovery from close-in ac-faults. Commutation failure and succeeding discharge of the cable into the inverter ac-network is often unavoidable if the ac-fault takes place electrically close to the inverter bus. Recovery of HVDC-schemes is usually more difficult and slower for weak ac-networks than for strong networks. Fast recovery is, however, more critical for weak networks in order to maintain stability, because they are less capable of withstanding the temporary deficit of power.

The recovery performance is investigated when a 50 ms single-phase and a three-phase to ground fault are applied at the ac inverter bus. The clearance time is mainly determined by the relay detection time and the breaker switching time. The latter obviously depends on the type of breaker used. The selected fault clearance (50 ms) requires state-of-the-art equipment both for relays and breakers. The recovery time is defined as the time from fault clearing to the instant at which 90 % of the pre-fault dc-power is restored.

The single-phase to ground fault is, in general, considered to be less severe than three-phase faults in terms of power system stability. This type of unbalanced fault can, however, be particularly critical for the CCC since it may generate imbalance in the series capacitor voltages resulting in performance deterioration during transients. The surge arresters will however limit any imbalances from reaching extreme values.

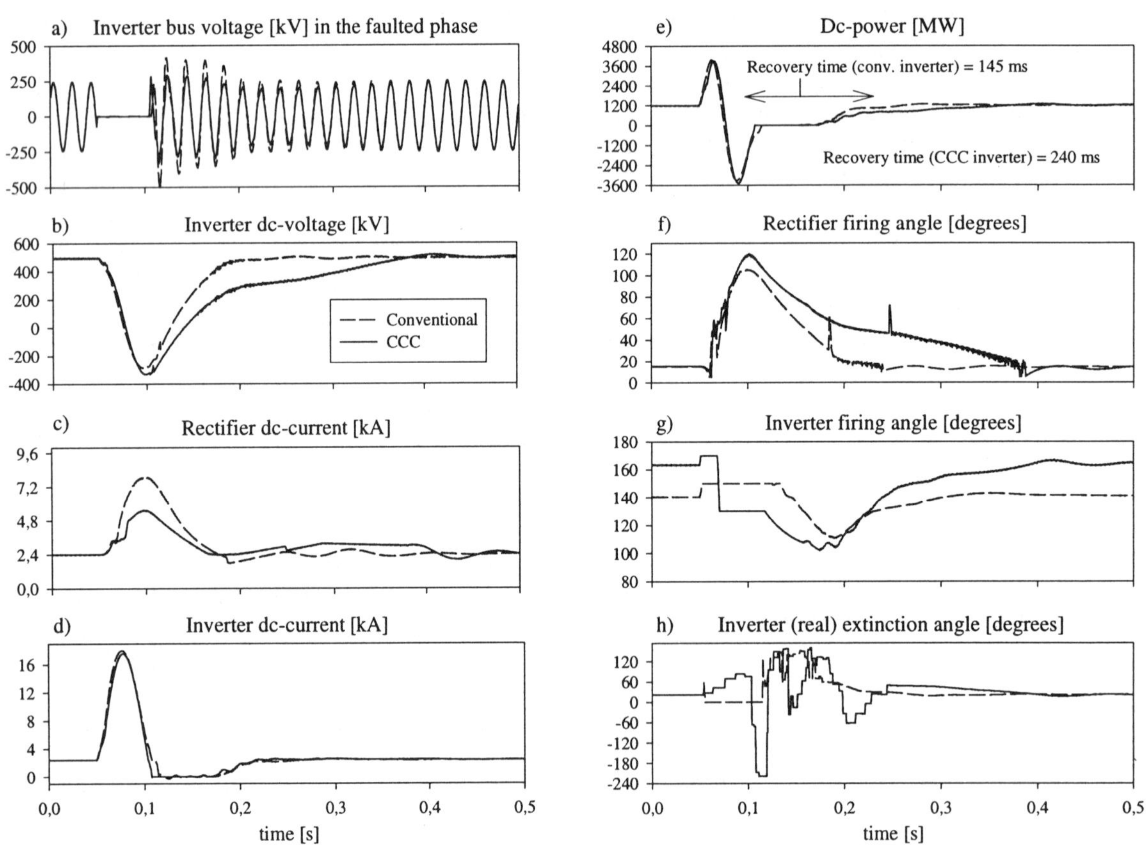

Fig. 3. Single-phase to ground fault applied at the inverter bus.

Figure 3 shows the recovery performance following the single-phase fault of both the conventional (dashed) and the CCC (solid-drawn) inverter based scheme. The results show that the CCC has a slower recovery (240 ms) than the conventional type (145 ms). The surge of current into the ac-network, given by the inverter dc-current, is close to identical (18 kA, i.e. 7.5 pu) for the two options. The negative power during recovery is a consequence of the voltage being negative. With a sustained line-to-line short circuit as happens during commutation failure, the ac-voltage gets applied to the dc-side resulting in this negative value.

The recovery performance following the three-phase fault show that the conventional alternative has a much higher temporary overvoltage after fault clearance in comparison to the CCC-alternative. The surge of current into the ac-network, given by the inverter dc-current, is reduced from 18.1 kA (i.e. 7.5 pu) in the conventional option to 10.6 kA (i.e. 4.4 pu) in the CCC-option. This is favourable since it reduces the current stress on the valves. The CCC-alternative demonstrates a quicker recovery (90 ms) than the conventional alternative (145 ms). This rapid recovery is quite remarkable, particularly in the view of the weak inverter ac-network. The dc-voltage, and thus the dc-power, becomes negative during recovery in the conventional option.

Both schemes demonstrate a recovery time that falls within 300 ms both for unbalanced and balanced faults. This is considered acceptable since most HVDC-schemes recover within 100 to 300 ms. The transient peak in the rectifier dc-current is lower in the CCC-option for both types of faults.

Setpoint-changes in the dc-controls. A step-change is made to the dc-current order at rated conditions to evaluate the robustness of the schemes to set-point changes in the controls. This essentially means a step-change in the ordered dc-power. A generally accepted performance level is that the HVDC-schemes should follow an instantaneous ±10% change in order. Both alternatives meet this specification, albeit with different margin. The maximum step-change that could be made without causing commutation failure is significantly larger (24 %) for the CCC than the conventional inverter (16 %). Both inverter options allow a 99 % reduction in the dc-current from rated conditions. The results are highly dependent on the settings in the controls, particularly the maximum allowable change in the Voltage Dependent Current Order Limit (VDCOL). With the settings selected here, it is evident that the CCC demonstrates better performance than the conventional scheme.

CONCLUSIONS

An analytical formulation was developed in order to investigate the steady-state behaviour of the Capacitor Commutated Converter and conventional HVDC converters. The validity of the formulation was confirmed by electromagnetic transient simulation. The steady-state results obtained by the formulation, show that the CCC is superior to the conventional converter in its voltage regulation, ability to work into depressed voltage systems, stability margin and ability to avoid commutation failure when operated in power control.

The transient analysis shows that the CCC does not demonstrate an uniformly favourable performance in a long cable HVDC transmission scheme when compared to the conventional converter type. It is true that the load rejection over-voltages are smaller for the CCC and its performance under balanced remote- and close-in faults is superior. The CCC also allows larger set-point changes in the controls. However, with unbalanced disturbances such as single-phase to ground faults, the conventional converter demonstrates the better performance.

The likely reason for this behaviour is the additional dynamics due to the energy storage in the CCC's series-capacitors. For unbalanced disturbances, each of the series-capacitor voltages impacts the system to differing degrees.

BIOGRAPHIES

Dr. M. Meisingset received a M.Sc. degree from the Norwegian Institute of Technology (NTH) in 1991 and a Ph.D. degree from the University of Manitoba at Winnipeg in Canada in 2000. He has since 1993 been employed at Statnett SF - The Norwegian Power Grid Company in Oslo.

Dr. A.M. Golé obtained his B.Tech (EE) degree from IIT Bombay in 1978 and the Ph.D. degree from the University of Manitoba in 1982. He currently serves in the capacity of Professor in the Department of Electrical and Computer Engineering at the University of Manitoba in Canada. He is a registered Professional Engineer in the province of Manitoba.

ACKNOWLEDGEMENTS

The authors acknowledge The Norwegian Power Grid Company (Statnett SF) of Oslo, Norway and the National Science and Engineering Council (NSERC) of Canada for their financial contributions to this research.

REFERENCES

1. J. Reeve et al, "A Technical Assessment of Artificial Commutation of HVdc Converters", IEEE Transactions on Power Apparatus and Systems, Vol. PAS-87, Oct. 1968, No. 10, pp. 1830-1840.

2. T. Jonsson and P. Björklund, 1995, "Capacitor Commutated Converters for HVDC", Stockholm Power Tech, Proc.: Power Electronics, pp. 44-51.

3. M. Meisingset and A.M. Golé, 2001, "Control of Capacitor Commutated Converters in long cable HVDC-transmission", IEEE PES Winter Meeting, Columbus, OH, USA.

4. C.V. Thio et al, April 1996, "Commutation failures in HVDC transmission systems", IEEE Transactions on Power Delivery, Vol. 11, No. 2, pp. 946-957.

APPENDIX

The analytical steady-state formulation for the CCC consists of 18 variables and 14 equations.

V_d, I_d — Dc voltage and -current.

$\alpha, \gamma_{app}, \mu$ — Firing-, apparent extinction- and overlap angle.

$\Delta v_1, \Delta v_2$ — Change in series capacitor voltage for incoming and outgoing valve during overlap interval.

B — Constant resulting from the solution of the overlap differential equations

n — Transformer turns ratio (primary/valve side)

V_{lp} — Transformer primary side ac-voltage

I_{ap}, ϕ_a — Magnitude and phase of ac-current in transformer.

I_f, ϕ_f — Magnitude and phase of the current in ac-filter.

I_s, ϕ_s — Magnitude and phase of the current in ac-network.

V_s, δ_s — Ac-network source voltage magnitude and phase.

Symbols:

L, C — Transformer leakage inductance, series capacitor.

Z_s, Z_f — Ac-network- and filter impedance.

ω — Ac-network angular frequency.

ω_0 — Angular frequency for com. circuit $(=1/\sqrt{LC})$.

The first six equations of the 14 required are obtained by applying Ohm's Law on the ac-side of the transformer.

$$V_{lp} - V_s = \sqrt{3}Z_s I_s \tag{1,2}$$

$$I_{ap} = I_f + I_s \tag{3,4}$$

$$V_{lp} = I_f \sqrt{3}Z_f \tag{5,6}$$

Each of these three equations in complex form is divided into two equations based on their real and imaginary part.

The other eight equations are related to the converter:

$$\frac{\Delta v_2}{2\omega_0 L} - \frac{V_{lp}\omega_0}{n\sqrt{2}(\omega_0^2-\omega^2)L}\sin\alpha - \frac{\pi I_d \omega_0}{3\omega} = B \tag{7}$$

$$\Delta v_1 + \Delta v_2 = \frac{I_d \mu}{\omega C} \tag{8}$$

$$n I_{ap} = \frac{\sqrt{6}}{\pi} I_d \tag{9}$$

$$\cos\phi_a + \frac{\cos\alpha + \cos(\alpha+\mu)}{2} = 0 \tag{10}$$

$$\alpha + \mu + \gamma_{app} = \pi \tag{11}$$

$$\left[\frac{I_d}{2} + \frac{V_{lp}\omega\cos\alpha}{n\sqrt{2}(\omega_0^2-\omega^2)L}\right]\cos(\frac{\omega_o}{\omega}\mu) + B\sin(\frac{\omega_o}{\omega}\mu) + \frac{V_{lp}\omega\cos(\alpha+\mu)}{n\sqrt{2}(\omega_0^2-\omega^2)L} + \frac{I_d}{2} = 0 \tag{12}$$

$$\left[-\frac{I_d}{2} - \frac{V_{lp}\omega\cos\alpha}{n\sqrt{2}(\omega_0^2-\omega^2)L}\right]\sin(\frac{\omega_o}{\omega}\mu) + B\cos(\frac{\omega_o}{\omega}\mu) + \frac{V_{lp}\omega_o\sin(\alpha+\mu)}{n\sqrt{2}(\omega_0^2-\omega^2)L} + \frac{\frac{2I_d\pi}{3\omega C} - \Delta v_1}{2\omega_0 L} = 0 \tag{13}$$

$$\frac{3}{\pi}\frac{\sqrt{2}V_{lp}}{n}\left[\frac{\cos\alpha + \cos(\alpha+\mu)}{2}\right] + \left[\frac{3}{\pi}(\Delta v_2 - \Delta v_1)(\frac{\pi}{3} - \frac{\mu}{4})\right] = -V_d \tag{14}$$

The <u>apparent</u> extinction angle γ_{app} is a measure of the inverter's power factor since it is related to the positive zero-crossing of the inverter bus voltage. The <u>real</u> extinction angle γ_{real} is related to the commutation voltage, and is hence a measure of margin to commutation failure. Based on the solution of the analytical formulation, γ_{real} is calculated by solving the following equation by an iterative technique.

$$\frac{\sqrt{2}V_{lp}}{n}\sin(\alpha+\mu+\gamma_{real}) + \frac{2\pi I_d}{3\omega C} - \Delta v_1 - \frac{I_d \gamma_{real}}{\omega C} = 0 \tag{15}$$

Figure 4 presents the CCC-inverter-based 1200 MW HVDC-scheme used for the electromagnetic transient simulation studies. Both transformers (YY and YΔ) have a turns ratio of 185/300 kV and a leakage reactance of 0.15 pu. The conventional HVDC scheme is similar in its design, but differs in transformer turns ratio, reactive compensation and control parameter settings.

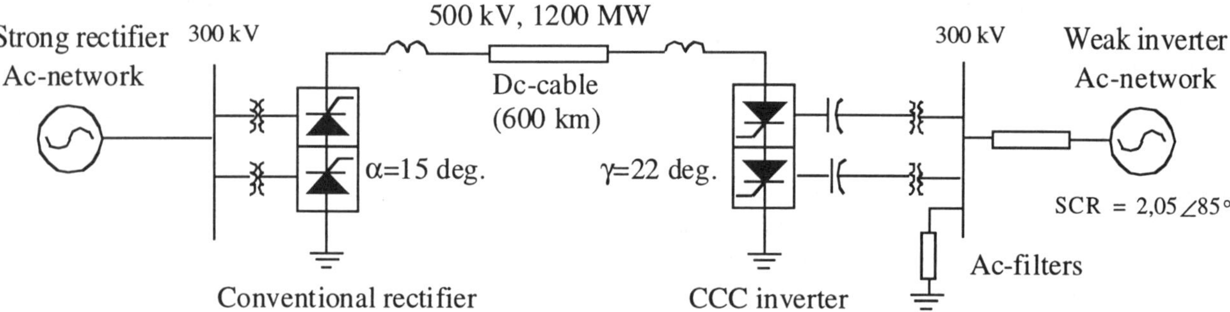

Fig. 4. The 1200 MW HVDC-scheme employing CCC-type inverters.

Modular back-to-back HVDC, with capacitor commutated converters (CCC)

Niclas Ottosson
Lena Kjellin

ABB Sweden

Keywords: HVDC, Back-to-back, Capacitor,
Comutated Converter, ConTune.

1.   SUMMARY

A back-to-back HVDC arrangement is used when two
asynchronous AC systems need to be interconnected for
bulk power transmission or for AC system stabilization
reasons. In an HVDC back-to-back station there are no
overhead lines or cables separating the rectifier and the
inverter, hence the transmission electrical losses on the
DC side can be neglected. In a back-to-back HVDC, the
DC current can be kept high and the DC voltage low.
The low DC voltage means that the air clearance
requirement is low, which is in favor of a compact
design of the valve housings. This enabled the modular
back-to-back HVDC concept to be developed.

The modular back-to-back converter station can be
made very compact, thus requiring a minimum of open
space. It can even be designed to fit into an existing
right-of-way of a typical 400 kV AC line. In
environmentally sensitive areas this is also considered a
great advantage from a permitting point of view. Our
ISO 14001 certification states that the design of the
equipment and the construction at the site are
performed in an environmentally acceptable way.

The Argentinean and Brazilian networks have been
interconnected via back-to-back HVDC with a rating of
1100 MW. A second interconnection of another 1100
MW is under construction. The back-to-back HVDC
converter stations are of the modular type with
capacitor commutated converters.

Fig.1 Overview of the 1100 MW modular back-to-back
HVDC station in Garabi, Brazil.

1. HVDC BACK-TO-BACK -ARRANGEMENT

A back-to-back HVDC arrangement is used when two
asynchronous AC systems need to be interconnected for
bulk power transmission or for AC system stabilization
reasons. In an HVDC back-to-back station there are no
overhead lines or cables separating the rectifier and the
inverter, hence the transmission electrical losses on the
DC side can be neglected. In a back-to-back HVDC, the
DC current can be kept high and the DC voltage low. In
order to optimize the costs of the thyristor valves as
well as the DC side equipment, the DC current is kept
as high as the standard thyristor rating and valve
cooling system can handle safely. This means that the
DC voltage can be kept fairly low, and is thus chosen in
relation to the rated DC power.

1.1 Applications

-    Asynchronous connection between AC networks

To stabilize a weak AC system, it might be desired to
do this by importing power from an AC system,
isolated from the receiving system. These two AC
systems could even have different frequencies. In such
situations, a DC transmission system is the only
possibility for power transmission.

-    Frequency control of AC systems.

The frequency of an AC system that becomes too
heavily loaded might drop, and when large loads are
suddenly disconnected, the frequency of the system will
rise. By interconnecting such an AC system to a
stronger system by means of a DC transmission, the
frequency deviation can be limited or controlled
through automatically imposed DC power modulations,
counteracting the frequency deviations.

-    Reduction of short circuit current in strong AC
     systems.

The short circuit power of a strong AC system with
several infeeds can cause a very high short circuit
current upon ground faults. At a certain point, the
number of infeeds added would cause the short circuit
current to exceed the level that the existing circuit
breakers are capable of handling, and the breakers in

the AC system might have to be uprated or replaced. A less costly way is to split the system up by means of one or several back-to-back converters, thereby enabling isolation of certain parts of the AC system at ground faults, thus avoiding the cost of replacing all the circuit breakers of the AC system.

2. THE CCC CONCEPT

The converter used in the CCC concept is characterized by the use of commutation capacitors inserted in series between the converter transformers and the converter valves. See Figure 2. Single line diagram. It makes it possible to operate HVDC in very weak networks and eliminates the need for synchronous compensators.

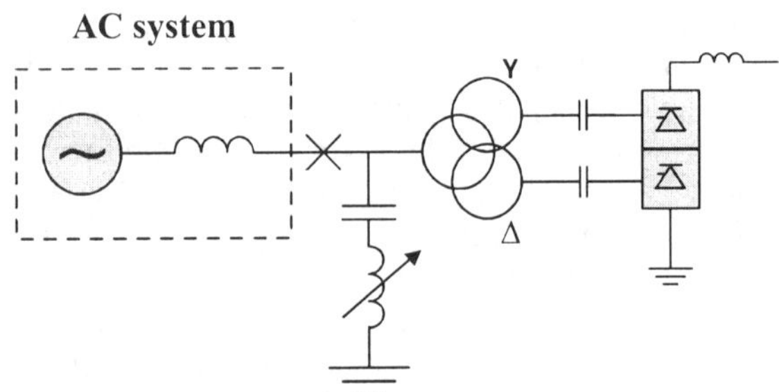

Figure 2 Single line diagram of a monopolar station with CCC and ConTune™ AC filter.

The ABB approach does not aim at achieving a self-commutated converter, instead, it provides reactive power compensation proportional to the load of the converter. The need of switchable shunt capacitor banks for reactive power compensation is thereby eliminated. Since the AC filters are necessary only from the point of view of filtering harmonics, the shunt connected reactive power generation can be minimized. In the ABB solution, the size of the commutation capacitor is chosen so that the full load reactive power consumption of the converter is compensated by the reactive generation of the small high performance AC filter. See Figure 3. Reactive power conditions.

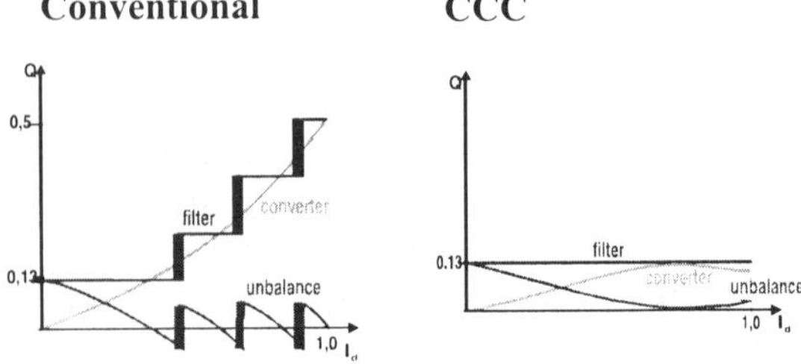

Figure 3 Reactive power conditions for a typical conventional converter and for a CCC.

2.1 The capacitor in the CCC concept

The commutation capacitors create the AC voltage to which the converter is commutating, hence these capacitors are called Commutation Capacitors, and the converter becomes a Capacitor Commutated Converter, a CCC. The steady state operating voltage of the commutation capacitor is defined by the direct current. The capacitors are protected against over voltages by parallel ZnO varistors. The voltage stresses on the

capacitors are relatively low compared to the installed capacity, and consequently the commutation capacitors can be of compact design.

Fig. 4 Commutation capacitors at Garabi

The commutation capacitors improve the commutation failure performance of the converter. Typically, a CCC can tolerate a sudden 15-20% voltage drop on the network voltage.

The contribution to the commutation voltage from the commutation capacitors results in positive inverter impedance characteristics for an inverter operating at minimum commutation margin control. An increase in direct current therefore results in a DC voltage increase rather than the opposite, which is the case for conventional inverters with commutation margin control. The dynamic stability of an inverter will thus be dramatically improved with a CCC.

Figure 5 shows the MAP (Maximum Available Power) curves for a conventional converter and a CCC for SCR = 2. As can be seen, the CCC is in a very stable situation while the conventional converter is close to the stability limit. The diagrams also show that the load rejection over voltage is reduced from 1.5 to 1.2 p.u. as a result of the small size of the shunt connected filters for the CCC.

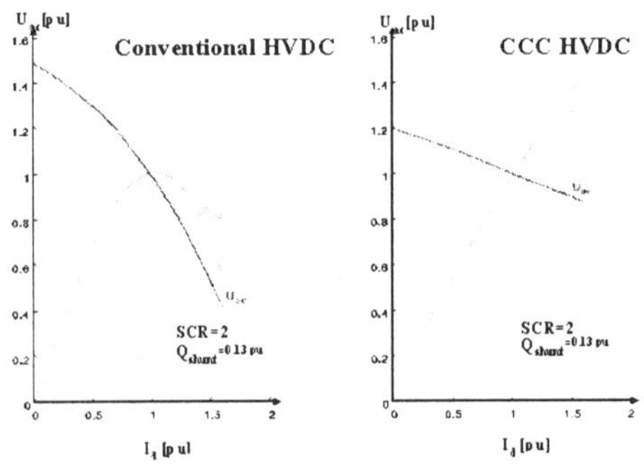

Figure 5 Maximum power curve for conventional and Capacitor Commutated Converters (CCC), SCR=2g=17°.

2.2 Impact on equipment and station design

The introduction of commutation capacitors results in different stresses on the other equipment compared to a conventional HVDC converter. The main influence from the capacitors is a considerable reduction of valve short-circuit currents. This is due to the voltage drop across the commutation capacitor varistors. On the other hand, a somewhat higher peak voltage across the valve, as well as higher extinction voltage steps, will be obtained compared to conventional HVDC. The voltage contribution from the commutation capacitors will support the commutation of the direct current from one valve to another; i.e., the overlap angle will be reduced compared to a conventional HVDC converter. The commutation capacitors reduce the rating of the converter transformer as the reactive power flow through the transformer is minimized.

The elimination of switched reactive power compensation equipment will simplify the AC switch-yard and minimize the number of circuit-breakers needed, which will reduce the area required for an HVDC station built with CCC.

ConTune AC filters:
The task for the AC filters is to filter out the harmonics, for which purpose quite small capacitor banks are needed, provided they can be tuned sharply enough. This is made possible by electronically controlled filter reactors included in the filter for the 11th and the 13th harmonics, which are the two dominant harmonics created by the twelve-pulse converter. The inductance of these reactors are continuously controlled or tuned, and hence such an AC filter branch is called ConTune. The adjustable reactors of the ConTune branches are continuously controlled by a DC current fed into a control winding mounted perpendicular to the main winding, enabling continuous adjustment of the inductance and thus continuous tuning of the filter branch.

The combination of the CCC and the ConTune offers several technical advantages. Firstly, operation is possible at very low short circuit power levels of the AC systems. A short circuit ratio of 1.0 is achievable without the need of synchronous machines added to the AC networks. Secondly, the immunity to AC system disturbances is improved, thereby maintaining constant DC power flow during severe disturbances in the AC systems. Thirdly, the reactive power balance between the HVDC converter and the AC networks can be kept within very narrow limits, since the commutation capacitors generate reactive power in accordance with the transmitted DC power, i.e. in accordance with the actual reactive power consumption of the converter.

Thyristor Valve Modules:
In order to optimize the costs of the thyristor valves, as well as the DC side equipment, the DC current is kept as high as the standard thyristor rating and the valve cooling system can handle safely. This means that the DC voltage can be kept fairly low, and is thus chosen in relation to the rated DC power. The low DC voltage means that the air clearance requirement is low which is in favor for a compact design of the valve housings. These facts enabled the modular back-to-back HVDC concept to be developed.

The thyristor valves are air insulated at atmospheric

pressure and installed in modular valve housings. Each valve module contains two or more single valves, which means up to six valve modules per twelve-pulse converter. The thyristor valves are suspended from the ceiling and are easily accessible for the maintenance crew. The housings also contain the surge arresters connected across the valves, and the valve control.

Fig.6 Outdoor HVDC Valves

Each modular housing is equipped with an independent fire protection system with detectors and inert gas injectors that are automatically initiated upon detection of overheated material. For a better and more reliable supervision of the thyristors, electrically triggered thyristors are used.

Valve Cooling module:
The valve cooling system for a modular back-to-back is located in a separate module, except for the cooling towers that are located outdoors. The thyristor valves are water cooled and preferably by means of a single closed loop system.

Auxiliary Power modules:
The low voltage part of the auxiliary system is installed in two separate modular housings, one for the distribution switchgear and one for the battery system. The auxiliary power can be provided either from a separate low voltage infeed, from a dedicated auxiliary transformer, or from tertiary windings of solidly connected AC line shunt reactors.

Control & Protection Module:
The MACH 2 Control & Protection System consists of a fully duplicated system, with commercially available computers including the control functions as well as the protection functions. The AC side protections are also included in MACH 2, in the same kind of hardware as the DC side protections. Operation can take place locally or remotely through serial protocols. The Control and Protection system is located in a dedicated module.

The whole back-to-back converter station can be operated from an Operator Workstation in the control and protection module, from the station control building in the converter station, or from remote dispatch centers.

HPL Compact breaker:
The HPL Compact switching module includes all functions performed by a traditional circuit breaker bay; HPL circuit breaker, pantograph disconnectors, earthing switches, digital optical current transducer, common base frame, prewired control cubicle. All these results in a substantial reduction of required space and facilitate ease and fast maintenance of the breaker bay.

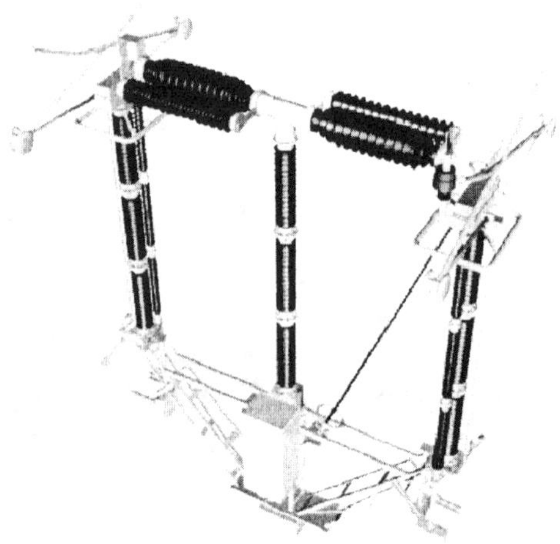

Fig.7 HPL Compact switching module

3. BENEFITS

3.1 Operation, maintenance, and availability benefits

The MACH 2 Station Control and Monitoring system, SCM, is a user friendly fully computerized operator tool. It has built-in facilities, such as operator language selector, self instructed operation flow charts, voice enunciator, sequence of events recorder, transient fault recorder, digital hierarchical documentation, etc.

The rest of the equipment, which is continuously supervised by the SCM systems, can be completely maintained from the fully duplicated MACH 2 Control and Protection systems. The system also includes the AC protections, integrated in the system in the same way as the DC protections are. Any HV, or LV equipment or measuring defect, as well as any internal MACH 2 malfunction will automatically be detected and recorded, thus eliminating the need for traditional maintenance by visual inspection and manual measurements.

Traditional maintenance like visual inspection, adjustments, lubrication, etc., is only needed for the moving parts, which are motors, pumps and fans used for the cooling equipment, and the converter transformer tap changer.

No AC breakers or disconnects, apart from the ones used for energizing purposes, are needed.

Since the number of moving parts are kept to a minimum, and the number of HV equipment is reduced, the availability of the facility is increased.

3.2 Economical benefits

The modular back-to back concept is time and cost saving with respect to permitting processes, as it can be designed to fit into an existing right-of-way.

As there is no need for a valve hall and a traditional control and service building, the civil design is reduced and consists primarily of design of the foundations for the different modules. The cost and time needed for the civil design is thus kept to a minimum.
The time and cost for construction is substantially reduced, as the construction of foundations for the different modules is easier than the construction of a valve hall and a control and service building.
The installation work can be reduced in time, because much of the installation work is factory-made in a more efficient way, and very little assembly work is needed at the site. The factory pre-assembly also detects any fitting problem at an early stage, which limits times and cost for trouble-shooting at the site. The factory pre-assembly of the modules can be done as a parallel activity to the construction at site and thus reduces the total project time.
Testing at site can be reduced as the modules are tested at the factory before shipping. As the different kinds of components included in the modular designed converter are limited substantially, the number of spare parts needed is also limited. This facilitates storage and handling of spares.

The CCC concept allows the location of stations at the edges of existing networks, avoiding costly transmission reinforcements. The modular concept also facilitates future changes in sub-station layout due to planned expansion. If the operating or the economic conditions change, relocation of the equipment might be desired. In a modular back-to-back the civil work is reduced to a minimum and therefor it allows for relatively easy relocation of the equipment.

3.3 Environmental benefits

The modular back-to-back converter station can be made very compact, thus requiring a minimum of open space. It can even be designed to fit into an existing right-of-way of a typical 400 kV AC line.

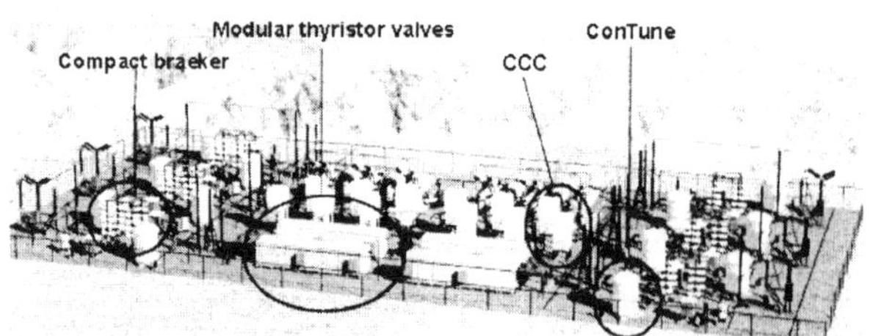

Fig.8 Modular back-to-back, side view.

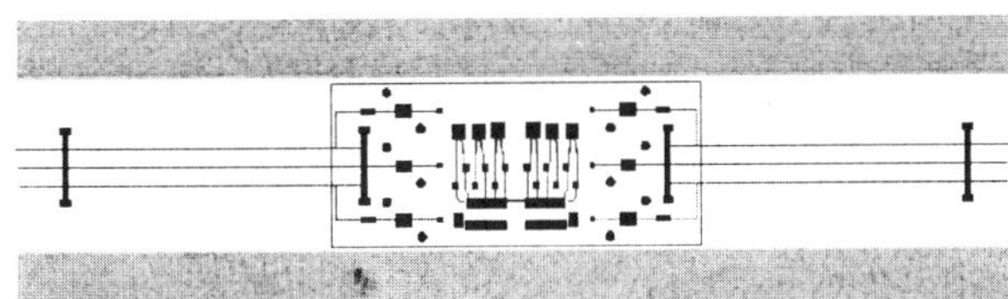

Figure 9 Modular back-to-back, overview. Fits in the right –of-way width of most AC lines.

In environmentally sensitive areas this is also considered as a great advantage from a permitting point of view.

REFERENCES

Hagman I, Jonsson T, "Verification of the CCC concept in a high power test plant", Cigré colloquium on HVDC and FACTS, South Africa, 1997.
Persson A, Carlsson L, Åberg M, "New technologies in HVDC converter design", SEPOPE, Recife, Brasil, 1996.
Jonsson T, Holmberg P, Tulkiewicz T, "Evaluation of classical, CCC and TCSC converter schemes for long cable projects", EPE99, Lausanne, Switzerland, 1999.

Capacitor Commutated Converters for High Power HVDC Transmission

G. Balzer, H. Müller

Darmstadt University of Technology, Germany

Abstract

Two concepts for the transmission with a high power capacity using HVDC technology are analysed in this paper. The technology of Capacitor Commutated Converters (CCC) is presented and the design of two system models is described. Their steady-state as well as transient performance is presented and the CCC is compared to the conventional HVDC transmission. The advantages of the CCC for high power transmission are shown.

Keywords: East-West interconnection, HVDC transmission, CCC, capacitor commutated converter, transient stability

I. Introduction

After the connection of the CENTREL to the UCTE power system and due to political approach of the countries in Eastern Europe towards the European Union, their power grids are in a process of new orientation. There is an effort of using the advantages of an electrical interconnection between the UCTE and the Russian power system.

A benefit of a high power East-West interconnection over the long distance from Germany to Russia of about 1000 km is the exchange of peak load due to the time shift. This results in the compensation of load peaks in the load curve profile of the networks. Moreover the grids can support each other with reserve power during faults and outages. In this way both systems can reduce their installed power capacities and there is a possibility of utilising the power plants more efficiently [1].

The maximum capacity level for the transmission depends on the available capacity on transmission lines and, in the case of total loss of the interconnection, the transient stability must be ensured. Consequently the maximum rated capacity of the transmission system is determined to be 4000 MW [1].

Since there are big differences in power system management and control a synchronous con-nection between the Eastern and Western networks can currently not be realised. So the only way of transmitting a large quantities of electrical power over long distances is the High-Voltage Direct-Current (HVDC) transmission. In the case of a later synchronous interconnection between the power networks, the HVDC system could support the power exchange further on and the stability of combined network can be improved [2].

The point of interconnection in Eastern Europe can be a problem for the transmission system. If the short circuit ratio (SCR) of the network is low, the operation of a HVDC converter can be difficult due to its consumption of reactive power.

A few years ago a new concept for HVDC systems was presented [3]. The Capacitor Commutated Converter (CCC) has capacitors connected in series between converter transformers and valves as shown in Fig. 1. These commutation capacitors (CC) provide an additional commutation voltage allowing to operate the rectifier at smaller firing angles respectively the inverter at smaller extinction angles [3, 4, 5].

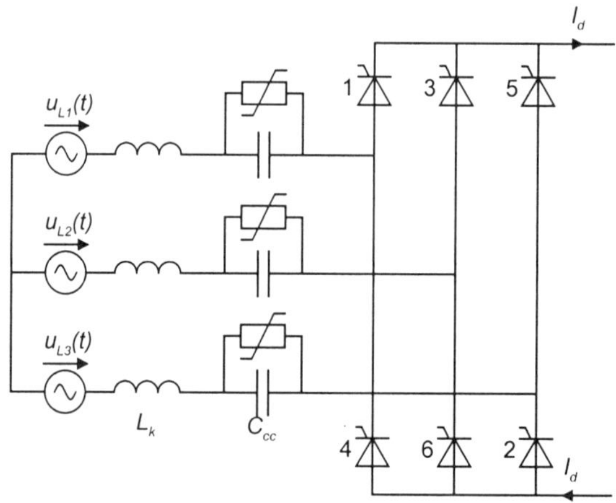

Fig. 1: Capacitor Commutated Converter (CCC)

Hence the consumption of reactive power and thus the size of the filter capacitance is well reduced. In this way the steady-state and the transient stability of the system is increased. The CCC can be operated at networks with lower SCR than the HVDC converter, which is shown in this paper.

AC-DC Power Transmission, 28-30 November 2001
Conference Publication No. 485 © IEE 2001

II. System Modelling

Using the conventional and the CCC-HVDC technology, design and modelling of a transmission system with a rated power capacity of 4000 MW over a distance of 1000 km is described in this chapter. The model should meet steady-state behaviour as well as the transient performance of both systems. With these models transient simulations are performed to analyse and compare their response to transient system faults. The simulation runs are realised using the transient simulation program EMTDC / PSCAD.

II.1. Basic System Characteristics

The maximum transmission capacity of the HVDC network is limited to 4000 MW due to the stability of the ac networks. To realise the dc transmission at this power level, the HVDC converter is designed as a bipolar system with a ground electrode. The rated dc voltage is ±600 kV, hence the rated dc current is I_{dr} = 3.33 kA.

The rectifier is operated in constant dc power control mode, whereas the inverter is controlled to a given minimum extinction angle γ_{min}. In the control of the converter VDCOL is included enhancing the stability of the system at transient ac voltage drops [6]. The design of control for the conventional as well as the CCC converters is basically the same, modelled on the control of the CIGRE benchmark model [7].

In Fig. 2 a schematic diagram of the whole system is shown. An ac network, a power plant and the reactive power compensation, i.e. ac filters and capacitors, are connected to the busbar at the rectifier side. The ac network is a relatively strong system with a SCR of 5. The generator supports a part of the power transmitted. The rated power output is controlled a constant value of 2600 MW and the generator bus voltage is controlled to be constant.

At the inverter side only the ac filters and compensation capacitors for supplying reactive power and the ac system are connected to the inverter bus. The short circuit ratio of this network is of special interest. It is assumed to be a high-impedance ac network. So the SCR is changed during the analysis of the systems.

II.2. Conventional HVDC System

The main converter parameters are specified in Table I. The ratings of the converter transformer can be seen as well as the minimum extinction angle γ_{min} and the installed reactive power Q_c at each converter bus. To filter the harmonics produced by the converter two ac filters for the 13th, for 11th harmonic and a high pass filter are installed. Each is designed to supply 500Mvar of reactive power at rated frequency.

TABLE 1 - Converter parameters

	γ_{min}	Q_c	S_{rT}	U_{rT}	u_{kr}
	°	Mvar	MVA	kV / kV	%
HVDC	18	2500	1212	400/257	18
CCC	2	850	1075	400/233	18

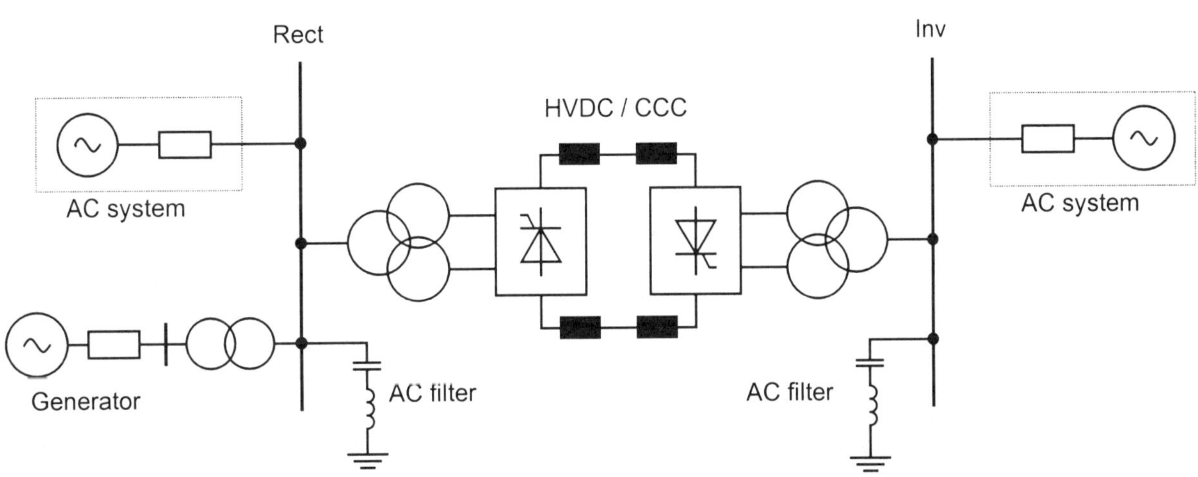

Fig. 2: Model system

II.3. CCC System

In contrast to the conventional HVDC transmission system the reduced extinction angle, due to the additional commutation voltage supported by the CCs, leads to a decreased consumption of reactive power. So the ac filter capacitors can be smaller and the quality of the filters can be improved.

It is practical to limit the size of the capacitors to a value allowing to extend the firing angle range at the inverter up to 180° [5]. The capacitance of the CCs used in this model is determined to $C_{cc} = 141{,}5\ \mu F$.

Fig. 3 shows the single line circuit diagram for the CCC inverter. The main parameters are also displayed in Table I. The values for the rated dc voltage and current are equal to the design of the conventional HVDC. The inverter bus voltage is 400 kV at the nominal operation point.

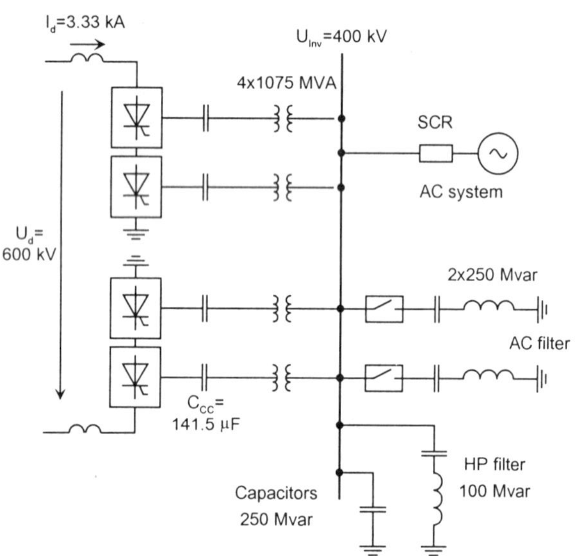

Fig. 3: Inverter station of the Capacitor Commutated Converter

Compared to the conventional converter design the rated secondary transformer voltage can be smaller and the rated power of the transformer can be reduced as well. For the 11th and 13th harmonics ac filters with 250 Mvar are installed. The high pass filter supplies 100 Mvar at nominal ac bus voltage. Additionally a capacitor bank with 250 Mvar is connected to the busbar. Therefore the filters can be installed in only one unit. The total installed reactive power Q_c is 850 Mvar, this is about 21 % of P_{rd}.

A disadvantage of the CCC concept is the increase in high ac harmonic currents caused by a reduced commutation time and thus a reduced overlap angle [3]. This effect can be met by increasing the quality of the ac filters. Improving the quality is possible since the filter capacitance is reduced by the factor of 2.

III. Steady-State Performance

In this chapter the steady-state performance of both HVDC systems is described and compared using MAP.

III.1. Reactive power consumption

The largest advantage of the CCC is the reduced consumption of reactive power. Hence smaller ac filter capacitors can be installed. During small load operation the filters don't have to be switched.

Fig. 4 a. and b. show the reactive power consumption of the conventional converter and the CCC as a function of the dc current referred to the rated dc power P_{rd}.

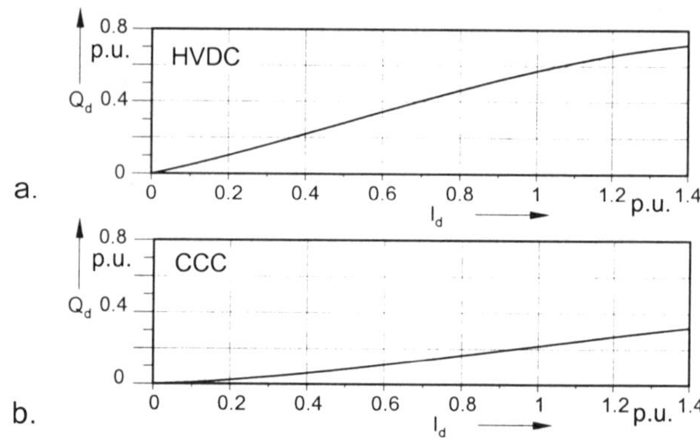

Fig. 4: Reactive power curves of conventional HVDC (a.) and CCC (b.)

The CCC consumes a reactive power of 0.23 P_{rd} at nominal operation point, whereas the conventional HVDC requires about 60 % of P_{rd} when operated at an ac network with a short circuit ratio of 3. Hence, at nominal operation point the reactive power consumption is reduced by nearly 2/3rd of its former value.

III.2. Maximum Available Power (MAP)

To analyse the steady-state stability of the two concepts, the Maximum Power Curves (MPC) are often used [6]. The theoretical curves show the maximum transmitted power when the inverter is operated in minimum extinction angle control mode at a certain SCR value of the ac network. All other values like tap changers, shunt capacitors and reactors are kept constant. The ac bus voltage is not controlled.

After a certain operation point an increase of dc current results in a decrease of transmitted power due to a higher consumption of reactive power. Hence the increase in dc current is smaller than the reduction of the ac voltage. The power transmitted at this operation point is called the Maximum Available Power (MAP). For a converter operated in constant power control mode the MAP point is the border between stable and instable operation of the system.

In the Fig. 5 b. and d. the MPCs are shown for both HVDC technologies at different ac system strengths. The corresponding ac bus voltages U_{inv} are presented in the graphs besides in Fig. 5 a. and c.

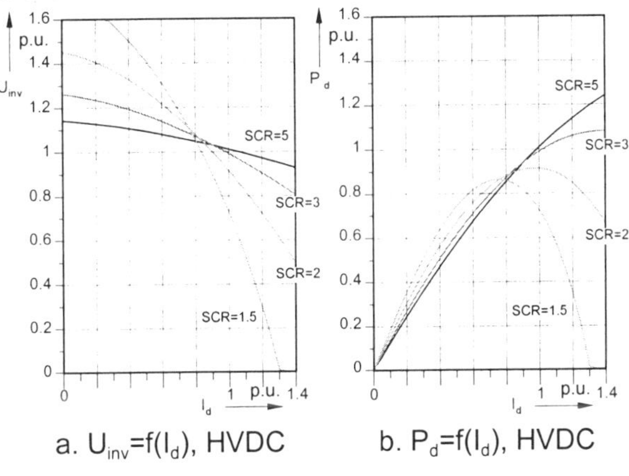

a. $U_{inv}=f(I_d)$, HVDC   b. $P_d=f(I_d)$, HVDC

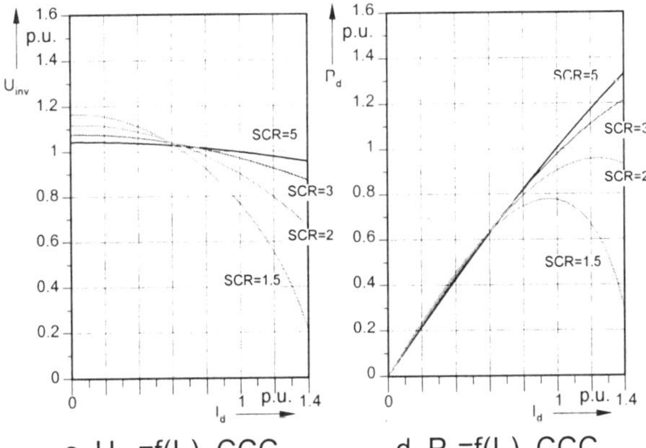

c. $U_{inv}=f(I_d)$, CCC   d. $P_d=f(I_d)$, CCC

Fig. 5: U_{inv} and P_d as a function of dc current depending on the SCR

When using the CCC concept the gradient of the ac voltage is smaller compared to the conventional HVDC converter. This leads to a more stable operation expressed in a MAP point at a higher value of dc current. The MPCs show this increase of stability.

Even at a SCR value of 1.5 the converter can be operated up to rated dc current without loosing stability. Yet the operation at a strong inverter network (SCR = 5) shows a difference between the two concepts. From the graph of the ac bus voltage can be seen that an operation without switchable filter components can be realised over the whole dc power range.

If the network strength is changed until the nominal operating point coincides with MAP, the corresponding short circuit ratio is termed the critical SCR (CSCR). The CSCR can be expressed as follows [8]:

$$\text{CSCR} = \frac{1}{U_{inv}^2}\left[-Q_d + P_d \, ctg\frac{1}{2}\left(90° - \lambda - u\right)\right] + \frac{Q_c}{P_{dcN}} \quad (1)$$

For the HVDC system the CSCR can be calculated to $\text{CSCR}_{conv} = 2.19$, whereas the CCC's CSCR is determined to 1.39. This suits to MPSs for SCR = 2 for the conventional converter respectively SCR = 1.5 for the CCC presented in Fig. 5.

IV. TRANSIENT PERFORMANCE

Conventional HVDC and CCC converter are now compared with respect to their transient behaviour. The reaction to a three phase ac bus fault and a reduction of the dc power setpoint are investigated.

IV.1. THREE PHASE AC BUS FAULT

Typical faults are applied to both system models designed in chapter II. The transmission systems are operating at the nominal operation point with $P_{rd} = 4000$ MW. For the simulation the transient simulation program EMTDC / PSCAD is used. To show the performance after a three phase ac bus fault at the inverter side of the converters, the dc current and voltage, inverter firing angle and the inverter ac bus voltage U_{inv} are presented in the Fig. 6 and 7.

After 0.1 s the fault occurs and is extinguished after 0.05 s which can be seen in the waveform of U_{inv}. After an initial overcurrent the dc current is brought to zero by the converter control. When the fault is extinguished, both systems recover to their nominal values.

The overcurrent at the beginning of the CCC is slightly smaller compared to the conventional converter. During the fault the waveforms of dc voltage and current of the CCC are more smooth due to better damping.

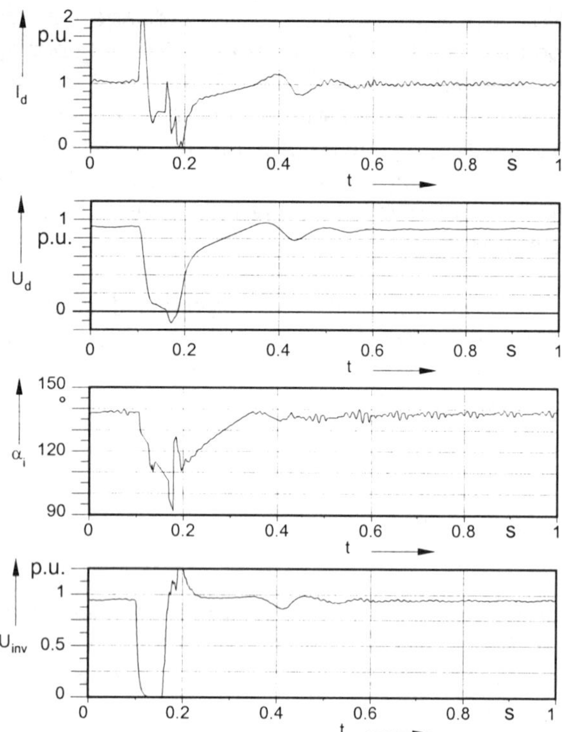

Fig. 6: Recovery from three phase ac bus fault at HVDC inverter

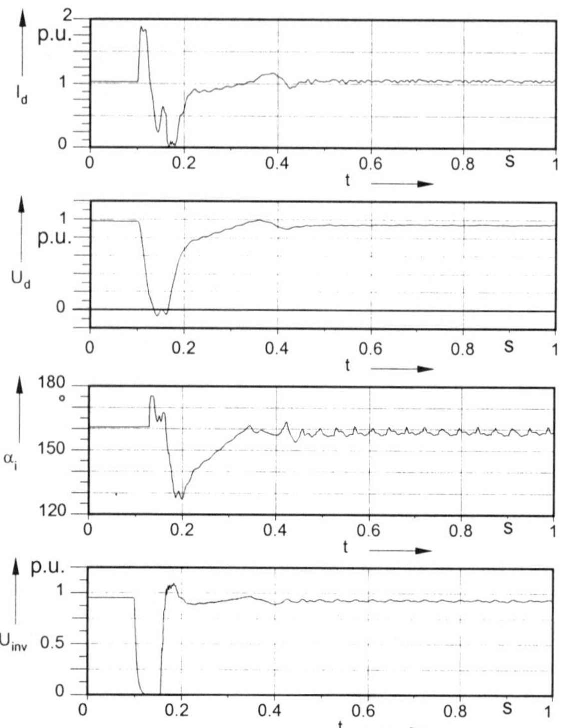

Fig. 7: Recovery from three phase ac bus fault at CCC inverter

The recovery times of the dc power of CCC and conventional system are nearly similar. 90 % of the power is restored after 130 ms.

After the fault is cleared the variables rise to the values of the former operation point. The inverter bus voltage U_{inv} overshoots its nominal value. The voltage overswing in the CCC system is lower. The transient response of the voltages and current is better damped than using the conventional HVDC technology.

IV.2. DC POWER REDUCTION

As a second transient event the response to a reduction of the power setpoint for the power controller is analysed. The power is decreased from nominal value to 0.75 P_{rd}.

In Fig. 8 and 9 the transient behaviour of the dc voltage, the dc current and the inverter bus voltage is shown. The duration of the simulation is one second, the event takes place at 0.1 s.

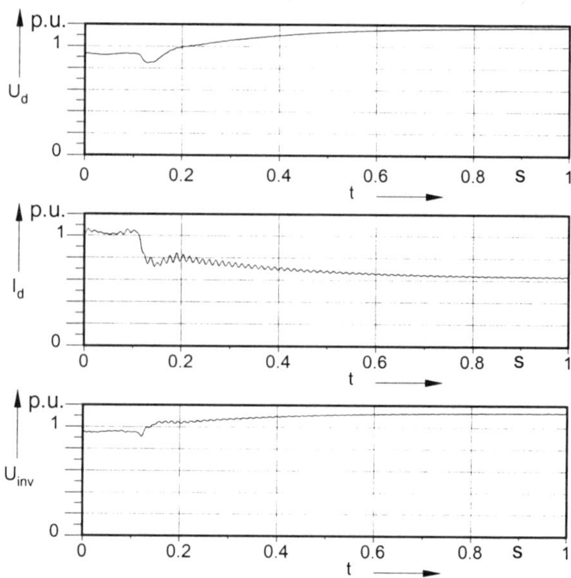

Fig. 8: Load reduction of 0.25 P_{rd}, HVDC

Due to the decrease in reactive power consumption the ac bus voltage at the inverter rises. Hence U_d is increased about 20 %. The power controller reduces the dc current to about 0.65 I_{rd}. Hence a reduction of 25 % in dc power is followed by a decrease of I_d of 35 %.

The waveforms of all graphs show a good damping characteristic. There is no overswing in the voltage waveforms. The dc current shows a high frequency oscillation. However U_{inv} is increased to a value that is not acceptable for steady-state operation, so the compensation capacitors or an ac filter has to be switched off.

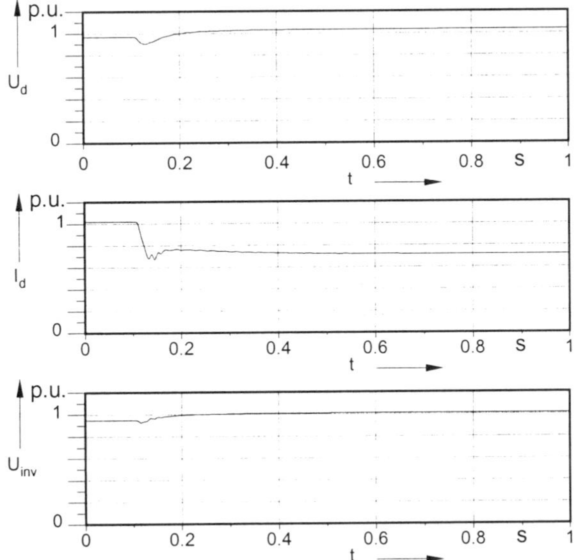

Fig. 9: Load reduction of 0.25 P_{rd}, CCC

Compared to the conventional HVDC the transient response of the CCC to the same event is shown in Fig. 9. The voltages U_d and U_{inv} are nearly kept constant. Their value is increased by less than 10 % indicating only a small reduction of the reactive power consumption.

The inverter bus voltage remains in the given voltage range for acceptable converter operation. The dc current is reduced to above 70 %. So the converter's behaviour results in an improved steady-state operating point.

Like the conventional HVDC system the CCC doesn't show any overshoot characteristics and the waveforms of the graphs are well damped.

V. CONCLUSION

In this paper two possibilities for the interconnection of the power systems of Eastern and Western Europe are shown. The CCC as a new concept of HVDC transmission system is presented.

The basic design parameters for a high power transmission system are introduced. The maximum rated dc power is limited to 4000 MW and the distance is about 1000 km. For both technologies a model system for steady-state and transient simulations is designed and described. The differences in the specification of the converter parameters are described. Here the design of the ac filters and the installed reactive power is important.

The steady-state performance is analysed by examining the reactive power consumption and the

Maximum Power Curves. The shift of the MAP and hence the smaller CSCR indicates the improved steady-state behaviour of the CCC system.

For investigating the transient performance of both HVDC technologies, a three phase ac fault at the inverter busbar and a reduction of the DC power are applied to the system models. The recovery time from the fault is nearly the same, but the waveforms show a better damping characteristic for the CCC. The response to the decrease of the DC power setpoint show a faster reaction. Here the waveforms are better damped, too. In contrast to the conventional HVDC, the ac bus voltage at the new operation point, reached by the CCC, is within an acceptable range. Therefore switching of compensation devices is omitted.

These investigations show advantages of the CCC in its steady-state and transient behaviour especially when feeding into a weak ac network.

VI. REFERENCES

1. Brumshagen H., Radke U., Berger F., 1996 "East-West European High Power Transmission System", 36th CIGRE Session

2. Tubandt H.-P., Focken B., 1995, "Future Development and Tasks of the Interconnected Network (German)", Elektrizizätswirtschaft, 25, 1663-1670

3. Jonsson T., Björklund P.-E., 1995, "Capacitor Commutated Converters for HVDC", PowerTech Conf., 44-51

4. Reeve J., Baron J. A., Hanley G. A., 1968, "A Technical Assessment of Artificial Commutation of HVDC Converters", Trans. on PAS, PAS-87, 1830-1840

5. Sadek K., Pereira M., Brandt D.P., Gole A.M., Daneshpooy A., 1998, "Capacitor Commutated Converter Circuit Configurations for DC Transmission", Trans. on Power Del., 13, 1257-1264

6. Kundur P., 1994, "Power System Stability and Control", McGraw-Hill Inc.

7. Szechtman M., Wess T., Thio C.V., 1991, "First Benchmark Model for HVDC Control Studies", Electra, 135, 56-73

8. 1997, "IEEE Guide for Planning DC Links Terminating at AC Locations Having Low Short-Circuit Capacities. Part I", IEEE Std 1204-1997

INNOVATIVE TECHNOLOGY IN THE MOYLE INTERCONNECTOR

I. Atkinson, M. Smith   P. Damgaard   M. Haeusler, M. Kuhn, P. Lips   G. Balog

Moyle Interconnector plc   Danish Power Consult   Siemens AG   Nexans Norge AS
United Kingdom   Denmark   Germany   Norway

INTRODUCTION

Power transmission by High Voltage Direct Current (HVDC) technology has been in commercial use for more than 40 years. Even though the basic concept has remained unchanged during the past decades, many technological improvements have been developed increasing system reliability and availability. Several recent examples of such innovative technology are implemented in the new HVDC interconnector between Northern Ireland and Scotland, the Moyle Interconnector, a dual monopole (2x250 MW) system connecting the transmission systems in Northern Ireland and Scotland.

This paper focuses on the direct light triggered thyristors and the modular redundant control system at the heart of the modern converter stations. It also discusses the use of cables with integrated return conductors, combining the high-voltage and metallic return cables into a co-axial design. Other new technologies as triple tuned ac harmonic filters and hybrid optical dc current measuring devices, which have already been discussed in other publications (e.g. Wild et al (1)), are only briefly mentioned.

THYRISTOR VALVES

Due to the excellent operating performance demonstrated in the Pacific Intertie, the modern technology of direct-light-triggered thyristors (LTT) including integrated overvoltage protection (Niedernostheide et al (2)) is used for the thyristor valves (Harvey at al (3)).

Valve Design

The HVDC thyristor valves for the Moyle Interconnector including direct-light-triggered thyristors with integrated overvoltage protection (LTT valves) are based on the same electrical and mechanical design features as electrically triggered valves used in earlier systems, including
- A hierarchical modular structure
- Water cooling of all heat producing components with parallel cooling circuits at all levels
- Wire-in water technology for the snubber resistors
- Exclusive use of fire retardant insulating material and wide spacing for thermal separation of components

Considering the transmission voltage of 250kV d.c. and the performance requirements, a series connection of 39 series connected 8kV LTT thyristors per valve was chosen, arranged in three valve sections of thirteen thyristor levels each. Two valve sections are mechanically combined into one modular unit, Figure 1.

Figure 1: View into an LTT thyristor modular unit including two valve sections during assembling.

The sizing and rating of the valve damping and voltage grading circuits ("snubber circuits") is not different to earlier designs using electrically triggered thyristors (ETTs), since the electrical parameters of 8kV LTTs and ETTs are practically the same with one exception: the increased di/dt capability of the LTT allowed the reduction of the number of standard reactors per valve section from four to three.

Triggering and Monitoring

Direct-light-triggered thyristors with integrated overvoltage protection do not require auxiliary energy nor electronic logic circuits for protection at high potential. Instead, the light trigger pulse generated at ground potential is applied directly to the thyristor gate, Figure 2.

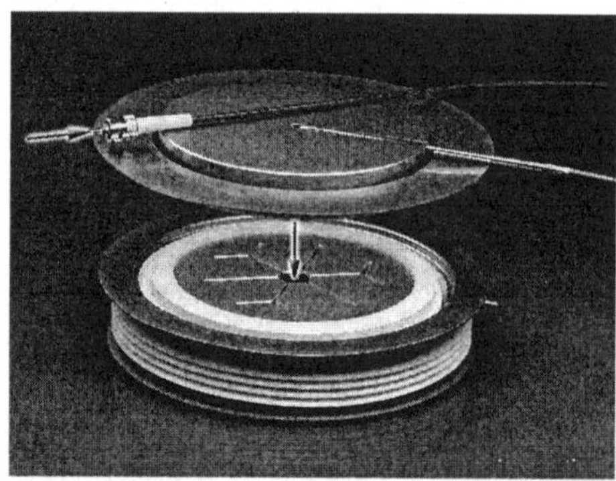

Figure 2: Artist's rendering of opened LTT with fibre optic cable illustrating the transmission of the firing pulse to the gate.

AC-DC Power Transmission, 28-30 November 2001
Conference Publication No. 485 © IEE 2001

Similarly, information on the status of the thyristor level as obtained from a voltage divider is transmitted to ground potential using an infrared diode without the need for auxiliary power. The information from each thyristor level is then evaluated in the valve base electronics and provides the same level of diagnostics as is known for ETT technology.

In conventional ETT technology, it is general practice to use one fibre optic cable per thyristor level to transmit triggering information in one step from ground potential to the thyristor electronics. For LTT valves, a different approach is more appropriate, Figure 3.

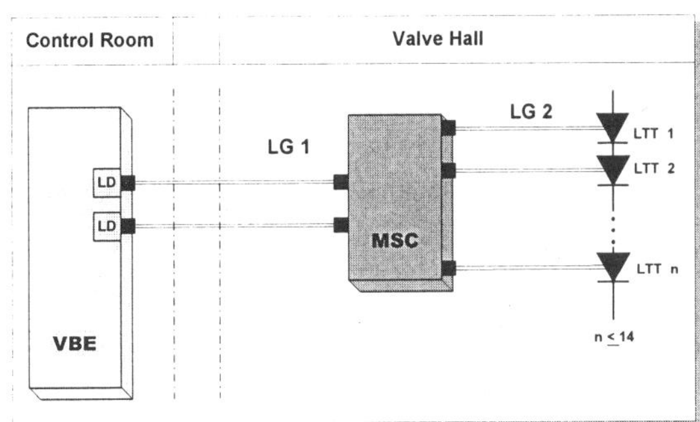

Figure 3: Schematic of LTT trigger pulse transmission (explanations in text).

The trigger pulses for the thyristors in a complete valve section are generated by redundant laser diodes and transmitted to the modular unit through separate fibre optic cables (LG1). At this point, the redundant channels are combined in a Multimode Star Coupler (MSC) and are then spread out to the individual thyristors of the section through separate fibres (LG2).

In this way, the number of active optical components in the triggering system as well as the number of long fibres from the control equipment to the high voltage potential in the valves is considerably reduced. Naturally, there still need to be individual fibres per thyristor level for monitoring purposes.

The multimode star coupler is a standard device in the communications industry when optics technology is used. Figure 4 shows a typical example with the cover removed.

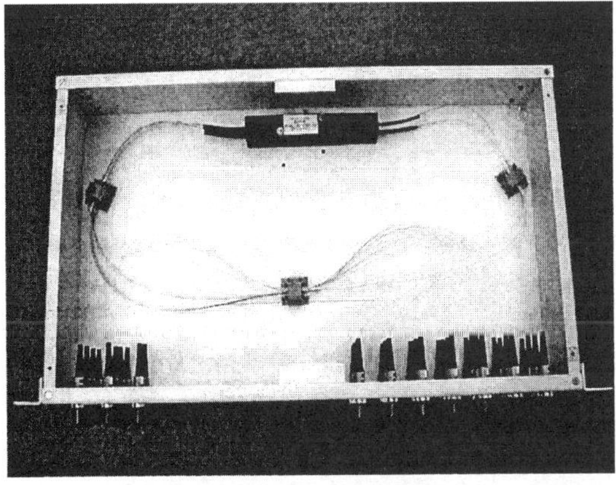

Figure 4: Inside of a Multimode Star Coupler unit showing the coupling device (top) and the internal fibre optic connections to the plug connectors at the bottom.

It is obvious that the MSC is a purely passive device that can best be compared to a soldered wire connection. The fibre optic cables from the VBE terminate at the plug connectors seen at the lower left, while the cables continuing to the individual thyristors are connected at the lower right.

Redundancy Considerations

For reliable operation of HVDC converter stations, parallel redundancy is included for all control and protection circuits at ground potential including the VBE and redundant series connected thyristor levels are included in the valve. Since each thyristor has its individual firing and monitoring, this is part of the series redundancy.

Somewhere between the VBE and the individual thyristor levels there needs to be a transition from parallel redundancy to series redundancy. There are basically three possibilities:

- At the output of the VBE in an optical interface using infrared diodes, that are each feeding a fibre optic cable which connects to the gate electronic of one thyristor level (ETT only). Here, the IREDs are part of the series redundancy of the valve; the transition from parallel to series redundancy is in the electronic switches for activating the IREDs and thus relies on active devices.

- The IREDs can be included in the parallel redundancy of the VBE by statistically mixing the individual fibres of a number of cables into a bundle at the ground end and then splitting the combined cable into two, which are connected to one light source each. With this method, the transition from parallel to series redundancy is now in the mixing point of the fibre optic cables - comparable to hard wiring and not depending on active components. One fibre optic cable per thyristor from the VBE is still required.

- An additional improvement and simplification is made in the LTT technology, Figure 3. With this method, the transition from parallel to series redundancy is left in the passive, "hard wired" fibre optic system. However, the number of long cables from ground to high voltage potential is greatly reduced, resulting in a lower risk of damage.

Testing

According to the project specification, type/design tests were to be performed according to IEEE 857, 1990. That document specifies dielectric tests on the valve base, a single valve, a multiple valve, and operational tests on valve sections. Figure 5 shows a complete quadruple valve including valve arrester housings set up for dielectric tests at the IEH test facility in Karlsruhe, Germany.

The operational tests were performed in the synthetic test circuit available in the factory which imposes stresses that are equal or higher than the worst conditions that can be expected in service. Throughout the whole test series, there was no failure or degradation of any kind.

Figure 5: LTT quadruple valves assembled on site.

CONTROL OF THE LINK

The control and protection systems in each converter station of the Moyle Interconnector are following the modern concept of a completely modular and redundant design, which provides the following benefits:
- Modular design which provides optimum redundant solutions
- Design structured for easy expansion in the future
- Application of well proven standard hardware and software systems
- High integration of all control and protection systems
- Satellite synchronised station master clock system (time base for fault and event recording)
- Modern redundant fibre optical field bus systems and local area networks
- Standardised telecommunication systems and protocols for inter station communication as well as for communication to control centres for remote control
- I/O Units with redundant field bus interface

Figure 6 shows the various levels of the HVDC control hierarchy for the two converter stations of the Moyle Interconnector.

Basically there are three main levels in each converter station:
- Operator Control Level
- Control Level
- Process Level including interfacing

The **Operator Control Level** in each converter station consists mainly of the fully redundant station initiation and monitoring system. The interfaces to the remote control facility at the higher network control level (e.g. the dispatch centre) and the telecommunication interface to the corresponding converter station for exchange of monitoring data are also part of that level. To have a common time base for all station equipment a central master clock system (GPS syncronised) is provided.

The **Control Level** comprises the station control system and the pole control system including the thyristor gate control system (valve base electronic). The telecontrol system, which exchanges fast control and protection data between the stations, is directly connected to the related pole control. The event recording is integrated within the controls. The transient fault recorder as an independent recording equipment is not explicitly shown in Figure 6. All the control systems at the control level are fully redundant.

The **Process Level** mainly consists of the HVDC converters (transformers and valves), the AC filters/shunt capacitors, the complete AC and DC switchyard and the auxiliary systems (auxiliary power supply, valve cooling, fire protection, ...). The respective I/O interfaces are also part of the process level.

Information exchange between the operator control level and the control level is provided by a redundant Local Area Network (LAN). The required information exchange between the control level and the process level is achieved via a redundant field bus system.

The measured dc values (dc current and dc voltage) are **redundantly** transmitted via **fibre** optic links from the

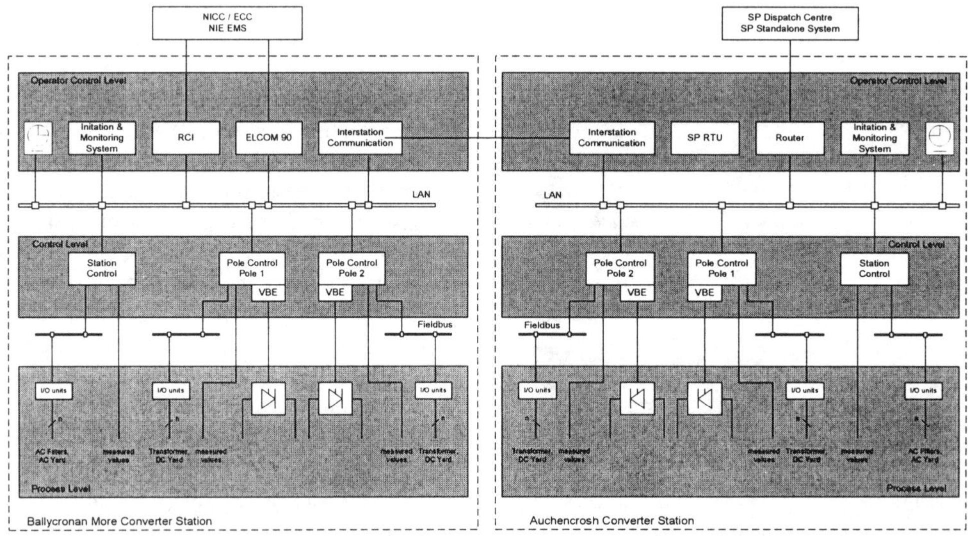

Figure 6: HVDC Control Hierarchy Moyle Interconnector (redundancy is not shown)

measuring point to the related control and protection cubicle. This fast digital (numerical) light-signal is insensitive to electromagnetic noise and therefore an excellent solution for data transmitting in the high voltage yard.

The **Protection Systems** are totally independent from the control systems and located in separated cubicles to avoid any interference (not shown in Figure 6).

Operation of the Moyle Interconnector

Operation of the Moyle Interconnector will be performed according to a clearly structured control location hierarchy with different control function groups under consideration of the two control levels (for HVDC control) system control level and station control level.

The control location hierarchy is based on the following control locations:

- field control (operation close to the process with highest priority with prevention of control from any distant control),
- local control from operator initiation and monitoring systems in the converter stations (Northern Ireland: Ballycronan More converter station; Scotland: Auchencrosh converter station),
- remote control from central control centres in Northern Ireland (from Northern Ireland Electricity's (NIE) control centre (NICC) near Belfast or NIE's emergency control centre (ECC)), respectively Scotland (Scottish Power's (SP) dispatch centre near Glasgow).

The control functions of the Moyle Interconnector are divided into the following 3 control function groups:

- AC Control Ballycronan More:

  Control of the line bays and bus coupler bay at Ballycronan More with control locations Northern Ireland (local initiation & monitoring system Ballycronan More, NIE control centre (NICC), or NIE's emergency control centre (ECC)).

- AC Control Auchencrosh:

  Control of the line bay at Auchencrosh with control locations in Scotland (local initiation &monitoring system Auchencrosh, SP dispatch centre via SP standalone system, or SP dispatch centre via SP RTU). Normally controlled by SP RTU.

- HVDC Control (Ballycronan More and Auchencrosh):
  The HVDC Control function group covers the remaining part of the converter stations i.e. filter bays, converter bays, HVDC control functions and auxiliary power. If inter station communication between the two converter stations is available, then the HVDC system is operated in the system control level, otherwise in station control level. The system control levels permits operation of HVDC related functions from only one of the converter stations. The controlling station is referred to as "master station", the other as "slave station".

Since the converter stations will be unmanned, the control systems are designed for fully remote operation of both converter stations. The remote control will primarily be from Northern Ireland Electricity's (NIE) control centre (NICC), but full remote control is also possible from NIE's emergency control centre (ECC) or Scottish Power's (SP) dispatch centre. The HVDC link can also be controlled from the operator control rooms at the converter stations, but this will only take place during commissioning and maintenance due to NIE's operational philosophy.

In Northern Ireland the remote control of the converter stations is fully integrated into the existing energy management system (EMS) at NICC and ECC. All data related to the AC control is transmitted from the Remote Control Interface (RCI) in Ballycronan More to the front ends of NIEs EMS in NICC and ECC using the IEC870-5 101 protocol. All data related to the HVDC control is interfaced via a gateway computer in Ballycronan More to the front ends in NICC and ECC. Data exchange is provided via ELCOM90 protocol.

The communication between the Ballycronan More Converter Station in Northern Ireland and the NIE's dispatch centre for RCI and ELCOM90 gateway are provided via two redundant 2Mbit/sec circuits, which are routed via diverse routes to improve the reliability. The two circuits are at the dispatch centre connected to the redundant front ends, which are then connected to a LAN of NIE's EMS system. The communication between Ballycronan More and the emergency control ECC centre is via one 2Mbit/sec circuit.

In Scotland the remote control from SP's dispatch centre is carried out from a stand-alone system (1 operator station). The standalone system is interfaced with a WAN connection via routers and telecommunication equipment to the LAN of Auchencrosh converter station. The communication between Auchencrosh and SP`s dispatch centre is provided via a 2Mbit/sec circuit and is also routed via diverse routes.

The inter station communication between the two converter stations is via two redundant 8Mbit/sec circuits over the optic fibres in the IRC cable.

CABLE CONCEPT

Up to the present many submarine DC links operated in the mono-polar mode, where a single cable or overhead line constituted the high voltage carrying component while the return current flowed through the soil or sea water. In the last decade, in connection with mono-polar DC transmission in some case stray currents were detected far from the electrodes on land. Expensive intervention had to be used to eliminate excessive corrosion on pipelines and other long metallic installations in such areas. It has been asserted that sea electrodes produce large amount chlorinated organic compounds and that the magnetic fields from the cable have negative influence on marine life. All in all the deployment of electrodes are time consuming, costly and meets large environmental resistance.

On this background the idea of integrating the return conductor into the cable was born. By applying the return conductor concentrically around the main

conductor, outside the lead sheath, the following major goals are satisfied:
- the core is a conventional massimpregnated cable core
- the return conductor insulation may be of different material than the main insulation
- there is no external magnetic field
- the laying properties are as for a conventional cable
- the return conductor is also part of the armouring

To keep the return current in the return conductor it has to be insulated from the ground.

Cable Design

The cable requirements for the Moyle project are transmission of 250 MW at 250 kV, across a total distance of 63 km, 3+5 km on land and greatest submarine depth of 160 m.

From the idea of integrating the return conductor (RC) into the cable there are several steps before the design is established.

The first step is to analyse the system in view of the probable stresses.

As the core is conventional mass-impregnated cable core, only the additional heating from the losses in the return conductor had to be taken into account for dimensioning the conductors.

As the return conductor may be earthed at one end only, the stationary DC voltage at full load was calculated to be approximately 1 kV on the other end.

To study the transient behaviour of the design the following equivalent diagram was established:

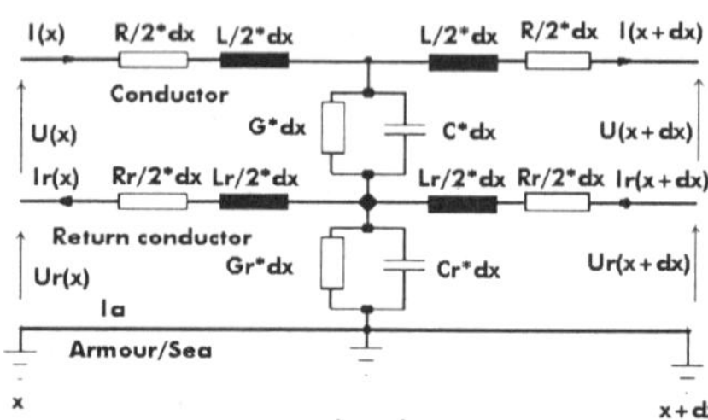

Figure 7: Unit length equivalent diagram for transient analysis

The transient analysis showed a low induced sheath voltage from atmospheric transients. However, in this case the end terminals are located indoors, so the only transients are the ones generated by the converters.

Most of the converter-generated transients occur rather seldomly. Commutation failure and bushing flashover type of faults are most common. The transients are of the dampened oscillating type and the voltage induced by them on the RC is even lower.

Based on the above the return conductor insulation was designed to satisfy the requirements in IEC 60229.

A layer of steel amour outside the RC insulation act both as a mechanical protection of the insulation and together with the return conductor as a torsionally balanced longitudinal load member.

The rest of the design is as for a conventional mass impregnated cable. The insulation thickness is 11,5 mm giving a maximum electrical stress at full load of 26,5 kV/mm.

In addition a FO element with six fibres was integrated into the power cable. The FO element was placed into the extruded plastic sheath, in contact with the lead sheath. The plastic sheath thickness is increased so it has the nominal thickness also over the FO part. The placing of the FO part in that position is the result of a project where both the properties of the fibres during cable handling and the jointing technique of the fibres were developed.

Cross-section of the resulting design is shown below.

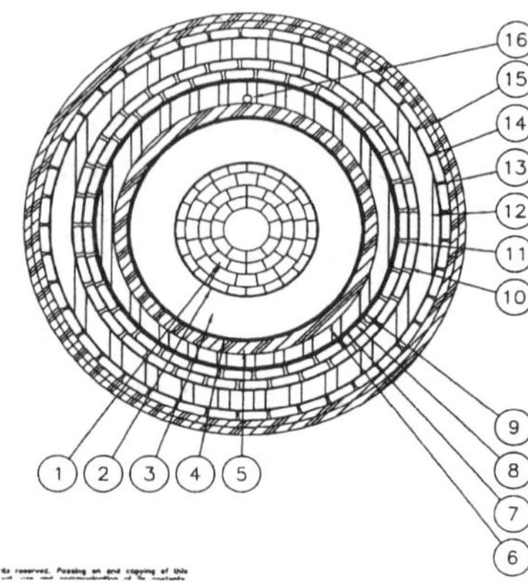

Figure 8: Cable cross-section

The salient points are: 1) Conductor 1000 mm^2, 3) Insulation 11,5 mm, 5) Lead sheath, 6) Plastic sheath, 9)&10) Return Conductor 1150 mm^2, 12) Return Conductor Insulation 5,5 mm, 16) FO cable 6 fibres

The cable outer diameter is 115 mm and the weight is 400 N/m. At full load the losses are 36 kW/km.

The design of land and submarine cables are identical. This reduces the spare cable necessary and allows long lengths to be pulled on shore.

Accessories

Joints. The jointing technique of mass-impregnated cables has been refined but it is a well-known technology. State of the art is 500 kV, flexible repair joints mechanically tested to 500 m water depth according to the CIGRE recommendations. In this case the development of the return conductor insulation that required most effort. The jointing method is replacing the insulation across the joint by longitudinally split tubes of the same material and welding the groves with a specially designed extruder. Thereafter welding the tube to the cable by a circumferential weld. The extra material is removed from the weld and the overall diameter is not increased in the process.

Cable Termination. The termination is the conventional pressurised type, to ensure internal over pressure. The lead sheath, return conductor and also the steel tube of the FO cable may carry potentially high voltage and they have to be terminated on an platform, isolated from earth and protected from unintentional touching. Also the steel tube of the FO cable and the pressurising pipe to the termination must be interrupted by a non-conductive stretch capable of withstanding stationary and transient voltages.

Testing

Type tests. The mechanical tests were performed according to the CIGRE Recommendations, as given in Electra No. 171, January 1997. The design depth was 165 m.

The electrical tests were performed according to CIGRE Recommendations in Electra No. 72, October 1980. The nominal voltage was 250 kV. The impulse voltage tests were conducted at $2,7*U_0$, superimposed on 250 kV DC with opposite polarity.

Two lengths with flexible repair joints were type tested. The return conductor insulation was tested according to IEC 60229. The DC withstand test was 25 kV for four hours, the impulse voltage level was 37,5 kV.

The FO parts were tested before and after each major test sequence with OTDR.

Performance tests.
- Bending fatigue test with 50000 cycles, moving the cable ±0,3 m out of the centre line while at the maximum test tension, was performed
- Drainage test at maximum conductor temperature for 8 days was also performed

Routine tests. Each production length was tested with $2*U_0$ high voltage, the conductor and return conductor resistance were measured.

On a 20 m length from each production length the loss angle and the capacitance were measured.

Factory acceptance testing. The complete delivery length was tested with $2*U_0$ on the main insulation for 15 minutes, 25 kV DC on the RC insulation for 15 minutes and OTDR FO fibres.

After laying tests. The take-over test will comprise the following items: Conductor continuity test, HVDC test on main insulation, - 338 kV for 15 min., DC test on IRC-insulation, 10 kV for 1 min., OTDR and insertion loss measurements on the fibres and checking the end termination pressurisation alarm functions.

Transport and Laying

Land cables. On both sides of the crossing there are long land cable sections. The route lengths are 3 km on the Northern Ireland side and 5 km on the Scottish side. There is an elevation difference along the routes, especially on the Scottish side where it is 160 m. To facilitate a short construction time and keep the trenches in the meadows open for only a short period, cable lengths were transported by sea and pulled onshore in approximately 1.5 km lengths. In this way the number of joints per length were kept to one on the Irish side and two on the Scottish side. The cables were transported and pulled ashore during autumn 2000. In order to minimise land damage trenching of the cables was started in the spring of 2001. The jointing on land started last year and was finished this summer.

Submarine cables. The submarine cables were transported in complete 55 km lengths and laid by C/S Skagerrak. It is one of the two ships capable handling and laying such heavy cables in the assigned cable corridors. The cable laying was completed in May.

Embedding. Following laying the cables were embedded with the Capjet. The Capjet utilises water jets for loosening the soil and keeping it fluidised while the cable sinks to the bottom of the trench. Propulsion is provided by a combination of water jets and thrusters. The jetting technique makes the Capjet light and responsive as compared to surface towed burial machines as ploughs. It is also a minimally invasive technique. This operation is ongoing at the time of writing this paper.

OTHER INNOVATIVE TECHNOLOGIES

Other new technologies are also implemented in the Moyle Interconnector Project. In the following only a brief summary is given since detailed information is available in (1).

Triple Tuned AC Harmonic Filters

In series to the well known double tuned filter arrangement another resonance circuit is placed. Such filtering at three harmonic frequencies provides as main advantages an improved harmonic filtering performance, optimised space requirements and minimum high voltage equipment (HV capacitor and circuit breaker). For the Moyle Interconnector the filters were tuned to 3/12/24 harmonic avoiding potential resonance problems at low order harmonics.

Hybrid Optical DC Measuring System

The DC current is measured using an ohmic shunt integrated in the DC Circuit. The measured signal is transmitted via a fibre optic link to ground potential. This technology significantly reduces weight, provides and inherent protection against electromagnetic interferences and increases the overall system availability.

ACKNOWLEDGEMENTS

The individual authors wish to acknowledge the partnership involved, not only in writing and presentation of this paper, but also during the execution of the project.

REFERENCES

1.	J. Ammon, H. Huang, A. Kumar, P. Lips, M. Pereira, K. Sadek, G. Wild, "New Developments in HVDC Technology", CEPSI 2000, Conference Paper

2.	F.-J. Niedernostheide, H.-J. Schulze, J. Dorn, U. Kellner-Werdehausen, and D. Westerholt, "Light-Triggered Thyristors with Integrated Protection Functions", Proceedings of the 12th International Symposium on Power Semiconductors and IC's, ISPSD'2000 (Toulouse, France) pp. 267 - 270.

3.	C. Harvey, K. Stenseth, M. Wohlmuth, "The Moyle HVDC Interconnector: Project Considerations, Design and Implementation.", Companion paper.

AN EXAMPLE IN SIZING A SERIES COMPENSATION SCHEME WITH FIXED AND VARIABLE CAPACITORS

H G Sarmiento[1]   E Estrada[1]   J Naude[1]   C Tovar[1] M A Avila[2]   C Fuentes[2]

Instituto de Investigaciones Eléctricas, México[1]   Comisión Federal de Electricidad, México[2]

INTRODUCTION

FACTS technology is slowly finding its way in the electric power industry. Although many FACTS controllers have been defined, some might never materialize, others are finding their place (1), and new ones will surely appear for specific applications.

This paper gives an example of a methodology to size a series compensation scheme comprised of a fixed capacitance and a thyristor controlled series compensation (TCSC). A base case for the 2004 winter peak demand in Mexico is used as reference. New generation is added in that year, but a new modified case base is formed, assuming no new lines are built to account for the added generation. Henceforth, the existing lines will have an increased power transfer, so studies are needed to design a compensation scheme that will carry this new loading.

BASE CASE DESCRIPTION

The base case used as reference corresponds to the winter peak demand of the year 2004 for the National Electrical System (NES). By this year, new generation will be added in the MMT hydro plant consisting of three 300 MW units for a total added capacity of 900 MW. The purpose of this paper is to present a structured methodology to specify a series compensation scheme with fixed and variable capacitors for a scenario in which no transmission lines are added, and only the existing transmission infrastructure is to be used to account for the new generation. In other words, a series compensation scheme is to be designed for the new increased power transfer in the existing transmission lines.

The transmission system used as an example of the study methodology described in this paper is shown in Figure 1. This 400 kV system transfers power from the main hydro plants in the southeastern part of Mexico to the central part of the country. Table 1 presents basic information on the transmission lines where the analysis is centered on.

AC-DC Power Transmission, 28-30 November 2001
Conference Publication No. 485 © IEE 2001

TABLE 1 – Base case information

Line	Length (km)	Voltage (kV)	Series Compensation (%)
MMT-JUI	243	3x400	25
JUI-TMD	155	1x400	0

Table 2 gives the power flow base case condition for the lines under study.

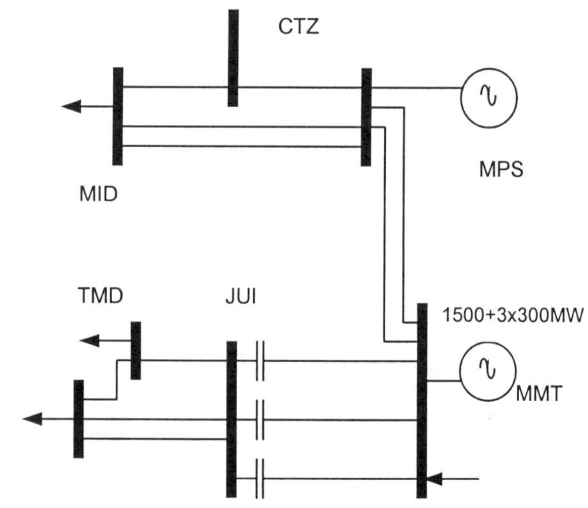

TABLE 2 – Power Flows during peak demand

Line	Transfer Level (MW)
MMT-JUI	3 x 635 = 1905
JUI-TMD	1 x 628 = 628

Figure 1 Base case transmission system

MODIFIED BASE CASE

In the modified base case, it is assumed that no new transmission lines are built in the system under study; and the existing transmission infrastructure will carry the added power at the MMT hydro plant. The present transmission system consists of two uncompensated lines from MMT to JUI, instead of three (refer to figure 1); and two lines from JUI to TMD (no third line).

Reinforcement to the 400 kV transmission system

The first activity is to reinforce the existing transmission system so a *steady state performance similar to the base case is achieved*.

The modified base case must meet the following requirements:

- The modified transmission system must exhibit steady state equilibrium, considering critical contingencies.
- Transmission voltage magnitudes must be kept within established limits.
- Transmission system components are loaded within their continuous operating limits.
- Transmission system loadability during steady state and contingencies does not show an important increment. This restriction translates into bus angles between voltages at section terminals to remain approximately the same as in the base case.

The above requirements were met with the

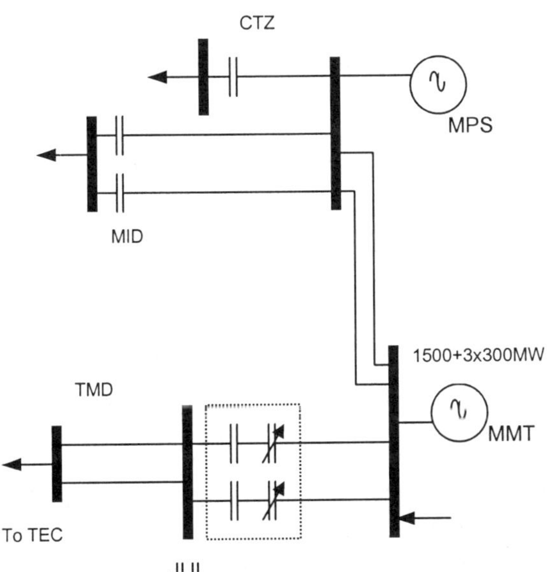

following reinforcements.

1. Upgrading existing series compensation.
2. Adding shunt compensation for voltage support.
3. Installing 60% series compensation in the two lines from MMT to JUI.

Figure 2 shows the transmission system under study with some of the reinforcements.
Figure 2 Modified base case
With reference to series compensation upgrading, it should be noted that the existing transmission network under study includes series compensation in other lines. Nevertheless, in the base case this compensation is bypassed, since its current rating is surpassed with the peak demand for 2004. When the new lines are removed, it becomes necessary to upgrade the existing series capacitor banks.

Upgrading consists of adding units in parallel to the existing capacitor bank, Kosterev, et al (2). This results in an increased current rating, although the degree of series compensation is somewhat lowered.

The equation used to simulate this upgrading is the following:

$$X_c^{nva} x I_{nom}^{nva} = X_c^{act} x I_{nom}^{act} \qquad (1)$$

Where:

X_c^{nva} = New series reactance, when upgrading the capacitor bank
I_{nom}^{nva} = New current rating of the capacitor bank
X_c^{act} = Present series reactance
I_{nom}^{act} = Present current rating of the capacitor bank

As an example, upgrading the series capacitor bank at CTZ will be shown. The present information for this bank is:

X_c^{act} = 0.00951 p.u.
I_{nom}^{act} = 844 A

The new power flow through this capacitor in the modified case is (including an additional 35%):

I_{nom}^{nva} = 1185 A

Therefore, the new series reactance is:

$$X_c^{nva} = \frac{0.00951 x 844}{1185} = 0.00677\, p.u.$$

The new degree of series compensation is 21%, whereas before it was 29%, but the capacitor bank has now the required current rating.

Table 3 shows the same information as Table 1, except for the new degrees of series compensation.

TABLE 3 – Modified base case information

Line	Length (km)	Voltage (kV)	Series Compensation (%)
MMT-JUI	243	2X400	60
JUI-TMD	155	2x400	
MPS-MID		2x400	21
MPS-CTZ		1x400	21

Table 4 shows the power flows for the modified case base.

TABLE 4 – Power flows for the modified base case

Line	Transfer level (MW)
MMT-JUI	2x856=1712
JUI-TMD	2X768=1536

Steady state analysis also included critical contingencies, mainly the following line outages: One line from MMT to JUI, one line from JUI to TMD, and one line from TMD to TEC (not shown in the figures).

From the complete steady state studies, additional reactive shunt compensation was identified, totaling 600 MVAR. The procedure followed is presented in Gao et al (3).

With the three types of reinforcements described above, the modified base case shows steady state equilibrium for all contingencies analyzed.

Bus voltage magnitudes

Additionally, the modified base case must satisfy bus voltage requirements. For buses shown in Figure 2, Table 5 presents voltage magnitudes in steady state and under one critical contingency.

TABLE 5 – Bus voltage magnitudes (modified base case)

Bus	Voltage Magnitude- kV (Steady state)	Voltage magnitude-kV (Outage TMD-TEC)
MMT	412	411
JUI	405	404
MPS	410	409
CTZ	399	398
MID	399	398
TMD	404	402

Load flow and transient stability analysis

As loading increases in the 400 kV transmission system under study in the modified base case, keeping bus angles with similar values as in the base case, prevents power flows from circulating and overloading transmission lines at lower voltage levels. Bus angle conditions are also of prime importance from the transient stability viewpoint, since they relate to the available steady state and transient stability margins.

Table 6 shows the bus angles between voltages for the modified base case and its two main outages (TMD-TEC and MMT-JUI). For reference, base case bus angle values are also shown. The installation of sixty percent degree series compensation in the modified case base, meets this restriction for the TMD-TEC outage as shown in Table 6.

TABLE 6 – Bus voltage angles

Line	Base Case	Modified Base case	TMD-TEC outage	MMT-JUI outage
MMT-MPS	4.2	4.8	4.9	6.3
MMT-JUI	15.4	12.1	12.1	**20.7**
MPZ-CTZ	12.9	10.7	10.7	12.3
JUI-TMD	13.0	16.5	16.6	13.7

As observed in Table 6, the MMT-JUI outage increases the bus angle by 71%. To reduce this angle, 15% additional series compensation in the form of TCSC (Thyristor controlled series compensation) is installed.

For the complete series compensation scheme (60% fixed and 15% TCSC), transient stability studies were carried out to assure good performance for all critical contingencies, see Appendix A.

SIZING THE SERIES COMPENSATION SCHEME

Current magnitudes and degree of compensation

Previously, series compensation was determined for the two transmission lines from MMT to JUI, in what has been called the 'modified base case', characterized by using only the existing transmission infrastructure to account for new generation at MMT. Table 7 summarizes these series compensation requirements.

TABLE 7 – Series compensation for the modified base case

Series compensation	Percent (%)	Ohms
Fixed – Steady State	60	56
TCSC – Outage	15	14

Table 8 shows current magnitudes in the MMT-JUI lines for the modified base case: Steady state and two critical contingencies. Values for the base case are also shown as reference.

TABLE 8 – Current magnitudes (A) in the MMT-JUI line

	Base Case	Modified Base case	TMD-TEC Outage	MMT-JUI Outage
MMT-JUI	890	1253	1250	2087

Continuous current rating of the fixed and variable compensation

Rated current I_{rat} of the TCSC bank must meet three requirements, according to the following inequality, Kosterev, et al (2):

$$I_{rat} \geq \max\{X_{ord}^{nor} I_{nor}; X_{ord}^{ot} I_{ot} / 1.35; X_{ord}^{sw} I_{sw} / 2\}$$
.. (2)

Where:

I_{nor} = Maximum normal line current in steady state.
I_{ot} = Maximum contingency line current in steady state.
I_{sw} = Maximum swing line current.
$X^{(nor,\ ot,\ sw)}_{ord}$ = Reactance orders of magnitude required for normal, contingency and swing conditions.

The lower limit of X_{ord} is established to prevent subsynchronous resonance in contingency and oscillation conditions, the recommended value is 1.1 (2). So:

$$X_{ord}^{ot} = X_{ord}^{sw} = 1.1$$

In normal conditions, a 15% margin is allowed in the variable capacitance for limited control in loop flows and power modulation:

$$X_{ord}^{nor} = 1.1(1.15) = 1.26$$

The following current magnitudes complete the information needed to substitute in Equation (2):

I_{nor}=1 253 A (from Table 8)
I_{ot}=2 087 A (from Table 8)
I_{sw}=2 654 A (from transient stability studies)

Substituting the X_{ord} and current magnitude values in Eq.(2) and evaluating the inequality results in:

I_{rat} = 1 700 A

It is also required to specify an upper limit for X_{ord}. This upper limit is dictated by the thyristor firing angle and typically should be less or equal to three. X_{ord} for a 30 min overload representing one of the critical contingencies has the following expression:

$$X_{ord} \leq 1.35 \frac{I_{rat}}{I_{nor}} = \frac{1.35x1700}{1253} = 1.83$$

Studies were also performed to identify power system oscillations in the modified base case for all critical contingencies. Although some damping ratios were identified with values below the recommended 0.05 (4), no power oscillation damping (POD) of the TCSC was considered necessary. In such cases, X_{ord} values for POD are in the range of 2 to 3.

For the fixed series compensation, X_{ord} = 1.0 and the rated current requirements are expressed in the following inequality:

$$I_{rat} \geq \max\{I_{nor}; I_{ot} / 1.35; I_{sw} / 2\} \qquad (3)$$

Substituting above, the rated current for the fixed series compensation is 1 546 A.

TCSC specification

Finally, other important parameters for the TCSC specification will be determined.

Apparent reactance.- This value is shown in Table 7 and equals 14 ohms.

Rated line current.- This value has already been determined and equals 1 700 A.

Rated TCSC voltage.-

$$V_{TCSC} = 14x1700 = 23.8kV / phase$$

Physical capacitive reactance.- With the upper limit of X_{ord}=1.83

$$X_C = \frac{X_{app}}{X_{ord}} = \frac{14}{1.83} = 7.65 \qquad \text{ohms}$$

Rated capacitance.-

$$C = \frac{1}{120\pi X_C} = \frac{1}{120\pi(7.65)} = 347 \qquad \text{µfarads}$$

Rated reactive power.-

$$Q_{t\,csc} = 3\frac{kV_{tcsc}^2}{X_C} = \frac{3x23.8^2}{7.65} = 222 \qquad \text{MVAR}$$

Inductive reactance and inductance.- To determine inductance 'L' of the TCSC, factor λ is used. This factor is the ratio between the resonant frequency of the LC circuit and the fundamental frequency (60 Hz). A value of $\lambda = 2.5$ is recommended. This results in lower thyristor currents. Using $\lambda = 3.0$ may cause problems, since the circuit is tuned to the 3rd. harmonic when TCSC is bypassed.

$$X_L = \frac{X_C}{2.5^2} = \frac{7.65}{2.5^2} = 1.224 \qquad \text{ohms/phase}$$

$$L = \frac{X_L}{\omega} = \frac{1.224}{120\pi} = 3.25 \qquad \text{miliHenrys}$$

TCSC bypass reactance.- This is the TCSC reactance when thyristor valves are not conducting.

$$X_{bp} = \frac{X_C}{\lambda^2 - 1} = \frac{7.65}{2.5^2 - 1} = 1.46 \qquad \text{ohm/phase}$$

MOV protection margin.- Normally, the continuous operation requirement dictates the protective voltage level of the Metal Oxide Varistors (MOV) connected in parallel with the capacitor bank. This protective voltage is the maximum instantaneous voltage appearing across the MOV in any fault condition. Typically, this protective level is about 2-2.5 times the peak voltage at continuous rating. That is,

$$V_{tcsc}x2.5 = 23.8x2.5 = 59.5 \qquad \text{kV}$$

Thyristor firing angle.- Traditionally, thyristors are described using the firing angle "α" (alpha) as the control variable. This angle indicates the delay in electrical angle from the first instant in which the thyristor receives the voltage signal until it fires, Angquist et al (6).

The apparent reactance characteristic at fundamental frequency vs. firing angle is shown in Figure 3.

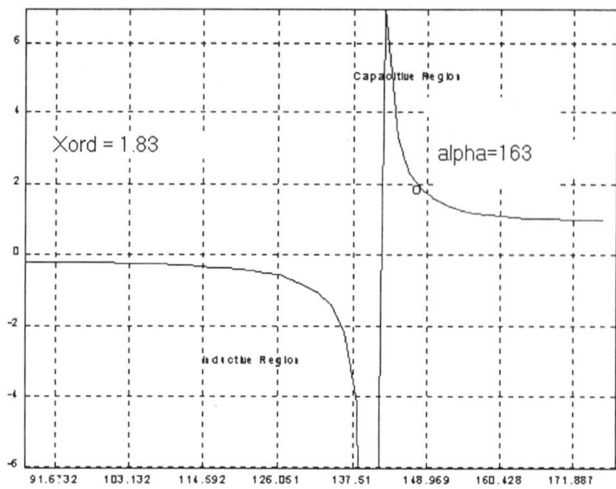

Figure 3 Firing angle vs. X$_{ord}$

CONCLUSIONS

1. A structured study methodology has been presented to size a series compensation scheme consisting of fixed and variable capacitors.
2. The example presented in this paper illustrates how conventional series compensation with a TCSC module can be used to defer the building of new lines when electricity demand increases.

REFERENCES

1. IEEE Power Engineering Society, 1996, "FACTS Applications", IEEE 96TP116-0, U.S.A.

2. Kosterev, D.N., W.A.Mittelstadt, R.R.Mohler, W.J.Kolodziej, 1996, "An Application Study for Sizing and Rating Controlled and Conventional Series Compensation", IEEE Trans. on Power Del., 11, 1105-1111.

3. Gao, B., G.K.Morison, P.Kundur, 1996, "Towards the Development of a Systematic Approach for Voltage Stability Assessment of Large-Scale Power Systems", IEEE Trans, on Power Sys. 11, 1314-1319.

4. CIGRE Task Force 07 of Study Committee 38, 1996, "Analysis and Control of Power System Oscillations", Technical Brochure 111, France.

5. Paserba, J., N.W.Miller, E.Larsen, R.J.Piwko, 1994, "A Thyristor Controlled Series Compensation Model for Power System Stability Analysis", IEEE 94 SM 476-2 PWRD.

6.Angquist, L., Ingestrom, H.A., 1996, "Dynamical Performance of TCSC schemes" CIGRE Session article 14-302.

APPENDIX A

A TCSC module consists of a series capacitor and a parallel path with thyristor valves in antiparallel and a reactor in series, Paserba et al (5).

Operation modes range from a condition in which the thyristor valves are blocked, resulting in conventional series compensation; to a condition where these valves are conducting continuously, resulting in a small inductive reactance. In between, a condition of partial conduction or 'Vernier' control exists, which can be used to increase the apparent capacitive reactance of the line.

Figure 4 illustrates the block diagram of the TCSC model used in transient studies, Paserba et al (5).

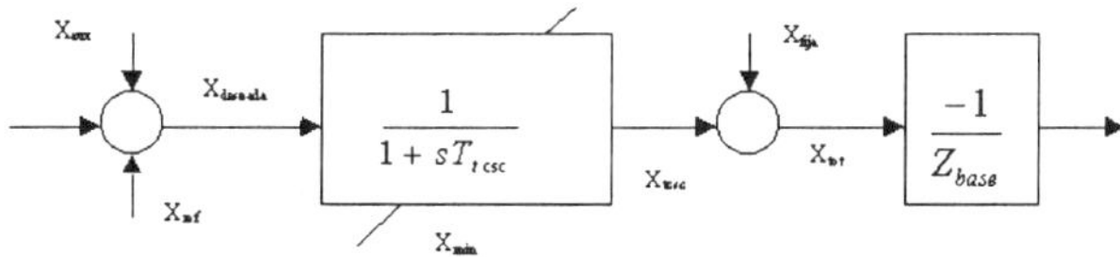

Figure 4 Block diagram of TCSC model

HVDC EAST–SOUTH INTERCONNECTOR II IN INDIA: 2000 MW, +/- 500 KV

V. K. Prasher and D. Kumar
Powergrid, India

C. Bartzsch, V. Hartmann and Dr. A. Mukherjee
Siemens, Germany

ABSTRACT

The world-wide electrical power consumption is projected to increase by over 70% during the next two decades. India is expected to enhance its electricity consumption at a rate of about 5% annually. That means enormous investments in power generation as well as in building up an interconnected national power grid. For an optimum utilisation of the electrical power available in the regional grids, they must be interconnected with each other by inter-regional lines. Particularly, the surplus energy available in one grid could then make up for the deficit in another, caused there e.g. by seasonal fluctuations in the availability of hydro-power in winter, or by major system breakdowns. The traditional AC power transmission technology with extra high voltages comes to its limits however, when asynchronous networks are to be interconnected with each other, or when the transmission distance is long. System stability problems may occur with AC interconnections in such cases.

In India there are presently six HVDC systems in operation which have contributed considerably to the linking of its dissimilar regional grids. In March 2000, Powergrid, the largest power transmission utility of India, awarded a new long distance HVDC transmission project to Siemens Germany. This project is being executed now. The HVDC system 'East-South' will transmit 2,000 MW of electrical energy from the power generation centre in Talcher in the State of Orissa – over a distance of about 1,400 km - to Bangalore in the State of Karnataka, a rapidly developing industrial high-tech area that needs this bulk power urgently to cope up with its increasing electricity demand.

The present paper highlights some key aspects of this mega project starting from design considerations for all vital components including converter valves, converter transformers and a state-of-the-art control and protection system. Furthermore, the paper introduces the design and performance criteria for the transmission system.

KEYWORDS

HVDC Systems - Design Criteria - Performance - Rating - HVDC Control – Converter valves – AC / DC filter

AC-DC Power Transmission, 28-30 November 2001
Conference Publication No. 485 © IEE 2001

1.0   INTRODUCTION

For the East-South link, the DC transmission technique was chosen not only because of the asynchronous eastern and southern grids, but also because of the vast transmission distance (approximately 1400 km) between the centres of power generation and consumption in Talcher and Kolar respectively (Fig. 1).

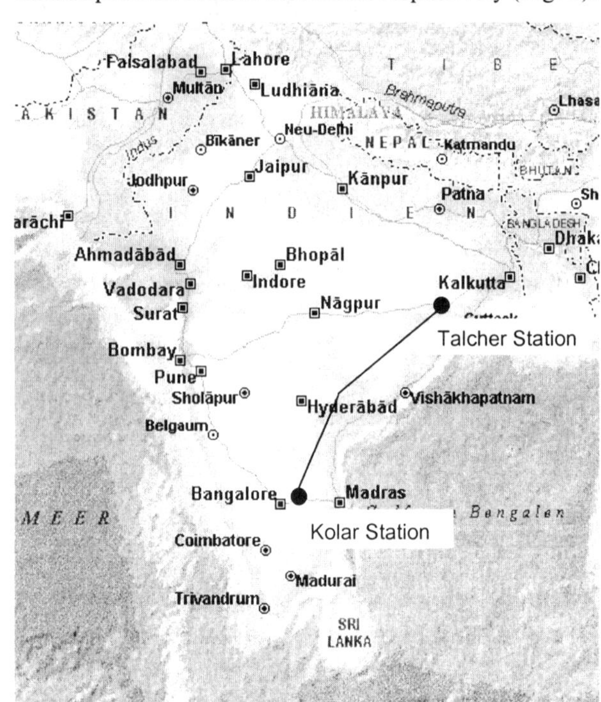

Fig. 1: The East-South Interconnector II Project will transmit power over a distance of 1400 km

Over such a long distance, the line losses with DC are only a fraction of the losses of a conventional AC transmission line, because DC transmission entails ohmic losses alone and no losses due to skin effect or the transport of reactive power. Further advantages of the HVDC transmission are:

- Very fast control of power flow with regard to both magnitude and direction, using a fully digitized control system. This characteristic will be effectively used to stabilise both AC networks.

- In addition, the DC system can intelligently control the reactive power exchange with the interconnected AC networks to control the AC system voltage.

- Narrow rights of way, combined with relatively low costs for lines and poles compared to a 3-phase HVAC transmission system.

- High reliability resulting from the initial design and built-in reliability for individual key components.

2.0 PROJECT SCHEDULE AND FINANCING

The contract came into effect in March, 2000. The scope of this turn-key project comprises design, manufacturing and testing of the electrical equipment and components, their delivery to site, installation and commissioning, architectural design of the buildings and execution of civil works, training of client's personnel, and numerous design studies. The AC switchyards - connecting the DC link to AC grids at both converter stations - are included too. Design and installation of both ground electrodes is part of the contract as well.

The project schedule foresees a stage by stage completion of the two poles of this bipolar scheme. The commercial operation of pole 1, thereafter pole 2, is planned 33 and 39 months respectively starting from the effectual contract date. The project is being financed by the KfW-Bank of the German government, except for the local portion of the project (approximately 20 percent of the total contract value), for which Powergrid itself has taken up responsibility.

3.0 DESIGN CRITERIA

3.1 Power Transmission Capacity

The HVDC interconnection between the 400 kV eastern and southern grids over a 1400 km long DC line is conceived as a $\pm$ 500 kV bipolar scheme. The single line power circuit diagram is shown in Fig. 2.

The bipolar system is rated for a continuous power of 2000 MW ($\pm$ 500 kV, 2000 A) at the DC terminal of the rectifier, i.e. Talcher converter station. The converter stations are designed to transmit full rated power up to a maximum ambient dry bulb temperature of 50°C without redundant cooling system in service. With redundant cooling, a two-hour overload of 2300 MW and a half-hour overload of 2400 MW is achievable. At low ambient temperatures, i.e. below 33°C, an overload capability of 2500 MW can also be utilized continuously. Additionally the DC transmission permits a five seconds overload of 1.47 p.u. power. This five seconds overload rating is valid under all ambient conditions and for all preceding operating conditions of continuous or two-hour overload. Such short time overload capabilities are important for some contingencies, like power modulation after AC system faults.

The HVDC interconnection scheme is capable of continuous operation at a reduced DC voltage of 400

kV (80%) from the minimum current up to the rated DC current of 2000 A with all redundant cooling equipment in service. In order to optimize the filter and converter transformer design, the capability of the valves to operate at high firing angles combined with an extended range of the tap-changer is used for this special mode of operation.

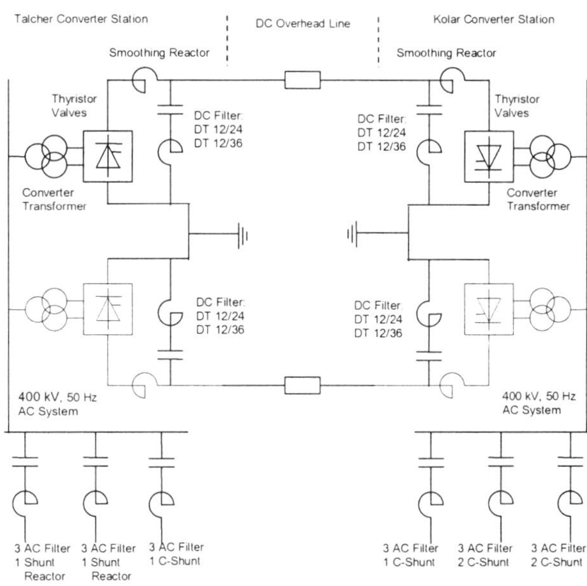

Fig. 2: Single Line Power Circuit Diagram

3.2 Special Features

The HVDC scheme can be operated in a number of operational modes, such as balanced bipolar mode, monopolar mode with ground return or metallic return as well as power reversal mode. Although the normal power flow is from eastern terminal at Talcher to the southern terminal at Kolar, the HVDC system can also transmit up to 1900 MW of power in the reverse direction.

In case a sequence of outages cause the eastern terminal to become islanded onto one or more generators connected to the 400 kV busbar, the HVDC system is designed to operate under this condition. The reactive power controller will adjust the filter configuration to prevent the harmonic current in the generators exceed the specified limits. In addition, a specific control logic is implemented to prevent self-excitation of the generators at Talcher. Under AC fault conditions resulting in an isolation of Talcher generators from the eastern region, the HVDC frequency controller can rapidly control DC power to prevent frequency deviation of the isolated generators beyond a certain limit. This frequency control feature can also be used to participate in the frequency regulation of the network on either side. Power modulation functions are integrated to increase the damping of the system following faults and improve the AC system stability as well as system recovery under multiple contingencies.

3.3 Performance Requirements

A high degree of energy availability was a major design objective. This design goal can be achieved mainly by minimizing the downtimes using fast fault detection as well as effective repair and maintenance strategies. Fault-tolerant control systems, redundancy, spare components and quality assurance ensures high component and system reliability. To provide the highest level of reliability and availability and hence quality of the HVDC control and protection system intensive off-side tests (e.g. functional performance test) will be performed. The energy availability for both stations together is guaranteed to be not less than 97 %, the number of forced outages should not exceed 10 (ten) per year.

The performance requirements for audible noise, electrical noise, radio and noise interference have been considered in the system design as per the limits specified in Owner's specification. Noise filter equipment is provided for the AC switchyard and the DC line to damp any interference between the thyristor valves and the PLC equipment.

Low loss design was of central importance for technical and economical optimisations. This resulted in converter station designs with total losses of approximately 14 - 15 MW per station at 2000 MW of transmission power. At rated transmission capacity the main loss sources are the converter valves and the converter transformers.

3.4 Reactive Power Requirements

Since the Talcher converter station is directly connected to the substation of the Talcher Power Station, the reactive power demand of the HVDC converters can be partly covered (up to 300 MVAr) by the power generators. In order to balance the reactive power flow to the AC system at rated power transfer, a total of 1077 MVAr shunt capacitance is required. This is configured as six switchable filters of 120 MVAr each, three filters of 97 MVAr each and one shunt capacitor of 66 MVAr rating. In order to limit the reactive power flow to the AC system to less than 20 MVAr at minimum load, two shunt reactors of 80 MVAr (at 420 kV) rating each are required. All reactive compensation equipment will be connected in form of banks and sub-banks to the AC station busbar as shown schematically in Fig. 3.

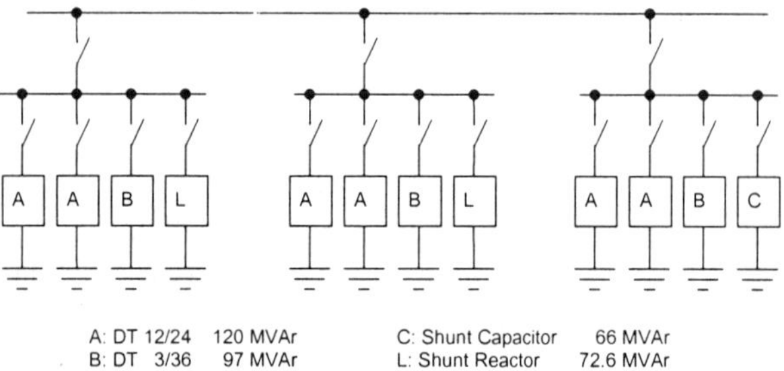

Fig. 3: Reactive Power Arrangement at Talcher Station

The reactive power compensation equipment to be installed is capable of regulating the reactive power exchange with the AC system up to rated power, with the largest sub-bank out of service.

Although the Kolar converter station is connected via seven 400 kV AC lines and one 220 kV interconnection transformer to the AC system, the minimum short-circuit level corresponds to a short-circuit ratio of 2.5 respectively about 5000 MVA only at rated power transfer. Since the AC system is relatively weak, increased capacitive compensation needs to be provided to support the system voltage at high power transfer levels. In order to supply 175 MVAr to the southern AC system at continuous overload a total of 1701 MVAr shunt capacitance is required. This is configured as six switchable filters of 120 MVAr each, three filters of 97 MVAr each and five shunt capacitors 138 MVAr rating each. All reactive power compensation equipment will be connected in the form of banks and sub-banks to the AC bus as shown schematically in Fig. 4.

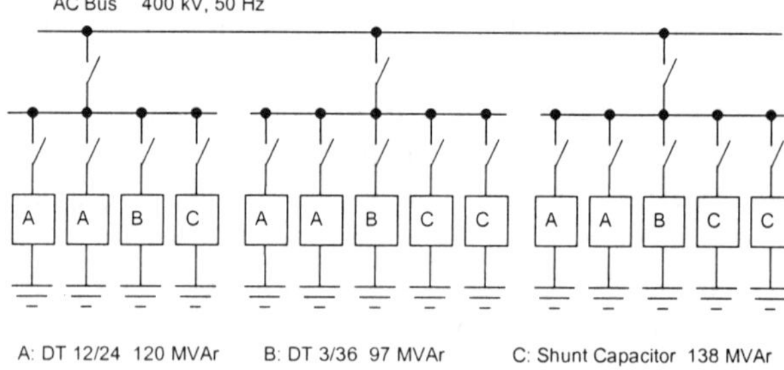

Fig. 4:   Reactive Power Arrangement at Kolar Station

Using a special control mode with increased firing/extinction angles the reactive power consumption of the DC converter can be increased in order to limit the reactive power flow to the AC system to 150 MVAr at 200 MW minimum load. Consequently the required exchange limit can be achieved without additional shunt reactors.

The compensation elements were subdivided into sub-banks also to comply with the limitation on AC voltage change due to Q-element switching.

However, the high degree of compensation especially at Kolar converter station combined with the high system impedance may lead to high temporary overvoltages after load rejection. Therefore a temporary overvoltage control of the HVDC system has been conceived to minimize the impact of fundamental frequency overvoltages (FOV) by restarting the DC system and restoring power transfer to pre-disturbance levels. If the DC system is not able to recover or can only recover to partial load levels, an additional logic is implemented to ensure that the FOV is less than 1.5 times 400 kV and is reduced to within the normal voltage range within 1 second following fault clearing. This strategy will ensure that only the minimum amount of reactive compensation necessary to meet the FOV requirements will be tripped to allow recovery to the maximum DC power transfer levels. Subsequent reduction of temporary overvoltages will be effected by extended control measures in order to prevent self-excitation of the Talcher thermal power generators.

4.0 MAIN EQUIPMENT AND MAJOR TECHNICAL FEATURES

4.1 Thyristor Valves

Continuous R&D and decades of experience using power electronics in HVDC, SVC, and similar applications in a high voltage environment have led to a design of the thyristor valves today which meets the highest requirements with respect to fire resistance, seismic robustness, efficient corrosion-free cooling as well as reliability and easy maintenance.

The engineering efforts focused on decreasing the number of components in a thyristor valve while using only equipment with a proven record to fulfil the above mentioned requirements. Keeping the number of components as small as possible without neglecting protection and monitoring aspects results in high reliability, as well as compact and economical thyristor valves with little maintenance requirements. Special care was taken to minimize the risks of destructive fires. An exceptional feature of the thyristor valves is the corrosion-free and extremely reliable performance of the cooling system.

Fig. 5: 'East-South' Quadruple Valve under High

Voltage Tests

Figure 5 depicts a fully assembled quadruple valve or multiple valve unit (MVU) set up for the high voltage tests. Each valve comprises 84 thyristor levels (three redundant) at Talcher station and 78 thyristor levels at Kolar station (three redundant), 24 non-linear reactors distributed to three thyristor modules. The three modules forming a valve are arranged in the twin tower. A thyristor level consists of a 4" 8 kV flatpack thyristor, a single snubber resistor and capacitor as well as the thyristor electronics. The latter includes a backup trigger circuit besides its primary task as a gating unit. Likewise, it comprises the electronic logic for individual thyristor supervision as well as for conversion of optical control signals received via fiber optics from the valve base electronics (VBE).

The MVU is suspended from the ceiling and all joints between the modules like suspension insulators, buswork, and piping are made flexible to withstand maximum seismic stresses. Cooling water and fiber optics are led from the top. The Aluminium frame of the modules and the large electrodes at the bottom act as corona shields. Three MVU´s are arranged side by side to form a 12-pulse group for one 1000 MW pole.

The use of a modular technique has simplified manufacture, testing, transport and installation of the valves and helped to reduce costs. In the unlikely event of an outage, replacement of faulty individual components within a module is also easy and fast.

4.2 Converter Transformer

The converter transformers at Talcher and Kolar are of identical design, so that on need the spare transformer from Kolar may be used at Talcher and vice versa.

Single-phase three-winding units (six plus one spare at each station) with following main electrical data have been designed (Fig. 6):

Rated power		(397 / 198.5 / 198.5) MVA
Rated voltage	- line side	400/√3 kV
	- valve side wye winding	210.3/√3 kV
	- valve side delta winding	210.3 kV
Leakage reactance		16.5 %
Tap-changer	- step size	1.25 %
	- range	-5% to +20%
Insulation level LIWL		
	- line side	1300 kV
	- valve side star winding	1550 kV
	- valve side delta winding	1050 kV

The leakage impedance was determined by taking several factors into consideration like permissible short-circuit current of the thyristor used, optimisation of rating and cost of the transformer etc. Selection of the on load tap changer range is adapted to the requirements about AC voltage variation range, operation modes, reduced dc voltage operation and valve capability of operating at high firing angles.

All the secondary bushings of the transformer will protrude directly into the valve hall. Both star and delta connections are made inside the valve hall, thereby eliminating the need for wall bushings as well as avoiding lightning surge stresses of the valves caused by direct strokes.

Fig. 6: 'East-South Interconnector II' Converter Transformer during Type Test

4.3 Smoothing Reactor

Taking different DC circuit configurations including DC filter outage as well as AC system impedance into account, the calculations indicated that a 250 mH smoothing reactor per station is an adequate size to avoid resonance at low order harmonics. This size of the smoothing reactor ensures fulfilment of further tasks as well, like limiting the transient overcurrents caused by DC side faults or commutation failures, avoiding discontinuous current operation at low DC currents, especially at 80 % DC voltage operation with high firing angles.

However, the most decisive design criteria for selection of a 250 mH smoothing reactor was the necessity to bring down the DC as well as AC harmonics to a permissibly low level. The filter arrangement, especially the concept of two DC filter branches per pole, associated with this reactor size, ensures that the specified requirements are met.

The smoothing reactor to be installed outdoor is of air-core dry type design and thus maintenance free.

The main design data of the smoothing reactor are as follows:

Inductance 250 mH (two coils @ 125 mH)

Rated voltage	500 kV dc
Rated current	2000 A dc
Insulation level	1425 kV LIWL to ground
	1425 kV LIWL across coil

4.4 AC Harmonic Filters

Double tuned damped passive filters are foreseen at both converter stations to meet the specified harmonic performance requirements:

- Individual harmonic distortion Dn ≤ 1 %
- Total harmonic distortion (rms) D ≤ 4 (3) %
- Telephone influence factor TIF ≤ 30
- Generator harmonic current content Ig ≤ 1 %
- Arithmetic sum of 5th and 7th harmonic currents in any generator Ig5/7 ≤ 0.6 %

The performance limits shall be met at any DC power transfer level from minimum up to the continuous rating with any complete AC filter bank (three filter) out of service, and from continuous rating to continuous / half-hour overload with any one AC filter sub-bank not available for service. Two types of filters - DT 12/14 and DT 3/36 - are used to provide the necessary filtering. In order to improve the harmonic performance, the current limiting reactors of shunt capacitors are chosen such as to build resonance circuits at 36th or 48th harmonic frequency.

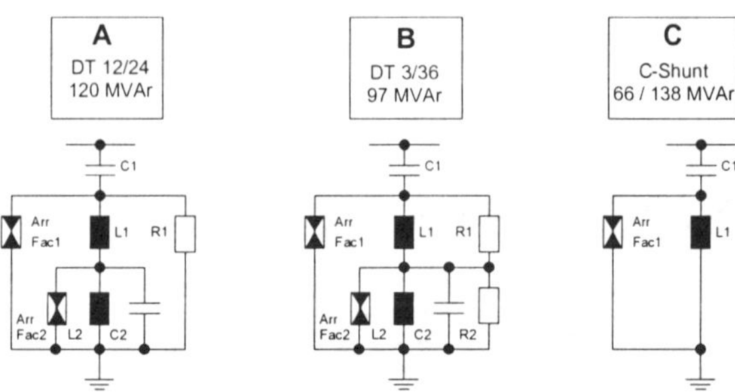

Fig. 7: Harmonic Filters and Shunt Capacitors

Figure 7 shows the single-line diagram and the corresponding AC filter circuits installed at both converter stations. The advantage of using double-tuned filter circuits combined with additional high-pass damping resistors is their excellent performance in the whole range of operation. In addition to the steady state harmonic performance, the filtering effect at low order harmonic frequency (3rd) provides also a good damping for the temporary low order harmonics generated by switching actions like transformer inrush.

4.5 DC Switchyard and DC Harmonic Filters

Specific equipment such as Metallic Return Transfer Breaker (MRTB) and Metallic Return Switch (MRS), are implemented in the HVDC system which are capable of switching from ground return mode to metallic return mode while in monopolar operation, without any reduction of the power transfer level or interruption of the DC transmission.

The measured value of the direct current is one of the most important system quantities used to control and protect the HVDC system. State-of-the-art technology today is the new Hybrid Optical Direct Current measuring system using ohmic shunt. The voltage drop across the shunt proportional to the DC current is measured and digitized by a light powered electronic circuit located in the same housing as the shunt at high voltage potential. The measured signal is transmitted as a serial telegram via a fibre optic link to ground potential.

DC filters are installed in both converter stations in order to reduce harmonic currents flowing in the DC transmission lines and electrode lines. Double tuned passive filters are foreseen to meet the specified requirements. The maximum allowable equivalent disturbing current Ieq [mA] during normal operation, with all filters in service, are shown below (electrode line limits in brackets):

Talcher station / Kolar station

- bipolar operation: 350 (250)   250 (150) mA
- metallic return: 700 (500)   500 (350) mA
- ground return: 1000 (700)   750 (500) mA

As per Owner's Specification, in case of loss of any one filter branch the maximum Ieq at balanced bipolar operation shall not exceed the limits for ground operation. Therefore, two filter branches per pole are foreseen: one tuned to the 12th and 24th harmonics and the other tuned to the 12th and 36th harmonics.

4.6   Control and Protection

The control and protection systems in each converter station of this HVDC transmission follow the most modern concepts. The control and protection components are all located in the control room. The communication between the controls and the microprocessor based I/O units is implemented using a redundant field bus. The central controllers and the operator control consoles – work station and MIMIC board - are all interconnected via a local area network.

The HVDC system can be operated

- either remote controlled via the Remote Terminal Unit, or locally via
- the Initiation and Monitoring Workstations
- the MIMIC boards.

The pole / converter related control functions are carried out by the redundantly configured (active / hot-stand-by) SIMADYN D pole control, the AC Switchyard control, whereas station related HVDC functions like reactive power control are tasks of the SIMATIC S5 station control system (also redundant) using active / hot-stand-by systems. The Valve Base Electronics (VBE) with integrated thyristor monitoring serves as the interface between pole control and the valves.

A tailor-made redundant protection system for HVDC applications using SIMADYN D properly co-ordinated with pole control and the AC system protection is responsible for a selective fault clearing, respectively prevention of damage to HVDC components. Diagnosis of faulty conditions is assisted by the

- Sequence of Events Recording and
- Transient Fault Recorder,

time synchronised via a GPS controlled Masterclock.

5. SUMMARY

The present paper introduces the general design criteria and performance requirements for the HVDC long distance project 'East-South Interconnector II' in India. It further highlights major technical features of the key components such as converter valves, converter transformers, AC and DC harmonic filters etc.

TYPE TESTING OF THE GTO VALVES FOR A NOVEL STATCOM CONVERTOR

M. L. Woodhouse and M. W. Donoghue

ALSTOM T&D Ltd., Power Electronic Systems, UK

M. M. Osborne

The National Grid Company plc, UK

INTRODUCTION

A relocatable Static Var Compensator (SVC), rated 0 to +225Mvar at 400kV, has been supplied to the National Grid Company (NGC) for its East Claydon substation in Buckinghamshire, England. The SVC comprises a fixed filter of 23Mvar, a conventional TSC of 127Mvar and a GTO-based STATCOM of ±75Mvar. All are connected, to a LV bus, at nominally 15.1kV, fed via the secondary winding of the compensator transformer. See Knight et al (1).

The STATCOM is a voltage-sourced convertor (VSC), using the novel chain-circuit topology as described by Ainsworth et al (2). In this, a valve comprises a number of individually switched, series-connected, single-phase bridges (Links), each link having its own dc capacitor. For East Claydon, there are 16 links in series per phase, each link incorporating four 4.5kV gate turn-off thyristors (GTOs), having 3000A turn-off capability.

The steady state, temporary and transient voltage and current stresses on the valves were determined during the design phase. Translating these stresses into a practical equipment design, supported by a meaningful programme of type tests was a challenge. This was more so as this is the first application of this technology and the existing International Standard for testing of thyristor valves for SVCs (3) is not strictly applicable.

DESCRIPTION OF THE CONVERTOR

Figure 1 shows the main components of a single-phase chain-circuit convertor, while Figure 2 shows the important components of a single link.

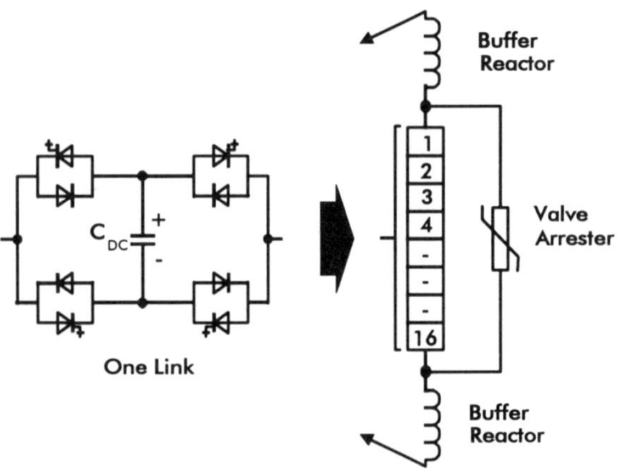

Figure 1: Single-phase Chain Circuit Convertor

AC-DC Power Transmission, 28-30 November 2001
Conference Publication No. 485 © IEE 2001

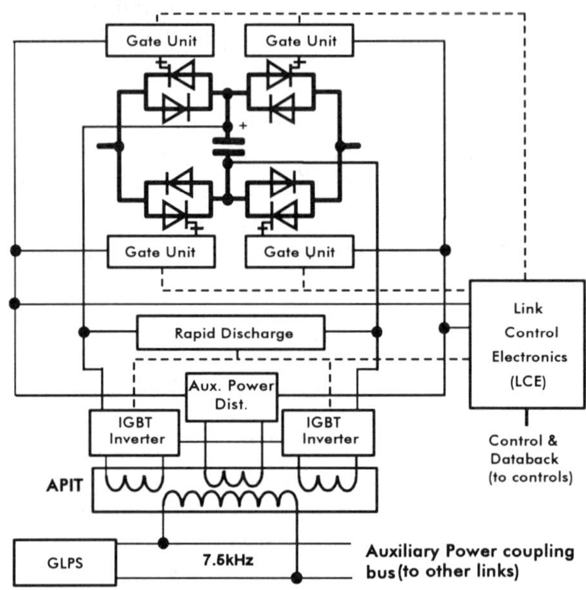

Figure 2: Important Components of One Link

Buffer Reactors: The inductance of the buffer reactors is chosen to ensure that the control system can provide fast and stable control of the valve over the full operating range. The reactors also limit the maximum voltage and current stresses on the valve in the event of an insulation fault on the 15.1kV bus. The insulation of the 15.1kV bus is designed to 17.5kV class while the GTO valve and all other equipment on the valve side of the buffer reactors is designed to 36kV class. 36kV class insulation ensures that the probability of an earth fault in this part of the circuit is made so low that it can be disregarded for design purposes.

Surge Arrester. A metal oxide surge arrester is connected between the terminals of the GTO valve to limit overvoltages. Because the natural frequency of the circuit formed by the buffer reactors and the dc capacitors of the links is very low (about 100Hz), the arrester is stressed only with very slow switching type overvoltages. System studies showed that a valve arrester with a switching impulse protective level of 49kV was adequate.

DC Capacitor Rapid Discharge. Overvoltage events can lead to trapped voltages on the dc capacitors that are above the overvoltage blocking level of the links (3kV). To prevent lockout, a dc capacitor rapid discharge circuit is provided in each link. This comprises an IGBT switch and a series resistor that can quickly reduce the link voltage to below 3kV. This circuit is also used to discharge the capacitors to below 50V, following shut down.

DC Capacitor Voltage Balancing. As with all multi-level convertors, there are systematic processes that cause the dc capacitor voltages to become unbalanced. Control action is used to counter this tendency and ensure that good voltage balance is maintained under all normal conditions. The chosen strategy achieves this without an increase in the switching frequency of the GTOs, thus maintaining high convertor efficiency.

Control action alone cannot always maintain the desired precision of steady state voltage balance. Provision is therefore made in the design to manage any residual unbalancing effects by exchanging energy between capacitors. This is accomplished using dual IGBT inverters at each link, which are connected to an auxiliary power transfer bus at earth potential via a fully insulated auxiliary power isolating transformer (APIT). The rating of the inverter/APIT combination is 5kVA at 7.5kHz

Auxiliary Power. When deblocked, in steady state, the net power flow on the coupling bus is zero and the links are self-sustaining with regard to auxiliary power for energising the GTO gate units and other electronic systems. However, during start-up, shut down and following certain faults, the links require an independent supply of auxiliary power. A ground level power supply (GLPS) provides this energy. The GLPS additionally provides energy for pre-charging the dc capacitors, prior to closing of the SVC 400kV circuit breaker.

STATCOM Controls. STATCOM Control provides all the high-speed control, protection and monitoring functions. This includes co-ordinated operation not only of the switching of the individual GTOs in each of the 16 links but also of the rapid discharge circuits and auxiliary inverter units. Further details may be found in the companion paper by Horwill et al (4).

Implementation. All equipment except the surge arrester and buffer reactors is cabin-mounted for ease of relocation (see Figure 3). Each cabin has two main rooms, one housing the links and their associated dc capacitors and one housing the STATCOM Controls, GLPS and other cabin auxiliary systems. Figure 4 shows the upper part of the valve room with the link power electronic assemblies in position, while Figure 5 shows how the principal components relate physically to one another. As can be seen, the links and dc capacitors are not free-standing items but are an integral part of the cabin assembly.

MAIN PARAMETERS

The main parameters for the GTO valve, arising from the design process and which underpin the test programme are summarised in Table 1.

Figure 3: STATCOM Cabin on the Move

Figure 4: Links Mounted in Cabin

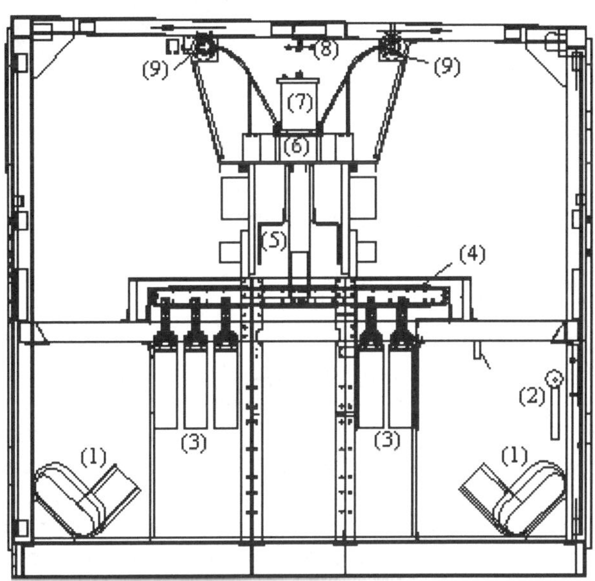

Key:

(1) Ventilation ducts   (6) IGBT inverter
(2) Earth switch   (7) APIT
(3) DC capacitor units   (8) Auxiliary power coupling bus
(4) Stripline busbar   (9) Coolant manifolds
(5) Link assembly

Figure 5: Cross Section of Valve Room

TABLE 1 – Main Parameters for the GTO Valve

Parameter	Value	Comments
Nominal valve voltage	15.1kV rms	
Maximum continuous valve voltage	19.6kV rms	At full leading output
System temporary overvoltage	1.3 pu for 1s	At up to full lagging current
Valve overvoltage protection level	35kV instantaneous	Action: block valve[1]
Link overvoltage protection level	3kV instantaneous	Action: block link (then valve), activate RD[1,2]
Valve arrester protective level	49kV	Slow switching impulse
GTO rated voltage	4.5kV	
No. of series links installed per phase	16	Cabin has capacity for up to 20
No. of redundant links per phase	2	
Maximum link voltage	2kV peak	Worst case steady state
Buffer reactance	4.8mH	Total between phases
DC capacitance	8000µF	5 x 1,600µF units per link, Un = 2.5kV
Nominal operating frequency	50Hz	Design has 60Hz capability
GTO switching frequency	50Hz nominal	Always same as ac system frequency
Maximum ambient temperature	40°C	
Maximum coolant temperature	50°C	Redundant coolers out of service
Maximum continuous valve current	1750A rms	
Overcurrent protection limit (soft)	2700A instantaneous	Control action
Overcurrent protection limit (hard)	2900A instantaneous	Action: block all links[1]
GTO safe turn-off current limit	3000A instantaneous	Guaranteed value
Fault current	12.9kA peak	Withstand, without blocking

Notes to Table 1:
(1) Valve electronic protection (self re-setting).　　(2) RD = Rapid discharge circuit.

TESTS AND TEST VALUES

NGC specified that any programme of tests should be based on CIGRÉ Guidelines for SVC valve testing (5). Using this document for reference, proposals for routine and type tests were worked-up in detail during the contract and agreed with NGC.

In accordance with the principles adopted by CIGRÉ (and, more recently, by IEC) for valve testing, the tests were broken down into two broad categories: "Operational Tests" and "Dielectric Tests".

Operational Tests

For many of the tests, two links, connected back-to-back via a load inductor, one operated in leading mode (GTOs turning off current) and one operated in lagging mode (GTOs turning on into current), proved the ideal way for replicating the critical service stresses (see Figure 6). Power losses in the circuit were replenished by a dc test supply connected to one of the dc capacitors. Two special test rigs were built to permit these tests to be carried out in as representative a manner as possible.

The dc capacitors, although an integral part of the test set-ups were not, themselves, the subject of these tests. The capacitors were independently type tested in accordance with BS EN 61071-1 (6).

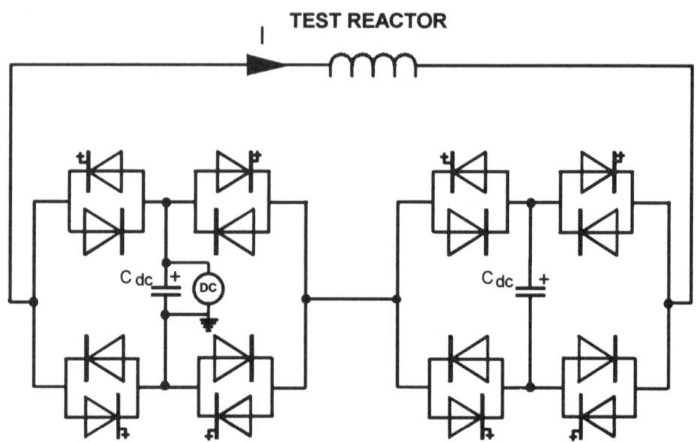

Figure 6: Circuit for Back-to-back Testing of Links

The following operational tests were carried out:

- Periodic Firing and Extinction Tests: a full programme of tests on 16 links as a type test, plus a sub-set as routine tests on all 48 links.
- Power Loss and Temperature Rise Tests: on 2 links as a type test.
- Overcurrent Tests: on 16 links as a type test.
- Rapid Discharge and Auxiliary Inverter Tests: a full programme on 16 links as a type test, plus a sub-set as routine tests on all 48 links.

Outside the agreed routine and type test programmes, the rigs were used also to:

- Explore dc capacitor thermal conditions at full load current when mounted as intended in the cabin and provided with a realistic air ventilation environment.
- Study internal fault scenarios.

The results were shared and discussed with NGC.

Periodic Firing and Extinction Tests. These were all performed at 50 Hz, using the design minimum coolant flow at the design maximum inlet temperature. During the tests, three parameters had to be achieved simultaneously: the rms current through the link (Irms), the instantaneous dc capacitor voltage at GTO switching (S-level) and the instantaneous current at GTO switching (Isw). Each of these parameters was set to be 5% above the maximum value in service. Table 2 details the specified test conditions. Figure 7 shows representative waveforms recorded during the tests.

Power Loss and Temperature Rise Tests. The power loss tests were aimed at determining the switching losses. The principle adopted was to accurately measure the total power fed into the circuit from the test supply and subtracting from it the low frequency voltage- and current-dependent losses of the main circuit components. The residue is the switching loss. Measurements were made over a range of S-level and Isw values that encompass service operation. Using methods similar to HVDC valves (7), the total valve losses were determined by a combination of measurement and calculation, with the switching loss component being derived from the results of these tests.

Many important components in the valve are liquid cooled but some are natural air cooled and others rely on a combination of both. For accuracy in temperature rise testing, the liquid coolant temperature was controlled to be in the same relationship with respect to the ventilation air temperature as in service. The reference temperature was the ambient temperature of the test laboratory. The operating conditions were generally as for the periodic firing and extinction tests but with revised S-level and Isw values that take into account the effects of voltage balancing control action. The temperature rise of critical components above the reference temperature was recorded and extrapolated to the maximum service temperature.

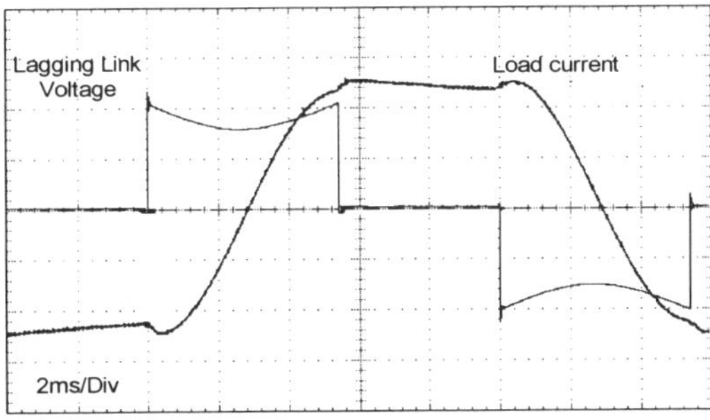

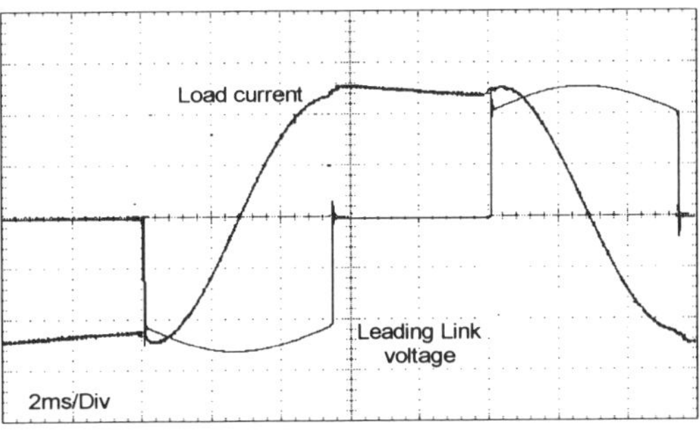

Figure 7: Waveshapes for Back-to-back Link Testing

Overcurrent Tests. The worst case overcurrents originate from the sudden release of the energy stored in the dc capacitors. The design reference case for this project was a phase-to-phase short circuit on the 15.1kV bus, on the transformer side of the buffer reactors, coincident with the convertor output being at the valve overvoltage protection limit (35kV). Two sub-cases exist: a) the normal case where overcurrent protection acts quickly enough to suppress the overcurrent by turning off the GTOs, before the instantaneous current rises above the safe-to-turn-off limit of 3kA; b) the abnormal case where current rises above 3kA before blocking can be applied and it is no longer safe to attempt turn-off. For case b), the dc capacitors discharge completely via the buffer reactors, with the complication that, at voltage zero (current peak), the free-wheel diodes in the links start conducting. Diode conduction prevents voltage reversal of the dc capacitors but leaves the peak of the overcurrent trapped as a slowly decaying dc component, with $\tau = L/R$ seconds. Opening of the main 400kV SVC circuit breaker has no effect on this current, which continues to circulate, causing considerable heating of the conducting GTOs and free-wheel diodes.

TABLE 2 – Summary of Periodic Firing and Extinction Test Conditions

Description of test	Irms	S-level	Isw[1]	Duration
Leading mode, continuous	1843A	2100V	2600A	30mins as a type test, 5mins routine
Lagging mode, continuous	1843A	2100V	2600A	30mins as a type test, 5mins routine
Lagging mode, temporary	2020A[2]	2750V	2850A	1s, only as a type test

Notes to Table 2:

(1)   Isw is a function of control angle, α, which is different for all links. Specified value is for $\alpha = 0°$ which corresponds to GTO switching at the peak of the load current.

(2)   Estimated value. The rms current during the temporary overload was not specified, this being subservient to achieving the technically more important S-level and Isw values.

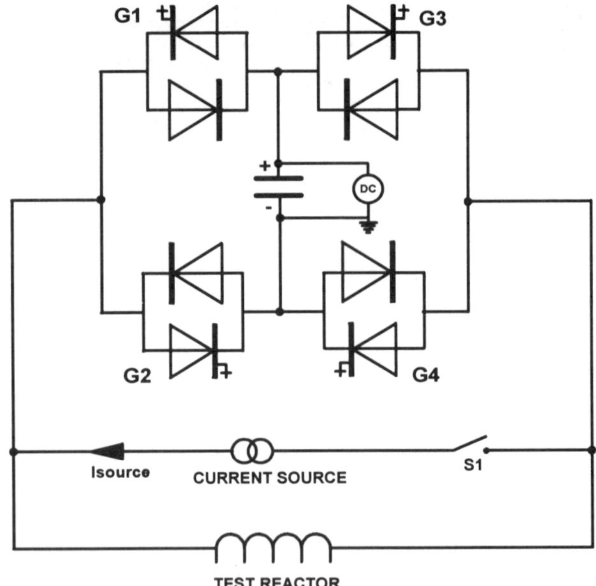

Figure 8: Test Circuit for Overcurrent Tests

The overcurrent tests were performed on one link at a time, using the circuit of Figure 8. The test reactor is proportionally scaled to one link and the GTOs and free-wheel diodes brought to operating temperature using 50Hz current from the low voltage current source, while toggling the current path between positive bypass (G1/G3 ON) and negative bypass (G2/G4 ON). When ready, the dc capacitor is charged to a voltage just above 2.5kV, the GTOs are blocked, S1 is opened and a chosen diagonal pair of GTOs, e.g. G1/G4, is turned ON to initiate the overcurrent. For the overcurrent with blocking test, the conducting GTOs are ordered OFF again as the current passes through 3kA, see Figure 9. For the overcurrent without blocking test, the GTOs are held in conduction, see Figure 10. The two tests are repeated for the other diagonal pair of GTOs (G2/G3).

During the overcurrent without blocking test, it was found that the L/R ratio of the test circuit was much smaller than for the real circuit, giving a faster decrement to the trapped dc component and less heating of the GTOs. To compensate for this, it was necessary to artificially increase the value of dc capacitance, reduce the test reactor inductance and allow the peak overcurrent to rise above the specified value of 12.9kA. The peak test current found necessary to obtain equivalent heating of the GTOs was 20kA.

Rapid Discharge and Auxiliary Inverter Tests. With the GTOs blocked, a 0.6A constant current dc test supply was connected to the dc capacitor of the link under test. The rapid discharge circuit was set to activate at 4.4kV (i.e. just below the voltage limit of the GTOs) and de-activate at 2.7kV (the reset level of the link overvoltage protection circuit). A single discharge from 4.4kV to 2.7kV, followed, within a few seconds, by a discharge from 2.7kV to zero, was applied to all links as a routine test. Links selected for type testing were given 3 successive discharges from 4.4kV to 2.7kV, followed by a discharge to zero.

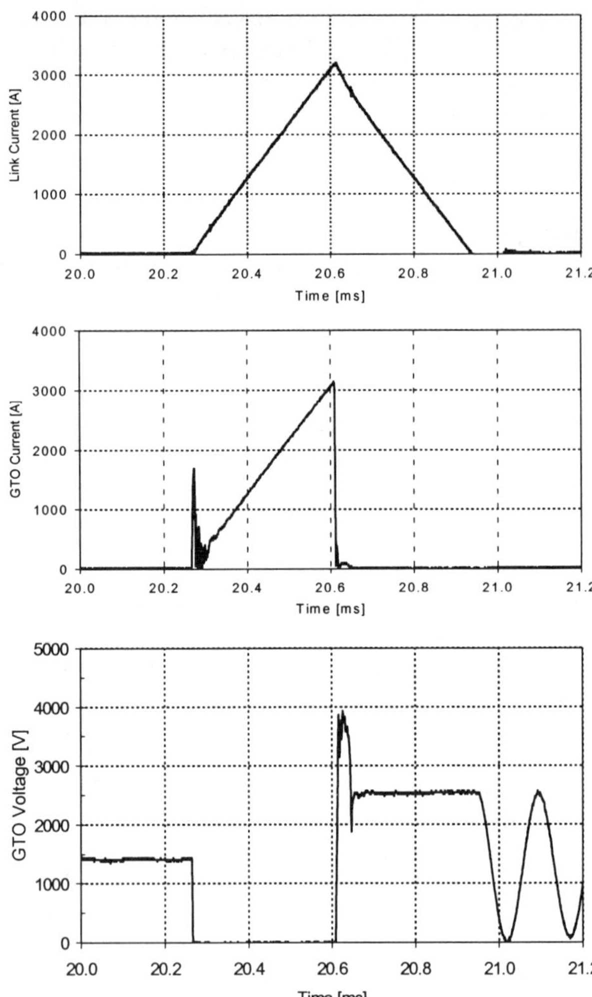

Figure 9: Overcurrent with Blocking

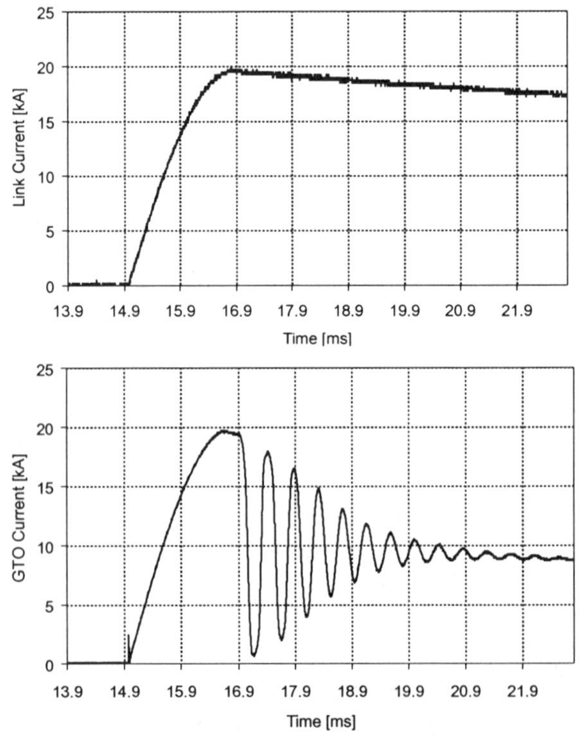

Figure 10: Overcurrent without Blocking

89

The auxiliary inverters were tested, two links at a time. The inverters of both links were connected to an auxiliary power coupling bus via APITs, as in service and the coupling bus energised from a test GLPS. Starting from an equilibrium voltage of about 1400V on the dc capacitor of each link, the inverter of one link was commanded to full import and the other to full export of power. When either the discharging link reached 750V or the charging link reached 2900V, the roles were reversed. The links then continued to exchange energy between themselves, the inverters alternately importing and exporting full power. Circuit losses were provided by the GLPS. The inverters of all links were tested this way for 5 minutes, as a routine test, while the inverters of the 16 type test links were operated for 30 minutes. NB: In service, the inverters normally operate near to float (power close to zero) with only temporary excursions to full power.

Dielectric Tests

Because the links and their associated dc capacitors form an integral part of the cabin assembly, the dielectric tests were performed on a complete valve assembly when mounted in its cabin.

The type tests comprised power frequency and lightning impulse tests to earth, appropriate to 36kV class insulation, plus power frequency and switching impulse tests between valve terminals, derived from worst case service conditions, with agreed test safety margins. One cabin was tested.

The special nature of the test object and aspects of its behaviour required ingenuity to arrive at a satisfactory programme. The dielectric type tests were carried out at the BSTS Clothier Laboratory, Hebburn.

Introduction to Dielectric Test Programme. The links and dc capacitors were routine and type tested as stand-alone items before assembly into the cabins. Only after assembly in the cabin did the whole valve come together and the proper means for energising and controlling it become available. Following installation, a comprehensive cabin routine test programme was carried out that culminated in a full open-circuit test of the convertor at 5% above the highest in-service voltage (see Figure 11). This is possible because, as a chain circuit VSC, it is only necessary to provide a charge on the dc capacitors for it to become usable as a 50Hz sinewave voltage generator, as in service. For East Claydon, the GLPS, in conjunction with the individual link auxiliary inverters, provides the means for off-line charging of the dc capacitors. Furthermore, it is within their ratings to supply the 50Hz no-load switching losses of the links. This ability to operate the complete convertor, off-line, in a realistic manner, is a major confidence-booster and proved to be an invaluable tool for rapidly checking the integrity of the complete system. It was especially useful during and following the dielectric tests and as part of site commissioning (4).

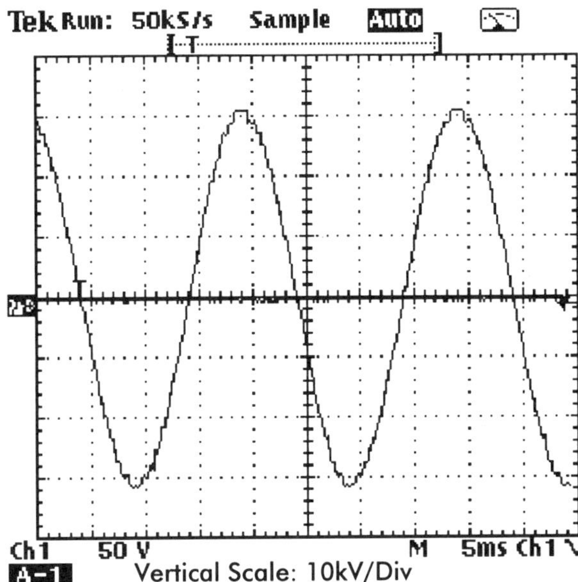

Figure 11: Open Circuit Output Voltage

Insulation Tests to Earth. For these tests, all the active components of the valve were bound out and the test voltages applied between the two cabin wall bushings, tied together, and earth. The tests comprised 70kV rms for 1 minute, followed by 3 impulses, in each polarity, of a standard lightning impulse voltage of 170kV peak. A partial discharge measurement was performed during the ac voltage test.

Voltage Tests between Valve Terminals. For these tests, the binding wire was removed and the valve electronics energised from the GLPS. Two of the 16 links were commanded into by-pass mode, to simulate loss of redundancy, while the others were held blocked. To reduce the burden on the test voltage sources, the dc capacitors at each link were disconnected, voltage sharing between links being enforced only by the internal circuits of the links.

A two-part ac voltage test was applied, in accordance with (5). The one minute test amplitude was 27.2kVrms, this being 1.1 x 35/√2kV, where 35kV is the valve instantaneous overvoltage protection setting. This was followed, for 10 minutes, by 23.5kVrms, this being 1.2 x 19.6kV, where 19.6kV is the maximum continuous convertor output voltage. Partial discharge tests were not performed at this time: supplementary tests being performed instead (see below).

Three impulses of a non-standard switching impulse voltage waveform were applied in each polarity. The specified waveshape was 2.5ms time to peak and 100ms time to half value. The peak amplitude was 53.9kV, this being 10% above the valve surge arrester protective level. The non-standard waveshape reflects the very slow natural response of the convertor to an externally applied overvoltage, as discussed earlier. Even though the dc capacitors were disconnected for this test, the test object presented a significant burden to the impulse generator. This needed a special impulse circuit set-up.

Supplementary Dielectric Tests. It was not meaningful to perform partial discharge tests during the ac voltage test between valve terminals. This is because the valve and its electronic systems are inherently noisy due to the presence of built-in switched-mode power supplies, choppers and auxiliary inverters etc. To verify that the insulation systems of the valve were sound and inherently discharge free at the required voltage, two supplementary tests were performed. In the first, the individual links of one valve were disconnected from one another and the insulation between each pair of links checked at the proportionally scaled voltage. In the second, a special dummy link assembly was constructed using representative contract material in the correct physical arrangement but with certain components disconnected and others replaced by discharge-free insulating blocks. The dummy link permitted the integrity of all the highly stressed insulation systems within a link to be verified at the proportionally scaled voltage, without interference from active electronics etc.

TEST RESULTS

At the highest level, the equipment performed in an exemplary way throughout the whole programme of tests, with no fundamental problems emerging.

As may be expected with a new product that was undergoing rigorous testing for the first time, a number of minor issues did emerge which required attention. These included:

- Higher than expected regulation in the auxiliary power supply system. Resolved by a reconfiguration of the windings on the APIT.
- Link Control Electronics firmware glitches and logic inconsistencies. Resolved by reprogramming.
- Marginally excess temperature rise on some link components that are predominantly air cooled: Resolved by adjustments to the ventilation air flow.
- Marginal lightning impulse voltage withstand of a few APITs at 170kV. Spare units used as replacements. Marginal units investigated and rectified.

No GTOs or free-wheel diodes or other valve components failed directly as a result of the test conditions. However, a number of incidents related to defective test technique did occur which led to device failures. These testing difficulties were overcome and the tests were then passed without incident.

CONCLUSIONS

- A novel STATCOM, based on GTO valves using the chain link VSC circuit topology, has been designed, built and tested.
- In the absence of any directly applicable International testing standards for VSC valves, the testing was based on an adaptation of available test guidelines for related conventional SVC valves.

- The interpretation of the existing test practices into a viable test programme, that was mutually acceptable to both purchaser and supplier, was achieved through the close collaboration and co-operation of both parties.
- To realise some of the tests required the separate design and construction of specialised test rigs and some ingenuity in test circuit design.
- The tests were carried out successfully without any major problems being encountered.
- The tests confirmed that all performance requirements were met.
- The successful outcome has firmly established the single-phase chain link VSC as a viable product for providing fast dynamic reactive power control for transmission systems.

REFERENCES

1. Knight R C, Young D J and Trainer D R, "Relocatable GTO-Based Static Var Compensator for NGC Substations", CIGRÉ 14-106, 1998 Session.

2. Ainsworth J D, Davies M, Fitz P J, Owen K E and Trainer D R, 1998, "Static Var Compensator (STATCOM) Based on Single-Phase Chain Circuit Converters", IEE Proc.- Gener. Transm. Distrib., Vol 145, No. 4, 381- 386.

3. International Standard, 1999, "Power Electronics for Electrical Transmission Systems – Testing of Thyristor Valves for Static VAR Compensators", IEC 61954.

4. Horwill C, Totterdell A J, Hanson D J, Monkhouse D R and Price J J, 2001 " Commissioning of a 225 Mvar SVC Incorporating a ± 75 Mvar STATCOM at NGC's 400kV East Claydon Substation", IEE Conf. Proc. AC-DC Power Transmission.

5. Working Group 14.01, Task Force 02, 1995, "Guidelines for Testing of Thyristor Valves for Static Var Compensators", CIGRÉ publication 93

6. British Standard, 1997, "Power Electronic Capacitors", BS EN 61071 Part 1

7. International Standard, 1999, "Determination of Power Losses in High-Voltage Direct Current (HVDC) Convertor Stations", IEC 61803, section 5.1

INTERACTIVE AC-DC HARMONIC AND INTERHARMONIC ANALYSIS

G.N. Bathurst,   N.R. Watson, J. Arrillaga,   B.C. Smith

UMIST   University of Canterbury   Transpower
United Kingdom   New Zealand   New Zealand

Abstract – Advanced frequency domain software is used, via graphical interfaces, to provide fast and accurate information of the ac-dc system harmonics, interharmonics and converter impedances on the interactive PC platform.

Keywords – Harmonics, interharmonics, converter impedances.

1. INTRODUCTION

Recent CIGRE documents provide guidelines for the modelling of the ac system impedances at harmonic frequencies and for the derivation of frequency transfers across hvdc converters, CIGRE (1), CIGRE (2). Putting these guidelines into practice is a difficult task, however, due to insufficient reliable system data, and lack of suitable software. In practice manufacturers rely heavily on their experience with previous schemes to derive the harmonic information required for filter design. This approach can sometimes cause unexpected behaviour at the commissioning stage.

The harmonic interaction between the converters and ac and dc systems can be assessed by electromagnetic transient simulation of the steady state condition followed by Fast Fourier Transform of the periodic waveforms. This solution has great flexibility in terms of detailed component and control system representation. Its main limitations are insufficient speed for interactive simulation on PC's and imperfect frequency dependence representation of the ac system.

This paper describes the use of advanced frequency domain software via graphical interfaces which, subject to the availability of reliable data, provides fast and accurate information of the ac-dc system harmonics and interharmonics on the interactive PC platform.

2. HARMONIC MODELLING REQUIREMENTS

With reference to ac-dc converters, an effective ac filter design requires accurate knowledge of the overall system resonant frequencies and of the converter harmonic current injections.

The resonant frequencies result from the parallel combination of ac system, filters and converter harmonic impedances. Only the first two, i.e. the ac system and filter impedances, can be derived from linear circuit analysis. The ac system harmonic impedances are largely influenced by the transmission system, which is well known in terms of layout, distances and conductor data; the main uncertainty being the earth resistivities.

Other ac system components, such as transformers, synchronous generators, VAR compensating plant and as much information as possible from the loads must be combined with the transmission system to determine the ac system harmonic impedances. Suitable programs are generally available to obtain this information as accurately as the reliability of data permits, Arrillaga et al (3).

The converter effective impedances, on the other hand, are the result of complex interactions between the ac and dc systems via the converter and its controls and can only be determined via iterative ac-dc system solutions, as explained in the following section.

3. HARMONIC DOMAIN SOLUTION

The commercial programs currently available for harmonic analysis only provide approximate information of the uncharacteristic converter harmonic behaviour. Moreover, the converter is usually represented as a perfect current harmonic source when analysing harmonic power flows. It has already been shown that the approximations made in these models (particularly the lack of harmonic modulation of the switching instants via the converter controls) can lead to inadequate filter designs, Arrillaga et al (4). Instead, an advanced algorithm, based on the Newton solution, and referred to as the Harmonic Domain, has been developed to provide accurate and reliable modelling of the converter harmonics. In this method the converter model must be differentiable with respect to the solution variables. This is achieved by the convolution technique, which calculates the transfers analytically using easily differentiable real-valued functions, (4). The advantage of describing the system in real-valued terms is that, both, electrical and non-electrical variables, such as control variables, can be solved simultaneously.

Each of the six-pulse bridges of a converter terminal can be viewed as a four-port circuit, i.e. of two inputs and two outputs. The inputs in this case are the ac phase

AC-DC Power Transmission, 28-30 November 2001
Conference Publication No. 485 © IEE 2001

voltage spectra and the dc current spectra. The outputs are the ac current and dc voltage spectra. The convolution technique approximates the transferred waveshapes (dc voltage and ac phase currents) to piece-wise waveshapes consisting of twelve distinct periods of conduction. These periods are defined explicitly by the switching angles of the converter. In the case of the dc voltage transfer, twelve analytic frequency domain expressions are derived from nodal analysis of the twelve commutation circuits, which describe steady state waveshapes valid for those circuits. The spectra calculated using these expressions are then convolved with those of band-limited rectangular windowing functions. The resultant sum of the convolved spectra is that of the total waveshape. Figure 1 illustrates one convolution for the dc voltage of an inverter as seen in the time domain.

A similar process is followed to obtain the harmonic spectra of the ac phase currents. These transfers are general for both inverter and rectifier, with only the direction of dc current flow and the resultant sign of the dc voltage across the bridge being different.

The harmonic transfers across a converter are largely dependant upon the accuracy of determining the switching instants, particularly the end of commutation angle. For the convolution model these angles are obtained from twelve single variable Newton-Raphson steps. The commutation angles are defined to be the crossover points of the commutation currents with the dc current (harmonics included). The firing angles are dependant upon the control strategy but are also solved using single variable iterative steps. As the full solution Jacobian includes the effect of modulating these switching instants, the main solution is still of a full Newton nature.

To achieve the standard twelve-pulse configuration two convolution models are used, one star/star and the other star/delta. The harmonic transfers from each model are added together to yield the full transfer characteristic.

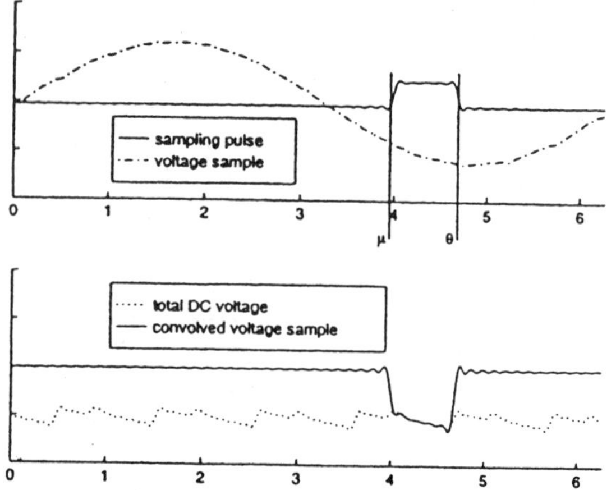

Figure 1. Convolution sampling waveforms for a six pulse converter.

Each frequency is ordered by phase components and is partitioned into real and imaginary parts. This leads to six variables for each harmonic at an ac busbar, and two at a dc busbar. Consequently for a study to the 50th harmonic, 300 variables need to be defined at the converter ac terminal and 100 at the dc terminal. The remaining required mismatches are those for the average dc current and the control variables. For a current controlled rectifier these are the firing angles, which are modulated by the dc current through a PI controller. The minimum gamma controlled inverter assumes equidistant firing but requires the commutating voltage zero-crossings as variables in order to calculate the extinction angles.

Variables	Mismatches	Description
1	A	Rectifier AC busbar
2	B	Inverter AC busbar
3	C	Rectifier DC busbar
4	D	Inverter DC busbar
5	E	AC voltage source busbars
6	F	System control variables

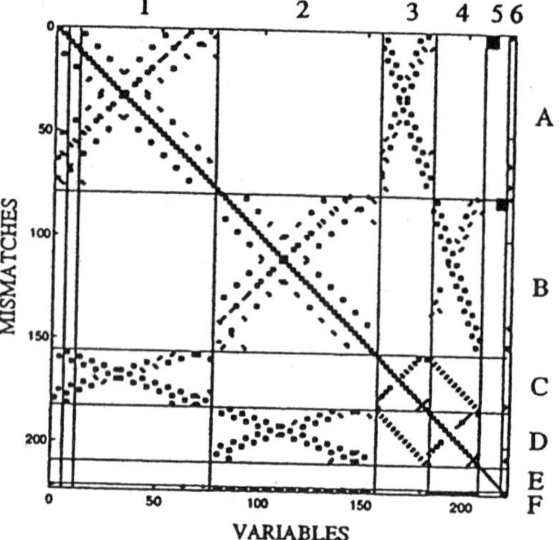

Figure 2. Jacobian of the CIGRE hvdc test system for 13 harmonics.

The Jacobian, in Newton's solution, only needs to be approximate, and therefore only the significant terms need to be retained. Thus the full harmonic Jacobian can be about 96% sparse without affecting convergence. The Jacobian elements are best calculated analytically, despite the significant effort required for their derivation, because the solution speed is increased by a factor of 50 with respect to using one obtained numerically.

Good variable initialisation is achieved in a two stage process. The first stage uses a positive sequence power flow estimation and classical converter equations. This information is used to start a three-phase ac-dc power-flow solution including the control variables; the results of the latter are then used to initialise the full harmonic solution.

As the two converters of the hvdc link interact non-linearly with each other, it is necessary to solve them simultaneously. At each converter busbar the variables represent the specified harmonics in ascending order.

The structure of the Jacobian is illustrated in Figure 2, although for simplicity it only includes 13 harmonics.

As the two ac systems are not directly connected, there is no direct coupling between the two ac busbars as represented by numbers 1 and 2. This can be seen in the matrix by the presence of zero blocks A2 and B1. The two dc busbars at 3 and 4 are directly coupled through the nodal analysis of the system; blocks C4 and D3 are the diagonal linear linking elements.

Some refinements have been made to the Harmonic Domain algorithm to make it suitable for the efficient computation of inter-harmonics, Bathurst et al (5). These refinements take advantage of the inherent sparsity present in the harmonic arrays, which provide several orders of improvement in solution time with no significant degradation in solution accuracy. As a result, the increase in computation is now almost linear with the number of power frequency cycles, as opposed to cubic with the original full method described in reference, (4).

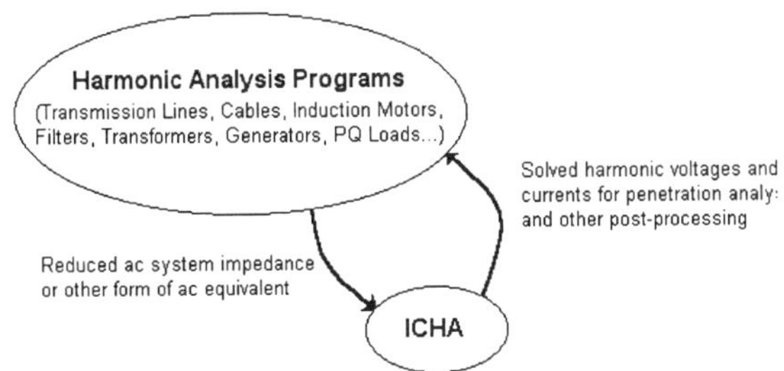

Figure 3. Data flow for harmonic analysis.

The converter impedances are calculated using a partial inverse of the system Jacobian. The Jacobian is calculated after the solution has been converged and so represents the system at that operating point.

4.   STEERING GRAPHICAL INTERFACE

Figure 3 illustrates how the Interactive Converter Harmonic Analysis (ICHA) program links with existing analysis software to derive and present harmonic information.

While the engine behind ICHA can perform the full system analysis, i.e. including the load flow and system reduction, Bathurst et al (6), the users may prefer to work with their own software to generate the external ac system equivalent. In such case ICHA will then load that information, or use a simple R-X model, and thus restrict the use of ICHA for the local loads, including any generators and filters in close proximity to the converters. To simplify the interface only a set of fixed configurations are available, although virtually any desired configuration is possible. If required, a special style can be easily formed and then loaded as an external style file.

A library of Forms facilitates the task of editing the system components. As an illustration of data editing, Figure 4 shows the complete screen with a window displaying the Form used to edit an ac filter branch. Similar Forms are used to edit the converter transformers, converter controls and ac system sources and impedances.

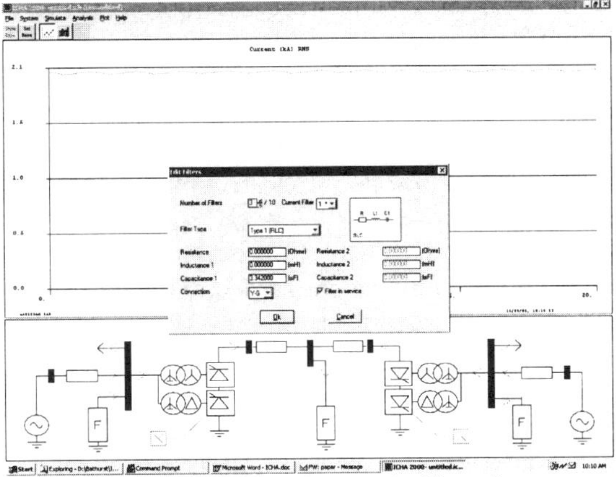

Figure 4. Editing harmonic filter data

The Simulate option is then selected and the Harmonic Domain iterative solution is carried out. During the iterative process a temporary window appears, giving information on the progress of the solution, as illustrated in Figure 5. When convergence has been reached, the Plot option is highlighted. The quantity to be displayed is then selected by directly clicking on the desired measurement in the schematic. Presently available measurements include network, filter, load and converter currents as well as the relevant ac and dc voltages. Toolbar buttons are used to decide whether these quantities are to be displayed as waveforms or bar charts. The bar charts can display either phase or sequence components.

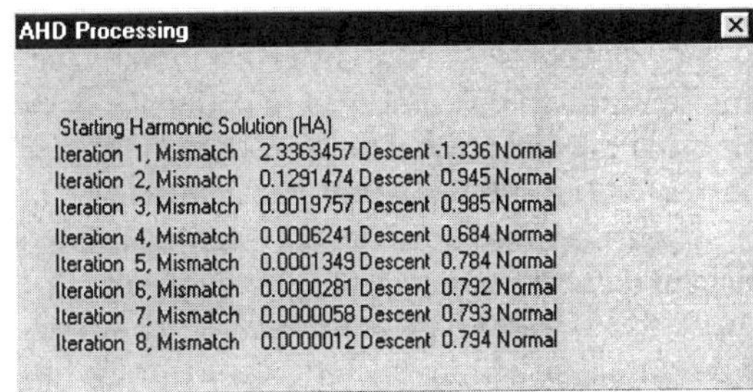

Figure 5. Solution progress.

The results can be displayed in either kV/kA, per unit or as a percentage of the power frequency in the case of the bar charts. The impedance can be shown on either a Log or a Linear scale, and a zoom function is also available. All displays can be copied to the Windows Clipboard along with project information for Quality Control so that they can be easily incorporated into associated documentation. Output data files can also be written for external post-processing.

It is possible to make system or component modifications, and compare them by displaying the new and base case results simultaneously. For instance to simulate the effect of asymmetrical operation, the reactance in one phase of the transformer delta winding may be altered from, for example, 0.18 (the base case) to 0.25 per unit, by clicking on the transformer in the circuit diagram to change the parameters in the Transformer Form and then resimulating.

By default ICHA autoranges to fit the minimum and maximum values of the data to be displayed, however ICHA also permits the user to specify any required magnitude range. Forms with standard harmonic indices (THD, TIF, etc.) and load flow data are also available in the ANALYSIS section.

Once the modifications have been made, the SAVE option can be used to place them on file. The cases can then be retrieved by OPENING the specified file containing the system data. There is also an option to use the default built-in data, which is based on the CIGRE benchmark model, Szechtman et al (7).

5. EXAMPLE OF APPLICATION

The capability of the Interactive Harmonic Domain to derive useful information not easily obtainable by other means, is demonstrated here by detailed analysis of the CIGRE benchmark model (7), a test system proposed for the comparison of alternative simulation algorithms and control systems. The parameters of this test system were calculated from conventional analysis to create ac and dc system resonant conditions at the second harmonic and fundamental frequency respectively.

The Harmonic Domain is used here to check the accuracy of the theoretical resonant frequencies. The test consists of the full CIGRE model under nominal operation, i.e. with 2 kA dc current (1000 MW). The harmonic solution is first converged to define the operating point using the Adaptive Harmonic Domain routine (5). Then the full analytic Jacobian is calculated to the desired frequency resolution and a partial inverse performed to obtain the equivalent system impedance.

The results are shown in Figures 6 and 7 for the positive and negative sequence impedance cases calculated with a 2Hz resolution over a window from 2Hz to 500Hz. The negative sequence plot (Figure 7) shows a step in the converter impedance at 450Hz. This is due to the converter frequency coupling to terms greater than 500Hz, which was the cut-off frequency in this case, and not the actual converter impedance. However the plot shows that the converter has negligible effect in this frequency region.

An important observation from the resonant plots is that the positive and negative sequence impedances are noticeably different when the converters are represented.

6. CONCLUSIONS

The Harmonic Domain, helped by a suitable steering graphical interface, constitutes an ideal steady-state simulator for interactive work involving ac-dc converters. As well as providing fast and accurate harmonic and inter-harmonic information, the iterative algorithm permits the derivation of direct relationships between any selected input and output variables. This is extremely useful to obtain the effective converter harmonic impedances and is therefore an invaluable tool in filter design.

Although the algorithm involved is rather complicated, the user does not need to become familiar with the mathematical formulation of the Harmonic Domain and related numerical techniques.

7. REFERENCES

1. CIGRE SC14, WG 14.03, 1996, "AC system modelling for ac filter design, an overview of impedance modelling", ELECTRA, 164, 133-151.

2. CIGRE SC14, WG 14.25, 1999 "Harmonic cross-modulation in hvdc transmission", ELECTRA, 184, 141-146.

3. Arrillaga, J, Bradley, D. and Bodger, P.S., 1985 "Power System Harmonics", John Wiley and Sons, London.

4. Arrillaga, J, Smith B.C, Watson, N.R. and Wood, A.R., 1997, "Power System Harmonic Analysis", John Wiley and Sons, London

5. Bathurst, G.N., Watson, N.R., Arrillaga, J., 2000, "Adaptive frequency-selection method for a Newton solution of harmonics and inter-harmonics", Proceedings IEE Gen. Trans. & Dist, 142(2), 126-130.

6. Bathurst, G.N., Smith, B.C. Watson, N.R. Arrillaga , J., 1998, "A modular approach to the three-phase harmonic power flow", ICHQP'98, 653-659.

7. Szechtman, M, Weiss, T. and Thio, C.V., 1991, "First benchmark model for hvdc control studies", ELECTRA, 135, 55-75.

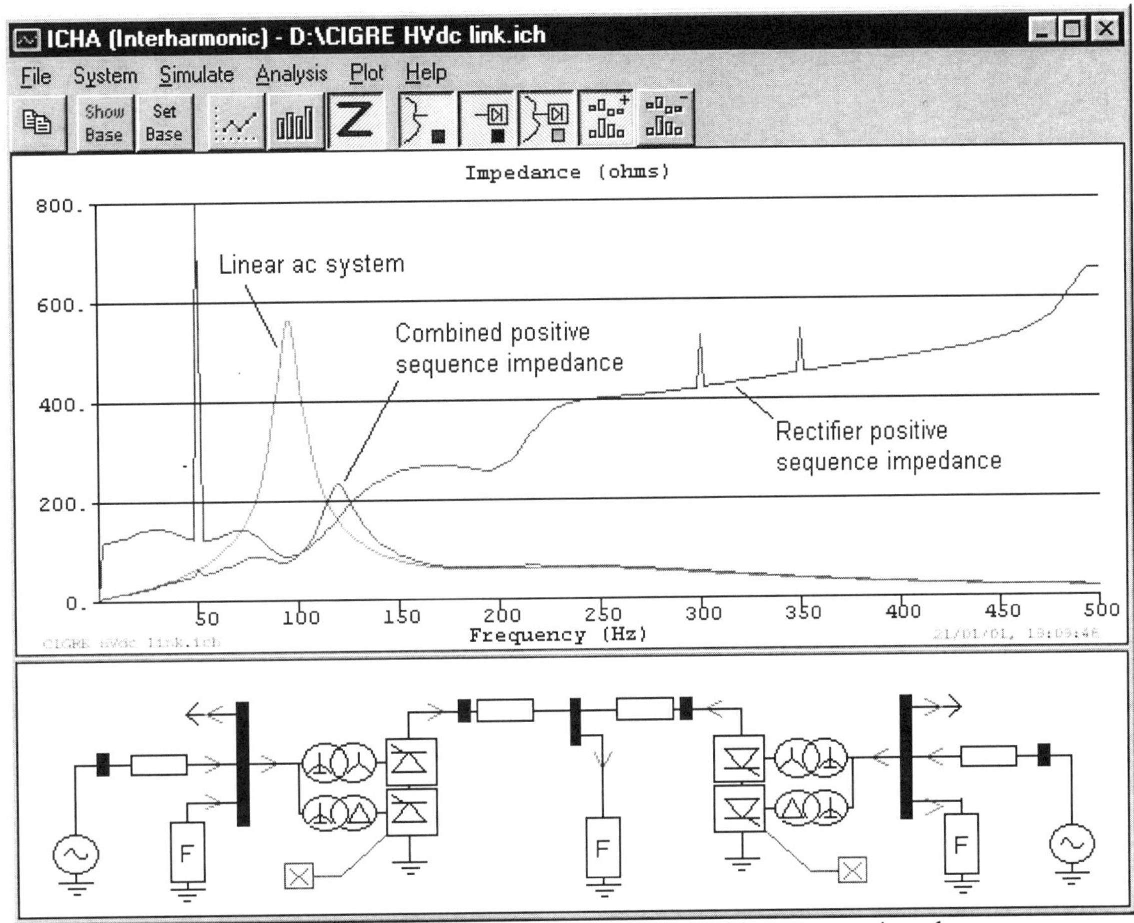

Figure 6.   Effect of the converter on the ac system positive sequence impedance.

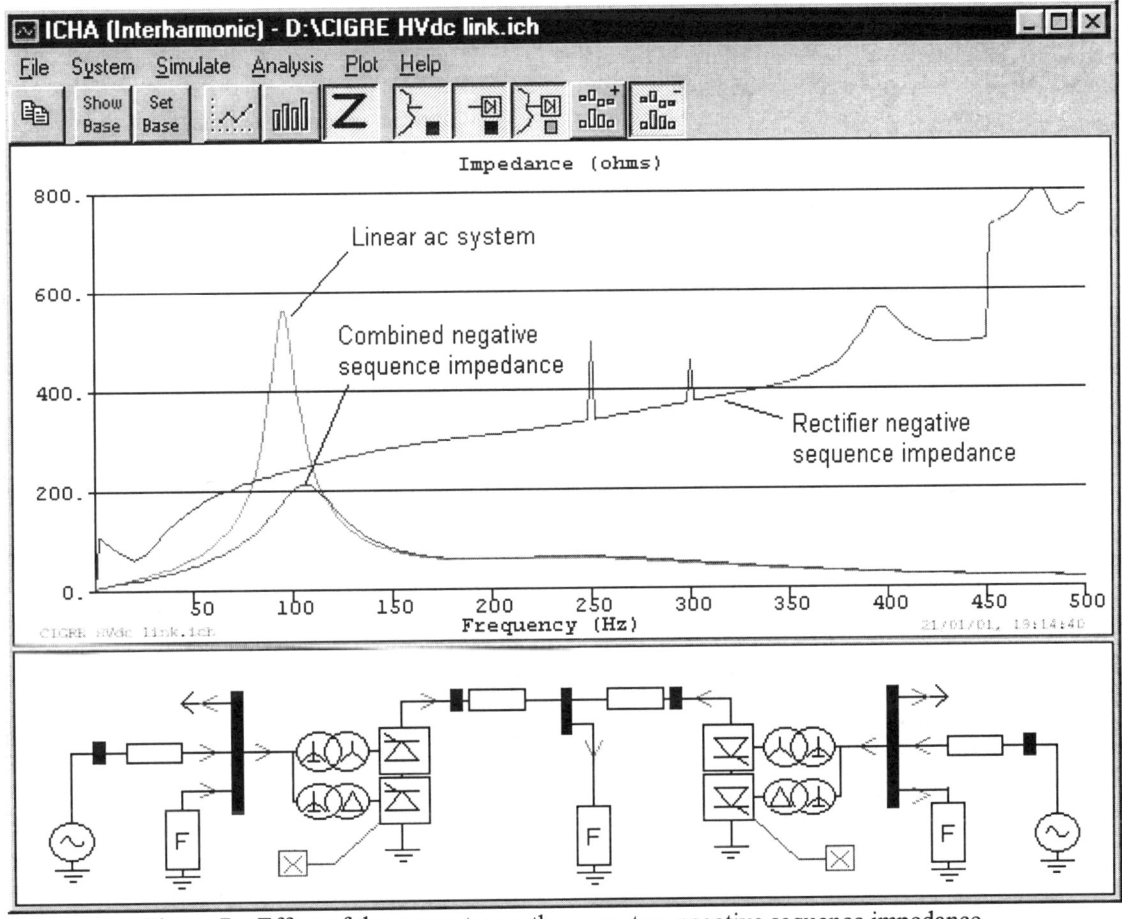

Figure 7.   Effect of the converter on the ac system negative sequence impedance.

EVALUATION OF THE RELATION BETWEEN VOLTAGE SOURCED CONVERTER PERFORMANCE AND DESIGN PARAMETERS

M Takasaki, N Gibo, T Hayashi

Central Research Institute of Electric Power Industry, Japan

ABSTRACT

To promote the application of HVDC system and FACTS devices with Voltage Sourced Converters, it is essential to attain performance improvement and cost reduction simultaneously through the rationalization of converter design. This kind of design coordination first requires comprehensive evaluation of the relation between converter performance and design parameters.

This paper discusses the effect of various design parameters on VSC performance such as loss, harmonics and maximum overcurrent at the system fault. The design parameters particularly include semiconductor device characteristics as well as PWM pulse number and converter control speed. Performance of a 300MW IGBT converter when applied to a HVDC system has been evaluated through numerical analyses. The results obtained make it possible to elaborate a rational converter design with satisfying performance requirements for the HVDC application.

INTRODUCTION

With the progress of deregulation, it is required to improve the utilization of existing power transmission and distribution networks. This can be achieved by providing high-speed controllability of power flow and system voltage by installing HVDC systems and FACTS devices to the network.

Voltage Sourced Converter (VSC), when applied to HVDC interconnection or FACTS device, enables coordinated control of network power flow and voltage by utilizing its independent controllability of active and reactive power[1],[2]. It can be designed to control short circuit current as well. Such controls devised to cope with power flow bottleneck and/or stability limit bring utilization improvement of the existing transmission lines up to the thermal limit.

City center infeed by underground cable is another promising application of HVDC with VSC. The capacity can be increased on existing power corridors. Multi-terminal configuration, which is easily made up with VSC terminals, can feed to any points even no generator exists in the receiving ac system.

VSC applied HVDC[1], STATCOM and UPFC[3] are now in operation emphasizing each particular advantage. Sufficient cost reduction of VSC with satisfying required performance promotes extensive applications of these equipments. The performance to be investigated in designing VSC includes converter loss and harmonics which affect operation or construction cost. Continuous operation capability at the ac system fault should also be discussed to estimate a contribution to stability enhancement that increases the network utilization.

The evaluation of the relation between VSC performance and various design parameters is indispensable to achieve a rational or possibly an optimal system design.

FRAMEWORK OF THE EVALUATION

VSC Performance

The relation between VSC performance and design parameters has been evaluated assuming the application to the HVDC system shown in Fig.1. VSC should be designed to minimize total cost, which consists of construction and operation costs of the HVDC system, with satisfying required system performance.

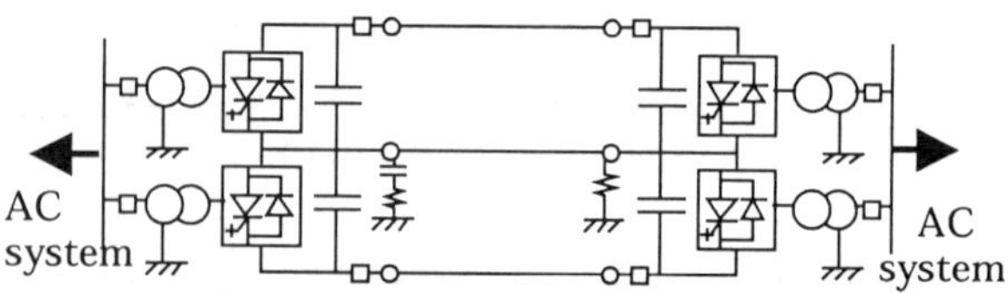

Figure 1: VSC applied HVDC system

The following performance items need to be considered when designing the VSC.

(a) Overcurrent suppression
(b) Overvoltage suppression
(c) Loss
(d) Harmonics
(e) Reliability
(f) Continuous operation at the ac system fault
(g) PQ controllability (speed, independency)
(h) Compactness
(i) Electromagnetic compatibility

In this study, items (a) – (d) are chosen for quantitative evaluation considering the effect pointed out below.

(a) Overcurrent suppression

The maximum overcurrent level at the system fault determines the current rating of a converter for a specific semiconductor device. Overcurrent suppression leads to increase converter rated current using the same device. Consequently cost reduction of a converter can be expected with providing continuous operation capability for the system fault condition.

(b) Overvoltage suppression

Both dc and ac side overvoltages force to increase

AC-DC Power Transmission, 28-30 November 2001
Conference Publication No. 485 © IEE 2001

converter insulation level, thus compactness is also obstructed. Overvoltage suppression can therefore make contribution to cost reduction for a converter as well as for a transmission line.

(c) Loss

Loss is an important item to be discussed, because it can be incorporated in the operation cost estimation. Higher switching frequency enhances converter controllability to suppress the overcurrent and reduces harmonic filter capacity, but on the contrary it increases converter loss. Loss estimation is indispensable when designing a converter to minimize the total cost. Loss of a line-commutated HVDC converter including converter transformer decreases to at around 1.2% in these days. Generally, it could be a target level for a VSC when applying to HVDC system.

(d) Harmonics

Harmonic voltage produced by a converter affects ac filter capacity required and consequently converter system cost. Coordination with loss is needed when determining, for instance, the total pulse number of a VSC.

Design Parameters

The performance items listed above are evaluated in relation with the following design parameters of a converter. A PWM and multi-stage type (constituted with multiple 3-phase converters connected in series or parallel) converter is chosen for the evaluation. It should be noticed that the switching scheme and the topology of a converter specified here are also the important design parameters.

i) DC voltage and DC current ratings

In the multiple converter structure, series connection of unit converters at dc side should be adopted for long distance transmission, parallel connection is generally used for BTB system. The combination of unit converter (3-phase bridge) capacity and stage number (unit converter number) is determined taking semiconductor device ratings and harmonics reduction effect into account.

ii) Total pulse number

The total pulse number is defined with multiplying PWM pulse number of a unit converter by stage number. It affects the control speed of converter output current and voltage as well as converter harmonics and loss.

iii) Converter circuit

Wiring inductance of a converter circuit affects energy to be disposed at every switching and consequently the snubber capacity required. Converter transformer impedance that relates to converter current at the ac system fault is also an important design parameter.

iv) Semiconductor device

IGBT was selected as a promising device. A 4.5kV class high voltage IGBT has already been developed. It can contribute loss reduction by attaining low inductance with a compact design and eliminating anode reactors. Device characteristics such as fall time and on-state voltage are also included as design parameters.

v) Converter controls

Fig. 2 illustrates a block diagram of a converter control based on the PQ vector control scheme. Control speed and functions are design items to be discussed.

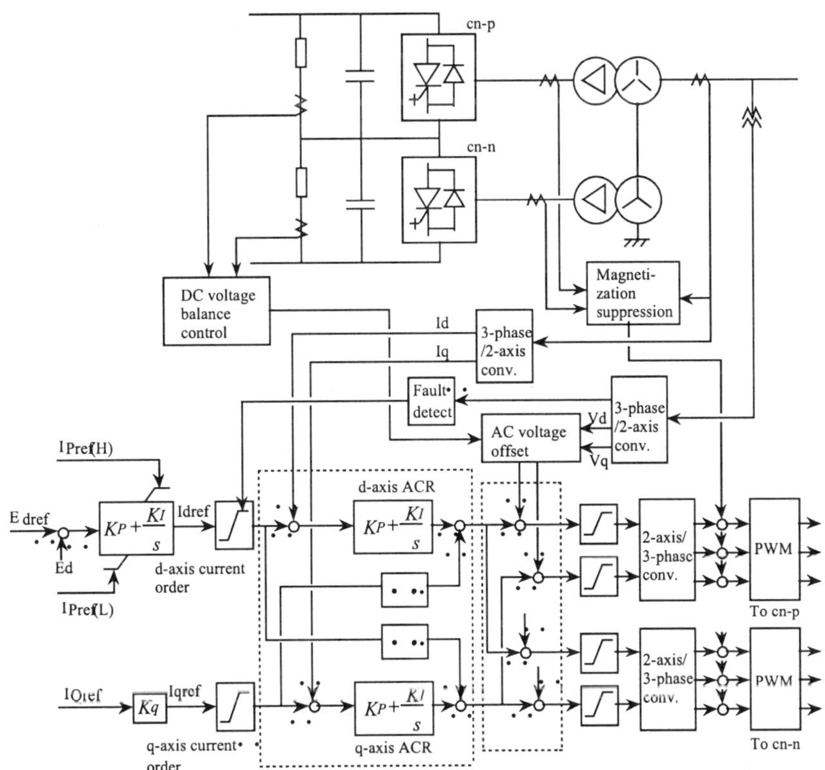

Figure 2: Converter control bock diagram

TRANSIENTS AT THE SYSTEM FAULT

Overcurrent Suppression

Overcurrents flowing through HVDC converter may occur both at ac and dc system faults. As for a VSC, ac system 3LG fault at the location adjacent to converter bus is a severe fault for a continuous operation. The overcurrent in this case is caused by converter control delay (dead time) and uncontrollable period between PWM switching events. The dead time of the converter control includes times for A/D conversion and PWM that generates firing pulse from converter control output. Overcurrent in case of 3LG fault at the inverter ac bus of Fig. 1 has been analyzed using EMTP. The main parameters are the total pulse number and the converter control dead time described above. Table 1 summarizes conditions for EMTP analyses. Converter ratings are based on a VSC model installed in CRIEPI s analogue simulator. The results however can be generalized because the same overcurrent in pu value is obtained when the pu impedance of converter transformer keeps constant.

Changes in peak overcurrent with the total pulse number and the dead time are shown in Fig. 3 and Fig.4 respectively. Fault timing is swept by 20μs step with the output ac voltage peak as the time center (t=0).

Table 1: Conditions for overcurrent evaluation

Converter capacity	50.3kVA (45kW, 22.5kVar)
DC voltage, DC current	1500V, 30A
Converter configuration (All in series connected at the DC side)	· 21-pulse, 2-stage · 15-pulse, 2-stage · 9-pulse, 2-stage · 9-pulse, 4-stage
Control dead time	360μs, 310μs, 270μs
Converter trans. imp.	X: 20%, R: 5%
Fault condition	3LG at inverter ac bus

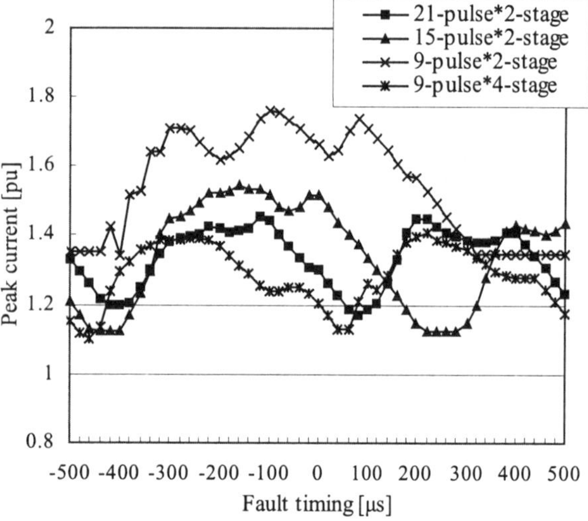

Figure 3: Change in overcurrent factor with the total pulse number

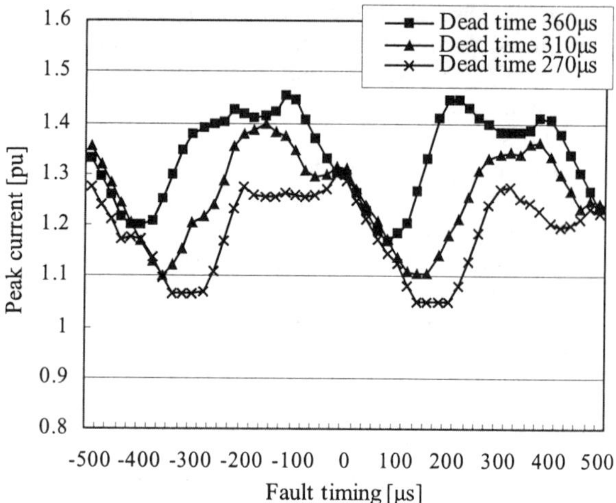

Figure 4: Change in overcurrent factor with the dead time of converter control

These results prove the followings.
1) The increase of the total pulse number and the shortening of the dead time are both effective for overcurrent suppression.
2) The total pulse number and the dead time can be determined for a specified overcurrent level. To suppress the maximum overcurrent 1.5pu or under, for example, 36(9*4) or larger in total pulse number and 360μs or shorter in dead time are needed.

In addition, for the unbalanced fault in the ac system such as 1LG, we have developed dead time compensation control to suppress the peak current (4).

Overvoltage Suppression

Overvoltages of a HVDC system with VSC primarily occur in the dc side. In case of 1 line to ground fault in the dc system of Fig.1, voltages of the neutral line and the sound pole line will be increased by 1pu, these voltages reach 1pu and 2pu to the ground level respectively. This kind of overvoltages are caused by the grounding resistance on the neutral line as shown in Fig.1, it is however needed to suppress overcurrent flowed into dc system through anti-parallel diode. We have proved by a simulator study that high-speed gate blocking of VSC in accordance with a dc line grounding fault detection gives sufficient overvoltage suppression.

LOSS EVALUATION

The converter loss consists of the following loss factors.
a) On-state loss
  1) On-state loss of the semiconductor device
b) Switching loss
  1) Turn-off loss
    i) Device loss in the turn-off period
    ii) Snubber loss
    iii) Anode reactor loss
  2) Turn-on loss
    i) Device loss in the turn-on period
    ii) Snubber loss

iii) Diode recovery loss

These losses can be estimated by calculating the loss energy [J] in the unit period (1/3-cycle or 120° period for a 3-phase converter), then the loss power [W] as time average value for respective loss factors.

Base conditions to evaluate the loss are summarized in Table 2. The effect of the design parameters can be estimated by changing the correspondent condition.

Table 2: Conditions for loss evaluation

Rated capacity	300MW
Converter configuration	9-pulse, 4-stage
DC voltage, DC current per 3-phase bridge	60kV, 1.25kA
AC current: RMS (peak)	2.08kA (2.95kA)
Rated modulation rate[*1]	0.8
IGBT ratings	4.5kV, 4.5kA
Number of IGBT series connection per arm	15
IGBT on-state voltage	3.0V/device
Diode on-state voltage	3.0V/device
IGBT turn-off time (fall time)	2µs (0.6µs): Data from a 3.3kV, 1.2kA IGBT
Wiring inductance	0.2µH/device
Snubber: C, R	0.4µF, 5Ω

*1: The ratio of the control voltage amplitude to the carrier voltage amplitude

On-state Loss

The current flowing pattern of a 3-phase bridge can be classified with on-state pattern for each phase. Each phase of a 3-phase bridge has an IGBT arm or a diode arm in on-state.

When the current flowing pattern j is defined in a period of $[\theta_{sj}, \theta_{ej}]$, the on loss energy of 120° period can be calculated with the following equation.

$$E_{on} = \sum_{j;\theta=\frac{\pi}{6},\frac{5\pi}{6}} n \left\{ \int_{\theta_{sj}}^{\theta_{ej}} V_{da}\cdot i_a\, d\theta + \int_{\theta_{sj}}^{\theta_{ej}} V_{db}\cdot i_b\, d\theta + \int_{\theta_{sj}}^{\theta_{ej}} V_{dc}\cdot i_c\, d\theta \right\}$$

(1)

where,

V_{da}, V_{db}, V_{dc}: On voltage drop for respective phase, that correspond to on-state voltage of an IGBT or a diode

n: Number of series connected devices

As the waveforms of device currents i_a, i_b, i_c are approximately sinusoidal, on loss power P_{on} [W] of the 3-phase bridge can be derived as:

$$P_{on} = E_{on}\cdot(3/2\pi)$$

(2)

On loss calculated for the condition of Table 2 is 1.012MW or 0.34% (1.012/300).

Turn-off Loss

Fig. 5 illustrates a model to simulate turn off behavior of one phase in a 3-phase bridge. Turn-off characteristic of IGBT is represented as a resistance linearly increased with time. The reactor represents the wiring inductance of converter circuit and no anode reactor exists.

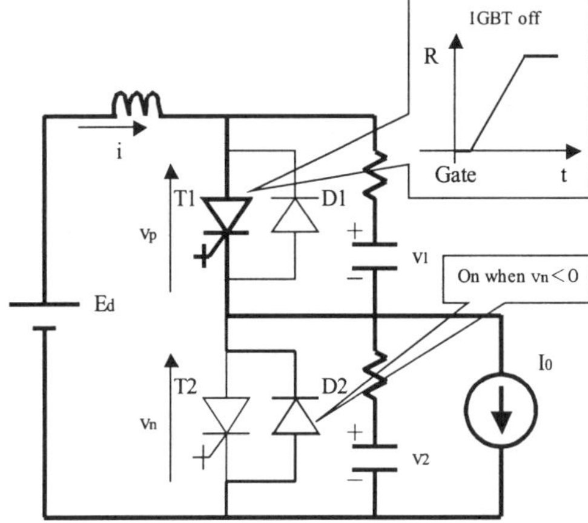

Figure 5: Simulation circuit for the p-arm turn-off action

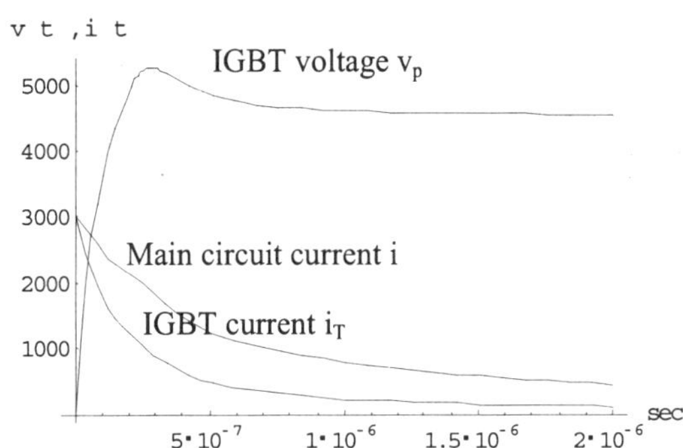

Figure 6: Transients in the turn-off period

IGBT loss. Fig. 6 shows a simulation result for a device (4.5kV) when cutting off the normal peak current of 2.95kA. Loss energy consumed in the IGBT device at turn-off can be derived by calculating the time integral of $v_p{}^*i_T$. The device loss therefore varies with the switching phase because of the change in current to be cut off.

Suppose the device loss is proportional to the cut off current level, the average loss for a switching is given by $(2/\pi)\cdot E_{dofm}$, where E_{dofm} is the device loss in Joule when cutting off the ac peak current in normal operation.

In a 3-phase bridge of pulse number p, turn-off actions occur p times in the 120° period. Thus the turn-off loss power P_{toff} [W] is obtained from the following equation.

$$P_{toff} = (2/\pi)\, E_{dofm}\cdot n\cdot p\times 3f$$

(3)

where n: Number of series connected devices

f: System frequency

Snubber loss. The snubber loss in the p-arm turn-off period is produced by the discharge of the n-arm snubber capacitor. It is consumed in the n-arm snubber resistor. When C_s is snubber capacity and E_{d1} is dc voltage per device, snubber loss per one turn-off action equals to $(1/2)\cdot C_s E_{d1}{}^2$. So the snubber loss power P_{soff} for a 3-phase bridge can be calculated as:

$$P_{soff} = (1/2)\, C_s\cdot E_{d1}{}^2\cdot n\cdot p\times 3f$$

(4)

Loss analysis for the turn-off action. Table 3 shows the losses in the turn-off period calculated using the above equations.

Table 3: Result of turn-off loss evaluation

IGBT loss	Loss energy when cutting off the peak current	3.1 [J]/device
	Loss power per bridge	0.159 [MW]
	Loss rate	0.053 [%]
Snubber loss	Loss energy per turn-off	4.1 [J]/device
	Loss power per bridge	0.332 [MW]
	Loss rate	0.11 [%]

Turn-on Loss

Fig. 7 illustrates a circuit model similar to Fig. 5 for simulating turn-on behavior. As the turn-on action of IGBT device itself is considerably fast, we can assume the turn-on of IGBT is completed before starting the diode recovery. The diode recovery action starting at t_r as shown in Fig. 7 therefore becomes crucial when analyzing the turn-on behavior. It can be characterized by the maximum reverse current I_r and the time from $I=-I_r$ to $I=0$.

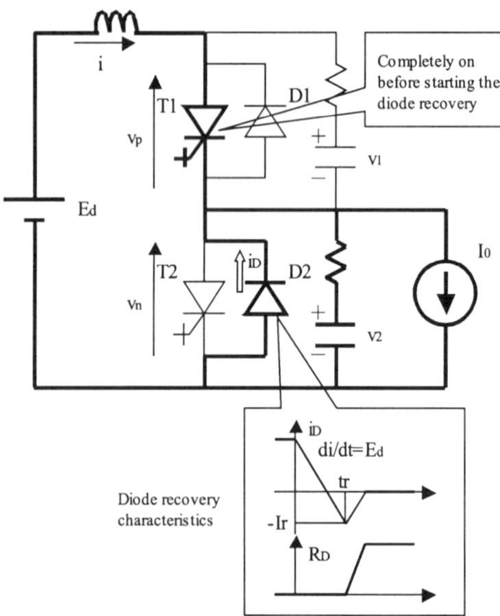

Figure 7: Simulation circuit for the p-arm turn-on action

IGBT loss. The loss produced in an IGBT device at the turn-on action can be obtained with the same manner as the turn-off period by calculating a time integral of $v_p * i_T$. The turn-on loss power P_{ton} [W] is obtained by substituting E_{donm} (IGBT loss in Joule when turning on at the ac peak current level) for E_{dofm} in equation (3).

Snubber loss. The snubber loss produced by the turn-on action becomes the same amount as that in the turn-off period. Because the p-arm snubber capacitor is discharged at the p-arm turn-on action and its energy is consumed in the snubber resistor.

Diode recovery loss. The diode recovery loss can be estimated by calculating a time integral of the product of diode current and diode voltage based on a transient simulation of the turn-on action. The coordination between characteristics of IGBT turn-on and diode recovery is also an important designing point.

Loss analysis for the turn-on action. Table 4 shows the loss rates for respective loss factors at the turn-on action. IGBT loss is estimated based on measured data of a commercial IGBT (3.3kV, 1.2kA). In this study, diode recovery loss is simply assumed to be the same amount as the IGBT loss at the turn-off action.

Table 4: Loss rate estimated for the turn-on action

IGBT loss	0.032 [%]
Snubber loss	0.11 [%]
Diode recovery loss	0.053 [%]

Summary of Loss Evaluation

The total loss estimated from the above quantitative analyses becomes 0.7 [%] in loss rate or 2.1MW for the 300MW converter. This value is rather theoretical when realizing:

1) Low on-state voltage of 3V for a 4.5kV, 4.5kA IGBT
2) Compact converter design that can reduce the snubber capacitor to 0.4μF/device or less

Fig. 8 shows the effect of the total pulse number on the loss divided into four factors. The switching loss increases proportionally with the total pulse number. For a converter with the total pulse of 36 or under, on loss reduction is quite important for reducing the total loss.

An application of high voltage and large current IGBT enables to reduce the converter system loss including transformer loss to about 1.2%, that is comparable to the total loss of a conventional line-commutated converter system.

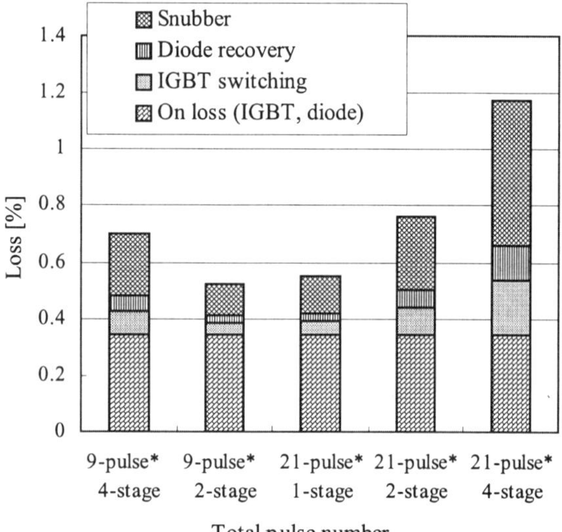

Figure 8: Effect of the total pulse number on converter losses

HARMONICS EVALUATION

The required ac filter capacity for reducing all harmonic voltage components to less than 1% has been analyzed. Short Circuit Ratio (SCR) was assumed to be 2.0 for the HVDC application shown in Fig. 1. The main parameter is the total pulse number. The effects of the pulse number per bridge and the stage number should be separately evaluated in harmonics analyses. PWM modulation rate was also changed within 0.6 – 0.94.

Fig. 9 depicts the required minimum filter capacity for the fundamental frequency expressed in the ratio to the converter capacity. The filter configuration is shown in Fig.9, and its constants are determined to minimize the capacity with satisfying the condition to reduce all harmonic voltage components to less than 1%.

This result proves that the required ac filter capacity is in inverse proportion to (pulse number per bridge)*(stage number)^2. Increasing stage number is more effective for reducing ac filter capacity.

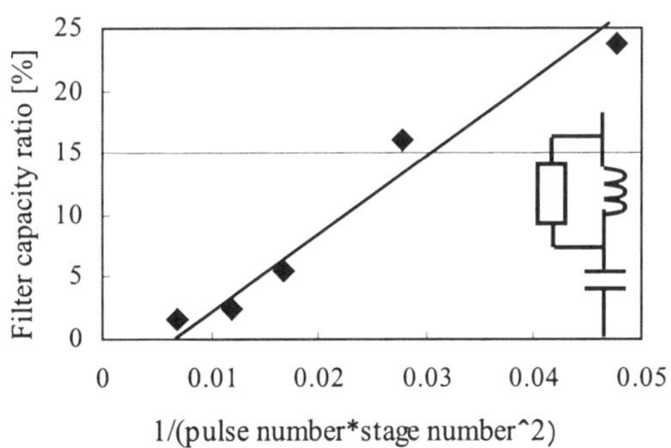

Figure 9: Filter capacity required to reduce all voltage harmonic components to less than 1%

RELATION BETWEEN VSC PERFORMANCE AND DESIGN PARAMETERS

Table 5 summarizes the relation between VSC performance and design parameters discussed in this paper. This kind of comprehensive evaluation is useful when pursuing design coordination between the performance improvement and cost reduction.

CONCLUSIONS

The effect of VSC design parameters on the principal converter performance, such as transient voltage and current control, loss and harmonics, has been evaluated with simulation analyses. The loss evaluation proved clearly the development of a large capacity IGBT has a great impact on the design improvement of VSC.

The results obtained in this research make it possible to clarify an adequate design procedure for attaining cost reduction with satisfying the required system performance.

REFERENCES

1. T.J. Hammons, et al, 2000, "Role of HVDC Transmission in Future Energy Development", IEEE Power Engineering Review, Vol. 20, No. 2, 10-25

2. M. Takasaki and T. Hayashi, 1996, "Effect of HVDC System with Self-commutated Converter for Enhancing Transmission Capability", IEE Conference Publication No.423, 411-416

3. A. Edris, 2000, "FACTS Technology Development: An Update", IEEE Power Engineering Review, Vol. 20, No. 3, 4-9

4. N.Gibo, K.Takenaka, M.Takasaki, T.Hayashi, H.Konishi, S.Tanaka and H.Ito, 1999, "Enhancement of Continuous Operation Performance of Self-commutated HVDC System", IEEE Transactions on Power Systems, PE062PRS

Table 5: The relation between the VSC performance and design parameters

Design parameters	Performance	Over-current	Over-voltage	Loss	Har-monics	Subjects for technical development
Rated dc voltage and dc current			**	**		(Cost minimization)
Total pulse number		***		***	***	(Cost minimization)
Main circuit	Wiring inductance		***	***		Compact stack design
	Snubber capacitor, resistance		***	***		Optimal snubber design
	DC capacitor		**			
	AC filter capacity	*	*	**	***	Minimization
	Converter trans. reactance	***			*	
Semi-conductor device (IGBT)	Voltage and current rating		*	***		Large capacity IGBT
	Fall time		***	***	**noise	Optimal characteristics
	On-state voltage			***		Minimization
	Active gate control	*	***	**		Control scheme
Converter control	Control speed (Dead time)	***				High speed processor
	Dead time compensation	***	***			[Developed (4)]
	High-speed gate block	*	***			[Developed]

***: Main design factor, **, *: Minor or less design factor

A CONSIDERATION OF STABLE OPERATING POWER LIMITS IN VSC-HVDC SYSTEMS

H Konishi
Hitachi Research Laboratory, Hitachi Ltd., Japan

C Takahashi, H Kishibe, H Sato
R & D Center, Tohoku Electric Power Co., Inc., Japan

Abstract - The stable operating power limits of a small scale HVDC system composed of voltage source converters (VSC-HVDC system) are analyzed with a simple model. The VSC-HVDC system could operate where the AC system must be somewhat larger in capacity than the VSC-HVDC system capacity. The stable operating power limits were between one and two times the SCR (short circuit ratio). When the inverter of the VSC-HVDC system was operated with lead reactive (capacitive) power control conditions, the stable operating limits were increased through AC voltage stabilization. If there were loads near the inverter, the VSC-HVDC system could transmit a little bit larger power than those without loads.

Keywords: HVDC, voltage source converter (VSC), operating power limit, SCR

INTRODUCTION

As voltage source converters (VSCs) consist of self-commutated devices such as IGBTs (insulated gate turn-off bipolar transistors), they can operate in areas which lack rotating machines or do not have enough power in the rotating machines. Additionally, they control active powers and reactive powers independently and rapidly [1]. A small scale HVDC system composed of voltage source converters (VSC-HVDC system) has been used world-wide to supply power to small islands, to get easy connection to wind power facilities or fuel cells, solar batteries, etc.[2]. To extend applications of the VSC-HVDC system, it is important to note the operational characteristics and limitations. The stable operating power limits of a small scale HVDC system composed of voltage source converters (VSC-HVDC system) are analyzed with a simple model in this paper.

CONFIGURATION OF VSC-HVDC SYSTEM

The configuration of the VSC-HVDC system studied is shown in Fig. 1 and specifications are listed in Table 1. Its converters consist of four VSCs with nine-pulse PWM, having a total capacity of 14 MVA (10MW for active power and 10Mvar for reactive power) and DC voltage of 20 kV. The control systems of the converters use conventional PQ vector control to control their active powers P and reactive powers Q independently. A rectifier controls the DC voltage to the rated values,

while an inverter controls the DC power to the reference values[3,4]. The active power vs. DC voltage characteristics of the VSC-HVDC system are shown in Fig. 2. The operating point is the crossing point of the lines. The active power of the inverter is adjusted by changing the active power reference of the inverter Pref while keeping the DC voltage constant with the rectifier.

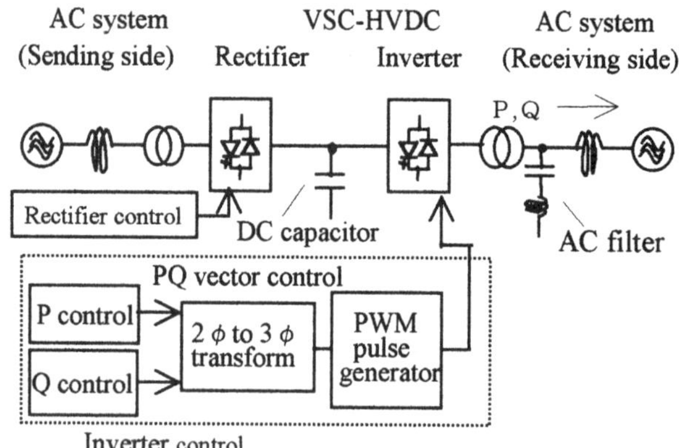

Fig. 1 VSC-HVDC transmission system model studied

TABLE 1-Specifications of VSC-HVDC

Items	Specifications
AC voltage	66 kV
Converter transformer	8.72 kV/66 kV、20%Z
Converter capacity	14 MVA(10 MW、10MVar)
DC voltage	20 kV
Converter configuration	4 multi-converter、9 pulse PWM
Control circuit	P Q vector control P control: Active power control Q control: Reactive power control
AC filter	30th, 0.5MVA
DC capacitor	2000 μ F/converter

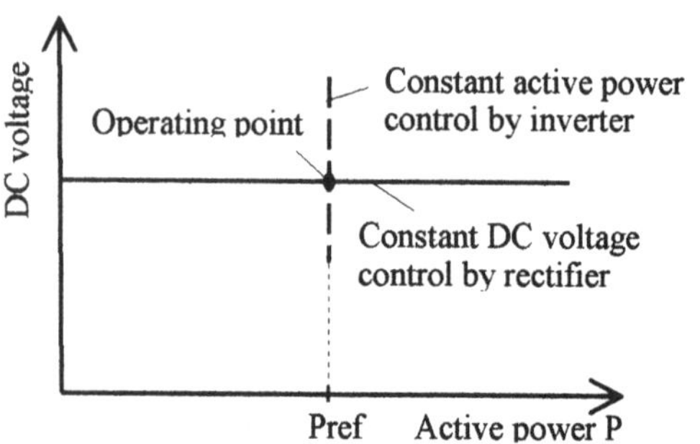

Fig. 2   Active power vs. DC voltage characteristics of the VSC-HVDC system

AC-DC Power Transmission, 28-30 November 2001
Conference Publication No. 485 © IEE 2001

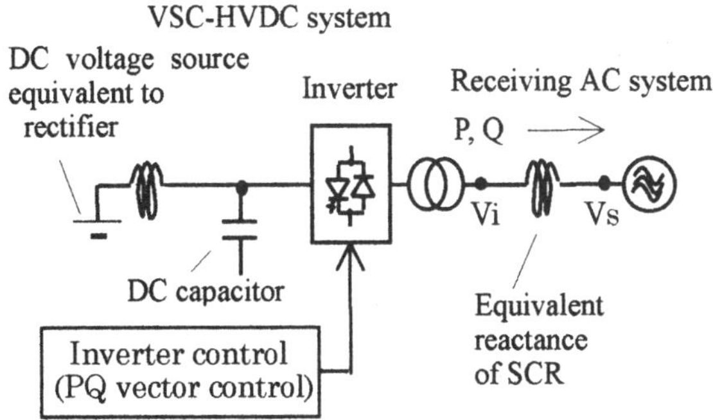

Fig. 3   VSC-HVDC transmission system model

ANALYSIS OF STABLE POWER LIMITS

Analysis Conditions

The VSC-HVDC transmission system model shown in Fig. 3 is used to analyze and evaluate stable operating limits. In the transmission system model, a constant DC voltage source is used in place of the rectifier. A constant AC voltage source and an equivalent reactance representing the SCR (short circuit ratio; the ratio of an AC system short circuit capacity to the inverter capacity) are set for the AC system of the inverter side. The converter rated capacity of 14 MVA and the AC system rated voltage of 66kV are selected for the pu base value. Analyses of VSC-HVDC system start-up operation while changing reactive control powers of the inverter are carried out for the SCR. The same analyses are done for STATCOM operation of the VSC-HVDC system, too.

Analysis results

(a) AC system via an equivalent SCR impedance

Table 2 summarizes analysis results for VSC-HVDC system start-up operation to rated active power and Fig. 4 shows stable operating regions vs. SCRs of Table 2. Figure 5 shows four of the analysis waveforms when reactive powers of the inverter are controlled as -0.8 pu (capacitive). The waveforms show AC voltages, active and reactive powers and phase angle difference θ between the inverter output voltage Vi and the AC system voltage Vs. In the figure, the VSC-HVDC system reaches a stable operating point when the SCR is greater than 1.2, but it is oscillatory when SCR is 1.0. Although the inverter of the VSC-HVDC system can operate where there are no rotating machines in the AC system [2], Table 2 and Fig. 4 show that for the AC power system including rotating machines, the system must be somewhat large to get stable operation. As the SCR of the AC system becomes smaller, stable inverter operation of the VSC-HVDC becomes harder. The stable operating limits are between one and two times the SCR. If the inverter operates with lead reactive power (capacitive) controls, the stable operating limits become larger because AC voltages are stabilized. Due to the non-linearity of the VSC-HVDC system, the

TABLE 2-Analysis cases and results
(VSC-HVDC start-up of P=1.0 pu)

SCR	Reactive power control Q(p u)					
	0.0	-0.2	-0.4	-0.6	-0.8	-1.0
2.0	△	○	○	○	○	○
1.5	×	×	×	○	○	○
1.2	×	×	×	△	○	△
1.0	×	×	×	×	△	×
0.9	×	×	×	×	×	×

-Q : lead reactive power control by VSC
○ : stable operation, △: nearly stable operation,
× : unstable operation

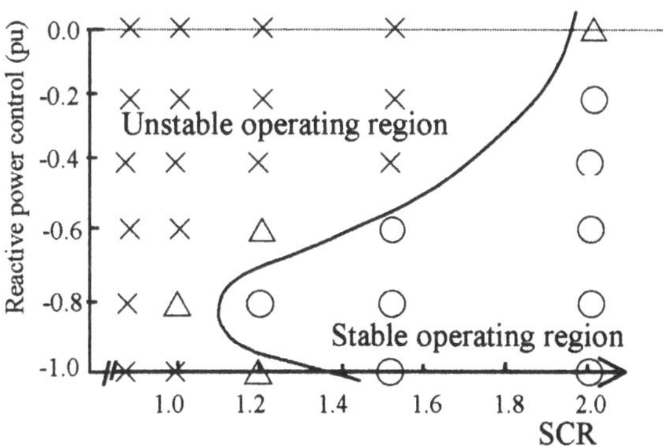

Fig. 4   Stable operating region of VSC-HVDC

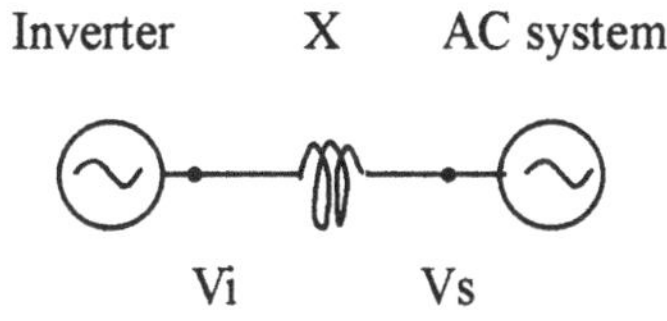

Fig. 6   Equivalent circuit

stable operating regions become a little bit large when the reactive power control of the inverter is –0.8 pu.

These results are investigated theoretically using a related equation of transmission powers and voltage phase angles. The equivalent circuit of the VSC-HVDC system is shown in Fig. 6. The AC voltage source Vi presents the inverter output voltage of the VSC-HVDC system, Vs shows AC system voltage and X shows reactance of AC lines representing the SCR. Then the transmission power P can be described by the following equation:

$$P = Vs \cdot Vi \cdot \sin\theta / X \qquad (1)$$

where θ is phase angle difference between AC system voltage and inverter output voltage. If it is supposed that Vs=Vi=1 pu, and $\theta = 90°$, X must be less than 1 pu to transmit the active power $P \geqq 1$ pu. This means the AC system short circuit capacity (a reciprocal of the reactance X) must be greater than 1. The simulation analysis results agree with the theoretically calculated results considering that the phase angle difference θ is

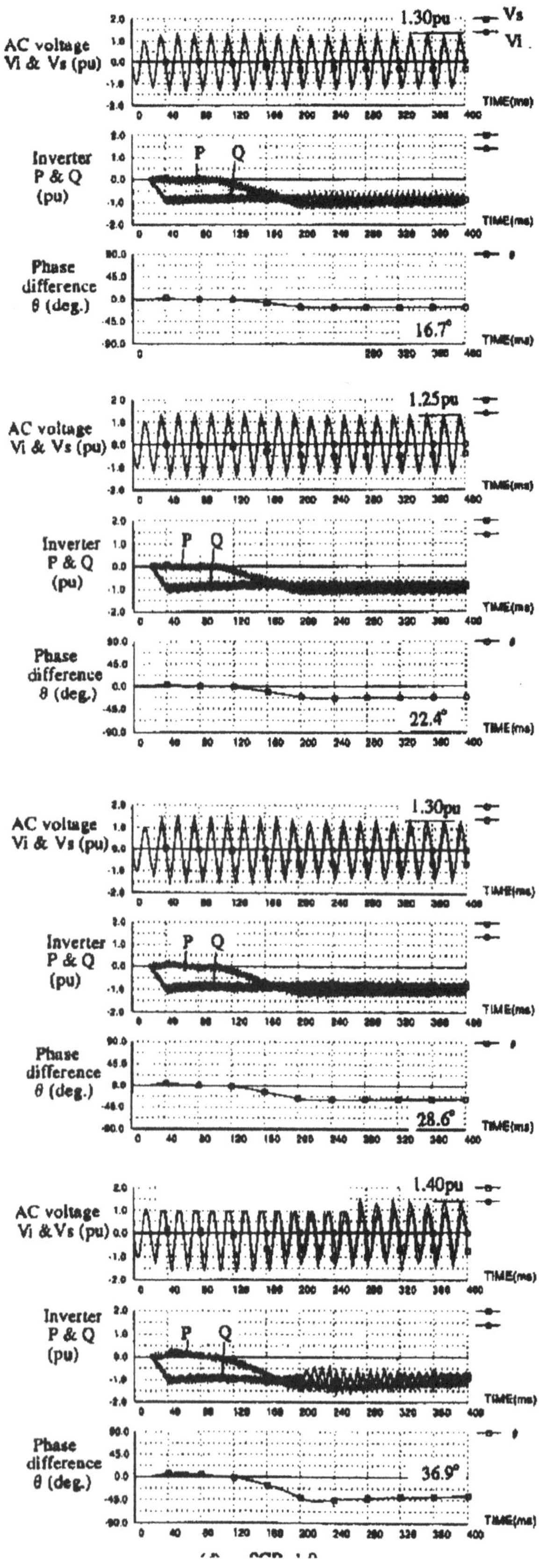

Fig. 5 Analysis results in VSC-HVDC start-up operation

generally much less than $90°$ to get stable operations dynamically.

TABLE 3- Comparison of inverter output voltage

SCR	Reactance X (pu)	Phase angle difference θ (deg.)	Inverter output voltage Vi (pu)		Fig.
			Measured	Calculated	
2.0	0.5	16.7	1.30	1.23	5(a)
1.5	0.67	22.4	1.25	1.24	5(b)
1.2	0.83	28.6	1.30	1.23	5(c)
1.0	1.0	36.9	1.40	1.18	5(d)

Table 3 compares both calculated and measured inverter output voltages Vi. The former are calculated using equation (1), inserting measured θ while setting P=1 pu, Vs=1 pu and X=1/SCR. The measured Vi values are good agreement with calculated ones when the SCR is greater than 1.2. So the analysis results are confirmed to be valid.

(b) STATCOM operation

Table 4 summarizes analysis results for STATCOM operation of the inverter in the VSC-HVDC system. The stable operating limits are independent of the SCR of the AC system. The STATCOM can not operate with lag reactive (inductive) power control Q=1 pu because of deterioration of AC voltages. Figure 7 shows analysis results when SCR=1.0 and reactor powers are (a) Q=1.0 (inductive), (b) Q=—0.8 (capacitive) and (c) Q=—1.0 (capacitive). The waveforms illustrate inverter output voltage Vi (pu), inverter current Iac (pu) and inverter active power P (pu) and reactive power Q (pu). The STATCOM oscillates with high frequency for Q=1.0 since the AC voltages decrease due to lagging currents.

TABLE 4 - Analysis cases and results (STATCOM operation (P = 0 pu))

SCR	Reactive power control Q (pu)			
	1.0	-0.8	-1.0	Fig.
2.0	×	○	○	
1.5	×	○	○	
1.2	×	○	○	
1.0	×	○	△	7

○: stable operation, △: nearly stable operation, ×: unstable operation

(c) AC system with load near the inverter

Another case with loads near the inverter is also studied. This VSC-HVDC system is shown in Fig. 8. In the simulation the equivalent reactance of the SCR setting is 0.5. Analyses of VSC-HVDC system start-up operation while changing reactive control powers of the inverter are carried out for the load and the analysis results are shown in Table 5.

When the SCR is 0.5, the VSC-HVDC system can transmit the maximum active power of 0.5 pu. If the loads are greater than 0.6 pu, it can transmit 1.0 pu by AC voltage control (reactive power control) of the

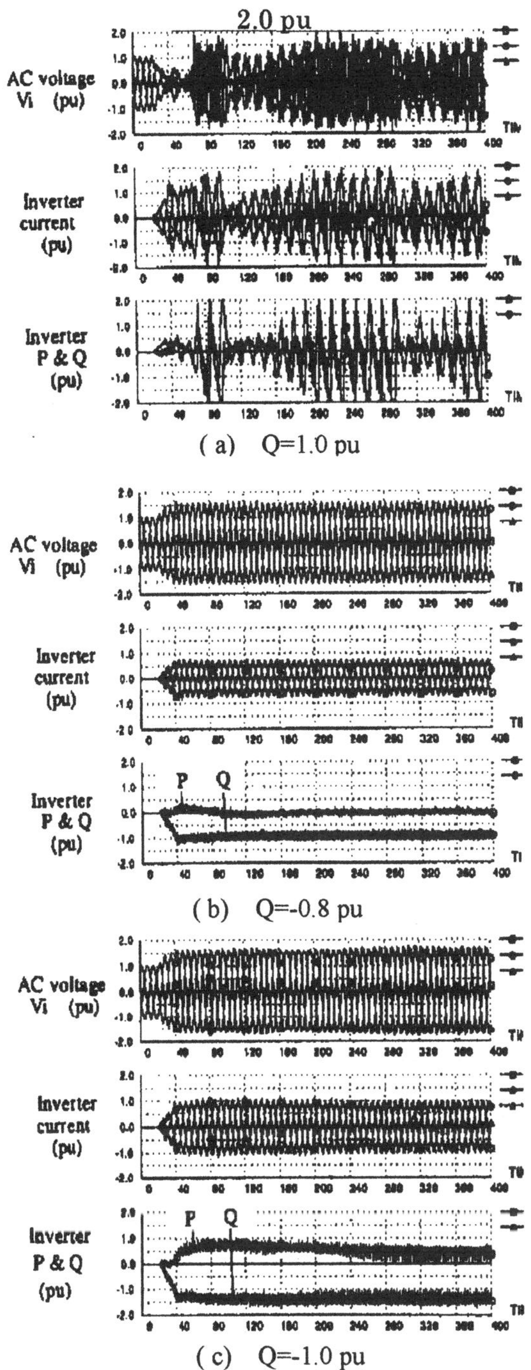

(a)   Q=1.0 pu

(b)   Q=-0.8 pu

(c)   Q=-1.0 pu

Fig.7 Analysis results of STATCOM operation

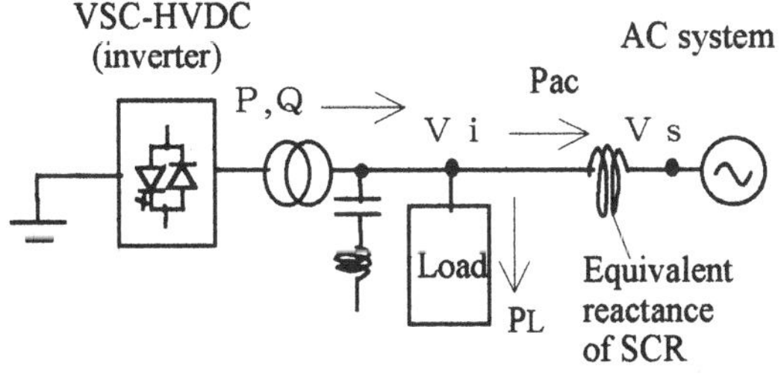

Fig. 8   VSC-HVDC model used for analysis

TABLE 5-Analysis cases and results with loads
(P=1.0 pu, SCR=0.5)

ZL(pu)	PL(pu)	Pac(pu)	Q(pu)	Vi(pu)	Operation
Non	-	0.5	<-0.3	1.2	×
0.3	0.3	0.7	<-0.5	>1.5 *	×
0.4	0.4	0.6	<-0.3	//	×
0.5	0.8	0.2	<-0.2	>1.4*	×
0.6	0.85	0.15	<-0.1	1.1	○ (Fig.9)
0.7	1.0	0	//	//	○
0.8	//	//	<0	//	○
0.9	1.1	-0.1	//	1.0	○
1.0	1.13	-0.13	//	//	○
1.5	1.17	-0.17	<0.1	0.9	○

ZL: impedance of the load
PL: measured active power of the load
Pac: measured AC active power flowing to the AC
system
Q: reactive power control of the inverter
Vi: inverter output voltage,   Vi*: not accept
○ : stable operation, △: nearly stable operation,
×: unstable operation

inverter. The active powers of the VSC-HVDC system
flow partly to the load and the rest to the AC system.
The sum of the load and the active powers flowing to
the AC system equal the active power output of the
VSC-HVDC system. The results show that the stable
operating power limits become larger than those without
loads due to AC voltage stabilizing effects of the loads.
Fig. 9 shows the analysis results when the load is set 0.6
pu with SCR=0.5. Figure 9(a) indicates at the reactive
power control of inverter setting 0 pu. The start-up
operation of the VSC-HVDC system does not succeed
due to the AC voltage instability. On the other hand, as
shown in Fig. 9(b), it reaches stable operating
conditions by the reactive power control of –0.1 pu. In
this case, the load consumes active powers of 0.85 pu
regardless of 0.6 pu loads and the rest of the active
powers of 0.15 pu flow to the AC system. The AC
voltages of the inverter settle down to 1.1 pu by both the
reactive power control and voltage stabilizing effects of
the load.

CONCLUSIONS

Stable operating power limits for the HVDC system
composed of voltage source converters were analyzed
and evaluated theoretically. The VSC-HVDC system
could operate where the AC systems must be somewhat
larger capacity than the VSC-HVDC system capacity.
The stable operating power limits were between one and
two times the SCR. When the inverter of the VSC-
HVDC system operated with lead reactive power
control conditions, the stable operating limits were
increased through AC voltage stabilization. When there

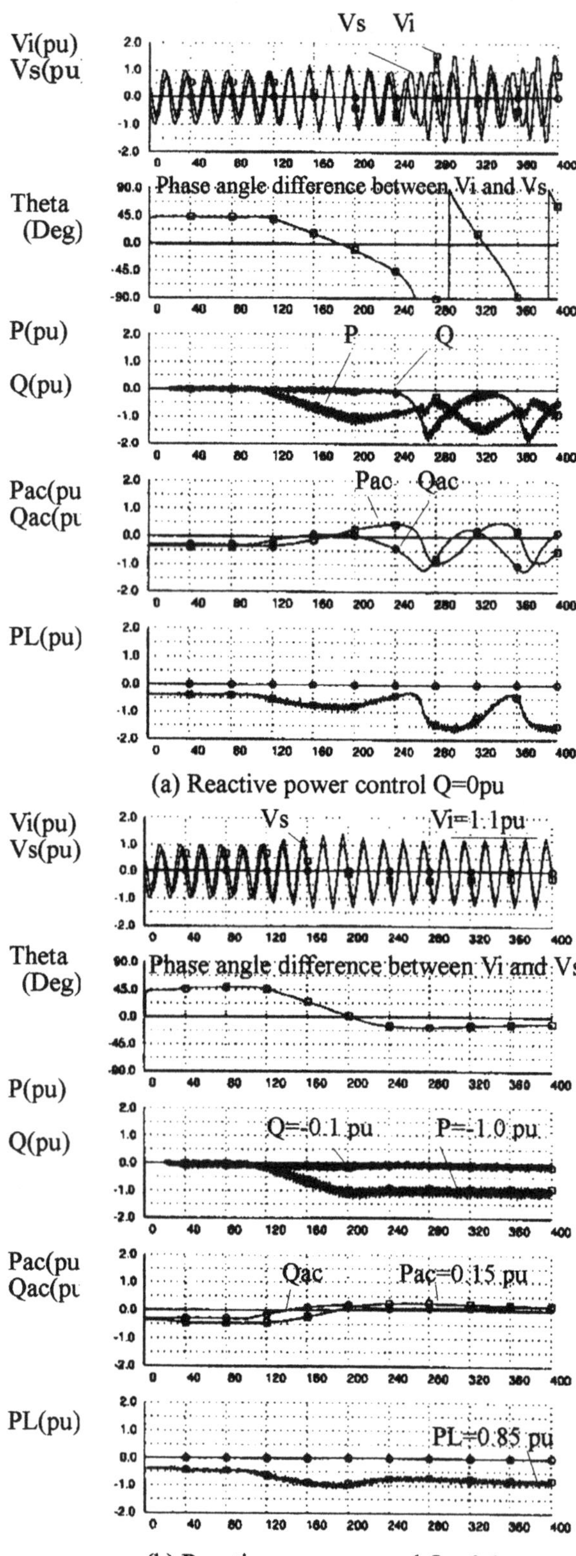

(a) Reactive power control Q=0pu

(b) Reactive power control Q=-0.1 pu

Fig. 9   Analysis results with loads
(PL=0.6 pu, SCR=0.5)

were loads near the inverter, the VSC-HVDC system could transmit a little bit larger power than those without loads by the reactive power control and due to AC voltage stabilizing effects of the loads.

REFERENCES

1.   Eriksson K, Jonsson T and Tollerz O, 1998, "Small scale transmission to AC networks by HVDC light", 12th Cepsi Conference

2.   Asplund G, Eriksson K and Svensson K, 1997, "DC Transmission based on Voltage Source Converter", CIGRE SC14 Colloquium in South Africa

3.   Iwata Y, Tanaka S, Sakamoto K, Konishi H and Kawazoe H, 1996, "Simulation study of a hybrid HVDC system composed of a self-commutated converter and a line-commutated converter", 6th International Conference on AC and DC Transmission, IEE No.423, pp381-386

4.   Tanaka S, Maeda T, Mitsuma H, Konishi H, and Kawazoe H, 2000, "Study on Control and Operational Characteristics of SCC Applied to Distribution Systems", IEEJ PSE-00-1, pp.1-6

BIOGRAPHIES

Hiroo Konishi was born in Tokushima Prefecture, Japan on February 6, 1948. He received the M.S. degree in electrical engineering from Osaka University in 1972. He joined Hitachi Research Laboratory, Hitachi Ltd., Japan the same year. Since then he has been working in research and development of control and protection systems of HVDC. He received the Ph. D degree from Osaka University in 1989. He is interested in FACTS devices and power system simulations.
Dr. Konishi is a member of the IEE of Japan.

Choei Takahashi was born in Iwate Prefecture, Japan, on November 12, 1967. He received the B.S. degree from Iwate University in 1990. He joined Tohoku Electric Power Co., Inc. the same year. Since then, he has been mainly engaged in research on power electronics application for power systems.
Mr. Takahashi is a member of the IEEE and IEE of Japan.

Hideto Kishibe was born in Akita Prefecture, Japan, on February 22, 1959. He received the B.S. degree from Akita University in 1981. He joined Tohoku Electric Power Co., Inc. the same year. Since then, he has been mainly engaged in research on power device application for power system equipment.
Mr. Kishibe is a member of IEE of Japan.

Hiromichi Sato was born in Hokkaido, Japan, on September 1945. He received the B.S. degrees from Hokkaido University in 1968. He joined Tohoku Electric Power Co., Inc. the same year. Since then, he has been mainly engaged in planning, operation, analysis and technical development as related to power systems.
Mr. Sato is a member of the IEE of Japan.

SMALL SIGNAL MODELLING OF HVDC TRANSMISSION SYSTEMS

A. R. Wood, C. M. Osauskas, D. J. Hume
University of Canterbury, Christchurch, New Zealand

ABSTRACT

An exact small signal linear model of the HVdc converter has been recently published, and this paper describes two applications of the model. The first application is the calculation of steady state harmonic levels in the presence of unbalanced ac systems. The method presented is general, and can be applied to asynchronous links of different operating frequencies, with different sources of waveform distortion. The second application is the calculation of the dynamics of HVdc systems, particularly within the frequency range from 2 to 200Hz. All interactions relevant to the dynamics and control of HVdc systems are accurately represented in the s-domain. Both applications are in excellent agreement with time domain simulations using PSCAD/EMTDC.

INTRODUCTION

Modern power systems are becoming more and more dependent on power electronic control, either for transmission purposes, or for interfacing a potentially wide range of alternative power sources to a network. The issue of how these power electronic circuits interact with each other through the system interconnection, either through steady state harmonics, or during system transients, is becoming more important.

Power networks are large and complex, as are many of the power electronic devices embedded within them. Complete modelling of such networks is impractical, and network reductions or simplifications must be made. One such simplification involves linearisation, which allows fast solutions for large networks, and also allows access to simple and powerful control design and optimisation techniques. While linearisation is accurate for many network components, for thyristor based power electronic circuits it is a small signal approximation. This approach has been used in the past for harmonic studies by Hu et al[1], Larson et al [2], Hume et al [3], and for control studies by Persson [4], Todd et al [5], Jovcic et al [6]. A recently developed exact small signal linearised model of an HVdc converter has been recently published by Osauskas et al [6], and this paper uses this model. The accuracy of the linearised approach for distortion levels between 2 and 5 % is demonstrated.

Part I briefly describes the linearised converter model, while Parts II and III describe the use of the model and the results for a harmonic and a control study respectively.

PART I: THE HVDC GRAETZ BRIDGE MODEL

In the steady state, free from system distortion, the HVdc converter operates at its base case operating point, producing characteristic harmonics. Converter waveforms and switching instants can be exactly defined.

When under system distortion or unbalance, the converter wave-shapes change, and the converter switching instants also change. The linearised model assumes a single source of distortion at a single frequency, and uses the principle of superposition to extend to multiple frequencies or sources of distortion. The effects of the applied distortion are calculated for two circuit configurations; direct conduction and commutation. There are three parts to the solution, being:

- The steady state solution, multiplied by a sampling function for the appropriate conduction period, to yield the Partial Steady State spectrum
- The transient solution, multiplied by the sampling function for the appropriate conduction period, to yield the Partial Transient spectrum
- The relationship between the switching instant variation and the applied distortion is linearised, and the resulting spectrum is written as a Pulse Amplitude Modulation Spectrum.

This process results in the relationships between waveform distortion around the HVdc converter, which can be written as follows:

$$\begin{bmatrix} \Delta I_p \\ \Delta I_n \\ \Delta V_d \\ 0 \end{bmatrix} = \begin{bmatrix} a & b & c & d \\ e & f & g & h \\ i & j & k & l \\ m & n & o & -1 \end{bmatrix} \cdot \begin{bmatrix} \Delta V_p \\ \Delta V_n \\ \Delta I_d \\ \Delta \alpha \end{bmatrix}$$

$$(Eqn.\ 1)$$

Transfers a to l are derived by the converter model, and m, n and o represent the converter control functions. Each transfer is a frequency dependent multiplier. All the variables are small signal variations from the base

AC-DC Power Transmission, 28-30 November 2001
Conference Publication No. 485 © IEE 2001

operating point, and each is a vector of frequencies, specified according to requirements. Subscript d means a DC side variable, subscript p denotes an AC side variable in positive sequence, and subscript n an AC side variable in negative sequence. All the transfers are, for small levels of distortion, exact. The model is reported in detail in Osauskas et al [7].

The model can be used for prediction of harmonic spectra around one or more interconnected HVdc converters, or can be restricted to the terms that describe the feedback paths that govern the transient response of an HVdc system. Both these applications are described in the following sections.

PART II: HARMONIC MODELLING

The steady state cross modulation that occurs in HVDC links can be split into two basic categories. Synchronous links experience no non-characteristic distortion, unless there is an independent distortion source or an AC system imbalance that includes a negative sequence component. In asynchronous links, the characteristic harmonics of each converter are non-characteristic for the converter at the opposite end of the link, and these are an additional source of non-characteristic distortion around an HVdc link. Although both cases can be modelled by the linearised method, this section focuses on the distortion resulting from an unbalanced AC system voltage at one end of a synchronous back to back HVdc link.

The method is based closely on that described by Larson et al [2], who used a linearised harmonic cross coupling matrix to represent the converter. This was later extended by Hume et al [3], who modelled the full HVDC link, incorporating the time variant nature of the transfers using tensor elements, described by Smith et al [8]. Both Larson and Hume perturbed the converter in the time domain before applying the FFT to acquire the linearised frequency coupling matrix transfers and the results were shown to be in reasonable agreement with full time domain simulation. The model used for the rectifier and inverter in this paper was developed by Osauskas et al [7]. A modified version of the benchmark CIGRE HVDC link (Szechtmann et al [9]) with minimal dc side impedance is used and all results are validated against PSCAD/EMTDC time domain simulation.

Linear steady state harmonic solutions of networks including power electronic devices require that the definition of linear is extended to include both frequency coupling and phase angle dependence. Frequency coupling means that a single frequency will generate a multitude of other frequencies, all frequencies linearly related to that single frequency. Phase angle dependence is the result of time-variance and is described later.

Each of the transfers a to l are linear frequency coupling matrixes. The matrices all have a sparse lattice like structure similar to that shown for transfer a in Figure 1.

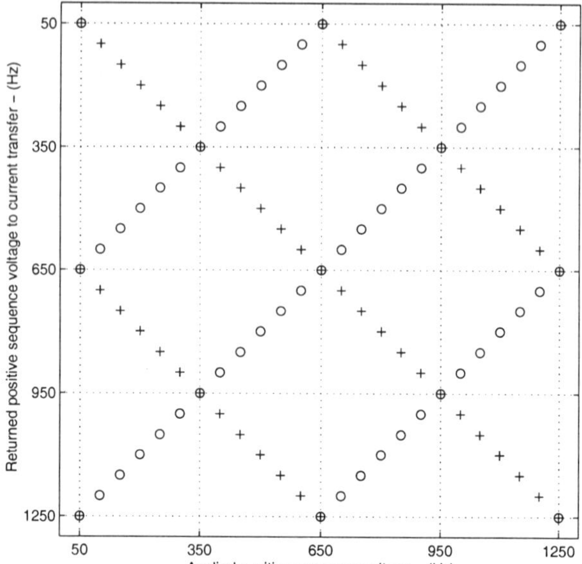

Figure 1. Structure of transfer 'a'.

The matrix size depends upon the frequency range and incremental step required. For the example used in this paper a 12 pulse synchronous back-to-back 50-50Hz link is used with a 50Hz negative sequence system distortion. The lowest common denominator for this system is simply the fundamental, or 50Hz, which is the base frequency and incremental frequency step.

Time variance and Tensor representation

As in all the transfers associated with the HVDC converter, the spectrum resulting from applied distortion consists of sum and difference frequencies. The difference frequencies are dependent on the complex conjugate of the applied distortion. This can be accommodated by using negative frequencies, or by writing each individual transfer as a 2 by 2 matrix, which relates the real and imaginary parts of each distortion phasor. These 2 by 2 matrices are called tensors.

HVDC system circuit analysis

Larson's equation describes the frequency cross coupling inter-relationships of the converter around a base operating point. Full modelling of the converter must include the time-invariant ac and dc system admittances.

Though linear and time-invariant, the admittances must be represented in the same manner as the full time-variant frequency cross-coupling inter-relationships. Hence, the ac and dc system admittances are represented in matrix form and are the same size as the converter frequency cross-coupling matrices, the diagonal representing the admittance frequency dependent vector, the matrix being zero elsewhere.

Nodal analysis provides a convenient way of combining each converter of the link together with its associated ac and dc systems. The converter itself can be described in nodal form by rearranging Larson's equation. Figure 2 represents the converter in nodal form.

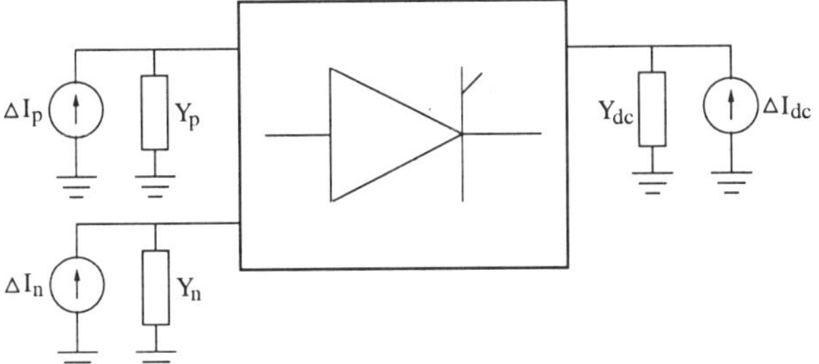

Figure 2. Nodal current injections for a single converter

Implicitly incorporating the control, and rearranging Larson's equation to nodal form gives

$$\begin{bmatrix} I_p \\ I_n \\ I_d \end{bmatrix} = \begin{bmatrix} Y_{11}+Y_p & Y_{12} & Y_{13} \\ Y_{21} & Y_{22}+Y_n & Y_{23} \\ Y_{31} & Y_{32} & Y_{33}+Y_{dc} \end{bmatrix} \cdot \begin{bmatrix} V_p \\ V_n \\ V_d \end{bmatrix}$$

(Eqn. 2)

where Y_{11} to Y_{33} are the converter frequency cross-coupling admittance matrix transfers, rearranged from Larson's equation and Y_p, Y_n and Y_{dc} are diagonal matrices representing the frequency dependent system admittances. This generalised nodal system approach means large systems of linearised time-variant power electronic devices can be built up with their time-invariant ac and dc systems easily and solved for the particular distortions of interest.

The nodal admittance matrix for the CIGRE HVdc link is built up in this way. For this study, the dc side system is made up with a large admittance between converter terminals and a small admittance to ground, to convert the link to a back to back link.

Solution of Nodal equation set

This paper investigates the distortion inter-harmonics on a synchronous HVDC link resulting from a 2 % negative sequence voltage distortion on the rectifier terminals. As distortion inter-harmonics only are investigated the effect of the base case harmonics is ignored as these are assumed to stay constant. Hence, the knowns of the nodal equation set are all current injections at all frequencies (except fundamental negative sequence on the rectifier side), as well as the fundamental negative sequence terminal voltage at the rectifier. The unknowns are the fundamental negative sequence current injection at the rectifier and all voltages around the link (except the negative sequence fundamental voltage at the rectifier).

The equation set must be ordered to solve for these unknowns. This can be achieved by restructuring of the nodal matrix into the knowns and unknowns and requires indexing of the tensor admittance matrix elements for partitioning into the form shown in equation 3.

$$\begin{bmatrix} \Delta I_{known} \\ \Delta I_{unknown} \end{bmatrix} = \begin{bmatrix} A & B \\ C & D \end{bmatrix} \begin{bmatrix} \Delta V_{unknown} \\ \Delta V_{known} \end{bmatrix}$$

(Eqn. 3)

where ΔI_{known}, ΔV_{known}, $\Delta I_{unknown}$, $\Delta V_{unknown}$ are vectors of the known and unknown currents and voltages around the link. Once the matrices A, B, C and D are partitioned from the original admittance tensor matrix they can then be rearranged, as shown in equation 4 and solved for the unknowns.

$$\begin{bmatrix} \Delta I_{known} \\ \Delta V_{known} \end{bmatrix} = \begin{bmatrix} A-BD^{-1}C & BD^{-1} \\ -D^{-1}C & D^{-1} \end{bmatrix} \cdot \begin{bmatrix} \Delta V_{unknown} \\ \Delta I_{unknown} \end{bmatrix}$$

(Eqn. 4)

This is the general method adopted for rearranging the matrix equations to solve for the unknowns.

The effect of negative sequence distortion on a back-to-back link

To validate the linearised method the effect of a 2 % negative sequence voltage distortion on the rectifier terminals is investigated. Comparisons of the returned system AC current spectra are made with the time domain and analytic model including and excluding the associated effects of the switching instant variation. This shows the effect SIV has on the resulting spectra around the converter. To do this the system is solved twice, once with the full transfers of equation 1, including the effects of SIV, and once without the effects of SIV. The resultant spectra are shown in Figures 3 and 4, for the resulting rectifier and inverter AC system currents respectively.

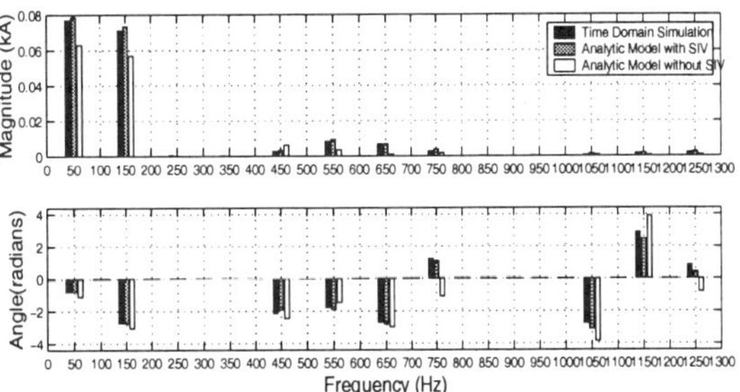

Figure 3. Rectifier AC current spectrum resulting from a 2 % negative sequence voltage distortion on the rectifier.

The returned AC current spectrum for both figures excludes the characteristic harmonics, showing only the generated spectra resulting from a 2 % negative sequence voltage unbalance at the rectifier. The unbalance results in approximately 80 Amps of negative sequence current in both rectifier and inverter AC systems, including filters. A positive sequence 3rd harmonic current of approximately 75 amps is present in the inverter AC system impedance, while only about 20 Amps flows in the rectifier AC system.

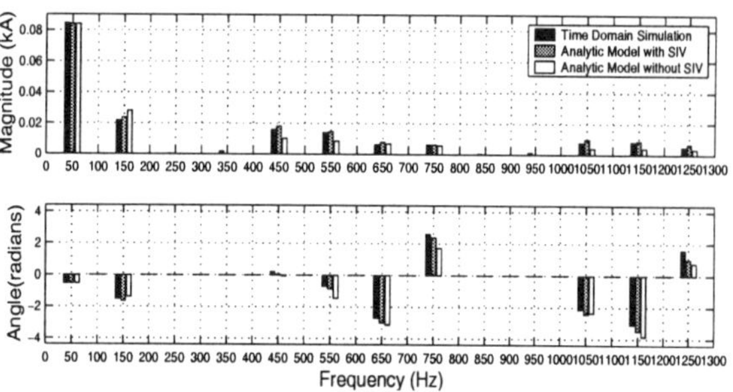

Figure 4. Inverter AC current spectrum resulting from a 2 % negative sequence voltage distortion on the rectifier

The current flow in the inverter results mostly from the transfer from DC side current to the AC side, whereas at the rectifier, the presence of a significant negative sequence unbalance (or distortion) results in an additional AC side current flow. In this case, the additional current flow results in a reduced AC side current magnitude.

It can be seen in Figure 4 that the switching instant variation has a substantial effect on the returned current spectrum at the inverter (up to over 50 % in some cases). This is because all the transfers to DC voltage strongly depend on switching instant variation, while transfers to AC current do not.

PART II: DYNAMIC MODELLING

System oscillations associated with power electronic devices are electromagnetic in nature and in the frequency range of approximately 2 to 200 Hz. The use of small signal linearisation methods based on analytical models is a very powerful method of analysis which provides analytical insight into system dynamics, and facilitates the use of classical and modern control design methods.

The model described here has similarities with the models previously presented by Persson [4], Sucena-Paiva et al [10], Todd et al [5], and Jovcic et al [6]. Also of relevance are the models of the STATCOM described by Padiyar [11] and Schauder [12].

General Description of the Model

The configuration of the HVDC transmission system model is shown in Figure 5, and is of the same form as the CIGRE benchmark model [9]. The purpose of the model is the representation of electromagnetic modes of oscillation, in the DC side frequency range from 2 to 200 Hz, which interact with the fastest level of converter controls. The electromechanical dynamics of the power system are not represented, as they are considered to below the frequency range of interest.

The HVDC system is modelled using 9 sub-systems, which are the rectifier and inverter AC systems, the rectifier and inverter AC filters and shunt capacitors, the DC system, the rectifier and inverter HVDC converters, and the rectifier and inverter Phase Locked Loops (PLL's).

State Model Formation

The formation of a state model is a straightforward process in the case of control sub-systems defined in the s-domain and linear electrical sub-systems, which are described using discrete RLC components. When a sub-system is described in terms of frequency response data it is necessary to fit an s-domain transfer function to the frequency response, which is then converted to a state model.

Once the state models of the sub-systems are formed they can be connected together to form a state model of the HVDC system. The state model can be used in conjunction with both classical and modern control theory, for steady state stability analysis and controller design.

AC System and Filters

The converter model is derived using AC variables represented in sequence components. To enable the direct connection of sub-systems without the need for variable transformations all AC electrical variables are also represented using sequence components. It is assumed that the positive and negative sequence admittances of the AC system Y_{ac} are the same, and that there is no coupling between the sequences.

The frequency conversion process of the converter is accounted for by frequency shifting the AC system equations. This is achieved by representing the admittance Y_{ac} in zero-pole transfer function form, and then adding $\pm j\omega_0$ to the values of the zeroes and poles, as described by equation 5.

$$Y_{acp}(s) = Y_{ac}(s + j\omega_0)$$
$$Y_{acn}(s) = Y_{ac}(s - j\omega_0)$$

$$(Eqn.\ 5)$$

It is these two frequencies, with the associated frequency on the DC side, that govern the transient response of the overall system.

As the transfer function zeroes and poles have been shifted in opposite directions, the poles of the sub-system still form complex conjugate pairs. The frequency shifted transfer functions are required to be converted into state model form.

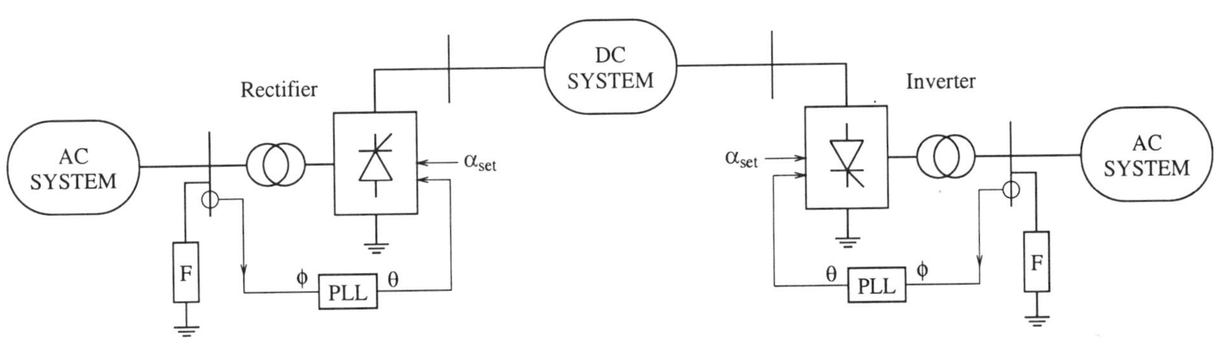

Figure 5: HVdc transmission system

DC System

The DC system has two electrical terminals which are connected to the DC terminals of the rectifier and inverter. The series inductive nature of the DC system (smoothing reactor and DC transmission line) means that the system is best described in admittance form. This representation is compatible with the converter DC terminal current input and voltage output variables.

$$\begin{bmatrix} I_1 \\ I_2 \end{bmatrix} = \begin{bmatrix} Y_{11} & Y_{12} \\ Y_{21} & Y_{22} \end{bmatrix} \cdot \begin{bmatrix} V_1 \\ V_2 \end{bmatrix} \qquad (Eqn.\ 6)$$

HVDC Converter

The state model of the HVDC converter used in this paper was obtained from the frequency domain model derived Osauskas [7].

An examination of the dominant transfers of the frequency domain model shows that, while all transfers exhibit some frequency dependence, this is largely above the frequency range of interest. The change in the HVDC system dynamics, resulting from the use of converter models which take into account varying degrees of frequency dependence in the transfers, indicates that a model where the transfers are approximated as constants is of sufficient accuracy. The constants are naturally chosen to be the value of the frequency domain transfers at zero frequency on the DC side, and are consistent with the differentiation of the standard steady state converter equations. Transfer k, from dc current to dc voltage, has the form of a zero, and is the only case where a constant approximation is inappropriate.

Phase Locked Loop (PLL)

The PLL is a negative feedback control system which tracks the changes in the angle of the positive sequence fundamental frequency component of the AC commutating bus voltage. The output of the PLL adjusts the position of the firing angle ramp references so that the thyristor firing instants are synchronised to the AC voltage.

The PLL system modelled is of the DQZ type, the major components of which are a frequency modulator, voltage controlled oscillator (VCO), and controller. The development of the PLL small signal model is described in Jovcic et al [6] and represented by the block diagram of Figure 6. The input to the model is the angle of the AC bus voltage, which is obtained from a sequence or d-q components representation of the AC voltage. The open loop transfer function consists of the series combination of a non-linear gain K, a PI controller, and an integrator which represents the operation of the VCO. The non-linear gain is the magnitude of the AC bus voltage, and is approximated by a constant using the operating point AC voltage of the system.

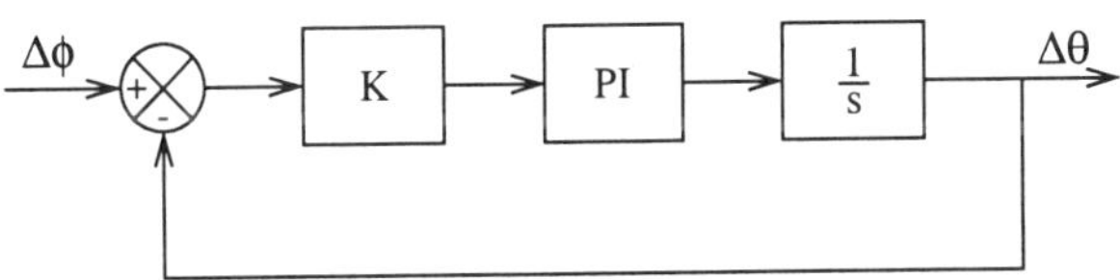

Figure 6. Small signal model of the Phase Locked Loop

The gain of the controller is normally set such that the bandwidth of the PLL is in the order of 5 Hz. This limits the possibility of the PLL interacting with the dominant electromagnetic modes of system oscillations which are typically at higher frequencies. The inclusion of the PLL improves the accuracy of system models in the low frequency range.

Validation

The formation of a state model of the CIGRE HVdc benchmark system results in a model of 39'th order. A significant number of the system eigenvalues are above the frequency range of interest and are associated with the AC filter sub-systems. The electromagnetic dynamics of the CIGRE HVdc system in the frequency range of interest is accurately represented by a state model of approximately 20'th order.

The accuracy of the state model is investigated by comparing the disturbance response of the state model with that obtained from PSCAD/EMTDC simulations of the CIGRE system operating with a DC current of 2 kA. A rectifier DC current controller sub-system is incorporated in the state model, while constant firing angle control at the inverter is assumed. The disturbance input is chosen to be a voltage source in series with the mid point capacitor of the DC line model. This disturbance input was especially chosen, as it has no effect on the operating point of the system, minimising the non-linear behaviour of the system.

Figure 7 shows the response of the rectifier DC current to a 50kV or 10 % step increase in the disturbance voltage at a time of 0.02 seconds. The agreement between the state model and PSCAD/EMTDC responses is very good, and suggests that the discrete switching nature of the converter can be modelled using a continuous model in the frequency range of interest. The response of other system variables such as the DC voltage and current, and the AC voltage and current, at both the rectifier and inverter have also been found to be in excellent agreement with PSCAD/EMTDC simulations.

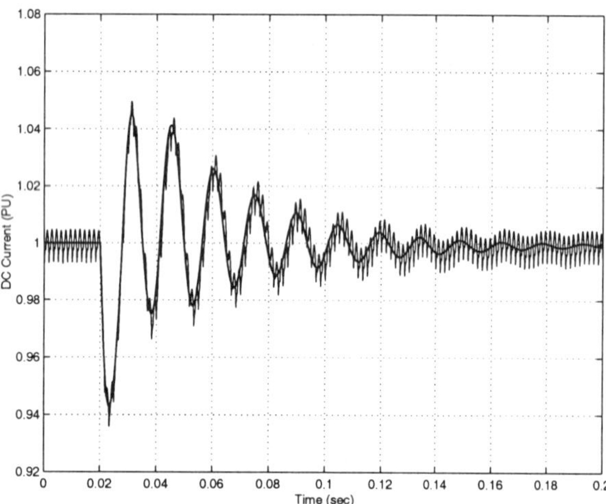

Figure 7 Rectifier DC current response to a 50kV step increase in DC capacitor voltage. Smooth line is linearised model response, and the rough line is the full time domain simulation.

CONCLUSIONS

This paper has utilised a recently reported small signal linear frequency domain HVdc converter model for a harmonic study on a back to back HVdc link, and a control study for the CIGRE HVdc transmission system. Harmonic current levels resulting from a 2 % negative sequence unbalance at the rectifier terminals of a back to back link have been calculated to good accuracy. The rapid calculation that this approach allows, and the fact that it incorporates any linear control contribution, makes it useful for examination of a range of system conditions. Harmonic distortion in a system is required by legislation to be small, so small signal studies are ideally suited to this application.

The control study verifies that a linearised modelling approach correlates well with time domain studies, for small disturbances. The linear model reveals system poles and zeroes, and allows fast and easy access to frequency and s domain control design and analysis techniques. As such it allows a systematic approach to the tuning of control gains, although for large signal disturbances, where controls are normally non-linear, this type of model is less useful.

REFERENCES

[1] L. Hu and R. Yacamini, 1992, Calculation of Harmonics and inter-harmonics in HVdc schemes with low DC side impedance, IEE Proc Pt C, 140(6):469-76.

[2] E. V. Larson, D. H. Baker, and J. C. McIver, 1989, Low order harmonic interaction on ac/dc systems. IEEE Transactions on Power Delivery, 4(1):493 to 501.

[3] D. J. Hume, A. R. Wood, B. C. Smith and J. Arrillaga, 1998, Linearised direct harmonic solution method for a back to back HVdc link. ICHQPS Conference, Athens.

[4] E. V. Persson, 1970, Calculation of transfer functions in grid controlled converter systems. IEE Proc 117(5):989 to 997.

[5] S.Todd, A. R. Wood, P. S. Bodger, 1997, An s-domain model of an HVdc converter. IEEE Trans on Power Delivery, 12(4):1723-29.

[6] D. Jovcic, N. Pahalawattha, M. Zavahir, 1999, Analytic modelling of HVdc-Hvac systems. IEEE Trans on Power Delivery, 14(2):506 to 511.

[7] C. M. Osauskas, D. J. Hume, A. R. Wood, 2001, A small signal frequency domain model of an HVdc converter, accepted for IEE proc Pt C.

[8] B. C. Smith, N. R. Watson, A. R. Wood and J.Arrillaga, 1998, Harmonic tensor linearisation of HVdc converters. IEEE Trans on Power Delivery, 13(4):1244 to 1250.

[9] M Szechtman, T. Weiss and C Thio, 1991, First benchmark model for HVdc control studies. Electra 135:55to 75.

[10] J. P. Sucena-Paiva and L. L. Freris, 1974, Stability of a DC transmission link between weak AC systems. IEE Proc, 121(6):508 to 515.

[11] K. R. Padiyar and A. M. Kulkarni, 1997, Design of reactive curent and voltage controller of static condensor. Electrical Power and Energy Systems, 19(6):397 to 410.

[12] C. Schauder and H. Mehta, 1993 Vector analysis and control of advanced static Var compensators. IEE Proc Pt C., 140(4):299 to 306.

CONTROL OF HVDC SYSTEMS OPERATING WITH LONG DC CABLES

Dragan Jovcic

University of Ulster, United Kingdom

Abstract- **This paper studies dynamic aspects of HVDC system operation with long DC cables. The eigenvalue analysis of CIGRE HVDC system with extended DC cable is presented at the beginning. It is concluded that significant aggravation in the system stability is present with cables of *1000km* and the system dynamic instability is expected at frequencies *2Hz<f<4Hz*. The design solutions based on smoothing reactors modifications or existing control loops re-tuning can not improve the system stability. Direct voltage feedback or AC current angle feedback at inverter side are the suitable control strategies. AC current angle is more favorable since the probability of commutation failure is reduced. The last section studies the operating conditions with low SCR inverter AC system and concurrently long DC cables. Eigenvalue analysis shows that this is particularly bad combination and resonance-like conditions rapidly aggravate system stability. Nevertheless, AC current angle feedback significantly improves the system stability. The analysis in the paper is accomplished using analytical system model and MATLAB software.**

Keywords: HVDC control, HVDC modeling, Power system dynamic stability,

I. INTRODUCTION

Traditionally, HVDC transmission is employed mainly into three categories: medium length underwater cables (below *100km*), long overhead transmission lines and connection of asynchronous systems (back to back schemes). The successful HVDC operation in these areas, over the last 5 decades, has signaled possible expansion of the HVDC application domain.

The recent restructuring of electrical energy market, and privatization of utilities, has brought forward many different concepts of power generation and different options for power transfers, including the transmission over long sea cables, Karlsson and Liss [1], Hammons et al [2]. HVDC appears as a viable candidate for long under sea transmission, however this technology is yet to be confirmed.

Reference [1] elaborates on a number of possible HVDC projects with sea cables of *500 – 700km*, and points that some projects with *1000km* cables are being discussed.

It is expected that prolonged DC cables will alter DC system dynamic behavior and that stability of interconnected AC systems may also be affected. Reference [1] further summarizes that main control issues stem from the high cable capacitance and they are reflected in:

- Inability of rectifier controller to "see" beyond the high capacitance cable, i.e. inverter system end, leading to different rectifier and inverter responses,
- The discharge of huge cable energy that would lead to AC system collapse.

The above conclusions are derived using static *Ud/Id* HVDC indicators, and observed employing digital simulators, whereas they are not confirmed and studied using dynamic system analysis.

This paper discusses the HVDC operation with long DC cables, from the dynamic pint of view and primarily in the frequency domain *1Hz<f<100Hz*. The primary aim of this research is to offer eigenvalue-based analysis of the above phenomena. Using the eigenvalue decomposition analysis and studying the movement and sensitivity of oscillatory modes, it is possible to predict

The influence of the existing control strategies can also be studied by analysing the affect that changes in controller parameters have on dominant eigenvalues. In a similar way, the influence of other parameters like smoothing reactors size is investigated in order to advice best design options with long DC cables. A new HVDC control method that can counteract the observed stability problems will further be discussed. In the last section, the paper studies the dynamic implications of combining long DC cables and weak receiving AC systems.

The dynamic HVDC system model employed is presented by Jovcic et al [3]. The standard CIGRE HVDC model, as presented by Szechman et al [4] is used in the analysis as the starting test system and modified accordingly to represent long DC cable.

The study and results from this paper will assist the system designers and operators because of the generic type conclusions on expected dynamic difficulties and possible instabilities with long DC cables. They are of further importance since they signal directions for the development of future HVDC control strategies, particularly at inverter side. The results present preliminary theoretical confirmation of technical feasibility of long DC cable HVDC projects.

II. TEST SYSTEM

The starting test system used in the analysis section is the CIGRE monopolar HVDC system given in [4], and

AC-DC Power Transmission, 28-30 November 2001
Conference Publication No. 485 © IEE 2001

it is modeled as shown in[3], implemented in MATLAB software. The controller uses PI rectifier current controller and gamma controller at inverter side as given in PSCAD/EMTDC HVDC Benchmark model.

To represent the operating changes with long DC cables, the following model parameters are varied:

- Cable capacitance $C = C_{cig} + C_i$, (1)

- DC line resistance $R_{dc} = R_{dc-cig} + C_i * 10^6 / 40$, (2)

where: $C_i \in [0,800 \times 10^{-6} \mu F]$, C_{cig}, R_{dc-cig} -original CIGRE value.

The above highest value for capacitance ($C_i = 800 \times 10^{-6} \mu F$ in (1)) is used in the system analysis as the expected capacitance for approximately *2000km* long cable. Although this length might not have practical significance, the study gives generic theoretical conclusions. In the controller design section, a more realistic value of $C_i = 400 * 10^{-6} \mu F$ is used, representing approximately *1000km* cable and this system is termed original system in the design section.

DC line resistance is assumed to vary linearly with the variation in cable capacitance, as given by formula (2). It is noted that formula (1) and (2), derived using PSCAD simulation, are a crude approximation of the actual values, nevertheless the validity of conclusions is confirmed in sections III and IV.

In testing the results with PSCAD, a detailed frequency dependent cable model is used and only the cable length is varied.

III. STABILITY ANALYSIS

III.1 Eigenvalue analysis of HVDC systems with long cables

In this section, the cable capacitance and cable resistance are varied in small steps as given by (1) and (2), using the analytical system model and the location of eigenvalues is observed.

Figure 1 shows the root locus where crosses denote original CIGRE model eigenvalues. Observing the movement of eigenvalues at higher frequencies (branches *a* and *b*) it is seen that long cables do not significantly alter higher frequency stability. These branches are short and stability is somewhat improved. In mid frequency range (branch *c*) the frequency of the original mode is altered and the system stability is somewhat degraded. Finally, at lower frequencies, a significant degradation in the system stability occurs, as seen by the root branch *d*.

The end value of eigenvalue branch *d* (location A1) is dynamically unstable, confirming that system becomes unstable with cable length around *1500km*. The eigenvalues move slower as the cable length is increased, and the eigenvalue location for

approximately *1000km* long cable (location A2) is very close to the instability.

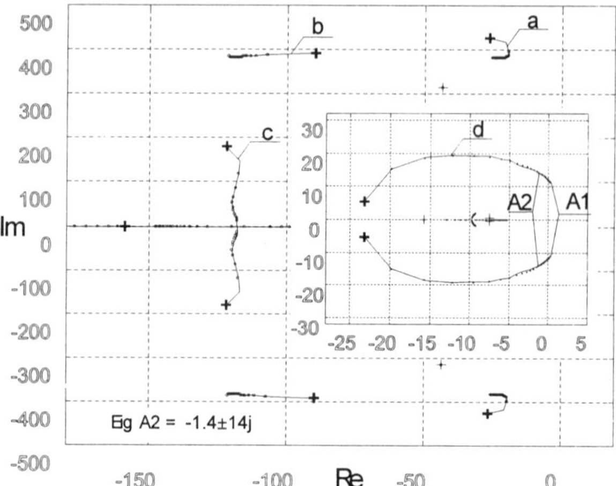

Figure 1. Root locus with the increase in DC cable capacitance and resistance. "+" – Standard CIGRE model.

The system responses are expected to be very poor with *200-300km* long cables since eigenvalues move very fast for lower cable lengths.

The frequency for possible instabilities is *2Hz<f<4Hz*. It should be noted that long DC cables do not significantly aggravate damping of oscillatory mode close to first harmonic (*50Hz*), and therefore composite resonance and second harmonic instability is not accelerated.

The above conclusions are confirmed with PSCAD/EMTDC simulation, as shown in Figure 2. A single dominant oscillatory mode is evident and it also seen that inverter side is much more prone to the oscillations. The figure further verifies the analytical model, pointing also that model can not perfectly match non-linear digital simulation.

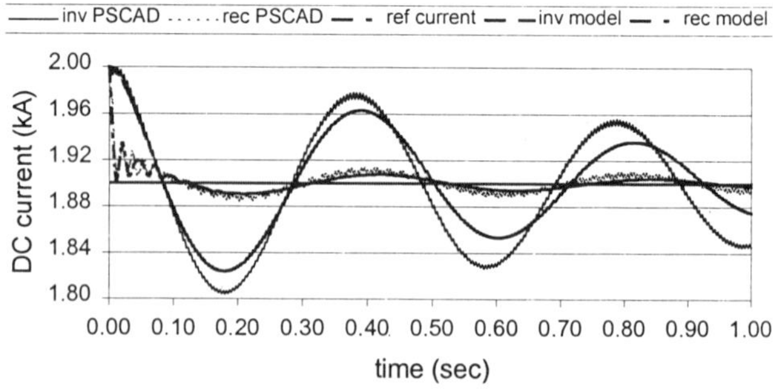

Figure 2. 1000km cable HVDC system, DC current (rectifier and inverter) response following a current reference step change.

III.2 Influence of Existing Control Loops

The existing rectifier controller gains (proportional and integral) are varied in order to improve dynamics of the system with long DC cables. The method of analysis follows the classical root locus method for HVDC systems as described by Jovcic et al [5]. The results are similar to those obtained for rectifier DC current feedback in this reference. In summary, with the existing HVDC control structure, it is not possible to improve the system stability without destabilizing other modes or reducing the speed of the response.

III.3 Influence of Smoothing Reactors

Smoothing reactors are normally added to attenuate DC harmonics and to limit the rate of DC current responses in cases of DC faults and DC side overvoltages. In line with the same design method, the value of smoothing reactors is further increased in order to compensate for large cable capacitance and to attenuate DC current responses.

The value of inductance of smoothing reactors is varied in the following manner:

$$L_{dc} = L_{dc-cig} + L_i, \qquad (3)$$

$L_i \in [0,1H]$, L_{dc-cig} -original CIGRE value.

Figure 3 shows the root locus, where the final eigenvalue locations (A1, B1, C1 and D1) correspond to $L_i = 1H$. The original system corresponds to $C_i = 400*10^{-6} \mu F$ (A2 location in Figure 1).

It is see that increase in smoothing rectors significantly slows system responses, as demonstrated by reduced frequency in c branch. However the damping of the dominant eigenvalues is also reduced and the system is further destabilised. Is is evident that soleley increase in smoothing recators can not improve dynamic stability of the systems with long DC cables. Reducing the smothing reactors marginally enhabces the stability.

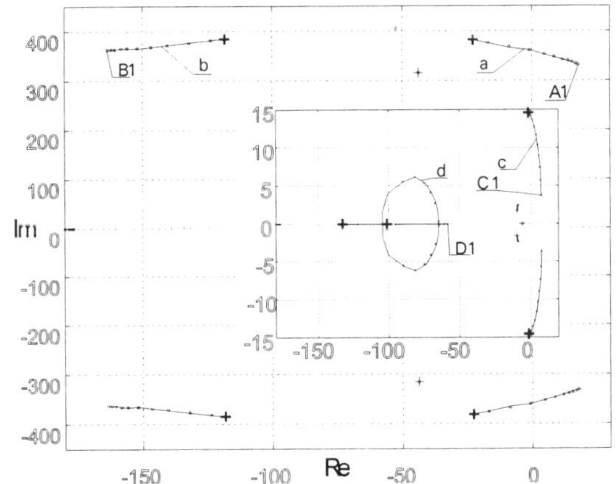

Figure 3. Root locus with the increase in DC smoothing reactance. "+" - Original system with long DC cable.

IV. CONTROLLER DESIGN

IV.1 Design based on stability requirements

In this section, various inverter control strategies are examined in an attempt to improve the system stability.

The original system corresponds to A2 eigenvalue location in Figure 1. The main design goal is to improve damping of eigenvalues A2 without stability deterioration at other frequencies (other modes). The method of analysis resembles the one used in [5].

Most inverter variables are studied as feedback candidate signals including: AC voltage, AC current angle, DC voltage, DC current. Only two feedback strategies are found to improve the system stability: Direct voltage feedback (DCV) and AC current angle (ACCA) feedback, and they are further examined and compared. It is noted that AC current angle feedback is earlier found to be effective inverter control strategy with weak receiving AC systems, as presented by Jovcic et al [6].

Figures 4 and 5 show the root locus with DCV and ACCA feedback respectively. It is seen that both control strategies improve damping at lower frequencies (branch c in Figure 4 and a in Figure 5) and system stability is not jeopardized at higher frequencies. More precisely, the negative influence at higher frequencies with DCV feedback is evident by observing branch a in Figure 4, however the system is still stable with high gain values and this negative effect can be compensated with a feedback filter. DVC will have faster responses since branch c moves towards higher frequencies. Therefore, both control strategies can be used for system operation with long DC cables.

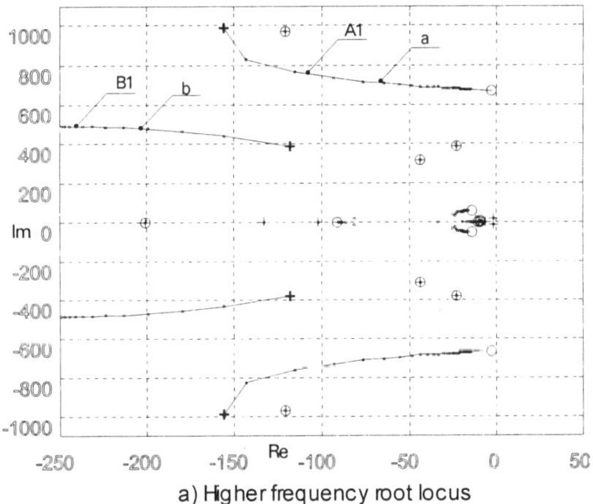

a) Higher frequency root locus

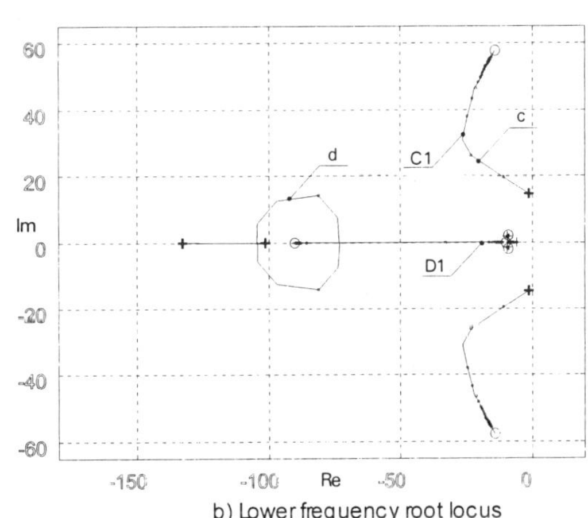

b) Lower frequency root locus

Figure 4. Root locus with direct voltage feedback. "+" – Original eigenvalues with long DC cable.

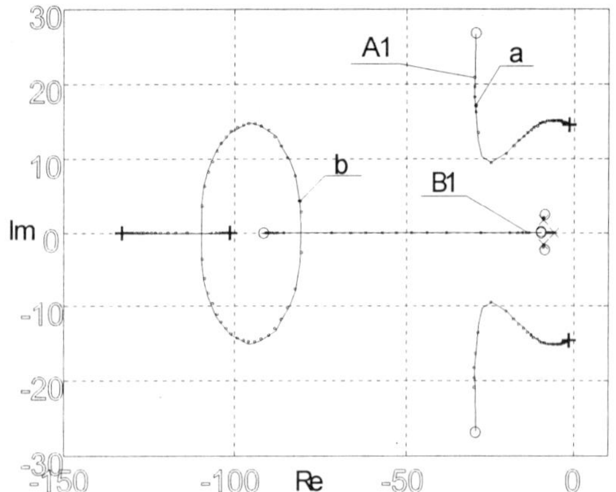

Figure 5. Root locus with AC current angle feedback. "+" – Original eigenvalues.

Figure 6 compares the time domain responses for the two feedback loops. The exact value of the feedback gain is selected such that the eigenvalue location corresponds to the position marked A1, B1, C1 and D1 in Figure 4 and A1, B1 in Figure 4. Using the classical control theory recommendations, DCV feedback would be selected from Figure 6 since the responses are somewhat faster and the overshoot is within 20-25% boundary.

However to fully reflect the particular requirements of the HVDC system, especially commutation failure considerations, the disturbance responses for AC and DC variables are studied as shown in the next section.

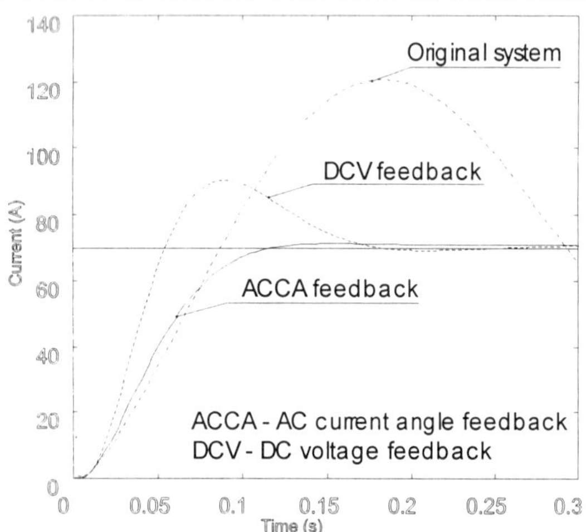

Figure 6. Inverter DC current response after current reference step change. Rectifier DC current responses are not influenced with feedback.

IV.2 Commutation failure considerations

This section compares the DCV and ACCA feedback by considering the probability of commutation failure.

It is known that commutation failure is caused by reduction in gamma values, which is primarily caused by AC voltage depression and DC current increase at inverter side, as shown by Thio et al [7].

Figure 7 compares the inverter DC current and AC voltage responses for the two feedback loops and considering the disturbance at inverter AC system (tap changer action). It is seen that in the interval *0.02-0.04sec* DC current values are higher and AC voltage values are lower for DCV feedback. This result leads to conclusion that probability of commutation failure is higher with DCV feedback than with ACCA feedback and consequently ACCA feedback loop would be more favorable.

It is therefore concluded that although both control strategies improve the system stability, considering the commutation failure probability, ACCA feedback is better option.

It should be noted that both feedback control strategies in the above analysis use simple feedback gain without any feedback filter. This is contrary to the ACCA control employed in [5], which uses *Hinf* controller in ACCA feedback loop to maximise the feedback control effect.

Figure 8 shows the PSCAD simulation of the above control method. Figure 8 a) shows that the ACCA feedback control gives quite acceptable responses (*25%* overshoot and *0.2s* settling time) with *100km* DC cable. Figure 8 b) confirms that AC current angle stabilises the *2000km* system, the responses are somewhat sluggish though (settling time of *0.5s*).

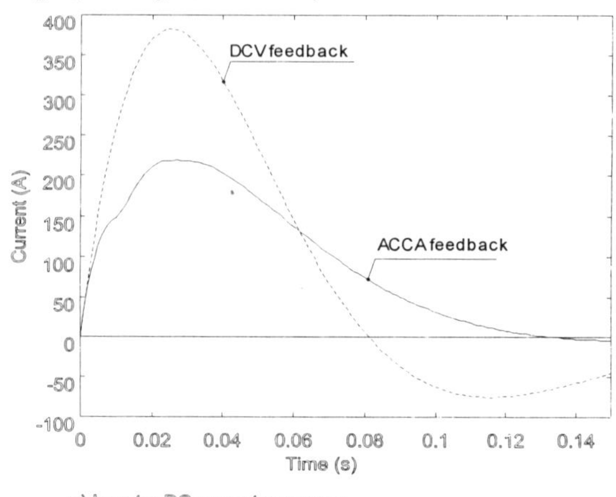

Figure 7. System response following disturbance on inverter AC voltage

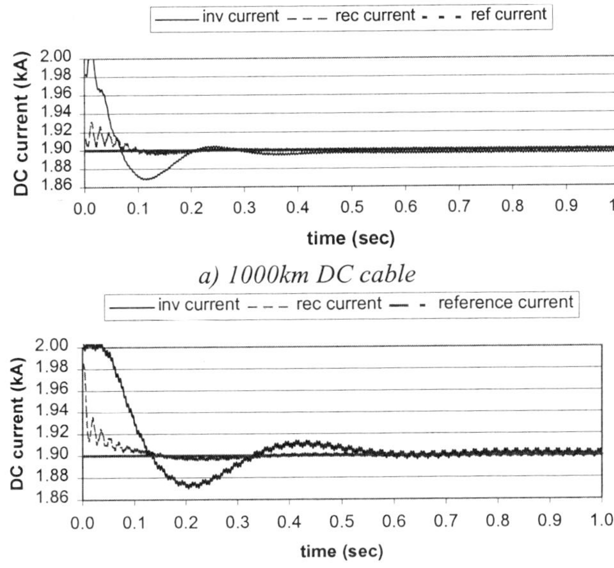

a) 1000km DC cable

b) 2000km DC cable.

Figure 8. PSCAD simulation of long cable HVDC system with supplementary inverter control. System response following a step change in reference current.

The simulation with PSCAD confirms the above generic conclusions, nevertheless because of the presence of noise and non-linear elements, the actual value of feedback controller gains is restricted. As a result, a slightly slower responses and larger overshoots, compared with responses with analytical model in Figures 6 and 7, are obtained.

V. VERY WEAK RECEIVING AC SYSTEMS AND LONG DC CABLES

V.1 System analysis

This section studies the case of long DC cables and simultaneously very weak receiving AC system. Using the static analysis and simulation, this operating condition is shown to be very difficult [1], however this reference does not offer dynamic system analysis to support the conclusion.

Reference [6] concludes that weak AC systems will cause dynamic instabilities in the frequency range $2Hz<f<7Hz$. Since this frequency range is close to the dynamic instability domain with long DC cables, as demonstrated by branch *d* in Figure 1, the concern is now raised that combination: long DC cables with weak AC systems will cause rapid worsening in the system stability.

Table 1 shows the dominant eigenvalues for very weak receiving AC system, and *600km* long DC cable case. In the first column, only the inverter AC system SCR is reduced to *SCR=1.3*, and the dominant oscillatory mode at frequency around *3Hz* is evident. In the second column, only the DC cable is extended (with original *SCR=2.5*) to approximately *600km*, creating a complex

eigenvalue pair with frequency very close to *3Hz*, i.e. close to earlier case with low SCR. Therefore, both operating conditions degrade system stability in the same frequency range, although the system is stable in both configurations. The last column in the Table confirms that the system becomes unstable if the system configuration with *SCR=1.3* and *600km* long DC cable is used.

It can be concluded that low inverter SCR and long DC cable is particularly difficult combination for system dynamic stability. This operating conditions can lead to specific "resonance" at frequencies around *3Hz*. The above mechanism for instability is very similar to the known HVDC composite resonance as shown by Wood and Arrillaga [8]. A specific condition on AC side (low SCR) combined with a corresponding condition on DC side (long DC cable) will lead to dynamic instabilities in particular frequency range.

V.2 AC current angle feedback

Figure 9 shows the use of ACCA feedback with the system having weak AC system and long DC cable. The original system is the configuration from third column, Table 1. It is seen that ACCA can stabilise the system and the step responses are satisfactory.

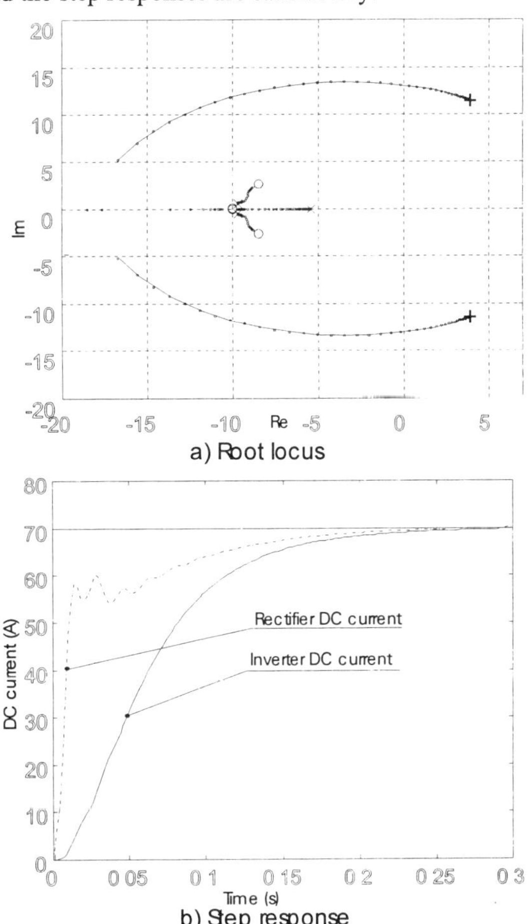

Figure 9. System with low SCR and long DC cables. Root locus and step response with ACCA feedback. "+" Original system.

TABLE 1 - DOMINANT EIGENVALUES FOR LOW SCR AND LONG DC CABLE

System Configur.	Inverter AC *SCR=1.3*	*SCR=2.5, 600km* DC cable	*SCR=1.3, 600km* DC cable
Dominant eigenvalue	-5.1±17.27j	-5.3±17.98j	+3.8±11.50j

It is further noted that in Figure 9 original proportional gain at rectifier side is reduced by half in order to improve damping of oscillatory mode close to first harmonic.

It should be observed in Figure 9 b) (also in Figure 8) that inverter DC current response is noticeably lagging the rectifier current response. This is a consequence of high DC cable capacitance that contributes decoupling rectifier from inverter. For this reason, solely used rectifier controller is not effective for long cable configurations.

VI. CONCLUSIONS

Eigenvalue analysis of HVDC system operating with long DC cables shows that significant aggravation in the system stability occurs with cables of around *1000km*. The system dynamic instability may occur at frequencies *2Hz<f<4Hz*. The analysis further shows that design solutions based on smoothing reactors modifications or existing control loops re-tuning can not improve the system stability. Direct voltage feedback and AC current angle feedbacks are suitable control strategies. The study of commutation failure probability verifies that AC current angle is more favorable since the DC current and AC voltage overshoots are reduced. PSCAD simulation in general confirms that AC current angle feedback can give satisfactory responses with *1000km* DC cable. The last section studies the operating conditions with low SCR inverter AC system and simultaneously long DC cables. Eigenvalue analysis shows that this is particularly bad combination and resonance-like conditions rapidly deteriorate the system stability. Nevertheless, AC current angle feedback at inverter side significantly improves the system stability. This feedback strategy will move the unstable eigenvalues into stable region and it will not influence other eigenvalues.

REFERENCES:

[1] T.Karlsson, G.Liss, 1995, "HVDC Transmission with extremely long DC cables control strategies" IEEE/KTH Stockholm Power Tech Conference, Stockholm pp 24-29.

[2] Hammons, K.O.Lee, K.H. Chew and T.C. Chua., 1998, "Competitiveness of renewable energy from Iceland via the proposed Iceland/UK HVDC submarine cable link". Electric Machines & Power Systems, vol.26, no.9, pp.917-33.

[3] D.Jovcic N.Pahalawaththa, M.Zavahir, April 1999 "Analytical Modeling of HVDC Systems" *IEEE Transactions on PD, Vol. 14, no 2, pp. 506-511.*

[4] M. Szechman, T. Wess and C.V. Thio, April 1991, "First Benchmark model for HVDC control studies", CIGRE WG 14.02 *Electra No. 135, pages: 54-73.*

[5] D.Jovcic N.Pahalawaththa, M.Zavahir, March 1999, "Stability Analysis of HVDC Control Loops" *IEE Proceedings Generation, Transmission and Distribution, Vol. 146, no 2, pp. 143-148.*

[6] D.Jovcic N.Pahalawaththa, M.Zavahir, May 1999, "Inverter Controller for Very Weak Receiving AC Systems" *IEE Proceedings - Generation, Transmission and Distribution, Vol. 146, no 3, pp. 235-240.*

[7] C.V. Thio, J.B. Davis, K.L. Kent, 1995, "Commutation Failures in HVDC Transmission Systems" Paper presented at IEEE/PES Summer meeting 1995. 95 SM 377-2 PWRD.

[8] A.R. Wood Arrillaga, Oct 1995, Composite resonance; a circuit approach to the waveform distortion dynamics of an HVDC converter, *IEEE PWRD, vol 10, no 4, pp 1882 to 1888.*

ANALYSIS OF A ROBUST DC-BUS VOLTAGE CONTROL SYSTEM FOR A VSC TRANSMISSION SCHEME

JL Thomas, S Poullain and A. Benchaib

Power Electronics Research Team
ALSTOM Technologies, France

ABSTRACT

This paper proposes a high dynamic digital control for IGBT based VSC transmission scheme, i.e. VSC-HVDC system, based on equivalent continuous-time state-space models for the rectifier/inverter actuators, using a nonlinear approach. In this paper, both tracking and regulation problems are presented in order to perform a robust DC-Bus voltage control. results illustrate the good behaviour of the system variables in response to positive and negative active power steps.

KEY WORDS

VSC-HVDC system, DC-Bus Voltage, State-Space Model, Nonlinear Control, Power Flow Control.

I. INTRODUCTION

The HVDC control laws, used for standard Thyristor based transmission schemes, are today well known. Nevertheless, these algorithms cannot be directly applied to the VSC-HVDC schemes. The freedom degrees offered by the VSC-PWM rectifier / inverter authorize new behaviours. Furthermore, the DC-Bus voltage control appears in standard VSC Transmission scheme as the main difficulty from the control point of view. A high performance and robust DC-Bus voltage control will result in a reduction of the DC-Bus voltage ripple, thus allowing a reduction of the DC capacitor banks. However, the switching frequency of the two actuators combined with closed-loop high-bandwith of controllers could be a source of excitation for the DC line, introducing instabilities and oscillations in the long distance cable.

Some control structures for VSC-HVDC Back-to-Back schemes have been proposed based on a set of standard Proportional-Integral controllers [3]. Basically, two PI controllers are dedicated to the current control loops, i.e. inner loops, for rectifier / inverter side. One additional PI controller is used for the DC-Bus voltage control. The others are reserved for active / reactive power control.
In this paper, an equivalent continuous-time averaged state-space model of the HVDC scheme is presented, introducing particular nonlinear properties of the transmission. The proposed modelling approach could be used for both Back-to-Back HVDC scheme or long distance VSC Transmission scheme. Then, the synthesis of nonlinear controllers are presented highlighting fast and slow dynamics respectively associated to the current control loops and the DC-Bus voltage control loop. The performance analysis is carried out on the base of simulation results obtained in Matlab/Simulink environment for various active power flow disturbances.

II. PROBLEM FORMULATION

The standard VSC-HVDC Transmission scheme is defined by Figure 1. The rectifier and inverter are respectively called "Station 1" and "Station 2" to simplify notations , i.e. indices, of the associated equations.

II.1 Control objectives

Voltage source converters could be viewed as two freedom degrees actuators. Thus, the VSC-HVDC scheme of Figure 1 is characterized by four control variables, i.e. magnitude and phase of the two voltage vectors $\vec{v}_1$ and $\vec{v}_2$. Then, because the system to be controlled must be square, only four independant control objectives can be satisfied. In this paper, "Station 1" is dedicated to the control of the reactive power Q_1 and the DC-Bus voltage u_{c1} , "Station 2" is used for control of the reactive power Q_2 and the active power P or DC-current i_{c1} ,if u_{c1} is considered as a constant variable. The active power flow must be also bidirectional, i.e. $P \geq 0$ and $P \leq 0$.

II.2 Assumptions

The three main assumptions of the proposed approach could be defined as follows:

- Three-phase line voltages are well balanced, i.e. no negative sequence of line voltages,
- Rotating reference frames (d,q) are well oriented, i.e. synchronized, using 2 robust PLL,
- "Station 1" & "Station 2" are ideal VSC actuators, i.e. symetricals and able to work in four quadrant operation modes.

AC-DC Power Transmission, 28-30 November 2001
Conference Publication No. 485 © IEE 2001

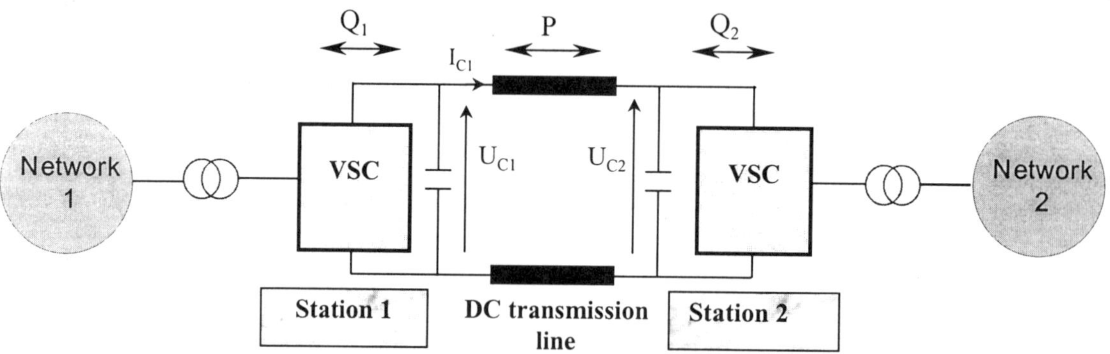

Figure 1 : VSC – HVDC Transmission Structure

III. CONTINUOUS-TIME EQUIVALENT MODEL

In order to perform the best control structure synthesis for the system illustrated in Figure 1, a full analytical control model is required. Furthermore, this model could be used directly for simulation in Matlab / Simulink environment.

Figure 2 presents a description of a VSC-HVDC scheme based on the connection of the Rectifier / simplified DC-line / Inverter through the variable u_{cc}. The building of equivalent continuous-time models of the actuators is based on the methodology detailed in [1]. Thus, a PWM VSC-converter is described by an average model defined on one "switching period".

III.1 Rectifier average model

Given the physical model defined by Figure 2, the average model of the rectifier is based on the following differential equations:

$$\begin{cases} L_{l1}\dfrac{di_{l1A}}{dt} + R_{l1}i_{l1A} = v_{l1A} - v_{1A} \\[2mm] L_{l1}\dfrac{di_{l1B}}{dt} + R_{l1}i_{l1B} = v_{l1B} - v_{1B} \\[2mm] L_{l1}\dfrac{di_{l1C}}{dt} + R_{l1}i_{l1C} = v_{l1C} - v_{1C} \end{cases} \tag{1}$$

Considering the following phasors associated to the three-phase balanced systems, and rotating at the pulsation ω_1, with $\theta_1 = \omega_1 t$:

$$\begin{cases} \vec{v}_{l1} = V_{l1}e^{j\theta_1}e^{j\gamma_1} \\ \vec{v}_1 = V_1 e^{j\theta_1}e^{j\psi_1} \\ \vec{i}_{l1} = I_{l1}e^{j\theta_1}e^{j\varphi_1} \end{cases} \tag{2}$$

and given the (d_1, q_1) rotating reference frame, initially oriented on $\theta_1 = \omega_1 t$, we can define:

$$\begin{cases} \overline{v}_{l1} = V_{l1}e^{j\gamma_1} = V_{l1}\cos(\gamma_1) + jV_{l1}\sin(\gamma_1) = v_{l1d} + jv_{l1q} \\ \overline{v}_1 = V_1 e^{j\psi_1} = V_1\cos(\psi_1) + jV_1\sin(\psi_1) = v_{1d} + jv_{1q} \\ \overline{i}_{l1} = I_{l1}e^{j\varphi_1} = I_{l1}\cos(\varphi_1) + jI_{l1}\sin(\varphi_1) = i_{l1d} + ji_{l1q} \end{cases} \tag{3}$$

Furthermore, given the duty cycle r_1, we can write:

$$\begin{aligned} \overline{v}_1 &= r_1\frac{u_{c1}}{2}e^{j\psi_{1w}} = \frac{u_{c1}}{2}r_1\cos(\psi_{1w}) + j\frac{u_{c1}}{2}r_1\sin(\psi_{1w}) \\ &= v_{1dw}\frac{u_{c1}}{2} + jv_{1qw}\frac{u_{c1}}{2} \end{aligned} \tag{4}$$

where v_{1dw} and v_{1qw} are the control variables of the rectifier. Finally, considering the power equality $u_{c1}i_1 = v_{1A}i_{l1A} + v_{1B}i_{l1B} + v_{1C}i_{l1C}$, on both sides of the rectifer, we can deduce the following expression:

$$i_1 = \frac{3}{4}(v_{1dw}i_{l1d} + v_{1qw}i_{l1q}) \tag{5}$$

Then, the state-space model of the rectifier is defined by the first four equations of the model (6). By analogy, the inverter side model is defined in (6) by the last four equations. The fifth represents the interconnection between "Station 1" and "Station 2".

$$\begin{cases} \dfrac{di_{l1d}}{dt} = -\dfrac{R_{l1}}{L_{l1}}i_{l1d} + \omega_1 i_{l1q} - \dfrac{1}{L_{l1}}v_{1dw}\dfrac{u_{c1}}{2} + \dfrac{1}{L_{l1}}v_{l1d} \\[2mm] \dfrac{di_{l1q}}{dt} = -\dfrac{R_{l1}}{L_{l1}}i_{l1q} - \omega_1 i_{l1d} - \dfrac{1}{L_{l1}}v_{1qw}\dfrac{u_{c1}}{2} + \dfrac{1}{L_{l1}}v_{l1q} \\[2mm] \dfrac{du_{c1}}{dt} = -\dfrac{1}{C_1}i_{c1} + \dfrac{1}{C_1}\dfrac{3}{4}(v_{1dw}i_{l1d} + v_{1qw}i_{l1q}) \\[2mm] \dfrac{di_{c1}}{dt} = \dfrac{1}{L_{c1}}u_{c1} - \dfrac{R_{c1}}{L_{c1}}i_{c1} - \dfrac{1}{L_{c1}}u_{cc} \\[2mm] \dfrac{du_{cc}}{dt} = \dfrac{1}{C_c}i_{c1} - \dfrac{1}{C_c}i_{c2} \\[2mm] \dfrac{di_{c2}}{dt} = -\dfrac{1}{L_{c2}}u_{c2} - \dfrac{R_{c2}}{L_{c2}}i_{c2} + \dfrac{1}{L_{c2}}u_{cc} \\[2mm] \dfrac{du_{c2}}{dt} = \dfrac{1}{C_2}i_{c2} - \dfrac{1}{C_2}\dfrac{3}{4}(v_{2dw}i_{l2d} + v_{2qw}i_{l2q}) \\[2mm] \dfrac{di_{l2d}}{dt} = -\dfrac{R_{l2}}{L_{l2}}i_{l2d} + \omega_2 i_{l2q} + \dfrac{1}{L_{l2}}v_{2dw}\dfrac{u_{c2}}{2} - \dfrac{1}{L_{l2}}v_{l2d} \\[2mm] \dfrac{di_{l2q}}{dt} = -\dfrac{R_{l2}}{L_{l2}}i_{l2q} - \omega_2 i_{l2d} + \dfrac{1}{L_{l2}}v_{2qw}\dfrac{u_{c2}}{2} - \dfrac{1}{L_{l2}}v_{l2q} \end{cases} \tag{6}$$

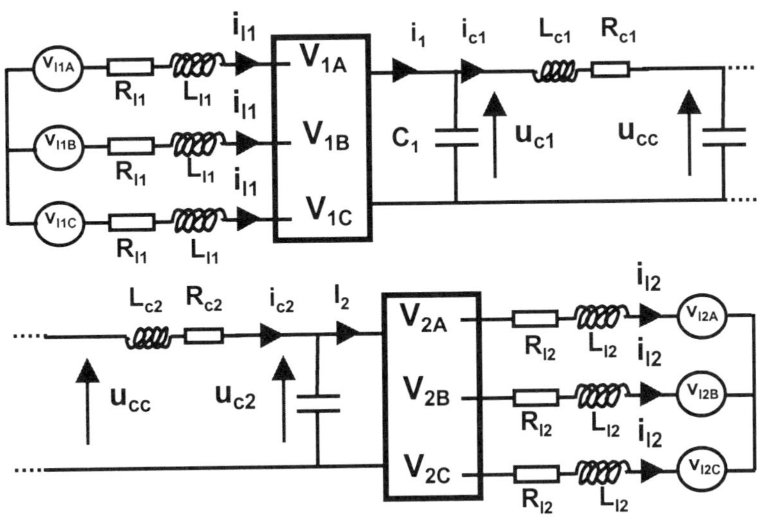

Figure 2 : Continuous-time equivalent VSC-HVDC model

III.2 Full state-space model

Given system (6) and the following definitions:

$$\begin{cases} x^T = \left[i_{l1d}, i_{l1q}, u_{c1}, i_{c1}, u_{cc}, i_{c2}, u_{c2}, i_{l2d}, i_{l2q} \right] \\ u^T = \left[v_{1dw}, v_{1qw}, v_{2dw}, v_{2qw} \right] \\ z^T = \left[v_{l1d}, v_{l1q}, v_{l2d}, v_{l2q} \right] \\ y^T = \left[P_{l1}, Q_{l1}, P_{l2}, Q_{l2} \right] \end{cases} \tag{7}$$

where the expressions of the instantaneous active power and reactive power on both sides are defined by:

$$\begin{cases} P_{l1} = v_{l1}i_{l1}\cos(\gamma_1 - \varphi_1) = v_{l1d}i_{l1d} + v_{l1q}i_{l1q} \\ Q_{l1} = v_{l1}i_{l1}\sin(\gamma_1 - \varphi_1) = v_{l1q}i_{l1d} - v_{l1d}i_{l1q} \\ P_{l2} = v_{l2}i_{l2}\cos(\gamma_2 - \varphi_2) = v_{l2d}i_{l2d} + v_{l2q}i_{l2q} \\ Q_{l2} = v_{l2}i_{l2}\sin(\gamma_2 - \varphi_2) = v_{l2q}i_{l2d} - v_{l2d}i_{l2q} \end{cases} \tag{8}$$

the full state-space model of the simplified scheme depicted in Figure 2 could be written by the bilinear system (9), characterized by a non usual nonlinear output equation:

$$\begin{cases} \dot{x} = [A]x + g(x)u + [R]z \\ y = h(z)x \end{cases} \tag{9}$$

IV. CONTROL STRUCTURE

In this paper the synthesis of the control structure will be restricted in a first approach to the DC-Bus voltage u_{c1} control, including current loops on both sides. Nevertheless, active power flow steps could be applied by action on i_{l2d}, with $i_{l2q} = 0$ to set $Q_{l2} = 0$.

IV.1 Slow and fast dynamics

For the "Station 1" optimal behaviour, i_{l1d} and i_{l1q} has to follow varying set points. This is a "Tracking" control problem. The voltage u_{c1} has to be maintained at a set point in order to compensate the performance during dynamic variations. This is a "Regulation"

control problem, i.e. the effect of $[R]z$ must be rejected.

Extracting from systems (9) or (6), the equations of the currents define the "fast dynamics" and the equation of the DC-Bus voltage represents the "slow dynamics".

$$\begin{cases} \dfrac{d}{dt}i_{l1d} = -\dfrac{R_{l1}}{L_{l1}}i_{l1d} + \omega_1 i_{l1q} + \dfrac{1}{L_{l1}}v_{l1d} - \dfrac{1}{L_{l1}}\dfrac{u_{c1}}{2}v_{1dw} \\ \dfrac{d}{dt}i_{l1q} = -\dfrac{R_{l1}}{L_{l1}}i_{l1q} - \omega_1 i_{l1d} + \dfrac{1}{L_{l1}}v_{l1q} - \dfrac{1}{L_{l1}}\dfrac{u_{c1}}{2}v_{1qw} \\ \dfrac{d}{dt}u_{c1} = -\dfrac{1}{C_1}i_{c1} + \dfrac{1}{C_1}\dfrac{3}{4}\left(v_{1dw}i_{l1d} + v_{1qw}i_{l1q}\right) \end{cases} \tag{10}$$

IV.2 Controller synthesis

Because the input matrix $g(x)$ is not square, the exact linearisation control, proposed in [2], is not applicable. Then, if we consider the "fast dynamics" of system (10) with the assumption of (d_1, q_1) frame right orientation, i.e. $v_{l1q} = 0$, the following nonlinear state feedback laws:

$$\begin{cases} u_d = -\dfrac{R_{l1}}{L_{l1}}i_{l1d} + \omega_1 i_{l1q} + \dfrac{1}{L_{l1}}v_{l1d} - \dfrac{1}{L_{l1}}\dfrac{u_{c1}}{2}v_{1dw} \\ u_q = -\dfrac{R_{l1}}{L_{l1}}i_{l1q} - \omega_1 i_{l1d} + \dfrac{1}{L_{l1}}v_{l1q} - \dfrac{1}{L_{l1}}\dfrac{u_{c1}}{2}v_{1qw} \end{cases} \tag{11}$$

performs a full linearized and decoupled closed loop control of the line currents components i_{l1d} and i_{l1q}.

$$\begin{cases} \dfrac{d}{dt}i_{l1d} = u_d \\ \dfrac{d}{dt}i_{l1q} = u_q \end{cases} \tag{12}$$

Nevertheless, closed-loop dynamics (12) is not acceptable in practice. Then, in order to achieve asymptotic tracking, the tracking error dynamics can be chosen to be a second order system. In this case an integral action will appear in the control laws.

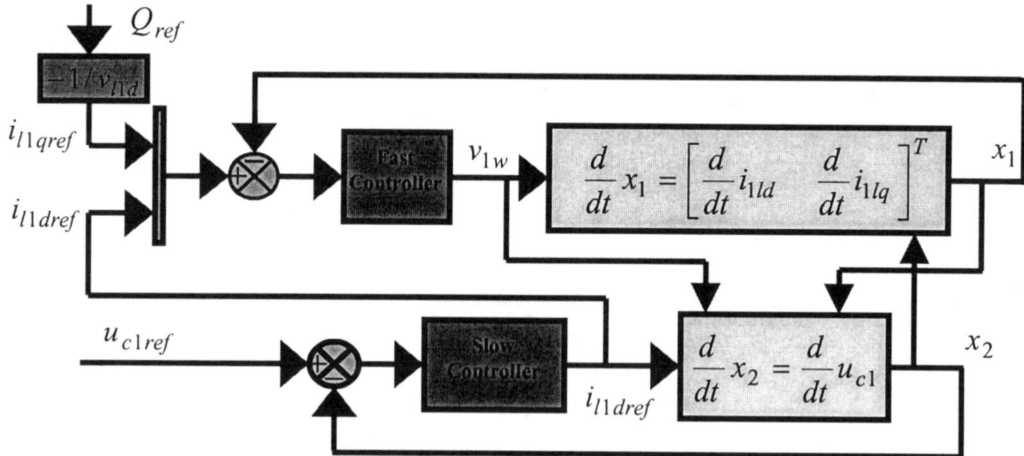

Figure 3 : Nonlinear control structure

Considering the "slow dynamics" of system (10) and a second order controller for the DC-Bus voltage regulation, the set point of i_{l1d} , noted i_{l1dref} , can be expressed by the following two equations:

$$i_{l1dref} = \frac{4C_1}{3v_{1dw}}\left(I_e + \frac{1}{C_1}i_{c1} - \frac{1}{C_1}\frac{3}{4}v_{1qw}i_{l1q}\right) \qquad (13)$$

$$I_e = K_p(u_{c1ref} - u_{c1}) + K_i \int (u_{c1ref} - u_{c1})dt + \frac{d}{dt}u_{c1ref} \qquad (14)$$

where K_p and K_i are respectively the adjustments for proportional and integral terms of the slow controller shown in Figure 3.

Notice that control variables v_{1dw} and v_{1qw} of the PWM rectifier appear in the slow nonlinear controller equation (13), in order to improve the DC-Bus voltage regulation.

Figure 3 defines the two time-scale nonlinear control structure of the "Station 1", highlighting the very simple open-loop control of the reactive power Q_{ref} .

The set point u_{c1ref} is the DC-Bus voltage reference to be kept constant at 300 kV in our example.

In this paper, the control of the "Station 2" is voluntary restricted to the linearizing and decoupling control of the line currents i_{l2d} and i_{l2q} components, based on the "Station 1" control structure.

## V.	SIMULATION RESULTS

One of the advantages of the equivalent continuous-time modelling of static converters, more particularly for PWM VSC-actuators, is the improvement of simulations in Matlab/Simulink environment by using only standard functional block diagrams. Obviously, switching effects resulting from PWM "Station 1" and "Station 2" are not introduced. Then, all waveforms are plotted without any ripple or noise. However, the

"fondamental" dynamics of the system in closed-loop can be well analysed.

In this paper, the simulation model of the DC-line is simplified (see Figure 2). It is of course possible to introduce easily a more complex cable model between u_{c1} and u_{c2} , as for example, the model proposed in reference [4].

### V.1	Direct power flow control

In this section, all simulation results are obtained for a 200 MW step disturbance of active power P , by action on i_{ld2} component, at the instant $t = 0.1s$. Maximum power ratings is 300 MW in our case study.

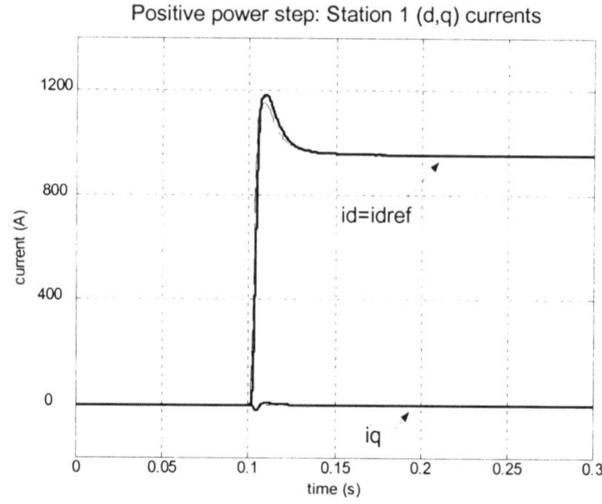

Figure 4: Tracking of current components Station 1 side – P>0

Both reactive power Q_{l1} and Q_{l2} are set up to zero, i.e. unity power factor. Figure 4 shows the resulting value of i_{l1d} which is the right value to maintain DC-Bus voltage at the 300 kV nominal value. More precisely, it is not possible to distinguish between i_{l1d} and the setpoint i_{l1dref} , highlighting fast dynamics of the rectifier current loop. The "good" decoupling between d and q components can be also observed.

The regulation behaviour of the DC-Bus voltage u_{c1} is illustrated by Figure 5.

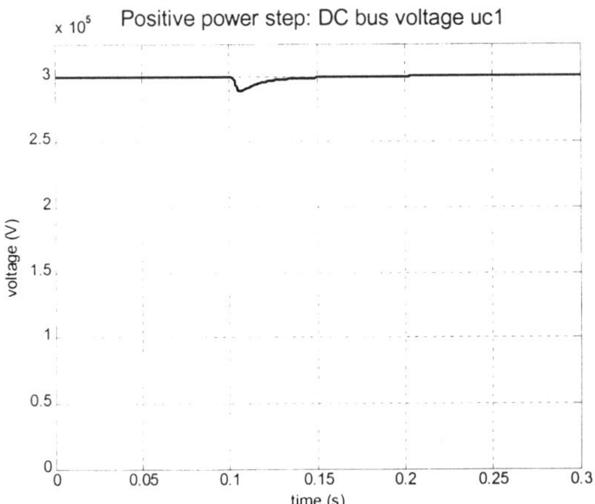

**Figure 5: DC-Bus voltage regulation
Station 1 side - P>0**

Figure 6 shows the behaviour of the associated DC-line current at the input of the cable, i.e. "Station 1" side. The response of the two d and q components of the "Station 2" is depicted in Figure 7, demonstrating the decoupled behaviour between active and reactive power components.

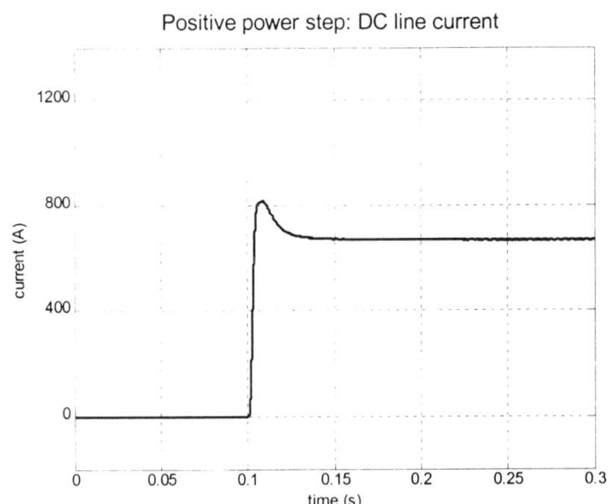

Figure 6: DC-Line current – P>0

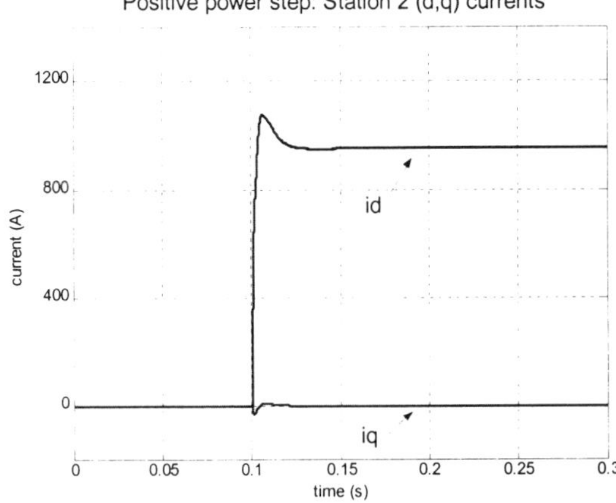

**Figure 7: Tracking of current components
Station 2 side – P>0**

The unity power factor of "Station 2" is also well satisfied by keeping i_{l2q} to zero.

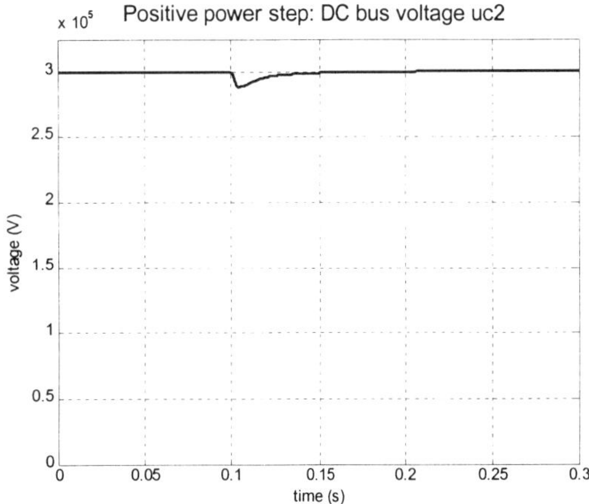

**Figure 8: DC-Bus voltage regulation
Station 2 side - P>0**

Figure 8 illustrates the "good open-loop" regulation of the DC-Bus voltage u_{c2} around 300 kV, without voltage loop on "Station 2" side. Obviously, the true value is slightly greater than u_{c1} value to ensure positive power flow, i.e. power transfer from "Station 1" to "Station 2".

V.2   Inverse power flow control

One of the major features of a VSC-HVDC transmission system is to be able to reverse the power flow without any dynamic modifications of the control structure, taking benefits of the 4 quadrant converters.

Figure 9 shows the "exact symetrical" closed-loop behaviour of the current components of "Station 1", compared to those of Figure 4.

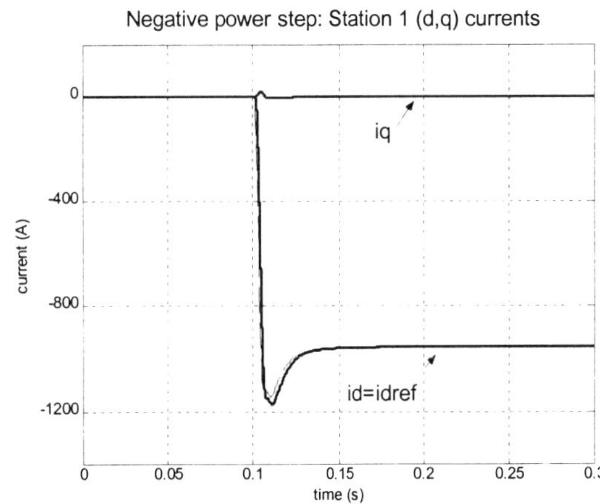

**Figure 9: Tracking of current components
Station 1 side – P<0**

Finally, in the same inverse power flow operating conditions the regulation of the DC-Bus voltage u_{c1} at 300 kV is illustrated by Figure 10, where the overshoot appears as the "inverse" of the overshoot obtained for positive power flow control (see Figure 5).

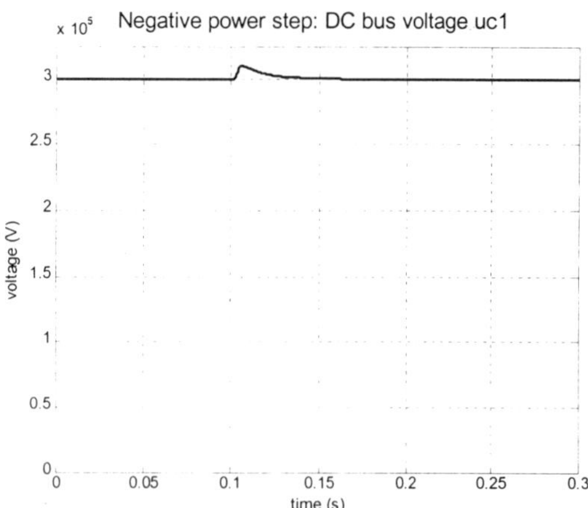

**Figure 10: DC-Bus voltage regulation
Station 1 side – P<0**

VI.   CONCLUSION

An equivalent continuous-time model of a standard VSC-HVDC tranmission scheme has been proposed in order to improve simulation performances and to perform the synthesis of a well adapted control structure. The DC-Bus voltage control problem, which is one of the critical aspects of a VSC-HVDC system, has been solved by nonlinear approach.

Performances of the proposed control are demonstrated through the simulation results of a 300 MW / 300 kV case study.

VII.   ACKNOWLEDGEMENTS

The authors would like to express their gratitude to B. Andersen of ALSTOM T&D in Stafford, for initiating and supporting this work.

VIII.   REFERENCES

1. F. Labrique, H. Buyse, G. Seguier and R. Bausière, "Les convertisseurs de l'électronique de puissance", Vol 5, "Commande et comportement dynamique", Tech. & Doc, Lavoisier, 1998, Paris, France.
2. Isidori, "Nonlinear Control Systems", Third Edition, Springer, 1995.
3. Lindberg A, "PWM and control of two and threelevel high power voltage source converters", PhD thesis, Royal Institute of Technology, Department of Electric Power Engineering, Division of High Power Electronics, Stockholm, SWEDEN, April, 1995.
4. A. Henni, F. Heliodore, S. Poullain, J-L. Thomas "Modelling of the dynamic characteristics of the DC line for VSC Transmission scheme", 7th International Conference on AC-DC Power Transmission, IEE, London, 28-30 November 2001.

HARMONIC MODELLING OF VOLTAGE SOURCE CONVERTERS FOR HVDC STATIONS

M Madrigal and E Acha

CERPD Centre for Economic Renewable Power Delivery
The University of Glasgow, Scotland, UK

Abstract— Interest in the use of VSCs for HVDC stations is increasing rapidly due to their economical and technical advantages. Most of the work reported in the open literature has been on PWM control and VSC topologies, and in HVDC commissioning, construction and operation. Thus far, little emphasis has been placed on VSC harmonic generation and its impact on the power system, since it is assumed that the use of PWM control yields low harmonic distortion and that there is no risk of interacting with other power plant components. But in real systems this preconceived assumption is not always met. The harmonic frequencies generated by the PWM control and the multiple resonant impedance points exhibited by transmission lines and cables, may lead to unwanted interactions between the HVDC stations and the power system causing resonance problems at harmonic frequencies. In order to predict such problems, a suitable HVDC station model for harmonic propagation is required. In this paper, a new and comprehensive harmonic model of a VSC is developed using first principles. The VSC model is used as the basic building block with which harmonic models for HVDC-VSC back-to-back and HVDC-VSC transmission station are assembled. The response of the newly developed harmonic models are in good agreement with the steady-state responses of time domain models implemented in PSCAD/EMTDC.

Index Terms— Voltage source converters, switching functions, PWM, harmonic analysis.

I. INTRODUCTION

Voltage source converters (VSCs) are employed in several of the power electronic-based controllers recently made available for improved power flow reliability in high-voltage transmission and power quality in low-voltage distribution systems. Examples of such controllers are the advanced static compensator, the unified power flow controller, the dynamic voltage restorer [1][2] and HVDC stations [3].

Thus far, most of the available literature on VSC applications on HVDC transmission relates to the all-important issues of operation, construction and commissioning. Theoretical studies of PWM control schemes and bridge topology have also received attention [4][5][6]. The following applications of HVDC transmission based on VSCs are discussed in [7]: (1) systems interconnection using an HVDC-VSC back-to-back tie; (2) grid-connection of offshore wind farms; (3) grid-connection of otherwise autonomous systems such as islands and oil platforms; (4) city centre infeed. The two converter stations are linked together by sea or underground cables. The HVDC transmission scheme is referred to as HVDC-VSC transmission station.

This paper presents models of HVDC-VSC stations for harmonic analysis. These models are developed using first principles and are based on the following key assumptions: (1) The VSCs use GTO or IGBT electronic switches; (2) the VSCs are represented as linear, time variant systems; (3) complex Fourier series may be used to determine the steady-state solution of the linear, time variant system. From the modelling point of view,

(2) is a reasonable assumption to make since GTOs and IGBTs have fully controllable turn-on and turn-off capabilities. Bearing these three considerations in mind, a VSC model is developed in the harmonic domain [8], where PWM switching functions are used to represent the electronic switches turn-on and turn-off characteristics. This is the basic three-phase VSC block with which the HVDC-VSC back-to-back and HVDC-VSC transmission station models are assembled.

II. VOLTAGE SOURCE CONVERTERS

A. Steady-state Operation of the VSC

Fig. 1 represents a six-pulse, three-phase VSC. It includes a star-delta connected transformer bank with its impedance transfered to the primary side with value Z_e. The transformation ratio is considered to be $1{:}1$.

The equation that relates the steady-state voltages from both sides of the VSC is given by

$$\mathbf{v}_{abc}(t) = \mathbf{p}_s(t)v_{\text{cap}}(t) \tag{1}$$

and for the currents,

$$i_{\text{cap}}(t) = \mathbf{q}_s(t)\mathbf{i}_{abc}(t) \tag{2}$$

The voltage and current vectors are

$$\mathbf{v}_{abc}(t) = \begin{bmatrix} v_a(t) \\ v_b(t) \\ v_c(t) \end{bmatrix} \qquad \mathbf{i}_{abc}(t) = \begin{bmatrix} i_a(t) \\ i_b(t) \\ i_c(t) \end{bmatrix}$$

and the voltage in the capacitor is

$$v_{\text{cap}}(t) = \frac{1}{C}\int_0^t i_{\text{cap}}(t)\mathrm{d}t + v_{\text{cap}}(0^+) \tag{3}$$

where $v_{\text{cap}}(0^+)$ is the charged capacitor voltage from an earlier period. For steady-state analysis the voltage $v_{\text{cap}}(0^+)$ may be taken to be constant, i.e. the capacitor is not charging or discharging [9]. This condition results in a zero dc-term of the current $i_{\text{cap}}(t)$.

The switching vectors are given by

$$\mathbf{p}_s(t) = \begin{bmatrix} s_{ab}(t) \\ s_{bc}(t) \\ s_{ca}(t) \end{bmatrix}$$

$$\mathbf{q}_s(t) = \begin{bmatrix} s_{ab}(t) & s_{bc}(t) & s_{ca}(t) \end{bmatrix}$$

AC-DC Power Transmission, 28-30 November 2001
Conference Publication No. 485 © IEE 2001

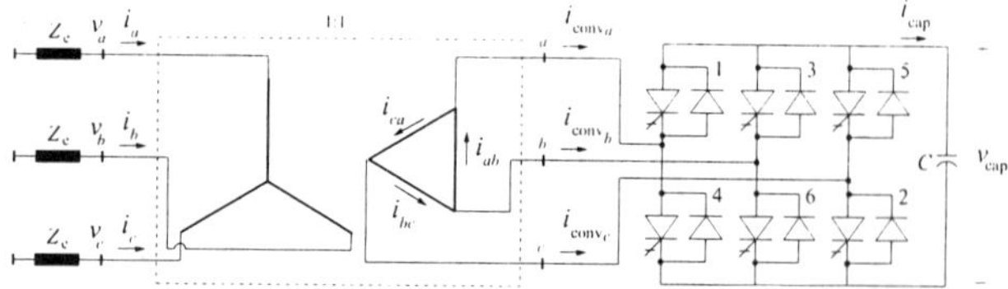

Fig. 1. Three-phase voltage source converter

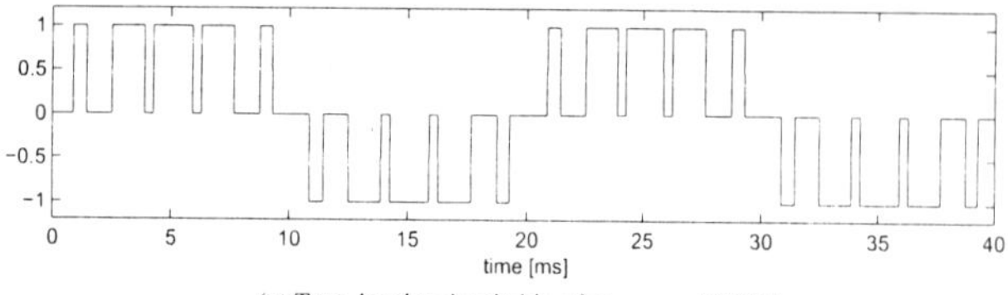

(a) Two-level and switching frequency of 250 Hz

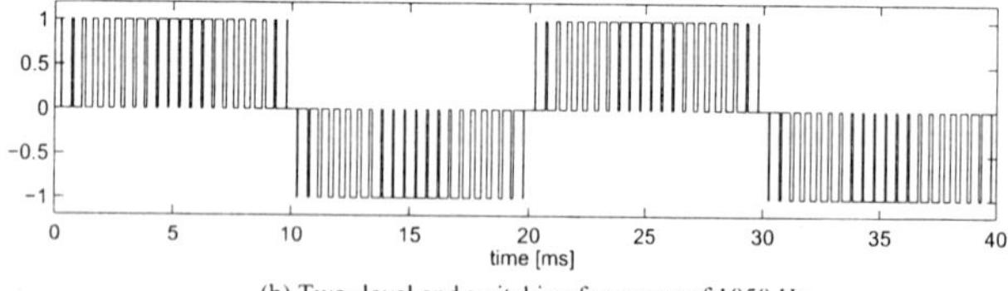

(b) Two-level and switching frequency of 1050 Hz

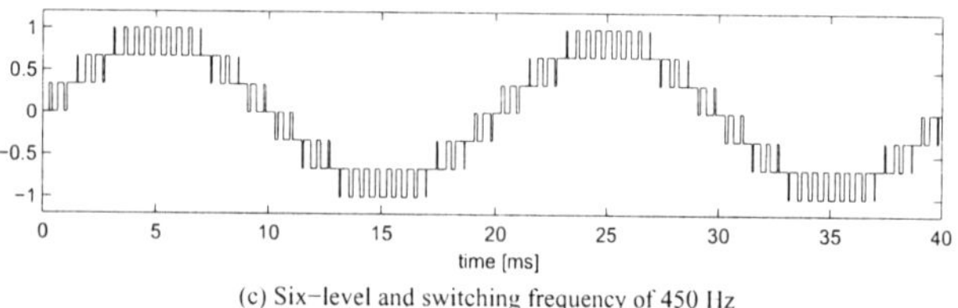

(c) Six-level and switching frequency of 450 Hz

Fig. 2. Switching function that represent different VSCs

where $s_{ab}(t)$, $s_{bc}(t)$ and $s_{ca}(t)$ are the PWM switching functions that govern the firing of valves 1-6, 3-2 and 5-4, respectively. In general, by using these three switching functions (one per line to line voltage), different switching functions can be represented, as shown in Fig. 2. It should be remarked that only the two-level switching functions in Fig. 2(a) and (b) correspond to the VSC topology shown in Fig. 1. The switching function in Fig. 2(c) corresponds to a six-level topology not shown in the paper. With the switching strategy, the magnitude, the phase angle and the frequency of the VSC output voltage can be controlled [10].

B. Steady-state Harmonic Model of the VSC

The Harmonic Domain (HD) is based on the complex Fourier series, where the periodic function $f(t)$ is given by the series,

$$f(t) = \sum_{h=-\infty}^{\infty} F_h e^{jh\omega_0 t}$$

and represented in the HD by the vector,

$$\mathbf{F} = \begin{bmatrix} \cdots & F_{-2} & F_{-1} & F_0 & F_1 & F_2 & \cdots \end{bmatrix}^{\mathrm{T}}$$

From eq. (1), the voltage on the DC side and its relationship to the AC phase voltages, are expressed in the HD as

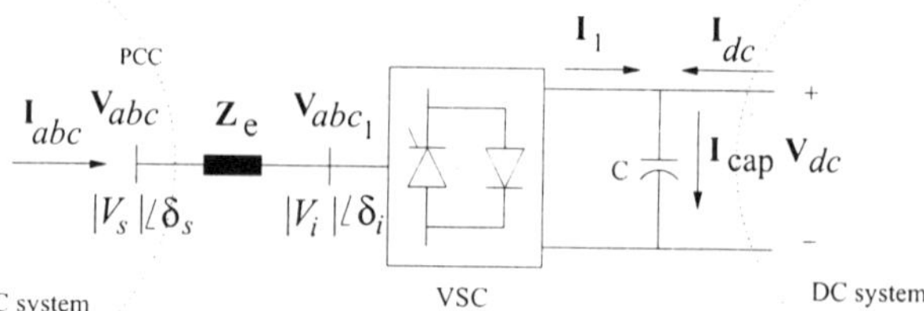

AC system   VSC   DC system

Fig. 3. Voltage source converter

$$\mathbf{V}_{abc} = \mathbf{P}_s \mathbf{V}_{cap} \tag{4}$$

Similarly, the relationship between currents is given by

$$\mathbf{I}_{cap} = \mathbf{Q}_s \mathbf{I}_{abc} \tag{5}$$

where the transformation matrices $\mathbf{P}_s$ and $\mathbf{Q}_s$ are

$$\mathbf{P}_s = \begin{bmatrix} \mathbf{S}_{ab} \\ \mathbf{S}_{bc} \\ \mathbf{S}_{ca} \end{bmatrix}$$

$$\mathbf{Q}_s = \begin{bmatrix} \mathbf{S}_{ab} & \mathbf{S}_{bc} & \mathbf{S}_{ca} \end{bmatrix}$$

$\mathbf{S}_{ab}$, $\mathbf{S}_{bc}$ and $\mathbf{S}_{ca}$ are Toeplitz matrices which contain the harmonic content of their switching functions. The switching functions are treated as general periodical functions, where their harmonic coefficients depend on the PWM technique used, the VSC levels and its operating point.

The harmonic voltage and current vectors are given by

$$\mathbf{V}_{abc} = \begin{bmatrix} \mathbf{V}_a \\ \mathbf{V}_b \\ \mathbf{V}_c \end{bmatrix} \qquad \mathbf{I}_{abc} = \begin{bmatrix} \mathbf{I}_a \\ \mathbf{I}_b \\ \mathbf{I}_c \end{bmatrix}$$

Also eq. (3) may be expressed by

$$\mathbf{V}_{cap} = \frac{1}{C}\mathbf{D}^{-1}(jh\omega_0)\mathbf{I}_{cap} + \mathbf{V}_0 \tag{6}$$

where $\mathbf{V}_{cap}$ gives the harmonic content of $v_{cap}(t)$. $\mathbf{D}(jh\omega_0)$ is a diagonal matrix of differentiation with entries $jh\omega_0$. Likewise, vector $\mathbf{I}_{cap}$ includes the harmonic content of $i_{cap}(t)$, and $\mathbf{V}_0$ contains only a constant dc-term equal to $v_{cap}(0^+)$.

Eq. (6) can be written as

$$\mathbf{V}_{cap} = \mathbf{Z}_{cap}\mathbf{I}_{cap} + \mathbf{V}_0 \tag{7}$$

where $\mathbf{Z}_{cap} = \frac{1}{C}\mathbf{D}^{-1}(jh\omega_0)$ represents the equivalent impedance of the DC capacitor at harmonic frequencies.

An alternative schematic representation of the VSC is given in Fig. 3 which includes a current injection from the DC side. This additional current gives the option to represent active power exchange with the AC system.

With the previous equations and some algebra, the equation that represents the VSC shown in Fig. 3 is obtained,

$$\begin{bmatrix} \mathbf{V}_{abc_{\mathrm{T}}} \\ \mathbf{V}_{dc_{\mathrm{T}}} \end{bmatrix} = \begin{bmatrix} \mathbf{A}_1 & \mathbf{B}_1 \\ \mathbf{C}_1 & \mathbf{D}_1 \end{bmatrix} \begin{bmatrix} \mathbf{I}_{abc} \\ \mathbf{I}_{dc} \end{bmatrix} \tag{8}$$

where

$$\mathbf{V}_{abc_T} = \mathbf{V}_{abc} - \mathbf{P}_s\mathbf{V}_0 \tag{9}$$
$$\mathbf{V}_{dc_T} = \mathbf{V}_{dc} - \mathbf{V}_0 \tag{10}$$

and

$$\mathbf{A}_1 = \mathbf{Z}_e + \mathbf{P}_s\mathbf{Z}_{cap}\mathbf{Q}_s \qquad \mathbf{B}_1 = \mathbf{P}_s\mathbf{Z}_{cap}$$
$$\mathbf{C}_1 = \mathbf{Z}_{cap}\mathbf{Q}_s \qquad \mathbf{D}_1 = \mathbf{Z}_{cap}$$

$\mathbf{Z}_e = R_e\mathbf{U}_I + L_e\mathbf{D}(jh\omega_0)$, where R_e and L_e are the equivalent resistance and inductance of the coupling transformer and $\mathbf{U}_I$ is the identity matrix.

The steady-state condition is given when the dc-term of $\mathbf{I}_{cap}$ is zero. It should be noted that in Fig. 3, $\mathbf{I}_{cap} = \mathbf{I}_1 + \mathbf{I}_{dc}$, $\mathbf{I}_1 = \mathbf{Q}_s\mathbf{I}_{abc}$ and $\mathbf{V}_{dc} = \mathbf{V}_{cap}$.

C. Phase Angle Control for the VSC

Since the angle δ_i of the VSC must be adjusted in order to obtain a zero dc-term current in the capacitor, an iterative process may be used to such end. Fig. 4 illustrates the iterative process used in this paper, the solid line is the current-angle characteristic, and the initial condition for the iterative process is given by the dashed line. The iterative process is explained below with relation to Fig. 4:

1) Calculate the dashed line slope M,

$$M = \frac{\Delta I_1 - \Delta I_0}{\delta_1 - \delta_0}$$

2) Calculate the new angle,

$$\delta_{\text{new}} = -\frac{1}{M}\Delta I_1 + \delta_1$$

3) Update the harmonics of the three switching functions, e.g. for $s_{ab}(t)$,

$$S_{ab_h} = S_{ab_h}e^{-jh\delta_{\text{new}}}$$

and a new current ΔI_{new} is obtained by solving the VSC model using eq. (8).
4) Update variables: $\Delta I_0 = \Delta I_1$, $\Delta I_1 = \Delta I_{\text{new}}$, $\delta_0 = \delta_1$ and $\delta_1 = \delta_{\text{new}}$ and go to step 1 until $\Delta I_{\text{new}} \leq \epsilon$.

D. VSC Test Model

The system has the following parameters:
Voltage source connected at PCC: 115 kV line to line, RL parallel equivalent impedance with $R_g = 100\ \Omega$ and $L_g = 0.01$ H.

Transformer: An equivalent series impedance $R_e = 0.1\ \Omega$ and $L_e = 0.05$ H.

VSC: Six-pulse, three-phase converter with IGBTs switched at 150 Hz. Capacitor: $300\ \mu$F to be charged at 115 kV.

Simulation: For the proposed model 100 harmonics and $\mathbf{I}_{dc} = 0$ were used, obtaining the steady-state condition in two iterations.

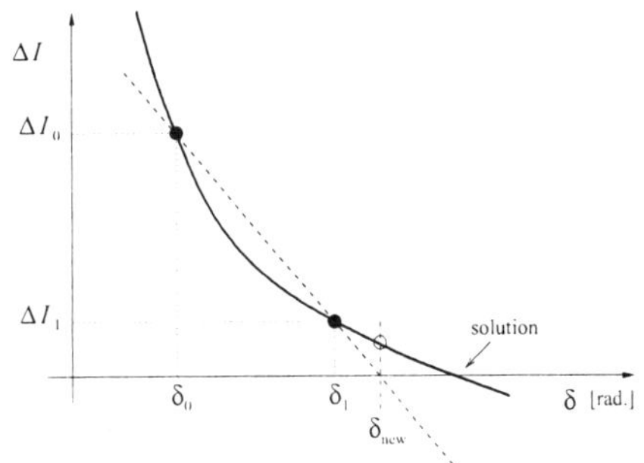

Fig. 4. Current-angle iterative process

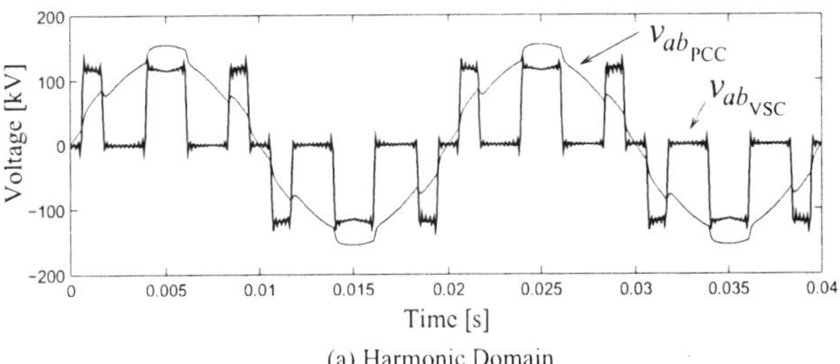

(a) Harmonic Domain

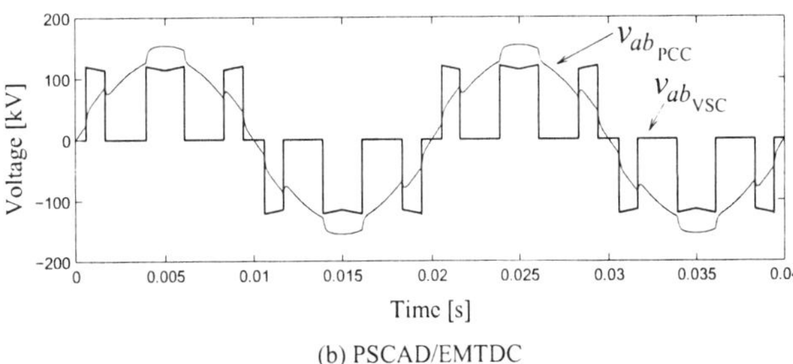

(b) PSCAD/EMTDC

Fig. 5. Line voltages in the PCC and the VSC

The proposed VSC model was compared against a model implemented in PSCAD/EMTDC [11], a simulation time of 1.7 seconds, with a time step of $20\ \mu$s, were used to obtain a suitable steady-state.

Fig. 5 shows the line voltages at the PCC and at the VSC's terminal. The results show that voltages obtained with both models match with each other. The rounded sections at the top and bottom of the VSC line voltage are due to the influence of the capacitor. The results in Fig. 5(a) were obtained with HD techniques and are presented in waveform only to enable comparison with the results generated with PSCAD/EMTDC shown in Fig. 5(b).

E. Response of the VSC

With reference to Fig. 3, the exchange of active power between the AC and the DC system can be controlled by adjusting the phase angle difference of the PWM converter output voltage and the voltage at the PCC, i.e. $\delta_{si} = \delta_s - \delta_i$. The reactive power exchanged between the AC system and the VSC is controlled by adjusting the voltage magnitude difference across the coupling transformer, i.e. $|V_s| - |V_i|$. Where the active and reactive power injected by the AC system are given by

$$P \approx \frac{|V_s||V_i|}{X_e} \sin \delta_{si} \qquad (11)$$

$$Q \approx \frac{|V_s|^2}{X_e} - \frac{|V_s||V_i|}{X_e} \cos \delta_{si} \qquad (12)$$

and X_e is the equivalent reactance of the coupling transformer.

The VSC of Fig. 3 is used to illustrate the response of the VSC for active and reactive power exchange with the AC system. The switching functions have been selected to represent a six-level VSC (as in Fig. 2(c)) with a switching frequency of 150 Hz. Also, the PWM control uses a variable modulation index to control the output voltage magnitude. A capacitor of $500\,\mu$F with a precharged voltage of 1.5 p.u. was considered. A three-phase AC balanced voltage source of 1 p.u. with $\delta_s = 0°$ were used and equivalent impedance system of $Z_e = 0.0133 + j0.6283\,\Omega$ at 50 Hz. The analysis was carried out with 15 harmonics. The results were obtained in the HD and shown in waveform representation. The powers are those injected into the VSC terminal.

Active power exchange. The active power exchange was obtained by selecting different values for the DC current injection in the range $-0.5 < I_{dc_0} < 0.5$. A constant modulation index of $m = 0.8$ was used.

Fig. 6 shows results only for phase a. These results show that for different DC current values, the direction of the active power is controlled by changing the direction of the current $i_1(t)$ while maintaining the voltage $v_{dc}(t)$ in the same polarity. The reactive power shows the same direction (absorbed by the AC system) with an almost constant value since it has more dependence on the voltage magnitude $|V_i|$ (regulated by m) than on the phase angle. From the harmonic point of view, $v_{dc}(t)$ shows an almost constant waveform distortion, and the VSC output voltage $v_{a_1}(t)$ presents the same waveform (because of the PWM technique used) with different phase shift angle respect to $v_a(t)$. The total harmonic distortion (THD) and the rms value of $v_{a_1}(t)$ are almost constant, meaning that the harmonics generated by the VSC does not have a strong dependence on the voltage angle. The maximum THD current is giving for minimum active power exchange.

Reactive power exchange. The reactive power exchange was obtained by selecting different values of the modulation index in the range $0.5 < m < 1$, and zero active power exchange, i.e. $I_{dc} = 0$.

Fig. 7 shows the results for phase a. Since no active power is exchanged between the systems, the current $i_1(t)$ does not have dc-term. The current $i_a(t)$ has a 90° phase shift with respect to the voltage but with different magnitude and direction. From the harmonic point of view, $v_{dc}(t)$ shows an almost constant waveform distortion, and the VSC output voltage $v_{a_1}(t)$ presents different waveform maintaining the same phase shift angle with respect to $v_a(t)$. The reactive power exchanged equals zero when $|V_i| = |V_s|$, a condition that occurs at $m \approx 0.665$ ($|V_i| = 0.665 \times 1.5 = 0.998$ p.u.). It should be noted that at zero reactive power exchange, the current THD

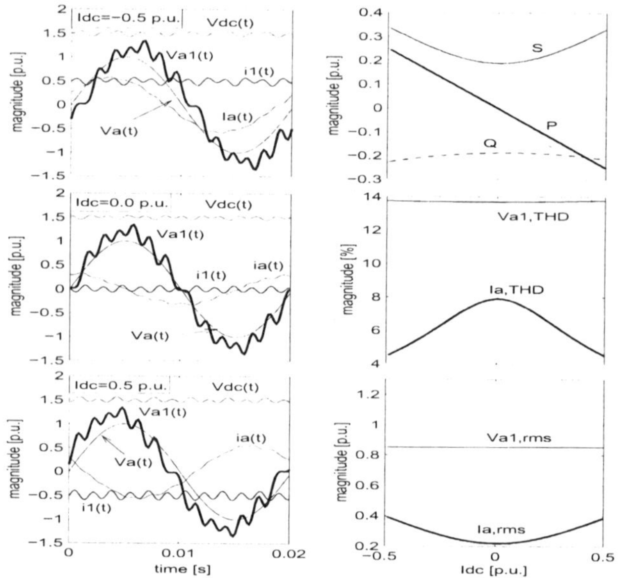

Fig. 6. Active power exchange between the AC and DC systems

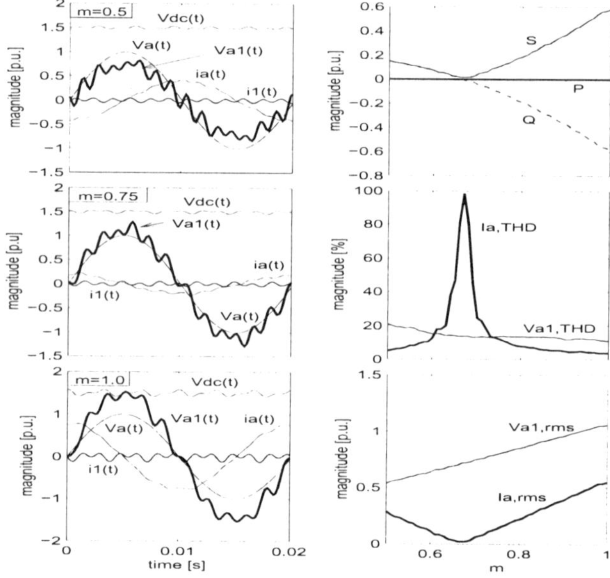

Fig. 7. Reactive power exchange between the AC system and the VSC

is maximum and the rms current is minimum. The rms voltage presents a linear response but not the THD, meaning that the harmonics generated by the VSC are also a function of the modulation index, which clearly is not the case for the voltage phase angle.

III. HVDC-VSC Back-to-back Station

A. Analytical Model

Basically, the HVDC-VSC back-to-back is built with two equal VSCs sharing a capacitor, as shown in Fig. 8. The model of the HVDC-VSC back-to-back is given by

129

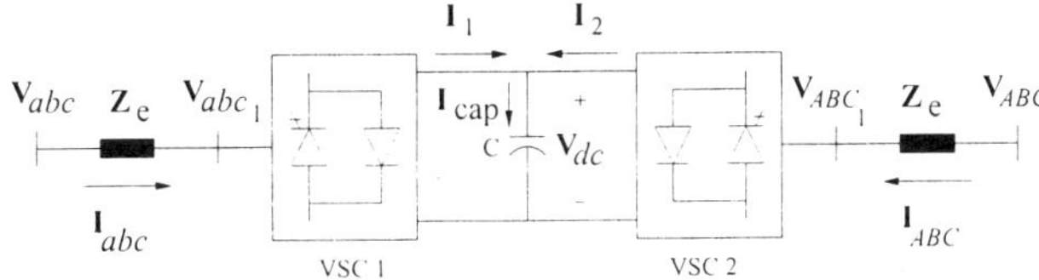

Fig. 8. HVDC-VSC back-to-back station

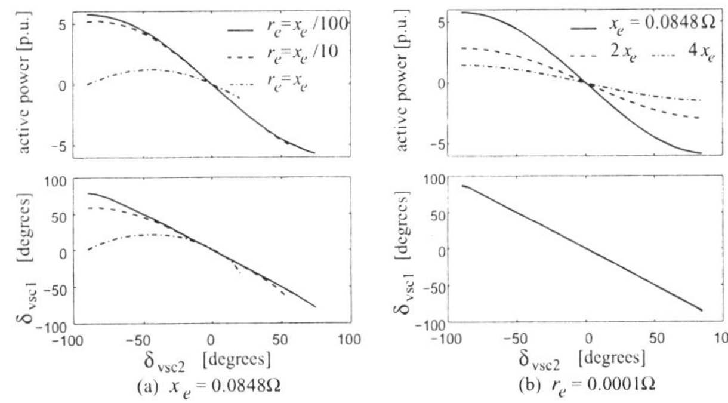

Fig. 9. HVDC-VSC back-to-back responses

$$\begin{bmatrix} \mathbf{V}_{abc_\mathrm{T}} \\ \mathbf{V}_{ABC_\mathrm{T}} \end{bmatrix} = \begin{bmatrix} \mathbf{A}_{12} & \mathbf{B}_{12} \\ \mathbf{C}_{12} & \mathbf{D}_{12} \end{bmatrix} \begin{bmatrix} \mathbf{I}_{abc} \\ \mathbf{I}_{ABC} \end{bmatrix} \tag{13}$$

where

$$\mathbf{V}_{abc_\mathrm{T}} = \mathbf{V}_{abc} - \mathbf{P}_{s_1}\mathbf{V}_0 \tag{14}$$
$$\mathbf{V}_{ABC_\mathrm{T}} = \mathbf{V}_{ABC} - \mathbf{P}_{s_2}\mathbf{V}_0 \tag{15}$$

and

$$\mathbf{A}_{12} = \mathbf{Z}_e + \mathbf{P}_{s_1}\mathbf{Z}_{cap}\mathbf{Q}_{s_1} \qquad \mathbf{B}_{12} = \mathbf{P}_{s_1}\mathbf{Z}_{cap}\mathbf{Q}_{s_2}$$
$$\mathbf{C}_{12} = \mathbf{P}_{s_2}\mathbf{Z}_{cap}\mathbf{Q}_{s_1} \qquad \mathbf{D}_{12} = \mathbf{Z}_e + \mathbf{P}_{s_2}\mathbf{Z}_{cap}\mathbf{Q}_{s_2}$$

Eq. (13) may be seen as two equivalent STATCOMs [8] with a coupling impedances between them given by $\mathbf{B}_{12}$ and $\mathbf{C}_{12}$.

B. HVDC-VSC Back-to-back Operation

Since the HVDC-VSC back-to-back is built with two VSCs, reactive power may be exchanged with both AC systems in a similar fashion as with the STATCOM, with the only major difference that active power can also be exchanged between the two systems. The active power flow during steady-state operation is achieved when no dc current is flowing in the capacitor. In order to achieve such an operation, one VSC is selected as a power dispatcher and the second VSC as a master DC voltage regulator [9].

Power dispatcher. The function of the power dispatcher is to supply a pre-specified active power demand P_{dmd}, which can be positive or negative and supplied by the power dispatcher in a rectifier or inverter mode. In this case, VSC 2 is selected to be the power dispatcher and the active power, $P_{\mathrm{vsc_2}}$, is measured and compared with the reference P_{dmd}. The error,

$$\epsilon = P_{\mathrm{dmd}} - P_{\mathrm{vsc_2}}$$

is used as a negative feedback signal to increase or decrease the output voltage angle $\delta_{\mathrm{vsc_2}}$ of the converter by means of the PWM control until the error is nulled.

Master DC voltage regulator. The function of the master DC voltage regulator is to maintain the voltage $\mathbf{V}_{dc}$ at a preselected constant value. Since VSC 2 was selected as power dispatcher then VSC 1 is the master DC voltage regulator.

The dc-term of $\mathbf{V}_{dc}$, i.e. V_{dc_0}, is measured and compared against the voltage reference V_{ref}. The voltage error,

$$\epsilon = V_{\mathrm{ref}} - V_{dc_0}$$

is used as a command in a negative feedback loop to adjust the voltage angle $\delta_{\mathrm{vsc_1}}$ by means of a PWM control to null the error. A constant DC link voltage in steady-state is obtained when the sum of dc-terms of the currents $i_1(t)$ and $i_2(t)$ become equal to zero, i.e. zero dc-term current in the capacitor,

$$I_{1_0} + I_{2_0} = 0$$

In this control scheme, $\delta_{\mathrm{vsc_1}}$ is adjusted until the power from the AC system satisfies the power demanded by the power dispatcher.

C. HVDC-VSC Back-to-back Model Response

Fig. 9 shows the response of the HVDC-VSC back-to-back under different values of phase angle $\delta_{\mathrm{vsc_2}}$ of the power dispatcher and AC equivalent impedance. VSC 1 is the master DC voltage regulator in these tests. The active power correspond to the input power at the terminal of the power dispatcher. The simulations were carried out for the range $-90° \leq \delta_{\mathrm{vsc_2}} \leq 90°$, and equal AC voltage sources with constant magnitude and zero phase angle.

The following features are observed from these results: 1) On the right, these curves are limited by the maximum angle $\delta_{\mathrm{vsc_1}}$ which can maintain constant DC voltage (lost of convergence in the method). 2) Active power transfer follows the same pattern of voltage angle control as in AC generation stations. 3) Active power reversal operation is achieved by changing the voltage angle from a negative value to a positive value.

For the case when $x_e = 0.0848\,\Omega$ and the value of r_e changes, the following key points are observed: 1) As r_e increases, the AC transmission losses increase. The maximum power transfer is defined by the extreme points in the curves or by the point where the curve reaches its maximum value. 2) The active power follows a non-symmetrical response from negative to positive voltage angles due to the losses. 3) The relationship between the angles $\delta_{\mathrm{vsc_2}}$ and $\delta_{\mathrm{vsc_1}}$ is non-linear, due to the losses.

For the lossless case ($r_e = 0.0001\,\Omega$) and different values of x_e, the following is noticed: 1) As x_e increases, the AC transmission capability weakens. The maximum power transfer is defined by the extreme points in the curves. 2) The active power follows a symmetrical response from positive to negative voltage angles. 3) The relationship between the angles $\delta_{\mathrm{vsc_2}}$ and $\delta_{\mathrm{vsc_1}}$ is a linear relation.

IV. HVDC-VSC Transmission Station

The HVDC-VSC transmission station operates under the same principle as the HVDC-VSC back-to-back configuration, with the only difference that the VSCs are connected through a DC cable. One VSC plays the role of master DC voltage regulator and the other the role of power dispatcher.

A. Model in Steady-state

A schematic representation of an HVDC-VSC transmission station is shown in Fig. 10. Following a similar procedure as for the HVDC-VSC back-to-back, the equivalent representation of the HVDC-VSC transmission station is given by

$$\begin{bmatrix} \mathbf{V}_{abc_{\mathrm{T}}} \\ \mathbf{V}_{ABC_{\mathrm{T}}} \end{bmatrix} = \begin{bmatrix} \mathbf{A}_{1c2} & \mathbf{B}_{1c2} \\ \mathbf{C}_{1c2} & \mathbf{D}_{1c2} \end{bmatrix} \begin{bmatrix} \mathbf{I}_{abc} \\ \mathbf{I}_{ABC} \end{bmatrix} \qquad (16)$$

where

$$\mathbf{V}_{abc_{\mathrm{T}}} = \mathbf{V}_{abc} - \mathbf{P}_{s_1} \mathbf{V}_{0_1} \qquad (17)$$

$$\mathbf{V}_{ABC_{\mathrm{T}}} = \mathbf{V}_{ABC} - \mathbf{P}_{s_2} \mathbf{V}_{0_2} \qquad (18)$$

and

$$\begin{aligned}
\mathbf{A}_{1c2} &= \mathbf{Z}_e + \frac{1}{2}\mathbf{P}_{s_1}\mathbf{A}_c\left(\mathbf{Y}_{\mathrm{cap}}\mathbf{A}_c + \mathbf{C}_c\right)^{-1}\mathbf{Q}_{s_1} \\
&\quad + \frac{1}{2}\mathbf{P}_{s_1}\mathbf{B}_c\left(\mathbf{D}_c + \mathbf{Y}_{\mathrm{cap}}\mathbf{B}_c\right)^{-1}\mathbf{Q}_{s_1} \\[4pt]
\mathbf{B}_{1c2} &= \frac{1}{2}\mathbf{P}_{s_1}\mathbf{A}_c\left(\mathbf{Y}_{\mathrm{cap}}\mathbf{A}_c + \mathbf{C}_c\right)^{-1}\mathbf{Q}_{s_2} \\
&\quad - \frac{1}{2}\mathbf{P}_{s_1}\mathbf{B}_c\left(\mathbf{D}_c + \mathbf{Y}_{\mathrm{cap}}\mathbf{B}_c\right)^{-1}\mathbf{Q}_{s_2} \\[4pt]
\mathbf{C}_{1c2} &= \frac{1}{2}\mathbf{P}_{s_2}\mathbf{A}_c\left(\mathbf{Y}_{\mathrm{cap}}\mathbf{A}_c + \mathbf{C}_c\right)^{-1}\mathbf{Q}_{s_1} \\
&\quad - \frac{1}{2}\mathbf{P}_{s_2}\mathbf{B}_c\left(\mathbf{D}_c + \mathbf{Y}_{\mathrm{cap}}\mathbf{B}_c\right)^{-1}\mathbf{Q}_{s_1} \\[4pt]
\mathbf{D}_{1c2} &= \mathbf{Z}_e + \frac{1}{2}\mathbf{P}_{s_2}\mathbf{A}_c\left(\mathbf{Y}_{\mathrm{cap}}\mathbf{A}_c + \mathbf{C}_c\right)^{-1}\mathbf{Q}_{s_2} \\
&\quad + \frac{1}{2}\mathbf{P}_{s_2}\mathbf{B}_c\left(\mathbf{D}_c + \mathbf{Y}_{\mathrm{cap}}\mathbf{B}_c\right)^{-1}\mathbf{Q}_{s_2}
\end{aligned}$$

where $\mathbf{Y}_{\mathrm{cap}} = \mathbf{Z}_{\mathrm{cap}}^{-1}$. The other elements correspond to the DC cable parameters given by the following diagonal matrices,

$$\begin{aligned}
\mathbf{A}_c &= \cosh(\gamma l/2) & \mathbf{B}_c &= Z_c \sinh(\gamma l/2) \\
\mathbf{C}_c &= \tfrac{1}{Z_c}\sinh(\gamma l/2) & \mathbf{D}_c &= \cosh(\gamma l/2)
\end{aligned}$$

where γ and Z_c are the wave propagation and characteristic impedance of the cable, respectively. l is the total cable length in km. At harmonic frequencies, the wave propagation and the characteristic impedance are given by

$$\gamma = \sqrt{(R' + j\omega_0 hL')\,(G' + j\omega_0 hC')} \qquad (19)$$

$$Z_c = \sqrt{(R' + j\omega_0 hL')\,/\,(G' + j\omega_0 hC')} \qquad (20)$$

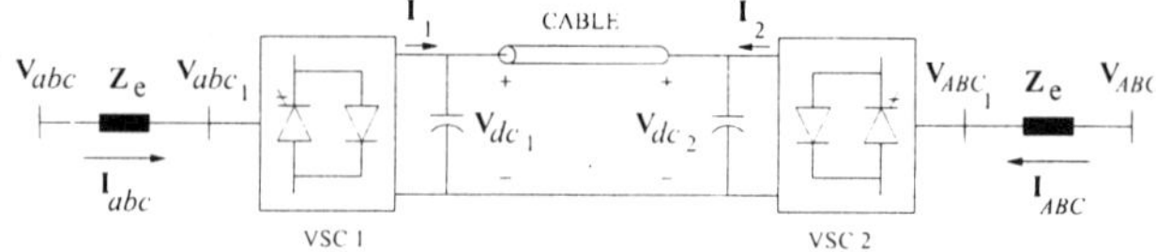

Fig. 10. HVDC-VSC transmission station

where R' in Ω/km, L' in H/km, G' in S/km and C' in F/km are the distributed series resistance, series inductance, shunt conductance and shunt capacitance of the cable, respectively. The harmonic analysis in frequency domain permits the use of the exact model of the cable; something which is more difficult to achieve when time domain simulations are used, where the common way to represent the cable is by using one $\pi-$section per km [12].

Assuming the same control rules for VSC 1 and VSC 2, the dc-term voltage in the capacitor of VSC 2 is given by $V_{dc2_0} = V_{dc1_0} - R_{\mathrm{cable}}I_{1_0}$. The steady-state condition of the HVDC-VSC transmission station is achieved when $I_{1_0} + I_{2_0} = 0$, which results in no dc-term current flowing in the capacitors.

B. HVDC-VSC Transmission Station Response

An HVDC-VSC transmission station with the following parameters was used:

AC voltage sources: 115 kV line to line, RL parallel equivalent impedance with $R_g = 1\,\Omega$ and $L_g = 0.03$ H.

Transformers: Equivalent series impedance $R_e = 0.30\,\Omega$ and $L_e = 0.15$ H.

VSCs: Two-level VSC with a switching frequency of 250 Hz and modulation index of 0.85. VSC 1 was selected as the master DC voltage regulator to maintain the DC link at 115 kV. VSC 2 was selected as the power dispatcher with the phase angle fix at $-45°$. Capacitors: 50 μF.

Cable: $R' = 0.068\,\Omega$/km, $L' = 2.422$ mH/km, $C' = 0.476\,\mu$F/km and $G' = 0.030\,\mu$S/km. Different lengths of the cable were used. 15 harmonics were used in the study.

Fig. 11 shows the impedance response as seen from the sending end of the arrangement capacitors-cable. This figure shows that resonances appear at different frequencies for different cable lengths. For example, resonances with the 6th harmonic exist for cable lenghts of 5 km and 54 km. Also, for cable lengths of 27 km, 52 km and 76 km resonances will appear with the 12th harmonic.

Fig. 12 and 13 show the response of the system for different lengths of cable. Fig. 12 show the injected power by each AC voltage source. It can be seen that the active power is flowing from the AC source ABC to the AC source abc. The losses in the whole system are given by $P_{losses} = P_a + P_A$. These results show that resonance points occur at cable lengths of 5, 27, 54 and 76 km.

The results in Fig. 13 show that the current $\mathbf{I}_1$ (DC side) contains the 6th and 12th harmonic, which interact with the DC system producing resonances for 5 km and 54 km long cables. It should be remarked that these resonances were previously predicted by inspecting the DC side impedance responses in

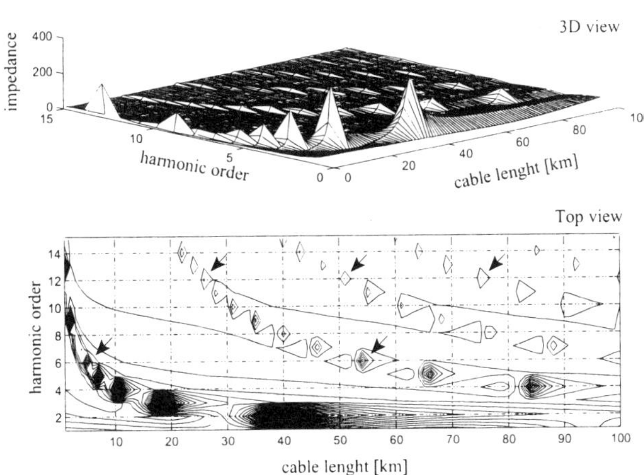

Fig. 11. DC side impedance response

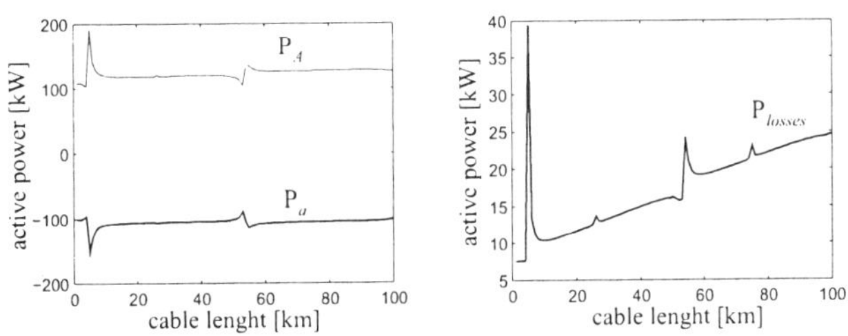

Fig. 12. Powers in the AC sides of the HVDC-VSC transmission station

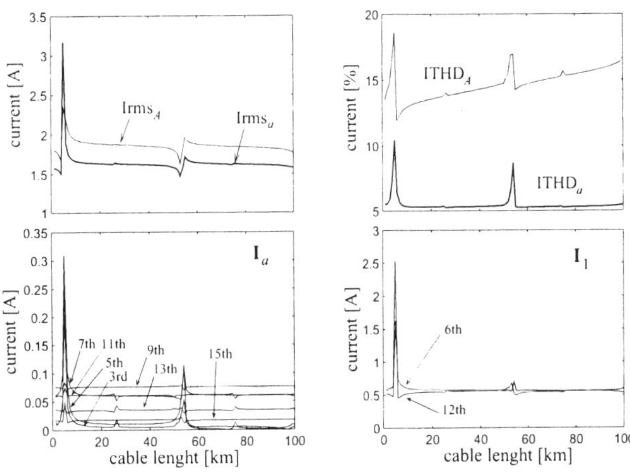

Fig. 13. Currents in the DC and AC sides of the HVDC-VSC transmission station

Fig. 11. Resonances in the DC side interact with the AC side currents producing resonance at other harmonic frequencies, as can be seen in the harmonic content of current $\mathbf{I}_a$. Resonance at other frequencies also exist and are due to the cross-coupling between harmonics in the impedance equivalent of the converters.

V. CONCLUSIONS

A flexible and comprehensive model for the periodic, steady-state operation of voltage source converters was presented in this paper. Switching functions based on complex Fourier series are used very effectively to represent the operation of the VSC. The modelling approach applies to a wide range of VSC configurations, e.g. single-phase, multi-phase, two-level, multi-level, single-converter and multi-converter. The response of the newly developed VSC model was compared against the steady-state response of a comparable VSC time domain model implemented in PSCAD/EMTDC. The VSC model was used as a building block to assemble harmonic domain models for the HVDC-VSC back-to-back and for the HVDC-VSC transmission station. The model of the HVDC-VSC transmission station goes on well with an accurate model of the cable since both models are in the frequency domain. Possible applications of the frequency domain models presented in the paper are: harmonic propagation studies, resonance prediction and harmonic stability analysis.

REFERENCES

[1] N.G. Hingorani, L. Gyugyi, *Understanding FACTS: Concepts and Technology of Flexible AC Transmission Systems*, The Institute of Electrical and Electronics Engineering, Inc., New York, 2000.

[2] Y.H. Song, A.T. Johns, *Flexible AC Transmission Systems (FACTS)*, The Institution of Electrical Engineers, England, 1999.

[3] B. Andersen, C. Barker, "A New era of HVDC?", *IEE Review*, March 2000, pp. 33–39.

[4] A. Lindberg, T. Larsson, "PWM and Control of Three Level Voltage Source Converters in an HVDC Back-to-back Station", *Proceedings of the AC and DC Power Transmission Conference*, 29 April–3 May 1996, pp. 297–302.

[5] Z. Zhang, J. Kuang, X. Wang, B.T. Ooi, "Force Commutated HVDC and SVC Based on Phase-Shifted Multi-Converter Modules", *IEEE Transactions on Power Delivery*, Vol. 8, No. 2, April 1993, pp. 712–718.

[6] J. Kuang, B.T. Ooi, "Series Connected Voltage-Source Converter Modules for Force-Commutated SVC and DC-Transmission", *IEEE Transactions on Power Delivery*, Vol. 9, No. 2, April 1994, pp. 977–983.

[7] G. Asplund, "Application of HVDC Light to Power System Enhancement", *Proceedings of IEEE Power Engineering Society Winter Meeting*, Singapore, January 23–27, 2000. paper 0-7803-5938-0/00.

[8] E. Acha, M. Madrigal, *Power Systems Harmonics: Computer Modelling and Analysis*, John Wiley & Sons, Chichester, 2001.

[9] B.T. Ooi, X. Wang, "Boost Type PWM HVDC Transmission System", *IEEE Transactions on Power Delivery*, Vol. 6, No. 4, October 1991, pp. 1557–1563.

[10] B.T. Ooi, X. Wang, "Voltage Angle Lock Loop Control of the Boost Type PWM Converter for HVDC Application", *IEEE Transactions on Power Electronics*, Vol. 5, No. 2, April 1990, pp. 229–235.

[11] Manitoba HVDC Research Centre, *PSCAD/EMTDC: Electromagnetic Transients Program Including DC Systems*, 1994.

[12] Z. Yao, B.T. Ooi, "Utilization of Cable Capacitance in GTO-HVDC Transmission", *IEEE Transactions on Power Delivery*, Vol. 13, No. 3, July 1998, pp. 945–951.

VI. BIOGRAPHIES

Manuel Madrigal was born in Purépero Mich., Mexico. Obtained his BSc and MSc from Inst. Tec. de Morelia and UANL, Mexico in 1993 and 1996, respectively. Since 1996, he has taught and conducted research in harmonics and power quality at Inst. Tec. de Morelia. At present he is carrying out PhD studies at University of Glasgow. Mr. Madrigal wishes to thanks the National Council of Science and Technology from Mexico (CONACyT) and the Inst. Tec. de Morelia for financial support to carry out PhD studies. e-mail: madrigal@elec.gla.ac.uk.

Enrique Acha was born in Mexico. He graduated from Univ. Michoacana, Mexico in 1979 and obtained his PhD degree from the Univ. of Canterbury, New Zealand in 1988. He holds the position of Reader at the University of Glasgow and is chairman of the inter-university Glasgow Strathclyde Centre for Economic Renewable Power Delivery. e-mail: e.acha@elec.gla.ac.uk.

FAULT PROTECTION OF METALLIC RETURN CIRCUIT OF KII CHANNEL HVDC SYSTEM

S Hara,
The Kansai Electric Power Co., Inc. Japan

M Hirose,
Shikoku Electric Power Co., Inc. Japan

M Hatano
Electric Power Development Co., Ltd. Japan

S Kinoshita,
Mitsubishi Electric Corp. Japan

H Ito,
Mitsubishi Electric Corp. Japan

K Ibuki
Mitsubishi Electric Corp. Japan

ABSTRACT

Kii channel HVDC transmission system (1400MW in the first stage) was commissioned in the year 2000 and transmits power generated in Shikoku Island to AC 500kV system in Honshu Island. The transmission line of approximately 100 km between the converter stations consists of two sections--undersea cable and overhead line sections. The transmission lines consist of 2 main lines and 2 return lines. The fault protection function was verified by making artificial grounding faults both for the main lines and return lines during a series of commissioning tests. The characteristics of the protection system of the return circuit including the MRTBs were evaluated from the tests. At some of the points along the line, the grounding arc extinguishes itself before closing MRTB and at other points, the grounding arc is cleared by closing MRTB. It was confirmed that the clearing time of the grounding arc was consistent with the system design parameters.

Keywords: *HVDC Link, Bipolar Metallic Return Operation, Fault Protection, Metallic Return Transfer Breaker, Grounding arc extinction*

INTRODUCTION

Kii channel HVDC transmission system [1] [2] (±250kV, 2800A in the first stage) was constructed and commissioned to transmit 1400MW power generated at Tachibana-Bay coal thermal power plant located in Shikoku Island to the AC 500kV transmission system in Honshu Island since June 2000. The bipolar double-metallic return system was employed for the reliable operation of the system that is shown in Fig. 1. Fig. 2 shows an outline this system particularly about the return circuit. The return wires is grounded at Anan Converter Station (which is located in Shikoku Island) and the undersea cable section spans 48.9 km. The cable is connected to the overhead lines at Yura Switching Station immediately after landing the Honshu Island. In most of the 50.9 km overhead line section, the return wires function as grounding wires. The overhead grounding wires are provided only in the 1 km section from the switching station and the converter station. At Kihoku C/S (which is located in the Honshu Island) each of the return wire is connected to a metallic return circuit breaker (MRTB) which is normally open.

When a grounding fault on the return wire takes place under the single pole operation, the return current is split into two parts--a part that goes through the ground return and the other part that goes through the return wire. When the grounding arc does not extinguish itself, the

MRTB that is located at Kihoku C/S is closed to assist the extinction of the grounding arc to avoid the continuous flow of the ground current. The time required for the extinction is dependent on various parameters such as the grounding resistance of the grid of the stations and the grounding arc characteristics.

Detail of the MRTB and its performance is to be discussed on another occasion and the confirmation of the characteristics will be briefly presented in the final chapter of this paper. Schematic circuit of the MRTB is given in Fig. 11.

The performance of the grounding fault clearing system was verified by making artificial grounding faults. The grounding faults were generated at Kihoku C/S and Yura SW/S during a series of commissioning tests. The protection sequence and the performance of MRTB were evaluated and the data were used to estimate the expected behaviour of the grounding arcs at an arbitrary point in the transmission lines. The record of an actual lightning fault that occurred last summer after commissioning was also investigated.

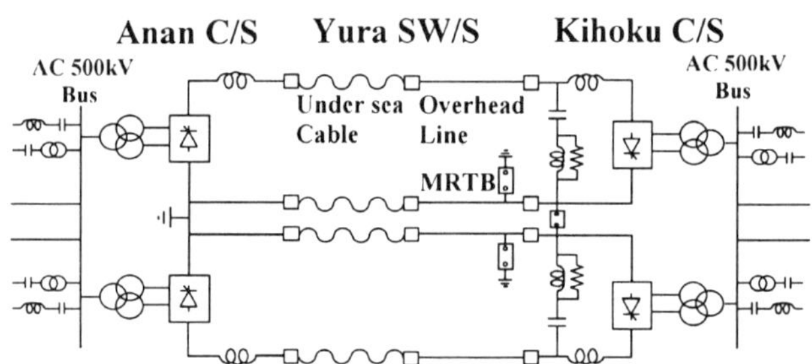

Fig.1 One-line diagram of Kii channel HVDC Link

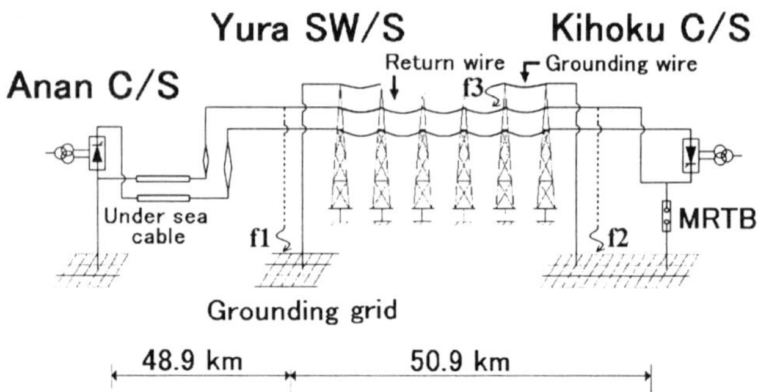

Fig.2 Return circuit of Kii channel HVDC transmission system. 3 points in the system which have records of grounding arcs are indicated by f1, f2 and f3.

AC-DC Power Transmission, 28-30 November 2001
Conference Publication No. 485 © IEE 2001

PROTECTION SYSTEM SPECIFICATION

Fig. 3 shows the operation sequence of the protection system. When the return line is grounded in the overhead line section, a part of the return current goes through the arc and the ground path while the lest of the current goes through the metallic return. The protection relay detects the fault and in approximately 300 ms from the initiation of the grounding arc, the closing command for the MRTB is issued. The MRTB is closed and the grounding arc extinguishes by transferring the return current to the MRTB and ground. A typical closing time of the MRTB is 100ms and a grounding fault will be extinguished in 30 ms or less after the MRTB's closing. After some time delay for the dielectric recovery of the arced point, the opening command is issued. The required time to recover the insulation of the line after arc extinction was estimated to be 50ms by a fundamental experiment[3] and the time delay from closing to opening of MRTB in this system was decided to be about 300 ms. Then MRTB is opened to commutate the return current back to the return circuit.

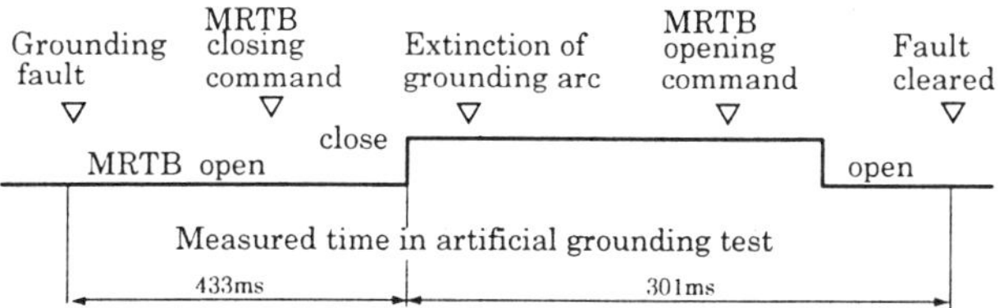

Fig.3. Protection sequence of the return circuit and the close/open position of the MRTB.

VERIFICATION TEST OF SYSTEM

Table 1 shows a summary of a series of artificial grounding fault tests conducted during the commissioning tests and an actual lightning fault recorded after commissioning. The fuse started grounding arcs were formed between the return wire and the grounding grid of the stations in the artificial fault tests. The grounding arcs in all the cases were cleared within the time expected by the system design specification. The artificial fault generated at Yura SW/S did not sustain itself and extinguished before the closing command to MRTB was issued in the tests in which the system current was less than 1680A.

Because of the change of the balance of resistance of the ground return and the metallic return the value of the grounding arc current as well as the time required for extinction changes depending on the distance from the Kihoku C/S.

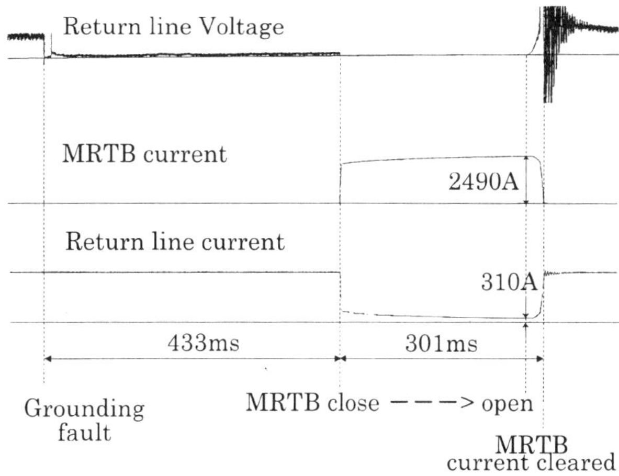

Fig.4 Oscillograms of an artificial grounding fault test. The grounding arc is generated at Kihoku Converter Station

Fig. 4 shows the voltage and current records of an artificial grounding fault generated at the Kihoku C/S. The current that flowed through MRTB after closing operation is measured to be 2490A at a system current of 2800A.

The value of the grounding current is estimated by the following relation.

$$\frac{i_g}{i_l} = \frac{r_l}{r_{gkihoku} + r_{ganan}}$$

,where r_{ganan} / $r_{gkihoku}$ show the grounding resistance of the grid at these stations and r_l shows the resistance of the return wire. From this result the grounding resistance of the grid at Kihoku C/S is estimated to be 0.1 Ohm. Fig.5 shows an oscillogram obtained when an actual lightning fault occurred at the transmission tower 18km away from Kihoku C/S. The grounding resistance of the grid is estimated in the same way was 0.2 Ohm in this case. In this paper, the expected grounding arc extinction was evaluated with a grid resistance of 0.1 Ohm, as this value gave a better fit to the artificial fault test results and the results with higher resistance showed little difference in the extinction time estimation.

It was not practical to measure the grounding arc current in the tests but it was quite critical to know the arc

Table 1 Brief description of the return circuit protection operations

Grounding fault position	System current	Results
Artificial fault at Yura SW/S	280A (0.1 pu)	Grounding arc current: 70A. Grounding arc cleared before the closing command is issued.
	1680A (0.6 pu)	Grounding arc cleared before the closing command is issued.
Artificial fault at Kihoku C/S	2800A (1.0 pu)	Grounding arc current: 2000A. The fault was cleared in 3 ms after MRTB is closed.
Actual lightning fault at a transmission tower 18km away from Kihoku C/S	2800A (1.0 pu)	Grounding arc current: 300A. The fault was cleared after MRTB is closed.

extinction time to have a clear understanding of the characteristics of the typical arc. It was possible to evaluate it through a detailed observation of the MRTB current, more specifically, the initial current build up process. Fig.6 shows the build up stage of the current through MRTB shown in Fig.4. The di/dt showed a stepwise change at around 2 ms of the closing of the MRTB. This is interpreted as the moment of the grounding arc extinction.

The behaviour of the grounding arc (particularly the extinction time at an arbitrary point in the system) that interferes with the grounding current through the MRTB can be simulated by assuming the grounding arc characteristics that is solved with the circuit equations of the return circuit.

The arc characteristics are usually given by a function of i_a and r_a as follows:-

$$\frac{dr_a}{dt} = f(i_a, r_a), \quad r_a : grounding\ arc\ resistance$$
$$i_a : grounding\ arc\ current$$

It was convenient to assume that the grounding arc is represented by a simple Mayr type arc that is usually described by 2 parameters--arc loss and arc time constant. The simulation work with a varied value of related resistance and inductance as well as the assumed characteristics of the grounding arc could define the typical arc characteristics to be used for the further evaluation work of the system behaviour.

Another critical parameter for estimation was the grounding resistance of transmission towers. Fortunately, the data of the actual fault that took place on a transmission tower was available for this estimation work. From the value of the arc current which was

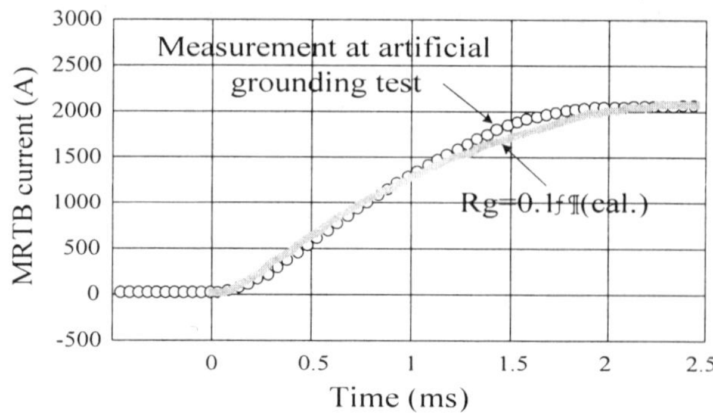

(a) MRTB current wave form

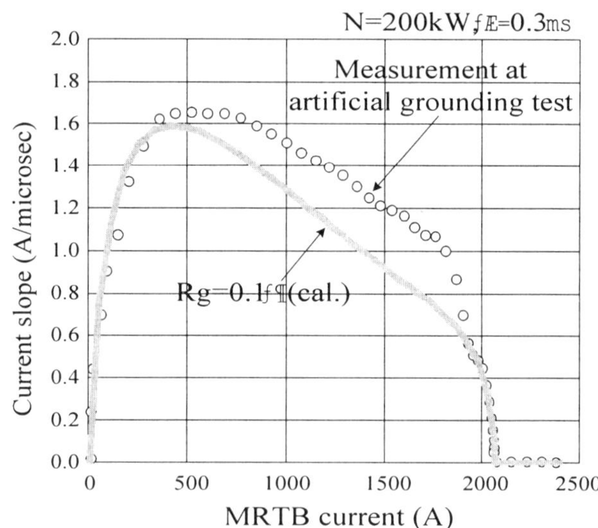

(b) MRTB current vs. di/dt

Fig.6 The detail of the current through MRTB of the oscillogram in Fig.4 together with a simulation result to estimate the arc characteristics and the related parameters.

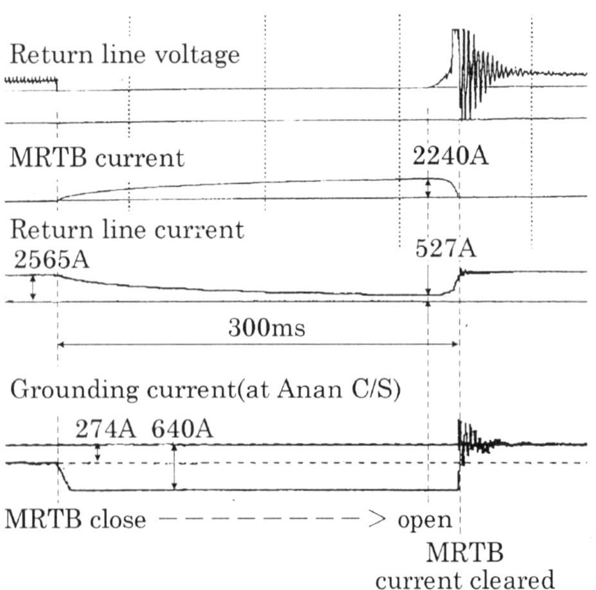

Fig.5 Oscillograms of an actual lightning fault occurred at a transmission tower.

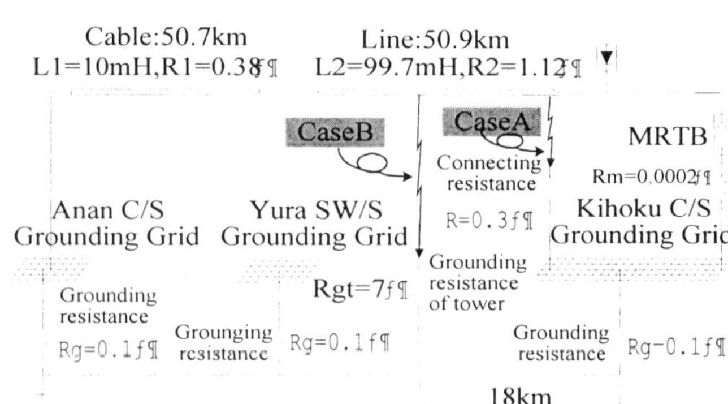

CaseA: Artificial fault at Kihoku C/S
CaseB: Lightning fault at a transmission tower 18km from Kihoku C/S

Fig.7 A calculation model circuit of the return circuit. Estimated values of grounding resistance and related circuit parameters are shown.

measured at Anan as 274 A, the grounding resistance was estimated as 5 to 7 Ohm.

These estimation works provided the necessary basis for the overall evaluation of the behaviour of the system. Fig.6 shows the equivalent circuit of the return circuit. The inductance and resistance of the overhead lines and cables are also included in it.

The simulation of the fault at Kihoku (which is shown in Fig. 8.) and at a tower 18 km away showed arc extinction time of 3.7 ms and 3.6 ms respectively. The extinction time of the arc at a tower was not estimated very clearly from the recorded oscillogram but 3.6 ms is a value that can convince to be consistent with it.

Rg=0.1 Ω, N=200kW, θ =0.3ms

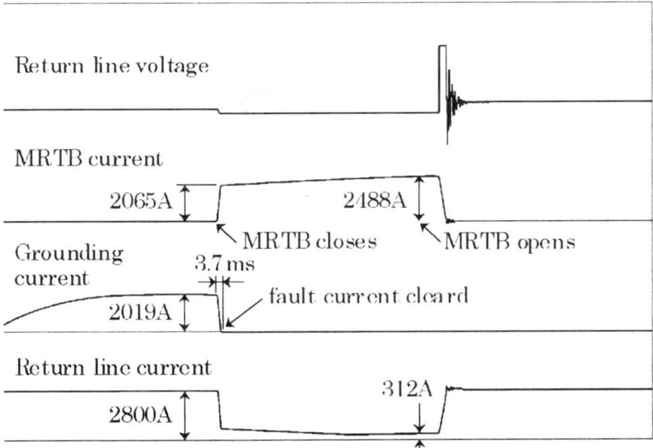

Fig.8 Voltage and current behaviour of a fault at Kihoku C/S by the circuit shown in Fig.7

ESTIMATION OF THE CHARACTERISTICS

The consistency of the extinguishing time of the grounding arc with the system design values were confirmed using circuit conditions and arc characteristics obtained through the experimental data by assuming the grounding fault at various points in the return circuit.

Fig.9 and Fig.10 shows the grounding fault current and the arc extinguishing time calculated at various fault points along the return line when a system current is 2800A.

The grounding arc current at Kihoku C/S is estimated to be 2019A and at the section where the overhead grounding wires are installed, the grounding current stays approximately at this value. The arc extinguishing time is estimated to be less than 10 milliseconds after closing operation of MRTB. The grounding current expected at Yura SW/S and the neighbouring section is around 1000A.

In the section where the return wires are to function as grounding wires, the grounding current decreases with an increased distance of a fault point away form Kihoku C/S. The grounding arc current reduces down to 500A or less because the grounding resistance of the

transmission

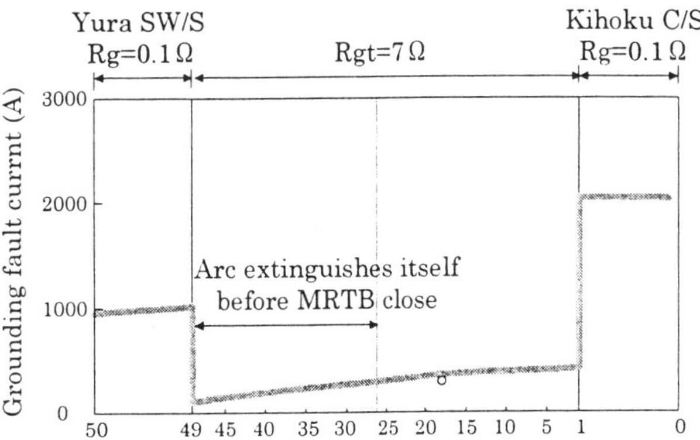

Fig.9 The grounding current as a function of the fault point away from the Kihoku Converter Station

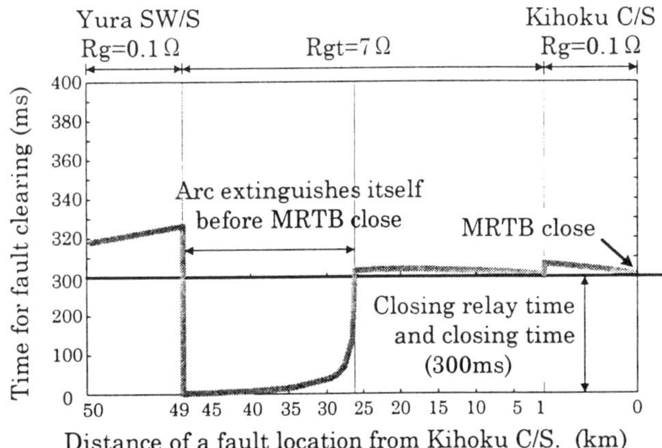

Fig.10 A fault clearing time as a function of the fault point from the Kihoku Converter Station

tower is high in comparison with that of the stations. As the result, the fault clearing time becomes relatively short typically 4 ms or less.

In the section far from the converter station, the expected grounding current decreases further and the arc is likely to extinguish by the time when the closing command for the MRTB from the protection relay is issued. This was expected in the section further than 26 km from Kihoku C/S. This estimation is consistent with the observation of the artificial fault test at Yura Switching Station.

The investigation work showed that the behaviour of the arc extinction can be classified into following three patterns.

1. The grounding arc is relatively high and more than 2/3 of the system current goes through the grounding arc. This type of arc is extinguished by closing operation of MRTB with relatively long arc duration.

This mode appears when the fault occurs at transmission towers located close to Kihoku C/S or Yura SW/S where overhead grounding wires are provided. The grounding current is relatively high because of a low grounding resistance of the stations. The fault clearing time from MRTB closing is around 10ms.

2. The grounding current of the arc is less than the value in the neighbouring sections of the station but it is extinguished by closing operation of MRTB with relatively short arc duration.

This mode appears when the fault occurs at transmission towers of the overhead line from 1km to 26km from Kihoku C/S. The grounding current is smaller than 500A as the grounding resistance of the tower is estimated as 5-7 Ohm. The fault clearing time from MRTB closing is a few milliseconds.

3. The grounding current is still lower and it extinguishes itself before closing operation of MRTB

This mode appears when the fault occurs at transmission towers of the overhead line more than 26km away from Kihoku C/S. The grounding arc does not sustain itself because the impedance of the cable is smaller in comparison with that of the grounding resistance of the tower. The arc extinguishing time ranges from a few milliseconds to 200 milliseconds which is less than the time of MRTB closing command delay.

PERFORMANCE OF MRTB

MRTB is an interrupter that is not covered by the standards for AC interrupter and it has two duties-- closing and opening. Closing duty is with the line drop voltage of the return circuit and the resulting current is expected to be the system current at its maximum. The opening duty requires to interrupt the current in the MRTB and commutate it back to the return wire. The MRTB shown in Fig.10 consists of a disconnecting unit and an interrupter unit paralleled by a reactor and a capacitor connected in series. A surge suppresser is also connected in parallel to the interrupter unit. A current interruption process is initiated by separating the arc contacts of the interrupter and a current oscillation continues to grow and eventually leads to a current zero.[4]

Since the interrupting process of the MRTB is not affected by external circuits such as the transmission line length and grounding resistance, the MRTB can be verified to interrupt the maximum overload rating of 3500A within a designed breaking time at factory. Commutation process is also relatively simple because the current is pushed back to the return circuit established in the capacitor parallel to the interrupter. Therefore, the main point to be checked about the operation of the MRTB is the consistency of the operation with the estimation in the factory evaluation stage.

This was confirmed mainly by comparing the dissipated energy by the arc that is directly related to the growth of the oscillation of the interrupted current and the

interruption time. Fig.12 shows the average arc energy evaluated in 3 final loops of the oscillation before interruption and the arcing time of an interrupter measured as a function of the breaking current. The arc dissipated energy (arc loss) was estimated to be 10MW at a breaking current of 3500A and the arcing time was around 30ms. The arcing time was around 18ms in the artificial fault test at Kihoku C/S. A typical oscillogram in Fig.13 is recorded at the tests and Fig.14 shows an artificial grounding arc in the test. From the data, the energy was estimated to be 4-5MW at a breaking current of 2490A.

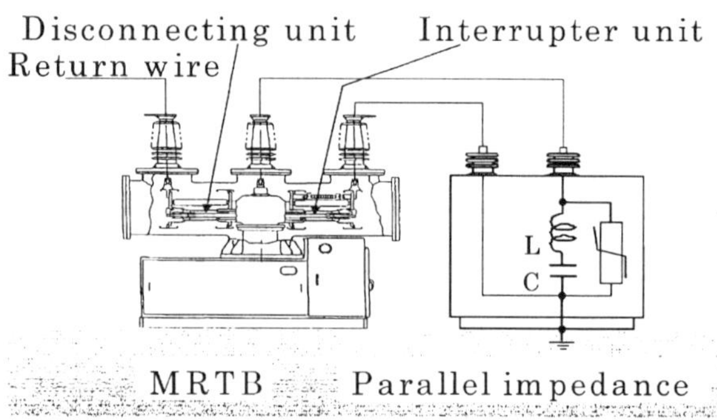

Fig.11 Construction of MRTB consists of an interrupter unit with a parallel impedance and a disconnecting unit

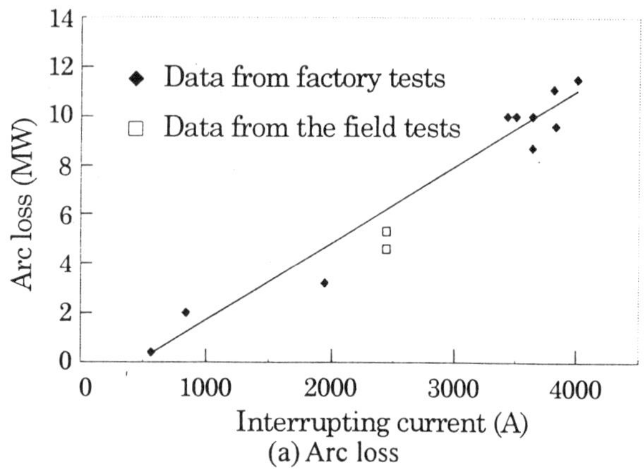

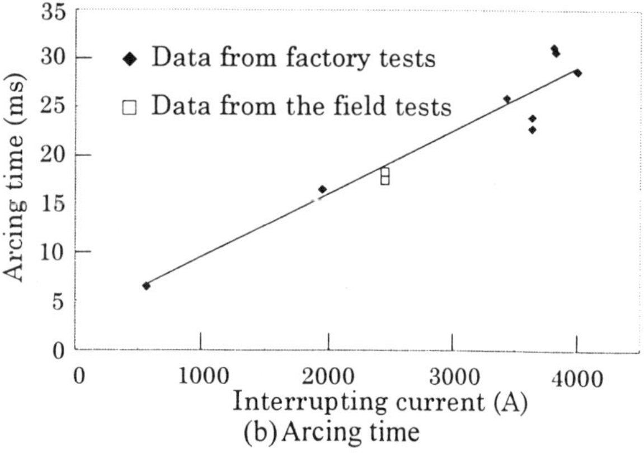

Fig.12 Arc loss and arcing time of an interrupter evaluated from data measured in factory test and artificial fault tests

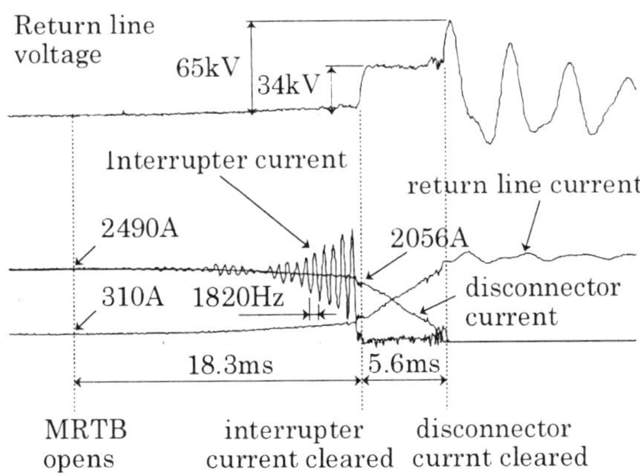

Fig.13 Voltage and current behaviour of MRTB obtained at an artificial fault test at Kihoku Converter Station

Fig.14 Grounding arc initiated by a fuse in the artificial grounding test at Kihoku C/S

CONCLUSIONS

The performance of the return circuit protection was verified by a series of the artificial fault tests and actual lightning fault. The effectiveness of the protection sequence and the performance of MRTB were evaluated from data and the system including the MRTB was confirmed that the fault is expected to be cleared wherever it may take place along the return lines. From the investigation of data the behaviour of arc extinction was classified into three patterns. It was concluded that the protection scheme and the equipment for the system are consistent.

References

[1] Y. Sekine et al., "Kii Channel HVDC link between Shikoku and Kansai Electric Power Companies by Submarine Cables", CIGRE Tokyo Symposium 1995

[2] J. J. Vithayathil et al., "HVDC Circuit Breaker Development and Field Tests", IEEE transactions on Power Apparatus and system, Vol.PAS-104, No.10, pp.2693-2705, 1985

[3] K. Sunabe et al., "Influence of starting voltage waveform on time sequence to restart DC transmission lines", T.IEEJ, Vol.119-B, No. 10, pp.1096-1101, 1999

[4] H. Ito et al., "Instability of DC Arc in SF6 Circuit breaker", IEEE transactions on Power Delivery, Vol.12, No.4, pp.1508-1513, 1997

Thyristor Protected Series Capacitor: Design Aspects

Lutz Kirschner[1], Jeff Bohn[2], and Dr. Kadry Sadek[1]

SIEMENS, Germany [1], Southern California Edison, USA[2]

1 Abstract

In spring 1998 Southern California Edison requires to refurbish some older series capacitors in the Vincent – Midway corridor located near Los Angeles, California, USA in the southwest of the pacific ac-system tie. Due to the heavy load and ac-fault current requirements it has been decided not to establish a conventional MOV and gap protected series capacitor but a new type of thyristor protection. In June 1999 SIEMENS company has been ordered to refurbish an existent capacitor by a Thyristor Protected Series Capacitor (TPSC) as a turnkey project. The project time schedule considers commercial operation mid April 2000.

2 Summary

Series capacitor installations have been established in high voltage ac-system networks to increase the power transmittal across existing ac-lines. Also they improve the ac-system steady state stability due to the line impedance reduction. In case of ac-system faults resulting in high ac-line currents the capacitor must be protected against overvoltages. The conventional way of protection is the use of Metal Oxide Varistor (MOV) across the device. Nowadays installations require a great amount of energy dissipation capability resulting in a huge number of MOV housings in parallel because of estimated future ac-system requirements. The paper describes a new approach to eliminate the MOV housings by thyristor valves being able to withstand ac-system fault currents and a control strategy for adaptive valve firing and blocking for capacitor protection on the basis of the TPSC Vincent scheme at Southern California Edison in California, USA.

Keywords

TPSC, Thyristor Protected Series Capacitor, TCSC, Thyristor Controlled Series Capacitor, LTT valves, MOV, external faults, internal faults

AC-DC Power Transmission, 28-30 November 2001
Conference Publication No. 485 © IEE 2001

3 SC Development

The development of series capacitor systems is always coupled with the development of the protective bypass-devices like gap or breaker. Semiconductor played no important role in this development until now because of the relative high investment costs. Nowadays semiconductor devices like thyristors became a "common" application and therefore also the price decreases.

Because the SC is connected to the line it is exposed to an ac-short circuit fault current resulting in a high overvoltage across the device. Therefore the development of FSC always has the goal to protect the series capacitor against overvoltage by bypassing.

In the beginning bypassing has been considered by mechanical switches. The elementary problem was that closing of the device was rather slow. Therefore a gap was connected across the breaker for fast bypassing. To limit the capacitor discharge current a damping reactor also was added in series to the gap. This design leads to a fast capacitor bypass and an adequate overvoltage protection but tends to inexact firing and poor maintenance due to arc erosion. To avoid excessive arc erosion the breaker closes after gap firing. But it was still not desirable to lock out the SC for longer times than the fault duration and need to be reinserted manually or automatically.

With improved MOV development the idea was born to limit voltage across capacitor by a MOV. The nonlinear clipping voltage characteristic of the MOV leads to an acceptable economic overvoltage across the capacitor which establishes at about 2.3pu in most applications. The bank is bypassed by a breaker or gap only in case of severe faults where the MOV tends to overload. Otherwise the capacitor is only bypassed for the fault duration by the MOV and smoothly inserted to the system when the fault disappears. It must be considered that the MOV dissipated energy is very sensitive on the ac-system short circuit level: Due to the non-linear VI-characteristics of the MOV an increase of short circuit level lead to a high increase of MOV energy dissipation. Due to the increasing power demand and high future short circuit levels the actual tendency for installing MOVs on the platform results in a huge amount of MOV housings. Space must be provided on the platform resulting in a inconvenient platform size. In addition long cooling down times of the MOV must be considered after a fault that stresses the MOV with a nominal energy discharge.

With decreasing costs for semiconductor devices it therefore became of interest to replace the MOV by a thyristor switch.

4 Benefits using TPSC

An ac-fault current flowing through a MOV always leads to a high energy dissipation of the MOV. The MOV heats up heavily. Due to an upper temperature limit the MOV must cool down before the next current stress can be absorbed. Cooling down requires a big amount of time, time constants of several hours are known. A thyristor provides a high current carrying capability in combination with a low conduction loss. Therefore heating as well as cooling down will be performed much faster than for a MOV. Also there is no stringent design limit for a maximum ac-short circuit level: SIEMENS high current thyristors provide a current carrying capability that withstands the highest ac-system short circuit currents in high voltage ac-networks. A closer look to the single line diagram of the TPSC installation implies the impression of a Thyristor Controlled Series Capacitor (TCSC). In fact the TPSC can be upgraded to full TCSC operation by adding the especial control features and valve water cooling equipment. For customers benefit this upgrade can be foreseen in the TPSC design stage. The upgrade itself can be carried out at a later stage during commercial operation. Compared to a conventional MOV solution the TPSC alternative requires less space on the platform. The thyristor switch is located in a container on the platform.

5 Single Line Diagram

The basic single line of a TPSC Installation is shown Figure 1. The system can be connected to the line by several disconnect switches and the bypass switch. The series capacitor is build up in H-connection to detect capacitor unbalances. The thyristor branch and reactor are connected in parallel to the capacitor, to provide the fast temporary bypass. For permanent lockout and bypass, a switch with an associated damping circuit is located across the capacitor. To protect the equipment on the platform, current transformers for measuring ac-line currents, capacitor currents unbalance currents and platform currents are provided.

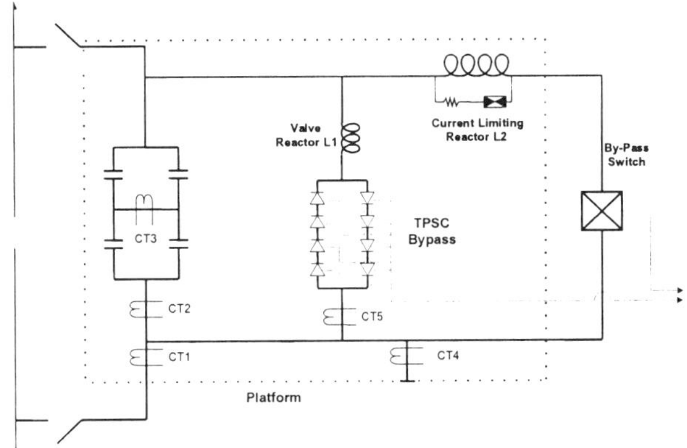

Figure 1: TPSC Single Line Diagram

6 AC-System Fault Cases

Depending on the severity of the faults, the fault locations in the ac-system are classified in two types. Depending upon the type of fault, different protection actions are initiated:

Internal Faults:

Internal faults are those occurring within the series compensated line terminated by breakers. If an internal fault is detected, the thyristor valves are fired and simultaneously the series capacitor gets bypassed by a switch.

External Faults:

External faults are those occurring outside the series compensated line terminated by breakers. In this case, only the valves receive a firing command. The thyristor valve and reactor must be designed to withstand external faults without damages, as the series capacitor will not be locked out by the switch.

7   TPSC Protection Strategies

In case of normal operation, the valves are blocked. The ac-line current flows through the capacitor and is processed by the protection system. If an ac-system fault occurs the capacitor has to carry the current until the thyristor bypasses the capacitor.

If the ac-line current exceeds a threshold value, a fault situation is identified by protection and a valve firing sequence is initiated.

After valve firing, the ac-line fault current splits between the capacitor and the thyristor branch. Due to the fact that the thyristor branch impedance is small compared to the capacitor impedance, most of the fault current flows through the thyristor branch.

The ac-line current peak value is used as a criteria to distinguish between an internal and external fault. Depending on this value, different reactions are initiated for external and internal faults.

As long as the fault is not cleared, the valves continue to get firing pulses. Fault clearance is detected as soon as the current is below a threshold value. In this case, the valve blocks and the capacitor is inserted into the ac-system. The resulting swing current flows through the capacitor and after the decay of this current it settles down to the overload value, following the disturbance.

8   Component Ratings for TPSC Vincent

For a cost and performance optimized design, various studies with different system conditions have been carried out. To determine the component ratings under worst case conditions, not only steady state voltage and current stresses need to be considered but also the transient stresses during severe ac-system faults have to be investigated in detail. The complete Vincent-Midway corridor has been modeled in the PSCAD/EMTDC program including all parallel lines and existing fixed series capacitors (Figure 2). A detailed fault current analysis has been carried out. Different fault locations as well as different ac-system conditions have been simulated. The point of fault insertion has been varied over a period of time to establish the worst case of current stress which the valves have to withstand.

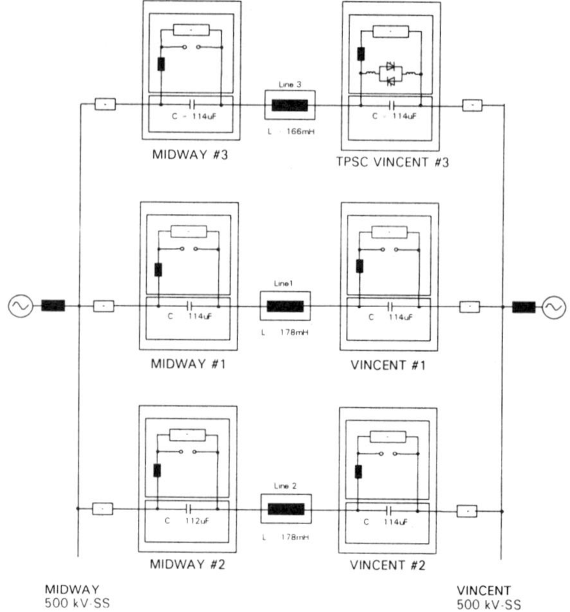

Figure 2: Vincent Midway-Corridor

8.1   Capacitors

The series capacitors have a value 23.23 Ohms and consist of one segment in H-configuration. Four stacks constitute such a configuration. The total number of installed capacitor units is 768 for all three phases. They are of outdoor fuseless type.

The rated voltage for the capacitor is 56 kV rms at a rated current of 2400 Arms. The overload requirements are 3500 Arms for 30min. The capacitor is designed for a protective level of 2.3pu = 180 kVpeak.

8.2   Thyristor Valves

8.2.1 Thyristor Main Design Features

The thyristor valve technology is based on the well proven design of water-cooled thyristor valves for SVC- and TCSC-applications which have been supplied since many years. Further improvements have been achieved by introducing thyristor valve technology with direct light triggered thyristors (LTT), which represent the most up-to-date technology. The main advantages compared to conventional valves with electrically triggered thyristors are:

- The entire individual thyristor protection is integrated in the thyristor itself by means of a forward overvoltage protection.

- The optical firing pulses are generated in the valve base electronics and transmitted directly to the gate. This results in an improved turn-on characteristic of the LTT.

- Firing pulses are available any time even at undervoltage conditions of the valve.

No thyristor electronics (gating units) with power supply and protective firing circuit on high potential are required, leading to a completely passive design.. Only a voltage divider for thyristor voltage monitoring is placed across each thyristor. This minimizes wiring and eliminates a potential source of partial discharge and electromagnetic interference (EMI) which leads to an inherently higher reliability of LTT valves.

8.2.2 Thyristor Requirements

Under steady state condition the valves are blocked. The valves have to withstand only voltage stresses. These no load voltage losses are dependant on the valve snubber circuit. Through optimized circuit design, snubber circuit losses are kept low thereby allowing an air cooling system.

However, it is possible to extend the air cooling to a water cooling system in case the TPSC is upgraded to a full TCSC operation.

In case of ac-system faults, the valves get fired . As a result, the thyristors have to withstand the discharge current of the capacitor as well as the superimposed power frequency ac-line current. Coordinating the line current with the swing current the valves will immediately fire after exceeding 6.8kA. The adaptive valve blocking sequencer will not block the valves before the line current falls below 4.8 kArms for a time of 50 msec. Simultaneously the ac-line current is observed and if the fault current exceeds 17 kApeak an internal fault is interpreted. In this case a signal to close the bypass switch is initiated and the switch closes within 50 msec.

Internal faults at the capacitor terminals leads to the highest current stress which the thyristors have to withstand. Figure 3 shows the current through the valves, voltage across the valves and the current through the bypass switch for such internal faults when the bypass switch closes 50msec after occurrence of the internal

fault. The current carrying capability of the thyristors is much higher than the currents through them under these worst case conditions.: The fault duration may last three times longer than stated in this figure without overstressing its capabilities. The surge forward current can be up to 110kApeak.

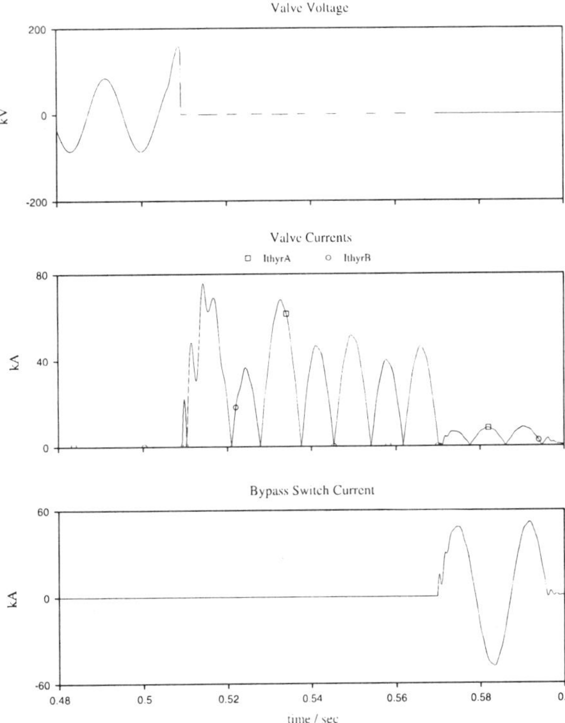

Figure 3: Internal fault

In case of an external fault, the number of thyristors connected in series must be so chosen that they can withstand the voltage appearing across them immediately after they block. While deciding on this number, one has to consider that the blocking capability of the valves is influenced by the losses., which are due to the valves conducting during external fault conditions..

Further, the TPSC Vincent is able to insert at an overload line current of 3500 Arms (Figure 5). To optimize the number of thyristors in series required for the TPSC Vincent,a small MOV has been used to limit the insertion voltage to 180 kVpeak. The MOV energy rating is fairly low.

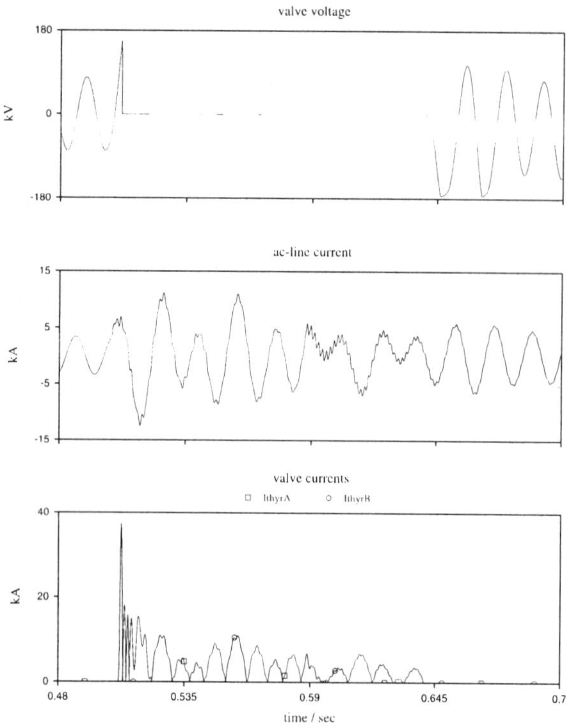

Figure 4: External fault

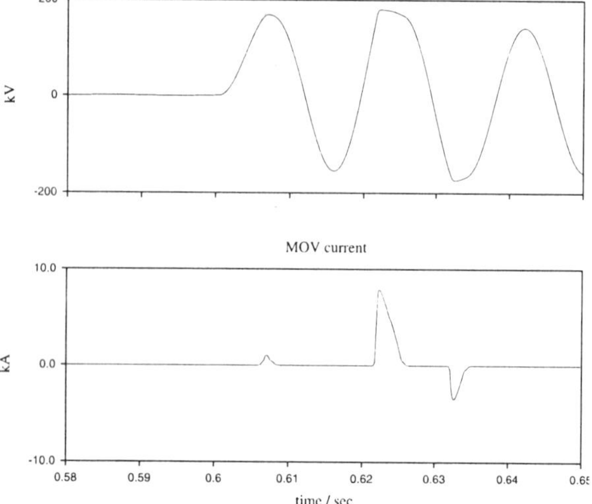

Figure 5:   Insertion at a line current of
            3500Arms

8.3   Reactor Requirements

The reactor is stressed by the valve currents in
both directions (Figure 6). A 2mH inductance is
chosen to limit the maximum rate of rise of
current acceptable to the thyristors. To dampen
the high frequency components occurring
immediately after valve firing, a quality factor of 4
has been chosen. Since the current through the
reactor under steady state conditions is

negligibly small, it need not be designed for this
condition. The reactor as well as the thyristors
are designed to withstand not only a worst case
internal fault until the bypass switch closes but
also for a duration which is three times longer
(ca. 150msec).

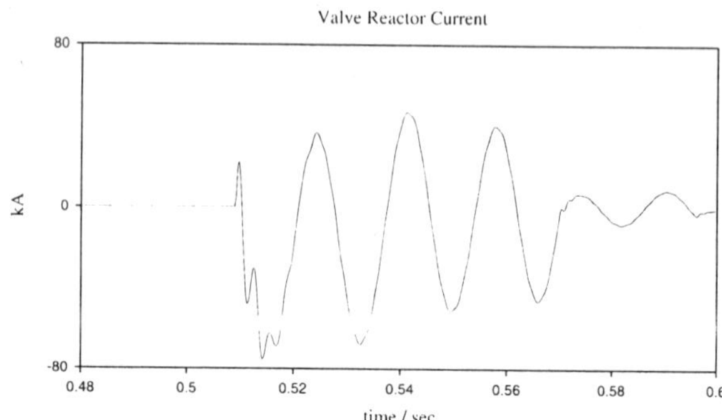

Figure 6: Reactor Current for Internal Fault

8.4   Bypass Switch Capabilities

The 550kV SIEMENS SF6-bypass switch is
designed as three single-phase units, one for
each phase with an interruptor voltage of 245kV
and a closing time of 50 ms. It consists of
porcelain insulator columns with a hydraulic
opening mechanism. The switch is also used for
insertion and disconnection of the bank. In case
of excessive valve or reactor duty cycle on an
external fault, the bypass closes as well as
during internal faults.

8.5   Damping Circuit Design

Current through the damping circuit flows when
the bypass switch closes on internal faults or
excessive valve or reactor duty cycle. It is
designed for a maximum capacitor voltage
discharge and a continuous current of
3500Arms.

To limit the current peaks during capacitor
discharge, a reactor of 500μH has been chosen.
This results in a discharge frequency of 666Hz.
The appropriate damping is achieved by a
resistor of 7ohms located in parallel to the
damping reactor. To avoid a steady state current
drain in the resistor, a small spark gap is also
included in series with the resistor: The small
gap fires for voltages exceeding the overload
requirements of the capacitor (Figure 1).
Therefore the resistor need not be rated for
steady state conditions. The impulse energy

rating of the resistor is sufficient for a duty cycle of two consecutive discharges.

9 Layout Considerations

The TPSC components are installed on a platform which is connected to the high potential of the ac-transmission line. This design leads to the lowest creepage distance and flashover voltages across the components. The platform itself is built up on reinforced structures of porcelain insulators to provide an appropriate line to earth creepage distance with a height of 6 meters. The capacitor is built up in one segment with four stacks in an "H"-connection to detect unbalances. The damping circuit resistor is located within the damping reactor. The valves are installed in a container located in the middle of the platform. The valve consists of 3 modules with 14 thyristors in each direction (Figure 8). The bypass switch is not installed on the platform but located beneath on the ground. Due to the optimized design, the platform size is only 15m x 8m and has a total weight of about 30 tons. Figure 7 show the Platform Layout and associated bypass switch for one Phase. The control cubicles are located in a container on ground near the platform.

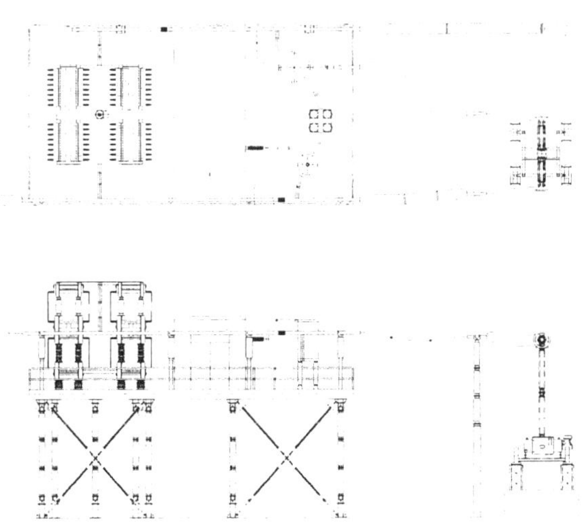

Figure 7: TPSC Vincent Platform Layout

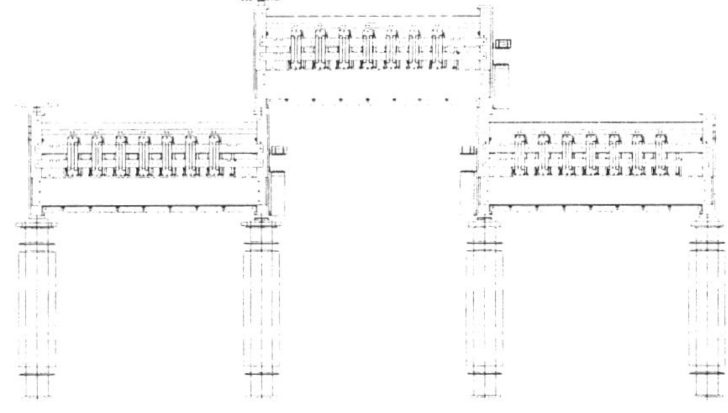

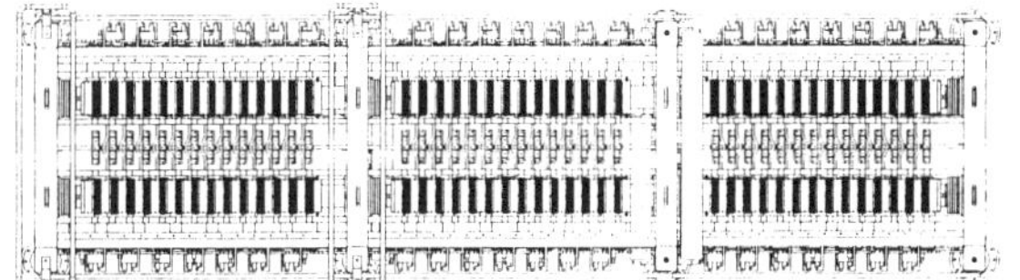

Figure 8: Valve Layout

10 SCE-TPSC Vincent Main Data

The 550kV TPSC Vincent has a three phase power rating of 401MVAr. The nominal impedance is 23 ohms at a current of 2400 Arms. Figure 9 shows a photograph of the installation in operation. The series capacitor is designed to operate safely at 3500 Arms for 30min. In case of an ac-fault the valves are fired immediately when the line current exceeds 6800 Apeak. The thyristor valve consists of 42 direct light triggered thyristors in series per phase and direction. The surge current capability of each thyristor is 110 kApeak. The valve damping reactor has an inductance of 2mH to limit the rate of rise of valve current.

Figure 9: TPSC Vincent Scheme

11 Control and Protection

The implemented control as well as the according tests and are documented in detail in a separate paper [1].

12 Conclusion

This paper discusses the design considerations of the components of a Thyristor Protected Series Capacitor installation (TPSC). The protection strategy for different system fault scenarios are discussed. and dictates the requirements for the component rating. Details of the TPSC Vincent scheme in commercial operation since April 2000, are presented. The major benefits of fast cooling down times resulting in faster re-insertion after ac-faults compared to MOV-protected schemes as well as the upgrade-capability to full TCSC functionality during commercial operation makes the TPSC a flexible transmission component and a competitive alternative to MOV-protected schemes. Additionally platform space can be reduced considerably by replacing MOV-banks by thyristors.

13 References

[1] Design Aspects of a Thyristor Protected Series Capacitor – Part 2 : Control and Protection Details :

To be published.

THE MOYLE HVDC INTERCONNECTOR
PROJECT CONSIDERATIONS, DESIGN AND IMPLEMENTATION

C. Harvey[1], K. Stenseth[2] and M. Wohlmuth[3]

(1) Northern Ireland Electricity plc, UK (2) Nexans Norway AS, Norway (3) Siemens AG, Germany

INTRODUCTION

Northern Ireland Electricity (NIE), the electricity transmission and distribution company for Northern Ireland, signed agreements in 1990 with ScottishPower, its counterpart in the south of Scotland, for the construction of a High Voltage Direct Current (HVDC) interconnector, known as the Moyle Interconnector, between the companies' transmission systems. After a protracted period in which all the necessary consents were obtained, contracts have been awarded and construction is in progress.

NIE is a member of Viridian Group PLC. In 1999, Moyle Interconnector plc (Moyle), another member of the Viridian Group was established to construct the link. The aim of the Moyle Project is to deliver to customers the benefits of an interconnector that is economically viable, technically feasible and environmentally acceptable. This required:

i. An appraisal of the engineering options available in order to optimise an electrical interconnection between the electricity systems of Scotland and Northern Ireland that meets the requisite reliability and availability.

ii. Minimising the environmental impact of the electrical interconnection by identification followed by elimination or acceptable mitigation of the impact.

iii. Going to the HVDC market to obtain a competitive price for an optimised solution that will stand the test of an economical appraisal by the regulatory authority, which allows a revenue stream to meet the capital investment in an evolving electricity market.

The Moyle Interconnector will transmit power between the electricity systems in Ireland and Great Britain from the end of 2001, see Figure 1. It will provide Northern Ireland with an important new source of electricity supply, promoting competition in the emerging markets in Northern Ireland and the Republic of Ireland and enhancing security and quality of supply. The overall project costs are approximately £150 million. The project is sponsored with a contribution of £52.5 million by the European Regional Development Fund.

The implementation of the Moyle Interconnector involves the construction of 64 km of 275kV a.c. overhead line in Scotland, HVDC converter stations in Scotland and Northern Ireland, underground d.c. cables from each station to the coast and the installation and protection of 2x55km of subsea d.c. cable (Figure 2).

The elements that make up the Moyle Interconnector project are required to be technically feasible, reliable and economically viable. They also require to be designed, constructed and operated taking into account environmental considerations. As imports over the interconnection will represent a significant proportion of the Northern Ireland system demand, the design of the link must ensure that quality of supply experienced by customers will be at least as good with the interconnection in service as prior to its introduction. A main criterion influencing the Interconnector design was that its performance in reliability and availability terms should be no worse than that of an equivalent sized onshore modern generating unit.

The implementation of the Moyle Interconnector requires it to be economically viable with a balance between mitigation of any potential environmental impact in conjunction with the requirements of statutory bodies and a technical solution that delivers the requisite reliability and availability.

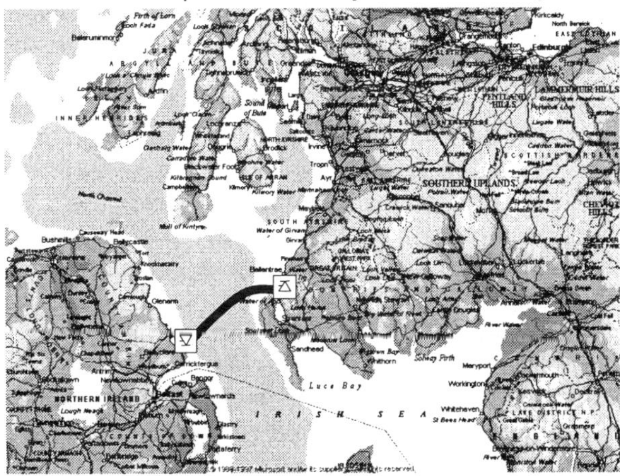

Figure 1: The Moyle Interconnector is connecting the island of Ireland to the British grid and, through it, to the central European grid

AC-DC Power Transmission, 28-30 November 2001
Conference Publication No. 485 © IEE 2001

PROJECT CONSIDERATIONS

The electricity system in Northern Ireland, serving a population of around 1.6 million, is a self-contained system for the generation, transmission and distribution of electricity. The North-South Interconnector between the systems of Northern Ireland and the Republic of Ireland was restored in March 1995 after being out of service for twenty years. The restoration enables the Northern Ireland system to become part of a system approximately three times the size when isolated. Therefore the restrictions determining the maximum capacity of a single infeed into an isolated system were relaxed. Future enhancement of the connections between Northern Ireland and the Republic of Ireland will further relax these restrictions. The electricity system in Scotland is part of the large interconnected British and Western European systems, UNIPEDE Group (1), Rainey et al (2).

Electrical interconnection between Scotland and Northern Ireland has been under consideration for more than thirty years. Early feasibility studies established that direct current interconnection was superior to alternating current interconnection from both the technical and the economic points of view. There have also been considerable technical advances in direct current technology over the intervening period so that present day advantages of d.c. interconnection outweighs those of a.c. interconnection.

The design for the Moyle Interconnector was influenced by both environmental and system considerations. From the viewpoint of the Northern Ireland system, the main consideration was obtaining the reliability and availability required by an island system dependent on the d.c. link for approximately 20% of its total demand at any time.

The location of the 275kV transmission systems and associated substations in Scotland and Northern Ireland determined the general area in which interconnector route options could be developed. Various undersea cable routes between Northern Ireland and Scotland were ruled out for different reasons, such as high sea currents, deepness of the seabed, exposed bedrock indicating a very hostile environment for undersea cables, post war dumping of munitions as well as for reliability and environmental reasons regarding the enormous length of overhead line sections required in Scotland and Northern Ireland. The overhead line route selection, the seabed route selection and the landfall evaluation, together with the underground cable route selection and converter station site selection were developed and co-ordinated towards selecting a preferred route for the interconnector. In order to meet the objective that the project should create the least disturbance to the environment, the effects of the total route on the environment and its users were studied and as far as possible minimised. The preferred route was further refined after consultation with local authorities, statutory consultees and affected parties.

An environmental consultant was commissioned to assess the environmental effects of all the elements of the Interconnector. The necessary detailed studies were undertaken by the companies' own experts or by outside specialists. Landscape architects using modern computer graphics techniques were deeply involved in the identification of route and site options, the assessment of the visual and landscape effects of such options and development of mitigative measures. Studies of ecology, archaeology, forestry and tourism and recreation, together with noise surveys and an assessment of electric and magnetic field strengths were undertaken by specialists. The undersea aspects of the Interconnector, including marine biology, war graves, shipwrecks and safety to mariners, were studied. The overall assessment was presented in three separate documents: Environmental Statement covering the Scottish elements, Environmental Appraisal covering the undersea elements and Environmental Statement covering the Northern Ireland elements. A compromise between environmental, technical and economic factors was achieved in order to find the best route for the interconnector, Harvey et al (4).

Applications for the various consents in Northern Ireland and Scotland were made in December 1993. Public inquires were held in the relevant jurisdictions with full consent granted in Scotland for the overhead line and outline consent for the converter station at Auchencrosh in October 1997. In Northern Ireland the application for the first converter station site was refused and after an application was made for another site on Island Magee, outline consent was granted for BallycronanMore converter station in June 1998. Detailed planning permission for both converter stations was not finally granted until March 2000. The main elements of the optimised interconnector could be determined (Figure 2).

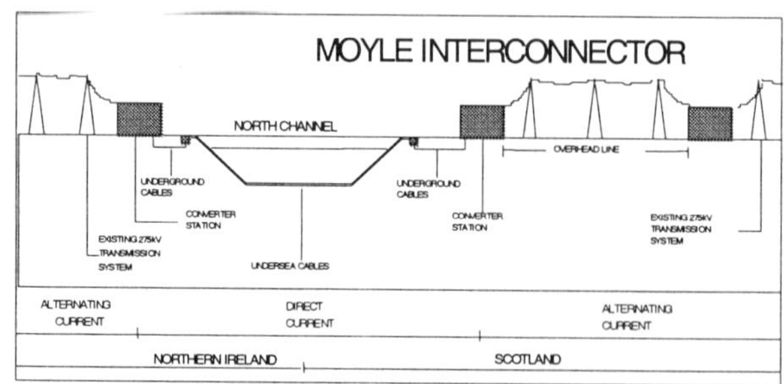

Figure 2: Elements of the Moyle Interconnector

DESIGN AND IMPLEMENTATION

The configuration of the Moyle Interconnector consists of two monopolar submarine HVDC cable links operating in parallel on the a.c. systems (Figure 3). The converters have a power rating of 2x250MW in either direction, referenced at the rectifier station. The d.c. operating voltage of each monopole is 250kV, the a.c. sides of the converters are connected to the 275kV networks of Northern Ireland Electricity and ScottishPower respectively. The nominal direct current is 1000A per monopole. Each d.c. cable system, which connects the two converter stations, consists of 55km of subsea cable and 8.5km of underground cable. The cable system is of the Integrated Return Conductor type (IRC), where the return cable is integrated into the HVDC cable, i.e. a metallic coaxial layer integrated in the cable forms the return path for the current.

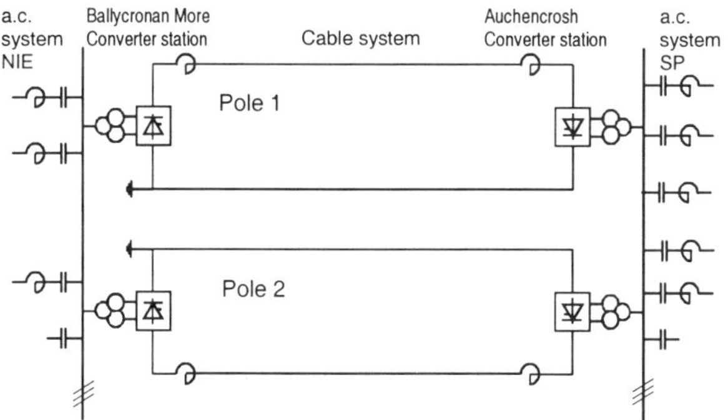

Figure 3: The Moyle Interconnector, configured as dual monopole with a capacity of 2x250MW

When going out to the market for the construction contracts, Moyle had sought tenders for alternative link capacities of 250MW and 2x250MW. Working closely with the leading tenderers, Moyle was able to obtain excellent value for money and to contract for a 2x250MW capacity link within the original budget with no additional environmental impact. Contracts have been awarded for all the elements of the project. At the end of 1999 Moyle awarded turnkey contracts to:

i. Siemens for the design, installation and commissioning of the HVDC converter stations linking the interconnector to the transmission systems in Northern Ireland and Scotland. The stations will be located at Ballycronan More Island Magee, Co Antrim, Northern Ireland and at Auchencrosh, Ayrshire, Scotland (Figure 1).

ii. Nexans Norge AS for the submarine and underground cables which connect the two converter stations. This includes 8.5km of underground cable route connecting the converter stations to the landfalls at Currarie Port in Scotland and Portmuck South in Northern Ireland and 55km of undersea cable route.

Beginning of year 2000 ScottishPower awarded a contract to Balfour Kilpatrick Ltd for 64km of 275kV single circuit overhead line from the existing Coylton substation to the converter station at Auchencrosh, using flat formation towers of 25m standard height and 360m nominal span.

Cable Systems

The cable, rated at 250MW, is of the well proven mass-impregnated paper insulated type but, in a development of the conventional design, includes an integrated return conductor (IRC). This innovative design by Nexans has a metallic coaxial layer integrated in the HVDC cable to form the return path for the current and also to work as a part of the torsion balanced armouring (Figure 4).

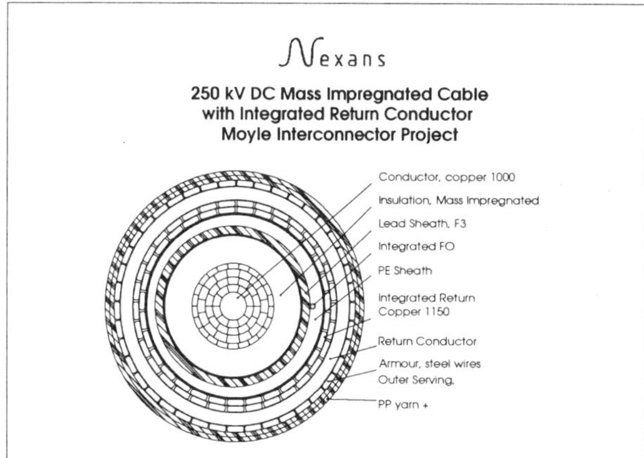

Figure 4: Mass impregnated 250kV d.c. cable with Integrated Return Conductor (weight 42kg/m)

The IRC design has a number of advantages:

- Power cable and return conductor in one cable, good transport and laying properties.
- Good mechanical properties, torsion free and higher tensile strength.
- No external magnetic fields.

The fibre optic cable for control and communications between the converter stations is integrated into the polyethylene sheath of the IRC.

Prior to manufacturing at Nexans' Norway facilities the cable has passed an extensive type test. The two undersea cables have been laid as continuous lengths, separated by one kilometre from each other. The laying was performed by the purpose built vessel c/s Skagerrak, whereafter the cables have been trenched into the seabed by Nexans's Capjet water jetting system, or otherwise protected by concrete mattresses or rock dumping to a nominal depth of one metre. At each shore landing the undersea cables are jointed to the land cables.

The land cable sections were installed from a separate vessel prior to the laying of the undersea cables. The land cables were pulled on shore in lengths of approximately 1.5km, jointed together and laid in a 1.2m deep trench from the shore line and up to the converter stations where the cables are terminated in sealing ends. The cables are also equipped with a single point temperature measurement system.

Converter Stations
Both converter stations have identical designs except for the a.c. switchyards and the a.c. filters, Wohlmuth (3). Each converter station consists of two valve halls - one for each monopole - with the control building in between (Figure 5).

Figure 5: BallycronanMore Converter Station

Adjacent to each valve hall is a d.c. hall, in which the d.c. switchyard including the measuring equipment, the smoothing reactor and the cable sealing end are located. The a.c. switchyard in BallycronanMore converter station is designed as a double busbar arrangement fed by the existing 275kV transmission system. Both outdoor switchyards are designed for a short circuit current of 31.5kA and a specific insulator creepage distance of 31mm/kV.

Thyristor Valves
The arrangement of the thyristor valves are three branches with a quadruple valve in each branch fed from a single-phase three-winding transformer. Each quadruple valve consists of four identical single valves connected in series. The quadruple valves are suspended from the ceiling of the valve hall with the high voltage connections at the bottom.
The converter stations are equipped with the latest high voltage semiconductor technology: direct-light-triggered thyristors with integrated overvoltage protection.
Conventional thyristor valves for HVDC converter stations are made from high voltage semiconductor devices with a peak blocking voltage of 8kV and a rated current of up to 4000A d.c. for transmission voltages up to 500kV d.c. This requires series connection of up to 80 thyristors to make a complete thyristor valve. Each thyristor requires its gate pulse at the same time but at a different electrical potential for the correct operation of the complete valve. So far, this has been achieved by using a rather complex hybrid system, including light emitting diodes and fibre optics for transmitting triggering signals from ground to each thyristor. An electronic board powered locally by an auxiliary energy circuit is used to generate the electric gate pulse and perform various monitoring and protection functions.

The new direct-light-triggered thyristor no longer requires the electronic board at thyristor potential. This thyristor only requires 40mW of light power for reliable turn-on. Therefore triggering is initiated by light pulses generated at ground potential and applied directly into the thyristor gate through a set of fibre optic cables. The light pulses are generated by laser diodes which have a life expectancy in excess of 40 years. In addition this new thyristor has an integrated forward overvoltage protection function, which makes the separate external circuits used so far for this purpose unnecessary. As a result, by introducing direct-light-triggered thyristors the electrical parts required for the HVDC thyristor valve is reduced to approximately 20%, resulting in better reliability with longer maintenance intervals. Furthermore, the spare parts requirement is substantially reduced.

The new technology has been used in the commercial operation of a complete valve for a period of four years. Bonneville Power Administration (BPA) of Portland, Oregon installed the valve in its Celilo Converter Station of the Pacific Northwest-Southwest HVDC Intertie, replacing one of the 36 mercury arc valves (Figure 6). Due to its excellent performance in service, BPA had opted to purchase the valve after one year of operation. Meanwhile BPA awarded a contract to Siemens to replace the remaining mercury arc valves at the Celilo converter station.

Figure 6: Direct-light-triggered thyristor valve rated 133kV, 2000A replacing a mercury arc valve in the Celilo converter station

Converter Transformers
The transformers are of the single-phase three-winding type rated 96/48/48MVA each. A tap changer with 8x1.25% steps is used to keep the

valve side voltage at the ideal value. The transformers are located directly adjacent to the valve hall with their dry-type valve side bushings penetrating the valve hall walls. It is not necessary to install separate d.c. wall bushings with this arrangement.

AC Filter Arrangement
Three triple tuned a.c. filters rated at 59MVAr each and tuned to the 3/12/24 harmonics in combination with one 59MVAr shunt capacitor cover the similar demands on reactive power and harmonic currents filtering of both stations. To compensate for the reactive power demand of the 64km transmission line in Scotland, two 59MVAr filters tuned to the 12 harmonic have been installed additionally in Auchencrosh. Each filter branch can be switched individually by means of circuit breakers with two interrupter units. This design satisfies both the maximum limit of switching overvoltage and the requirements on the harmonic currents suppression. Since both converters are directly connected to the d.c. cable system, there is no need for a d.c. filter.

Smoothing Reactor
A smoothing reactor rated 200mH is installed in each d.c. hall to smooth the direct current and to limit the overcurrent in the cable system during faults. The smoothing reactor will be of the air insulated type. In combination with the dry type bushings of the converter transformers and the carefully selected fire-resistant materials in the converter valves this reduces the risk of catastrophic damage by fire considerably.

Direct Current Measuring System
A reliable HVDC transmission system requires an adequate measurement of the direct current at several locations in each station. The measured direct current is used in the fast converter firing angle control loop and for protection functions. The hybrid-optical system uses an ohmic shunt, which is integrated into the d.c. circuit. The voltage drop across the shunt is proportional to the direct current. This signal is measured and digitised by an electronic circuit located at high voltage potential. The measured signal is transmitted as a serial telegram via a fibre optic link to ground potential. The main advantages are no electromagnetic interference to the measuring signal due to the optical system and an excellent dynamic performance.

Control System
The control system is fully redundant in order to meet the very high demands on availability and reliability of the installation. The control system combines the functions for control, supervision and protection of the link. The normal operating mode of the link is absolute power control. In addition other control functions which are typical for modern HVDC transmissions are available, including delta power control, emergency power control, direct current control, frequency limit control, stability control functions and automatic reactive power control. Since the converter stations will be unmanned, the link is designed for fully automatic remote operation from NIE's dispatch centre including automatic load scheduling operation.

Operation and Maintenance
The main design objective of the Moyle Interconnector is to establish an electricity interconnection with low losses and a very high availability and reliability combined with a low maintenance requirement throughout the expected life time of more than 30 years. This is reflected in a state-of-the-art design using a high degree of redundancy and a combination of the latest HVDC technology and components with a long time record of operation experience. The high performance of the converter stations is reflected in the guaranteed losses of less than 1.35 % and a guaranteed value for the energy availability of more than 99.6 %. The converters are designed for a biannual scheduled maintenance. A safe and reliable operation of the Interconnector will be achieved by experienced and skilled operating and maintenance personnel well trained in Siemens' training program.

CONCLUSION
Public concern about the environmental impact of major infrastructure projects continues to increase and can be compounded by other concerns. Therefore the lead times for such projects must allow for a protracted consents process, which can span the introduction of new technology. Major changes to the electricity market and transmission systems being connected may also occur. This can affect initial study results and capacity requirements.

REFERENCES
1. UNIPEDE Group of Experts on Island Systems, 'Island Systems', UNIPEDE Kobenhavn Congress June 10-14 1991, Report 91 en 40.2, 1991.

2. Rainey, A. et al, 'Interaction between an HVDC Interconnection and an Island System - Impact on the Design of the Moyle Interconnection', CIGRÉ 1994 Session Paper 37-105, 1994.

3. Wohlmuth, M. Dr., 'The Celtic Connection', Power Engineering International Magazine, PEI June 2000.

4. Harvey, C., Stenseth, K., Wohlmuth, M., 'The Moyle Interconnector – Implementing an infrastructure project', Queens UPEC 2000 Conference, September 2000.

AUTONOMOUS LOOP POWER FLOW CONTROL FOR DISTRIBUTION SYSTEM

N. Okada

Central Research Institute of Electric Power Industry, Japan

Abstract: This paper describes an autonomous control for loop power flow in a distribution system. The purpose of this study is to show that an autonomous method using local voltage information is able to distribute feeder power flow and to maintain the correct voltage with stable operations. The result of our experiment clearly shows that the proposed autonomous method for loop power flow control balances the power flow and regulates the voltage with stable operations.

INTRODUCTION

In Japan, 6.6kV overhead distribution systems are currently constructed as radial networks. In these systems each feeder voltage is controlled to within 101±6V for the standard voltage 100V and 202±20V for the standard voltage 200V. The legal range for 100V is narrower than for 200V, therefore, voltage remains within 101±6V for each customer. A line voltage compensator (LDC) maintained proper voltage for a bank of about six feeders. The load imbalance of each feeder therefore had a negative impact on voltage stability.

In the near future, many distributed generators (DG) will be connected to the power distribution system. The reverse power flow of these DG systems will rise to the upper limit of the proper voltage range. On the other hand, disconnecting the DGs with faults causes line-over load or under-voltage. In this case, it has been proposed that loop or mesh distribution systems be designed to balance the power flow and regulate voltage. However, this will increase short-circuit current in the distribution system, and fault location detection methods for loop distribution systems have not yet been established. A loop power flow controller (LPC) using a pulse wide modulation (PWM) AC-DC-AC converter is able to control loop distribution systems without increasing short-circuit current. LPC loop distribution systems can be used in present protection method.

The question here is whether the LPC is aware of the current status of the distribution system. A loop power flow control method using feeder power flow information has been proposed (1). It must be pointed out that data communication cannot be used in every case, therefore it is necessary to estimate the state of the distribution system autonomously.

DISTRIBUTION SYSTEM

System model

In this paper we use a simple model for the distribution system consisting of one substation transformer and two-feeders. Fig. 1 shows the distribution system model. The feeders have 5km of distribution line, where load is assumed to be uniformly distributed along the feeder, and the load power factor is assumed to be 0.85, which shows that the current lags the voltage. An LDC installed on the substation transformer in the distribution substation regulates line voltage. The LDC estimates the voltage drop of the lines from the line current and controls the voltage rate of the transformer. The line voltages remain within a proper range. Pole transformers convert 6.6kV to 100V, and conversion ratios of 6750V/105V and 6600V/105V are used between 0km and 2km and 2km and 5km respectively. The sectionalizer is always open in the radial distribution system. Table 1 shows the specifications of the 6.6kV distribution system model. Power base and voltage base are defined as 10MV and 6.6kV respectively.

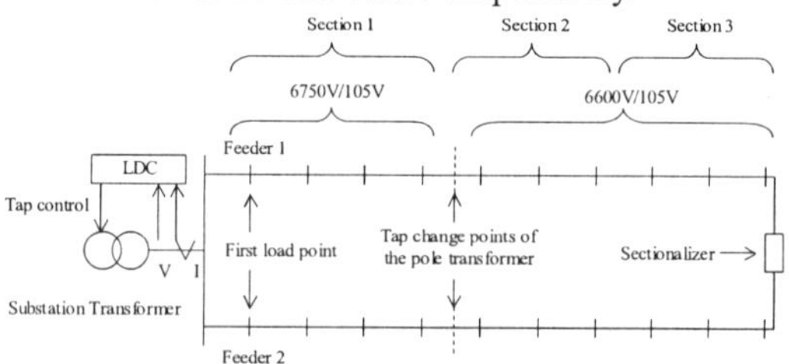

Figure 1: Distribution system model with two feeders

Table 1 Specifications of model distribution system

Item	Specification
Substation Transformer	
Capacity	2 pu
Secondary voltage	1.076 ~ 0.970 pu
	Interval 0.015 pu
Number of feeders	2
LDC	
Reference voltage	1 pu
Impedance	$0.062 + j\,0.079$ pu
Feeder	5km
Impedance	$0.069 + j\,0.087$ pu/km
Load	
Power factor	0.85
Distribution	uniformly distributed
Fixed capacitor	0.05 pu/feeder

Basic characteristics

In Japan, generally one LDC performs voltage regulation for about six feeders. In this model the voltage of two feeders is adjusted using one LDC. Therefore, if load between feeders becomes unbalanced, neither feeder is able to maintain a suitable voltage. Fig. 2 shows the voltage characteristics of each feeder assuming loads of 0.3 pu and 0.075 pu on feeders 1 and 2 respectively. The upper limit of suitable voltage is based on 107V. The

AC-DC Power Transmission, 28-30 November 2001
Conference Publication No. 485 © IEE 2001

lower limit is based on 101V including pole transformer voltage drop and low voltage line drop. In a radial system, the voltage difference is produced from the two feeders by an unbalanced load. On the other hand in the loop distribution system, the voltage of a terminal corresponds to the other terminal and the voltage difference between feeders is small compared to the radial distribution system. Here we must note that short circuit capacity increases and the influence range is expanded at the time of failure by the simple loop.

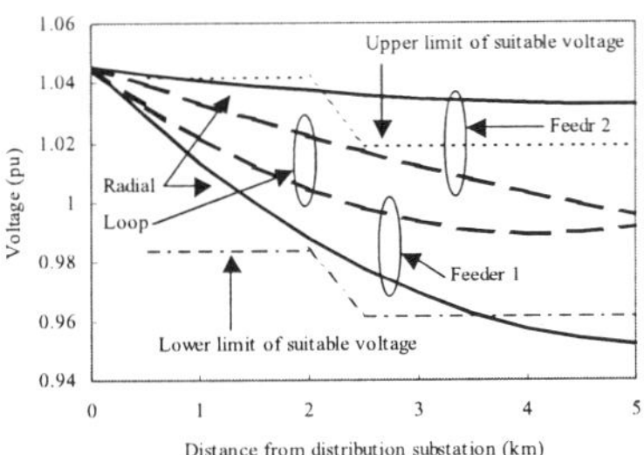

Figure 2: Simulated voltage profile of distribution lines

BASIC CONCEPT OF LOOP DISTRIBUTION SYSTEM WITH LOOP POWER FLOW CONTROLLER

The basic concept of the loop distribution system using the loop power flow controller we propose is as follows.

- Aim is free access to a distributed power supply.
- System responds flexibly to unbalanced load between feeders, and makes effective use of equipment.
- To enable this, the system is constructed in the shape of loop from radial.
- A loop distribution system is provided without altering existing systems such as the protection system, except for loop points.

The equipment to achieve these aims is called as a loop power flow controller (LPC), which replaces the sectionalizer at the loop points. In this paper we show a method for controlling voltage and balance of power flow when a PWM AC-DC-AC converter is used as the LPC.

PROPOSAL FOR AUTONOMOUS LOOP FLOW CONTROL

One idea is to use local voltage for loop power flow control and voltage regulation. Theoretically, the voltage of the power system is controlled by reactive power. Here however I would like to emphasize line resistance (R) close to line inductive reactance (X) in Japanese 6.6kV distribution lines.

Rules of control

Voltage at each end of the LPC was set at V1 and V2 respectively. The maximum and minimum values of suitable voltage used for autonomous control were set at VH and VL. Specifically, when suitable voltage is maintained, the maximum and minimum values of the voltage were the voltage at minimum load and maximum

load respectively. Fig. 3 shows the terminal voltages of LPC and reference voltage of the suitable voltage range.

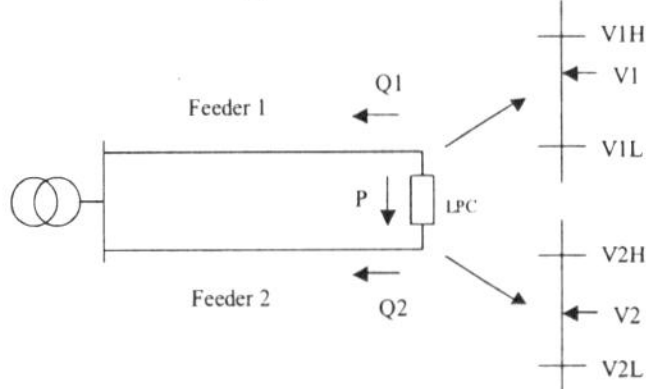

Figure 3: Terminal voltages of LPC and reference voltage of suitable voltage range

If a distribution line removes effective power from a power line, voltage will fall, and if effective power is injected into the line, voltage will rise. Moreover if the inductive reactive power is injected into, voltage will rise, and if the capacitive reactive power is injected into, voltage will fall. From the relationship between the self-terminal voltages V1 and V2 and reference voltages V1H, V1L, V2H and V2L, we see the status divided into nine domains, and as shown in Table 2, P, Q1, and Q2 of LPC are controlled by the rules of each domain.

Table 2 Autonomous flow control table

	V2>V2H	V2H≥V2≥V2L	V2L>V2
V1>V1H	Q1 and Q2 are decreased.	Q1 is decreased.	P is increased.
V1H≥V1≥V1L	Q2 is decreased.	Absolute value of P, Q1 and Q2 are decreased.	Q2 is increased.
V1L>V1	P is decreased.	Q1 is increased.	Q1 and Q2 are increased.

Loop power flow control

To make the LPC efficiently requires combining two sets of PWM voltage source inverters (VSIs). In order to perform the above-mentioned loop flow control operation, both sets of terminal voltage information are needed. We propose a voltage flow control that exchanges voltage information between units using DC voltage of VSIs. Each VSI has DC voltage control, and assumes that the setting voltage value of each DC voltage control is derived using the following formula from the current self-terminal voltage V and effective power P.

$$\frac{V-1}{dV} = \frac{V_{dc}-1}{dV_{dc}} - \frac{P}{K} \tag{1}$$

dV is 1/2 of the suitable voltage range and dVdc is 1/2 of the DC voltage range of fluctuation. K is a coefficient that determines the loop effective power corresponding to voltage difference. In inverter unit 1, at self-terminal voltage V1, the effective power output P1 and the common DC voltage Vdc, equation (1) is as follows.

$$\frac{V_1-1}{dV_1} = \frac{V_{dc}-1}{dV_{dc}} - \frac{P_1}{K_1} \tag{2}$$

Similarly, in inverter unit 2, at self-terminal voltage V2, the effective power output P2, and the common DC voltage Vdc, equation (1) is as follows.

$$\frac{V_2-1}{dV_2} = \frac{V_{dc}-1}{dV_{dc}} - \frac{P_2}{K_2} \tag{3}$$

An effective electric power output has the relation in the following equation, when losses of each inverter unit are assumed to be zero.

$$P_1 + P_2 = 0 \qquad (4)$$

The following equation can be derived from equation (2), (3) and (4).

$$\left(K_1 + K_2\right)\frac{V_{dc}-1}{dV_{dc}} - K_1\frac{V_1-1}{dV_1} - K_2\frac{V_2-1}{dV_2} = 0 \cdot \qquad (5)$$

This gives us the relationship between V1, V2, and Vdc. In inverter unit 1, the other terminal voltage V2 of inverter unit 2 may be obtained from the self-terminal voltage V1 and the DC voltage Vdc, which are known. The control characteristics of inverter 1 and inverter 2 are shown in Fig. 4. The self-terminal information V1 and V2 on each inverter is the section at which the control characteristic intersects the V axis. Because the relation of P1=P2 consists of the equation (4), the intersection of the central line (broken line) and V axis gives the DC voltage Vdc.

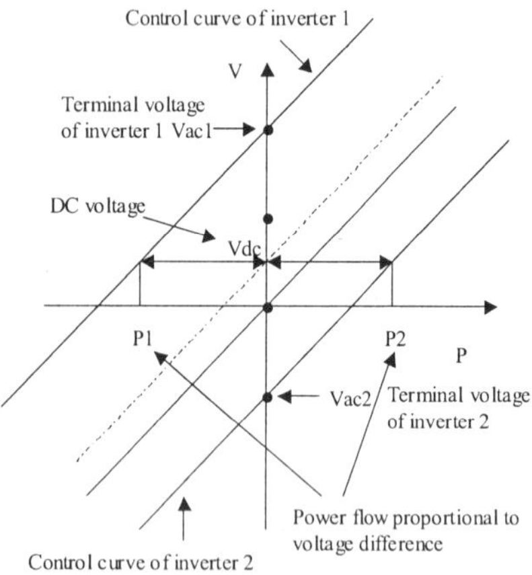

Figure 4: Power flow control characteristic proportional to voltage difference for LPC

Vdc of equation (1) is placed with DC voltage reference Vdcref, and DC voltage is controlled by the following formula.

$$V_{dcref} = dV_{dc}\left(\frac{V-1}{dV} + \frac{P}{K}\right) + 1 \qquad (6)$$

Reactive power control

Reactive power control generates reactive power for self-terminal voltage in accordance with Table 2, as compared with the maximum voltage VH and the minimum voltage VL. Voltage control reference values may be derived from self-terminal voltage V passes to the limiter with the maximum VH and the minimum VL. The difference between the reference voltage and self-terminal voltage becomes zero within the limits. When the usual PI control is used and reactive power generated so that a voltage rise may occur at once and be controlled, the integration element holds this generating reactive power value. Then when the voltage returns to within its suitable limit, the reactive power value possessed by the integration element will continue being generated. When this enters the suitable limit, we use a low path filter instead of the usual integration element so

that the integration value may reach zero gradually.

$$G(s) = \frac{K_{IQ}}{s + K_Q} \qquad (7)$$

KQ (=2πf) is the cut off frequency of a low path filter. KIQ is the gain of the integration region higher than the cutoff frequency. Fig. 5 shows a control block of the LPC.

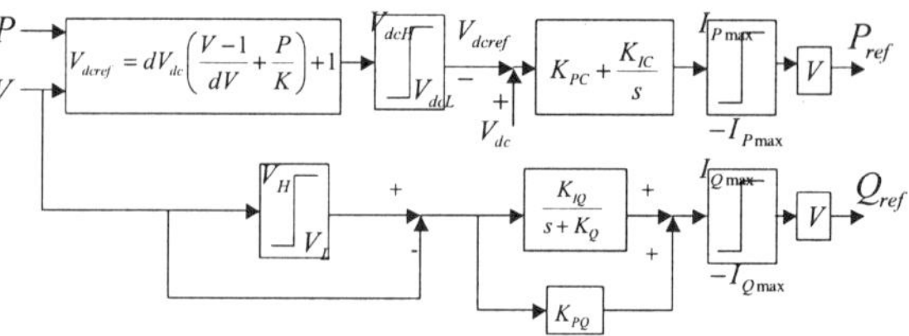

Figure 5: Block diagram of autonomous power flow and reactive power control

Here, VdcH and VdcL are the maximum and minimum values of DC voltage. Ipmax and Iqmax are the current limits of the d axis and q axis respectively. When current capacity is set to 1, the limitations of Ipmax and Iqmax are as follows.

$$Ip\max^2 + Iq\max^2 \le 1 \qquad (8)$$

When Ipmax and Iqmax are taken equally:

$$\begin{aligned} I_{p\max} &= 1/\sqrt{2} \\ I_{q\max} &= 1/\sqrt{2} \end{aligned} \cdot \qquad (9)$$

Fig. 6 shows the composition of the LPC. Current controllers are PI controller designed on synchronous frame (2).

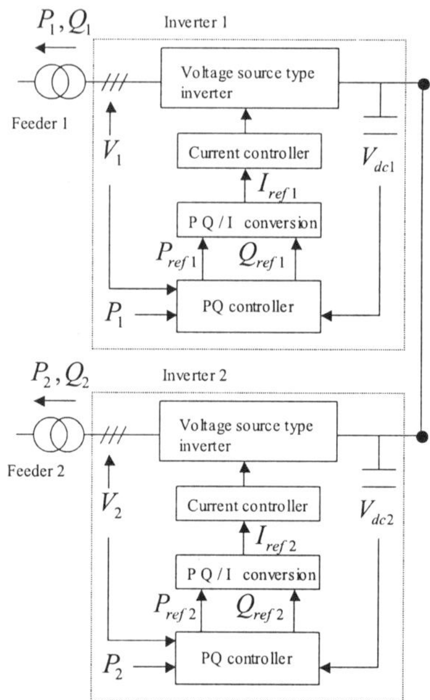

Figure 6: Composition of LPC

STEADY STATE VOLTAGE AND FLOW ANALYSIS

Set up control parameters

Fig. 7 shows the maximum and minimum voltage along

the feeder when the load changes from 0.075 pu to 0.25 pu in the model system shown in Fig. 1. The solid curve and broken curve in the figure represent voltage at power factors 1 and 0.85 respectively. In this paper, the maximum and minimum voltage assuming the power factor of 1 are set to VH and VL respectively. DV is derived as follows.

$$dV = \frac{V_H - V_L}{2} \qquad (10)$$

The setting value of each installation point is shown in Table 3.

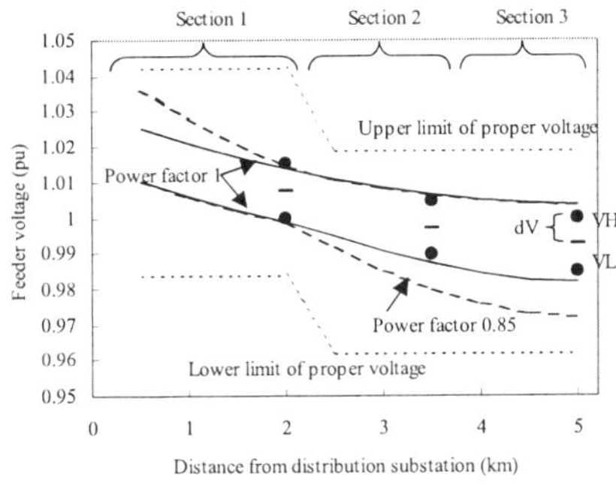

Figure 7: Maximum and minimum voltage of distribution feeder

Table 3 Setting at each location

	Section 1 (n4)	Section 2 (n7)	Section 3 (n10)
VH	1.015	1.000	1.000
VL	1.000	0.985	0.985
DV	0.00758	0.00758	0.00758

Voltage characteristics

The supply voltage to low-voltage customers is restricted to within the limits of 101±6V. If we assume a 2V drop in the pole transformer and 4V drop in the low-voltage, the voltage adjustment width becomes 101V-107V. Evaluation of the suitable voltage uses the following ranges by each node.

$$101V \leq Vn \leq 107V \qquad (11)$$

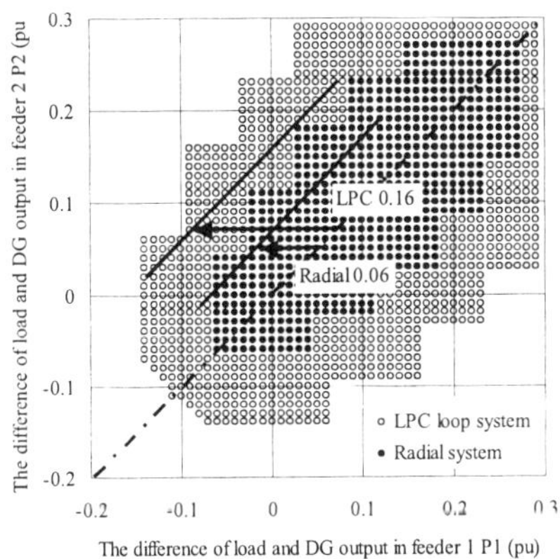

Figure 8: Proper voltage range of two feeders for radial distribution network and loop distribution system

The domain that satisfies the conditions of equation (11) to each feeder load is shown in Fig. 8. In this evaluation K is 4. If we assume that the two feeders have balanced, the minimum value with which DG connects with feeder 1 and each feeder is satisfied by equation (1) is 0.06pu. On the other hand, the value is set to 0.16pu in the loop distribution system by the LPC.

Power flow balance

Here we consider equalization of power flow. Fig. 9 shows the feeder sending effective power difference over the demand difference between feeders. The following equation defines the demand difference LU between feeders.

$$L_U = \left| (P_{L1} - P_{G1}) - (P_{L2} - P_{G2}) \right| \qquad (12)$$

The following equation defines the feeder sending effective power difference FU.

$$F_U = \left| P_{F1} - P_{F2} \right| \qquad (13)$$

PL1 and PL2 are the loads connected to feeders 1 and 2. PG1 and PG2 are the output of DG connected to feeders 1 and 2. PF1 and PF2 are the sending effective power of feeders 1 and 2. In this analysis, the ends of the feeders were connected by a LPC of capacity 0.05pu. The power flow difference FU with the LPC is half that without the LPC. The theoretical performance limit of the LPC is determined by LPC capacity.

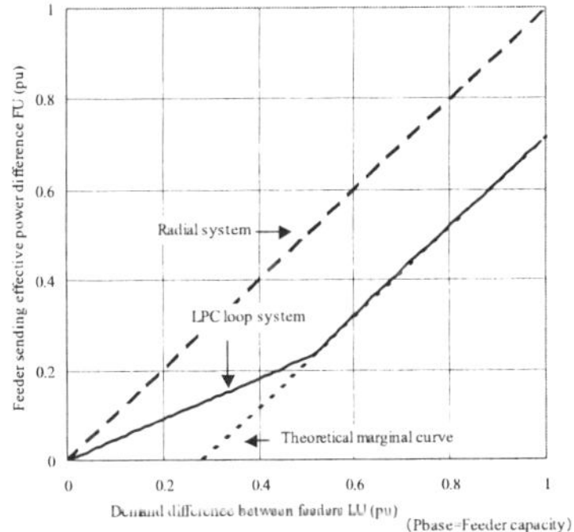

Figure 9: Simulation of feeder power flow balance for unbalanced feeder load

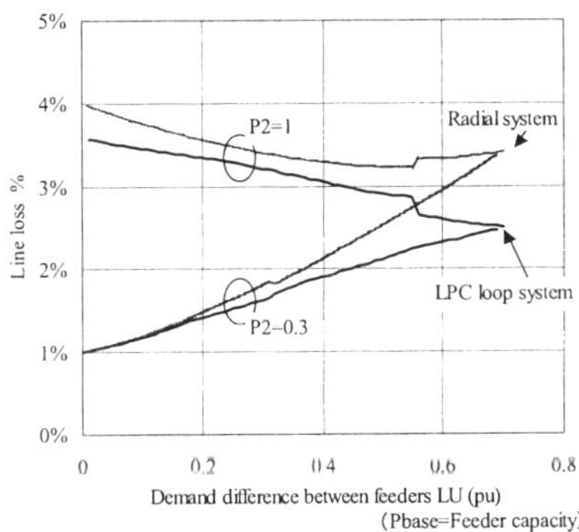

Figure 10: Estimated line loss for unbalanced feeder load.

Line loss

Fig. 10 compares line loss in a LPC loop distribution system with that of a radial distribution system. The feeder of the LPC system becomes low loss as the demand difference LU between feeders becomes greater.

TEST RESULTS FOR SMALL LPC

Test circuit

The test circuit consisted of two three-phase voltage regulators imitating a substation, two line impedance, a load and a small testing LPC as shown in Fig. 11. Table 4 lists the parameters of the test circuit. Table 5 shows the components of the experimental small LPC.

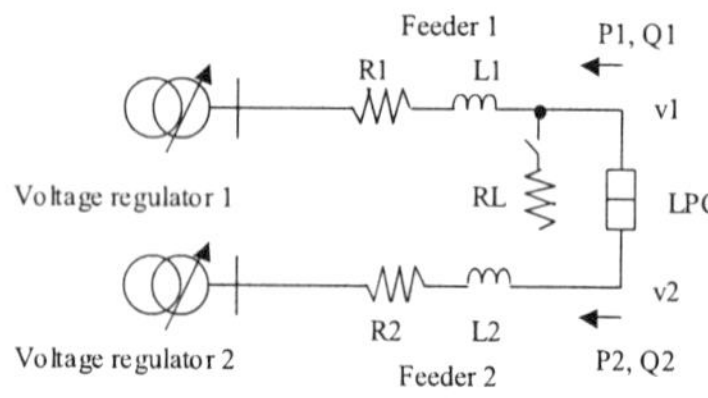

Figure 11: Experimental small loop flow controller with modeled autonomic flow control and test circuit

Table 4: Test circuit parameter

Items	Specifications
Three-phase voltage regulators 1, 2	200V, 10kVA
R1	0.5· ·
L1	3.8mH
R2	0.6· ·
L2	4.3mH
RL	1.5kW

Table 5: Components of experimental small LPC

Items	Specifications
Inverter #1, #2	MWINV9R122A (9.1kVA/200V) IPM: PM75RSA060
DC capacitor Cdc#1,Cdc#2	3000· F/450V
AC inductor L#1,L#2	1.3mH/15A
AC filter capacitor C#1,C#2	10· F/250V (Δ connection 30· F)
Controller #1, #2	MWPE2-P DSP TMS320C32, 50MHz
Compiler and development tool	TI DSP C compiler Ver.5.1 Tool: MWPE2 -DP

Basic characteristics

A half LPC unit and the characteristics of terminal voltage V and the DC voltage Vdc are shown in Fig. 12. By replacing P with 0 in equation (1), this equation becomes

$$V_{dc} = \frac{V-1}{dV} dV_{dc} + 1. \tag{14}$$

The experiment result was in agreement with the theoretical characteristics. In this investigation, dV is 0.1 and dVdc is 0.0555.

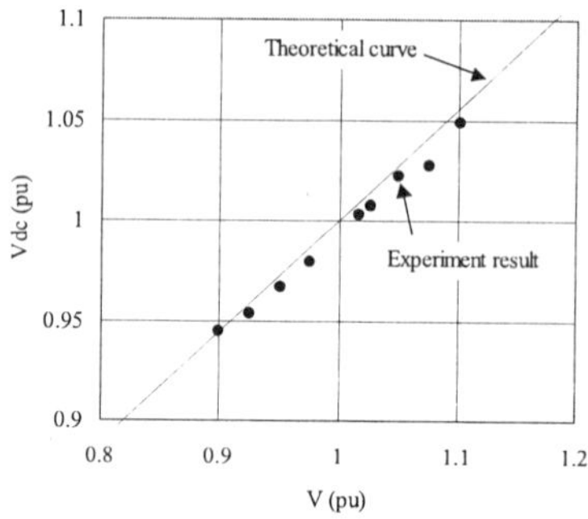

Figure 12: V-Vdc curve of half LPC unit

The results of investigating of the relationship between voltage difference and power flow are shown in Fig. 13 for the test circuit of Fig. 11. We fixed the sending voltage of feeder 2 to 1pu, and changed the sending voltage of feeder 1. By replacing K1 with K, K2 with K, dV1 with dV and dV2 with dV in equation (5), eliminating the terms in Vdc from equations (2) and (5), these equations become

$$P_1 = -\frac{V_1 - V_2}{2dV} K. \tag{15}$$

Similarly, equations (3) and (5) become

$$P_2 = \frac{V_1 - V_2}{2dV} K. \tag{16}$$

The value in the experiment agrees totally with equations (15) and (16), K is 2 in this test.

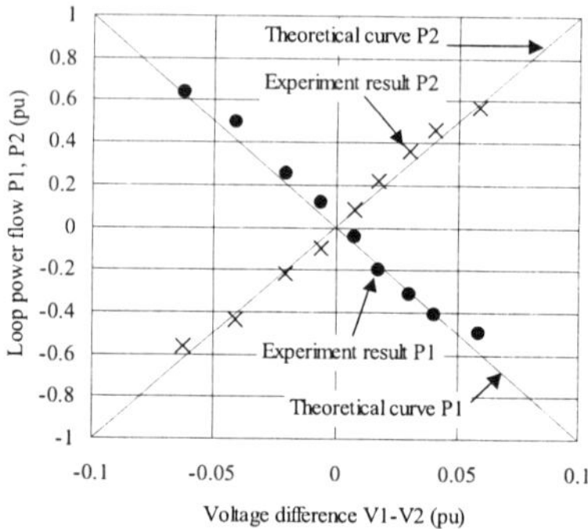

Figure 13: Voltage difference of two feeder and loop power flow on LPC

Dynamic response

Fig. 14 shows the voltage regulation characteristics of the half LPC unit. Since the effective power control does not work, the half LPC controls the voltage rise by about 0.2 seconds with reactive power, and is rationalizing voltage. This is shown in (a). Here the sending voltage was set at 1.11pu, and the load 0.015pu of feeder 1 disconnected to generates the voltage rise. From the state of the voltage rise control by reactive power, operation when voltage enters the proper range is as shown in (b). The voltage enters the proper range and reactive power decreases as in equation (7). The settings of the controller are shown in Table 6.

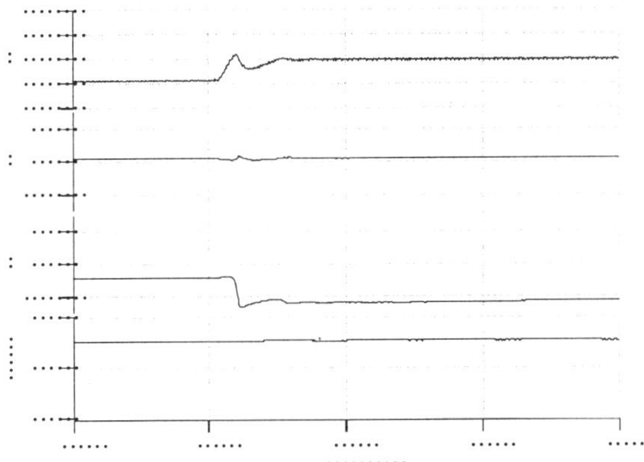

(a) From the proper range to maximum voltage

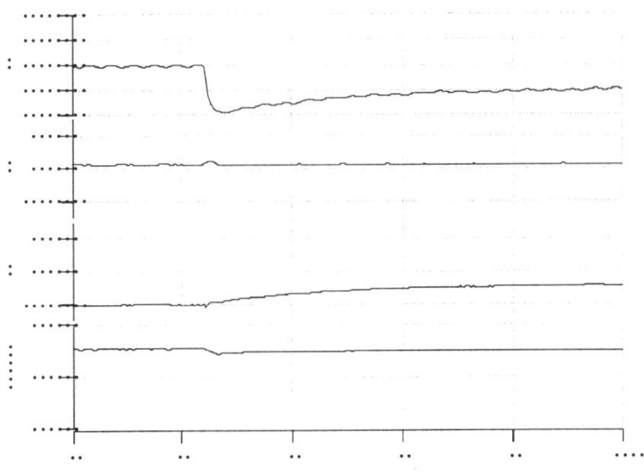

(b) From maximum voltage to proper range
Figure 14: Voltage regulation of half LPC unit

Table 6: Settings for controller

Items	Specifications
KPQ	160
KIQ	4000
KQ	8
KPC	3.37
KIC	14.3
K	2
dV	0.1
dVdc	0.0555

Fig. 15 shows the voltage and power flow control characteristics. Initially the sending voltage of feeders 1 and 2 was set to 1pu. We then changed the sending voltage of feeder 1 from 1pu to 1.14pu. Each feeder had no-load at this time. First, the start of a rise in voltage generates the voltage difference between V1 and V2. Next, DC voltage rises. Then the loop effective power flows into feeder 2 from feeder 1. Voltage is raised further and V1 becomes 1.1pu i.e. the maximum voltage. Then the injecting of reactive power performs a rationalization of the voltage. The effective power increases gradually due to the voltage difference, and reactive power decreases and remains steady state. Stable operation of autonomous control was confirmed in this way using a small test circuit.

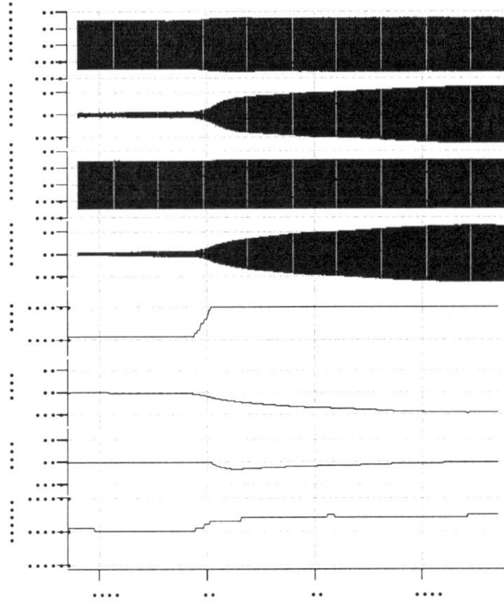

Figure 15: Voltage regulation and power flow characteristics of the LPC

CONCLUSIONS

The basic concept of the loop distribution system using LPC that we propose consists of free access to a distributed power supply in a flexible response to an unbalanced load between feeders, utilizing distribution system effectively. A loop distribution system is created without altering existing systems such as the protection system. The equipment for achieving this is called a LPC, replacing sectionalizer at the loop points. In this paper, we show the method of controlling voltage and the balance of power flow when a PWM AC-DC-AC converter is used as the LPC.

The question here is whether the LPC is aware of the current status of the distribution system. It must be pointed out that we cannot use the data communication for every case, therefore it is necessary to estimate the status of the distribution system autonomously. We proposed a loop power flow control method for this using the voltage difference between circuits.

A simulation was used to gauge the basic performance of this system. In the loop distribution system using the LPC, robustness for the load unbalance increases in comparison with the radial distribution system.

Using a small test circuit, basic characteristics and dynamic response demonstrate the stable operation of autonomous control.

REFERENCES

1. Okada, N., 2000, "Control of Loop Distribution Network and Result", <u>IEEJ PSE</u>, <u>00</u>, 7-12
2. Kim, H., S., et al., 1998, "Design of Current Controller for 3-Phase PWM Converter with Unbalanced Input Voltage", <u>Proceedings of PESC</u>, 503-509

ELECTRICAL STABILITY OF LARGE, OFFSHORE WIND FARMS

L. Holdsworth, N. Jenkins, G. Strbac

UMIST, United Kingdom

ABSTRACT - Transient and steady state modelling of large MW capacity, induction generator, wind turbines used for offshore wind farms is presented. The possibility of network voltage instability is investigated. Results are presented from simulations using predicted turbine and network data for UK offshore wind farm installations. Possible techniques to improve dynamic stability, applied to both the electrical and mechanical systems of the wind turbine, are discussed.

Keywords: Offshore wind farms, wind turbines, induction generator models, voltage collapse, dynamic stability.

1. INTRODUCTION

Offshore wind energy has been identified, by the DTI consultation document on New and Renewable Energy, as one of the technologies needed by 2010 in order to meet the targets of 10% of UK electricity from renewables, DTI (1). The prospects for offshore wind energy throughout Europe has been well documented, BWEA (2), with the first installations being constructed within the enclosed seas of the Netherlands and off the coasts of Denmark and Sweden. The recent 4 MW wind farm installation at Blyth Offshore, comprising two 2 MW wind turbines, represents the UK's first venture into the development of offshore wind farms and the worlds first multi-megawatt wind farm in a severe wave climate, Grainger et al (3). The success of this installation, together with the government policy of support for sustainable development via the Renewables Obligation, will encourage the future development of large MW capacity UK offshore wind energy.

On the 5th April 2001 the Crown Estate announced 18 potential sites for UK offshore wind farms. The proposed capacity of each site is likely to be some 60 MW consisting of thirty 2 MW induction generator based wind turbines. Elsewhere in Europe plans exist for wind farms of up to 1000 MW capacity. Thus, investigating the reliable operation of large wind farms and obtaining solutions of maintaining dynamic stability of the power system is critical.

This paper demonstrates that faults on the a.c. network, which depress the voltage at the wind farm, may lead to turbine overspeed. It is shown that the electrical instability of both the turbine and network voltage becomes more significant the weaker the a.c. network is relative to the windpower capacity. The effect of

machine parameters upon the stability of the system is discussed together with the magnitude and X/R ratio of the network source impedance. Various network fault conditions and different network parameters have been simulated. Results are presented from simulations using predicted turbine and network data for offshore wind farm installations. The usual steady state and classical fifth order (transient) induction generator models have been used. A doubly fed induction generator model is discussed, as variable speed induction generator based wind turbines have recently become a viable industrial option. Techniques for improving stability margins through voltage control and reactive compensation strategies are also presented.

2. MODELLING CONSIDERATIONS

Initial investigations have illustrated the potentially major issue of voltage stability resulting from the continued increase of the wind farm generation capacity to network short circuit level, Jenkins et al (4). It has also been presented that a transient short circuit in the connecting networks with large wind farms may result in voltage collapse, Akhmatov et al (5). A depressed wind farm terminal voltage resultant from faults on the a.c. network can lead to overspeed if the generating capacity to network short circuit level ratio is insufficient. This leads to the induction generator drawing a high level of reactive power that depresses the voltage further and instability results.

The current technology with the majority of land based wind farms is stall regulated wind turbines with conventional fixed speed induction generators which are connected directly to the power system. Due to the possibility of overspeed the wind turbines are automatically disconnected from the power system in the event of short circuit faults. However, in the case of offshore wind farms it is required, due to the large generating capacity, that the voltage stability be maintained throughout failure events in the connecting power system without disconnection of the wind farm, Akhmatov et al (6).

Steady state and transient models of the induction generator based wind turbines are implemented for a 60 MW offshore wind farm. As the first UK wind turbine installations are likely to be implemented using fixed speed squirrel cage induction machines, the conventional steady state equivalent circuit model and the classical 5th order (transient) model, Kundur (7), are

AC-DC Power Transmission, 28-30 November 2001
Conference Publication No. 485 © IEE 2001

implemented for the simulations. Operating experience has found that in practise it is possible to connect relatively high capacities of induction generating plant to rural distribution networks, Jenkins et al (4). The stability margins are consequently observed for strong and weak network connections, for short circuit levels of 3600 MVA and 360 MVA respectively, and for X/R ratios of 10 and 2 representing typical sub-transmission and distribution network impedance parameters.

3. STEADY STATE STABILITY MODELLING

The effect of network parameters, such as short circuit level, X/R ratio and machine power factor correction (PFC), upon the steady state stability margins of the wind farm can be observed through a simple network connected to the induction generator equivalent circuit, as shown in Figure 1. The induction machine, transformer reactance and PFC parameters used for the simulation are given in Table 1. The cable charging capacitance is not included for these simulations. All results are given in per unit and network parameters are calculated using a 60 MVA base.

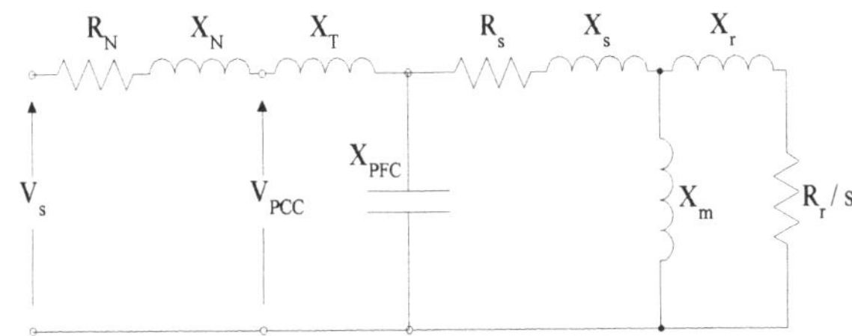

Figure 1: Steady state Wind Farm/Network model

Rating	2 MW
Voltage	0.69 kV
Stator resistance (R_s)	0.00488 p.u.
Stator reactance (X_s)	0.09241 p.u.
Magnetising reactance (X_m)	3.95279 p.u.
Rotor resistance (R_r)	0.00549 p.u.
Rotor reactance (X_r)	0.09955 p.u.
Transformer reactance (X_T)	0.05 p.u.
Power factor correction (X_{PFC})	3.33 p.u.

Table 1: Simulation parameters

The torque-slip and reactive power-slip curves may be used to graphically represent the steady state behaviour of an induction generator. Steady state instability occurs at the peak pull-out torque and the reactive power absorbed by the wind farm increases. The torque-slip characteristic of the 60 MW induction generator, shown in Figure 2, illustrates the influence of network source impedance upon the stability of the wind farm. For normal generating operation of −1 pu torque the induction generator operates in both network impedance cases with a slip of approximately -0.6%. However, the peak-pull out torque is obviously reduced as the network impedance increases. The large peak-pull out torque of the machine with a wind farm capacity to the short circuit level at the point of connection ratio of 60, representing a strong network connection, indicates that the system should have a good dynamic stability

capability. Whereas a short circuit ratio of 6, for the weak network connection, illustrates a rather asymmetrical torque-slip curve with much reduced peak-pull out torque. The pull out torque is approximately 1 pu in the motoring region and 1.25 pu in the generating region.

This indicates that although the wind farm will be operable at 1 pu generation connected to a weak network, it may have dynamic stability problems resultant from variations in the network voltages. Figure 3 shows the dependence of the pull-out torque on the terminal voltage. Varying the network voltage from 0.9 to 1 pu, to represent outage conditions, shows possible machine instability with a network voltage of 0.9 pu if the 1 pu generative torque is maintained.

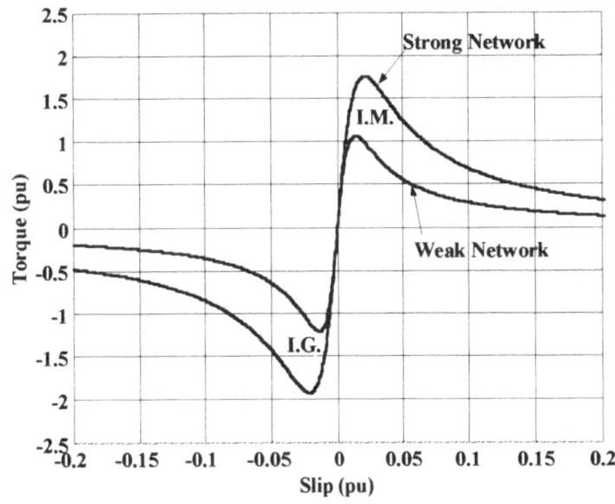

Figure 2: Torque-Slip curves for 60 MW wind farm

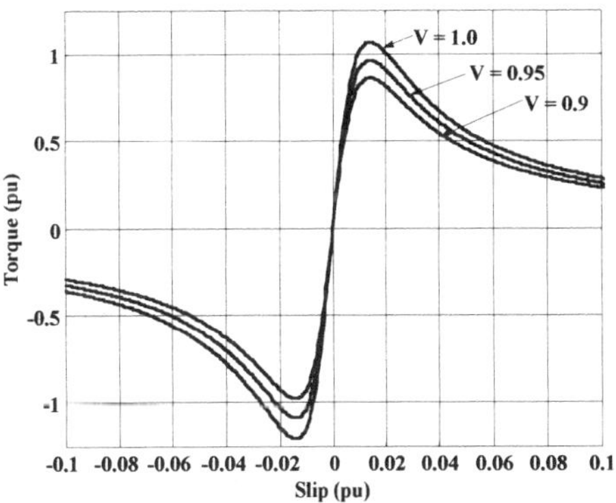

Figure 3: Effect of network voltage variations

Figure 4 shows the variation of wind farm reactive power (Q), drawn from the network, to slip. In both strong and weak network cases it can be observed that if the wind farm induction generators run above their pull-out torque then large amounts of reactive power would be absorbed from the network. This would clearly lead to voltage collapse in the network. With the influence of network short circuit level (SCL) and voltage variations upon the stability of the wind farm identified, the next case study considers the effect of the X/R ratio of the cable used for connecting the offshore wind farm to land. Although typical sub-transmission lines characteristically have an X/R ratio of 10, the submarine cable to be implemented for the offshore installation will typically be closer to a ratio of 2. The effect of the X/R ratio upon the torque-slip characteristic is negligible for a strong network. However, a characteristic difference can be observed for X/R ratios of 10 and 2 when connected to a weak network, as shown in Figure 5.

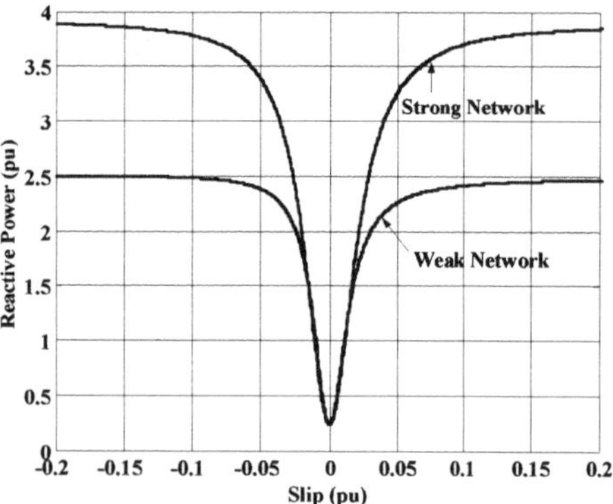

Figure 4: Q-Slip curves for 60 MW wind farm.

A large asymmetry between the motoring and generating pull out torque is now exhibited for the X/R ratio of 2, typical of the cable proposed for offshore installation. The final steady state case study is to observe the improvement upon the stability margin through implementation of machine terminal PFC. All characteristics up to this point have been studied without any PFC. It is typical in wind farm installations to connect 30% PFC of the wind farm capacity. Therefore for a 60 MW installation, the effect of 18 MW of PFC is studied. Figure 6 shows that for weak network connection scenario the PFC increases the pull out torque and thereby assists in improving the overall stability. It may be noted that any large submarine cable network will require careful control of reactive power.

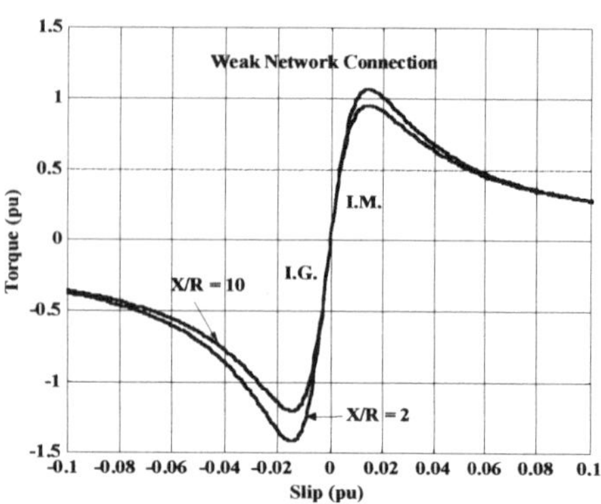

Figure 5: Weak network, X/R ratios of 10 and 2

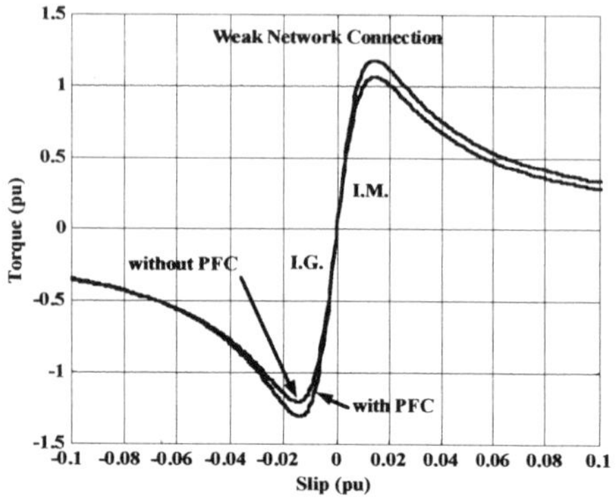

Figure 6: Effect of machine terminal PFC.

4. WIND TURBINE & NETWORK STABILITY

Using a classical 5th order transient model for the induction generator the stability of a 60 MW offshore wind farm is observed. The generator parameters given in Table 1 were implemented. A machine inertia constant of H = 3 s, is used for the model to represent the lumped turbine and generator inertia. The phenomena of network voltage instability, resulting from potential turbine overspeed due to machine terminal voltage variations, is investigated for various network and fixed compensation parameters. Balanced and unbalanced system faults are simulated to predict the response of the 60 MW wind farm for typical voltage sags and fault clearance times.

To investigate the transient stability of the wind turbine and the voltage stability of the connecting network, a typical sub-transmission connection of the fixed speed induction generator based wind turbine was used for preliminary simulations, as illustrated in Figure 7. With an infinite source at B1 and a 60 MW wind farm connected at busbar B3, the network voltage depressions were implemented by a network fault applied at the mid point between busbars B1 and B2. The breaker timing control and the resistance controlled the duration and magnitude of the voltage sag. For the following investigation voltage sags of approximately 0.5pu with clearance time of 150ms were implemented.

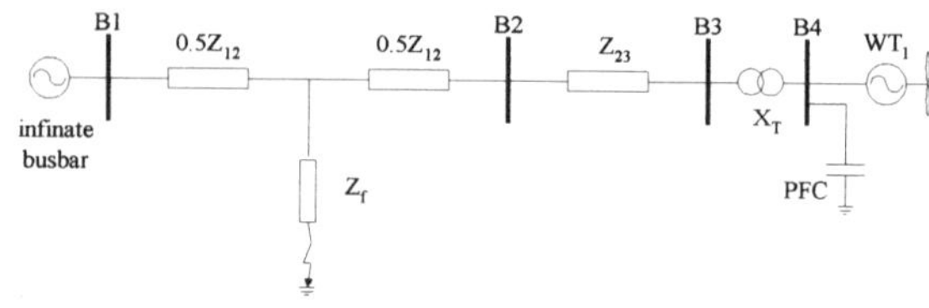

Figure 7: Model to investigate wind farm response to connecting network voltage fluctuations.

The system was initially simulated for 3-phase balanced network faults using an X/R ratio of 10, for section B2-B3, and a SCL of 3600 MVA, for the 60 MW wind farm point of common coupling (PCC), at busbar B3. This investigates the transient stability of the wind farm when connected to a strong network (short circuit ratio of 60). The connecting transformer has a leakage reactance of 5% and a typical value of 30% PFC (18 MVAr) is applied at the machine terminals. Under normal operating conditions the terminal voltage at B4 is 0.986 pu. Figure 8 illustrates that post fault the turbine and the network maintain stability after voltage variations.

To observe the stability of the wind farm when connected to a weak network the PCC SCL was reduced to 360 MVA (short circuit ratio of 6). Under normal system conditions the operation is satisfactory. However, during a voltage variation in the network the electrical power produced by the generator drops but the rotor, which is driven by the prime mover, accelerates.

The increased generator speed results in large quantities of reactive power being absorbed by the induction generator, as illustrated in Figure 4. This causes a significant voltage drop at the terminals and the generator continues to overspeed, as shown in Figure 9.

in Figure 9, a single-phase fault is applied. Figure 12 shows that the wind farm and connecting network maintain stability with this fault condition.

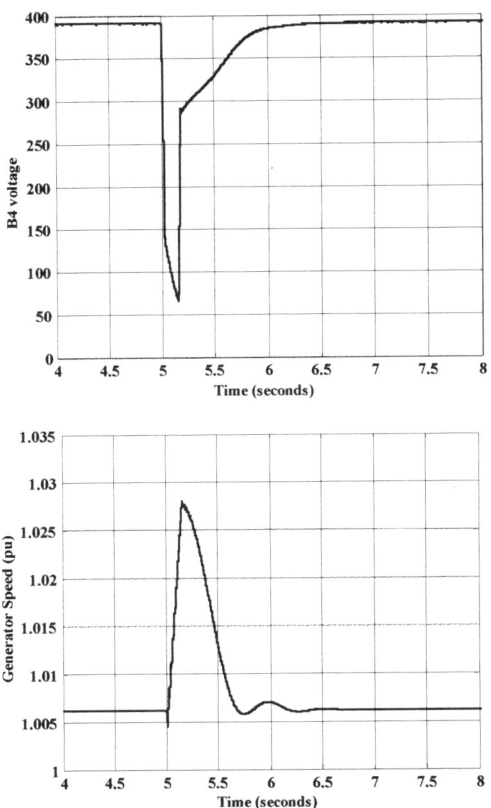

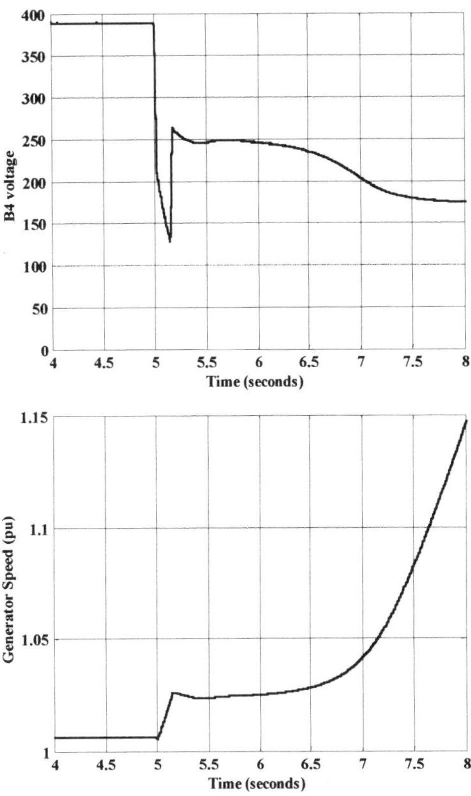

Figure 9: Terminal voltage and generator speed behaviour for a weak network connection, X/R = 10

Figure 8: Terminal voltage and generator speed behaviour for a strong network connection, X/R= 10

Maintaining the weak network connection but implementing an X/R ratio of 2, typical for submarine cables, the stability of the offshore installation was investigated. Figure 10 illustrates that a previously unstable installation has recovered stability due to the electrical parameters of the connecting cable. This reinforces the stability observations of Figure 5 that predicted an increased pull out torque in the generation mode for this X/R ratio. Continuing the parametric influence of the typical wind farm installation the previously stable case was now simulated with a reduced quantity of terminal PFC. With 15% PFC applied at the machine terminals (9 MVAr), under normal operating conditions the terminal voltage at B4 is now rather low at 0.952 pu. Figure 11 shows voltage instability occurs at B4 with insufficient wind farm fixed PFC.

Up to this point the simulation of the power system and the applied 3-phase fault is symmetrical. However, the effect of a single line-to-ground fault between B1 and B2, which will result in an unbalanced system, is now considered. Applying network parameters for the previously unstable weak network scenario, illustrated

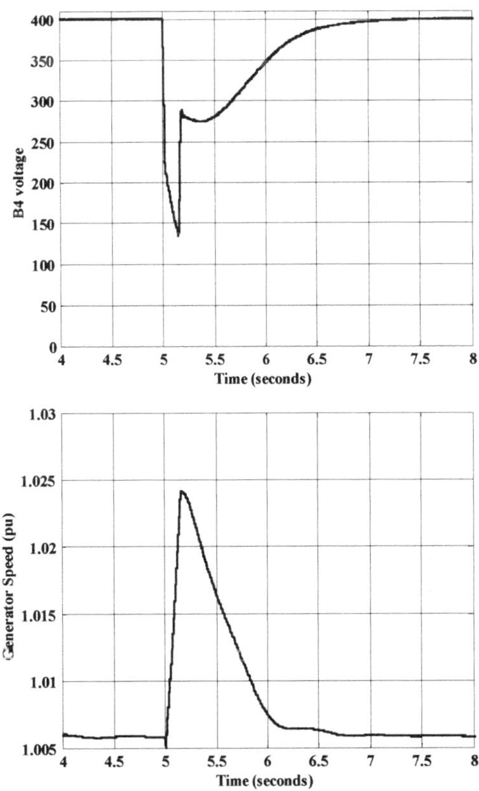

Figure 10: Terminal voltage and generator speed behaviour for a weak network connection, X/R = 2

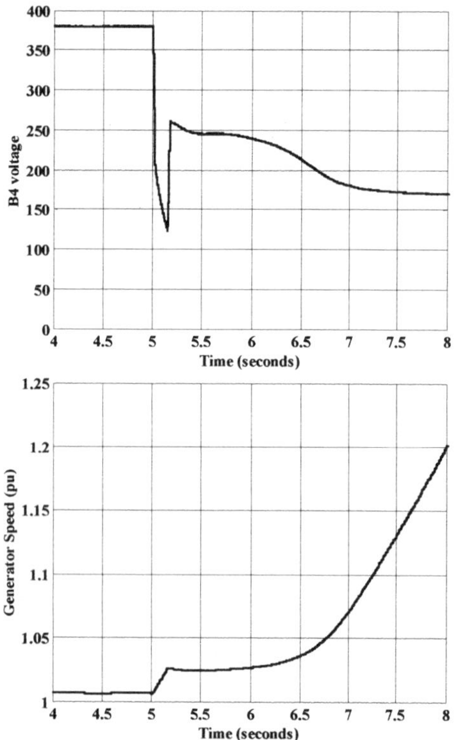

Figure 11: Weak network connection, PFC = 15%

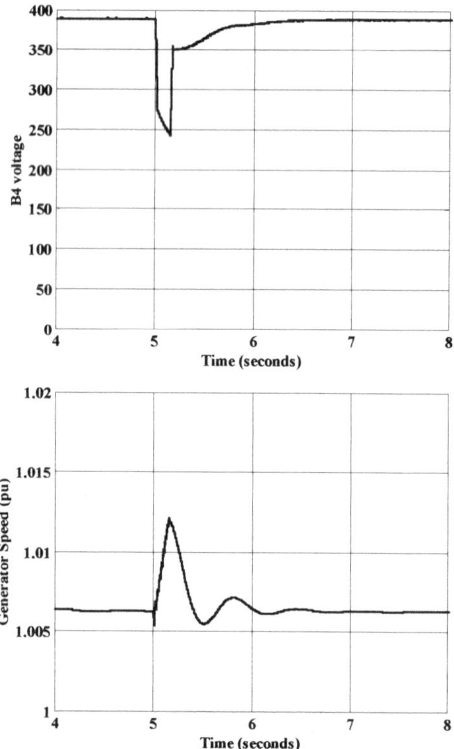

Figure 12: Weak network connection and single line-to-ground fault.

5. DYNAMIC STABILITY IMPROVEMENTS

Similar to the synchronous machine equal area criteria for operating angle, the wind farm can maintain stability if the slip of the induction generator does not exceed its critical value. A dynamic stability limit criteria for induction generators to demonstrate this has been presented by Akhmatov et al (5). Therefore, possible solutions for improving the stability may be implemented by increasing the pull out torque (increasing the electrical torque) or reducing the applied mechanical torque during voltage variations in the network.

Initially considering the electrical parameters influence on the stability, Figure 2 has shown that a strong network connection leads to a large pull out torque. Therefore, reinforcing the connecting network should assist in the stability of the system. The use of voltage source converter systems to enable control of the power flow, which may improve system stability, is also being considered.

The influence of the induction generator electrical parameters can also be taken into consideration at the design stage to improve stability. If the critical slip of the turbine is increased the short-term voltage stability will be improved. This can be obtained by reducing the values of R_s, X_s, X_m and X_r, or increasing the value of R_r, as shown in Figure 13. The possibilities of implementing a dynamic rotor resistance, Akhmatov et al (6), for increasing the critical slip and thereby stability offers control similar to partially variable speed induction generators.

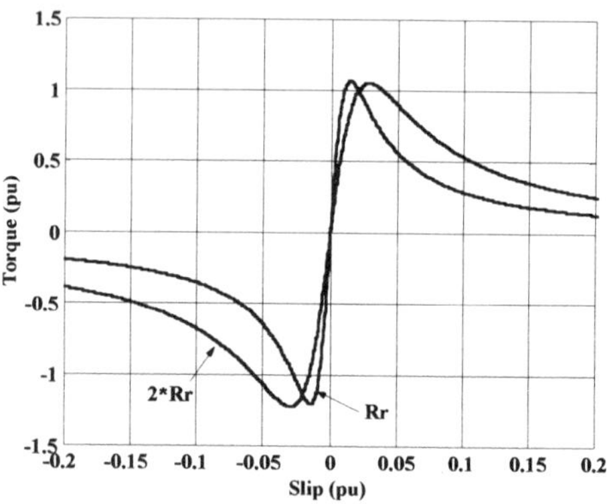

Figure 13: Variable rotor resistance influence on stability

Due to the possible improvements in dynamic stability the variable speed wind turbine has recently attracted much interest, Feijoo et al (8), and appears to be the trend for future years. The stability of the doubly fed induction generator used for large offshore wind farm installations is currently under investigation.

Another possibility for improving dynamic stability with conventional technology is the turbine mechanical construction and mechanical torque control. It is obvious that a more stable operation is obtained, in network post fault conditions, if the wind turbine inertia is high. Although the inertia value will not directly influence the critical slip value the dynamic stability will be improved by slow acceleration.

If fast blade angle control techniques are implemented for the turbines the mechanical torque of the system can be reduced in the event of network failures causing the electric torque of the generator to drop. An imbalance in the rotor swing equation, shown in equation (1), results in machine acceleration.

$$\frac{d(s)}{dt} = \frac{1}{2H}\left(T_e - T_m\right) \qquad (1)$$

With a reduction in mechanical torque the rotor speed can be controlled such that stability is maintained when the network fault is cleared.

Dynamic Reactive Compensation

The use of dynamic reactive compensation, in the form of voltage control through shunt connected static VAr compensator (SVC) devices, for improving the dynamic stability has been identified as a possible technique for large wind farm installations. As shown by the steady state and transient simulations, the necessity for accurate and dynamic voltage control at the terminal bus of the installation is critical for maintaining stability. Used in conjunction with the electrical and mechanical controls discussed above, an appropriately rated SVC may be effective.

Preliminary simulations utilising a shunt connected STATCOM for dynamic reactive compensation of a 60MW wind farm have been conducted. The response of the wind farm to a network short circuit fault has been investigated using the same parameters as presented previously for a weak network. The critical clearance time to maintain system stability without reactive compensation is t = 0.14 s, this has been increased to t = 0.37 s with the inclusion of the STATCOM at the wind farm terminal bus.

Controlling the Active Power

Another methodology for improving wind farm dynamic stability, which is currently under investigation at UMIST, is a technique of controlling the active power output of the wind farm under network disturbances. The implementation of a shunt connected STATCOM with an appropriate energy source, such as a battery, enables the power output of the wind farm to be absorbed during an imbalance in the rotor swing equation.

By a technique similar to the electric brake resistance system, shown in Figure 14, the STATCOM can absorb the active power to restrict overspeed of the generator and improve stability. Simulations have shown that using the STATCOM, for a combined voltage and power control of the wind farm, the critical clearance time can be increased to t = 0.41 s.

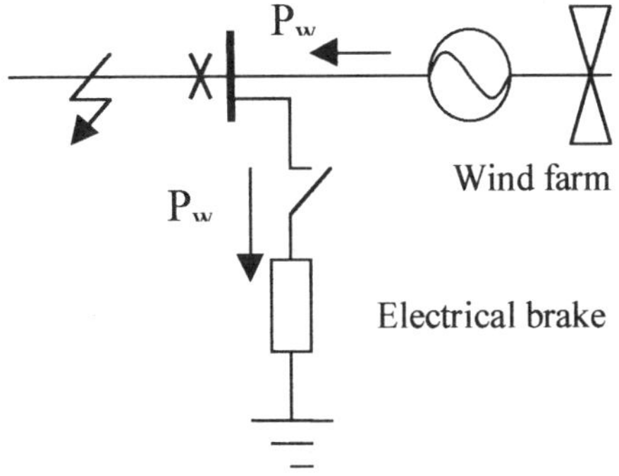

Figure 14: Electric brake for controlling active power

6. CONCLUSIONS

Steady state and transient modelling of a 60 MW wind farm has shown that relatively high ratios of wind farm capacity to network short circuit level can be accommodated successfully if conditions are favourable. However, voltage variations within the connecting network can lead to generator overspeed and network voltage instability. Dynamic stability techniques to improve the system stability in the event of voltage fluctuations have been discussed. The generator overspeed and network voltage collapse phenomena is definitely an important limiting factor to the future development of offshore wind farms unless sufficient dynamic stability improvement or voltage control technologies are implemented.

7. REFERENCES

1. DTI Consultative Document: NEW & RENEWABLE ENERGY – Prospects for the 21st Century, 30th March 1999.
2. "Prospects for Offshore Wind Energy", A report written for the EU by The British Wind Energy Association(BWEA),www.britishwindenergy.co.uk
3. Grainger, B., Thorogood, T., 2001, "Beyond the harbour wall" IEE Review.
4. Jenkins, N., Strbac, G., 1997, "Impact of embedded generation on distribution system voltage stability." IEE Colloquium on Voltage Collapse.
5. Akhmatov, V., Knudsen, H., et al, 2000, "A dynamic stability limit of grid connected induction generators" Power and Energy Systems Int. Conf. Proc.
6. Akhmatov, V., Knudsen, H., et al, 2001, "Modelling and Transient Stability of Large Wind Farms" 2nd Int. Workshop on Transmission Networks for Offshore Wind Farms.
7. Kundur, P., 1994, "Power System Stability and Control" Book, McGraw-Hill, Inc.
8. Feijoo, A., Cidras, J., Carrillo, C., 2000, "A third order model for the doubly-fed induction machine", Electric Power Systems Research, Vol. 56, Issue 2, pp 121-127.

HVDC TRANSMISSION FOR LARGE OFFSHORE WINDFARMS

N M Kirby, M J Luckett, L Xu,
ALSTOM T&D Ltd – Power Electronic Systems, UK

W Siepmann
ALSTOM Power Conversion GmbH, Germany

ABSTRACT

Renewable energy generation, in the form of both onshore and offshore windfarms, is a rapidly growing, high-tech industry on a worldwide basis. The development of larger, more efficient turbines is opening up new frontiers in wind energy generation in the form of large offshore windfarms. This paper describes how the use of High Voltage Direct Current (HVDC) technology can be used to fully realise the potential of these developments, overcoming a number of major technical challenges facing traditional AC solutions and providing the project developers and utility receiving the renewable energy supply with an efficient, economic and reliable solution.

INTRODUCTION

In recent years pressure has been applied to the international community to promote greater environmental responsibility. For the first time, in December 1997 at Kyoto, the industrialised countries agreed to reduce emissions of greenhouse gases.

One of the points of agreement was that

Parties to the protocol are to be encouraged to achieve emissions reductions through a series of policies and measures, such as improving energy efficiency, reforming energy and transport sectors, protecting forests and other carbon sinks, promoting renewable energy, and phasing out "inappropriate fiscal measures".

The German government has promoted the use of wind energy for many years as the multitude of onshore windfarms across northern Germany demonstrates. However to gain the maximum from this technology large, offshore windfarms are required. A crucial timescale is determined by the prices per kilowatt-hour laid down by the German government within the scope of the German Renewable Energies Act. This act proscribes that all offshore windfarms, which are commissioned by the end of 2006, can obtain the maximum kilowatt-hour price over a period of 9 years. Calculating back from this deadline and considering the different phase of such a project - i.e. prototype, prototype field tests, pilot plant, field experience with the pilot project and start of the final extension stage it can be seen that commencement of such a project is required imminently. Subsequently, wind farm developers depend on reliable, existing technologies.

AC-DC Power Transmission, 28-30 November 2001
Conference Publication No. 485 © IEE 2001

It is obvious that a total investment of approximately 2 billion Euros for a large offshore windfarm triggers particularly critical questions related to the project risks as well as the anticipated profit and payback periods. However, the developer will also bear in mind that the project will necessarily have minimal environmental impact and the utility concerned will define the level of electrical impact that the project may have on the existing ac network in terms of voltage fluctuation levels, harmonic distortion and reactive power exchange limits.

To maximise energy transmission, reliability and stability, HVDC technology is the ideal choice.

WINDFARM TECHNOLOGY

Experience with electrical energy generated by windmills is available since the eighties. But commercial use started in the early nineties, in particular in Denmark, USA, The Netherlands, Spain and Germany. Up to about 1 MW unit power, most of the windmills have been supplied with asynchronous generators, fixed speed, direct coupled to the grid and with stall control of the blades. This solution has some disadvantages because fixed speed, driven by the fixed grid frequency, causes variation of the output power according the wind speed. Consequently the grid voltage is very unstable especially in grids with poor capacities. This and overload protection were the reasons to change the windmill technology from fixed speed to variable speed when the unit size has been increased to the Megawatt class, and in the meantime to the Multi Megawatt class.

The design of a windmill with variable speed needs a converter between the generator and the grid. This converter is designed in most cases as IGBT active front-end converter with voltage source. There are two different solutions available:

- Fully fed synchronous generator, optionally with permanent magnets excitation

- Doubly fed asynchronous generator with slip ring rotor

The converter for each solution is different: Fully fed means, the converter size is equal to the generator size and it is connected at the stator windings. Doubly fed means, the converter size is only about 25 % of the generator size and it is connected at the rotor windings via the slip rings. But with respect to the line side behaviour there is no difference when we see the generator-converter-system as one unit. What are the main important characteristics of windmills with variable speed by using converters and pitch control?

1. According to specification the unit is designed for a power factor between 0,9 capacitive and 0,9 inductive and this can be controlled in a dynamic way. Reference values have to be supplied by the energy management computer. This design of the generator-converter-system makes a reactive power control range of ± 50 % of the full load active power available which can be used for cabling optimisation and grid voltage control/stabilisation.

2. The waveform of voltage and current are nearly sinusoidal, no extra filters outside the converter.

3. The converter limits the output power to the rated power.

4. Each windmill uses a line transformer for connection to the local grid, so an existing grid voltage level can be adopted.

5. The pitch controller controls the active power of the windmill by varying the angle of the blades.

Owing to the high cost of the foundations and other offshore facilities for an offshore wind farm with many fixed costs to be taken into account, wind farm developers will install the largest windmills possible. The blade material technology presently limits the power outputs to 5MW. This 5 MW is available at the terminals of the primary transformer windings.

All electrical equipment has to be supplied to protection class IP54, able to withstand salty atmosphere and ambient temperature between –20° and +50°C. The lifetime should be calculated for 20 years. Because maintenance costs in offshore wind farms are very high the design should be made for a minimum of maintenance.

In Germany there are concrete projects developed for offshore wind farms in the North Sea as well as in the Baltic Sea with capacities between 200 and 1500 MW, however, for the time being no project is approved. The distances from offshore to onshore vary up to about 150 km. Several windmill manufacturers are involved in the development of a 5 MW unit, the procedure being to start with a prototype to get experience during operation with it onshore. The second stage is a pilot plant with about 10 units 5 MW each installed at site. Experience with this pilot plant will be the basis for all decisions regarding the final stage of the planned wind farm with full capacity. For offshore wind farms 3600 full load hours per year are expected.

To reduce the impact from windmill to windmill by wind turbulence so much as possible the distance between the units should be about 700m. That means a wind farm for 500 MW with 100 windmills 5 MW each needs an area of 7 times 7 km. All windmills are connected to a local 30 kV AC grid in star or chain configuration or a combination of them. A substation with transformer and HV switchgears will be the local point of common coupling for the energy transmission line to onshore. AC- or HVDC-systems can do this. The simplest one appears to be AC because the offshore wind farm needs auxiliary energy (also when the wind is not blowing) and all necessary equipment is well known and already existing. Regarding existing cables the capacity of a 3-core AC submarine cable is 200 MW at 145 kV. These values influence the total number of cables. Because all cables have to cross the German National Park around the North Sea coastline (Wattenmeer) the number of cables have to be reduced as low as possible. Therefore, to get approval from the authorities involved, HVDC energy transmission technology should be chosen. That means using the well-known conventional HVDC technology for 500 kV dc. However, it has to be taken into account that the offshore converter needs AC voltage for commutation purposes. This could be supplied by a parallel AC cable if this is already installed during the pilot plant stage or could be provided by a synchronous compensator. For the offshore HVDC station a "second hand" oil platform would form an ideal base on which to mount the new convertor building.

USE OF HVDC WITH WINDFARMS

HVDC Technology Application

As windfarms become larger and more distant from shore, the justification of using HVDC to transmit the power to the onshore network becomes easier, particularly at power levels of 500 MW or more. The cost of the converter stations, offshore and onshore, are significant in themselves, but when put in the context of the complete project cost including the cables and the wind turbine generators, they feature less prominently. It is not presently economically feasible to use VSC Transmission technology at this power level due to the high cost of the multiple converters and cables that are required. The choice is therefore more logically between the use of AC or conventional HVDC.

Using conventional HVDC transmission offers many advantages to a direct AC inter-connection:

- Frequency at the sending end can be variable – the ability to de-couple the multiple turbines, and avoid synchronising them with the onshore network has a major advantage, as each end of the link may be allowed to operate according to its own control strategy, largely independent of the other end.
- Transmission Distance is not a technical limitation – the impact of cable charging current in AC cable interconnections is significant, even dominant, but in DC applications it is negligible.

- Offshore Installation is isolated from Mainland Disturbances – faults on one network will have a limited effect on the other, and in some cases the DC link is able to assist in the recovery from the disturbance.
- Power flow is fully defined and controllable – unlike AC interconnections the power flow through the transmission link is controllable by a range of strategies to suit system conditions during both normal and fault situations.
- Proven and Reliable Equipment – conventional HVDC is well-established and has a proven track-record in transmission links around the world, achieving the high availability required for such system interconnections as this.
- Low Power Loss – the ability to optimise the cable and converter ratings based on losses, by removing the effects of cable charging current means that overall the operating losses are lower than an equivalent rated AC interconnection.
- High Power Transmission Capability for each cable – in the same way as a pair of HVDC lines may carry 1.7 times that of a similar sized 3ph AC line.

Although the HVDC link is conventional in that it uses thyristor technology, in offshore situations there are many aspects which demand special attention, in order to achieve the necessary performance and reliability, such as provision of auxiliary supplies and a commutating voltage source.

Offshore Installation

The main constraints which must be addressed in any offshore electrical installation include:

- Limited space – in order to minimise the cost of the offshore installation the equipment should be as compact as possible, to reduce the overall size and weight.
- Extremely harsh and variable environment – constant exposure to the salt air, wind and water means that as much equipment as possible must not only be located indoors, but in many cases sealed to prevent moisture ingress.
- Auxiliary supplies – the auxiliary loads are graded such that those critical to the operation and safety of the equipment and personnel are always available. This requires some form of UPS, generator, battery, etc. or a combination of these, to ensure that even when there is no wind these supplies continue.
- Limited maintenance access – having reduced the size of and space between equipment, the accessibility for maintenance is obviously reduced. Therefore equipment must be as maintenance-free and reliable as possible, with appropriate redundancy incorporated to continue operation in the event of single (and in some cases multiple) faults.

These concerns lead to a set of design criteria for the offshore installation:

- Equipment should be as simple as possible with long maintenance intervals, or preferably no maintenance at all
- High reliability, which means that redundancy should be included in critical areas
- AC Voltage as low as possible to reduce AC harmonic filter and switchgear size
- Indoor equipment means that isolation levels and clearances may be reduced
- Multi-storey structure to reduce the base area to a manageable size for platform legs and supports
- Extensive use of automation, with remote control and monitoring of as much offshore plant as possible, with remote diagnostics

The control system is designed with the ability to control the offshore grid voltage and frequency, and to handle disturbances in both the sending and receiving end systems.

Auxiliary power is a problem for the offshore installation, and essential loads for the 1000 MW link are expected to be in the order of 1 MW for the converter station and 6 MW for the windmills. The loads are categorised according to:

- Un-interruptible - supplied via batteries, including control, protection, monitoring and communication
- Other essential loads - AC loads for cooling, heating, navigation lights, excitation, switchgear, emergency lighting
- Non-Essential – heating, lighting and air conditioning for manned areas, cranes, lifts, etc.

Normally the auxiliary power is supplied from that generated by the windfarm, however during conditions of low or no wind the auxiliary power may be taken from several sources:

- a parallel AC cable from onshore (if installed as part of a trial)
- reverse power from onshore through the HVDC link - if necessary supported by a synchronous compensator
- from a diesel generator

Onshore Installation

The onshore HVDC converter station is a conventional installation with thyristor valves, converter transformers, AC filters, and will provide the extra facilities required for remotely controlling the offshore end of the link. A typical converter station layout is shown in Figure 4

Cable Effects

Figure 1 shows the effect of cable charging current on a range of different cross-sectional areas (CSA) of AC cable. As the cable CSA increases, the capacitance increases, reducing the distance at which the charging current exceeds the rated current of the scheme. On the basis of charging current alone the breakeven distance could be as large as 180 km in this case. However the losses increase with resistance, which decreases with increasing CSA. In order to minimise the losses one would select the other extreme cable size shown in Figure 1, leading to a breakeven distance of 80 km.

This fundamental conflict between requirements of larger CSA to minimise losses and smaller CSA to minimise charging current requires a compromise solution when using AC cable. In the case of a DC cable the charging current may be removed from the calculations, so the cable may be optimised on the basis of conductor losses, copper cost and insulation ratings.

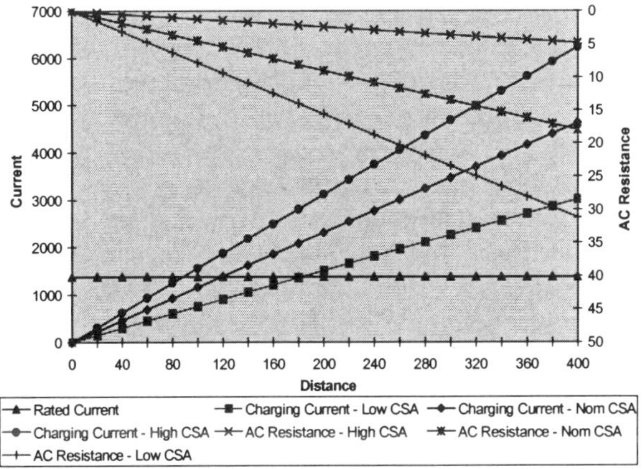

Figure 1 Cable Charging Current

Example Scheme

To illustrate these points, the example scheme shown in Figure 2 is described below.

The windmills are arranged in clusters of around 16 x 5 MW units, each of which is individually controllable in terms of real and reactive power. These are marshalled on to a common 30 kV busbar, and this feeds to 30 / 145 kV step up transformers. The 145 kV busbars are linked and cabled to the offshore converter station platform. The platform is illustrated in Figure 3. The main points to note in the platform layout are:

- converter transformers located outdoors and accessible for maintenance, configured as 3ph 2-winding to minimise the overall size, with no exposed bushings
- thyristor valves are arranged as bi-valves, which gives a good compromise on height/area
- use of a relatively low AC voltage reduces the clearances for AC filters and switchgear, and therefore the overall volume of the indoor filter area

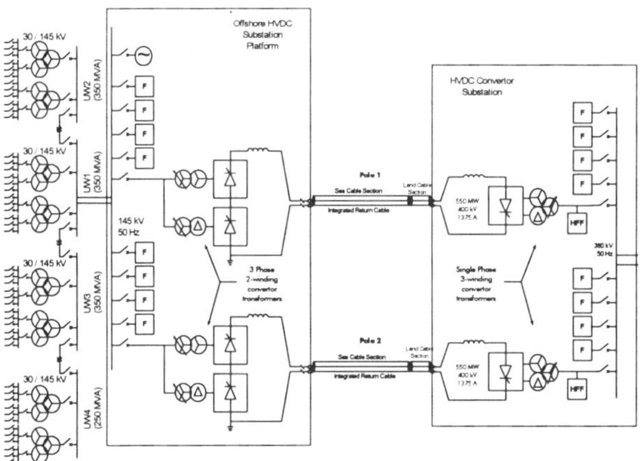

Figure 2 Single Line Diagram

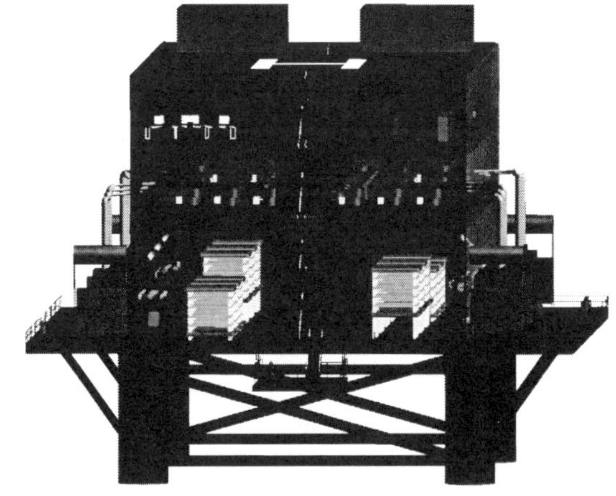

Figure 3 Offshore Converter Station Platform

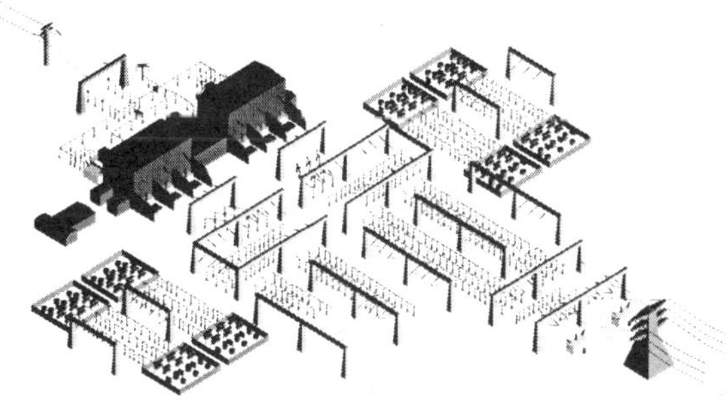

Figure 4 Onshore Converter Station Layout

ALTERNATIVE SOLUTIONS

For large, offshore windfarms with power ratings above about 200 MW it can be seen that thyristor-based conventional HVDC technology is the optimum solution. However, for smaller power ratings and offshore distances other viable solutions present themselves.

AC Transmission

Current AC technology can be economically used for the lower rated schemes over short distances. Limits of AC cable power ratings over longer distances will probably not be improved enough to allow utilisation of this technology on larger windfarms. Connection of windfarms via AC technology does not mitigate voltage and power variation effects due to the windfarms. Individual control of the voltage, power and reactive power of each windmill, combined with an overall energy management system to coordinate them, are essential.

Voltage-Sourced Convertors

Voltage-sourced convertor (VSC) based HVDC technology (VSC Transmission) is a fairly new development that may be suitable for the lower power rated schemes and over smaller distances. VSC technology allows flexibility in reactive power control and alleviates some of the offshore low voltage conditions. The reliability of this technology is yet to be fully proved by significant in-service time and can produce higher overall losses than conventional HVDC technology due to the nature of the switching devices. However as this technology becomes more widely used for conventional applications and the maximum power rating of the technology improves, the high reliability and low loss requirements for windfarm projects should cease to be a barrier to this approach.

Technology Combinations

Combinations of technology may produce useful hybrid solutions. For example, use of VSC offshore and conventional HVDC on-shore may produce an efficient and economic solution.

Conventional HVDC may be combined with Static Var Compensator (SVC) or STATCOM to provide effective var and voltage control where this is deemed necessary by AC network requirements.

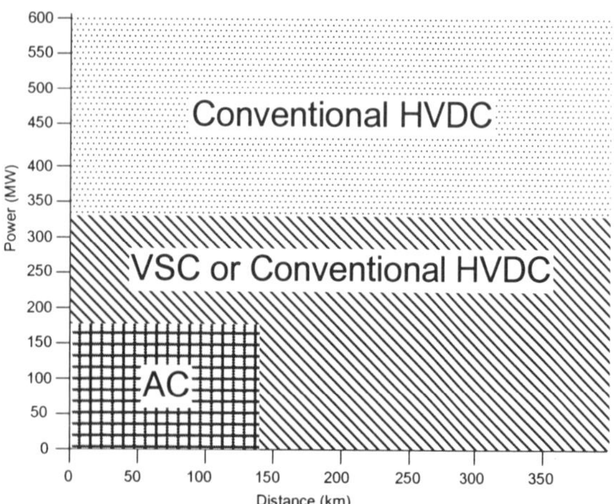

Figure 5 - Solutions with power range vs distance

Comparison

An estimated comparison of the different technologies is as shown in Figure 5. The figures in the graph assume the use of a single 3 conductor cable in the case of an AC solution, and a single 2 conductor cable in the case of a DC solution.

SYSTEM STUDIES AND AC NETWORK INTERACTIONS

The single-line diagram of the offshore system considered is shown in Figure 6. As shown, it contains 1000MW windfarms, a 100MVar synchronous compensator, 850Mvar capacitor banks and a 12-pulse HVDC transmission system rated at 1000MW. The 1000MW windfarm model contains 200 induction generators with each rated at 5MW. The operational power factor for the induction generator is assumed to be 0.9 at rated power. The synchronous compensator controls the AC voltage on the common bus by providing or absorbing reactive power. The rectifier of the HVDC converter controls the DC current whilst the inverter uses Gamma control. The frequency derivation on the rectifier side generates the DC current order. The relatively large synchronous compensator is used because there is no reactive power control by the HVDC converter for this preliminary study and no switched capacitor banks. In reality, due to the switched capacitor banks together with the reactive power control capability of the HVDC converter, a relatively small compensator can be used. It may also be possible to remove the synchronous compensator altogether through the use of a single AC cable suitably rated in parallel with the DC circuit. Furthermore, the use of doubly fed induction generators would remove the need for the synchronous compensator for providing reactive power support to the network. System studies using these technologies are currently being carried out in ALSTOM.

First, steady state and variations of wind speed are studied and simulated results are shown in Figure 7. From Figure 7 it can be seen that the windfarms generate 0.65pu of rated power and absorbs approximately 0.35pu reactive power. The synchronous compensator provides 0.67pu (67Mvar) of reactive power to the network. When the wind speed drops at the time of 2 seconds, the active power from the windfarm, active power to the inverter AC side reduce accordingly. As the reactive power taken by the HVDC converter reduces, the synchronous compensator changes from providing reactive power to absorbing reactive power.

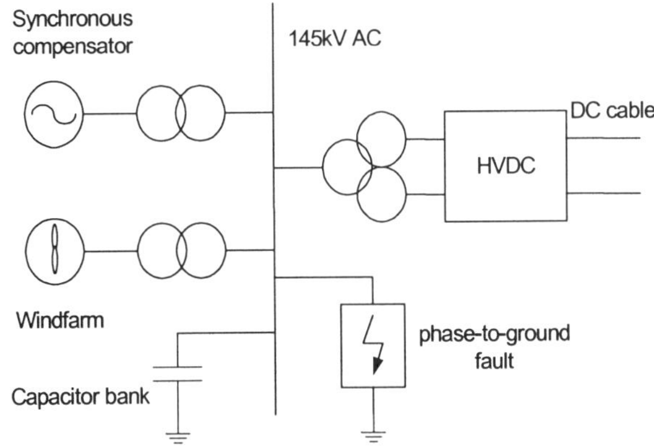

Figure 6 Single-line diagram of the system simulated (offshore)

Studies at fault conditions have also been carried out. Figures 8 and 9 show the simulation results when single phase to ground faults were applied directly on the rectifier and inverter AC buses, respectively. As can be seen when the fault is applied on the rectifier side, because of the drop of the AC voltage, the synchronous compensator, which controls the AC voltage, provides reactive power support to the network. However, because of the DC connection, there is only a small impact on the AC network at the inverter side. After the fault is cleared the offshore ac system returns to normal in about 2 seconds. When the fault is applied on the inverter side, the DC voltage reduces to zero and therefore, the transmitted active power goes down to zero. Consequently, the frequency on the rectifier increases rapidly. After the fault is cleared the offshore ac system returns to normal in about 1.5 seconds.

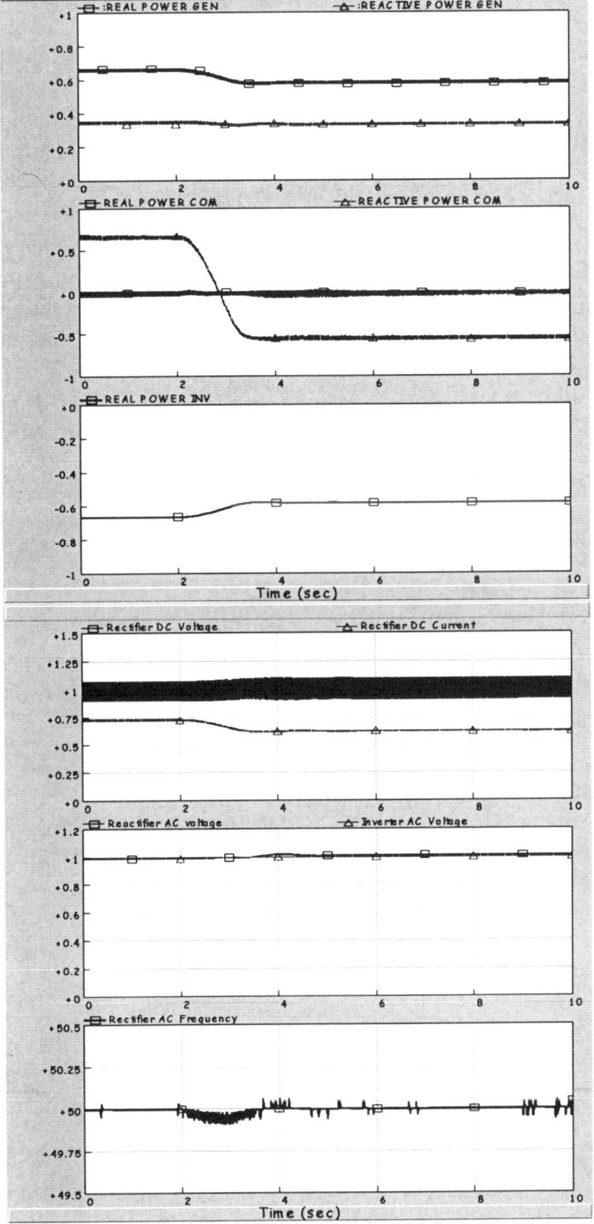

Figure 7 Operation with variation of wind speed

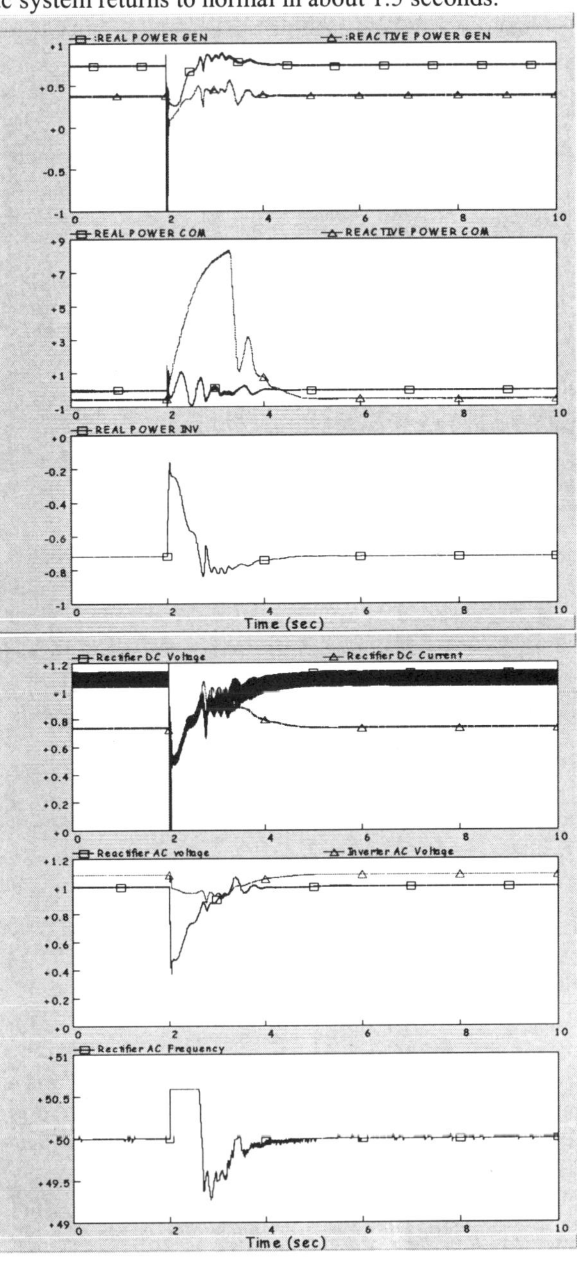

Figure 8 Single-phase to ground fault on the rectifier side (offshore)

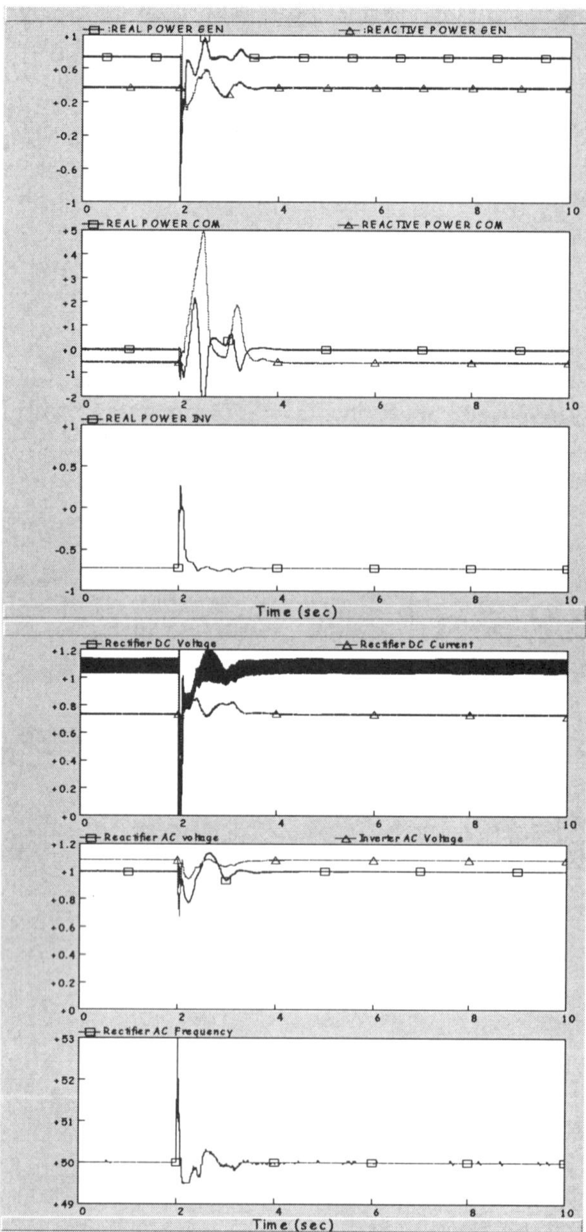

Figure 9 Single phase-to-ground fault on the inverter side (onshore)

CONCLUSIONS

Increasing use is being made of wind power generation through political and public pressure on a worldwide basis, and the limited availability of onshore sites is driving windfarms to offshore locations. This inevitably means that the power generated has to be transmitted over longer distances in order to make a connection to an AC network for onward transmission and distribution to customers. The lower power and shorter distance schemes may be most economically provided through use of AC cable interconnections, and at the other extremes of high power and long distance there is little alternative but to use conventional HVDC. In between these extremes there is no simple rule to apply, and the result for each particular scheme will be different based on its own merits, taking into account the issues described in the paper. In general however we expect the well established and proven technology of conventional HVDC will be the optimal solution in cases where:

- the distance offshore is greater than approximately 100 km
- connection into the AC network is at a weak point
- wind farm size is greater than approximately 350 MW and constraints are imposed on the quantity of cables laid
- where multiple or long AC cables are necessary, equivalent transmission through a DC cable and converter equipment may be a more economical alternative
- to reach a suitable AC network connection point requires a significant length AC transmission line onshore
- AC network studies and simulations under different fault conditions show unstable behaviour, and HVDC control of power transmission may assist in recovery, limit the effect, or even correct the instability

In order to operate most efficiently the individual windmills should each have their own power, reactive power, and voltage control, with an overall coordinating energy management system to control overall energy transmission.

REFERENCES

1 Kyoto Protocol to the United Nations Framework Convention on Climate Change, December 1997.
2 BICC Cables, Blockwell Science, London, 1977, "Electric Cables Handbook", third edition
3 L.E Jones, T. Ackermann, July 2001, "Transmission Networks for Offshore Wind Farms: Special Report", <u>European Wind Energy Conference</u>, Denmark

Real-Time Implementation of a HVDC-VSC Model for Application in a scaled-down Wind Energy Conversion System (WECS)

K.H. Chan   J.A. Parle   N. Johnson   E. Acha

Centre for Economic Renewable Power Delivery (CERPD)

Department of Electronics and Electrical Engineering, University of Glasgow

Abstract

This paper presents a time domain model of a HVDC-VSC system suitable for real-time simulations and its application to a Wind Energy Conversion System (WECS). The HVDC-VSC model comprises two Voltage Source Converters (VSC) linked together by a DC-link cable, their control systems and interfacing transformers. The operations of the VSC are based on Pulse Width Modulation (PWM). The WECS is represented by a scaled-down system comprising a three-phase induction machine driven by a controllable DC servomotor running at a speed that corresponds to a full-size asynchronous wind generator.

The mathematical model of the HVDC-VSC system is implemented in the real-time simulation environment using a commercially available Real-Time Station (RTS). The power output profile of the scaled-down WECS is entered into the HVDC-VSC model residing in the RTS and fed to an external load using the Digital-to-Analogue Converters (DAC) of the RTS.

The effect of fluctuating wind speed is simulated in the scaled-down WECS and the response of the HVDC-VSC system to maintain voltage regulation to the load is assessed. The detailed implementation and configuration of the test environment is described in the paper.

Keywords

HVDC-VSC, scaled down WECS, real-time simulation, wind speed, voltage regulation.

1   Introduction

The application of digital computer based simulation programs for analysing different aspects of power system network operations and developments has seen an exponential increase over the previous decades. However, these simulation tools do not provide a platform suitable for real-time test environments where mathematical models and physical electric devices can be analysed in the same instance. The mismatch between the CPU time taken by conventional simulation tools and the actual physical system's response makes this a difficult task.

In recent years, the development of modern fully digital dynamic simulators with real-time capabilities have made this approach possible [1][2]. The digital simulator or RTS executes the simulation of mathematical models and continuously produces output signals in real-time which accurately represent the real system under investigation. The RTS can run real-time tests on any power system network operations based on mathematical models. The output from the RTS is connected through the DAC, and can be used as control signals, for example, to any physical device, creating a complete hardware-in-the-loop (HIL) environment.

The RTS used in this work is a commercially available simulator developed by Applied Dynamics International (ADI). The main components of the RTS are the computing engine (CE3/CE4), the simulation processor (SP) and the VMEbus Interact Manager (VIM). The development of the mathematical model is undertaken within a GUI-based software package called EASY5.

In this paper, a HVDC-VSC model is implemented for real-time applications. The HVDC-VSC is a new type of HVDC system making use of the latest development in high speed power electronic switches with gate turn-off capabilities such as GTOs and IGBTs [3][4][5]. The model of the HVDC-VSC is implemented in the time domain suitable for any real-time simulations.

The main objective of this paper is to apply the real-time HVDC-VSC model in a scaled-down wind energy conversion system (WECS) to supply a load. The scaled-down WECS is created using a controllable DC servomotor running at speeds that correspond to a full-size asynchronous wind generator driving an induction machine. The power output profile of the scaled-down WECS is entered into one terminal of the HVDC-VSC model residing in the RTS. The output from the HVDC-VSC is then fed to an external load through the RTS.

2   Time Domain HVDC-VSC Model

The HVDC-VSC is based on the operation of the voltage source converter (VSC) technology. The basic building blocks of a HVDC-VSC are as shown in Fig. 1 [3][6],

AC-DC Power Transmission, 28-30 November 2001

Conference Publication No. 485 © IEE 2001

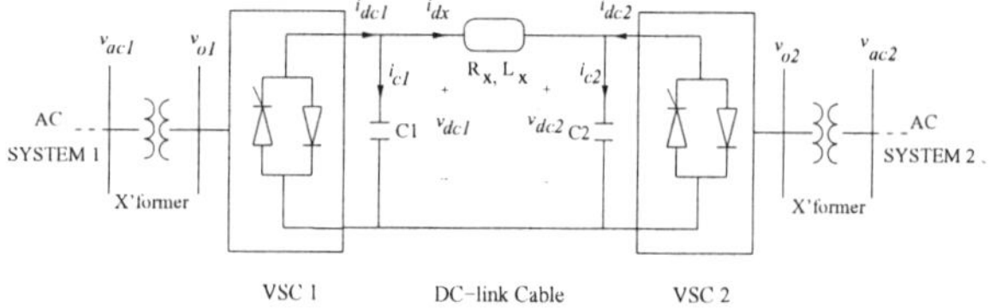

Figure 1: HVDC-VSC Configuration

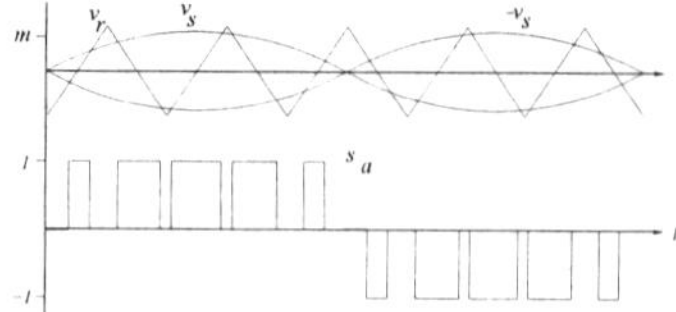

Figure 2: PWM Switching

where two VSC stations are linked together by a DC-link cable. As mentioned earlier, the switching elements used in the converters are high speed switches capable of turn-on and turn-off by a gating signal at a frequency rate higher than the fundamental frequency. The gating signals are controlled by a PWM technique. The AC system is connected to the terminal of the HVDC-VSC station through the interfacing transformers.

2.1 Mathematical Model

2.1.1 Voltage Source Converters (VSC)

The VSC stations are modelled based on the switching functions of a PWM control scheme [7]. The per-phase switching process of the VSC's output is as shown in Fig. 2, where, v_r represents the carrier signal with a frequency of f_r and v_c the modulation signal with a frequency of f_s. The signal v_c, which is in phase with the fundamental frequency at the terminal of the interfacing transformer has a magnitude which is determined by the modulation index, m. The switching function s is equal to ± 1 when $|v_c| > |v_r|$, and zero otherwise.

Harmonic reduction is possible by increasing the number of levels of the VSC. This can be achieved by adding another PWM switching function with the angle of v_r shifted while v_c remains the same [7]. A three level converter can be obtained by having three individual PWM switchings with v_r shifted by $0°$, $120°$ and $-120°$ respectively. The average of the three PWM switchings will be the output from the three level VSC.

The per-phase voltage output for the VSCs are given by,

$$v_{o1} = s_{a1} v_{dc1} \qquad (1)$$

$$v_{o2} = s_{a2} v_{dc2} \qquad (2)$$

The magnitude of the fundamental frequency of the VSCs output is a function of the modulation index, m. The voltage angle can be controlled by shifting the angle of v_c with respect to the AC voltage at the terminal of the interfacing transformer.

2.1.2 The DC-link Cable

The DC-link connection between the two VSC stations can be represented by an equivalent DC network resembling a π-circuit as in shown Fig. 1. As mentioned earlier, there exists an instantaneous power balance between the AC and DC system, giving rise to the following equations for VSC1,

$$P_{ac1} = P_{dc1} \qquad (3)$$

$$v_{oa1} i_{a1} + v_{ob1} i_{b1} + v_{oc1} i_{c1} = i_{dc1} v_{dc1} \qquad (4)$$

Substituting (1), with the other phases represented in similar way, into (4), i_{dc1} can be obtained,

$$i_{dc1} = s_{a1} i_{a1} + s_{b1} i_{b1} + s_{c1} i_{c1} \qquad (5)$$

Similarly the DC current for VSC2 can be obtained,

$$i_{dc2} = s_{a2} i_{a2} + s_{b2} i_{b2} + s_{c2} i_{c2} \qquad (6)$$

The following network equations are obtained from Thevenin's anaylsis of the DC network for VSC1,

$$i_{dc1} = i_{dx} + i_{c1} \qquad (7)$$

$$i_{c1} = C_1 \frac{dv_{dc1}}{dt} \qquad (8)$$

With (8) substituted into (7),

$$\frac{dv_{dc1}}{dt} = \frac{1}{C_1} \left(i_{dc1} - i_{dx} \right) \qquad (9)$$

For the DC-link of VSC2,

$$i_{dc2} = i_{c2} - i_{dx} \qquad (10)$$

$$i_{c2} = C_2 \frac{dv_{dc2}}{dt} \qquad (11)$$

With (11) substituted into (10),

$$\frac{dv_{dc2}}{dt} = \frac{1}{C_2} \left(i_{dc2} + i_{dx} \right) \qquad (12)$$

The current flowing in the DC-link cable is given by,

$$\frac{di_{dx}}{dt} = \frac{1}{L_x}\left(v_{dc1} - v_{dc2} - R_x i_{dx}\right) \qquad (13)$$

It can be observed that the variables are the DC voltage across the capacitor of the two VSCs and the current in the cable. In order to solve and simulate the HVDC-VSC model in the RTS, a discretised representation must be used. Hence, equations (9), (12) and (13) are discretised using the trapezoidal rule of integration,

$$v_{dc1} = \frac{\Delta t}{C_1}\left(i_{dc1} - \frac{i_{dx}}{2}\right) + hist_{dc1} \qquad (14)$$

$$v_{dc2} = \frac{\Delta t}{C_1}\left(i_{dc2} + \frac{i_{dx}}{2}\right) + hist_{dc2} \qquad (15)$$

$$i_{dx} = \left(1 + \frac{R_x \Delta t}{2L_x}\right)^{-1}\left(\frac{\Delta t}{2L_x}v_{dc1} - \frac{\Delta t}{2L_x}v_{dc2} + hist_{dx}\right) \qquad (16)$$

The term *hist* represents the variables known from preceeding time steps. The DC-link network equations are solved simultaneously with the equations describing the VSCs to obtain the output in a discretised form suitable for digital simulation.

2.2 Operations and Control of the HVDC-VSC

The power exchanged between the AC and DC system in a HVDC-VSC system is defined by the fundamental voltage across the VSC. The active power flow between the converter and the AC system is controlled by the voltage angle and the reactive power flow by the magnitude of the voltage. The application of PWM techniques enables the control of the voltage angle and magnitude of the VSCs to be changed almost instantaneously.

There are two required functions of the HVDC-VSC in order to ensure that the AC and the DC system connect robustly. The required functions are the active power dispatch control and the DC-link voltage regulation [4]. The individual VSC stations can be selected to control either of the assigned functions.

The control of the active power dispatch involves maintaining a reference point for active power in the converter station. The selected VSC must vary its output voltage angle to function as an inverter or rectifier based on the direction of power flow. The error between the reference power and the actual power of the VSC must be minimal. The control block is as shown in Fig. 3 where the VSC1 is selected for power dispatch control.

The exchange of active power between the AC system and the VSC will result in variation of the DC-link voltage of the converters. If the AC system provides more real power than the load demand and the converter losses,

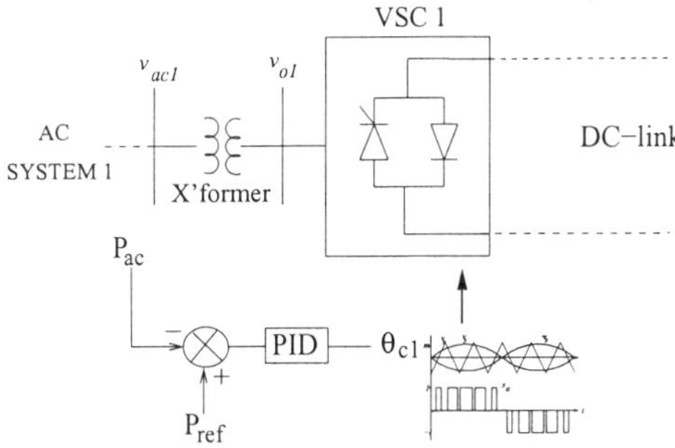

Figure 3: Power Dispatch Control

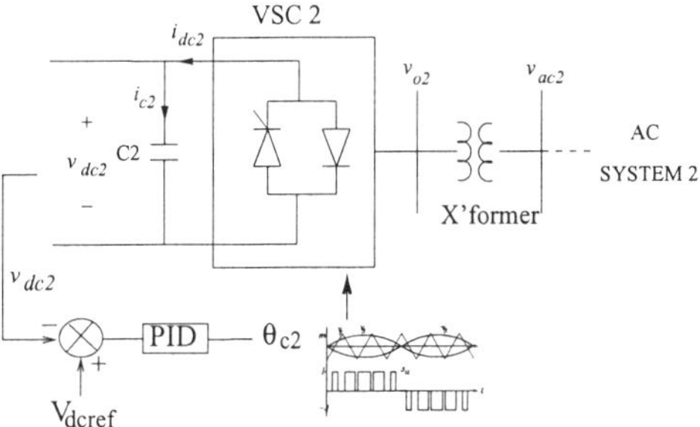

Figure 4: DC Voltage Regulation Control

the excess power will be absorbed by the VSC resulting in the DC-link capacitor voltage to increase. If this real power is less than the load demand and the converter losses, the DC-link capacitor voltage will decrease. It is important to maintain the voltage across the capacitor for HVDC-VSC to ensure continuous power flow.

Hence, one of the VSC stations will be needed to regulate the DC-link voltage of the HVDC-VSC station at a certain reference point. The DC voltage regulation control loop is as shown in Fig. 4. The voltage angle of VSC2 is adjusted to make sure the power flow is enough to maintain a charge across the DC-link capacitor at a desired level. In other words, the voltage regulator control ensures that the right amount of AC power is converted in one VSC from an AC system to satisfy the power requirements to be dispatched onto another AC system. This simultaneous control allows the direction of power flow to be alternated with ease.

3 Scaled-down Wind Energy Conversion System (WECS)

The scaled-down model of the WECS is constructed using a three-phase induction machine coupled to a DC servomotor. The speed of the servomotor which drives the induction machine can be adjusted by its excitation voltage. Hence, this excitation voltage is assumed to be equivalent to the wind speed. The induction machine runs at synchronous speed until the servomotor driving

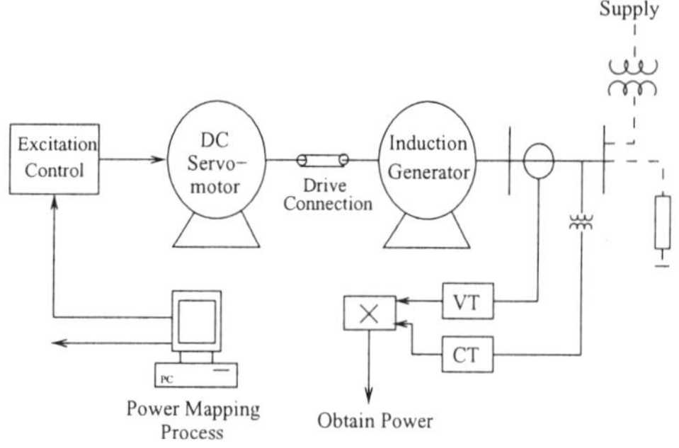

Figure 5: Schematic of scaled-down WECS

it reaches a cut-in speed where it will then run as a generator. The schematic of the scaled-down WECS setup is as shown in Fig. 5. The power output from the induction generator is monitored by acquiring the voltage and current generated using the respective transformers.

In order to study the behaviour of an actual wind generator using this scaled-down modelling approach, the output of the generator must reflect the actual characteristic of the wind generator. This can be achieved by first studying the power vs. speed characteristics of the scaled-down model [8] where the speed corresponds to the servomotor excitation volatge and the power is the output from the induction machine. A transfer function is developed to illustrate the relationship between this input excitation voltage and the output power. The characteristics of the scaled-down model can then be obtained as shown in Fig. 6. The value of the power is scaled to the actual characteristic in the range of kW.

It is observed that a mismatch occurs between the characteristics of the scaled-down model and the actual system. This can be overcome by mapping the characteristics of the scaled-down model to the actual system as shown in Fig. 6. With this mapped characteristic, which correctly represents the actual system, a new set of excitation voltages for the servomotor can be obtained using the transfer function approach.

With the flexibility in the transfer function approach to characterise the input and output, the scaled-down model can be constructed with any type and rating of DC servomotor and induction machine. The mapping process allows the actual generator characteristic of any system to be represented accurately with this transfer function approach.

4 Real-Time Simulation Approach

In this real-time simulation, the HVDC-VSC model is implemeted in the RTS. The WECS is considered to be connected directly to the terminal of VSC1 in the HVDC-VSC model. The general block diagram of the test setup is as shown in Fig. 7.

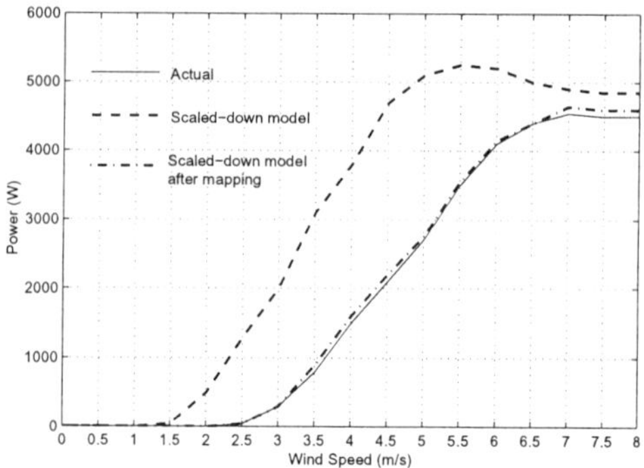

Figure 6: Scaled-down WECS Power Characteristic

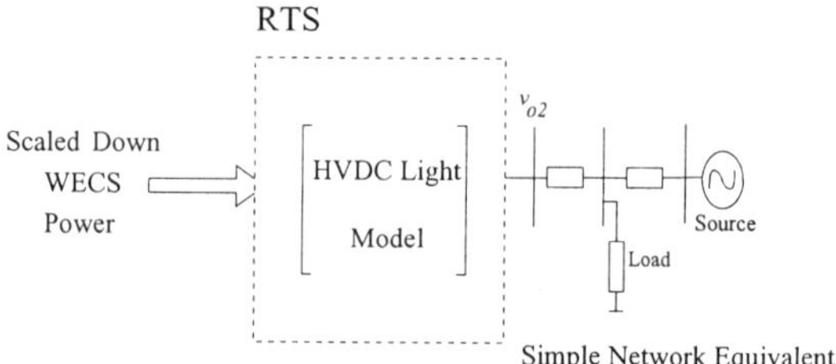

Figure 7: Test Setup

In (3)-(6) of the DC-link cable model, it is realised that the DC current, i_{dc1} and i_{dc2}, is obtained from the derivation of the power balance equation between the AC and DC system. Hence, it can be assumed that the DC current of the two VSCs corresponds to the instantenous power of the AC systems connected to them. In this case, the power into the terminal of VSC1 is the power generated by the WECS. With this consideration, the current equation for i_{dc1} is equal to the power from the WECS. The remaining DC-link cable model is unchanged in conducting the real-time simulation.

The terminal of VSC2 in the HVDC-VSC model is connected to a simple physical network representation through the DAC of the RTS. The voltage profile at the load point of the network is monitored.

5 Simulations and Results

Simulations are conducted to show the real-time capability of the HVDC-VSC model in the study of the WECS connection. The ability of the HVDC-VSC to transfer power generated is demonstrated by starting the WECS from stand-still through the cut-in period to full generation. The assessment of the HVDC-VSC to maintain voltage and frequency regulation is performed by oscillating the power output from the WECS.

In these simulations certain assumptions are being made on the system. The power of the WECS is assumed to be at the rated value at full generation. The power is represented in the equivalent p.u. value. The power vs. speed characteristic in Fig. 6 is scaled proportionally into an equivalent time frame of 2 s. Finally, DC values of the HVDC-VSC model are represented in p.u. assuming that an accurate base value exists for both the AC and DC system. The detailed rating and component values of the HVDC-VSC can also be included. These assumptions are made in the interest of the current work, since the main objective here is to demonstrate the ability of the HVDC-VSC model to operate in real-time with physical interactions.

5.1 WECS Run-Up to Full Generation

The power output from the WECS is as shown in Fig. 8. The initial period represents the wind speed where the WECS is not generating any power. The cut-in instant is at 0.8 s. The power transferred to the load network side or the output power, P_{ac2}, from VSC2 is slightly less than the WECS power input due to the losses in the cable.

It is observed that as the power from the WECS increases, the power transfer across the DC-link cable also increases as shown by the current in the cable. The DC voltage of VSC1 increases slightly as a result of this power input. The DC voltage of VSC2, where the voltage regulation control is taking place, is maintained at an almost constant level. The difference in the DC voltage level is to facilitate the current flow in the DC-link cable. As the voltage level at the terminal of VSC2 is dependent on the DC voltage, it will also maintain a constant profile throughout the transfer of power. This will in turn keep the voltage at the load point constant ensuring constant supply.

5.2 Oscillations in WECS Generated Power

In wind generation, the characteristics of the power depend on the wind turbine driving. It has been known that pulsations or oscillations of the power output can occur when the blades of the wind turbine passes its tower [9]. In this simulation, the output of the WECS is assumed to contain low frequency oscillations. This can be achieved by changing the power characteristics of the WECS which involves adding an oscillation signal to the excitation voltage of the servomotor.

It is observed from Fig. 9, that the DC voltage at both the terminals is not very much affected by the oscillations in the power input and output. Hence, from the previous explaination, the voltage profile at the load point also maintains its level. This shows the effectiveness of the HVDC-VSC in maintaining a constant voltage profile at its point of connection with the AC system.

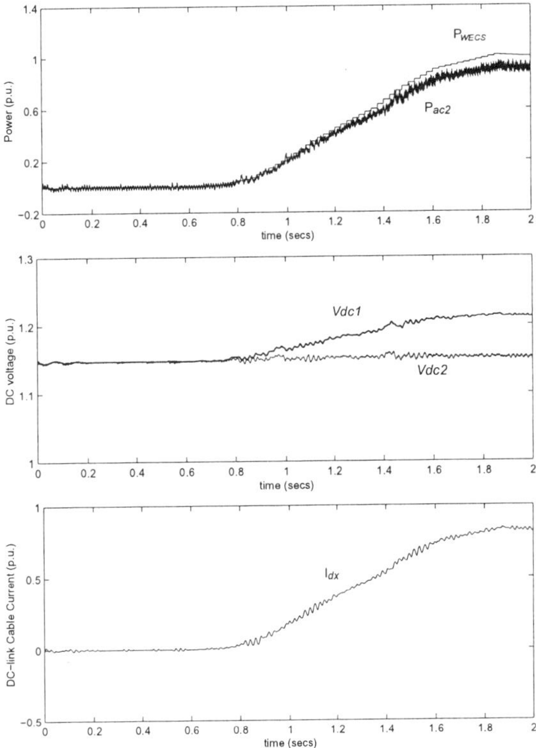

Figure 8: HVDC-VSC responses during WECS run-up to rated power

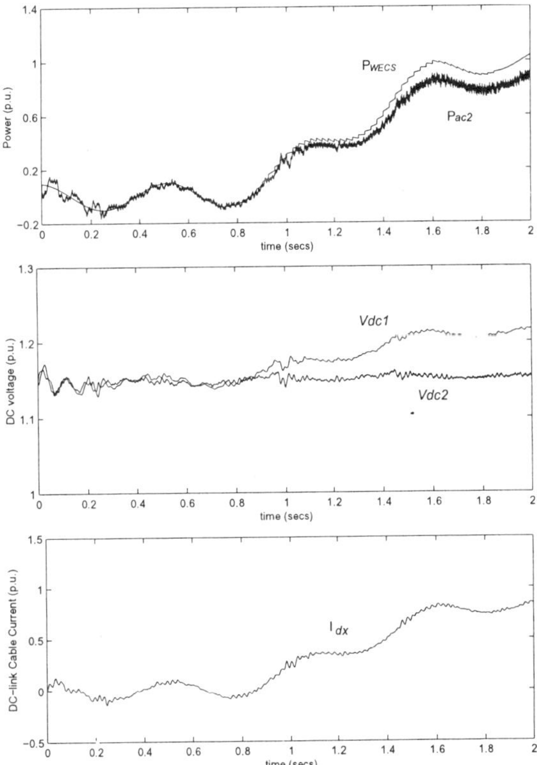

Figure 9: HVDC-VSC responses when transferring oscillating power

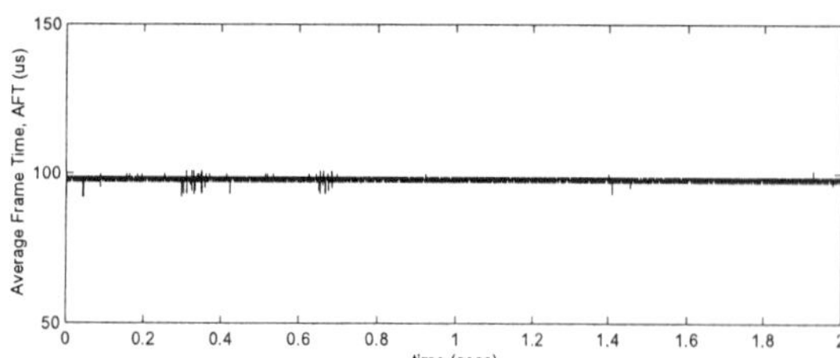

Figure 10: Average Frame Time (AFT) of simulations

5.3 Frame time of simulations

The main parameter that needs monitoring when performing real-time simulations is the Actual Frame Time (AFT) of the RTS. The AFT allows the user to observe the time taken by the RTS to compute each time step of the simulation. The AFT must not exceed the Desired Frame Time (DFT) at any point in the simulation. The DFT is always given by the simulation time step.

In the study cases, the simulations were undertaken with a time step of 100 μs. It is observed from Fig. 10, that the AFT of both the study cases are below the DFT of 100 μs. Hence, the real-time capability of the simulations is demonstrated.

6 Conclusions

This paper has presented a HVDC-VSC model in the time domain suitable for conducting real-time simulations. The model is applied in a practical situation [10][11] linking a scaled-down WECS to a load in the real-time environment. The scaled-down WECS model developed has the ability to represent the characteristics of any actual wind generation system. The distinctive feature of this paper is that studies are performed in an environment where the response of the HVDC-VSC mathematical model are simulated with interaction to physical devices. The simulations in this environment have indeed been performed in real-time as shown by the RTS frame time collected.

Results presented show that the HVDC-VSC has the ability to transfer fast changing power from one terminal to the other without affecting the voltage level of the AC system connected to it. This is true even for the case when oscillations exist in the power that is being transferred.

References

[1] J.A. Parle, *Phase Domain Transmission Line Modelling for EMTP-Type Studies with Application to Real-time Digital Simulation*, PhD Thesis, University of Glasgow, 2000.

[2] R.P. Wierckx, "Fully Digital Real-Time Electromagnetic Transients Simulator", *IERE International Electric Research Exchange, Workshop on New Issues in Power System Simulation, pp. 1-11, Caen, France, March 1992.*

[3] G. Asplund, "HVDC Using Voltage Source Converters - A New Way to Build Highly Controllable and Compact HVDC Substations", *Cigre SC 23 Symposium, Paris, France, August 1997.*

[4] B.T. Ooi and X. Wang, "Boost Type PWM HVDC Transmission System", *IEEE Trans. on Power Delivery, vol. 6, N0. 4, pp. 1557-1563, Oct. 1991.*

[5] B.T. Ooi and X. Wang, "Voltage Angle Lock Loop Control of Boost Type PWM Converter for HVdc Application", *IEEE Trans. on Power Electronics, vol. 5, N0. 2, pp. 229-235, April 1990.*

[6] G. Asplund, "Application of HVDC Light to Power System Enhancement", *IEEE PES Winter Meeting 2000, vol. 4, pp. 2498-2503.*

[7] M.Madrigal, O. Anaya, E. Acha, et al "Single-Phase PWM Converters Array for Three-Phase Reactive Power Compensation. Part 1: Time Domain Studies", *in Proc. 2000 International Conference on Harmonics and Quality of Power, ICHQP 2000, pp. 541-547, Oct. 2000.*

[8] N.P. Johnson, T.L. Tan, P. Miller and E. Acha, "A Laboratory Model for Harmonic Measurements in Windpower Generators", *IEEE Power Engineering Review, pp. 48-49, March 2000.*

[9] A. Feijoo and J. Cidras, "Analysis of mechanical power fluctuations is asynchronous WECS", *IEEE Trans. on Energy Conversion, vol. 14, N0. 3, pp. 284-291, Sept. 1999.*

[10] L. Weimers, "HVDC Light : A New Technology for a Better Environment", *IEEE Power Engineering Review, pp. 19-20, August 1998.*

[11] G. Asplund, K. Eriksson and O. Tollerz, "Land and Sea Cable Interconnections with HVDC Light ", *CEPSI Conference 2000, Manila, Philipines, October 2000.*

A GENERALISED CURRENT INJECTION APPROACH FOR MODELLING OF FACTS IN POWER SYSTEM DYNAMIC SIMULATION

W Freitas   A Morelato

State University of Campinas – Brazil

Abstract: This paper addresses the problem of modelling FACTS devices for simulation studies concerning power systems stability and control. In the approach describes here the dynamical behaviour of power electronic devices, such as STATCOM, SSSC and UPFC, are modelled based on current injection models. The performance of these FACTS models is reported during the utilisation of such FACTS devices to implement a new method of improving the transient stability margins of power systems, which consists of controlling directly the electrical torque of synchronous generators, just after the occurrence of a disturbance, by means of ac machine vector control.

1 INTRODUCTION

Flexible Alternating Current Transmission Systems (FACTS) devices have been increasingly introduced in power systems around the world. Among the various configurations of FACTS devices, those based on the Voltage-Sourced Converter (VSC) concept have several attractive features, such as to work together with energy storage devices, allowing both real and reactive power to be simultaneously exchanged with the ac system, Song and Johns (1). The devices of the SVC family covered here are the Static Synchronous Compensator (STATCOM), the Static Synchronous Series Compensator (SSSC) and the Unified Power Flow Controller (UPFC). In power system transient stability studies, FACTS equipments should be taken into account because their great influence in the system dynamics and, therefore, very reliable models are required to represent such devices into simulation programmes. Of the available models so far, the shunt current source model of FACTS appears to be the most adequate for stability simulations. However, even this kind of modelling approach has shown convergence issues, mostly related to the device-ac system interface. This fact occurs mainly due to the small reactances of coupling transformers, which causes high sensitivity of the converter current injections to changes in control variables.

In order to overcome these issues Padiyar and Rao (2) proposed a UPFC specific model for transient stability studies, but in their model is mandatory to compute the open circuit impedance matrix of the network external to the UPFC terminals, at every time step of simulation, for calculating the UPFC current injections. Mak *et al*

(3) presented a FACTS device-ac network interface calculation method based on the sequential solution approach, which has been successfully used for representing ac-dc interfaces in the ac/dc load flow and transient stability analysis. However, in his approach the bus admittance matrix must be reduced to the generator internal buses, except the FACTS ac terminal buses that should be retained. A similar approach is used in Huang *et al* (4), whereas Arabi *et al* (5) have introduced an auxiliary capacitor and a decoupling mechanism in the UPFC converters, aiming to smooth the convergence process.

In this paper the convergence process during the simulation of FACTS models is significantly improved by including an additional differential equation that represents the energy exchange between the FACTS device and the ac network. This procedure allows the FACTS and network equations to be solved as one block, reducing the interface mismatch. Additionally, the models proposed here avoid the usage of fictitious buses for representing series components in the SSSC and UPFC, preserving the topological structure of the original network and keeping constant the dimension of the network admittance matrix.

The proposed FACTS models are tested here to implement a vector control strategy applied to synchronous generators to enhance in real-time the transient stability margins of power systems, as described in Freitas and Morelato (6). This method can be seen as a new potential application of power electronic equipments for power system control, where the FACTS devices act as a controller that modulates the stator currents delivered by the generators just after the detection of a disturbance.

2 FACTS MODELS

A Voltage-Sourced Converter (VSC) is the building block of the STATCOM, SSSC and UPFC. The three-phase output voltages produced by a VSC can be considered sinusoidal in the fundamental frequency, assuming convenient design and the usage of pulse-width modulation, multi-pulse and multi-level techniques. Consequently, a VSC may be represented by voltage sources or equivalent current sources. The model of FACTS devices based on VSC is developed as follows.

AC-DC Power Transmission, 28-30 November 2001
Conference Publication No. 485 © IEE 2001

2.1 STATCOM

The STATCOM, which is schematically depicted in Fig. 1, consists of a VSC connected in shunt to the transmission network through a coupling transformer. Therefore, it can be represented by a current source expressed as:

$$\bar{I}_{sh} = \frac{\bar{V}_{sh} - \bar{V}_i}{z_{sh}} = I_{sh}^p + j.I_{sh}^q \qquad (1)$$

where $\bar{V}_{sh}$ is the output voltage phasor of VSC, $\bar{V}_i$ is the nodal voltage phasor and z_{sh} is the shunt transformer impedance. Moreover, $^{(p)}$ and $^{(q)}$ denote the current components corresponding to the real and reactive power injections, $i.e.$ the current components in phase and in quadrature with the nodal voltage, respectively. In the case that there are no energy storage devices in the dc side, the real component of the current injection is null.

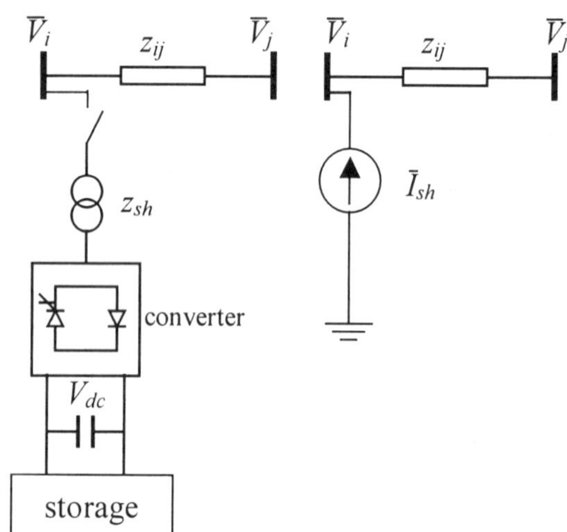

Fig.1 – The STATCOM current injection model.

2.2 SSSC

The SSSC also consists of a VSC connected to the power system through a transformer. On the contrary to the STATCOM, the transformer is connected in series with the transmission line. Consequently, the SSSC can be represented by a series voltage source, which can be replaced by a pair of equivalent current sources as shown in Fig. 2.

The symbols z_{ij} (y_{ij}) and z_{ij}^{sh} (y_{ij}^{sh}) in Fig. 2 and equations (2), (3) and (4) represent the impedance (admittance) of the line π-section and z_{se} (y_{se}) the series transformer impedance (admittance). Moreover, $\bar{I}_{ij}$ is the line current and f is a fictitious bus normally created to permit direct accessibility to the line side of the SSSC. However, it is convenient to eliminate this fictitious bus, utilising Gauss-elimination, in order to

preserve the original matrix structure and to prevent numerical problems caused by the representation of the bypass mechanism. The values of z_i^{sh}, z_j^{sh} and z_{eq} are obtained after the Gauss-elimination procedure. As a result, the current injected into the network by the SSSC can be given as:

$$\bar{I}_{se} = \frac{\bar{V}_{se}}{z_{se}} = \frac{V_{se}^p + j.V_{se}^q}{z_{se}} \qquad (2)$$

$$\bar{I}_{se}^i = \bar{I}_{se} - \frac{y_{se}}{y_{ij} + y_{ij}^{sh} + y_{se}}.\bar{I}_{se} \qquad (3)$$

$$\bar{I}_{se}^j = - \frac{y_{se}}{y_{ij} + y_{ij}^{sh} + y_{se}}.\bar{I}_{se} \qquad (4)$$

where $^{(p)}$ and $^{(q)}$ denote the real and imaginary voltage components, $i.e.$ the voltage components in phase and in quadrature with the line current, respectively. Again, if there are no energy storage devices in the dc side, then the real component of the series voltage is null.

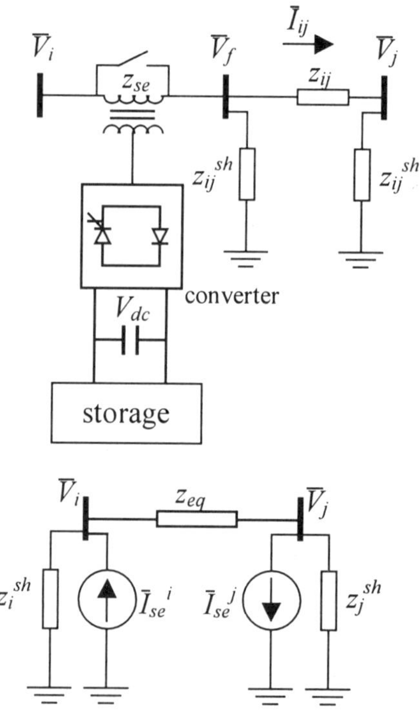

Fig. 2- The SSSC current injection model.

2.3 UPFC

The UPFC is a combination of the STATCOM and the SSSC. Hence, the UPFC model is obtained by combining the previous models, as depicted in Fig. 3, where the impedances (admittances) and current injections are exactly the same of previous devices. The UPFC configuration allows real power compensation to be made without any energy storage device in the dc side. This compensation is carried out by the dc link installed between the converters, $i.e.$ the real power injected at the transformer terminals is taken from the ac network by the other converter.

Since the UPFC does not absorb or generate real power at all, it is necessary to meet the active power balance requirements. The balance equation can be expressed mathematically, neglecting device losses, as follows:

$$\Re(\overline{V}_i . \overline{I}_{sh}^* + \overline{V}_{se} . \overline{I}_{ij}^*) = 0 \qquad (5)$$

where * denotes conjugated complex and $\Re$ is the real part of a complex number. This equation can be satisfied by adjusting the real component of either the shunt current or the series voltage.

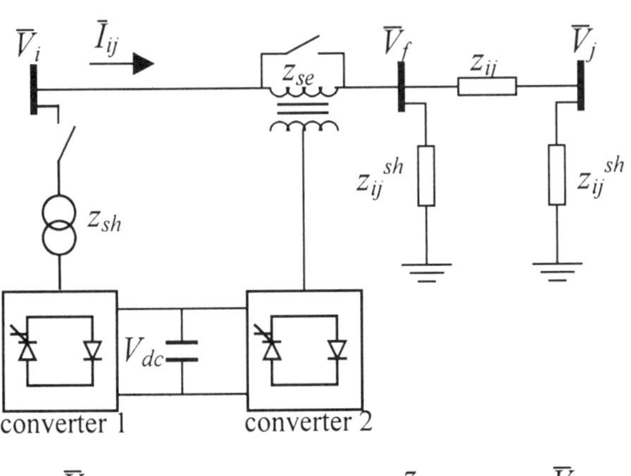

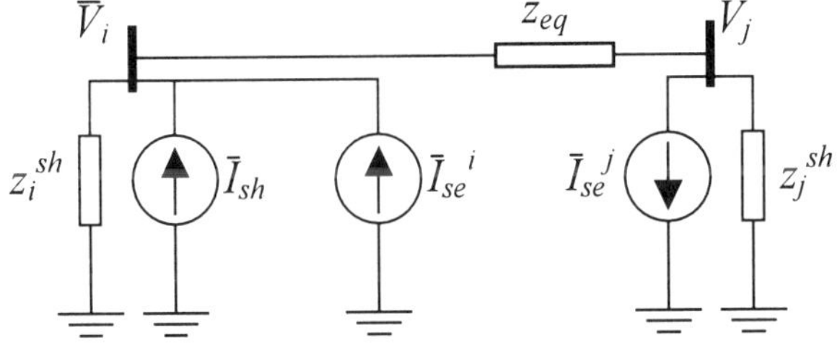

Fig. 3 – The UPFC current injection model.

2.4 Limits

Since the FACTS models use only current injections, the operating limits of each device can be imposed by limiting the magnitude of the current injected into the network, as follows:

$$\left|\overline{I}_{sh}\right| \leq I_{\max sh} \quad \text{and} \quad \left|\overline{I}_{se}\right| \leq I_{\max se} \qquad (6)$$

2.5 Energy differential equations

In general, the convergence issues frequently observed in FACTS model simulations are related to how the interface between the FACTS device and the ac network is modelled. Here this interface is represented by differential equations that describe the energy variation of the converters expressed as:

$$\frac{dE_{sh}}{dt} = \Re(\overline{V}_i . \overline{I}_{sh}^*) \qquad (7)$$

$$\frac{dE_{se}}{dt} = \Re(\overline{V}_{se} . \overline{I}_{ij}^*) \qquad (8)$$

where E_{sh} and E_{se} denote the energy of the shunt and series converters, respectively. The main advantage of this kind of approach is that these energy equations can represent very precisely the coupling between the FACTS device and ac network algebraic variables, reducing the number of iterations or even allowing convergence to be attained where the sequential solution approach fails.

2.6 Interface solution approach

In transient stability studies, power systems are usually modelled by an algebraic-differential equations set, as follows:

$$\dot{\mathbf{y}} = f(\mathbf{x}, \mathbf{y}) \qquad (9)$$

$$g(\mathbf{x}, \mathbf{y}) = 0 \qquad (10)$$

where $\mathbf{x}$ and $\mathbf{y}$ are the state variables of the algebraic and differential equations, respectively. The solution scheme adopted here is the so-called partitioned solution approach, in which the algebraic and differential equations are solved separately and then combined alternatively, Kundur (7). In this solution method, the algebraic equations are normally represented by the $\mathbf{I} = \mathbf{Y}.\mathbf{V}$ form, what it is very convenient for including FACTS models based on current injections. On the other hand, the energy differential equations of FACTS are included in the differential equation set. The block diagram of the partitioned solution scheme is shown in Fig. 4.

It should be noted that the most significant interface variables between the FACTS device and ac network are mixed into the energy differential equations. As a result, it is possible to execute only one iteration of the algebraic loop within each iteration of the differential loop, as usual in the partitioned scheme, and even so to obtain convergence towards the correct values of the algebraic variables. In addition, it was observed that the usage of the energy differential equations decreases the total number of the algebraic set solutions during the entire simulation.

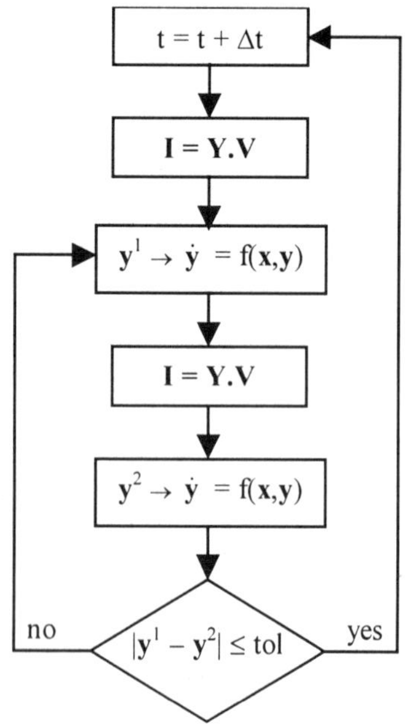

Fig. 4 – Block diagram of the partitioned solution.

3 VECTOR CONTROL BY FACTS

Vector control of ac machines is a technique that allows both the electromagnetic flux and torque to be controlled in an independent way, through control actions in both the amplitude and phase of currents, coining the term *vector control*, Novotny and Lipo (8). Usually, the vector control approach has been applied to synchronous or induction motors. Here this technique is applied to control the torque production of synchronous generators aiming at the enhancement of the power system transient stability.

Considering a synchronous generator with a field winding in the d-axis and a damper winding in the q-axis, the stator flux linkage and torque equations, in the rotor reference frame, are:

$$\psi_d = -X_d \cdot i_d + X_{ad} \cdot i_{fd} \tag{11}$$

$$\psi_q = -X_q \cdot i_q + X_{aq} \cdot i_{1q} \tag{12}$$

$$T_e = \psi_d \cdot i_q - \psi_q \cdot i_d \tag{13}$$

The above expressions are written according to the notation used in (7). In general, the machine torque production is determined by the components in both d-axis and q-axis of the stator current vector. However, a special situation can be created (by an external vector controller) where the stator current is all in the q-axis and therefore the d-axis current component is null. In this case (called field-orientation) the field current in the d-axis and the stator current in the q-axis are 90° apart and, as a result, the machine torque can be controlled by either the d-axis flux or the q-axis stator current in a decoupled manner, similar to what happens in a

separately excited dc machine. Setting $i_d = 0$ in equations (9) and (11) and controlling the torque through i_q, the control law can be expressed as:

$$i_q^{set} = \frac{T_e^{set}}{\psi_d} = \frac{T_e^{set}}{X_{ad} \cdot i_{fd}} \tag{14}$$

where i_q^{set} is the input control variable and T_e^{set} is the controlled variable.

3.1 Control strategy and implementation

Even though several control strategies to improve the first-swing stability of multimachine power systems would be implemented, the performance of the FACTS models proposed here was tested using the strategy described as follows.

When the electrical torque of the machine under control reaches the first maximum value after fault elimination, the vector controller starts aiming at keeping constant the electrical torque at this maximum value. This control action remains until the speed deviation of the synchronous generator becomes null. The objective of this procedure is to increase the capability of generators to produce decelerating energy after the fault clearance.

The vector control is carried out by modulating the stator current of the synchronous generator. Here this modulation is implemented by installing FACTS devices nearby the generator, as illustrated in Fig. 5.

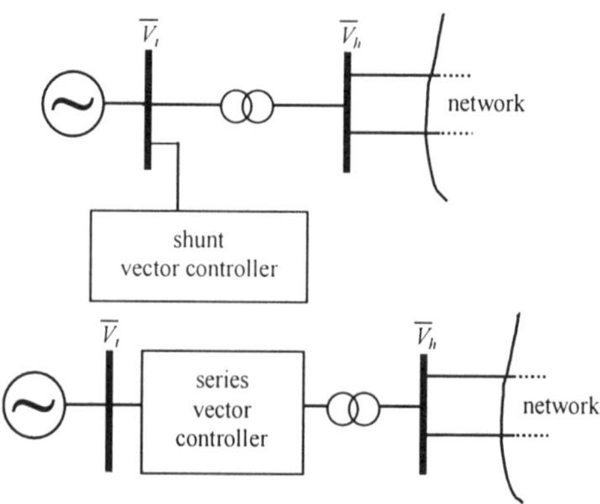

Fig. 5 – Placement of vector controllers.

It should be noted that the FACTS devices are bypassed before and after the vector control action, causing additional difficulties to the convergence process.

4 RESULTS

In this section simulation results are presented showing the performance of FACTS devices acting as vector controllers on a multimachine transient stability study using the **New England system**, with 39 buses, 46 branches and 10 machines, which on-line diagram is

179

depicted in Fig. 10. The complete system data can be encountered in Pai (9). All machines are represented by a 4th order model and its AVR by IEEE-Type 1 model. The simulated contingency is a three-phase fault at the bus 29, which is cleared after 50 ms by tripping the line 29-28.

In Fig. 6 are shown the relative angle responses of the machines in the absence of any vector controller. It can be verified that the system is unstable, since the machine 9 accelerate faster than the others machines. In this case, the machine 9 is eligible to have installed a vector controller, which may be implement by the STATCOM, SSSC or UPFC. In all circumstances the FACTS shunt components are connected to bus 38 and the series components are installed in the branch 38-29. In Fig. 7 to 9 the machine relative angle responses corresponding to each type of device are presented. In all cases the system becomes stable and damped after the control action to be active about 60 milliseconds.

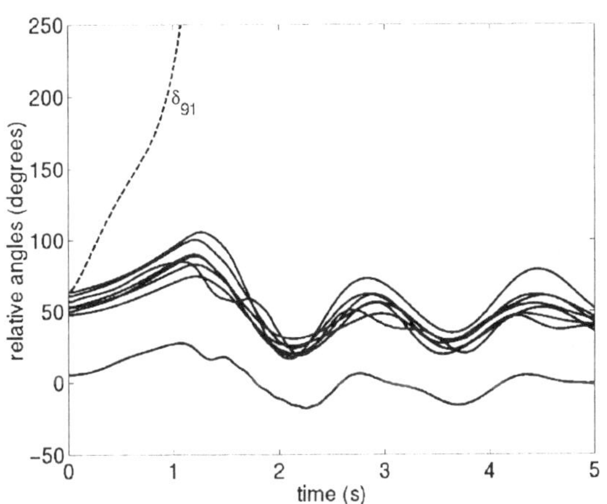

Fig. 6 – Machine responses without vector control.

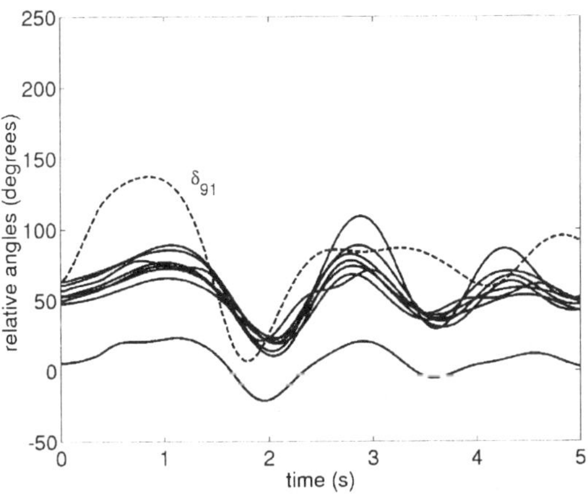

Fig. 7 – Machine responses with STATCOM controller.

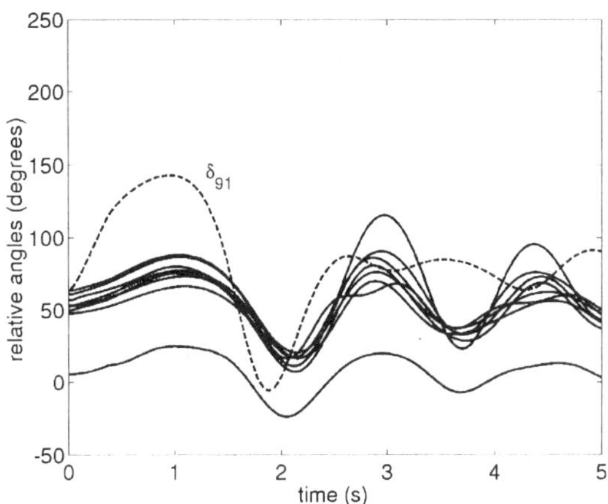

Fig. 8 – Machine responses with SSSC controller.

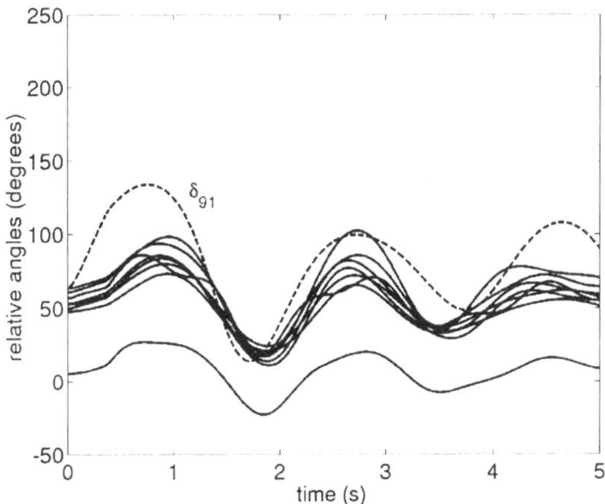

Fig. 9 – Machine responses with UPFC controller.

Table 1 compares the performance of the FACTS models proposed here and the traditional sequential solution. In the SSSC and UPFC cases, the proposed models converge whereas the sequential solution fails. Furthermore, in the STATCOM case both methods converge, however the number of algebraic solutions required during the vector control action is significantly reduced using the proposed model.

TABLE 1 – Performance of the FACTS devices models.

FACTS	Number of $\mathbf{I} = \mathbf{Y}.\mathbf{V}$ solutions.	
	Proposed model	Sequential solution
STATCOM	133	374
SSSC	570	no convergence
UPFC	734	no convergence

5 CONCLUSIONS

The proposed FACTS models, including the energy equations, can effectively smooth the convergence process in power systems dynamic studies performed under stringent test conditions as those required by real-time vector control applications. Furthermore, the synchronous generator vector controlling implemented by FACTS devices could be an effective technique in the enhancement of power system transient stability margins, establishing a new application for electronic power devices.

6 ACKNOWLEDGEMENTS

The authors would like to acknowledge the financial support of Conselho Nacional de Desenvolvimento Científico e Tecnológico (CNPq), Brazil (Proc. 141429/1997-6).

7 REFERENCES

1. Song, Y.H. and Johns, A.T., 2000, "Flexible ac transmission systems (FACTS)", The Institution of Electrical Engineers, London, UK

2. Padiyar, K.R. and Rao, K. U., 1999, "Modeling and control of unified power flow controller for transient stability", Electrical Power and Energy System, 21, 1-11

3. Mak, L.O., Huang, Z., Ni, Y., Shen, C.M., Wu, F.F., Chen, S. and Zhang, B., 1999, "A versatile interface between FACTS devices and ac power systems", 13th PSCC, 533-538

4. Huang, Z., Z., Ni, Y., Shen, C.M., Wu, F.F., Chen, S. and Zhang, B., 2000, "Application of unified power flow controller in interconnected power systems – modelling, interface, control strategy and case study", IEEE Trans. PWRS, 15, 817-824

5. Arabi, S., Kundur, P. and Rambabu, A., 2000, "Innovative techniques in modelling UPFC for power system analysis", IEEE Trans. PWRS, 15, 336-341

6. Freitas, W and Morelato, A, 2000, "Improvement of power system transient stability based on synchronous generator vector control", IEEE Power Engineering Review, 20, 64-66

7. Kundur, P., 1994, "Power system stability and control", McGraw-Hill, New York, USA

8. Novotny, D.W. and Lipo, T.A., 1997, "Vector control and dynamics of ac drives", Oxford University Press, Oxford, UK

9. Pai, M.A., 1989, "Energy function analysis for power system stability", Kluwer Academic Publishers, London, UK

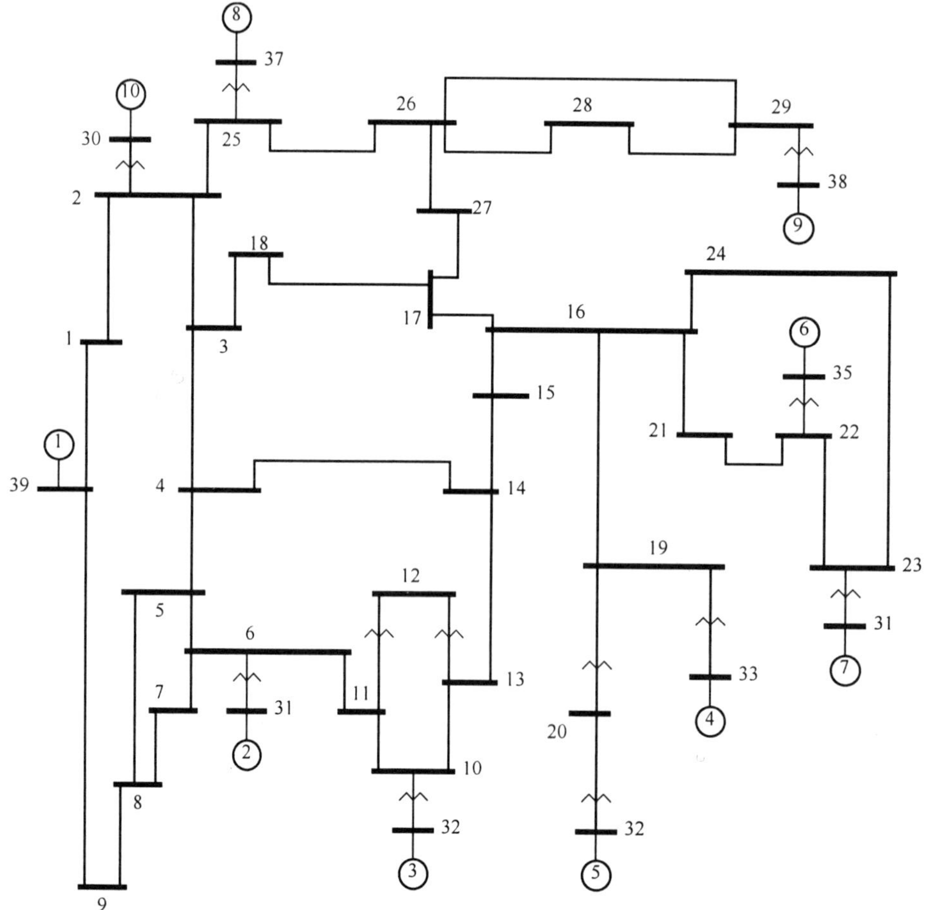

Fig. 10 – One-line diagram of the New England system.

A NOVEL CONTROL STRATEGY FOR TCSC TO DAMP SUBSYNCHRONOUS OSCILLATIONS

S Subhash[1], B N Sarkar[1] and K R Padiyar[2]

CPRI, INDIA[1] and IISc. INDIA[2]

1.0 INTRODUCTION :

Thyristor Controlled Series Compensators (TCSC) is the first FACTS controller under the new generation to have reached mature stage of development. The importance of adding series capacitor to ac transmission lines for increasing line loadability is known for a long time. But the potential risk of SSR oscillations had made it undesirable to widely use them in the system. Adding a thyristor controlled series compensator Larsen et al (1) is however a fairly recent phenomenon and provides greater flexibility in power transmission, There are several advantages in using TCSC and reducing the torsional oscillations caused by SSR is one among them.

In this paper a novel discrete control strategy to mitigate SSR oscillations is proposed for TCSC having discrete type of control over compensation level, since it behaves like a conventional series compensation under normal conditions. The control strategy is based on the Phase unbalance concept and is simple and easy to implement in the hardware.

This paper is organised as follows. The concept of Phase unbalance originally proposed by A.Edris (2) is described followed by theoretical analysis to support the design for practical implementation. An expression for the damping torque coefficient is given, which is useful to assess the damping effectiveness quantitatively, over a wide variety of system operating conditions and for arriving at an optimal design.

AC-DC Power Transmission, 28-30 November 2001
Conference Publication No. 485 © IEE 2001

Extension of this concept to TCSC with discrete control is subsequently explained.

Studies carried out on the IEEE second benchmark system for SSR demonstrate the utility of the damping torque coefficient derived at sub-synchronous frequencies and also the control strategy applied to TCSC. Results are validated through EMTP simulations and discussed in the end of the paper.

2. ANALYSIS OF PHASE IMBALANCE

As per IEEE (3), SSR is defined as a condition wherein the electrical system (transmission lines and series compensation) exchanges energy with the mechanical system (turbine generator set) at one or more of the natural frequencies of the combined system below the synchronous frequency of the system. The positive sequence currents flowing in the stator winding at subsynchronous frequency, ω_e can lead to severe torsional interactions when its complement frequency (ω_o-ω_e) is equal to one of the torsional mode frequency ω_m.

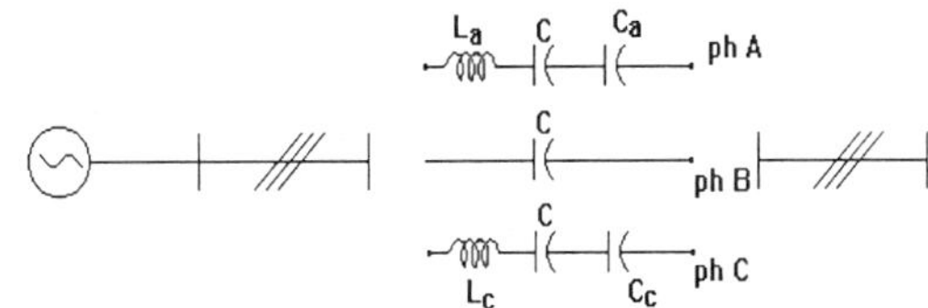

Fig 1. Series resonant compensation scheme

In the well known phase imbalance method (2), the network is modified by adding passive LC circuits in one or two phases as shown in Fig 1. The unbalance so created detunes the circuit, away from the SSR condition and damps the subsynchronous oscillations. The effectiveness of this method largely depends on the choice of LC circuits and the unbalance created.

3.0 ANALYSIS OF NETWORK WITH ASYMMETRY

In SSR phenomenon, the torsional interaction is more important than the induction generator effect. It is convenient to ignore the flux decay, damper circuits and transient saliency   in the generator. The synchronous machine is represented as a positive sequence voltage source behind a transient reactance, which is given by,

$$e_{pos} = \varpi E' \sin(\omega_o t + \delta),\qquad(2)$$

where,   E' – induced voltage proportional to the flux linkage, and ϖ – pu speed.

Assuming that the generator rotor oscillates about a constant speed sinusoidally, $\varpi = \varpi_o + A\sin(\omega_m t)$, we can show that the induced emf   in the stator $e_{pos}(t)$ is given by Padiyar (4)

$$e_{pos}(t) = \varpi_o E' \sin(\omega_o t + \delta_o) - \frac{AE'}{2\omega_m}(\omega_o - \omega_m)\cos[(\omega_o - \omega_m)t + \delta_o]$$

$$-\frac{AE'}{2\omega_m}(\omega_o + \omega_m)\cos[(\omega_o + \omega_m)t + \delta_o]$$

$$(3)$$

The emf has three sinusoidal components, viz., one of frequency ω_o and other two components of frequencies $\omega_o \pm \omega_m$, for small   amplitude (A) of oscillations. When these voltages are applied to an unbalanced transmission network, we have all the three sequence currents flowing in the network at these frequencies. as given by,

$$\begin{bmatrix} i_{pos} \\ i_{neg} \\ i_{zero} \end{bmatrix} = [Y_{seq}] \begin{bmatrix} e_{pos} \\ 0 \\ 0 \end{bmatrix} = \begin{bmatrix} Y_{11} & Y_{12} & Y_{13} \\ Y_{21} & Y_{22} & Y_{23} \\ Y_{31} & Y_{32} & Y_{33} \end{bmatrix} \begin{bmatrix} e_{pos} \\ 0 \\ 0 \end{bmatrix} \qquad(4)$$

where the $[Y_{seq}]$ is admittance matrix.

Among the three sequence currents flowing in the stator only the positive sequence currents produce torques at the subsynchronous frequency ω_m, and lead to the possibility of negative damping effect on SSR oscillation . Eqn (4) is further simplified as,

$$i_{pos} = Y_{11}\ e_{pos}\qquad(5)$$

Computation of i_{pos} from eqn(5) is equivalent to computing the positive sequence current flowing in an equivalent balanced network, whose admittance is equal to Y_{11} and we can write the expression for the damping coefficient torque T_D[3] as,

$$T_D = -\frac{(E')^2}{2\omega_m}\left[(\omega_o - \omega_m)G_{sub} - (\omega_o + \omega_m)G_{sup}\right]\qquad(6)$$

$$=\ -T_{Dsub} + T_{Dsup}$$

where, $G_{sub} + j\,B_{sub} = Y_{11}\,(j(\omega_o\text{-}\omega_m))$ and

$$G_{sup} + j\,B_{sup} = Y_{11}\,(j(\omega_o + \omega_m)),$$

are the admittances computed at the subsynchronous and supersynchronous frequency respectively. If the network impedance is close to SSR resonance condition, then the network impedance is purely resistive at the subsynchronous frequency and inductive in the supersynchronous frequency range. Moreover as G_{sub} is very large compared to G_{sup}, the damping coefficient is highly negative. It is therefore important to look into the effects of network unbalance on the factor G_{sub}. For simplicity the mutual coupling between phases is neglected and we define resonant frequency and impedances in phases a,b &c as,

$$\omega_{ea}^2 = \frac{1}{L_a C_a},\ \text{with}\quad Z_{aa} = R + j\omega L_a + \frac{1}{j\omega C_a}\qquad(7)$$

$$\omega_{eb}^2 = \frac{1}{L_b C_b},\ \text{with}\quad Z_{bb} = R + j\omega L_b + \frac{1}{j\omega C_b}\qquad(8)$$

$$\omega_{ec}^2 = \frac{1}{L_c C_c} \text{ , with } \quad Z_{cc} = R + j\omega L_c + \frac{1}{j\omega C_c} \qquad (9)$$

The expression of G_{sub} is given for two cases.

Case 1 : All three phase impedances are equal and close to the resonance frequency

$$\left(G_{sub}\right)_{bal} = \frac{1}{3R} + \frac{1}{3R} + \frac{1}{3R} = \frac{1}{R} \qquad (10)$$

Case 2 : Unbalance in phases a & c impedance and phase b impedance is close to its resonance frequency w_{eb}, ie., $(\omega_o - \omega_m) \sim \omega_{eb}$, ,

$$G_{sub} = \frac{R}{3\left[R^2 + \omega_{eb}^2 L_a^2 \left\{1 - \frac{\omega_{ea}^2}{\omega_{eb}^2}\right\}^2\right]} + \frac{1}{3R} + \\ \frac{R}{3\left[R^2 + \omega_{eb}^2 L_c^2 \left\{1 - \frac{\omega_{ec}^2}{\omega_{eb}^2}\right\}^2\right]} \qquad (11)$$

We can observe that G_{sub} is less than $(G_{sub})_{bal}$ at a particular torsional mode frequency ω_m, due to unbalance in phases a & c. Hence it is possible to quantify the reduction in G_{sub} in terms of degree of unbalance caused by the impedance imbalance and in turn the damping torque coefficient T_D.

4.0 EXTENSION OF UNBALANCE CONCEPT TO TCSC

TCSC with discrete control, is made up a number of capacitor modules connected in series to the transmission line, as shown in Fig 2 . The phase imbalance concept discussed above can be readily extended to TCSC, by introducing asymmetry under transient conditions in the system. Instead of adding passive LC resonant circuits, an unbalance is created in

TCSC by either inserting/bypassing capacitor modules in individual phases during the disturbances and by

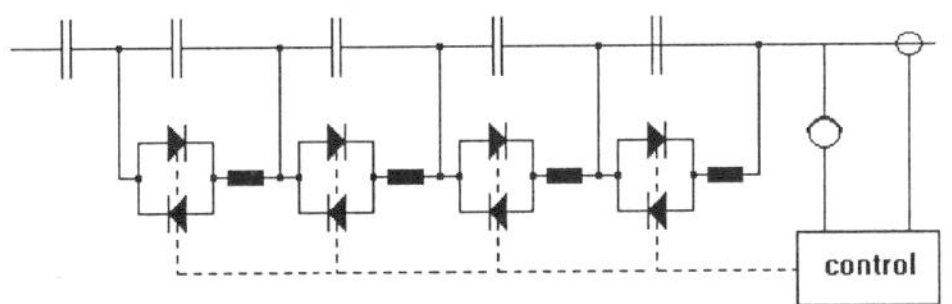

Fig 2. TCSC configuration

removing it when there is no SSR. This is possible due to the thyristor control, which otherwise would not be possible using mechanical breakers. Asymmetry so created reduces the torsional interactions. A control strategy for TCSC is discussed in the following section.

5. 0 TEST SYSTEM

The IEEE Second Benchmark model for SSR studies IEEE(5) as shown in Fig 3 is chosen as the test example. The system consists of a synchronous generator

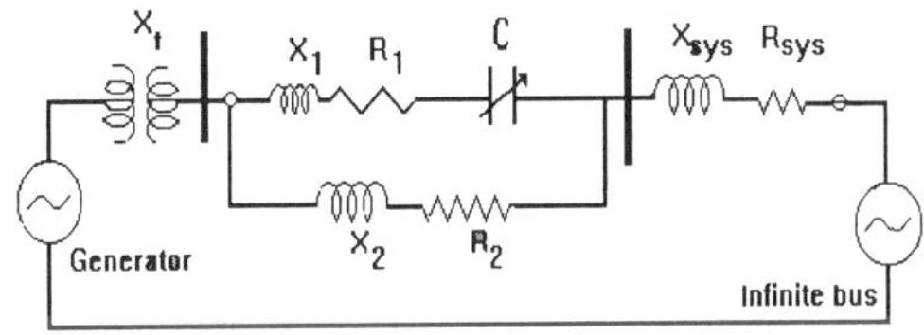

Fig 3. IEEE second benchmark system

connected to an infinite bus over two parallel transmission lines, one of which is capacitor compensated. The generator is represented as a four mass spring system:- high pressure turbine(HP), low pressure turbine(LP), the generator(GEN), and the exciter(EXC), coupled on a single shaft. The modal frequencies for the mechanical system are 24.65Hz , 32.39Hz and 51.10Hz [5]. It is reported in [5] that the torsional modes at 24.65Hz (mode 1) and 32.39Hz (mode 2) go unstable over certain range of compensation. Particularly the mode 1 is vulnerable at higher levels of compensation. The damping torque coefficient is calculated for two cases as mentioned above – one with balanced compensation on all phases and the other with unbalance in two phases.

Case 5.1 Analysis of T_D with balanced compensation level.

The variation of damping coefficient T_D is shown in Fig.4 for exciting the rotor at different modulating frequencies ω_m - 0 to 50Hz and having different balanced levels of compensation - 5% to 85%. We can observe that with increase in the level of compensation, the peak value of T_D increases and also the frequencies at which it occurs shifts to the lower

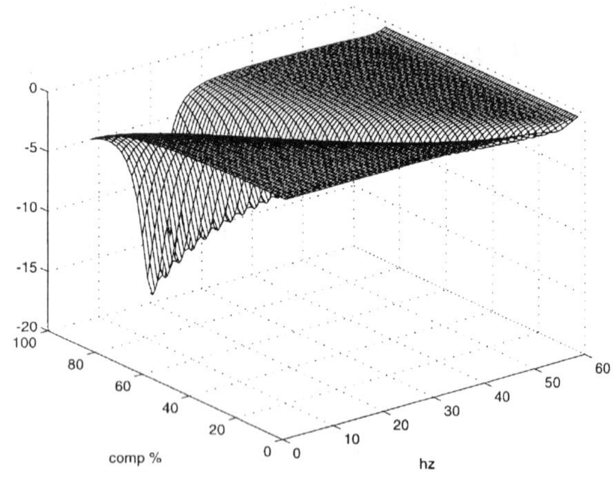

Fig 4. Variation of T_D for different compensation levels

side of the frequency range. Since we are interested in the damping effect of the torsional modes (at 24.6Hz and 32.39 Hz for the system), the comparison of T_D at 24.6Hz is shown in Fig 5. We can see that

- the magnitude of T_D at 55% level (3.55)is greater than the magnitude of T_D at 25% (0.5) and 75% (0.66)levels of compensation, which implies that there is poor damping in the case of 55%.

- the worst case is at 48% (T_D = 6.46)compensation level

Time domain simulation on EMTP is carried out with both the generator and the electrical network modeled in three phases, including the dynamics of the generator-turbine and the network. The disturbance

considered for the study is a three phase to ground fault on the high voltage side of the generator step-up transformer, with one cycle clearing time (fault duration is 17ms). At 55% compensation, we observe growing oscillations in the shaft torque as shown in Fig.6 which agrees with the analytical prediction.

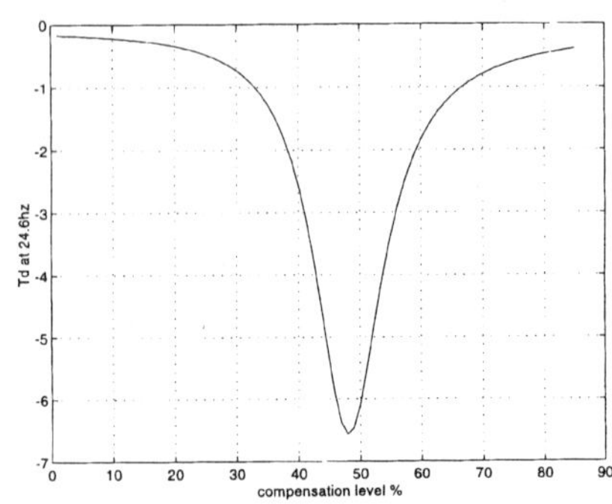

Fig 5. Variation of T_D at 24.6Hz for different compensation levels

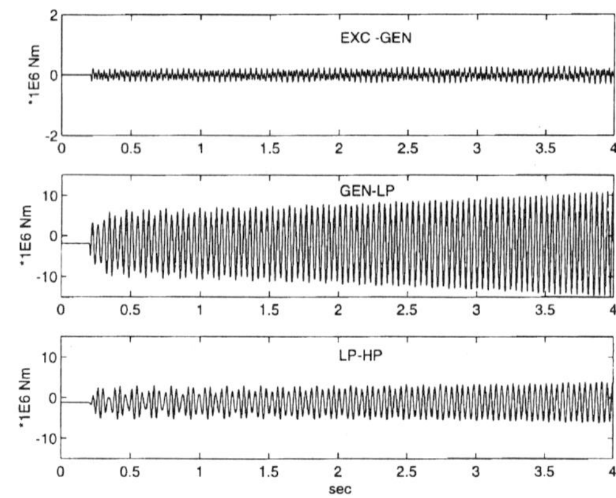

Fig 6 Shaft torque variation for a 3-ph to ground fault, at 55% compensation level.

Case 5.2 Analysis of T_D with unbalance in compensation level.

The damping torque coefficient T_D is computed for varying compensation levels in phases a & c in the range 10%-85%, keeping the compensation in phase-b constant at 55% compensation. Table 1 gives the maximum value of T_D and the frequency at which it occurs. It also shows T_D at 24.6Hz(one of the torsional modes). We can observe the following ,

Table 1 Maximum value of T_D for 0%   in ph-a and 55% in ph-b

Phase -c comp. (%)	T_D at 24.6Hz ,pu	Maximum	
		T_D ,pu	Freq. (Hz)
1	-2.03	-2.533	25.96
5	-2.11	-2.575	25.91
10	-2.24	-2.638	25.77
15	-2.39	-2.713	25.624
20	-2.58	-2.805	25.433
25	-2.795	-2.917	25.194
30	-3.02	-3.032	24.86
35	-3.2	-3.211	24.49
40	-3.22	-3.39	24.05
45	-2.96	-3.595	23.427
50	-2.47	-3.811	22.76
55	-1.925	-4.036	21.90
60	-1.497	-4.28	20.99
65	-1.211	-4.54	20.04
70	-1.04	-4.82	18.99
75	-0.937	-5.15	17.94
80	-0.88	-5.53	16.84
85	-0.85	-5.96	15.74

- an increase in the maximum value of T_D from –2.533 to –5.96 and a decrease in the frequency at which it occurs from 25.96Hz to 15.74Hz resp., as the compensation in phase-c increases.

- The critical levels are 30% to 35% because the maximum T_D occurs around the torsional frequency of 24.6Hz .

- the value of Td at 24.6Hz (shown in Table 1) for 35% compensation is lower than for the case of having balanced 55% compensation in all phases at 24.6Hz. (see Fig 5)

So the effect of detuning away from the torsional mode frequency can be observed with having unbalance compensation levels. Fig 7 shows the variation of T_D at 24.6Hz for varying compensation levels in phases a & c in the range 10%-85% keeping the compensation in phase-b constant at 55% compensation.   We can observe the following,

- The maximum value of T_D is –5.41, for having 48% and 47% compensation levels in phases a and c respectively. This value is less than the value of T_D for the balanced case of 48% compensation level (see Fig 5).

- There is symmetry between phases a and c . The change observed in T_D when we increase or decrease compensation in phase 'a', is also seen when we do a similar change in phase 'c' compensation.

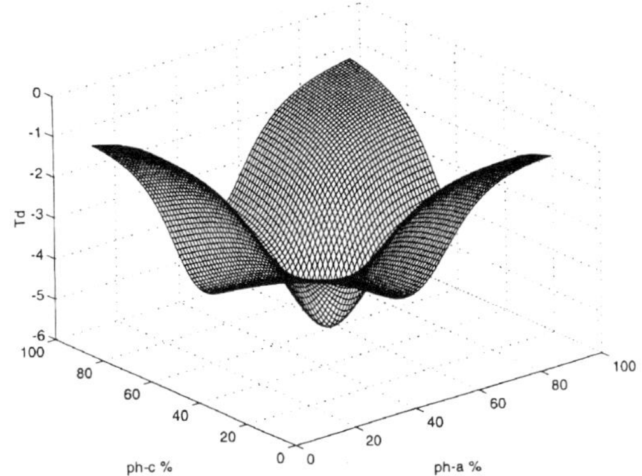

Fig 7. Variation of T_D at 24.6Hz for 55% in phase-b and different compensation levels in phases a & c

- Minimum negative damping is observed when we have either maximum or minimum compensation on both the phases or maximum compensation in one phase and minimum in the other phase.

5.3 Extension to TCSC -Validation using EMTP

EMTP simulations are carried out to demonstrate the control strategy adopted for TCSC. The nominal compensation provided by TCSC is 55%. When there is a disturbance the torsional oscillations are undamped as shown in Fig8. Five cases a-e of asymmetry (as listed below) are considered The desired asymmetry in the level of compensation is made in the phases at the time of fault application The change is effected either by switching in or switching out the capacitor modules. The disturbance is a three phase to ground fault lasting for 17ms duration.

Case (a)   10% in phase 'a', 55% in phase 'b' and 55% in phase 'c'

Case (b)   75% in phase 'a', 55% in phase 'b' and 55% in phase 'c'

Case (c) 10% in phase 'a', 55% in phase 'b' and 10% in phase 'c'

Case (d) 75% in phase 'a', 55% in phase 'b' and 75% in phase 'c'

Case (e) 0% in phase 'a', 55% in phase 'b' and 75% in phase 'c'

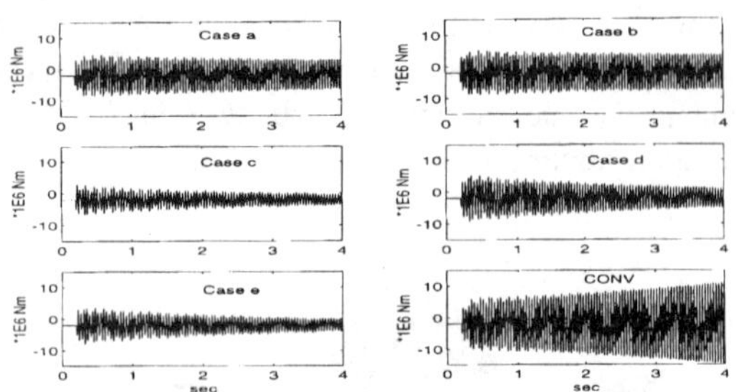

Fig 8. GEN-LP shaft transient torque for a 3-ph to ground fault for cases a-e. CONV- for 55% compensation in all phases

The oscillations in the GEN-LP shaft torque are shown in Fig.8 for all these cases. These oscillations are damped when there is a change in the compensation during transient period. The damping is more in cases c & e when compared to cases a, b & d. Though the damping is found to be almost same for the cases (c) and (e), the case (e) is better of the two for two reasons. One is that we can maintain more power flow (for three phases) during the transient in the case (e), compared to the normal (balanced) condition. But in case (c) the power flow is considerably reduced due to reduced compensation in both the phases. Secondly, 10% module may not be available for implementing, as it depends on the discrete steps adopted in the TCSC system. Instead complete bypass can be resorted to as in case (e) and easy to implement. Thus EMTP results confirm the damping of torsional oscillations by the proposed scheme.

6.0 CONCLUSION :

A theoretical analysis is presented to quantitatively assess the damping of torsional   modes under unbalanced network conditions due to asymmetric LC resonant circuits.. An expression for the damping torque coefficient is given in terms of equivalent network admittance evaluated at sub and supersynchronous frequencies. This coefficient is a useful measure in assessing the degree of asymmetry required for a given system.

The application of the unbalance concept to TCSC with discrete control is described for mitigation of SSR. The studies on the test system show that the change in the compensation level is to be made on both phases for better damping, instead of changing in only one phase. The best choice is found to have maximum in one phase and complete bypass of compensation in the other phase, during transients for damping transient torques. The strategy proposed is simple and easy to implement in the hardware.

7.0 References :

1. E.Larsen, C.Bowler, B.Damsky, S.Nilsson, 1992, "Benefits of thyristor controlled series compensation", CIGRE Paper no. 14/37/38-04.

2. A.Edris, 1990, "Series compensation schemes for reducing the potential of subsynchronous resonance", IEEE Trans on Power systems , 5, 219-226.

3. IEEE Committee Report 1992: "Reader's guide to subsynchronous resonance IEEE Trans on Power systems , 7, pp 150-157.

4. K.R.Padiyar , 1996, "Power system dynamics stability and control" Interline Publishing and John Wiley, Bangalore & Singapore.

5. IEEE SSR working group, 1985, "Second benchmark model for computer simulation of subsynchronous resonance", IEEE Trans. on PAS, 104, 1057-1066.

ANALYTICAL INVESTIGATION OF LARGE-SCALE USE OF STATIC VAR COMPENSATION TO AID DAMPING OF INTER-AREA OSCILLATIONS

A. R. Messina[a], D Olguin S[b], C A Rivera S[c], D Ruiz-Vega[d]

[a]Cinvestav, Mexico, [b]IPN, Mexico, [c]UAM, Mexico, [d]University of Liege, Belgium

ABSTRACT

This paper presents the results of analytical studies undertaken to assess the effectiveness of multiple Static VAR Compensation (SVC) in aiding damping of inter-area oscillation in the Mexican interconnected system. A frequency-domain methodology is adopted to coordinate controllers as well as to minimise the potential for adverse interaction between control loops. The effectiveness of SVC voltage support for damping inter-area oscillations is verified by non-linear time domain simulation studies.

INTRODUCTION

Dynamic shunt compensation by means of Static VAR Compensators (SVCs) is increasingly being utilised in the bulk Mexican interconnected system (MIS) to enhance voltage control and overall system dynamic performance. This paper presents the results of analytical studies undertaken to examine the application of extensive voltage support to damp power swings in the MIS. The paper examines several analytical procedures for estimating the capability of SVC voltage support to add damping to system electromechanical modes and discusses the experience in the modelling and simulation of these devices. Particular emphasis is being placed on the study of the effects of several large SVCs recently installed in the 400 kV network.

A 130-machine, six-area model of the MIS that includes the operation of 8 SVCs is used to assess the co-ordinated application of multiple controllers on dynamic system behaviour. Special efforts are devoted to quantify the influence of SVC dynamic support to enhance the damping of three critical inter-area modes in the MIS that involve the interaction between machine in the north and south systems of the network. To this end, a systematic approach based on modal analysis and decentralised control theory is proposed to analyse and design FACTS controllers.

Time-domain analyses are finally conducted to check the validity of the analysis as well as to assess the impact of dynamic voltage support on transient system behaviour. Study results include the identification of inter-area modes that are more controllable by existing SVCs and the use of supplementary modulation controls to enhance the damping of critical system modes.

AC-DC Power Transmission, 28-30 November 2001
Conference Publication No. 485 © IEE 2001

OVERVIEW OF THE STUDY SYSTEM

A simplified one-line diagram of major system elements is shown in Fig. 1. The MIS constitutes a huge 400/230/115/138 kV network extending from the border with Central America up to its interconnection with the USA. This transmission system is described in detail in Ref. (1). System studies in this research are based on a six-area model of the MIS that comprises 351 buses, 470 branches, 130 generators and 8 major SVCs. Dynamic and transient stability problems are a major source of concern to this network.

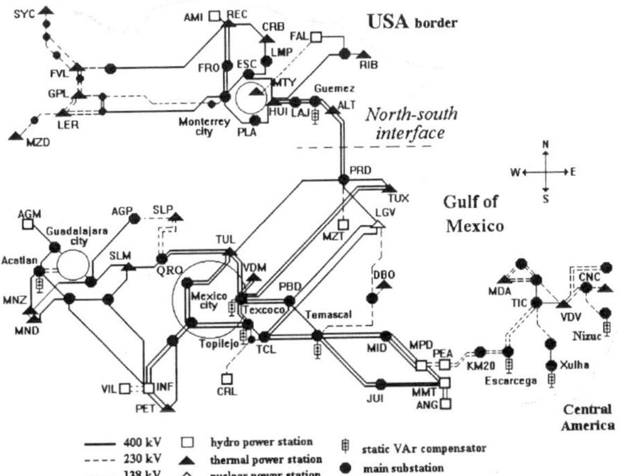

Fig. 1.- One-line diagram of the study system showing existing SVCs and major transmission elements

Dynamic voltage support in the MIS

Static VAR compensators are installed at several key locations on the 400/230 kV network of the MIS to provide voltage support as well as to enhance system dynamic performance (refer to Fig. 1) (2). Table 1 synthesises the main characteristics of SVCs considered in this study. Major control facilities include a SVC at the Temascal susbation on the 400 kV transmission network of the south-estern system, a SVC at Guemez substation on the north-south interface, and two SVCs at the Topilejo and Texcoco substations in the neighbourhood of the Mexico City metropolitan area.

The nature of these devices is briefly described below. Other existing devices were found to have a rather marginal impact on overall system damping and are therefore not included in the following analyses.

Table 1
Main characteristics of SVCs in the MIS

SVC location	Type	Rating (MVAr)	Main purpose/ Year of commissioning
Temascal	TSR/ TSC	±300	Steady-state and transient stability enhancement (1982)
Acatlan	TCR	-200	Local voltage Control (1982)
Texcoco	TCR/ TSC	-90/300	Dynamic voltage Control (1999)
Topilejo	TCR/ TSC	-90/300	Dynamic voltage Control (1999)
Guemez	TCR/TSC	-90/300	Dynamic voltage control and Power damping (1999)
Escarcega	TCR/ FC	-50/150	Dynamic voltage control and Power damping (1998)
Xulha	TCR/ FC	-20/40	Dynamic voltage control (1998)
Nizuc	TCR/ FC	-30/100	Dynamic voltage control (1999)

The SVC at the Temascal substation

The SVC at the 400 kV Temascal substation consists basically of two thyristor-switched capacitor (TSC) groups, one thyristor-switched reactor group (TSR) and one thyristor-controlled reactor group (TCR). Each of these groups has three-phase rating of 75 MVAr, thus producing an overall dynamic range of ±300 MVAr (refer to Fig. 2). The primary voltage control loop includes a measuring device (MD), a six-pulse rectifier and a low-pass filter with a time constant of 1.5 ms to reduce fast transients in the bus voltage. No modulation controls are presently used. This SVC is normally used to provide maximum reactive power reserve to prevent loss of stability following major contingencies on the south-eastern network.

The SVCs at the Topilejo, Guemez and Texcoco substations

The SVCs at the Topilejo, Guemez and Texcoco have similar control and operating characteristics. Each SVC has a dynamic range of –90 (inductive) to 300 Mvar (capacitive). Mechanical switched capacitors are also used to allow for the control of slow voltage variations at the SVC location. The normal operating mode for the SVCs at Guemez and Texcoco is voltage control. The SVC at the Topilejo substation, on the other hand, incorporates a supplemental modulation control loop. A single-line diagram of these devices is shown in Fig. 3.

Detailed models of SVC power equipment and control systems provided by the manufacturers were used in the simulations. Further, in addition to the normal SVC voltage control loop, a Power System Damping Control (PSDC) function, to allow modulation of the SVC voltage was also represented.

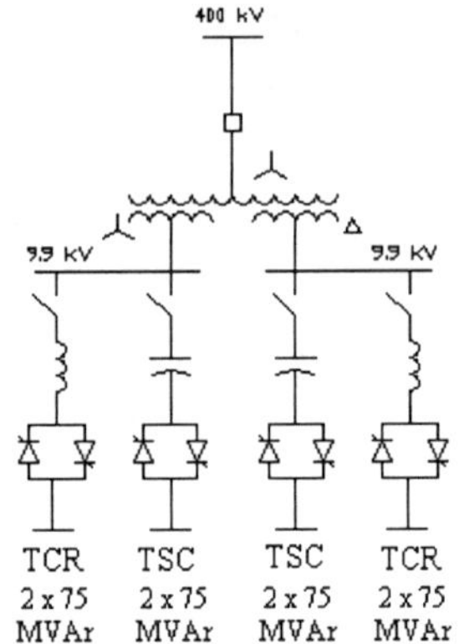

Fig. 2. Single line diagram of the SVC at the 400 kV Temascal station

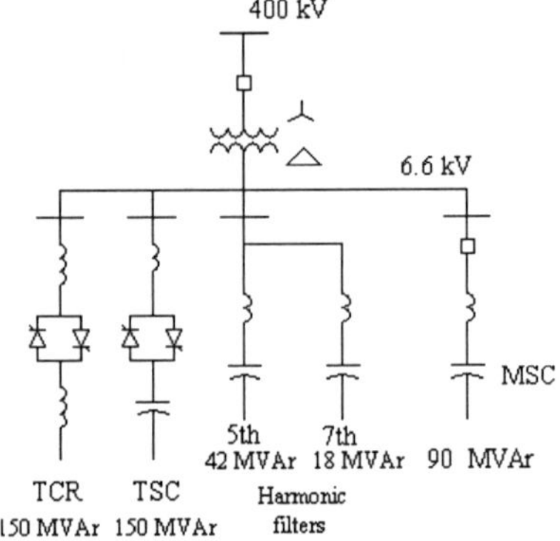

Fig. 3. Single-line diagram of the SVCs at Guemez, Topilejo and Texcoco substations

STUDY APPROACH

Consider a linear dynamical system described by the linear state representation

$$\dot{\mathbf{x}} = \mathbf{Ax} + \mathbf{Bu}$$
$$\mathbf{y} = \mathbf{Cx} \tag{1}$$

where $\mathbf{A} \in \mathfrak{R}^n$, $\mathbf{B} \in \mathfrak{R}^m$, and $\mathbf{x} \in \mathfrak{R}^n$ is the state vector; $\mathbf{u} \in \mathfrak{R}^m$ is the vector of control inputs and $\mathbf{y}$ is the vector of system outputs. The input-output transfer function of the system can be expressed in the form

$$\mathbf{y}(s) = \mathbf{C}\boldsymbol{\Phi}\left(s\mathbf{I} - \boldsymbol{\Lambda}\right)^{-1}\boldsymbol{\Psi}\mathbf{Bu}(s) = \sum_{i=1}^{n} \frac{\mathbf{R}_i}{s - \lambda_i} \tag{2}$$

where $\boldsymbol{\Lambda} = \text{diag}\{\lambda_1, \lambda_2, .., \lambda_n\}$ and $\boldsymbol{\Phi}, \boldsymbol{\Psi}$ are the corresponding matrices of right and left eigenvectors; the residue matrix $\mathbf{R}_i$ is used to compute transfer function residues as well as determine controllability and observability factors.

Controller design methodology

In dealing with interacting power system controllers, two major design aspects must be addressed: the study of the extent of interaction between the control loops and the nature of this interaction. The design task is divided into three parts (3):

1. Identification of modes that can be damped using output feedback
2. Selection of feedback signals for the controllers, or pairing of inputs and outputs, and
3. Decentralised control of PSDC controllers

In the following sections, a systematic approach for the selection of input signals for FACTS controllers is described.

Interaction measures

To gain insight into the nature of the interaction problem, consider the multiloop control system in Fig. 4. Let $\mathbf{G}(s)$ be the open-loop transfer matrix with elements $[g_{ij}(s)]$, for inputs u_i and outputs y_j, $i,j = 1,..,m$. In decentralised control theory each system input is expressed in the form $u_i = g_{ci} y_i$ where g_{ci} is the transfer function of the ith system controller. Clearly, if $G_{ij}(s)=0$, then each controller can be designed without any loss of performance.

Hence, the closed loop transfer function can be expressed as

$$\mathbf{G}_{CL}(s) = \left[\mathbf{I} - \mathbf{G}(s)\mathbf{G}_C(s)\right]^{-1} \mathbf{G}(s) \qquad (3)$$

where $\mathbf{G}_C = diag\{g_{c1}(s),..,g_{cm}(s)\}$.

Following Huang et al. the generalised dynamic relative gain (GDRG) is defined as (4)

$$GDRG_{ii}\{\mathbf{G}(s)\} = \frac{[\mathbf{G}(s)]_{ii}}{[\mathbf{G}_{CL}^i(s)]_{ii}} \qquad (4)$$

This coefficient provides a measure of loop interaction of a multiloop control system. For a given frequency of interest ω_h, the minimum interaction for the ith control loop with other controllers is obtained if

$$|GDRG(\omega_h)| = 1 \qquad (5)$$

Therefore, a practical measure of dynamic interaction is obtained from the GDRG coefficient

$$GDRG\{\mathbf{G}(j\omega)\} = \sum_{i=1}^{k} |GDRG_{ii}\{\mathbf{G}(j\omega)\} - 1| \qquad (6)$$

Given a set input-output, the procedure to determine the control structure for decentralised control is as follows: i) for each critical oscillation mode of interest, compute associated residues for all pairings of inputs and outputs, and ii) calculate the GDRG and select the control structure that presents minimum interaction.

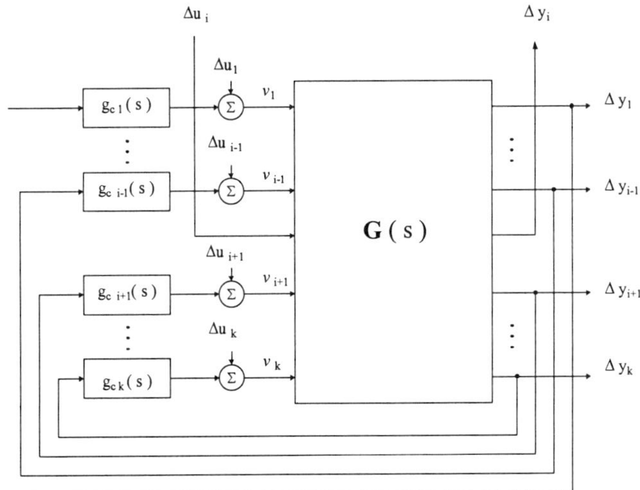

Fig. 4. Conceptual representation of the power system including the connection of supplementary controls

Damping controller design

A frequency-domain methodology based on the analysis of residues was used for coordination of controllers. Let the transfer function of a conventional power system control function be given by

$$M_k(s) = K_{psdc}(s)Q_k(s) = K_{psdc}\left(\frac{1+sT_{1k}}{1+sT_{2k}}\right)^{m_k}$$

where K_{psdc} is the gain of stabilizer and m_k is the number of compensation stages. The proposed iterative control design procedure consists of two major steps:

1. Compute the transfer function residues at each FACTS location and input signal. For a given mode of interest h, the time constants T_1k y T_2k are adjusted such that (5)

$$\arg\left\{R_{jk}^h \, Q^k(\lambda_h)\right\} = \pm 180°$$

2. Assuming ΔK_{psdc} is small, determine the control setting that minimises the real part of the mode of interest. A first order approximation for any system eigenvalue is then calculated from

$$\Delta \lambda_h \approx R_{jk}^h \, Q_k(\lambda_h) \, \Delta K_{psdc}k$$

DAMPING CHARACTERISTICS OF THE MIS

Small signal stability studies were directed to assess the potential benefits to be derived from the coordinated application of SVC controllers. The base operating condition with no SVC voltage support has over 1000 states. A root-loci of the dominant system eigenvalues for the base operating condition is shown in Fig. 5. The MIS model is characterised by six weakly damped inter-area modes:

- Two critical inter-area modes at 0.40, and 0.72 Hz. These modes are referred to as the 0.4 Hz North-south inter-area mode and the 0.7 Hz East-west mode

- Two higher frequency modes at 1.12 and 1.13 Hz

Damping studies in this research concentrate on the analysis of inter-area modes 1 through 3 and mode 9 that exhibit damping ratios below 5%. Table 2 synthesizes the main characteristics of these modes showing their modal damping ratios and swing patterns.

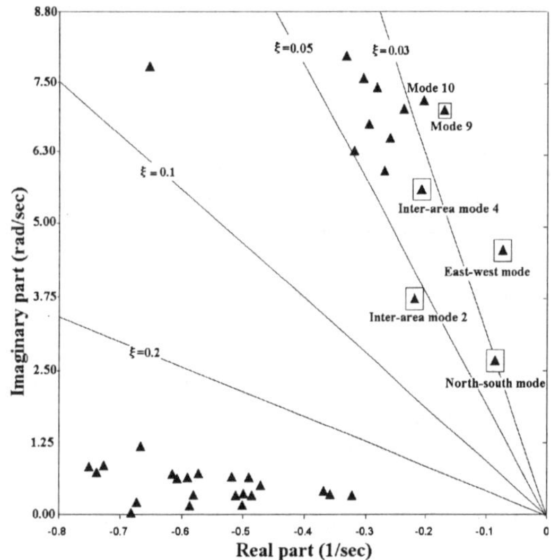

Fig. 5. Root-loci of dominant system eigenvalues

Table 2

Open-loop system eigenvalues

Mode	Eigenvalue	Freq. (Hz)	Swing Pattern*	Dominant machine
1	-0.0673±j2.52	0.40	North systems vs. South systems	MTY FAL RIB
2	-0.2175±j3.67	0.59	Machines in the north systems	MTY MZD GDL
3	-0.0510±4.49	0.72	W and C system vs. SE system	SLM MNZ ANG
9	-0.1694±j7.06	1.12	W systems Vs. C systems	SLM INF SLP

*C- Central, W-western, SE-south-eastern

COORDINATED APPLICATION OF SVCs

In the present study, several compensation alternatives to enhance dynamic performance were considered. In order to assess the effect of voltage support on system stabilisation, transfer function residues between the SVC susceptance and the SVC reference voltage were calculated for each of the critical modes. Table 3 shows the residues from the voltage reference inputs of SVCs to local signals for the modes of interest. Two classes of signals were chosen as input signals to the modulation controllers: tie-line power and current.

Study results suggest that that the SVC at the Guemez substation is more effective in damping both the 0.4 Hz North-south inter-area mode 1 and the 0.6 Hz mode. The SVCs at the Temascal and Topilejo substations, on the other hand, are expected to provide stabilization to modes 3 and 9, respectively. Other SVCs were found to have a negligible effect on the damping of critical modes.

Table 3

Top residues for the transfer function $\Delta\beta_{svc} / \Delta V_{ref}$

Mode	Input signal*	SVC location	Residue
1	Guemez-HUI (P)	Guemez	**5.257**
	Guemez-ALT (P)		4.858
	Guemez-ALT (I)		4.504
2	Guemez-HUI (P)	Guemez	**2.019**
	Guemez-HUI (I)		1.931
	Guemez-ALT (P)		1.862
3	JUI-Temascal (I)	Temascal	**5.674**
	Temascal-TEC (I)		4.506
	MND-Temascal (I)		4.007
9	Topilejo-SNB (P)	Topilejo	**0.271**
	Topilejo-CRU (I)		0.134
	TEC-Topilejo (P)		0.132

*P-tie-line power; I- tie-line current

From Table 3, four sets of pairings of SVC locations and input signals were selected for enhancing damping of critical system modes. Table 4 summarises the main characteristics of these sets indicating the control objective and selected input signals to the controllers.

Table 4

The pairing of locations and input signals

Set	Inter-area mode 1	Inter-area mode 3	Inter-area mode 9
	Input signal to SVC at Guemez	Input signal to SVC at Temascal	Input signal to SVC at Topilejo
1	Guemez-HUI Line current	MND-Temascal Line current	TEC- Topilejo Line power
2	Guemez-HUI Line current	Temascal–PBD Line power	TEC-Topilejo Line power
3	Guemez-ALT Line power	Temascal PBD Line current	Topilejo-CRU Line current
4	LAJ - Guemez Line current	Temascal–PBD Line current	Topilejo CRU Line current

Analysis of interaction among controllers

The analysis of dynamic relative gains was applied to assess the potential for adverse interaction among controllers. Fig. 6 depicts the magnitude of the GDRG for the frequency range of interest. Examination of these results shows that sets 3 and 4 exhibit the smallest degree of interaction for the frequency range of interest.

This procedure ensures the minimum adverse interaction between SVC controllers, necessary for decentralised design of controllers.

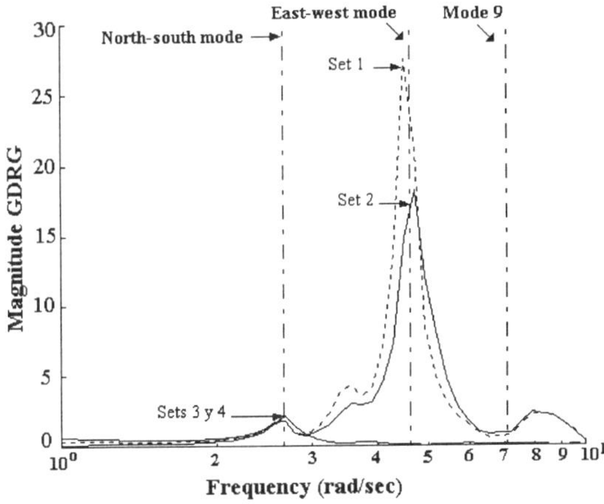

Fig. 6.- Magnitude of GDRG as a function of frequency

Based on the above analysis, a three SVCs alternative was investigated to aid damping of critical inter-area modes, namely:

- The SVC at Guemez substation tuned to damp the 0.40 Hz North-south inter-area mode 1 and the 0.59 Hz inter-area mode 2.

- The SVC at Temascal tuned to damp the 0.72 Hz East-west inter-area mode 3

- The SVC at Topilejo tuned to damp mode 9

Table 5 shows the closed-loop eigenvalues. It is seen that the improvement in the modal damping ratio of the critical system modes is noticeable. Note, however, that the 1.12 Hz mode is not observable with the tie-line input to the SVC at Topilejo.

Table 5
Closed-loop system eigenvalues

Inter-area mode	Eigenvalue	Damping ratio (%)	Freq. (Hz)
1	$-0.3695 \pm j2.58$	14.17	0.41
2	$-0.3175 \pm j3.63$	8.71	0.57
3	$-0.4635 \pm j4.62$	9.98	0.73
9	$-0.1788 \pm j7.07$	2.52	1.12

Further, several ac/dc transmission alternatives to removing operational constraints were investigated. Each alternative was simulated using the same base case system following the proposed design approach.

Coordinated application of SVC and TCSC

This alternative compares the application of SVC voltage support at Guemez with the application of a TCSC on the 400 kV North-south interface. A TCSC with a rating of 10% of the line reactance is installed in one of the circuits. The chosen input signal was the local active tie-line power through the TCSC.

Coordinated application of SVC and PSS

This alternative assumes the coordinated application of two major SVCs and PSSs at several dominant machines. It should be noted that each controller was designed to damp a specific system mode. Table 6 summarises the results of this control alternatives.

Table 6
Effect of different control actions on system damping

Control strategy	Eigenvalue
SVC at Temascal with PSDC & TCSC in line ALT-Guemez	$-0.3158 \pm j2.69$ $-0.1766 \pm j4.65$ $-0.1656 \pm j7.03$
SVCs at Temascal and Guemez with PSDC and PSS at SLM	$-0.3276 \pm j2.66$ $-0.4651 \pm j4.58$ $-0.3135 \pm j6.96$
SVCs at Temascal, Guemez and one 500 kV HVDC link between MMT andTemascal	$-0.3727 \pm j3.03$ $-0.2318 \pm j4.43$ $-0.2537 \pm j6.75$

TIME-DOMAIN SIMULATIONS

Detailed time-domain studies were finally performed to assess the effectiveness of SVC modulation for damping critical inter-area modes. Paramount to system operation, the various contingency cases simulated to excite these modes are:

A) Tie-line tripping without fault of the SLM-QRO 400 kV line in the Western system. This fault is known to strongly excite the 0.7 Hz East-west mode.

B) Tine-line tripping without fault of one circuit of the north south interface (ALT-PRD 400 kV line). This contingency is found to strongly excite both the 0.4 Hz North-south mode and inter-area mode 2

Simulation results with and without SVC voltage support for the cases A and B are shown in Figs. 7 through 10. As shown, outage of SVC voltage support leads to sustained or increased oscillations along major interconnectors and machines. For case A, the critical contingency results in growing power and voltage oscillations having a frequency near 0.7 Hz on the 400 kV of the south-eastern system. Further, loss of one circuit of the ALT-PRD 400 kV line on the north-south interface leads to growing power oscillations characterised by a frequency of 0.4 Hz. Rotor angle deviations of dominant machines indicate that loss of stability would occur. SVC modulation at Temascal and Guemez is seen to stabilise the system. Use of multiple SVC voltage support with supplementary modulation allows achieving satisfactory transient performance.

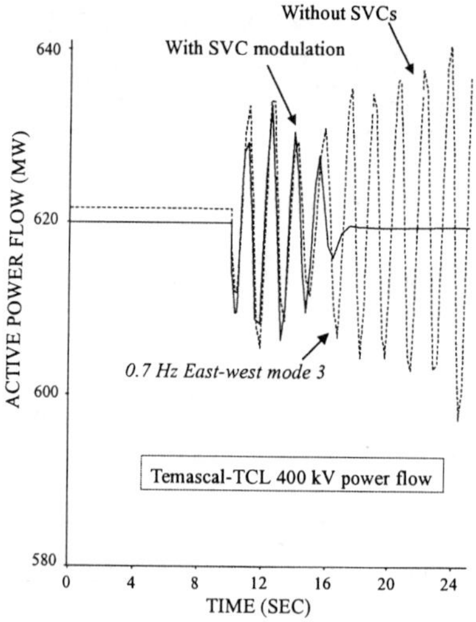

Fig. 7.- Comparison of active power flow with and without SVC voltage support for case A.

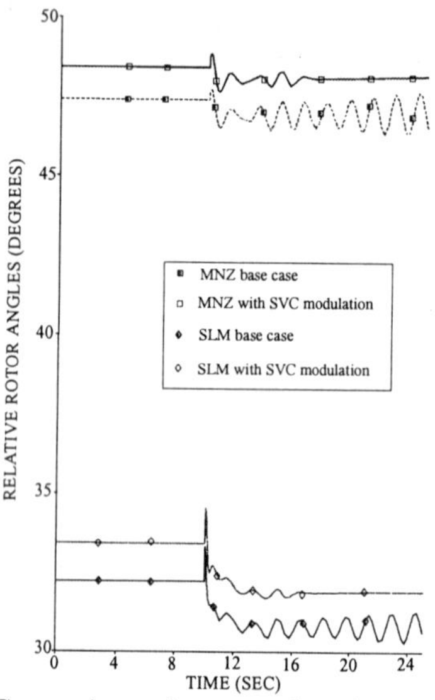

Fig. 8. Comparison of rotor angle swings with and without SVC voltage support for case A.

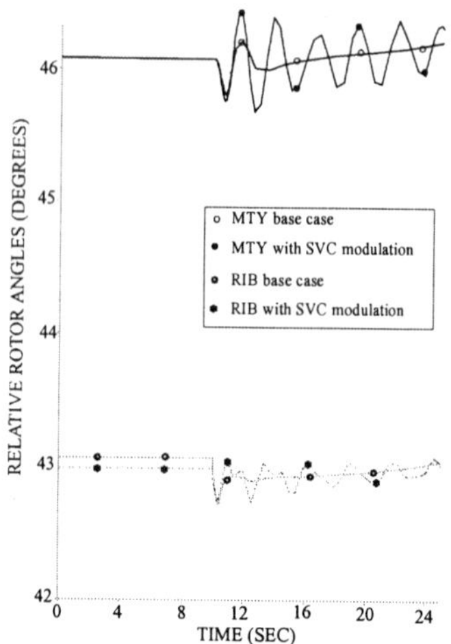

Fig. 9.-Comparison of rotor angle swings with and without SVC voltage support for case B.

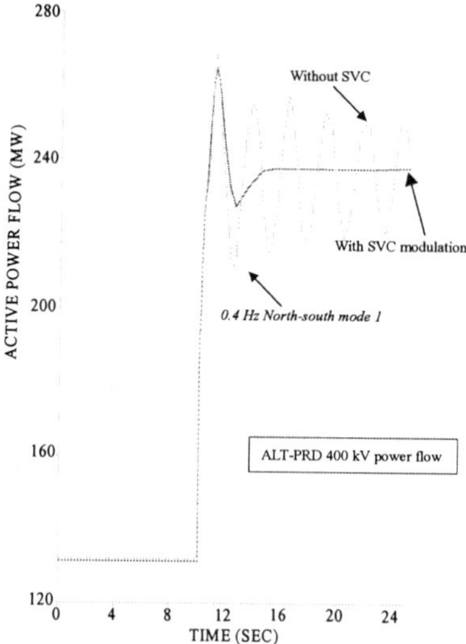

Fig. 10.- Comparison of active power flow with and without voltage support for case B.

CONCLUSIONS

This paper presents the results of analytical studies undertaken to assess the effectiveness of multiple SVC voltage support on system damping. Study experience with a complex system shows that existing SVCs in the MIS can be effectively used to aid stabilization, especially under the most stringent operating conditions. The selection of supplementary signals of the SVCs, was found to be crucial to the damping contribution to ensure minimum interaction with other controllers. Time-domain studies correlate well with small signal analyses showing the appropriateness of the developed models.

REFERENCES

1. Messina AR, Ramírez JM, and Cañedo JM, 1998, "An investigation on the use of power system stabilizers for damping inter-area oscillations in longitudinal power systems," <u>IEEE Trans. Power Systems, vol. 13</u>, 552-559, .

2. Jesus Gonzalez Flores, Cesar Fuentes Estrada, Miguel Angel Avila Rosales, Rolf Grunbaum, 1999, "Mexican grid uses FACTS for greater flexibility," <u>Modern Power Systems, June</u>, 1-4

3. Zhang P., Messina AR, "Selection of Locations and Input Signals for Multiple SVC Damping Controllers in Large Scale Power Systems", 1998 IEEE/PES Winter Meeting, paper IEEE 0–7803–4403–0.

4. Huang, Hsiao-Ping, Ohshima, Masahiro, Hashimoto, Iori, 1994, "Dynamic interaction and multiloop control system design", <u>J. Process Control, 1</u>, 15-27.

5. Yang Ning, Qinghua Liu, McCalley James D, 1998, "TCSC controller design for damping interarea oscillations", <u>IEEE Trans. On Power Systems, 14, No.4</u>, 1304-1310

AKNOWLEDGEMENTS

The authors wish to thank the support of Conacyt on grant 31840-A.

STUDY ON SSR CHARACTERISTICS OF POWER SYSTEMS WITH STATIC VAR COMPENSATOR

Hongtao Liu Zheng Xu Zhi Gao

Zhejiang University, P.R. China

ABSTRACT

Using the complex torque coefficient method realized by time domain simulation, the subsynchronous resonance (SSR) characteristics of power systems with static VAR compensators are studied in depth in this paper. The studied model is modified from the first IEEE Subsynchronous Resonance benchmark system added by a static VAR compensator (SVC), consisting of a thyristor-controlled reactor (TCR) and a thyristor-switched capacitor (TSC). A number of cases are studied by simulation through varying some key parameters of the model. By comparing the electric damping in different cases, the subsynchronous resonance characteristics of power systems with SVCs are identified.

INTRODUCTION

Static VAR compensator (SVC), a kind of dynamic shunt compensator, is applied extensively to the power system. Its applications include effective voltage control, transient stability margin enhancement and improvements on the damping of power oscillation. Although SVC can greatly improve system performance, whether it will bring about the same potential problems of subsynchronous resonance as series capacitor compensator does still remain an issue under study. It is necessary to quantify the potential effects of SVC on SSR characteristics of power system. Devices in FACTS such as SVC are composed of switching parts. In studying the SSR problems caused by such devices, numerous analysis methods based on analytical model meet serious problems in reality. In order to solve these problems, a newly developed approach is adopted in this paper. Based on the complex torque coefficient method, this approach is realized by the simulation in the time domain. The power system is modified from the first IEEE Subsynchronous Resonance benchmark system, and the SVC consists of a TCR and a TSC with their detail time domain simulation model. By the proposed approach, the SSR characteristics of power system with SVC are studied in detail. By varying some key parameters of the tested system and SVC, the electric damping characteristics under different conditions are compared, and some important conclusions are obtained.

AC-DC Power Transmission, 28-30 November 2001
Conference Publication No. 485 © IEE 2001

THE COMPLEX TORQUE COEFFICIENT APPROACH REALIZED BY TIME DOMAIN SIMULATION

In Xu and Feng (1), the applicability of the complex torque coefficient approach is analyzed in detail and the realization method of the complex torque coefficient approach by time domain simulation is also established. Here, some valuable results derived from paper Xu and Feng (1) will be applied to analyze the SSR characteristics of power system with SVC. The realization method of the complex torque coefficient approach by time domain simulation is concluded below.

For a system with a single machine and a number of fixed frequency sources. The prerequisite for the validity of the complex torque coefficient method is that the increment of the electromagnetic torque of the machine under small disturbance can be represented as

$$\Delta T_e = K_s \Delta \delta + K_D \Delta \omega \qquad (1)$$

In (1), $K_s \Delta \delta$ is called the synchronous torque, $K_D \Delta \omega$ is called the damping torque, K_s and K_D is called the synchronous torque coefficient and damping torque coefficient respectively, $\Delta \delta$ and $\Delta \omega$ are increment of the power angle and the angular velocity of the machine relative to the synchronously rotating frame. Generally, in (1), ΔT_e, $\Delta \omega$, K_s and K_D are in pu, while $\Delta \delta$ is in radian. Further, $\Delta \delta$ and $\Delta \omega$ have the following relationship:

$$\Delta \omega = \frac{1}{\omega_0} \frac{d\Delta \delta}{dt} \qquad (2)$$

where ω_0 is the synchronous angular velocity, t is in second.

Suppose that the machine rotor oscillates in small magnitude at frequency of $\lambda \omega_0$ ($\lambda < 1$), when the system is in steady state, each quantity can be represented by a phasor. According to (1) and (2), the following equations can be derived:

$$\Delta \dot{\omega} = \frac{1}{\omega_0} (j\lambda \omega_0) \Delta \dot{\delta} = j\lambda \Delta \dot{\delta} \qquad (3)$$

and $\quad \Delta \dot{T_e} = K_s(\lambda)\Delta \dot{\delta} + K_D(\lambda)\Delta \dot{\omega}$ (4)

where a variable with a dot over it represents a phasor. So the following expression can be deduced:

$$\frac{\Delta \dot{T_e}}{\Delta \dot{\delta}} = K_s(\lambda) + j\lambda K_D(\lambda)$$ (5)

$$\frac{\Delta \dot{T_e}}{\Delta \dot{\omega}} = K_D(\lambda) - j\frac{1}{\lambda}K_s(\lambda)$$ (6)

The objective of time domain simulation of the complex torque coefficient method is to obtain $\Delta \dot{T_e}$, $\Delta \dot{\delta}$ and $\Delta \dot{\omega}$, hence to obtain $K_s(\lambda)$ and $K_D(\lambda)$.

TIME DOMAIN SIMULATION MODEL OF POWER SYSTEM AND SVC

For subsynchronous resonance study, the tested power system is modified from the first IEEE SSR benchmark model, which is shown in IEEE subsynchronous resonance task force (2). The modification includes canceling the original fixed series capacitor, dividing the original transmission line into two segments, and installing a SVC on the joint of these two segment lines. The circuit diagram of test system is shown in Figure 1. The function of the SVC is to maintain the voltage of the bus connected at a constant value. It is supposed that the rated frequency is 50Hz and the rated voltage is 400kV.

On the basis of SVC configuration discussed in Bergman et al (3), the SVC model under study is established and its single line block diagram is shown in Figure 2. The main circuit consists of a 225MVAr TCR in delta connection, a 300MVAr TSC in delta connection and a 32MVAr filter to eliminate the 5th, 7th and 11th order harmonics. The voltage control loop includes a proportional-integral (PI) type controller with the ability to control voltage quickly and accurately.

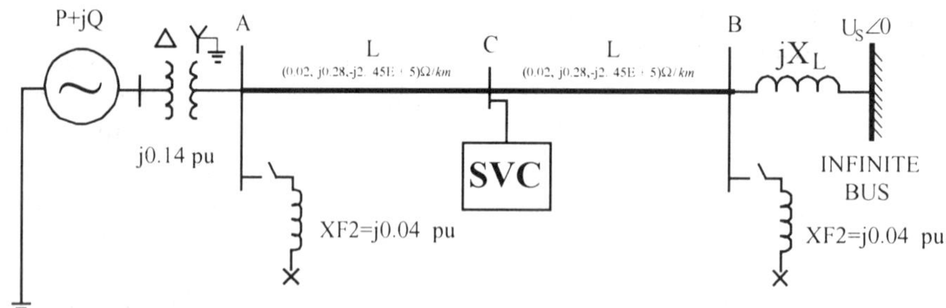

Fig.1 Circuit diagram of tested system

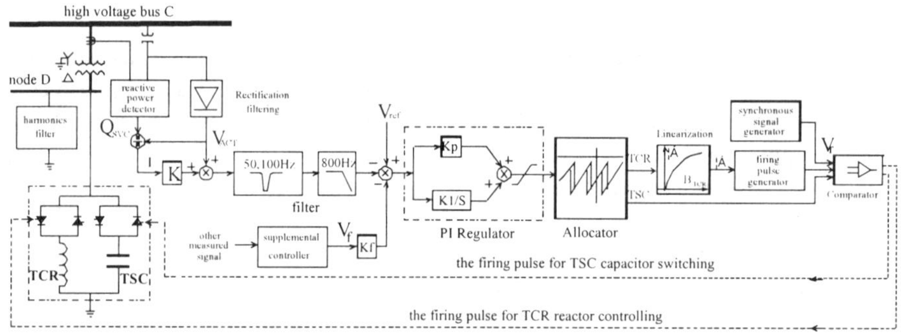

Fig.2 Single line diagram of SVC

SIMULATION RESULTS AND ANALYSIS

The simulation and analysis of SSR characteristics of power system with SVC will be divided into two parts and discussed in detail. First, by varying some parameters of SVC controller, their influences upon the SSR characteristics are analyzed by comparing the electric damping of the entire system operating with different parameters. Second, by varying the parameters of the power system, the corresponding electric damping characteristics of the entire system are analyzed.

Analysis with Different Parameters of SVC

Different slopes of SVC voltage-ampere characteristics curve. In Figure 2, for the curves depicting voltage-ampere characteristics of the SVC control system at fundamental frequency, the slope of them is decided by the constant K. Simulate the system with different K such as 0, 0.03 and 0.12, we obtained the curves of $K_D(f)$, which are shown in Figure 3. From Figure 3, the damping torque coefficient declines as K increases while the frequency ranges from 3 Hz to 45 Hz.

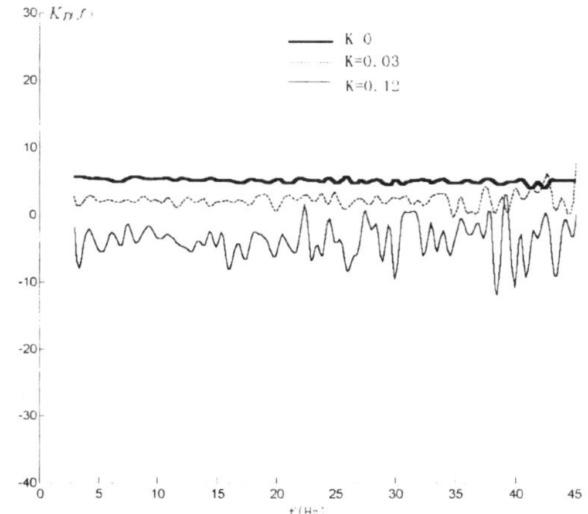

Figure 3   The damping torque coefficient vs for frequency different K

Supplemental reactive power controller. The role of supplemental reactive power controller in the power system is discussed in Romegialli and Beeler (4). In some cases, SVC must be configured to work with other reactive power sources. Their individual steady state reactive power output ought to be determined with respect to each other. To realize this goal, a supplemental reactive power controller with a relatively slow response can be added to the control system. When the reactive power reference Q_{ref} is assigned the value of -200MVAr, 0 and +250MVAr, then $K_D(f)$, the response of the system in frequency domain, is shown in Figure 4. Comparing these three curves in Figure 4, we reach the following conclusion: When Q_{ref} is negative, the damping torque coefficient of the system will increase. The system damping torque coefficient will decrease when Q_{ref} has a positive value. This is more obvious at a higher frequency. The block diagram of the supplemental reactive power controller for above analysis is illustrated in Figure 5.

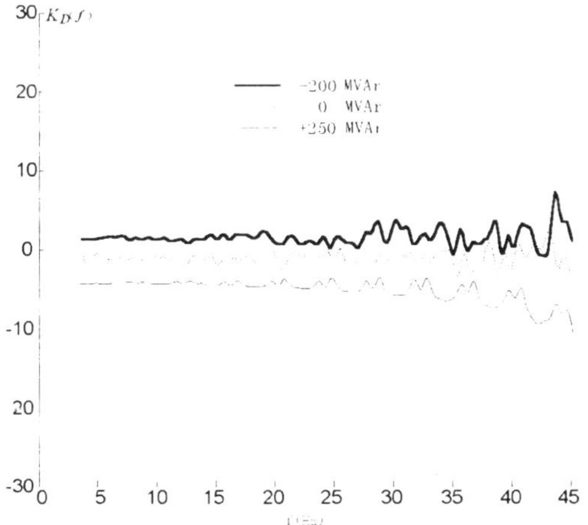

Figure 4 The damping torque coefficient vs frequency for different Q_{ref}

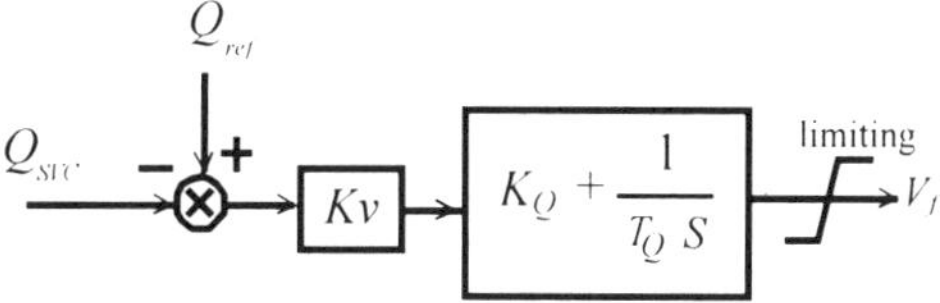

Figure 5   Block diagram of supplemental reactive power controller

Supplemental power oscillation damping controller. Supplemental power oscillation damping controller and its related functions are discussed in Wang and Swift (5). With this controller installed, SVC is able to damp the low frequency power oscillation in the system. According to the analysis, the current flowing through line AB (ΔI_{AB}) is the most sensitive to the control signal of power oscillation. So it is desirable to use ΔI_{AB} as the input for the supplemental power oscillation damping controller. The block diagram of the controller is shown in Figure 6. $K_D(f)$, the response in frequency domain when SVC has and has no such controller, is shown in Figure 7. From these two curves in Figure 7, we conclude that the inclusion of supplemental power oscillation damping controller can greatly enhance the damping torque coefficient of the system at low frequency.

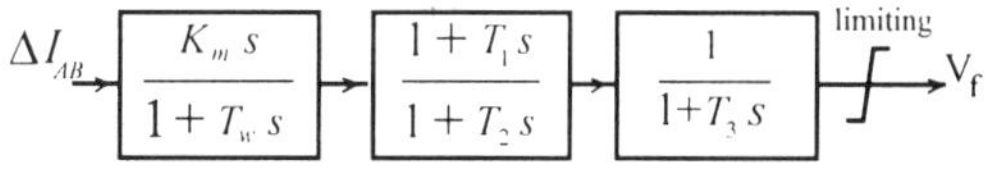

Figure 6   Block diagram of supplemental power oscillation damping controller

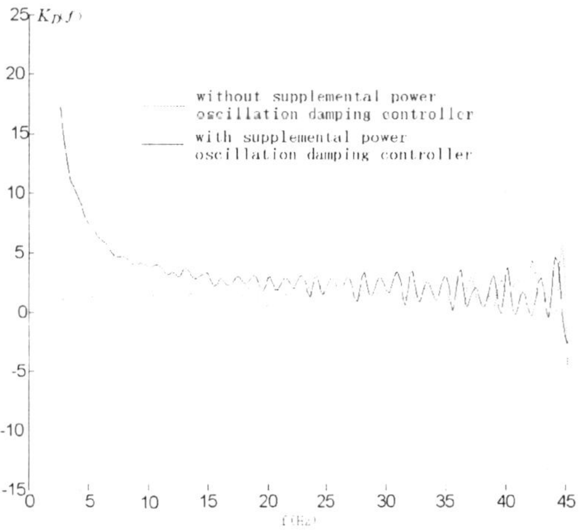

Figure 7   The damping torque coefficient vs frequency with and without supplemental power oscillation damping controller

Different synchronous methods. Different synchronous methods have quite different influences on the damping torque coefficient. Basic synchronous methods used in SVC are zero-crossing detection of individual line voltage and phase-locking loop (PLL). The former one uses the zero-crossing moment of line voltage of TCR and TSC as the starting time for the gating delay angle. The latter utilizes an oscillator, which is capable of tracing the system frequency and phase, to generate synchronous reference signal. The block diagram of PLL is shown in Figure 8. V_i is the line voltage of node D. Figure 9 shows the damping torque coefficient $K_D(f)$ in frequency domain resulting from these two methods. Comparing the curves in Figure 9, the fluctuation of the damping torque coefficient while using PLL is relatively smaller than that while using zero-crossing detection method.

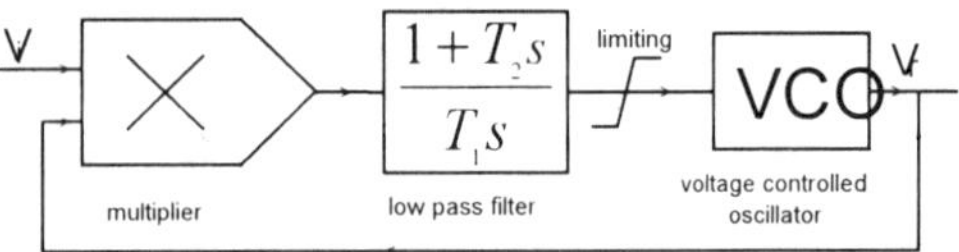

Figure 8   Block diagram of PLL

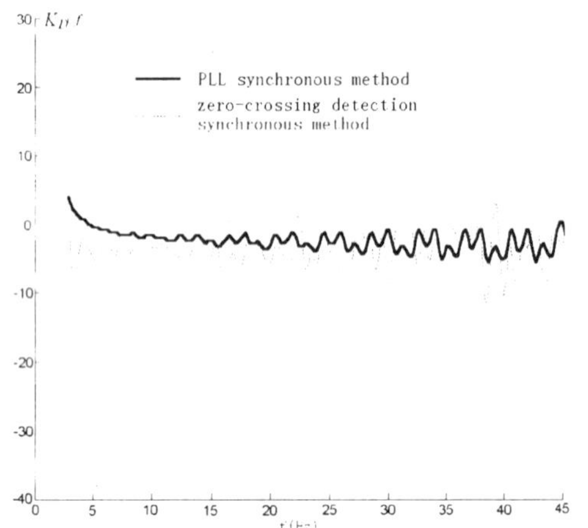

Figure 9   The damping torque coefficient vs frequency for different synchronous methods

Analysis with Different Parameters of the Test System

Different operating points of SVC. When the voltage on the bus C where SVC is connected to the system differs with the reference value V_{ref} (here V_{ref} =1pu), the reactive power that the SVC generates or absorbs will differ, and thus the operating point of SVC will fluctuate. In order to study the damping characteristics of SVC when it is operating at different points, we can change the magnitude of the voltage on the bus of the infinite system. By means of this, the magnitude of voltage on bus C is changed. When U_s=0.937pu, the magnitude of voltage on bus C is 1pu. Figure 10 is the result of the simulation for $K_D(f)$. For all U_s with value above 0.937pu, the SVC will absorb reactive power, and at this time the damping torque coefficient of the system oscillates not far above 0. On the other hand, when U_s<0.937pu, the SVC will generate reactive power, and at this moment the damping torque coefficient of the system declines significantly within the entire subsynchrounous spectrum.

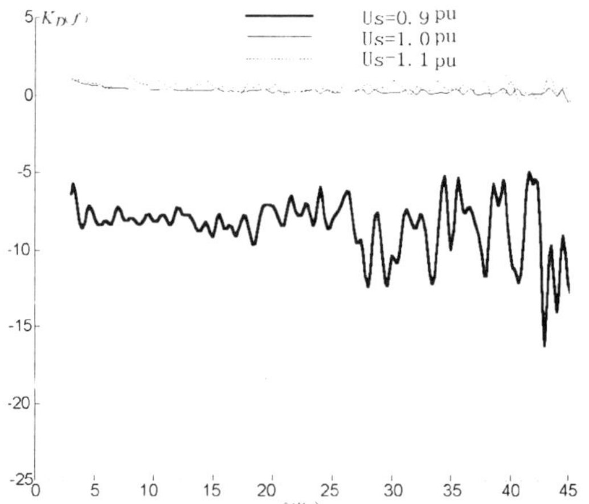

Figure 10 The damping torque coefficient vs frequency for different SVC operating points

Variation of the generator's load. Assume that the active power output of the generator (P) is 803 MW, 700 MW and 500 MW respectively, Figure 11 can be drawn by simulation. Analysis of the curves shows that damping torque coefficient declines as the load of the generator increases.

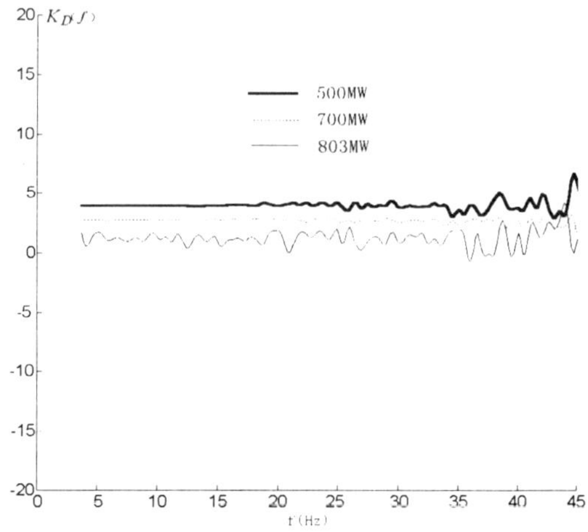

Figure 11 The damping torque coefficient vs frequency for different generator load

Variation of the system strength. X_L, the reactance of the network, indicates the electric strength between the generator and the infinite system. Simulate the system with X_L =3.36, 16.8 and 50.43 respectively and we get Figure 12. The figure shows that the damping torque coefficient of the system increases as its electric strength increases.

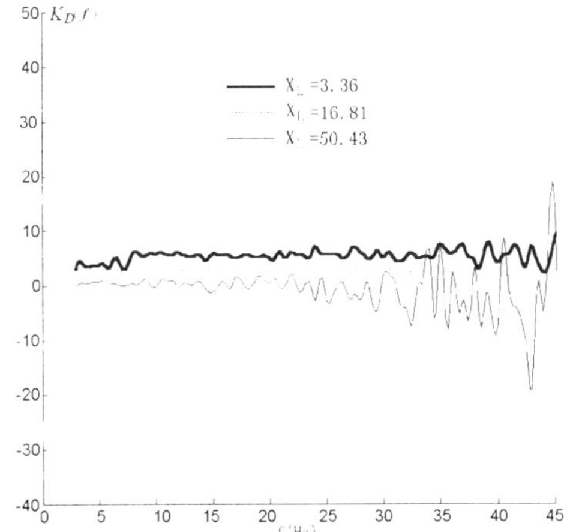

Figure 12 The damping torque coefficient vs frequency for different system strength

CONCLUSIONS

The analysis made in this paper is based on the model of the single generator to the infinite power system, and a relatively large SVC capacity is assumed here. From above analysis, the subsynchronous resonance characteristics of such a power system with SVC can be summarized as following:

1. For different voltage-ampere characteristics K of the SVC at steady state, the same amount of change in voltage will bring forth different amount of change in reactive power output, hence the response of the SVC toward the SSR signals will also be quite different. From above analysis, the damping torque coefficient decreases as constant K increases.

2. The supplemental reactive power controller has certain impact on the reactive power output of the SVC at steady state, and this will in turn influence the SSR characteristics of the whole power system. For example, in this paper, we conclude from the digital simulation that if the supplemental reactive power controller dictates SVC to absorb a certain amount of reactive power, the damping torque coefficient of the system will increase. The system damping torque coefficient will decrease if the SVC is configured to generate reactive power. This is especially obvious at high frequency.

3. The inclusion of a supplemental power oscillation damping controller will greatly enhance the damping torque coefficient of the system at low frequency. Generally speaking, the supplemental power oscillation damping controller makes SVC a device quite effective at curbing the low-frequency power oscillation. However, since the voltage and reactive power filters of the SVC are not ideal in reality, this controller also has some side effects in subsynchronous frequency range, as indicated in Figure 7.

4. According to the above analysis, the synchronizing strategy using zero-crossing detection of individual line voltage may incur a relatively large fluctuation while the one using PLL can reduce this fluctuation. This is due to the reason that the PLL can be designed to track the fundamental frequency signals while the zero-crossing detection of individual line voltage is not, thus the former one is relatively more insensitive to the subsynchronous frequency signals.

5. With no supplemental power controller present, the system components of single generator to infinite power system are primarily inductance. Working under such condition, the SVC is more capable of managing the damping torque coefficient of the system by interacting with those system components. In this paper, the simulation indicates that while supplemental power controller is absent and SVC absorbs reactive power, the damping torque coefficient of the system oscillates not far above 0. If SVC generates reactive power, then the damping torque coefficient of the system declines significantly.

6. The change of generator's active power output will lead to the change of the line current flown out of the generator, which will in turn cause the voltage at the SVC installation site to change. The damping torque coefficient of the system will be influenced by the SVC's response toward this voltage change. Digital simulation presented above reveals that the damping torque coefficient of the system declines as the generator's active power output increases.

7. The shorter the electric distance from the SVC installation point to the infinite power system is, the more stable the voltage input for the SVC is, thus the more insensitive the SVC is toward the subsynchronous frequency signals, and then the larger the damping torque coefficient of the system is.

REFERENCES

1. Xu Zheng, Feng Zhouyan, 2000,"A novel unified approach for analysis of small-signal stability of power systems", Proceedings of IEEE/PES Winter Meeting, 2, 963-967.

2. IEEE subsynchronous resonance task force, 1977, "First benchmark model for computer simulation of subsynchronous resonance", IEEE Trans, Power Apparatus and Systems, 96, 1565-1572.

3. Klaus Bergman, Keith Stump and William H. Elliott, 1993, "Digital simulation, transient network analyzer and field test of the closed loop control of the EDDY county SVC", IEEE Trans, Power Delivery, 8, 1867-

1873.

4. G. Romegialli, H. Beeler, 1981, "Problems and concepts of static compensator control", <u>IEE PROC.</u>,128, 382-388.

5. H.F.Wang, F.J.Swift, 1996, "Capability of the static Var compensator in damping power system oscillations ", <u>IEE Proc.-Gener. Transm. Distrib.</u>, 143, 353-358.

AUAS SVC, PERFORMANCE VERIFICATION BY RTDS AND FIELD TESTS

S. Boshoff   **C. van Dyk**   **L Becker**   **M Halonen, S Rudin, J Lidholm**   **Dr T Maguire**

S Boshoff Consulting   TAP   NamPower,   ABB Power Systems, Sweden   RTDS, Canada
South Africa   South Africa   Namibia   Sweden   Canada

INTRODUCTION

The new 400 kV interconnection between Namibia and South Africa was successfully commissioned in the last quarter of 2000. The 890 km single circuit 400 kV AC transmission line interconnects the two systems, ESKOM and NamPower, at Aries substation near Kenhardt in South Africa and Auas substation near Windhoek in Namibia. With the new interconnection, the NamPower system is strengthened but the new 400 kV line is also very long with a large charging capacitance which aggravates the inherent problems in the NamPower system; namely voltage stability and near 50 Hz resonance. The charging capacitance shifts the existing parallel resonance very close to 50 Hz and makes the network more voltage sensitive during system transients such as 400 kV line energisation or recovery after clearing of line faults.

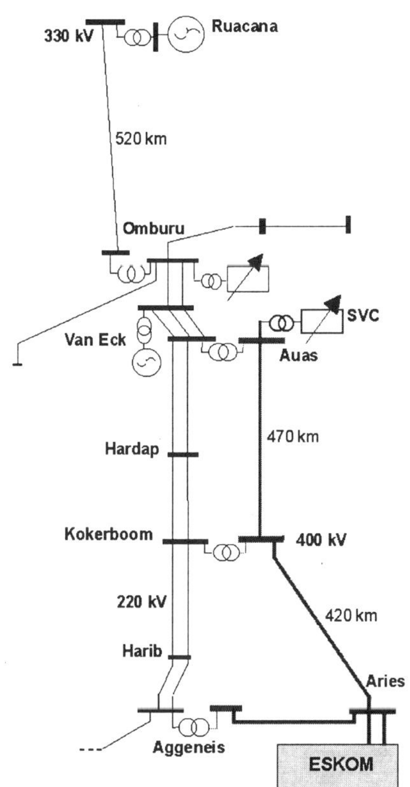

Figure 1 – NamPower network

BACKGROUND

The NamPower network consists of a radial network with its main generation at Ruacana (hydro) in the north and interconnected to Eskom in the south. At Auas substation, a SVC (80 MVAr capacitive to 250 MVAr inductive) was installed as part of the new interconnection with the primary function of controlling the system voltage and in particular the extreme (up to 1.6 p.u.) overvoltages expected due to the near 50 Hz resonance[1]. The Auas SVC is geographically located in the middle of the NamPower network, near one of its major load centres.

The use of this SVC (330 MVAR dynamic range) is unique in that it is installed in a system with very long lines, little local generation and low fault levels (from 1500 MVA to less than 300 MVA). Low frequency system eigen-frequencies (resonances), well below the second harmonic, is a result of that configuration.

This required careful design considerations and verification methods for the Auas SVC. Extensive digital simulations were carried out during various phases of the project as classical steady state load flow methods were proven to be inaccurate and unable to verify the near 50 Hz resonance. The SVC control system was tested thoroughly on a real time simulator. This was required due to the difficult and different network conditions that could occur, making traditional testing methods or simplifications impossible. Detailed modelling was required to ensure that the NamPower system (the hydro generators including controls, long transmission lines, distributed loads) as well as the SVC control system were correctly modelled to give the correct behaviour.

SYSTEM DESIGN AND SIMULATION STRATEGY

General design approach

The 330 MVAr dynamic range of the Auas SVC makes it one of the dominating devices in the NamPower network with the ability to control or blackout the network. The key role that the Auas SVC has in the 400 kV interconnection between the NamPower network and Eskom, placed the emphasis on correct modelling of the SVC and network, to minimise risk to the NamPower network.

For this project, the following system design, simulation and verification strategy was used. A transient network study was done to determine the impact that the near 50-Hz network resonance will have on the network equipment. With the introduction of a SVC to control the resonance, a typical textbook type SVC was used to derive the values to compile a SVC specification. After the contract was awarded to the manufacturer, a detailed design of the SVC was undertaken. At this stage, a unique resonance controller was developed and patented. To verify that the SVC could control the resonance, a digital simulator, RTDS, was used for the controller verification tests. Before the controller was send to site, a thorough verification of the interaction of the SVC with the network was conducted on the digital simulator. On site, commissioning verification was the final step in the process to prove that the installed SVC is operating as required.

Closed loop control

The closed loop control and the control and protection functions of the SVC are realised with MACH2. MACH2 is a microprocessor based control system and it is one of the most advanced and highest performance control and protection systems for high voltage applications on the market. The control system, MACH2, covers a wide range of applications (HVDC, SVC, SC and FACTS).

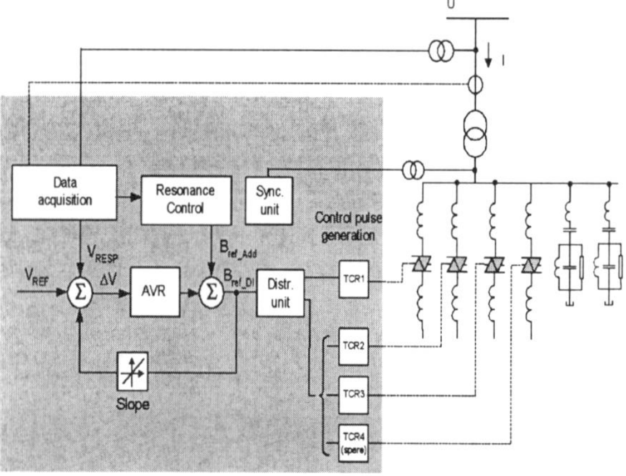

Figure 2 - Blockdiagram of the SVC closed loop control

Figure 2 gives an overview of the closed loop control of the Auas SVC. The normal operation mode of the

SVC is steady state voltage control. The conventional voltage controller used by ABB, for this application, is of an integrating type and the positive sequence component of the fundamental frequency voltage is used as a control signal. As a supplementary function to the conventional voltage controller, a new type of resonance controller is implemented which operates during resonance conditions, i.e. during energisation of the 400 kV line and recovery after line faults[2]. The resonance controller is the most important feature in the control system and optimally uses the complete inductive range of the SVC, from 0 to –250 MVAr.

Eskom network representation

One of the limitations of time domain simulation software is the size of the network that can be simulated. From the NamPower network's perspective, the Eskom network looks like an infinite source compared to itself. The Eskom network was therefore modelled as a fixed source behind an equivalent impedance. In order to retain the frequency response behaviour of the Eskom network, the frequency response is represented with a combination of series and parallel connected RLC components. This complex source impedance is derived from the Eskom network frequency over a range from 5 to 600 Hz. With the frequency dependant model, the interaction between the SVC controller and the rest of the network for both sub-system resonant frequencies and low order harmonics can be modelled.

NamPower network fault level considerations

For the NamPower network with low fault levels (around 300 MVA), the fault current or system strength is only provided by Eskom which, as described above, is modelled as an infinite source behind and complex impedance. Normally, when a SVC controller is stable during low fault level conditions, it will also remain stable for high fault levels in the network. For the NamPower network under high fault level cases (1000 to 1500 MVA), the generators at Ruacana and synchronous condensers (SC's) at Van Eck substations are modelled as dynamic machines. This included a full representation of their field and damper windings as well as controls such as the power system stabilisers (PSS), automatic voltage regulator (AVR) and governor. This ensured that the interaction between the SVC and other dynamic equipment in the network could be simulated correctly. This proved to be very important in determining worst case conditions for the network and the SVC. For the Auas SVC, the system with the lowest fault level did not produce the most onerous condition for the voltage controller, as is normally the case in traditional SVC applications.

Digital simulations on EMTDC

During the initial development phases of the interconnection, it was established that the near 50 Hz resonance phenomena could not be studied using classical load flow and dynamic simulation software. To overcome this, the rest of the project focussed all design and system studies on simulation software utilising electromagnetic transient principles. The flexibility of EMTDC made it easy to model the NamPower Transmission network. The dynamic behaviour of the hydro generation at Ruacana, the SVC at Omburu and the SC's at Van Eck substations were all modelled. EMTDC is not only capable at solving the network equations in the time domain, but it is well suited for control applications. A detailed model of the Auas SVC and its controller were developed to incorporate the exact SVC configuration that was eventually implemented on site.

EMTDC uses an interpolation technique to minimise inaccurate thyristor firing due to digital simulation timesteps. When a 50 Hz network is simulated using a 50 µs timestep, the firing accuracy is in the order of 1 deg. With the interpolation technique, exact firing is achieved.

SVC CONTROLLER AND SYSTEM VERIFICATION ON RTDS

The sensitivity of the NamPower network as well as the risk of damage to the complete network including the end-users, required extensive measures in order to verify that the Auas SVC would perform consistently correctly during various network conditions. This necessitated an approach were the optimum verification process can be followed before the actual SVC was to be tested and commissioned on site.

During the initial EMTDC system studies, it was established that the system had a large number of different configurations. Each configuration resulted in different type worst case contingencies with respect to the overvoltages, the frequency and duration thereof as well as the behaviour of the SVC controller. A simple network reduction of the NamPower network could not be considered. This resulted in a high number of critical conditions being identified. For the verification of the Auas SVC controller, it was important that the NamPower network was adequately represented over a wide range of frequencies. It was also established that

other dynamic devices in the network such as generators, SVC as well as the long stretched out network of transmission lines and distribution networks played an important role.

The standard controller verification methods utilising TNA (Transient Network Analyser) systems were not considered adequate due to the following reasons:

1. Correct representation of network components such as loads, transmission lines, generators, SVC's etc. are not easily facilitated in TNA systems.
2. Damping of the network plays an important role in level of the overvoltages.
3. Setting up different networks in TNA is very time consuming.
4. Optimisation of control strategies, which requires different network conditions to be evaluated in parallel, cannot be facilitated in TNA systems.

For the real time digital simulator, the configuration of the RTDS was optimised but it was important that the behaviour of the whole network was not compromised. The RTDS system used for the Auas study comprised of six racks, mostly made up out of TPC cards. Three 3PC cards was used for the SVC and Ruacana hydro generator model. The Tandem Processor Card (TPC) uses two NEC processors whereas the Triple Processor Card (3PC) is a more advanced card that uses three SHARC processors.

The frequency dependent source impedance that was used in EMTDC to represent the Eskom network could not be implemented directly in the RTDS. The RTDS had only a limited amount of nodes available for the source impedance and RTDS cannot implement negative resistance values. A reduction of the source impedance used in EMTDC was made to give the same response over a frequency range of 5 to 150 Hz.

The NamPower network model was further optimised in order to utilise the available RTDS capacity. This was achieved through reduction of the radial systems supplying small loads into single lumped branches. The RTDS controls compiler was used to create an exact model of the Ruacana PSS, AVR and governor.

All the electrical components of the SVC were modelled in the RTDS. The SVC control system used for the RTDS tests consisted of the MACH2 control system identical to the hardware implemented at site. The actual controller that was connected to the RTDS had access to the same VT and CT signals as implemented on site and returned the firing pulses to the RTDS. The ±10 V analogue output signals of the RTDS are amplified to 110 V input signals to the actual MACH 2 control. The firing pulses from the

Valve Control Unit (VCU) are connected to a DITS card on the RTDS. The DITS card enables the RTDS to model the switching of the thyristors digitally with almost the same accuracy as an analogue system.

The SVC controller was able to switch its 400 kV breaker in the RTDS. This made it possible to do comprehensive SVC energisation and de-energisation studies to verify that the start-up and shutdown sequences of the SVC are correct.

The RTDS proved itself very accurate. During the control verification phase, a problem with the synchronising PLL (phase lock loop) for the valve firing system was detected and rectified. This problem could not be detected with an analogue simulator running at 49.99 Hz or 50.01 Hz while the RTDS can run at exactly 50 Hz.

Using the RTDS made it possible to investigate a large number of cases despite the compressed time schedule. Extensive network simulations were done to ensure that the controller is operating correctly. A large number of fault cases and system conditions were tested, many of which can not be performed or would not be permitted on the real NamPower system. Various control irregularities were detected and improved well before the commissioning tests began which resulted in a fast, effective and successful commissioning.

FIELD TESTING

The new 400 kV interconnection was commissioned in October year 2000, only after the SVC had been successfully commissioned and tested. The critical nature of overvoltages on the NamPower system made it impossible to conduct system tests without the SVC. At the end of the SVC commissioning phase, in addition to the normal commissioning tests, a number of stringent acceptance tests were carried out in order to prove the effectiveness of the Auas SVC. Of particular importance was the resonance controller. The following system performance test were carried out with the Auas SVC and the NamPower transmission system:

1. Voltage step response test. An external 100 MVAr 400 kV busbar reactor was switched at Auas substation in order to determine the step response of the SVC.
2. Reactive Power control.
3. Black start of the SVC.
4. Maximum reactive power output.

5. Staged fault tests. Various phase to earth faults were applied in different branches inside the SVC such as the TCR, filter and auxiliary.
6. Staged fault tests on the SVC control and measurement system.
7. Simulated transmission line trips and re-closure.
8. Line energisation of the 400 kV line from Auas to Kokerboom and vice versa.

The most onerous condition for the SVC and the system is energisation of the 400 kV line from the northern section (Auas substation) of the 400 kV line. Energising the 400 kV line from the north forces the NamPower system into the critical 50 Hz resonance. This extreme test was eventually performed in the field, based on the high degree of confidence that the simulator studies on RTDS established with respect to the effectiveness of the resonance controller.

RESULTS

For comparison between the simulation results in EMTDC, RTDS and the results obtained from the system performance test performed in October 2000, the following cases have been selected:

1. EMTDC and RTDS, energisation of the 400 kV system from north to south, low fault level. The simulations were performed without the resonance controller.
2. EMTDC and RTDS, line energisation from north to south, high fault level (Ruacana generators in service) - without the resonance controller.
3. RTDS and Field, energisation of the 400 kV system from north to south, (Ruacana generators in service). The results are obtained with the new resonance controller in operation.

Results from EMTDC vs. RTDS

In Figure 3 and Figure 4 the same line energisation is shown for a very low fault level case (no generation in the NamPower system) and a higher fault level case (Ruacana generators in service) respectively. The RTDS model shows slightly less damping than the EMTDC model. This is due to the more reduced network that is used in the representation of the NamPower system on the RTDS.

It is important to note here the difference between the low and high fault level cases. For the low fault level condition (weak system), the resonance is very close to 50 Hz which was difficult for the SVC to control the voltage effectively. Under these conditions, the first voltage peak was not too critical (1.2 p.u.). However, for conditions when the fault level was higher due to the Ruacana generators or Van Eck SC's being in service, the first voltage peak was significantly higher (1.6 p.u.).

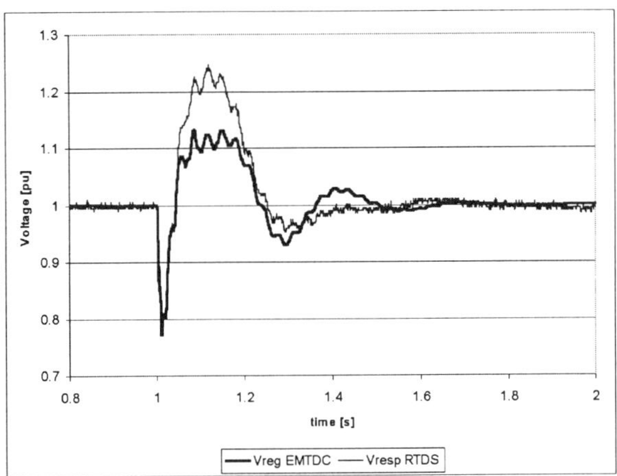

Figure 3 – Low fault level case comparison between EMTDC and RTDS

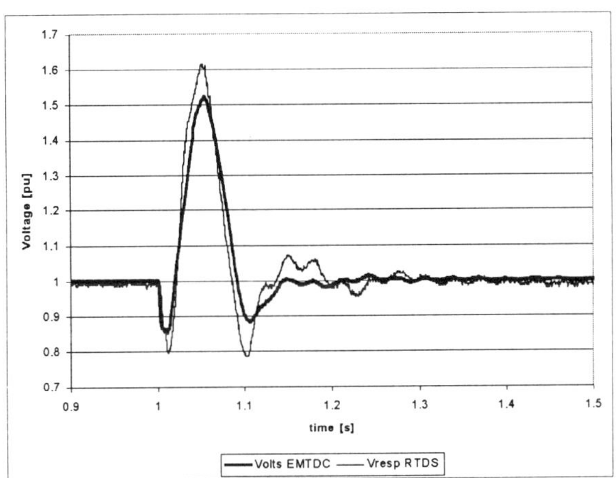

Figure 4 - High fault level case comparison between EMTDC and RTDS

Results of RTDS vs. Field testing

To compare the RTDS results with the field test results, the most onerous condition is presented here: line energisation from north (Auas) to south (Kokerboom), with Ruacana generators in service and with the new resonance controller active. Figure 5 and Figure 6 show the voltage response at Auas substation, the SVC controller output (Bref_DI) and the contribution from the resonance controller (Bref_Add).

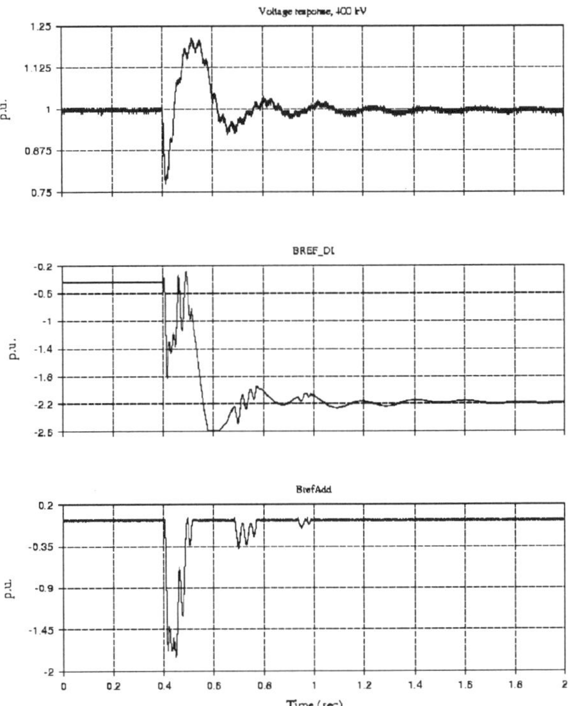

Figure 5 - RTDS, high fault level case for comparison between RTDS and field tests

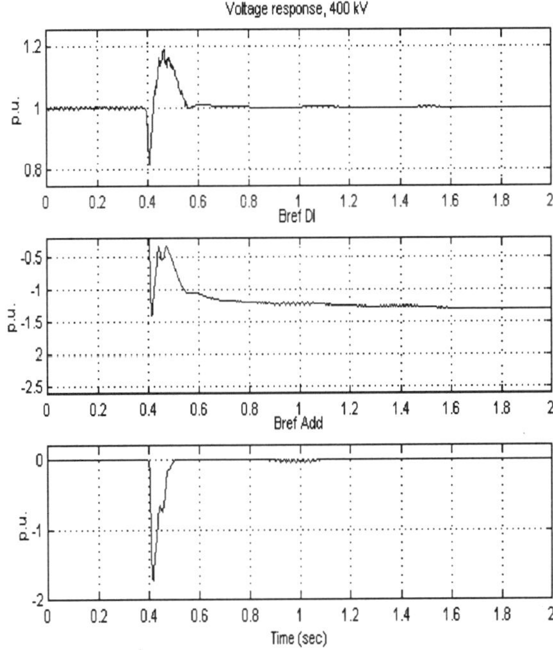

Figure 6 - Field, high fault level case for comparison between RTDS and field test

From these figures, it can be seen that there are 2 resonance frequencies, 56 Hz and 81 Hz, corresponding to the first and second pole in the system. The resulting overvoltage at Auas during energisation of the 400 kV line is below 1.21 p.u. It can be seen that the additional contribution from the resonance controller is rapidly forcing the SVC to go inductive. Comparison of results from the RTDS test and the system performance test shows good agreement and illustrates the improvement capability of the new resonance controller under resonance conditions. It was found that the results from RTDS have less damping than the results obtained from the system performance tests. However, the frequencies and time constants showed very good corresponding results.

The most significant factor here is the remarkable reduction in the first voltage peak to around 1.2 p.u. compared to the results obtained as shown in Figure 4. This is attributed to the effectiveness of the new resonance controller.

CONCLUSION

The Auas SVC project created an opportunity to investigate and compare alternative methods in the design and verification of SVC systems to be applied in transmission and distribution systems. A major advantage in the process followed, was the ability to verify the initial detailed EMTDC studies via RTDS with either identical or similar complexity. Classical methods utilising TNA would have necessitated large reductions in network in order to get the flexibility. However, this would have reduced the level of confidence and introduced more errors. In a project of this nature, it is necessary that a high level of confidence be achieved, with the minimum risk, while under the constraint of cost and time. The utilisation of EMTDC and RTDS provided the optimal solution.

The ability of the RTDS simulator to produce repetitive identical results instils a high level of confidence in the results obtained but more so in the SVC and its controller. This in itself created the ideal situation to optimally tune the SVC controller as well as fast and effective fault tracing. A major advantage in the RTDS testing was the effective utilisation of time through teamwork (ability to work in parallel between the operation of the RTDS and the analysis of results) and the ease in the documentation of results.

This process ultimately lead to a project being completed with a very successful performance testing without any surprises!

In conclusion, the project demonstrated that the technology is available to achieve effective and optimal solutions utilising systems such as RTDS. It also pointed the need to remain focused due to the almost endless flexibility that new technology provides. This requires a sensible and experienced engineering approach to ensure that time and cost constraints are always kept in mind and to avoid endless investigations just for the sake of it.

REFERENCES

1. Hammad A., Boshoff S., Van der Merwe W.C., Van Dyk C.J.D., Otto W.S. and Kleyenstüber U.H.E., "SVC for Mitigating 50 Hz Resonance of a Long 400 kV ac Interconnection", 1999, CIGRE Symposium Singapore

2. Halonen M., Rudin S., Thorvaldsson B., Kleyenstüber U.H.E., Boshoff S. and Van der Merwe W.C., "SVC for resonance control in NamPower electrical power system", 2001, IEEE Summer Meeting Vancouver.

OPERATIONAL EXPERIENCE OF HVDC LIGHT™

Kjell Eriksson

ABB, Sweden

Keywords: HVDC Light™, Electric power transmission Operation experiences, Voltage Source Converters, Power Control.

1. SUMMARY

HVDC Light™ is an electric power transmission technology based on Voltage Source Converters with Pulse Width Modulation and modern HVDC cables. This type of converters with modern power electronics has interesting characteristics based on separate and fast-acting controls of active and reactive powers. The technology is well suited for connection of networks that otherwise are difficult or impossible to inter-connect. It's suitability is based on accurate control of the transmitted active power and independent control of the reactive power in the connected AC networks.

HVDC Light™ is designed at standard units between 10 and 300 MW and are built in transportable housings. This together with the above technical characteristics make them suitable for various applications of power transmission, such as exchange of power between networks, infeed of wind power to a network or as a feeder to an isolated load.

There are four HVDC Light™ transmissions in operation in the world: one each in the countries of Sweden, Denmark, Australia and US. The HVDC Light™ links so far in operation or under construction have been justified for network interconnection or infeed of windpower. The speed and accuracy of both active and reactive controls have been used in customised ways in each of the above projects and is a key to success for HVDC Light™ projects.

As the first project, the Gotland Light in Sweden started operation in November 1999 and the others came along during 2000. The operational experiences for the Gotland Light, Tjaereborg, Directlink and Eagle Pass projects regarding the use of the controllability of HVDC Light™ projects is presented.

2. VOLTAGE SOURCE CONVERTERS WITH PULSE WIDTH MODULATION

In industrial drives the PCC (Phase Commutated Converter) technology, which is used in HVDC, is now almost totally replaced by VSC (Voltage Source Converter) technology. The fundamental difference is that in a VSC the current can be switched on and off by controlling the semiconductor valves. Thus there is no need for a network to commutate against.
By use of high switching frequency components e.g. IGBT's it is possible to use Pulse Width Modulation (PWM).

Thus it is possible to create any phase angle or amplitude (within limit) by changing the PWM pattern, which can be done almost instantaneous. Hereby PWM offers the possibility to control both active and reactive power independently.

Reactive power generation and consumption of an HVDC Light™ converter can be used for compensating the needs of the connected network within the rating of a converter. The combined active/reactive power cap-abilities are most easily seen in a P-Q diagrams, like the one below.

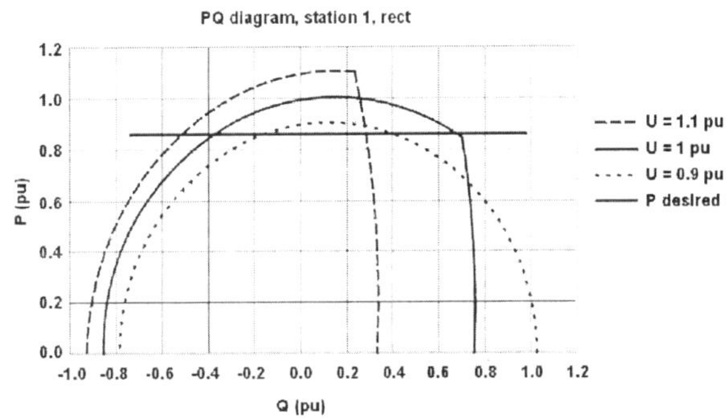

Figure 1   Shows a possible P/Q characteristics of an HVDC Light™ station (positive Q is fed to the AC network).

3. HVDC LIGHT™ DESIGN ASPECTS

The HVDC Light™ design is based on a modular concept with a number of standardized sizes, 10-300 MW. Most of the equipment is installed in enclosures at the factory. They can be tested there, which makes the field installation and commissioning short and efficient. The standardized design allows for delivery times of around 18 months.

The stations are compact and need little space, a 65 MVA station occupies an area of approx. 800 sq. meters as can be seen from the below Gotland Light station layout. A 250 MVA station would require around 3000 sq.metres.

AC-DC Power Transmission, 28-30 November 2001
Conference Publication No. 485 © IEE 2001

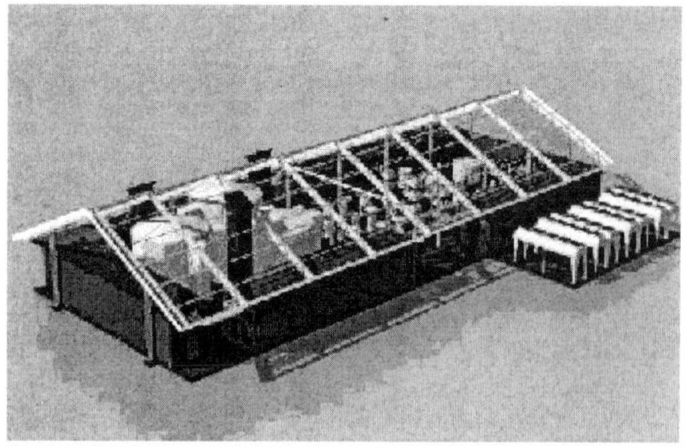

Figure 2 Model of the Gotland HVDC Light™ station 18x45 m.

The appearance can easily be adapted to local environmental requirements for easy permitting. The HVDC Light™ technology itself is designed to be environmentally friendly. Since power is transmitted via a pair of underground cables there is no visual impact along the transmission. The balanced voltage to ground eliminates the need of an electrode. Thus there is no ground current and consequently. There is no electromagnetic field from the cable pair.

The stations are designed to be unmanned and are in principle maintenance free. Operation can be carried out remotely or could even be automatic based on needs of the interconnected AC networks.

4. HVDC LIGHT™ TRANSMISSIONS IN OPERATION

March 1997 the Hellsjön transmission, rated 3 MW, +/- 10 kV over 10 km entered into operation to demonstrate the feasibility of the technology to operate in a network. It used merely setting controls for active and reactive currents. Yet it showed to have quite interesting characteristics for supporting the networks.

The Gotland HVDC Light™, 50 MW over 70 km is bringing wind power from southern Gotland to the center of the island. It is in operation since November 1999. The basic controls are changed to power controls of active and reactive powers. In the Gotland HVDC Light™ transmission the reactive power capabilities are used to control the AC voltages of the networks connected to the converter stations.

Directlink is a 180 MVA HVDC Light™ project that links the regional electricity markets of New South Wales and Queensland. Directlink will be a non-regulated link, operating as a generator by delivering energy to the highest value regional market and directly participating in the spot market. The ability to control power flow over the facility means that the capacity rights required for fully commercial network service are readily defined. The VSC stations can act independently of each other to provide ancillary services (such as var support) in the weak networks to which Directlink connects. The system has been in commercial operation since June 2000.

A DC feeder, using an 8 MVA HVDC Light™ link with VSC is in operation since September 2000 at Tjæreborg in Denmark to investigate and demonstrate how a DC feeder may be used to transmit power from wind farms to a receiving AC grid. It can operate as rectifier or inverter and at the same time absorb or supply reactive power to the AC network and is therefore suitable for connection of wind farms with induction generators.

The Eagle Pass HVDC Light™ Back-to-back, 36 MW controls the reactive power at each converter independently. In Voltage Control Mode, the AC voltage control may use all capability of the VSC regardless of the set power order. One of the converters may be connected to a passive network, i.e., a system without synchronous machines. The transmission has been in operation since November 2000.

5. HVDC LIGHT™ IN THE NETWORK

The HVDC Light™ links in operation have a common basic control design and then each one of them has its specific control features to act in its specific network environment. Thereby each one of them can give ample support to the connected AC networks.

5.1 Active power

An HVDC Light™ transmission can control the active power transmission in an exact way, so that the contracted power can be delivered when requested. The power transmission can be combined with a frequency controller that varies the power in order to support the network frequency controller. Spinning reserve in one network combined with free transmission capacity can be valid also for the network in the other end of the transmission.

This can be exemplified from operation characteristics of already operating HVDC Light™ transmissions.

The Hellsjön transmission was entered into the Swedish grid to demonstrate the feasibility of the technology to operate in a network and was not provided with any sophisticated controls, but used merely setting controls for active and reactive currents. Testing was however extensive. During testing it proved to be able to operate with variable frequency and also to provide frequency control, so that the power needed to keep the frequency control could be delivered.

On the Gotland network the introduction of an HVDC Light™ link permits the active power flow in the network to be controlled. Continuous calculation is considering both reactive compensation and active power flow and the optimization are made for the losses in the entire Gotland system. This makes the HVDC Light™ a very powerful tool for power flow control for the Gotland scheme.

Directlink is a 180 MVA HVDC Light™ project that will link the regional electricity markets of New South Wales and Queensland. Directlink will be a non-regulated project, operating as a generator by delivering energy to the highest value regional market. The HVDC Light™ technology advantages include that the flow of energy over the link can be precisely defined and controlled, so that the capacity rights required for fully commercial network service are readily defined.

Tjæreborg in Denmark is an 8 MVA HVDC Light™ link with Voltage Source. The VSC converter's ability to change the stator frequency of the induction generator within 30-65 Hz gives the possibility to optimize the power output from the wind turbine by adjusting the frequency in relation to the wind

5.2 Reactive power

The HVDC Light™ converters can provide reactive power and combined with a master controller, provide AC voltage control to the networks connected to the converter stations. Such an AC voltage control can also be used for improving the power quality by inclusion of flicker control. The following can be referred to the existing installations.

In the Gotland HVDC Light™ transmission the reactive power capabilities are used to control the AC voltages of the networks connected to the converter stations. With the verified speed of response the AC voltage control will be able to control transients and flicker up to around 3 Hz and other disturbances and keep the AC bus voltage constant. It is thus capable to relieve a considerable part of the wind power generated flicker from the AC bus.

The normal operation of the Tjaereborg link will be to use frequency and voltage control in the converter connected to the wind farm. The HVDC Light™ has the capacity to provide the reactive power necessary for the operation of the induction generators predominant in wind power generators.

The control scheme of the Eagle Pass HVDC Light™ back-to-back, 36 MW will control the reactive power at each converter independently. In Voltage Control Mode, the AC voltage control may use all capability of the VSC converter regardless of the power order set point.

5.3 Black start

In case of connection to a passive network, the HVDC Light™ transmission can provide control functions for active and reactive power, so that both voltage and frequency can be controlled from the converter station. This especially this provides possibilities for black start by control of voltage and frequency from zero to nominal. A transmission to such a passive network will give the same control possibilities as the connection of a generator. The use of black start capability has been tested and/or foreseen in some cases.

In Tjaereborg the black start capability will be used to start up the windmills by providing the reactive power necessary to the induction motors. In Eagle Pass one of

the converters may be connected to a passive network, i.e., where no synchronous machine is connected to the network. In this case, the magnitude of the AC voltage will be controlled, as long as the converter valve current is below the permissible value and the station has black start capability.

6. EXPERIENCES FROM COMMISSIONING AND OPERATION

6.1 Gotland

Experiences from testing and early operation included measurements at steady state and a staged fault in the 10 kV network. The behaviour was analysed during fault cases and for several regulation mode shifts executed manually and automatically. The wind power generation has been analysed during faults including the possibility to trip wind power units when necessary. The unique power flow controllability within the AC network is utilised by means of the network master control including set point values calculated from an overall losses minimisation algorithm. The loss minimisation utilises reduced DC voltage at reduced power, which gives as a drawback a reduced reactive power capability. The voltage control consisting of the master control set point values, the flicker control due to wind power tower shadowing around 1.5 Hz and the transient voltage control during faults were analysed. It was observed that the voltage control of the HVDC Light™ reduced voltage and angle variations so that the wind power did not synchronise to the flicker and a separate flicker control may be unnecessary.

In a conventional AC system the asynchronous generation from windmills and with very low short circuit power, ratio of only 1.75 would have made the system basically impossible to operate. Changes in the power flow from the windmills would strongly influence the voltage in the windpower infeed station and result in a very sensitive and unstable system. With the HVDC Light™ it proved to be easy to operate and control. Operation showed a smooth handling of the power flow, control functions for voltage levels, reactive and active power settings in an optimum way that should have been difficult for an operator to achieve in a conventional AC application. Operation can be automatic even at the onset of the windmills with no significant stress to the system The experience shows that this operation is much simpler than utilising capacitor banks to control reactive power balance. One experience from the early operation has been a difficulty to restart the system from remote in connection with some failures that requires manual resetting in the control.

In the middle of 2000 an extensive test program was executed including various realistic operational modes. Dedicated measurements were performed during this period both within and close to the HVDC Light™ stations, in important network nodes and along the HVDC Light™ cables. Measurements of harmonics, flicker, EMC, voltage dips and power flows were recorded. Harmonics were measured at the 70 kV AC network side in both stations and also at lower voltages at the sending end station. The following values were recorded at 70 kV:

THD (%)	THD excl. h5 (%)	TIF	Station
1.5	0.6	26	Bäcks
1.3	0.9	12	Näs

where   THD is Total Harmonic Distortion
TIF is Telephone Interference Factor

There are strong indications that the 5:th harmonic comes from the network and is not generated by HVDC Light™. Both THD and TIF tended to be rather independent of whether the HVDC Light™ was in operation or not.

Measurements of RI were performed according to ENV50121:1996 at the frequency range 9 kHz – 30 MHz, and CISPR11 at the frequency range 30 MHz – 1 GHz. The measured interference levels were found to be within the specified limits in the whole range except for a peak in the frequency interval 270-300 MHz, where the limits were slightly exceeded in two measuring points.

A staged fault test was made in the 10 kV system at the Garda substation to examine the real effects on the behaviour of the HVDC Light™ link on a three-phase failure in the network. The fault was initiated by closing a 10 kV breaker to a solid three-phase short circuit to earth during 50 ms. During the fault voltage measurements were made in 10 different places. Voltage dips and overvoltage amplitudes defined and evaluated as 20 ms RMS values are given in the diagram below for some of the nodes during one of the short circuits with HVDC Light™ DC voltage of 155 kV. In figure 3 the corresponding values from the SIMPOW simulations for the Garda 10 kV, 50 ms three phase short circuit to ground, are compared with the measurements. The test results showed, that the voltage dips were smaller than at the conditions before the HVDC Light™ installation and the late increase of windpower.

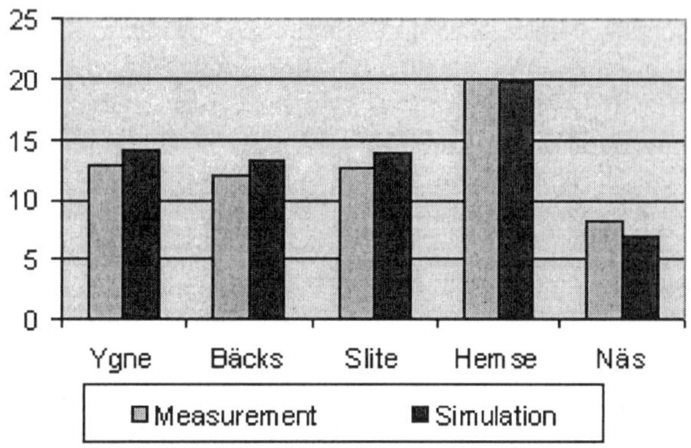

Figure 3.1 Voltage dips.

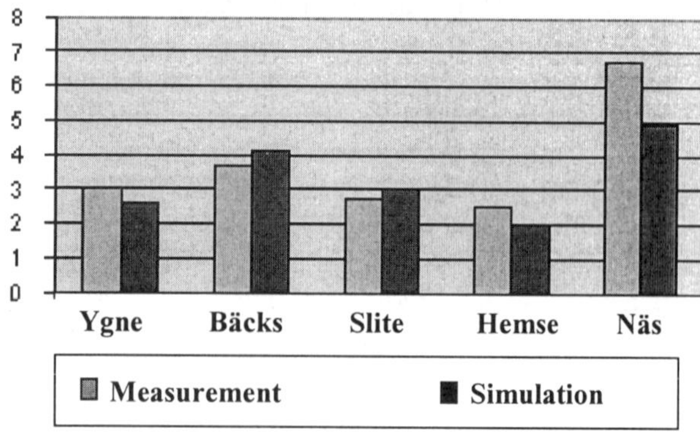

Figure 3.2 Overvoltage amplitudes.

6.2 Tjæreborg

Tests to investigate the behaviour and the co-operation between the HVDC Light™ and the wind farm were carried out on the Tjæreborg HVDC VSC project. Tests performed during commissioning are commented below.

At start/stop of wind turbines changes to transmission power was noticed only as corresponding active power changes without affecting the voltages in the connected networks. This was independent of the level of production at the event.

Start against black network is interesting for example at isolated power plants. When starting the wind farm as a black network the DC feeder was energised from the AC network via Tjæreborg substation, normally operating as inverter. Thus also the converter at Tjæreborg wind farm is energised via the DC network.. At the start it was shown that it is possible to determine both the voltage amplitude and frequency at the wind farm from the converter. The AC voltage can be ramped up smoothly by the VSC and transient overvoltages and inrush currents at energisation are avoided. The wind turbines were automatically connected to the 10 kV bus after seeing the correct AC voltage for 10 minutes. No difference between being solely connected to the HVDC Light™ or to the AC network could be noted.

During isolated operation the Tjæreborg HVDC Light™ connection has been designed to vary the AC voltage frequency between 30 and 65 Hz. With connected wind turbines the frequency was varied between 47 and 51 Hz during commissioning. Outside these frequencies the wind turbines are presently tripped by their abnormal frequency protection. A separate test with disconnected wind turbines was performed. The test demonstrated the capability of the VSC to vary the frequency between 30 and 50 Hz. When lowering the frequency the voltage amplitude is proportionally decreased to maintain the same flux in the generators and thus avoid saturation.

Simulation of three-phase faults in the receiving network of 180 ms and 250 ms were successfully simulated during commissioning by blocking the inverter temporarily. This possibility of the HVDC Light™ converter to isolate the wind turbines from the undervoltage in the receiving AC network, is important in preventing the wind turbines from tripping at network disturbances.

EMC performance of the HVDC Light™ has been verified by measurements against applicable standards. The measured TIF value is 6.2, which is well below the requirement of TIF 50. The harmonic distortion the measured THD value is 2.1 %. All individual harmonics are well below 1 % except the 5th harmonic, which is generated outside the HVDC Light™ link. The recorded flicker value was Pst < 0.23 which is well below the normal limit of 1.0.

6.3 Eagle Pass

During the BTB tie (HVDC Light™ link) commissioning, measurements were made for different phases of operation to verify the performance. Initial energization of the link was made by deblocking the VSC versus the stronger CFE transmission system, establishing a SVC Light mode of operation. The other side of the BTB was then off-line, i.e. disconnected from its line. The sequence involved charging of the DC link and the harmonic filters on the low voltage AC bus. Before the opposite VSC was deblocked, a steady state no load SVC Light condition was established on the CFE side. The complete start-up sequence including also the energization and start of the AEP-CPL side, was performed within 0.5 seconds.

The transient response of the BTB was checked by switching on one of the capacitor banks at the AEP-Eagle Pass substation. In the actual case, the reactive power from the converter was equal to 18 Mvar before capacitor switching. As the capacitor bank was switched on, the converter responded fast reducing its output with 15 Mvar, corresponding to the size of the fixed bank, settling to a new steady state in a few cycles.

The Black Start function was checked against an artificially created small-islanded "network" mainly consisting of the AEPstep-up transformer. Any larger islanded network was not possible to create, as this would have forced a power outage in parts of the city of Eagle Pass. Black Start was then initiated and the AEP-138 kV voltage was ramped from 0 kV to its nominal value at 138kV. The ramping ended without disturbing the CFE side voltages and currents. During operation of the islanded "network", the AEP 138 kV bus was supplied by the BTB at almost no load condition. The only load consisted of the harmonic filters rated totally 6 Mvar. The system was anyhow observed to perform well.

Finally, during a remote fault, the BTB was in operation at zero transfer power. Lightning conditions in a remote area caused a voltage dip in the AEP-CPL network. During stabilization of the voltage in Eagle Pass the BTB current (capacitive) during the fault condition was increased to almost 1 p.u. to support the bus voltage at Eagle Pass.

The Eagle Pass BTB tie has successfully demonstrated an unprecedented capability of providing reliable asynchronous interconnection between the AEPsystem and the Mexican interconnected network, this is relevant considering the difference of short-circuit power at the interconnection points. Furthermore, the use of VSCs with PWM technique has enabled the flexibility in providing independent voltage control, at the two ends of the BTB, and a bi-directional active power transfer between the interconnected power systems.

6.4 Directlink

The action of the Directlink transmission is shown referred to a fault in the NSW AC system during a storm, December 7, 2000. The three systems were in operation, each one with about 50 MW, power direction North. During the fault the transmitted power decreased to around 30 MW per system, as shown in the oscillogram from the Mullumbimby station (AC voltages, phase currents and DC power in one converter). At clearing of the fault temporary blockings of the converters were initiated. This was followed by a normal recovery, with a smooth increase of the power.

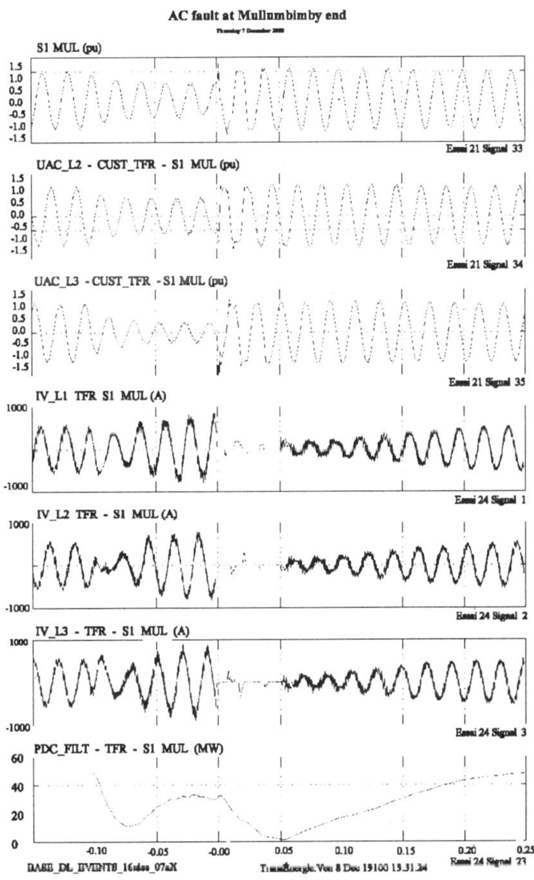

Figure 4 AC network fault on the Mullumbimby side.

It is interesting to see how the Directlink is operated when it comes to active power transmission and referring this to the intentions with the link as undertaken as a generator delivering energy to the highest valued market. The figure below shows how this is exercised in reality. Power is transmitted to the amount and in the direction where it earns money. This shows Directlink as a full commercial player in a deregulated environ-ment directly participating in capacity rights and the spot market. This has been made possible by the HVDC Light™ facility to control in an exact way the flow of energy over the link.

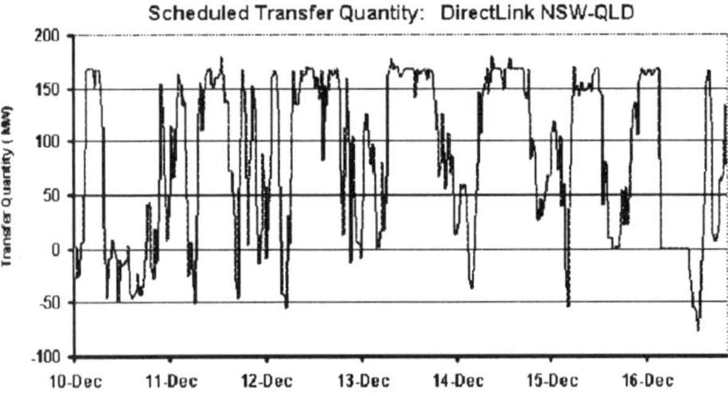

Figure 5 Power transmission by the Directlink.

Initially operating the three links with a weak network (disconnection of lines or shunt banks at one of the stations) showed tendencies to instabilities. This could be counteracted by retuning of the AC Voltage Controls of the links.

7. CONCLUSIONS

Theoretical investigations and simulations have repeatedly shown that HVDC Light™ gives a number of advantages for transmitting power and also for the operation of the surrounding networks. Four HVDC Light™ projects have now been finalized and taken into operation. The experiences from commissioning, trial operation and early operation confirms all expectations Regarding the behavior of an HVDC Light link both with regard to transmission of power and the possibilities to control it. The independent control of reactive power has shown to be very valuable for voltage control and keeping up network operation in connection with disturbances in the network and recovery after disturbances. Features such as black start and variable frequency operation have shown to work well.

8. REFERENCES

1. Eriksson K, Graham J,:HVDC Light™ - a transmission vehicle with potential for ancillary services; SEPOPE2000 Curitiba, Brazil, May 2000.

2. Axelsson U, Holm A, Liljegren C, Åberg M, Eriksson K, Tollerz O,: The Gotland HVDC Light™ project – experiences from trial and commercial operation; CIRED 2001, Amsterdam, Netherlands, June 2001.

3. Skytt A-K, Holmberg P, Juhlin L-E,: HVDC Light™ connection of wind farms, II Int. Workshop on Transmission Networks for offshore wind farms; Stockholm March 2001.

4. Soebrink K, Soerensen A, Jensen J, Holmberg P, Eriksson K, Skytt A-K,: Large-scale offshore wind power integration and HVDC using Voltage Sourced Converter; CIGRE Fourth Southern Africa regional conference, Cape Town, South Africa, October 2001.

5. Larsson T, Peterson Å, Edris A, Kidd D, Haley R, Aboytes F,: Eagle Pass Back-to-Back Tie: a dual-purpose application of Voltage Source Converter Technology; IEEE Summer Meeting, July 2001.

SwePol HVDC Link

Bernt Abrahamsson[1], Leif Söderberg and Krzysztof Lozinski[3]

ABB, Sweden[1], Swedpower, Sweden[2] and PPGC, Sweden[3]

Keywords: HVDC, Cable transmission, Sea electrodes, MACH 2, ConTune filter.

1. SUMMARY

SwePol Link is the latest and certainly not the last of the High Voltage Direct Current (HVDC) links between the Nordic countries and the continent. SwePol Link ties the 400 kV grids in Poland and Sweden together by means of 245 km long cables. SwePol Link's most unique feature is the absence of sea electrodes, for the first time cables are used to carry the return current rather than using earth return.

2. BACKGROUND

In November 1995 Vattenfall the Swedish State owned power producer signed a Power Purchase Agreement (PPA) with its Polish counterpart Polskie Siece Elektroenergetyczne SA, the national Polish power Grid Company, PPGC. The PPA's intention was to optimize utilization of the generating facilities the two countries possess. Sweden, like the other Nordic countries relies heavily on hydropower while, Poland almost entirely uses coal to generate electricity. Being hydro based, the weather strongly influences the power production and price in Sweden, which can be balanced by the steady fossil, fueled production in Poland. Also, there are peak load differences that may be accommodated by power transmission between the two countries.

SwePol Link AB a dedicated company, was formed by Vattenfall, PPGC and Svenska Kraftnät, the national Swedish power grid authority, to construct, maintain and operate a 600 MW HVDC link between Karlshamn in Sweden and Slupsk in Poland. On July 23, 1997, two contracts were signed with ABB to supply, install and commission the submarine cable in the Baltic Sea along with two converter stations. The transmission link entered commercial operation in June 2000.

AC-DC Power Transmission, 28-30 November 2001
Conference Publication No. 485 © IEE 2001

Fig. 1 SwePol Link map.

3. TRANSMISSION SYSTEM

SwePol Link was planned as a monopolar HVDC system with earth return. In spite of the very small environmental impact shown at similar systems already in operation, local opinion persuaded the owner to add a metallic return path consisting of two cables in parallel making SwePol Link the first monopolar system with metallic return.

The link connects the 400 kV grids in the two countries by means of a 245 km long mass impregnated cable operating at 450 kV DC, plus two polymer insulated cables rated 20 kV (in parallel) for the return current. The cables are extended from the coastline all the way to the converter stations therefore eliminating the need for overhead lines. The system is grounded at one point - in the valve hall at Stärnö, the Swedish converter station.

Fig. 2 Installation of seacable.

Under sea the cables are buried approximately one meter to avoid damage caused by anchoring or trawling. During the cable laying water jet technique was used wherever possible, (85% of the total length). The high voltage cable was laid separately with a physical separation of 3 to 43m to the two return cables. The high voltage cable has an outer diameter of 140mm and has double crosswind armoring to withstand the mechanical stresses during cable laying. Due to the length of the cable it was manufactured in four sections and joined on board the laying ship.

The rated capacity of SwePol Link is 600 MW, with 20% overload capacity at low ambient temperature and 10% overload capacity at any ambient temperature when redundant cooling is available. The converter stations use ABB´s classic and well-proven current stiff technology, which has proven its reliability since the introduction in 1954. Today performance has been largely improved through the introduction of different features (see below) and advanced control functions.

The converter valves are located inside the 20 m high valve halls. The transformers are located just outside the halls with the valve side bushings protruding through the walls.

Single-phase three winding transformers are used. There are 66 thyristors in each single valve. The valves are cooled using a water/ glycol mixture in the single circuit system.

Figure 3 Valve hall.

4. ENVIRONMENTAL CONCERNS

Earlier HVDC transmissions across the sea have used seawater and earth to carry the return current by an electrode connected to the station low voltage bus. The impact on the environment has been thoroughly discussed in the past and has been concentrated to the following issues:

- Magnetic field in the vicinity of DC cables
- Chlorine generation at the positive electrode
- Corrosion on metallic objects in the water or buried in the ground

Through the introduction of a metallic return path there are no longer concerns regarding chlorine formation or corrosion. While the magnetic field around DC cables is still present, it has been reduced. With the cable laying technique employed it was not possible to lay the return cables at the same time as the high voltage cable. Therefore physical separation had to be kept, which is also necessary because of thermal consideration during operation.

At a distance of 10m from a single DC cable, the intensity of the earth's natural magnetic field is greater than the field of the cable. With the introduction of the two return cables the magnetic field on the surface is typically reduced by 20% in shallow waters and 50% at a depth of 100m. While the magnetic field is still there, no adverse impact on marine life has been found, nor is modern navigation dependent on magnetic compasses.

5. TECHNICAL FEATURES

To avoid combustible material in the valve hall ABB has developed a dry type, high voltage transformer bushing. One SwePol link transformer unit is equipped with two such bushings, as a reference installation. The insulating medium used on the valve side of the bushing flange is sulphur hexaflouride (SF6). The operating experience to date is excellent, promoting the use of this bushing type for future projects.

One 95 Mvar filter meets the requirements on harmonic filtering with four branches. There are two branches tuned to the 11th and 13th harmonic plus two branches of high pass type tuned to the 24th and 36th harmonic. To satisfy the demand for reactive power there are also two 95 Mvar shunt banks. In addition there is a 117 Mvar shunt reactor on the Polish side.

The two sharply tuned filter branches employ the ConTune^{TM} reactor. The control winding of this reactor allows continuous adjustment of the reactor inductance. Thereby perfect tuning of the filter is maintained at all times, irrespective of network frequency excursions and variations in ambient temperature. The reactor has no moving parts permitting a high quality factor and low filter loss.

The control system continuously monitors the filter current and AC bus voltage for the harmonic of interest. The regulating direct current fed to the control winding of the reactor is adjusted to minimize the impedance of the filter.

The Compact switchgear is employed to save space in the switchyard. The switchgear is used for switching the reactive banks and consists of conventional switchgear units mounted on a common platform. The number of units may vary between applications. The Compact units for SwePol Link comprises of one SF6 breaker, one disconnector, one earthling switch and one Optical Current Transducer (shunt capacitor banks only). The OCTs are used for differential protection purposes. A core-and-coil assembly mounted on high potential monitors the bank current. The signal is digitized and transferred to ground with fiber optics. The measurement is fast and accurate enough to avoid protective actions at bank connection, which cause transients of high frequency.

Thyristors of high voltage rating, 9 kV, are used in the converter valves. With the introduction of gas insulated snubber capacitors and PVDF pipes in the water-cooling circuit, no combustible material exist in the valve halls.

Fig. 4 Interior of valve hall.

ABB´s recently developed MACH 2 control system is used to control, monitor and protect SwePol Link. This highly integrated system uses commercially available hardware and software to a great extent. Running under Windows NT, the main computers handle the operator's interface, event recording and transient fault recording along with controlling the converter process itself. Industry standard serial busses (CAN, TDM) are used for the communication between the main computers and the main circuit I/O cubicles.

The normal control mode is power control. Power modulations such as frequency control and emergency power control (EPC) are included. The settings of these modulations can easily be changed by the utilities. The frequency control is typically activated outside the range 50 ±0.1 Hz. There are a number of activation criteria for the emergency power function, typically a frequency drop below 49.5 Hz results in an EPC support, which at present is limited to 300 MW.
Telecommunication between stations is operated through one dedicated leased line and a back-up dial-up channel. During loss of telecommunication the link will continue to operate in back-up synchronous mode. Then the actual current is used as current order by the power regulator in the inverter.

5. OPERATION

The converter stations are normally unmanned, the operation is controlled by dispatch centers at Stockholm and Bydgoszcz. However, local control can be taken from the control rooms in the converter stations.

It is anticipated that the link will be primarily used by Vattenfall to export power to meet the growing demand in northern Poland, where consumption is expected to rise by 10 percent within the next five years. Vattenfall plans to export 1.5 percent of the annual Swedish production to Poland using SwePol Link. To date the main power direction has been Sweden to Poland. The link has been in almost constant use during weekdays, while at weekends, the link is mainly used during daytime.

Any transmission capacity in excess of the PPA is available to the market. There are no restrictions from the owners on which party may purchase available transmission capacity from SwePol Link AB. Power deals on short, medium or long-term basis may be contracted as complement to the PPA.

COMMISSIONING OF THE KII-CHANNEL HVDC LINK
- APPLICATION OF NEW HVDC TECHNOLOGY AND OPERATING EXPERIENCE –

Y Makino[1], S Hara[2], M Hirose[3]

Electric Power Development Co., Ltd. Japan[1], The Kansai Electric Power Co., Inc. Japan[2] and Shikoku Electric Power Co., Inc. Japan[3]

ABSTRACT

The Kii-channel HVDC Link is one of the world's biggest HVDC links that utilizes submarine cables and it has been in commercial operation since June 22, 2000. The purpose of the link is a bulk power transmission from a coal-fired thermal power plant to a load center. It also contributes to stabilize and reinforce the power systems in the western part of Japan. Many new technologies, such as DCGIS and large diameter LTTs, were applied to this project. Several coordination controls were also applied to the HVDC. Operation data for a year since the commissioning of the HVDC proves its high reliability, showing very high energy utilization of more than 90%.

Keywords: HVDC, DCGIS, LTT, metallic return, power modulation

INTRODUCTION

The Kii-Channel HVDC link is one of the world's biggest HVDC link which utilizes submarine cables, connecting Shikoku power system and Kansai power system in Japan. Main purpose of the link is to transmit electricity, which is generated by Tachibana-bay coal-fired thermal power plant (1050MW-2units and 700MW-1unit) in Shikoku island, to Kansai area where heavy energy demand exists. The HVDC link also contributes to stabilize and reinforce the power systems in the western part of Japan [1].

Figure 1 shows the geographical location of the link and related power systems. On the east coast of Tokushima prefecture in Shikoku island, there is Anan converter station that is connected to Tachibana-bay plant via a.c. 500kV transmission line (5km). The power converted from a.c. to d.c. is delivered to main·land of Japan (Kansai side) through the underground and the submarine cables (49km) across the Kii-Channel. On the shore in Wakayama prefecture, Kansai side, Yura switching station (cable terminal) locates which connects cables with 51km of overhead lines. At the end of the d.c. overhead lines, there is Kihoku converter station in inland Wakayama prefecture. From here, a.c. 500kV overhead lines are connected, then the electricity delivered through d.c. lines are passed to the Kansai power system network. Then a hybrid loop system is formed by the HVDC and existing interconnecting a.c. lines (Honshu-Shikoku interconnecting line), enabling more reliable and flexible power system operation. In figure 2, pictures of the both converter stations and Yura switching station are depicted.

The project had been jointly promoted by three electric power companies: the Kansai Electric Power Company, Shikoku Electric Power Company and Electric Power Development Company. Under the cooperation of these three companies, R&D and field test of new converter equipment, development and simulations of new concept control and protective relay systems had been carried out to achieve lower loss, more compact, lower cost and more reliable HVDC system. Results of the R&D were fruitful and many new technologies were applied to the HVDC link.

Since these new technologies were widely acknowledged as a great contribution to the development of industry, the link received several technical awards in Japan.

This paper presents the outstanding features of the HVDC, applied new technologies and the operating experience for a year since the commissioning of the HVDC.

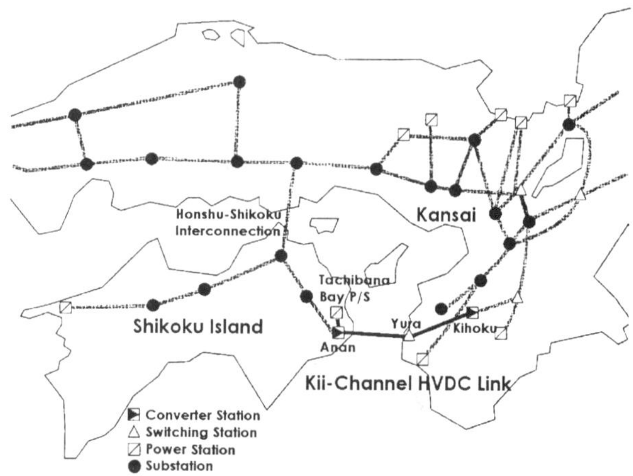

Fig. 1 Location of the Kii-Channel HVDC link

SYSTEM CONFIGURATION

Figure 3 shows the schematic diagram of the Kii-Channel HVDC Link. Installed capacity of the link is 1,400MW (+/-250kV, 2,800A) and it will be up-rated to 2,800MW (+/-500kV, 2,800A) in the future. The system configuration and the layout of the converter stations were taken into account of the up rating, including purchase of the installation area. Also, some of the installed equipment, such as submarine cables,

AC-DC Power Transmission, 28-30 November 2001
Conference Publication No. 485 © IEE 2001

(a) Anan converter station (Shikoku side)

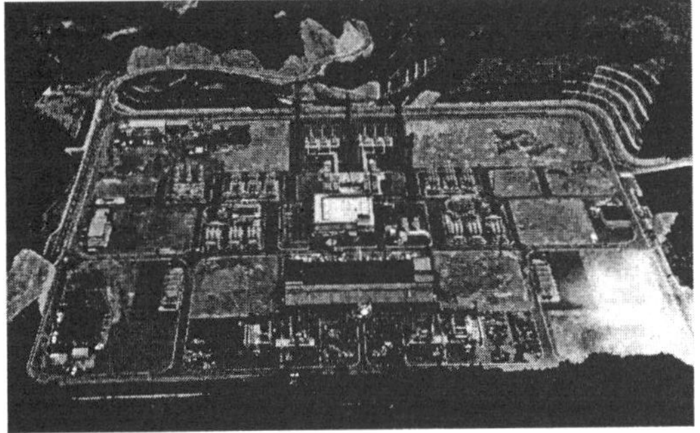

(b) Kihoku converter station (Kansai side)

(c) Yura switching station (Kansai side)

Fig. 2 Pictures of converter stations
and switching station

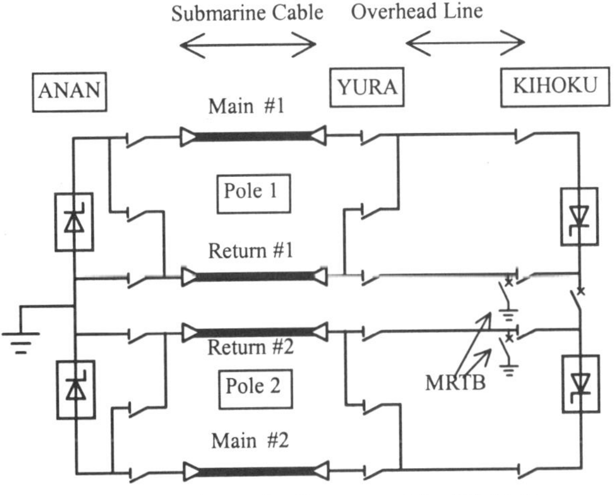

Fig. 3 Schematic diagram of the HVDC

smoothing reactors and DCGIS were already designed for d.c. 500kV operation.

The system configuration is bipolar, metallic return scheme. As in the case of the Hokkaido-Honshu HVDC Link (600MW, +/-250kV), metallic return scheme was selected to avoid electrolytic corrosion in the adjacent area. Each pole is independent of each other for basic concept of the HVDC; therefore, one pole outage doesn't affect the other pole.

Since the specification of the return cables are the same as the main cables, even if one of the main cables were damaged, it would be immediately substituted by one of the return cables for main cable. This could be achieved by switching the disconnecting switches of Anan and Yura. Then, the system could be back in normal operation again.

FEATURES OF THE HVDC AND APPLIED NEW TECHNOLOGIES

Thyristor Valve

For the thyristor valves, high voltage large capacity light-triggered thyristors (LTTs; 150mm in diameter, 8kV-3500A) were used and assembled in 6 modules/arm configuration (6 stages for quadrivalve). This enabled much more compact valves than before and excellent aseismatic structure.

LTTs has been used for many years at existing HVDC projects in Japan, such as Hokkaido-Honshu link, Shin-Shinano Frequency Converter and Sakuma Frequency Converter. Thyristors of 100mm diameter, 6kV-2500A are used for such projects and they exhibit excellent reliability. However, for achieving more compact and lower loss converter required further advance in thyristor device technology. But it is very difficult to fulfill the both conditions: large capacity and excellent switching property. They are in a trade-off relationship. To overcome this dilemma, proton or gamma ray radiation process were applied to make carrier lifetime within a limited area of the device shorter. Application of the new device and unique valve structure enabled compact valves of excellent aseismatic characteristics. Figure 4 shows the thyristor valve installed in Anan converter station. At Hokkaido-Honshu link, valve structure is 8 stages for quadrivalve (8 modules/arm) and the height is about 12 meters. In Kii-channel HVDC, it is 6 stages for quadrivalve (6 modules/arm) and about 9 meters high at Anan converter station (3/4 of the Hokkaido-Honshu valves).

Parts of the thyristor valves are not inflammable but the incipient fire detectors were installed in each valve hall. Primary cooling system for the thyristor valve is using deionized water, but the secondary system is different from each other in both converter stations. In Anan, open loop system is used because industrial water supply is available, but in Kihoku, closed loop type is used.

Fig. 4 Thyristor valve of Anan converter station

Fig. 5 DCGIS installed at Anan converter station
(Behind the DCGIS is the smoothing reactor)

Figure 5 shows the DCGIS installed in Anan. Due to the above-mentioned new technologies, the HVDC link has been operated stable since its commissioning.

D.c. Gas-Insulated Switchgears (DCGIS)

In Anan converter station and Yura switching station, d.c. gas insulated switchgears (DCGIS) were installed because of the heavy salt contamination in coastal area. They are the world's first large scale DCGIS which are used in the commercial operation. All of the energized conductors are enclosed in a tank, and therefore installation space was greatly saved.

In the course of developing the DCGIS, absolute elimination of metallic particles in the tank had been critical. If such particle existed in the GIS, they were to be lifted up towards the main conductor along with the electric field. Particles would move to-and-fro between conductor and the tank, or move around the conductor discharging on the tip of their own. On such condition, insulation level of the GIS is drastically decreased, then, if some switching surge came in, there would be a flash over. To avoid such tragedy, several types of particle traps were set in the tank. Also, improvement of quality control during the assembly, conditioning tests after installation were carried out. Because the charging on the insulating spacers would also bring about reduction of the insulation level, semi-corn type insulating spacers were developed to mitigate charging.

In Kihoku, metallic return transfer breakers (MRTBs) and pole separating breaker were installed. When the lightning strikes on the overhead line of the return circuit, MRTB is used to eliminate arcing at the fault point.   Both types of breakers use self-excited oscillation to achieve current zero point.

Filters and Reactive Power Compensation

For the same reason of the DCGIS, dead tank type a.c. filters are used in Anan converter station. D.c. filters are also dead tank type in Kihoku converter station, aiming compactness and higher reliability.

500kV dead tank type a.c. filters were developed since the highest voltage rating of the filter was a.c. 275kV before this project. Precise analyses of electric field in the capacitors of the filters had been carried out to optimize internal insulation structure. As a result of the study, compact and reliable filters were completed. Each pole has 11th, 13th order and high pass filters, and one 5th order filter, which is needed to dump non-theoretical harmonics emerging on some special system conditions, is installed at both terminals.

D.c. filters are 12th order, high pass type, and are installed only at Kihoku. On Anan side, since cables are directly connected to the d.c. circuit, their capacitance component can play roles of the d.c. filters.

Reactive power compensation can be done by switching the shunt capacitor banks. For Anan converter station, there are less shunt capacitor capacity because it can expect reactive power supply from nearby thermal power plant. Shunt reactors are also equipped at Anan, for floating or low load operation.

Converter Transformer and Smoothing Reactor

Converter transformers are three-phase three-winding type in both converter stations. Smoothing reactors are designed for d.c. 500kV from the beginning. And it is directly connected with DCGIS in case of Anan converter station.

During the design process, analyses of electric field including the polarity reverse event, had been carried out to optimize internal insulation structure. Consequently, 20% reduction of insulation length could be achieved.

The inductance of the reactor was decided so that the resonant frequency between submarine cables never coincided with fundamental frequency (60Hz). This prevented from the harmonic instability between a.c. system and the d.c. system.

Submarine Cables

For submarine cables, records making large capacity, high voltage (d.c. 500kV design) OF (Oil-Filled) type were used. The conductor size is 3,000mm^2. Two main cables and two return cables were laid on the seabed. These specifications are the same, so the return cable can be used as a spare for the damaged main cable.

There were two types of cables as candidates: mass impregnated type and oil-filled type. However, oil-filled type was preferable than mass impregnated type, because it could be used higher conductor temperature. For large-scale transmission, conventional paper insulation is no longer the best solution, because it tends to be thick and prevents heat radiation affecting on the mechanical strength. Applying the PPLP (PolyproPylene Laminated Paper) was the solution for that, and it was the very first time to use it for d.c. cables in the world. PPLP has been used in a.c. cables and the electrical property is excellent.

Optical fibers were installed in the cables as thermal sensors, in addition to the telecommunication purpose. These sensors can monitor the cable temperature, all along with the cable for about 49km.

Environmental Consideration

A.c. 500kV XLPE (Cross-Linked Polyethylene) cables are used to connect among converter station's equipment. They connect between a.c. buses and transformers, a.c. buses and a.c. filters. These minimize the exposure of energized parts, prevent salt contamination and save installation space of the equipment. Especially in the Anan, most of the energized conductors are enclosed except for the a.c. line parts before the GIS bus.

Thyristor valves are installed in the valve hall, where dust count is kept under 100,000, and the humidity is controlled less than 70%. The hall itself is electro-magnetically shielded to prevent radio noise emission. Converter transformers and smoothing reactors are enclosed in sound wall to mitigate audible noise to the surrounding residents.

In Kihoku, the color of the equipment, steel structures and transmission towers are coordinated with the surrounding nature, because it locates close to the historical buildings.

Each of the neutral point of the transformers is earthed through 5 ohms of resistors. These limit stray d.c. current to the a.c. network when d.c. line fault occurs. Effectiveness of the neutral grounding resistor was confirmed during the artificial d.c. line fault test.

New technology for the d.c. transmission line is the "Y" suspension assembly. Porcelain insulators used in the coastal area tend to be very long. Then, the insulators are formed not in the "V" suspension, but in the "Y" structure. This realized the compact d.c. line towers.

Control and Protection

In terms of control and protective relay systems, higher operability is required because the capacity of the HVDC is large and its influence to the existing a.c. network is serious. Then, a continuous operation control during and after the interconnected a.c. line faults, had been newly developed. Details of this control will be discussed in following section. Also, several damping controls had been developed to stabilize connected a.c. power system since they can be easily done by modulating the HVDC power in the a.c.-d.c. hybrid loop system. Power Modulation (PM) control is used to stabilize power swing, which is initiated by line faults within the power system. There are many types of system controls: Emergency Power Preset Switches (EPPS) and Emergency Frequency Control (EFC), depending on the system disturbance condition. They may be used to maintain and stabilize system frequency and to prevent overload of certain transmission lines.

In addition to that, coordination controls with nearby thermal power plant were applied. One of the controls is Supplemental Subsynchronous Damping Control (SSDC), and is used to avoid interaction between d.c. current control and turbine shaft (SubSynchronous Torsional Interaction : SSTI). Also, in the case of the transition to the islanded operation, in which the turbine generator and d.c. system is separated from Shikoku a.c. network, there are other coordination controls: one is EFC, which maintains frequency of the islanded system and the other is EPPS, which balances the d.c. power to the turbine generators output power during the transition.

COMMISSIONING TEST

HVDC system performance test had been carried out in the condition of connecting with commercial power system from January to June 2000. In addition to the verification of fundamental control, operation and protection functions, several power system control functions were also tested. These include PM, EFC and EPPS functions.

Shikoku Islanded system mode test was also carried out, in which interconnection between Shikoku system and main land (Honshu) system was done by the Kii-channel HVDC link only (a.c. ties were disconnected). In this test, the EFC for the Shikoku Islanded mode operated properly, maintaining the systems frequency stable.

Another islanded mode, which consists of Turbine generator and the d.c., was also tested. Manual tripping of the tie lines at the Anan converter station created this mode. In this test, SSDC and other coordination control functions were confirmed.

Artificial d.c. line fault tests, in which grounding faults on d.c. overhead lines were artificially triggered at both end of the lines, were also carried out and proper performance of protection and control systems were confirmed. Observed over voltages, over current and stray current were within the design criteria.

Due to these intensive tests, the HVDC link has shown outstanding performance and contribution to the systems stability.

OPERATION SYSTEM

Operation of the HVDC link can be done at either of the converter stations. The master control is to be exchanged by turns monthly. This allows systems redundancy. Thus, if operation system of one terminal broke down, the master control would be switched to other side operation system, then the link could be operated without interruption.

OPERATING EXPERIENCE

Energy Utilization, Availability

The Kii-Channel HVDC Link has experienced no major trouble and has operated very smoothly since the commissioning on June 22, 2000.

Monthly data of transmitted energy and energy utilization of the link are shown in figure 6. The energy utilization factor exceeds 90 % from August 2000 to February 2001. The number is still high as 83% even if it is averaged for a year. This clearly states that the link is really used for bulk power transmission. In terms of the availability, there are some influences of initial troubles, but if they are eliminated, the availability of the link ranges around 97%. The number is fairly good, but as many of the initial troubles were already repaired, a higher availability will be expected from now.

It should be emphasized that the transmitted energy through the link has already reached to 10 TWh within a year. The number is almost equal to the total energy transmitted through the Hokkaido-Honshu HVDC Link, which has been operated since 1979. Because the main purpose of the Hokkaido-Honshu link is emergency

power exchange and frequency control, these figures cannot be compared. However, the achievement done by the new HVDC link, and that is with submarine cables, is outstanding in the world.

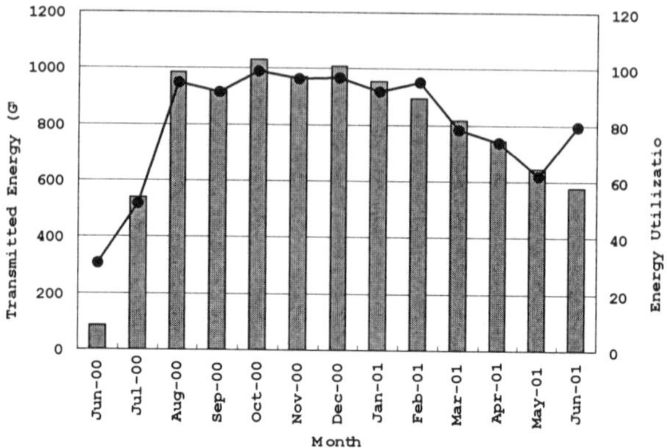

Fig. 6 Transmitted energy and energy utilization

Transient Reliability

When lightning strikes on the a.c. lines, which are close to the converter, station, existing HVDC project, such as Hokkaido-Honshu link, has to stop its operation during the fault. Then, after the a.c. side fault is cleared, it restarts the operation. For this scheme, improvement of the system reliability is limited. Then, a continuous operation control during and after the interconnected a.c. line faults, had been newly developed and applied to the Kii-Channel HVDC Link.

To achieve this operation, some new technologies were incorporated: new type of Phase Locked Loop (PLL) with phase memory function, high speed open loop gamma control with correction function for harmonic distortion, or gamma detecting type closed loop gamma control. Such new controls optimizes the gamma, so that both of the prevention of commutation failure and rapid power recovery after the a.c. side fault clearing can be accomplished simultaneously.

There were 12 a.c. side lightning strike faults in Kansai side, and the converter could be continuously operated for all the time. Figure 7 shows one of the example of the continuous operation during the a.c. line fault.

One lightning strike occurred on overhead return circuit for pole 1, near Kihoku converter station. At that time, MRTB operated successfully to extinguish the arc then cleared the fault.

Another example of the real operation of new control was PM. When the thermal power plant of the Shikoku side tripped, the PM could detect the disturbance and operated to damp the local oscillation. It was confirmed that the damping control was properly operated.

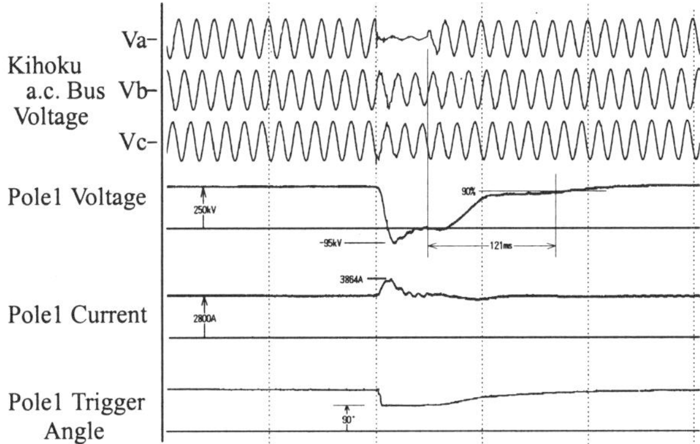

Fig. 7 Continuous operation during the a.c. fault

SUMMARY

The Kii-Channel HVDC Link has incorporated many new technologies, such as large capacity LTTs, DCGIS and advanced system controls. These new features contribute to achieve lower loss, compact, lower cost and reliable facilities. High energy utilization and availability of the link proves itself of the excellent reliability. Therefore the new HVDC link greatly contributes to the improvement of the system stability in western part of Japan.

REFERENCES

1. Y. Sekine, S. Kato, T. Motoki, S. Ito, "Kii Channel HVDC Link Between Shikoku and Kansai Electric Power Companies By Submarine Cables", Cigre Tokyo Symposium on Power Electronics in Electric Power Systems, 220-04, May 22, 1995

DYNAMIC PERFORMANCE OF THE EAGLE PASS BACK-TO-BACK HVDC LIGHT TIE

Å Petersson[1] and A Edris[2]

ABB Power Systems AB, Sweden[1] and EPRI,USA[2]

INTRODUCTION

Eagle Pass Back-to-Back (BtB) Tie is a Voltage Source converter (VSC) -based tie interconnecting the transmision grid of Texas with the Mexican Power system, at the AEP's Eagle Pass Substation in the State of Texas. The Tie was put in operation in July 2000.

As the VSC takes advantage of Gate-Turn-Off type of power semiconductors, the dynamic performance and capability are significantly improved compared to the conventional thyristor based technologies. This paper focuses on highlighting the performance of the VSC-based technology, using representative computer simulations and comparable field tests. The presented performance covers dynamic voltage support, operational mode changes and energizing of islanded networks. The influence on the ratings of the VSC main components is discussed, as well as the engineering tools used for simulations and validation of design.

KEYWORDS: HVDC, Back-to-Back, Voltage Source Converters, Pulse Width Modulation, Dynamic Performance.

PROJECT BACKGROUND

The Eagle Pass substation is located in a peripheral area of the Texas transmission grid, it is operated by AEP-CPL. Figure 1 illustrates the location on the US – Mexico border. The Eagle Pass load area is supplied over two 138 kV transmission lines. The grid structure and the relatively large distance to significant generation, results in weak voltage support to the Eagle Pass area.

Eagle Pass also has a 138 kV transmission line that cross-border ties into the Mexican transmission grid, operated by CFE. This tie is normally open and is mainly used in emergency conditions to transfer load from the USA to the Mexican grid. However, load being transferred needed firstly to be interrupted since it was not possible to interconnect the two asynchronous power systems. With a significant foreseen load growth for the load area it was eminent that voltage stability problems would in the near future be encountered at single contingencies.

Studies performed indicated that a facility providing 36 Mvar of dynamic reactive power support at Eagle Pass substation would provide years of relief from single contingency situations. A VSC type of installation, i.e. a STATCOM, would provide needed reactive support instantaneously and could maintain full reactive output at even lower voltage levels than what could result from the single contingency outage. Installation of a VSC would be ideal for weak systems whereas the reactive support provided by shunt capacitors is not very effective as it decreases with the square of the voltage. Expanding the installation to a Back-to-Back would enable also uninterrupted bidirectional active power transfer between the USA and Mexican grids, enhancing the reliability of the power supply. Thus a two-fold mission is accomplished by the dual VSCs. Additionally the tie can be used to energize the Eagle Pass load area from the Mexican grid as the VSC technology has an inherent black-start capability.

TECHNOLOGY APPLIED

A simplified one-line diagram of the BTB tie in Eagle Pass is shown in Figure 2.

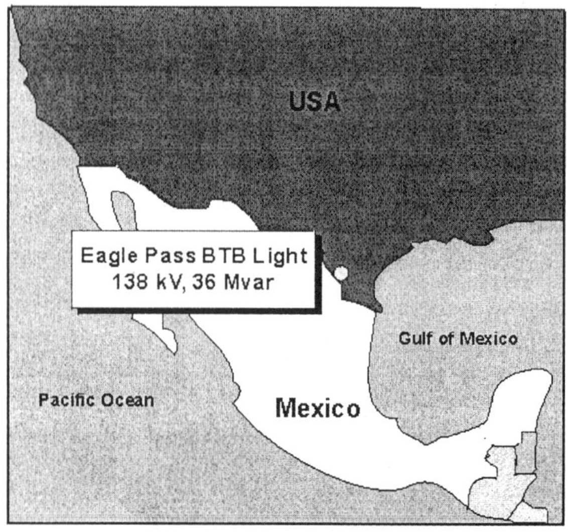

Figure 1. Eagle Pass location.

AC-DC Power Transmission, 28-30 November 2001
Conference Publication No. 485 © IEE 2001

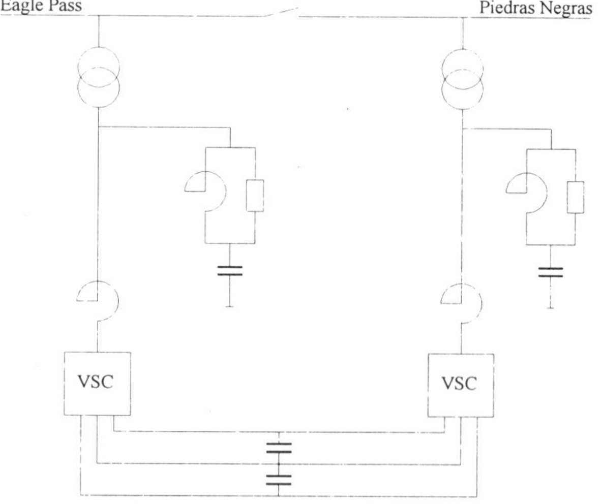

Figure 2. BTB simplified single line diagram.

The BTB scheme comprises two 36 MVA Voltage Source Converters coupled to a common DC capacitor bus. The VSCs are of the NPC, Neutral Point Clamped, type. The VSCs are equipped with IGBTs, Insulated Gate Bipolar Transistors, operated with PWM, Pulse Width Modulation. The power semiconductor equipment and the DC capacitor bank are of indoor design. The IGBTs are water cooled, utilizing a closed loop cooling system with water-to-air heat exchangers.

Each VSC is then on the respective ac terminal connected to phase reactors, which each in turn is connected to a conventional step-up transformer, interfacing the BtB to respective grid. The ac output voltage from the VSCs has a nominal value of 17.9kV, which is stepped up to 138kV. Harmonic filters, tuned to the 20th and 39th harmonic respectively, and with high pass characteristic, are shunt connected to the 17.9kV bus. Total filter rating is approximately 6 Mvar on each side. Due to the high frequency PWM switching applied, the filters can be kept small in rating and their respective tuning is not critical.

The whole BtB installation is located on the US side at the Eagle Pass substation. The substation is located in a residential area and specific consideration has been taken to control audible noise produced by some of the components. The layout of the BtB installation is shown in Figure 3.

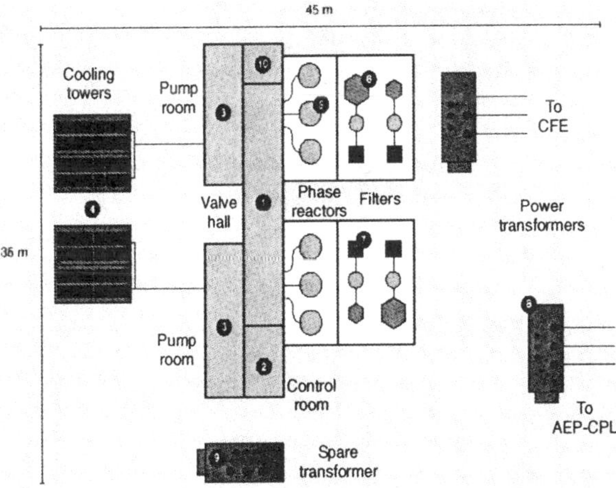

Figure 3. Eagle Pass BtB layout.

The BtB is designed to operate in two control modes. The primary mode is reactive power support for voltage control of the Eagle Pass bus and the secondary mode is active power flow across the tie. The BtB has the capability, within its MVA rating, of providing independent control of the voltages at the US and Mexican grids, as well as the real power transferred between the two grids. In voltage control mode, the BtB operates with a control slope to supply reactive power for voltage support. To transfer power in either direction the BtB operates in power control mode, it will continue to transfer power if the voltage at Eagle Pass is within a dead-band. For conditions outside the dead-band the BtB switches into the Voltage control mode. The dead-band is designed so that local capacitor switching or changes in remote generation which causes slight voltage swings do not cause the BtB to switch to voltage control mode.

VERIFYING DYNAMIC PERFORMANCE

The final verification of the dynamic performance characteristics was realized as part of the BtB tie commissioning. Then using adequate data acquisition equipment, real time recordings are taken of both main circuit parameters and internal tie controller variables. The Eagle Pass tie controller includes a built-in powerful Transient Fault Recording module with sub-millisecond resolution, which is capable of capturing also the inner control loop dynamics. This functionality also enables stored data to be digitally transferred to a remote user, e.g. the design engineer can have direct "just-in-time" access to the tie controller.

In the design phase of the BtB installation, extensive digital simulations were performed, as part of the development of the tie controller, and specifically the tuning of its important parameters. These simulations also to some extent verify the configuration of the main power circuit, e.g. the interaction between the main VSCs and the harmonic filters can be studied. The simulations have covered a large number of cases and events, out of which only a few could be conducted in the field.

When modeling electrical systems involving power semiconductor converters and extensive networks, the user interface becomes a critical factor. User friendly interfaces built around functional block representation is today a proven technique. For the Eagle Pass project a comprehensive model was developed using the EMTDC program. The complete model was built up with a number of functional blocks, such as: ac network, step-up transformer, phase reactor, harmonic filters, VSC bridge, DC capacitor and tie controllers. The ambition with the model was to represent the transmission grids accurately with respect to prospective resonances in the lower frequency range. The circuits within the BtB was modeled such that magnetic saturation phenomena could be identified. The representation of the IGBT converters was accurate with respect to the chosen NPC-topology and PWM switching at 1260Hz. The digital tie controller was modeled with sampling rates of critical control loops exceeding 7kHz. When the individual models had been validated, the actual simulations were performed with different parameter set-ups for different cases. Using a high performance personal computer, and an on-screen "control panel", very effective execution of the simulation can be achieved although the model becomes extensive. The design engineer can observe the behavior of the model as the simulation progresses and initiate actions accordingly.

There are typically a large number of dynamic events needing analysis and verification. Among those the following are illustrated below in an attempt to highlight the BtB tie unique and promising performance:

1. BtB power reversal
2. Capacitor bank switching
3. Black start capability
4. Nearby fault on the US side.
5. Remote line fault on the US side

For some of the cases below, the EMTDC simulations are presented together with field recordings with similar set-up.

1. BtB Power Reversal

The VSC converter technology applied in Eagle Pass, allows for the transferred active power to be dynamically controlled over the full available range. As an example of this, Figure 4 illustrates the performance after ordering a power reversal. This case is illustrated in the shape of a plot from an EMTDC simulation and the power order is changed from close to 36MW import to US, to 36MW export. Although locally significant, this transient is absorbed by the respective power systems representations. Looking at the traces of Figure 4, the sub-cycle reversal of active power appears well controlled without any overshoot, an example of the very high performance that can be achieved with PWM controlled VSC installations. The active power reversal is executed without any action needed by the voltage control.

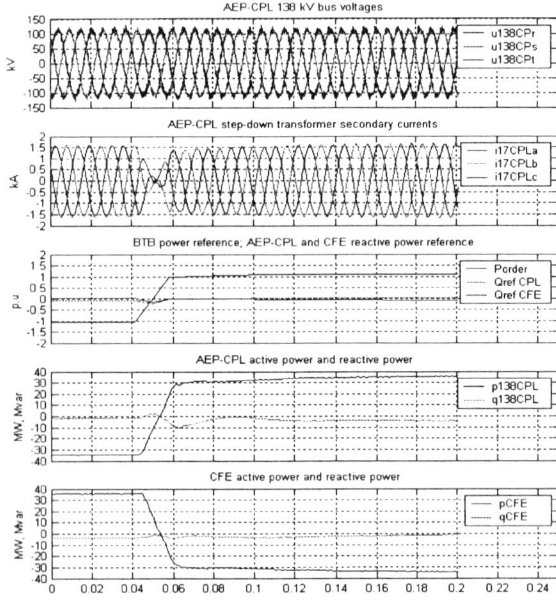

Figure 4. Power reversal at Eagle Pass (EMTDC simulation)
Subplot 1. AEP-CPL 138 kV voltages.
Subplot 2. AEP-CPL step-down transformer secondary currents.
Subplot 3. Back-to-Back power reference; AEP-CPL and CFE reactive power reference
Subplot 4. AEP-CPL active and reactive power.
Subplot 5. CFE active and reactive power

2. Capacitor Bank Switching

One possibility to check the transient response of the BtB was to switch on one of the capacitor banks at the Eagle Pass substation. In the actual case, the reactive power from the BTB before the switching was equal to 18 Mvar. As the capacitor bank was switched on, the BTB had to reduce its output by 15 Mvar, corresponding to the capacitor bank rating, down to 3 Mvar. The BtB controller responded in a well damped manner within a cycle as shown in subplots 2, 3 and 7 of Figure 5.

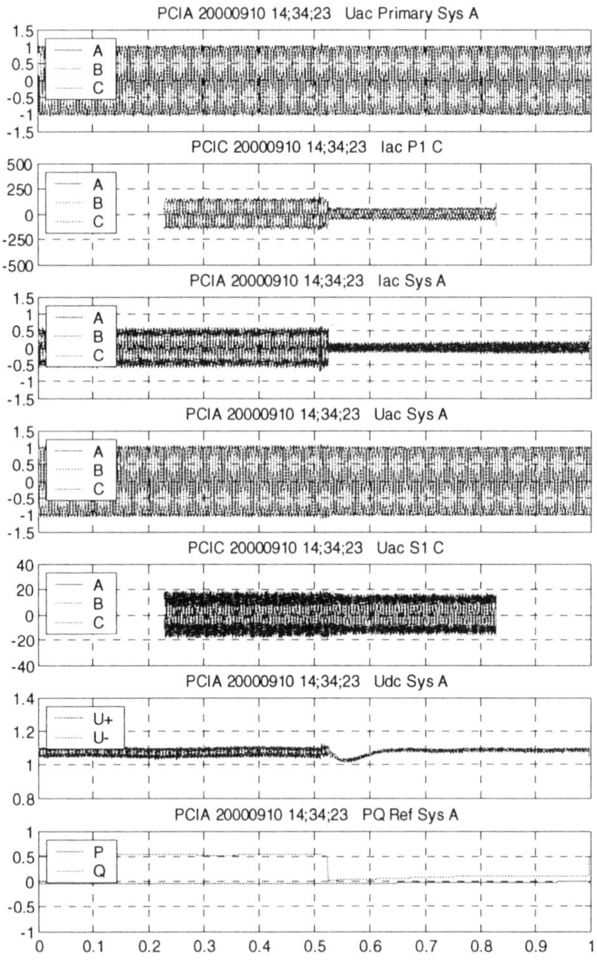

Figure 5. Capacitor bank switched on at Eagle Pass
Subplot 1. AEP-CPL 138 kV voltages.
Subplot 2. AEP-CPL step-down transformer secondary currents.
Subplot 3. AEP-CPL Phase reactor currents.
Subplot 4. AEP-CPL 17.9 kV voltages.
Subplot 5. AEP-CPL 17.9 kV phase-to-ground Voltages in kV.
Subplot 6. DC voltages.
Subplot 7. AEP-CPL converter Active (bottom curve) and Reactive Power.
It should be noted in Figure 7 that the switching causes also a 100ms transient in the active power. Although not significant, the switching produces a small angular

deviation in the bus voltage, relative the VSC output voltage. As a consequence a transient in the active power/DC voltage appears. On the other hand it is obvious that the BtB tie completely isolates the Mexican power system from the switching transients.

3. Black Start Capability

Next, to illustrate the Black Start function, the CFE side of the BTB tie was first energized as during normal energization. On the AEP-CPL side the line breaker, connecting the BtB tie to the AEP-CPL network, was open but was in the control system simulated as if it had been closed. In this way, a small islanded "network" consisting mainly of the AEP-CPL transformer was created. Any larger islanded network was not possible to create, as this would have forced a power outage in parts of the city of Eagle Pass. Black Start was then initiated.

For comparison Figure 6 shows the result of an EMTDC simulation where the black start is performed against a load of approximately 25MW.

Both the EMTDC simulation and the field recording (in Figure 7) show that the BtB energizes and picks up load in a controlled way without noticable transients.

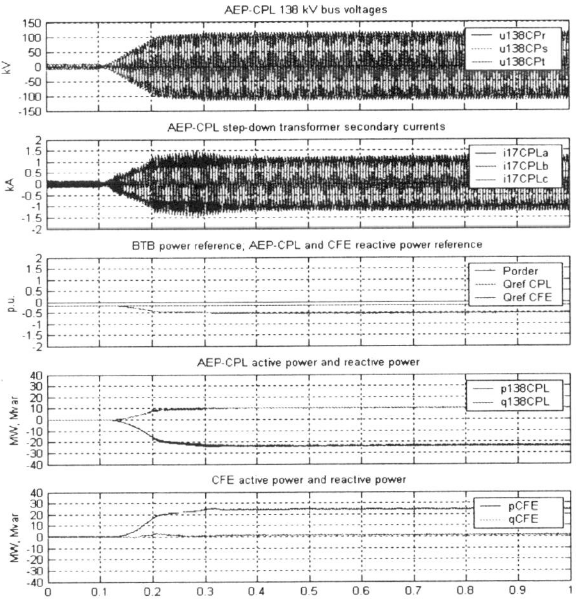

Figure 6. Black Start with 25 MW load (EMTDC simulation)
Subplot 1. AEP-CPL 138 kV voltages.
Subplot 2. AEP-CPL step-down transformer secondary currents.
Subplot 3. Back-to-Back power reference; AEP-CPL and CFE reactive power reference
Subplot 4. AEP-CPL active and reactive power.
Subplot 5. CFE active and reactive power

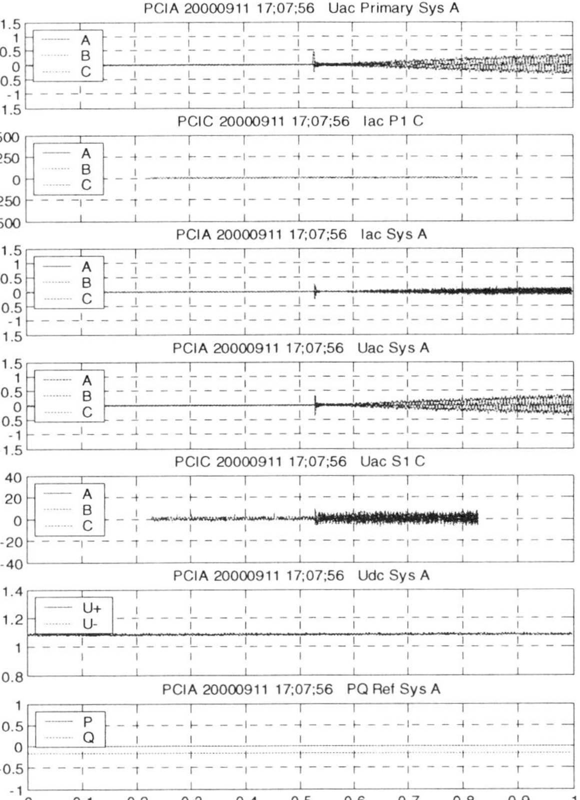

Figure 7. Black start.
Subplot 1. AEP-CPL 138 kV voltages.
Subplot 2. AEP-CPL step-down transformer secondary currents
Subplot 3. AEP-CPL Phase reactor currents.
Subplot 4. AEP-CPL 17.9 kV voltages
Subplot 5. AEP-CPL 17.9 kV phase-to-ground Voltages
Subplot 6. DC voltages.
Subplot 7. AEP-CPL converter Active (top curve) and Reactive Power (bottom curve) Reference.

As shown in Figure 7, The AEP-CPL 138 kV voltage was ramped from 0 kV, at a slower pace than in the digital simulation (Figure 6). The ramp ended at 138 kV although the plot, due to TFR limitations, does not show more than the part up to about 60 kV. On the CFE side, the voltages and currents remained undisturbed during the ramping phase. The only load on the AEP-CPL side consisted of the harmonic filters rated totally 6 Mvar. The system was anyhow observed to perform well. The modulation on the DC voltage in Subplot 6 is due to the frequency difference between AEP-CPL side, having constant frequency, and the CFE side. On the CFE side, the system operated as normal.

4. Nearby Fault on the US side

Near-by 1-phase to ground faults on the US side was simulated in EMTDC. For such a fault the primary voltage to be controlled, drops significantly. In order to avoid raising the voltage at fault clearing, the through-fault capacitive action is inhibited and the BtB stays at close to no load conditions during the fault. The post fault overvoltage reaches 1.3 p.u. in one phase, however it is practically eliminated in one cycle. Some current transients are seen in Figure 8 below, these being the result of oscillations excited in the harmonic filters. For this severe unbalanced fault it is worth noting that the transients seen on the Mexican side is practically negligible. The remaining disturbances are caused by the need to stabilize the DC voltage against the fluctuating active power caused by the unbalanced fault.

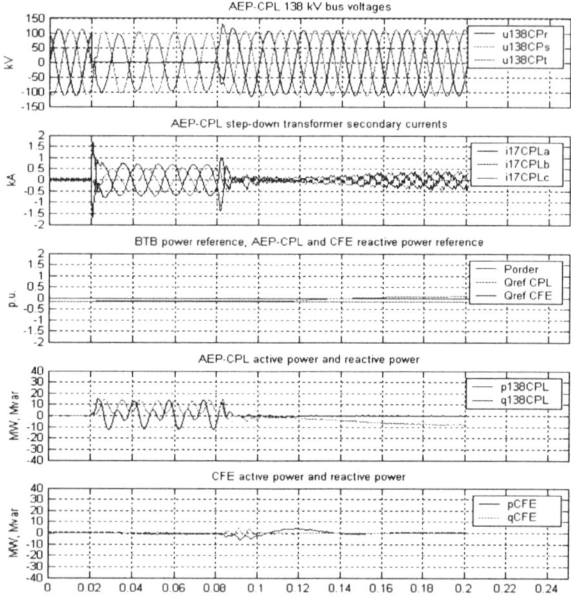

Figure 8. Near-by fault on US side (EMTDC simulation)
Subplot 1. AEP-CPL 138 kV voltages.
Subplot 2. AEP-CPL step-down transformer secondary currents.
Subplot 3. Back-to-Back power reference; AEP-CPL and CFE reactive power references
Subplot 4. AEP-CPL active and reactive power.
Subplot 5. CFE active and reactive power

5. Remote Fault on the US side

The final case to be highlighted is a remote fault on the US side. Where initially the BTB was in operation at close to zero active power. Lightning conditions in a remote area caused a voltage dip in the US transmission grid. This case was simulated also in EMTDC as is illustrated in Figure 9. The field recording is shown in figure 10 for comparison. Following the initial voltage drop in Eagle Pass, the BtB current during the fault condition was increased to almost 1 p.u. (capacitive) to support the bus voltage at Eagle Pass. When comparing the digital simulation with the field recording it can be seen that the reactive power transient in the simulation appears more exponential, indicating that the simulated fault creates a truly stepwise change in voltage. Otherwise the simulation and the field recording show good correlation.

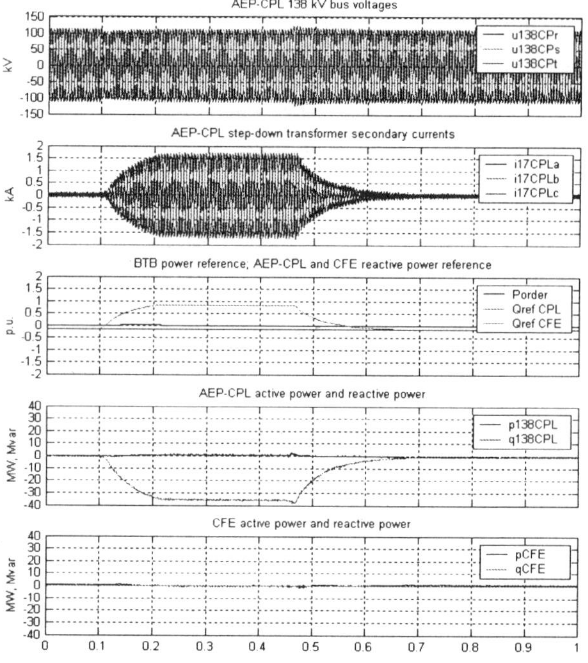

Figure 9. Remote fault on US side (EMTDC simulation)
Subplot 1. AEP-CPL 138 kV voltages.
Subplot 2. AEP-CPL step-down transformer currents.
Subplot 3. Back-to-Back power reference; AEP-CPL and CFE reactive power references
Subplot 4. AEP-CPL active and reactive power.
Subplot 5. CFE active and reactive power

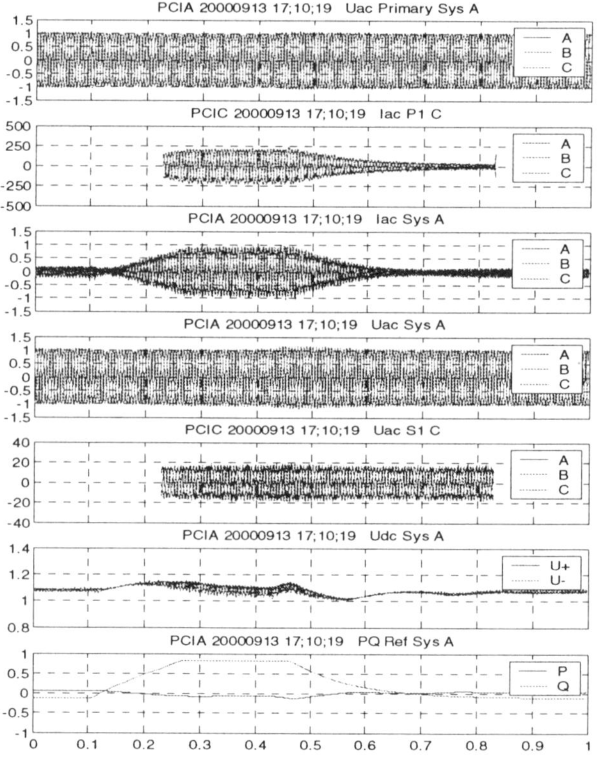

Figure 10. Remote fault case
Subplot 1. AEP-CPL 138 kV voltages.
Subplot 2. AEP-CPL step-down transformer secondary currents in Amps.

Subplot 3. AEP-CPL Phase reactor currents.
Subplot 4. AEP-CPL 17.9 kV voltages.
Subplot 5. AEP-CPL 17.9 kV phase-to-ground Voltages in kV.
Subplot 6. DC voltages.
Subplot 7. AEP-CPL converter Active (bottom curve) and Reactive Power (top curve)

It is worth noting from figure 9 that the BtB tie isolates the Mexican transmission grid from the fault, as is indicated by the reactive power reference being stable through the simulation.

CONCLUSIONS

The BtB HVDC Light tie at Eagle Pass is a first-of-its-kind installation, where VSC technology is used for the dual purpose of transferring active power between two asynchronous transmission systems and providing dynamic voltage support to the respective grids. The dynamic performance and characteristics of the BtB tie have been illustrated using digital simulations and field recordings. Among the features highlighted are instantaneous voltage support, black start of islanded networks and isolation of the grids with respect to switching transients and faults. State-of-the-art simulation software enables the design engineer to perform efficient simulations of the BtB tie on a standalone personal computer.

Acknowledgements

EMTDC is a simulation software package owned and developed by the Manitoba HVDC Research Center, Winnipeg, Canada.

The Eagle Pass BtB tie project was a joint effort between AEP, EPRI and ABB Power Systems. CFE of Mexico also played an important role providing assistance in many engineering matters. Today AEP operates the tie and an operation coordination with CFE is established.

CONTROL REFURBISHMENT PRINCIPLES APPLIED TO FACTS CONTROLLERS

J C Hinton, R S Whitehouse and M H Baker

ALSTOM T&D Ltd – Power Electronic Systems, UK

1.   INTRODUCTION

FACTS (Flexible AC Transmission Systems) controllers are now widely available and they demonstrate many years of service experience. Their contribution to the control of a transmission system will become more valuable as the demands and varying applications of the system increase.

The typical life expectancy of the main-circuit components of a FACTS controller is at least 30 years. The control system components however, though reliable and robust and often duplicated, are necessarily drawn from a technology that changes in a shorter period. Owners have shown increasing interest in the replacement or reassignment of control equipment to give the Controller further years of service or a change of application. Designers must therefore incorporate strategies to allow for advances in technology and owners will require the option to change control interfaces, for example to a new generation of system controller.

Of the FACTS controllers in service today, the majority are Static Var Compensators (SVCs), of which over 1000 give satisfactory service worldwide. Reviewing the available control packages and the experience of refurbishing an SVC, this paper will offer some guidelines on the principles of design for replaceable control components that can deliver a renewable and therefore ongoing performance for a FACTS controller.

2.   CONTROL SYSTEM REQUIREMENTS AND LIFE EXPECTANCY

Though a FACTS-controller may initially be installed to perform a specific function, the changes in ownership structures and circumstances may change its use. Some equipment can be relocated, for example see Figure 1, a useful concept given that reactive power does not easily travel and compensation tends to cover only a localised

region of a transmission system. To ensure relocatability, the top-level control functions should therefore be adaptable for change.

Figure 1: Valves and controls for a relocatable SVC are readily moved.

It is necessary at the outset to distinguish the demands placed upon the various levels within a control system, which can be considered as having three levels as shown diagrammatically in the table below:

Application or project specific		Increasing speed and dependence on supplier

High-level Controls	Low-level Controls	Valve Interfaces
Operator interface; SVS system	TCR & TSC Controls	Valve-base electronics
Millisecond response	10 micro-seconds to 1ms	Micro-second encoding to valve firing
Commercial desktop computer, PLC	Fast CPU/DSP, either commercial or proprietary	Specialised and proprietary

AC-DC Power Transmission, 28-30 November 2001
Conference Publication No. 485 © IEE 2001

To the left of the table, the controls are dependent on specific SVC actions whilst to the right, the prime requirement is speed. The various levels of control within a typical SVC are shown in block diagram form in Figure 2.

designed for the topology of the valve rather than the use of the controller, flexibility is not the prime requirement.

3. HIGH-LEVEL CONTROL EQUIPMENT

At this level within the control system, software

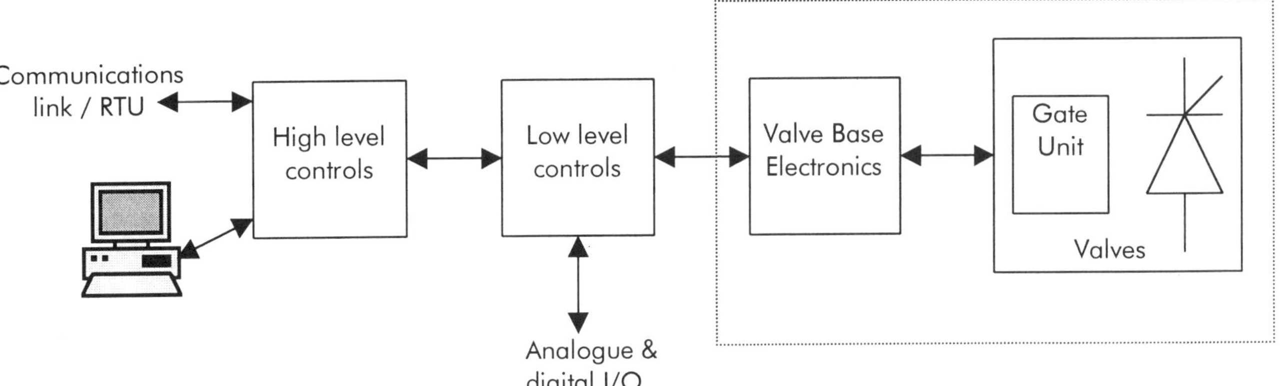

Figure 2: Block diagram of controls hierarchy

High-level controls refer to the control-system set points, start up and shutdown, and co-ordination of the controller with the power system, including in some examples, dynamically adjustable settings to assist system stability or the switching of other reactive power elements. Status monitoring feedback to the operator (the Human – Machine Interface) and to station Supervisory Control and Data Acquisition (SCADA) systems are also in this category. Ready adaptability of these functions is an advantage so that the task of altering settings is easy when system requirements change. Life expectancy can be justifiably short for this equipment, provided that replacement is straightforward.

Low-level controls refer to those parts of the system operating with cycle times in the order of tens to hundreds of microseconds, for example deriving the firing angles, and directing instructions to valves and breakers. Valve protection functions, usually independent of the control functions to maintain the robust performance required of a power transmission system, call for equipment with a similar speed. Such equipment must either be readily changeable or have a long expectancy of life.

Valve-Base Electronics (VBE) equipment, more fully defined in Section 6, provides the very fast and accurate generation of firing signals for individual valves and devices and the interpretation of data received from each valve level. Its prime requirements are speed and reliability and, being

applications are typically running with cycle times of milliseconds or slower and include parameter settings, the operator interface and the interface with SCADA. High-level control equipment can be realised by a combination of commercial and specialist software running on commercial equipment such as a personal computer or a programmable-logic controller. Since it is desirable that such equipment be replaceable, both the software and the physical interfaces must be readily removable. Equally, the physical interfaces and communication protocols must be of a standard likely to remain compatible with any new platform. On this basis, the long lifetime that is desirable can be achieved through re-use of software and interfaces whilst accepting that other hardware components are modular and plug-in, for which a shorter life is acceptable.

This does have the advantage that subsequent hardware technology advances can be utilised, which may result in improved performance. Equally, adaptation to new uses with new software can readily be deployed.

This also offers the opportunity for the owner to modify the application software and thus redefine the usage of the FACTS device in line with the changing requirements of his ac system. Although it may be advantageous for him to work with the original supplier of the equipment, the use of standard tools and languages will allow the use of competent third party vendors to provide solutions to changing system requirements.

228

4   LOW-LEVEL CONTROL EQUIPMENT

Low-level control equipment needs to operate at a speed that is beyond the capabilities of many commercial products. At this level we include fast control loops which require deterministic performance in the range from hundreds down to tens of microseconds. Standard microprocessors have difficulty in coping with such demands and typically high-speed digital signal processors (DSP) are used in conjunction with programmable logic support hardware. The designer must be aware of the significant risk of obsolescence with such hardware and the possible incompatibility with new equipment, which could make future changes difficult.

Commercial equipment may be able to satisfy the requirements of the low level control system. Commercial products that have adequate performance capability will come and go according to market demands and technology trends and the customer will be exposed to this, for example in 1994, a different technology was prominent [ref 1].

The alternative is custom-built hardware, but that implies that sufficient spares are also built (or remain obtainable) for the equipment lifetime. The full cost of such spares holding should be part of the assessment when this choice is considered.

The approach favoured by the authors is to use a combination of commercial and custom designed products to obtain the best compromise between cost and long term support. The benefits of commercial processor products with their standard hardware and software interfaces are combined with custom DSP and programmable logic support.

A complex control and protection system will be based on distributed processing and the individual processing elements need a common hardware base coupled with an efficient communications mechanism. The authors consider that the industry standard VMEbus platform is ideal for this purpose, supplemented as necessary by other industry standard technologies such as fieldbus interfacing and Ethernet communications. Using such industry standards, and eliminating as far as possible custom interfaces and protocols, ensures maximum protection against obsolescence and, furthermore, offers great flexibility in terms of expansion and interfacing potential with third party equipment. This clearly benefits both owners and suppliers.

These are the basic principles behind the ALSTOM Series V control system platform which is designed to utilise up-to-date technology whilst providing long term support and maintenance, apparently conflicting goals. The main processor board within the system is an industrial grade VME single board computer with an 850MHz Pentium processor, 64/128MB main memory and standard interfaces. Inevitably technology moves on and by the time the reader receives this paper we will probably be offering our customers over 1GHz processors (probably at a lower price too!). Most importantly, only minor changes will be needed to accommodate this change.

5.   SOFTWARE

Application software for the Series V control system is based on the industry standard C programming language. The front end design may be approached from the viewpoint of a software engineer, performing analysis and design in Unified Modelling Language (UML) for example then either hand writing code or autocoding from a Computer Aided Software Engineering (CASE) tool. Alternatively, the design may be approached from a control engineer's point of view by utilising a simulation tool such as Simulink and its code generation facilities. A third approach is to use one or more of the suite of languages defined by IEC1131-3 [2] including the graphical function block diagram (FBD).

All of these methods produce code that is easily interfaced with the system software and operating system described here.

The software will, in most cases, last for the lifetime of the product. The owner may wish to make enhancements but typically these are relatively minor and it would be unusual to need changes involving a fundamental rewrite. However, whilst the software may last, the hardware platform may become obsolete several times over during the lifetime of the system. Software portability is therefore a key issue to consider when designing a system with longevity and future refurbishment in mind.

The application of the controller can change too. As an example, a SVC was delivered to a steel-maker in the US in 1990 where it operated for 5 years as a flicker controller. The factory closed and the SVC was sold. The SVC was found a new role, but this time in another country and as a phase-balancer for a single-phase traction load.

Flexibility of the control system was clearly an important factor in this.

The key to software portability is to use a layered software design where the lowest layers insulate the upper layers from the real hardware (see Figure 3). The Application software sits at the top of the hierarchy and is supported by the System software and the real-time operating system (RTOS). System software supplements the operating system by providing high level services, for example standard communications interfaces, monitoring routines, standard Input and Output (I/O) interface etc. The hardware sits at the lowest level of the hierarchy and typically comprises processor boards together with I/O and communications boards. The only direct interface that the software has with the hardware is via the operating system, the processor Board Support Package (BSP) and other drivers. Intelligent processor boards are supplied with a BSP that provides a standard software interface with the board, regardless of the processor technology, clock speed, manufacturer etc. This is the mechanism by which processor boards can be replaced without requiring significant changes to software. A future replacement unit will have its own BSP which conforms to the standard software interface and thereby hides the physical details of the hardware. In fact, this philosophy can be taken a stage further by porting the system to run on another hardware platform such as PC/104 or Compact PCI, thereby removing reliance on the VME platform.

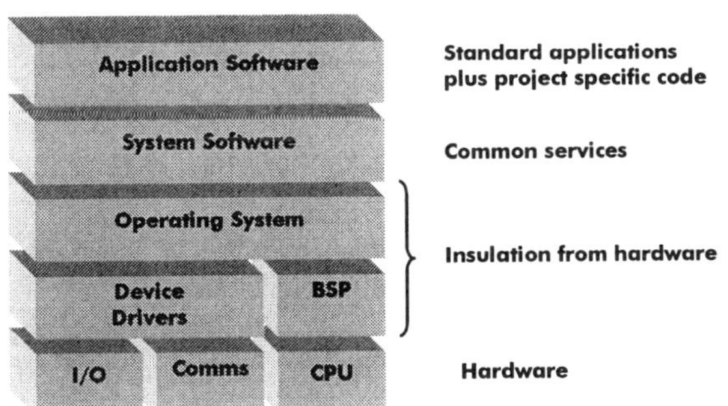

Figure 3: Layered approach to software design

As described earlier, the main processor board within the Series V control system is an industrial grade VME single board computer which conforms to the VME hardware standard and the operating system software interface standard. This ensures that the board can be substituted as necessary without impacting on the application software.

The custom hardware within the system is also largely based on VME bus and has been carefully designed to provide an upgrade path as the inevitable obsolescence occurs. The approach here is twofold: firstly, to design using carefully selected components (ideally available from multiple sources), and secondly to accept the inevitable and plan for it. At some point it will not be technically or economically viable to continue with the product. When this point is arrived at, the use of standard interfaces is then the key to allowing the obsolete equipment to be replaced by another that has the same standard interfaces. In essence, the core of the control hardware can be removed and replaced without disturbing existing peripheral equipment such as I/O, which significantly reduces the complexity and cost of the refurbishment.

6.   VBE

Valve-base electronics is, as its name implies, closely tied to the valve. All manufacturers of valves for high-power electronics applications have their own designs and VBE will therefore be supplier-specific.

The basic requirement of VBE is to provide a fast and very reliable distribution of signals to and from each power-electronic device. In addition, VBE may provide ground level protection and other features depending on the individual design. Clearly there will be design principles that each developer of FACTS controllers will maintain within his proprietary equipment and thus replacement during the lifetime of the main equipment, except by the original manufacturer is undesirable. On the plus side is the fact that VBE needs robust and simple equipment and long-life is compatible with this aim.

7.   PARTITIONING OF THE CONTROL SYSTEM

Valves and gate electronics are generally inseparable and are best considered as a single unit. At the ground level sits VBE with its highly specialised interface with the valve electronics and which is unique to the type of valve. It is logical therefore to consider the valve and VBE equipment as one package. Further up the control hierarchy, the low level and high level control system blocks can be defined as separate elements as shown in Figure 2 and standard interfaces can be defined between them. It is probable however that this distinction will become blurred with time as technology will eventually advance to the point where there is no justification for separating these functions architecturally.

The simplified system therefore consists of a controls block plus the valve/VBE package. The longevity of the system will therefore be highly dependent on the VBE and valve package as this represents a major part of the investment and is less easily replaced. Whilst accepting that the control system hardware may need replacing during the lifetime of the system, it is essential that the VBE and valve package remains functionally separate and maintainable.

Referring to Figure 2, VBE has two primary interfaces: at its 'input' is the interface with the control system and at its 'output' is the interface with the valves. The interface with the valves is entirely individual to each supplier and is determined by the data sent to and received from the valves. This is highly specialised and timing is critical. In comparison, whilst the 'input' interface with the control system will also carry timing information, the format of this data is much more flexible.

As microprocessor technology advances in terms of both speed and performance, we can confidently predict that commercial products will be able to perform more of the low-level functions associated with the control system. On the other hand, VBE equipment will remain part of the valve, diverse and proprietary. The time may be coming when it is appropriate to consider a standard definition of the interface between VBE and the low-level controls.

The purpose of such a standard interface would be to allow the owner flexibility in modifying the control system hardware, possibly after many years in service, without dependence on the original supplier. This could be in order to take advantage of new control hardware, to replace obsolete hardware or to add new functionality. In addition to the examples quoted above, it may be that software changes can extract in-built additional capacity from the main circuit components under certain conditions. One example of this would be to take advantage of seasonal variations in temperature and therefore cooling capacity, allowing extra capability of the FACTS controllers when the ambient temperature is low and the corresponding demand for energy is high. The authors propose that a standard interface between the controls and VBE should be based around a serial connection running an existing, industry

standard protocol. There are many examples of suitable candidates for the physical and data link layer standards. Whether the standard interface should go beyond this, and include a definition of the types of message that can be exchanged, is left open to debate.

The advantages of this approach to the owner are clear but what about the supplier of the equipment? A few of the advantages are:

- Development of controls and VBE/valves can be decoupled.

- The use of a standard interface, which is maintained in the long term across many products, will reduce the amount of new development effort required with each generation of control systems.

- Backwards compatibility between new control equipment and old valve/VBE packages is guaranteed

Equipment providers who may have the capability in high or low-level controls, but not the power electronics, would have a way into the FACTS market by partnership with equipment suppliers for new schemes or refurbishment of existing installations.

8. CONCLUSIONS

The general trend for power systems throughout the world is one of decentralisation and multiple independent ownership and yet, perhaps perversely, greater demands for control and reliability. This will mean that the original reasons and motivations for equipment procurement and installation may change. In addition the increasing use of standard commercial products for control makes future availability of spares from the original supplier uncertain. To avoid the untimely and costly replacement of perfectly serviceable power equipment, equipment today should be adaptable for uses that are impossible to specify at the initial contract stage and control implementation should not be specific to a particular hardware platform. Furthermore these trends infer that the control and power equipment should be de-coupled to the point that supplier dependency for particular hardware is removed.

To achieve the above requires the adoption of recognised industrial definitions of equipment functionality and standard rules of interfacing between such functions. To take advantage of advances in technology clearly benefits the power utilities and through them their customers, therefore it is these utilities, perhaps working with existing suppliers, which should promote the move towards standardised control and function boundaries.

REFERENCES

1. C A Brough; Transputers and the control of Power Electronic Systems; Parallel Computers and Transputers conference, Brisbane 1994

2. IEC1131-3 (1993); Programmable Controllers, Part 3, Programming languages.

COMMISSIONING OF A 225 Mvar SVC INCORPORATING A ± 75 Mvar STATCOM AT NGC's 400kV EAST CLAYDON SUBSTATION

C Horwill, A J Totterdell, D J Hanson

ALSTOM T&D Ltd , Power Electronic Systems, UK

D R Monkhouse, J J Price

National Grid Company plc, UK

INTRODUCTION

The privatisation of the electricity supply industry in the UK in 1990 is well documented (1). The unbundling of the generation and transmission functions into separate shareholder-owned companies has inevitably resulted in far less predictability in terms of generator siting and closure. NGC, as the sole transmission company in England and Wales, is required to plan and respond quickly to changing system patterns to maintain both security and power quality standards. This has been achieved by an extensive programme of investment in reactive compensation, comprising both the 'slow acting' (coarse control) mechanically switched capacitor (MSC), and 'fast acting' (dynamic) thyristor based static var compensator (SVC) types.

In view of the many system planning uncertainties faced by NGC, there is a clear need for readily relocatable reactive compensation systems with enhanced system performance. To date 12 relocatable SVCs (RSVCs) have been installed as part of a planned programme of work to meet these changing system needs. Developments in the rating capabilities of gate turn-off (GTO) thyristors have enabled power electronic converters using GTOs (2) to be proposed as serious competitive alternatives to the well proven thyristor controlled reactor (TCR) and thyristor switched capacitor (TSC). GTO-based STATCOM convertors have a number of benefits which are of value to NGC, such as reduction in required site area and ease of relocation. The East Claydon project is viewed as an important step towards long-term objectives which could see the deployment of a range of compact power electronic devices on NGC's system.

NGC's NEED AND BASIC REQUIREMENTS FOR AN SVC

During the mid 1990s NGC predicted a further build-up of low cost CCGT generation in the North which would have resulted in increased Midlands to South power flows with a consequent large potential deficit of Mvars in the South. The provision of a SVC at East Claydon substation, with its strategic location on the 400kV Grid System, therefore forms an important part of an overall scheme to provide reactive compensation in the South.

To enable design optimisation of the SVC, in particular the STATCOM element, NGC permitted some flexibility in the chosen Mvar rating. Tenderers were permitted to submit designs with a minimum range of 0-150Mvar (capacitive) and up to a maximum range of 0-225Mvar (capacitive). Adjudication of designs was based on an assessment of the cost per dynamic Mvar provided. To avoid the possibility of 'stranded assets' arising from future changes to the network rendering its position less optimal, the SVC must also be capable of subsequent relocation to other substations with either 400kV or 275kV connection voltage, and possessing similar site area. Up to 3 relocations are envisaged over an equipment lifetime of forty years.

BASIC DESIGN OF THE SVC

The SVC is designed to provide a smoothly variable reactive output of 0-225 Mvar (capacitive). No inductive capability was specified. The specified reactive power output is available at 0.95 per unit system voltage, giving a maximum SVC current of 225/0.95 = 2.37 per unit (on 100MVA base). The SVC equipment is connected to LV busbars operating at nominally 15.1kV fed via the secondary winding of the compensator transformer.

Design optimisation led to a configuration that employs a STATCOM with a dynamic range of 150 Mvar (±75 Mvar) which, in conjunction with a small fixed filter, provides outputs between zero and about 100 Mvar. A 127 Mvar TSC is also provided to deliver the additional Mvar required for an output of 225 Mvar. The TSC provides a rapid and reliable means of delivering additional reactive output when system conditions require the maximum available compensation. Figure 1 shows the SVC configuration. See also Knight et al (3).

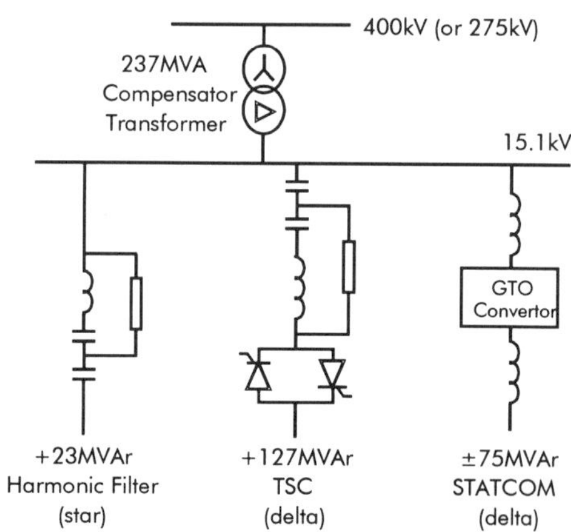

Fig. 1 SVC Configuration

AC-DC Power Transmission, 28-30 November 2001
Conference Publication No. 485 © IEE 2001

BASIC DESIGN OF THE STATCOM GTO CONVERTORS

STATCOM technology is based upon the principle of using power electronic switches to derive an approximately sinusoidal output voltage from a dc source. There are two main methods of implementing this principle in a way that gives realistic voltage ratings and ensures an acceptable level of residual harmonics. Simple voltage-sourced convertors can be combined magnetically to achieve an acceptable output voltage, although unbalanced system conditions can result in poor waveshapes. Alternatively the acceptable voltage waveform can be derived within the convertor using a multi-level circuit with a distributed capacitance. This approach has been adopted for the STATCOM at East Claydon which has three single phase multi-link convertors connected in delta. Each link consists of a single phase bridge voltage-sourced convertor using four GTOs and associated freewheel diodes with a dc capacitor as its voltage source. In order to satisfy the worst-case operating point each phase of the convertor requires a minimum of 14 series connected links. Two additional links are provided in each phase as redundancy, giving a total of 16 links per phase. Each convertor phase is connected to the LV busbars via a small buffer reactance. See also Woodhouse et al (4).

OVERVIEW OF SVC CONTROL SYSTEM

The control system for the SVC is implemented on a high-speed digital platform incorporating industry standard processors and using EPLD based controllers. A top-level varvolt controller is used to define the SVC operating characteristics in much the same way as for a typical TCR application. The aim of the controller is to produce a current order for the STATCOM (I.STM.order) and a block/deblock signal for the TSC. See Figure 2 below.

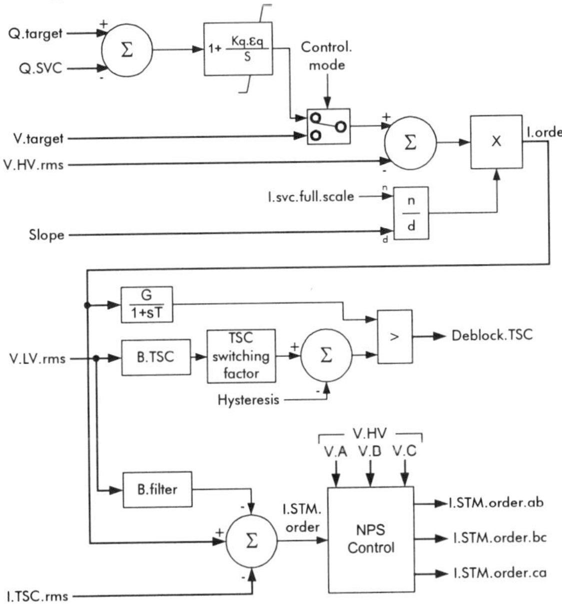

Fig. 2 Varvolt Control

As the STATCOM comprises three single-phase convertors connected in delta, a separate current order can be given to each phase to match prevailing system conditions. These *current orders* result from I.STM.order being modified to reduce the negative phase sequence voltage on the ac system.

STATCOM CONTROL SYSTEM

As discussed above, the STATCOM comprises three identical single-phase devices. The STATCOM control system for a single-phase is described here.

Figure 3 shows the STATCOM connected to the LV busbars via a small buffer reactance. By appropriate control of the STATCOM's output voltage (V_{STM}) to be higher or lower than V_{LV} the STATCOM will draw a capacitive or inductive current from the system.

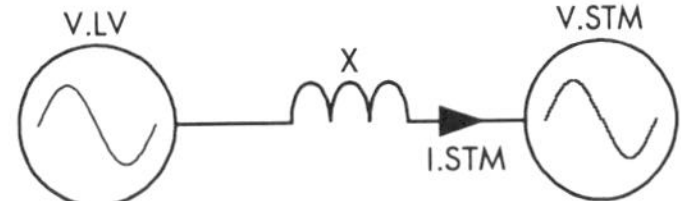

Fig. 3 Single line equivalent of a STATCOM connected to an ac system

In steady-state, the mechanism of control is to vary the phase shift (δ) between V_{LV} and V_{STM}, causing real power flow, which then charges or discharges the STATCOM link capacitors, thus varying the magnitude of V_{STM}. In-turn, this controls the flow of reactive current.

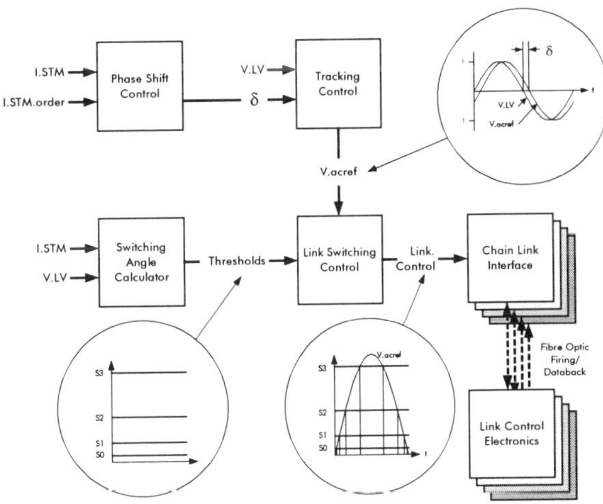

Fig. 4   STATCOM Control System

Figure 4 shows the five main blocks of the STATCOM control system. The *Link Control Electronics* are mounted at link-level and have been shown for information.

Phase Shift Control is at the heart of the closed loop current control, defining the phase shift (δ) of the converter waveform using a PI controller. *Tracking Control* constructs a digital representation of the required converter waveform, referred to as *V.acref.*

This waveform is then compared with *threshold levels (S(n))* calculated by the *Switching Angle Calculator* based upon preset harmonic cancellation criteria, the existing convertor current and V_{LV}. The combination of *V.acref* and the *threshold levels* results in a *Link.Control* signal, which defines when the links are switched in (and thus out). These signals are determined using high-speed DSP functions with a jitter of less than 2 microseconds. The *Chain Link Interfaces* then send the firing information to the links via fibre optic cables.

During system disturbances, the sinusoid *V.acref* is scaled to ensure that the convertor delivers full leading or lagging current, as appropriate, without needing to charge or discharge the dc capacitors. For a depression in system voltage, *V.acref* is modified to bypass links and for high system voltages *V.acref* is modified to widen the conduction period of the available links. The STATCOM is therefore able to provide a very fast response to system disturbances.

SITE INSTALLATION AND LAYOUT

To fit within the limited available site area, a compact arrangement of SVC equipment has been implemented occupying less than $1400m^2$ (or $15,000ft^2$). Each phase of the STATCOM is self-contained in a road-transportable cabin (located in the top-right part of Figure 5). The TSC valve, associated controls and other ancillary equipment are mounted in a similar cabin shown in the centre of the picture. The cooling plant and heat exchangers are located on the right of the picture. All other switchyard-type equipment is mounted, with associated connections, on metal frames to group items together for ease of transport.

Fig. 5. East Claydon SVC Site Layout

COMMISSIONING PROCESS

Following installation of the STATCOM and TSC cabins, harmonic filter skid and other SVC equipment, SVC commissioning was undertaken in two distinct phases. Firstly, pre-commissioning was performed to ensure that the equipment had not been damaged in transit or during installation and to prove that installation had been carried out correctly. Each item of

plant was thoroughly tested in order to prove its correct operation before energised commissioning could proceed.

The design of the STATCOM convertor, its auxiliary power supply equipment and its control system allows an "open circuit" sine wave to be synthesised by each phase of the convertor prior to making the final connections to the LV busbars. Each cabin essentially acts as an open-circuit voltage source able to generate a fundamental frequency output voltage but without any load current flowing. This test was an important part of the pre-commissioning process proving the operation of the STATCOM valve and its control system up to a peak terminal-to-terminal voltage of 24kV, equivalent to a voltage on each link capacitor of 1500V. The valve losses (including switching losses) were supplied from the substation 415V supply via a ground level inverter-based power supply (GLPS). A typical waveform of the STATCOM generated voltage is shown in Figure 6.

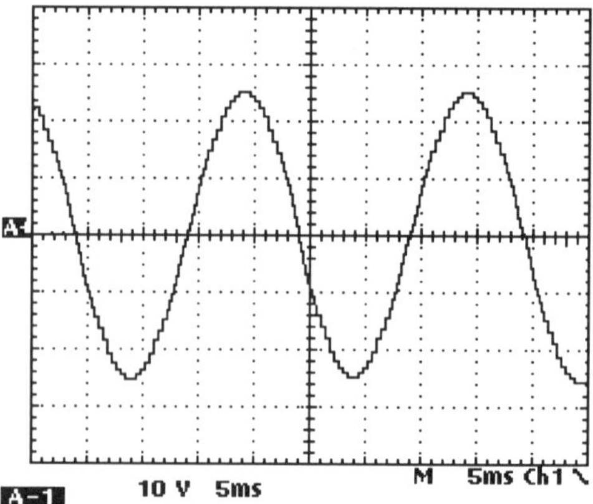

Fig 6. Open-circuit testing – STATCOM Voltage

Before commissioning commenced, measurements were taken to assess the magnitude of background harmonic voltage distortion present on the 400kV system. The first stages of testing energised and commissioned the compensator transformer, the 15.1kV LV busbars and the fixed filter. Harmonic measurements were then compared to the initial background readings to assess whether any 'magnification' had occurred. Subsequent tests then proved that the automatic start-up and shut-down of the valve cooling plant operated correctly. Once energised, the SVC is able to derive auxiliary power from the 415V secondary winding of an earthing transformer connected to the LV busbars. The correct changeover of the SVC's auxiliary supply from the substation to the earthing transformer was also verified.

Testing continued with the energisation and commissioning of the TSC. At this stage, TSC block and deblock commands were applied manually via the SVC controls. Various sequences of tests proved the correct operation of the valve and the relay-based and control-based protections associated with the TSC. Again, measurements of 400kV harmonic voltage distortion were taken and compared with background measurements.

The first stage of STATCOM commissioning was to deblock each phase of the STATCOM in turn whilst disconnected from the LV busbars. This test proved the correct STATCOM start-up sequence whereby each link dc capacitor is charged to 1500V. With the STATCOM still disconnected, the 400kV circuit breaker was closed to energise the LV busbars and generate the VT reference "tracking" signal for the STATCOM control system. With the STATCOM current order manually set to zero, a check of both the magnitude and phase relationship between the STATCOM generated voltage and the LV busbar voltage was performed. This test was then repeated on the other two STATCOM phases.

The final busbar connections from the STATCOM cabins to the buffer reactors were made, linking the STATCOM to the LV busbars. With the TSC manually blocked and the STATCOM overcurrent protection initially set to around 20% of nominal, a manual current order of zero was set and each STATCOM phase deblocked in turn. Once correct operation was verified, each STATCOM phase was selected to automatic control, and the STATCOM was operated as a three phase unit for the first time.

The SVC current order was adjusted in steps in order to exercise the STATCOM over its full (capacitive and inductive) current range. Figure 7 shows traces from the site transient recorders for full leading and lagging operation.

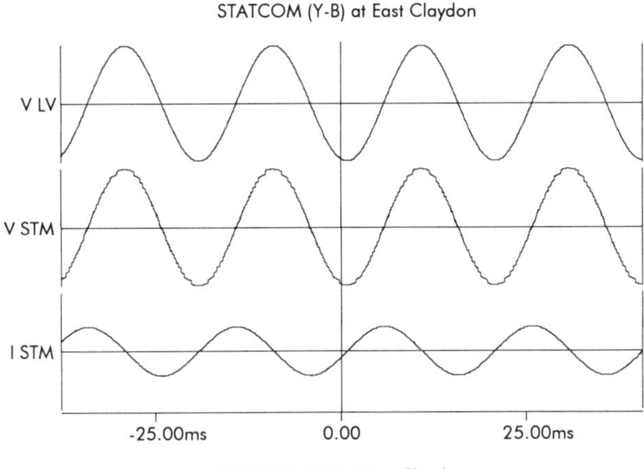

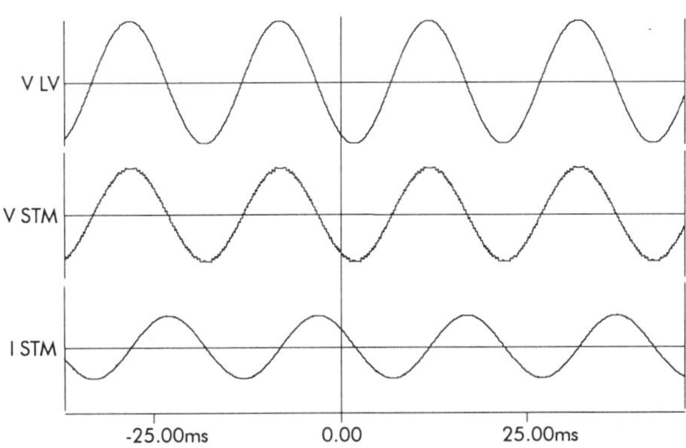

Fig. 7. STATCOM Waveforms at Full Leading and Lagging Current Operation

Note the voltage ripple appearing on the individual capacitor voltage steps due to current flow through the link capacitors, positive in leading and negative in lagging. Measurements of STATCOM current and voltage outputs were taken and the STATCOM electronic and conventional protections verified. Individual link voltages were monitored in some detail at various output levels to ensure that the dc capacitor voltages were balanced across the links of each phase.

Measurements of individual and total 400kV harmonic voltage distortion were taken across the entire STATCOM current range and comparisons made with 'background only' measurements. These comparisons indicated that, for the system conditions present during commissioning, the STATCOM's contribution to harmonic distortion was negligible. The results of the harmonic measurements are shown in Figure 8. Measurements were taken for approximately one hour prior to energising the SVC to assess background conditions (from 10:38).

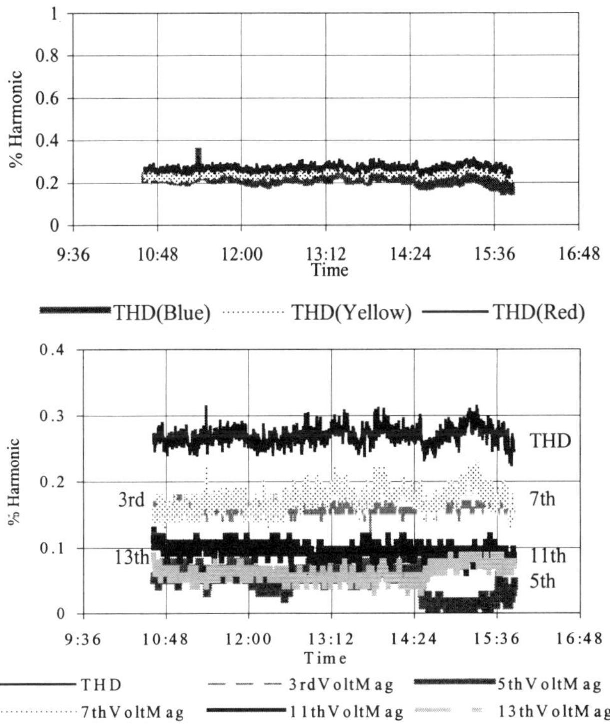

Time	Fixed Filter (Mvar)	STATCOM (Mvar)	Total SVC Output (Mvar)
10.38	0 (Off)	0 (Off)	0 (Off)
11.23	23	0	23
12.06	23	-23	0
12.41	23	22	45
13.08	23	47	70
14.39	23	67	90
15.06	23	82	105
15.47	0 (Off)	0 (Off)	0 (Off)

Fig. 8. 400kV Harmonic Voltage Distortion During STATCOM Operation

One of the benefits of the STATCOM is its improved speed of response and this was verified by performing a number of current order step change tests. From these tests, adjustments were made to the proportional and integral gains of phase control to achieve the optimum response and an adequate speed of response. Figure 9 shows a typical waveform demonstrating a smooth and rapid response from the STATCOM.

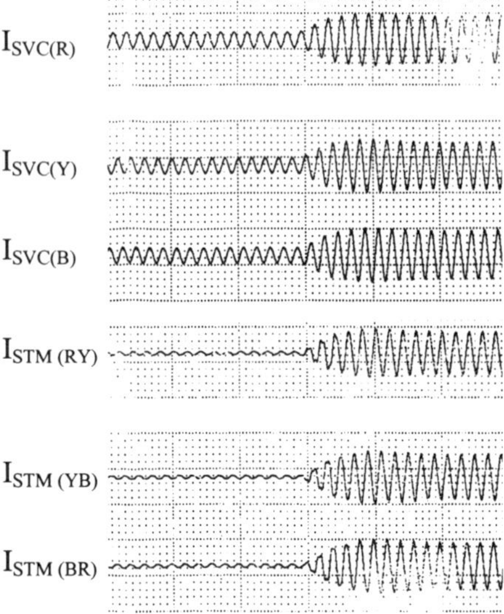

Fig. 9 Step of STATCOM Current Order

Automatic control of the SVC was checked, both in Target Voltage control and Constant Mvar control and correct automatic switching of the TSC was verified. A comprehensive range of harmonic voltage and current measurements was carried out at several different SVC currents. The measurements of 400kV harmonic voltage distortion were carried out by NGC and ALSTOM and both sets of measurements demonstrated that the increase in distortion as a result of connecting the STATCOM, TSC and harmonic filter was negligible.

The final response tests were performed by applying various step changes to the SVC's target voltage to cause automatic switching of the TSC. When the TSC is blocked or deblocked a transient "kick" is applied to the phase shift setting for the STATCOM to rapidly change its output from maximum capacitive to maximum inductive and vice versa. Various TSC block/deblock tests were carried out to determine the optimum settings for the "kick" function to minimise the disturbance to the SVC current. It was found that a much larger magnitude of "kick" is required for deblocking the TSC compared to blocking and its necessity, at all, for blocking appears to be marginal. Figure 10 shows the waveforms for deblocking and blocking the TSC and the corresponding operation of the STATCOM at the optimum "kick" settings.

These response tests verified that the SVC (STATCOM and TSC) met all the control performance criteria specified by NGC.

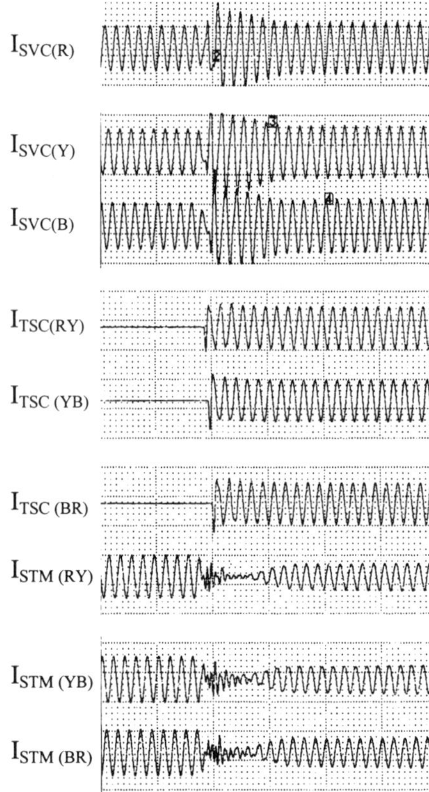

Fig. 10a   TSC Deblock

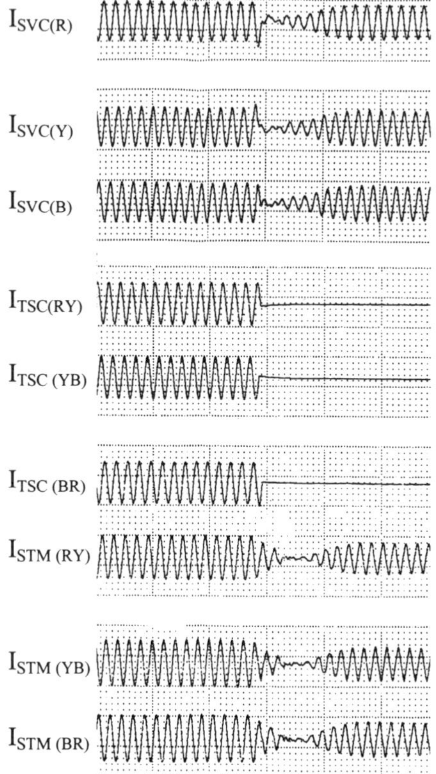

Fig. 10b TSC Block

The SVC can be operated via PC-based control points installed locally at the SVC and in the substation control room although the SVC is generally operated remotely by NGC's National Control Centre (Wokingham) via GI74 communication links. Operation from these local and remote control points was tested to verify that the communication between the SVC control and the control points operated correctly.

At the end of the commissioning process, two temperature rise tests were performed, one with the STATCOM at maximum leading current, lasting 12 hours, and one with the STATCOM at maximum lagging current with the TSC deblocked. This test lasted 24 hours. All results were satisfactory.

COMMENTARY ON SOME UNEXPECTED TECHNICAL CHALLENGES

The STATCOM GLPS is a source of power for the link electronics and dc capacitors prior to closing the SVC circuit breaker. In particular, the GLPS is used to pre-charge the dc capacitors to 1500V. Once operating, the STATCOM's nominal power requirements are satisfied from the system via phase shift control and the demand from the GLPS should reduce to virtually zero. Testing showed however that this interaction did not occur "naturally" and, with the circuit breaker closed, the GLPS was left supplying considerable load whilst the STATCOM phase shift operated at a relatively low value. Additional control was designed, tested and implemented to servo the load on the GLPS to a low value once the circuit breaker is closed thereby forcing the STATCOM phase shift control to supply the majority of the load from the system.

Control and protection of the chain-link GTO convertors requires rapid and high frequency communications between the ground level controls and the links. The link switching data and databack information is exchanged with the ground level interface via fibre optic cables using an in-house designed communications protocol. Testing in the factory showed the communications to be reliable under a range of operating conditions. At site however very small timing differences between links and the high transmission rates conspired such that the target data error rate was not achieved and this occasionally led to mal-operation of links. The data transmission protocol and its protections were examined and tested in detail and a number of improvements have been implemented. Extensive testing and STATCOM operation since these improvements has identified no further problems.

SUMMARY OF COMMISSIONING RESULTS

- The operation and performance of the STATCOM has been proven over its full leading and lagging operating range.

- The response of the STATCOM has been optimised for normal current order variations and the transient changes associated with switching the TSC.
- Compliance with the specified requirements for control performance and speed of response has been demonstrated.
- Measurements have shown that the total and individual harmonic contributions from the STATCOM (and remainder of the SVC) are negligible.
- Automatic and remote control of the SVC has been proven.

CONCLUSIONS

- The results of the commissioning programme demonstrate that the STATCOM operates correctly and satisfies NGC's specified requirements.
- The STATCOM provides a rapid and stable response throughout its operating range.
- The STATCOM can achieve very low levels of harmonic distortion.
- This project has provided the first application of ALSTOM's STATCOM technology and its performance during commissioning has exceeded expectations.
- The now demonstrated capabilities of this equipment offer many exciting and valuable possibilities to other utilities worldwide.

ACKNOWLEDGEMENT

The authors would like to acknowledge the contribution and commitment of A Bates (ALSTOM) and I Hemsley (NGC), as well as many others within NGC and ALSTOM, without which the commissioning of this unique project would not have been possible.

REFERENCES

1. Jefferies D G: 'Developing the highway of power', IEE Power Engineering Journal, Vol 11, No. 1.

2. Ainsworth J D, Davies M, Fitz P J, Owen K E and Trainer D R, 1998, "Static Var Compensator (STATCOM) Based on Single-Phase Chain Circuit Converters", IEE Proc.-Gener. Transm. Distrib., Vol 145, No. 4, 381-386.

3. Knight R C, Young D J and Trainer D R, "Relocatable GTO-Based Static Var Compensator for NGC Substations", CIGRÉ 14-106, 1998 Session.

4. Woodhouse M L, Donoghue M W, Osborne M, "Type Testing of the GTO Valves for a Novel STATCOM Convertor", AC-DC Power Transmission, Nov. 2001.

Aspects of Series Controlled Flexible AC Transmission System Planning and Operation

K H Kuypers[1], R E Morrison[2] and S B Tennakoon[3]

AEA Technology, UK[1], Monash University, Australia[2] and Staffordshire University, UK[3]

Abstract: Planning and operational aspects of a static synchronous series compensator (SSSC) applied to increase the utilisation of a transmission system are discussed. It is shown that planning of flexible AC transmission systems (FACTS) requires detailed and comprehensive system modelling and analysis. A process structure to design and specify a series FACTS device illustrates the relationship between the major analysis activities and the data, which are derived from operational needs. In accordance with the design process, a metal oxide varistor (MOV) based SSSC protection scheme and a control strategy to regulate the current flow in a transmission line is introduced. Test results demonstrate the application of the proposed control strategy, which has been implemented on a digital signal processor system (DSP) to control an experimental model of a 24-pulse SSSC. Power quality aspects are also considered.

1. INTRODUCTION

Circuit breakers and tap changing transformers are the main means of control in a traditional transmission system [1], [2]. Operation of these mechanical devices may over-correct operating points or excite transient events. Both of these actions may cause temporary or transient over-voltages, over-currents or mechanical oscillations. Closing or opening transmission lines may redistribute power, cause local voltage depression and overload transmission lines of parallel power flow paths [3], [4], [5]. The slow response of the mechanical equipment might result in temporary operation above voltage and current rating and there may be concerns of instability or hunting [6]. Generally, due to this lack of dynamic control, traditional power systems may be over designed above the maximum required steady state rating and utilised below the peak transmission capacity to ensure reliable and secure operation [3], [4]. The introduction of dynamically and continuously controllable elements into power systems is expected to improve the utilisation of power systems, reduce component stress and improve reliability [3]. In addition to improvements to steady state, contingency and transient problems [4], devices able to control power flow dynamically, may be applied to realise power transport contracts such as contract path arrangements in a deregulated electricity market [7].

In early 1990 the concept of Flexible AC Transmission Systems (FACTS) was introduced to enhance dynamic control in power systems and to improve system utilisation [3].

This concept is based on the application of high power, power electronic shunt and series devices. FACTS shunt devices like the static var compensator (SVC) and the static shunt compensator (STATCOM) are widely used today to compensate reactive power and support and regulate area and load voltage levels. Series FACTS devices, such as thyristor switched series capacitors (TSSC), thyristor controlled series capacitors (TCSC) [7], [8], [9], [10] and static synchronous series compensators (SSSC) have a broad range of application to increase the load capacity of power systems. They can be deployed in long distance transmission systems to improve transient stability. The application of series FACTS controllers in meshed networks enables power flow control and redistribution of load power flow. However, compared to shunt FACTS controllers, few series controllers are in use [6], [11], [12]. The majority of installed series devices are deployed to improve stability and transfer capacity on long distance transmission systems [11], [12]. In the steady state their behaviour is similar to a series capacitor without the risk of resonance posed by capacitors. The particular dynamic characteristics also allow them to be used for damping load angle oscillations. The lack of dynamic and continuous power flow control with tap changing transformers, phase angle regulators and phase shifting transformers has been identified to reduce power transfer capacity severely in meshed networks due to loop flows, wheeling and the limitation of options during contingency operation [5].

This paper considers the application of a Static Synchronous Series Compensator (SSSC) in a meshed transmission system to enhance the transfer capacity of an existing system particularly during contingency operation. In contrast to traditional power systems, networks employing FACTS are expected to operate closer to the thermal or theoretical transfer capacity, provide additional control freedom and more flexible operation. As demonstrated in this work, FACTS planning is associated with comprehensive analyses and modelling processes [4], [10]. The importance of through faults, power quality and control strategy analyses for the device specification, design and operation is illustrated. It is shown that for the co-ordination of analyses and modelling tasks, a structured analysis and planning approach is essential to realise FACTS and to devise operating regimes which optimise the FACTS operation.

AC-DC Power Transmission, 28-30 November 2001
Conference Publication No. 485 © IEE 2001

A SSSC is a FACTS device able to supply and absorb reactive power. It consists of a high power, voltage source inverter (VSI) with the dc terminals connected to a capacitor and the ac terminals connected in series with the transmission line via an interface transformer. By changing the firing angle of the inverter the capacitor voltage can be increased or decreased to control the amplitude and direction of the SSSC voltage.

As a result of a fault analysis a metal oxide varistor (MOV) based SSSC protection scheme is introduced and discussed. The power quality requirements are also considered. Moreover criteria for the SSSC control requirements are outlined and the structure of a SSSC control system is proposed. Experimental test results are presented demonstrating the application and practical relevance of the power quality requirements and the proposed control system.

2. INVESTIGATIONS AND ANALYSES FOR FACTS PLANNING AND OPERATION

Investigations and analyses required to plan and to operate series controlled FACTS are discussed in relation to the transmission system given in fig. 1 as a case study. The three phase transmission lines (fig. 1) have been represented as π-sections with parameters given in table 1. Transmission lines 2 and 3 have a thermal current limit of 5.0 pu whilst transmission line 1 has a current limit of 8.0 pu. The load data are given in table 2. All pu values correspond to a base voltage of 275 kV and a base power of 100 MVA. The feeding transmission system has a fault level of 50 pu at nominal voltage. It is represented by a constant voltage of 1.085 pu connected to a reactance of 0.02 pu.

TABLE 1: Parameters of the 275 kV transmission lines

R' * Ω km^{-1}	L' mH km^{-1}	C' μF km^{-1}	l_1 km	l_2 km	l_3 km
0.04	1.0154	0.01162	60	24	38

* For transmission line 1 R' = 0.03 Ω km^{-1}

TABLE 2: Load data for the test system

	S / pu	PF
Load 1	5.96	0.98
Load 2	6.08	0.94
Load 3	1.90	0.93

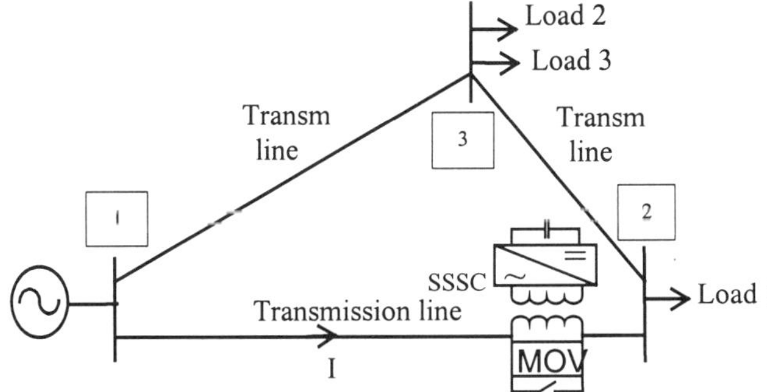

Fig. 1: Transmission system used to study planning aspects

Under normal conditions the system without the series controller feeds load 1 and load 3 only (table 3). In this case a load flow analysis reveals that the busbar voltages are close to 1 pu and line 3 operates at 84 % of its thermal current limit. A contingency situation arises, when load 2, which is normally supplied from a neighbouring transmission system is connected to busbar 2. Under the contingency condition the natural load balance is disturbed such that line 3 is over loaded and line 1 and line 2 are not fully utilised (table 3). In order to avoid the overloading during contingency operation the spare power transfer capacity of transmission line 1 is to be used. By connecting a SSSC in series to transmission line 1 the load can be diverted from transmission line 3 to transmission line 1. Table 3 summarises the load flow results for the contingency.

TABLE 3: Parameters for various system conditions

	Normal	Contingency in non-FACTS	Contingency in FACTS
Load 1	On	On	On
Load 2	Off	On	On
Load 3	On	On	On
I_{line1} / pu	3.6	4.99	8.0
I_{line2} / pu	2.3	0.56	2.53
I_{line3} / pu	4.2	7.62	5.0
V_1 / pu	1.032	0.966	0.977
V_2 / pu	0.998	0.911	0.933
V_3 / pu	1.005	0.909	0.928
V_{SSSC} / pu	0.0	0.0	0.160

The location of the series device within the selected power flow path has been determined on the basis of the contingency. Besides an auxiliary supply, road accessibility for maintenance and repair, criteria such as fault and power quality levels have to be considered before the position of the series device in transmission line 1 can be finalised.

Power quality levels depend on harmonic distortion, voltage flicker and voltage fluctuation introduced by the FACTS device. As the level of harmonic distortion depends on the harmonic resonance characteristic of the system under normal and contingency loading, it is necessary to analyse this characteristic. Voltage fluctuation and flicker could be caused by rapid control action of the SSSC. This type of distortion may not only deteriorate the quality of load voltages, but could also propagate to neighbouring transmission systems [10]. A thorough investigation of control strategy options within the frame of the SSSC control task helps to prevent the FACTS device posing a risk to voltage quality.

Often, positioning the series device at a particular location on the controlled transmission line can reduce the effect of harmonic distortion on the load and adjacent systems. However, a low fault level and the possibility of protecting the device effectively may outweigh the benefits of the reduction of harmonic distortion. Therefore the fault level is generally the dominant factor, which determines the SSSC location.

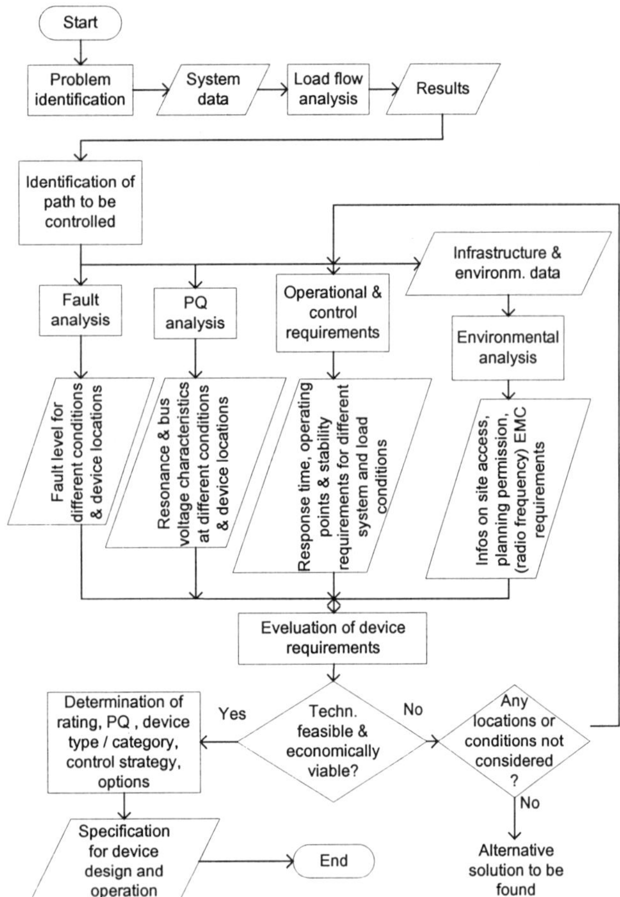

Fig. 2: Methodical overview of analyses and FACTS planning processes to specify the FACTS device

The harmonic distortion emitted by the series device may have to be mitigated by a more sophisticated device or harmonic filters.

Fig. 2 summarises the relationship between key-activities and information relevant for the specification of series FACTS controllers. The meaning and detailed background of the activities presented in the rectangular boxes are discussed in the following sections.

3. FAULT ANALYSIS, SSSC PROTECTION AND RATING

The maximum fault current at the feeding end of line 1 is 55 pu (11.55 kA); at the load end of transmission line 1 it is 16 pu (3.36 kA). Circuit breakers can not be deployed to protect the SSSC, because they are too slow to divert the fault current or disconnect the device. Compared with the feeding end of transmission line 1, the fault level at the load end is lower by a factor 3.4. Therefore it is expected that protection measures would be significantly cheaper when the device is located at the load end. One option to protect the series device against faults is to divert the fault current via a very fast responding crow bar. Based on the maximum fault current at the load end of transmission line 1, it is technically feasible to deploy a thyristor crow bar. However to be able to bypass the SSSC reliably at all

times, fail safe power thyristors and gate drive circuits are needed. Thus a thyristor crow bar has to be designed with sufficient redundancy and several backup systems. Therefore a fail safe crow bar may not be the most economical solution. Rather than pursuing a SSSC protection system, which relies on power electronic switches, it is proposed to investigate the use of a MOV surge arrester. MOV arresters have the advantage of being passive devices and their application is well established for the protection of series capacitors in transmission systems [13], [14].

A MOV arrester is connected across the transmission line side transformer winding, which connects the SSSC inverter with the transmission system (fig. 1). When a fault occurs and the instantaneous current limit of a GTO thyristor is exceeded, the GTO thyristors are blocked. Thus the transformer winding on the inverter side can be considered to be open circuit and the voltage across the transformer rises very steeply until the MOV limits the voltage. Referring to the MOV arrester current-voltage characteristic in fig. 3, it can be seen that when the voltage V_{SSSC} exceeds the MOV arrester protective voltage level the current is immediately diverted from the transformer.

In parallel with the MOV arrester a bypass circuit breaker is also connected (fig. 1). This bypass circuit breaker is needed, if a fault persists. Due to high fault currents, MOV arresters may exceed their energy absorption level in the case of a persistent fault. The circuit breaker across the MOV arresters is closed, when the energy absorption level is reached. The energy absorption level has to be rated such, that the MOV arresters withstand the maximum fault current for the time it takes to close the bypass breaker. A triggered air gap may be used to reduce the duty of the MOV arrester, if fault currents cause the MOV arrester to reach its energy absorption level before the bypass circuit breaker can be closed [13], [14], [15].

If the fault is cleared prior to the MOV arresters reaching their energy absorption level and the current falls below the maximum current limit of the GTO thyristors, the SSSC inverter resumes normal operation. As soon the SSSC starts operating, the voltage across the MOV arresters falls below its protection level and the transmission line current flows through the SSSC inverter (fig. 3).

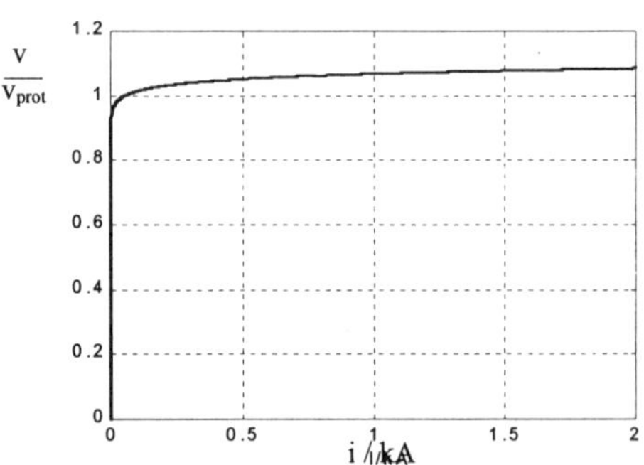

Fig. 3: Instantaneous voltage vs. instantaneous current of a MOV arrester

It must be noted that the protective voltage level v_{prot} should be specified to be as low as possible. Firstly a low v_{prot} enables the MOV to divert the current from the SSSC transformer as soon as possible after blocking the GTO thyristors. Secondly a low v_{prot} reduces the voltage across the GTO thyristors during the blocking state of the inverter and allows lower rated GTO thyristors. Unfortunately a low specified protective voltage level increases the active power losses of the MOV arresters during normal system operation [15]. In general, the protective voltage level of a MOV is specified at about 2.3 times the voltage peak across the MOV during normal system operation [13], [15]. This means for the MOV protecting the SSSC in fig. 1 with a maximum rms voltage of 0.16 pu (table 3) or 25.4 kV, that v_{prot} must be 83 kV. Simulation of a three phase to ground short circuit showed a fault current of 9.75 pu (2.05 kA) flowing through the MOV. Assuming a pessimistic bypass circuit breaker closing time of 450 ms, which includes a failed attempt to open the circuit breakers of transmission line 1 and the relay response times, the required energy absorption level of the MOVs is 2 pus or 67 MWs per phase.

Despite the advantage of not relying on auxiliary systems such as thyristor crowbars, the disadvantage of the MOV protection scheme is that the SSSC has to be rated for 2.3 times the normal operating voltage. However, as there is no current flow through the SSSC at v_{prot}, the power rating of the SSSC coupling and interface transformers as well as the dc capacitor has to correspond to the VAr required for the operating condition given in column 3 of table 3.

4. HARMONIC ANALYSIS AND HARMONIC PERFORMANCE SPECIFICATION

In order to specify the harmonic performance of the SSSC the resonance characteristic of the system was analysed. Fig. 4 represents the system harmonic admittance at the terminals of the series controller in fig. 1 versus frequency for normal operation and the contingency condition (c.f. table 3). The high admittance in the vicinity of the 12th harmonic is likely to cause unacceptable harmonic distortion, if the series device is realised with a 12 pulse voltage source inverter, which generates characteristic 11th and 13th order harmonic voltages (fig. 4). Since usually the harmonic components introduced by a power electronic inverter decay with frequency, the admittance peaks occurring at higher frequencies are generally associated with a lower harmonic distortion level.

Test and validation of the harmonic performance was carried out by implementation of the system given in fig. 1 on a hardware simulator. A DSP controlled 24-pulse SSSC (fig. 5) was built and integrated into the power system simulator. The measured harmonic voltage spectrum at the terminals is shown in fig. 6. How the harmonic voltage spectrum at busbar 2 changes under normal and contingency load is shown in fig. 7.

The correlation between the harmonic admittance at the SSSC terminals and the voltage spectrum for the normal and contingency load condition indicates that the system loading affects the resonance characteristic and consequently the harmonic distortion. Hence the harmonic analysis has to be carried out taking into account the load variation and changing network conditions.

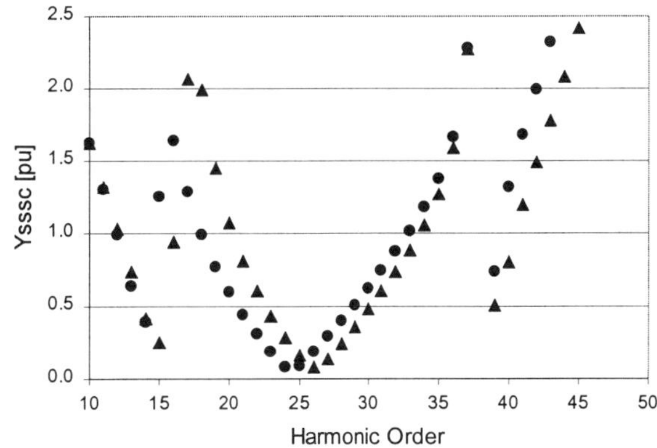

Fig. 4: Y(f) as seen from the series controller terminals for normal (•) and contingency (▲) operation

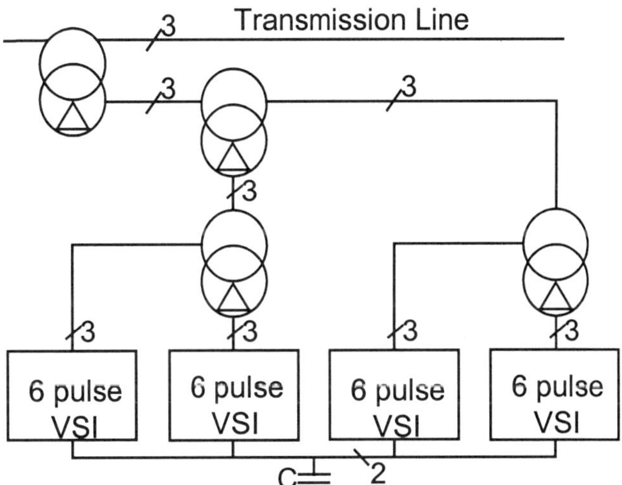

Fig. 5: 24 pulse SSSC selected based on the harmonic analysis

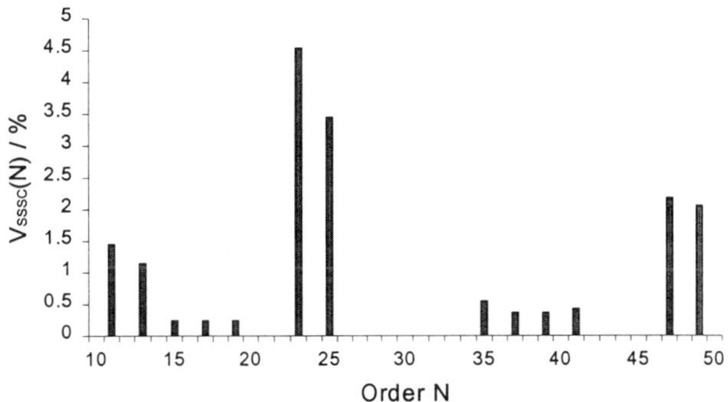

Fig. 6: Harmonic spectrum of the SSSC voltage as a percentage of the fundamental component

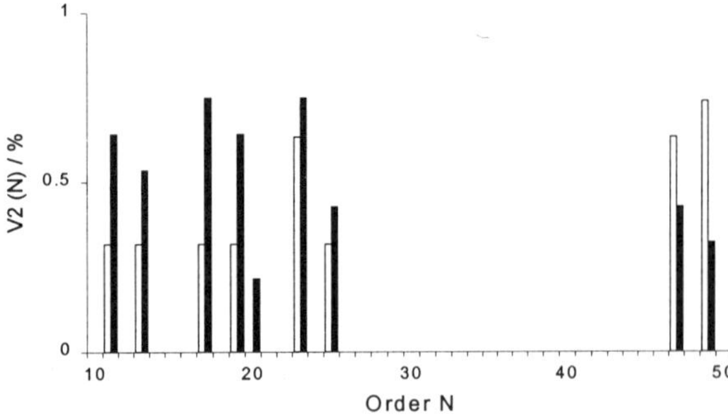

Fig. 7: Harmonic voltage spectr. (% of the fundamental freq.)
at bus bar 2, normal (bright) and contingency (dark)

5. OPERATIONAL AND VOLTAGE QUALITY REQUIREMENTS

Besides increasing the transfer capacity of the transmission system under contingency operation it is possible to use the SSSC for the enforcement of power flow contracts. Two control strategies denoted by X- and I-control were considered to investigate the effect of their control action on load voltage fluctuation and possible propagation of flicker towards feeding systems. With the X-control strategy the SSSC voltage is regulated to be proportional to I_{SSSC}. Hence the series controller acts as a capacitive reactance (Xc) with respect to its fundamental frequency voltage. The I-control strategy regulates the current of transmission line 1 to maintain a constant level equivalent to the rated current of transmission line 1.

Fig. 8 indicates how the control strategy can effect the voltage variation at the load and feeding busbars. It can be seen that the I-strategy results in less load voltage variation than the Xc-strategy. Also control of power flow is expected to be easier with a feedback controlled I-strategy than a feed forward implemented Xc-strategy. When the SSSC is controlled with Xc-strategy, the SSSC voltage-current-ratio would need to be re-calibrated each time a network parameter, e.g. load is changed. In large networks, this re-calibration may become unpractical, due to the long computation time to derive the V_{SSSC} to I_{SSSC} ratio. Due to the anticipated advantages and flexibility, the I-strategy is the preferred control mode and considered for further discussion of planning and operational aspects.

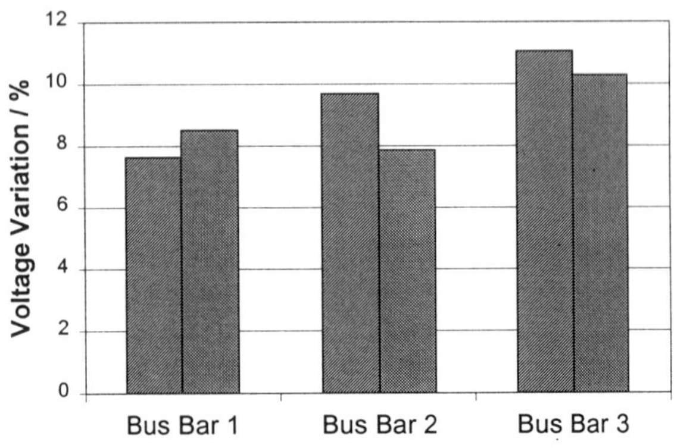

Fig. 8: Voltage Variation: Xc strategy (left column), I strategy
(right column)

The I-control strategy was implemented on a DSP board as digital control algorithm. With the SSSC employed to increase the power transfer capacity during contingency operation and to enforce power flow contracts under normal conditions, the following requirements must be met:

- The SSSC must operate such, that the power flowing across transmission line 1 follows a reference value with acceptable accuracy.
- The regulation must be tolerant to changes in the network configuration and load condition.
- The SSSC control must be stable and withstand disturbances.
- The response time to either a changing reference value, changing network configuration or changing load condition must be short, i.e. of similar duration as the response time of automatic voltage regulators and governor systems.

A control system was designed which consists of two cascaded control loops. The inner loop is used to control the inverter voltage; the outer loop controls the current of transmission line 1 (fig. 9).

Fig. 10 illustratess the performance of the control algorithm implimented. While connecting load 2 to the system, the controller was arranged to regulate the SSSC current to 8.0 pu, equivalent to the rating of transmission line 1.

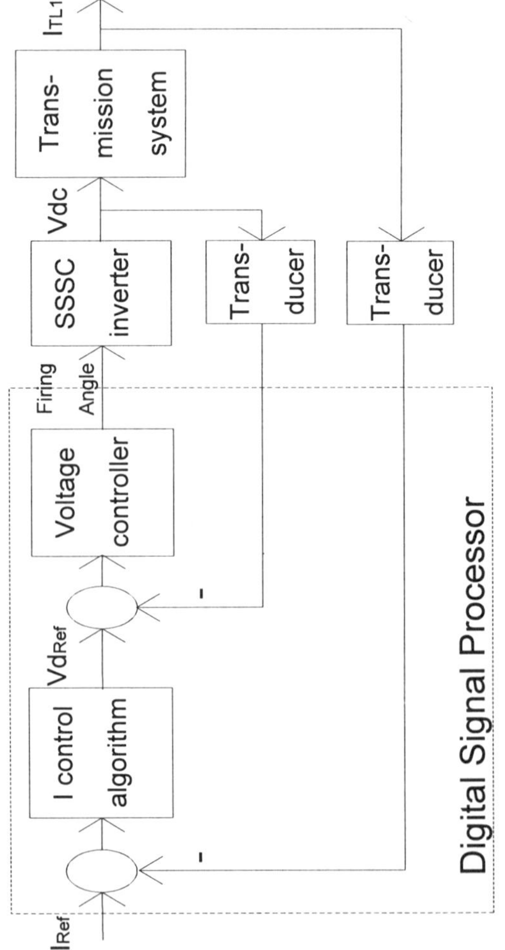

Fig. 9: Structure of the control system

Immediately after connecting load 2 the current in transmission line 1 rises suddenly, but the control system brings I_{Line1} back to the reference value of 8.0 pu within about 250 ms. Since the thermal current rating is 8.0 pu, an occasional occurrence of current pulses such as shown in fig. 10 do not compromise the operational safety of the regulated transmission line.

The test result shown in fig. 10 validates the feasibility of the I-control strategy. Bearing in mind, that the load increases by about 77 % on the change from normal to contingency operation fig. 10 also indicates that the algorithm is stable and tolerant to changes in network condition. Moreover the response time of 250 ms is in the order of fault clearing times and shorter than the reaction of automatic voltage regulators and governor system, which are usually larger than 0.5 s.

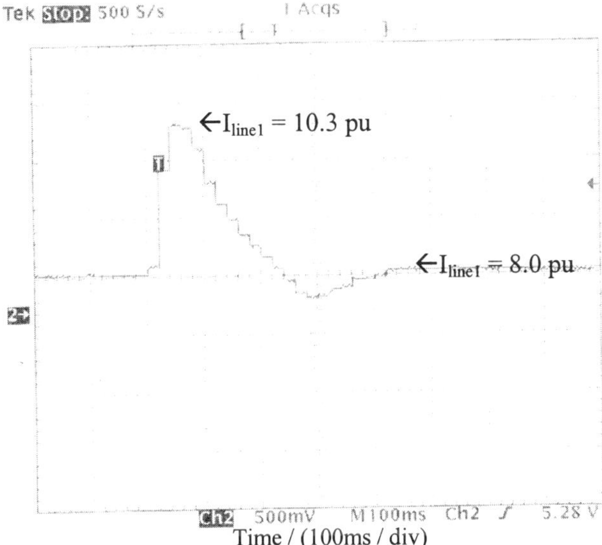

Fig. 10: RMS current of the SSSC following disturbances caused by the connection of load 2

6. CONCLUSION

The work presented illustrated that detailed fault, harmonic and dynamic control analyses have to be undertaken in order to plan FACTS, to specify the series device including protection and to develop operating and control regimes.

A passive MOV arrester based SSSC protection scheme has been proposed and the required rating for the MOV arresters and SSSC components has been discussed. It has been indicated that
- the positioning of the series FACTS device is tightly linked with the system short circuit levels,
- standard protection equipment like circuit breakers can not be applied to protect SSSC,
- the design of the protection system depends on the worst case fault condition.

By means of a case study it has been shown that the specification of the harmonic SSSC performance requires harmonic analyses, which take into account the variation of the resonance characteristics with changing load and network configurations.

It has been demonstrated that the SSSC control action can result in voltage variations at load and feeding busbars. Moreover the comparison of two different control strategies illustrated the importance of investigating alternative control strategies to achieve the SSSC control tasks and to provide acceptable power quality. For the system considered, tests results obtained by using a DSP controlled SSSC model indicate that the regulation of the transmission line can be one option to increase the transfer capacity in a meshed transmission system.

7. REFERENCES

1. Nelson R.J., 1994, "Transmission power flow control: Electronics vs. electromagnetic alternatives for steady state operation", IEEE Trans. Power Delivery, Vol. PWRD 9, No. 3, pp. 1678-1684

2. Gyugyi L., 1994, "Dynamic compensation of ac transmission lines by solid state synchronous voltage sources", IEEE Trans. Power Delivery, Vol. PWRD 9, No.2, pp.904-911

3. Hingorani N.G., 1991, "FACTS - flexible ac transmission systems", 5th Int. Conf. AC & DC Power Transm., London, UK, pp. 1-8

4. Lemay J., 1994, "A Planners view of FACTS: Meeting the needs of modern networks", Power Technology International, pp. 109-111

5. IEEE Committee Report, 1991, "Operating problems with parallel flows", IEEE Trans. Power Systems, Vol. 6, No. 3, pp. 1024-1033

6. Rahman M., Ahmed M., Nelson R.J., Bian J., 1997, "UPFC Application on the AEP system: Planning considerations", IEEE Trans. Power Systems, Vol. 12, No. 4, pp. 1695-1701

7. Weedy B.M., Cory B.J., 1998, "Electric power systems", John Wiley, 4th Edition, Chichester, New York, p. 515, p. 337

8. IEEE FACTS Terms & Definitions Task Force, 1997, "Proposed terms and definitions for flexible ac transmission system (FACTS)", IEEE Trans. Power Delivery, Vol. 12, No. 4, pp. 1848-1853

9. Mihalic R., 1998, "Power flow control with controllable reactive series elements", IEE Proc.-Gener. Transm. Distrib., Vol. 145, No. 5, pp. 493-498

10. Kuypers K.H., Morrison R.E., Tennakoon S.B., 2000, "Power quality issues associated with a series FACTS controller", Proc. 9th Int. Conf. Harm. & Power Quality, Orlando, Florida, pp. 176-181

11. Keri A.J.F., Chamia M., 1992, "Improving transmission system performance using controlled series capacitors", CIGRE Paper 14/37/38-07, Paris

12. Urbanek J., Piwko R.J., Larsen E.V., 1993, "Thyristor controlled series compensation prototype installation at the Slatt 500 kV substation", IEEE Trans. Power Delivery, Vol. 8, No. 3, pp. 1460-1469

13. Lee G.E., Goldsworthy D.L., 1996, 'BPA's Pacific AC intertie series capacitors: Experience, equipment & protection", IEEE Trans. Power Delivery, Vol. 11, No. 1

14. Hamann J.R., Mieske S.A., Johnson I.B., Conts A.L., 1981, "A zinc oxide varistor protective system for series capacitors", IEEE Trans. Power App. & Systems, Vol. PAS 100, No. 3, pp. 929-936

15. Pilvelait B., Ortmeyer T.H., Maratukulam D., 1993, "Advanced series compensation for transmission systems using a switched capacitor module", IEEE Trans. Power Delivery, Vol. PWRD 8, No. 2, pp. 584-596

TRANSIENT STABILITY AUGMENTATION BY PROGRAMMED POWER ANGLE RELATIONSHIP USING UNIFIED POWER FLOW CONTROLLER

K R Padiyar and S Krishna
Indian Institute of Science, India

ABSTRACT

Improvement in transient stability can be achieved by adequate system design and discrete supplementary controllers. The emerging Flexible AC Transmission System (FACTS) controllers are considered to be suitable for this purpose due to their speed and flexibility. The Unified Power Flow Controller (UPFC) is a voltage source converter based FACTS controller which injects a series voltage and a shunt current. In this paper, a control strategy is developed to achieve maximal improvement in transient stability using UPFC. It is shown that for a single machine infinite bus system, maximal improvement in transient stability can be achieved by maximizing the electrical power output of the generator w.r.t. the control variables. This result can be extended to multimachine systems and maximizing power flow on a critical line can improve transient stability. The control strategy is evaluated by a simulation study on the 10 generator 39 bus New England system.

INTRODUCTION

There are several discrete supplementary controllers [1,2] which can be initiated following a large disturbance. A comprehensive review of angle stability controls is presented in [3]. Braking resistors and switched series capacitors were among the earliest controllers used to enhance transient stability by changing the network parameters. In recent years, Flexible AC Transmission System (FACTS) controllers are considered to be viable solution to the problem of transient stability, due to their speed and flexibility.

FACTS controllers based on voltage source converters use turn off devices like Gate Turn-Off Thyristors (GTO) [4]. The magnitude and angle of the fundamental frequency voltage injected by the converter is varied by controlling the switching instants of the GTO devices. These type of FACTS controllers have the advantages of reduced equipment size and improved performance compared to variable impedance type controllers. Unified Power Flow Controller (UPFC) is a voltage source converter based FACTS controller which injects a series voltage and a shunt current. The series and shunt branches can generate/absorb reactive power independently and the two branches can exchange real power; therefore UPFC has three degrees of freedom.

UPFC is a versatile controller which can perform the functions of Static Synchronous Compensator (STATCOM) and Static Synchronous Series Compensator (SSSC). Padiyar and Kulkarni [5] propose a control strategy for UPFC to control real power flow through the line, while regulating magnitudes of the voltages at its two ports. Padiyar and Uma Rao [6] present a control scheme for the series injected voltage of the UPFC to damp power oscillations and improve transient stability.

Mihalic *et al* [7] propose maximization of power using UPFC for improvement in transient stability of a single machine infinite bus (SMIB) system. Bian *et al* [8] propose a control strategy to increase power transfer between two large systems during a contingency, using UPFC, while considering the operational constraints.

Padiyar and Uma Rao [9] devised a discrete control strategy for a Thyristor Controlled Series Compensator for transient stability improvement, using the concept of potential energy in a line. They have shown that under certain assumptions it is possible to express the potential energy of a system represented by classical model, as sum of energies in the lines belonging to a cutset. This result is applicable even if the generators are represented by detailed (1.1) model [10]. It is further shown in [10] that the system kinetic energy can be expressed as a function of the rate of change of phase angle across a line belonging to the cutset.

In this paper, a control strategy is derived for UPFC for maximal improvement in transient stability. The control strategy is based on the idea of maximizing energy margin which is a measure of transient stability. The control strategy is derived for a SMIB system. The extension to the multimachine system is based on the energy function given in [10].

CONTROL STRATEGY

The SMIB system with UPFC shown in fig. 1 is used to derive the control strategy. The generator is represented by the classical model. The series and shunt branches of the UPFC are represented by voltage and current sources $(V \angle \phi$ and $I \angle \psi)$ respectively.

The transient energy function for the SMIB system is defined as

AC-DC Power Transmission, 28-30 November 2001
Conference Publication No. 485 © IEE 2001

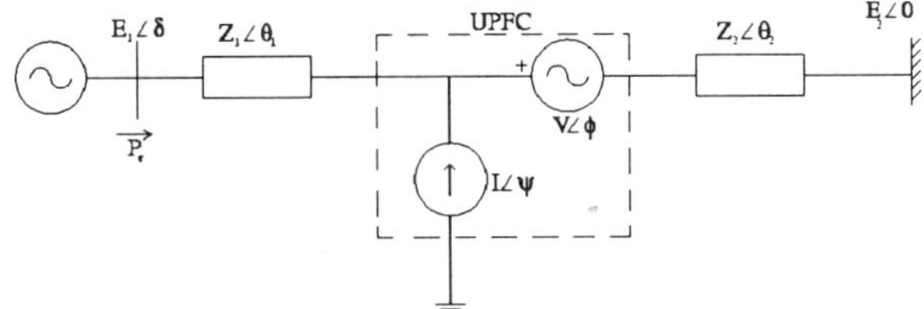

Fig. 1 SMIB system

$$W = W_1 + W_2 \tag{1}$$

W_1 is the kinetic energy and W_2 is the potential energy given by

$$W_1 = \frac{1}{2} M \omega^2 \tag{2}$$

$$W_2 = \int_{\delta_o}^{\delta} (P_e - P_m)\,\mathrm{d}\delta \tag{3}$$

where

δ : rotor angle

ω : rotor speed

M : inertia constant

P_e : electrical power output of generator

P_m : mechanical power input to generator

δ_o : initial steady state rotor angle

P_e is a function of δ and the control variables V, I, ϕ and ψ. P_e is given by

$$P_e = cV\cos\phi + dV\sin\phi + fI\cos\psi + gI\sin\psi + l \tag{4}$$

The real power constraint on the UPFC is given by

$$mV^2 + nI^2 + oV\cos\phi + pV\sin\phi + qI\cos\psi + rI\sin\psi + sVI\sin(\psi - \phi) = 0 \tag{5}$$

The expressions for the coefficients c to s are given in the appendix. The following constraints are imposed on the ratings of the series and shunt converters.

$$0 < V < V_{max} \tag{6}$$

$$0 < I < I_{max} \tag{7}$$

The energy margin W_{em} given by the difference between the critical energy and the energy at the instant of fault clearing, is a quantitative measure of transient stability. The critical energy is the energy at the controlling unstable equilibrium point (UEP) (δ_u,0).

$$W_{em} = \int_{\delta_{cl}}^{\delta_u}(P_e - P_m)\,\mathrm{d}\delta - \frac{1}{2}M\omega_{cl}^2 \tag{8}$$

where the subscript cl indicates quantities at the instant of fault clearing. Maximal improvement in transient stability can be achieved by maximizing the energy margin. The second term on the RHS of (8) is independent of control. The rotor angle at the controlling UEP δ_u depends on control. By applying Pontryagin's principle [11], maximization of the functional on the RHS of (8) requires maximizing the Hamiltonian **H** defined as

$$\mathbf{H} = P_e - P_m \tag{9}$$

Since P_m is a constant, maximization of energy margin implies maximization of P_e w.r.t. control variables

subject to the constraints (5), (6) and (7). Since P_e does not depend on any derivative of the control variables w.r.t. δ, the problem is reduced to maximizing a function for a given value of δ.

EXTENSION TO MULTIMACHINE SYSTEM

When a power system becomes unstable, it initially splits into two groups. There is usually a unique cutset consisting of series elements (connecting the two areas) across which the angle becomes unbounded. The system can be represented by two areas connected by the critical cutset as shown in fig. 2. The two areas can be assumed to be coherent in order to neglect the oscillations within the areas and account only for interarea oscillations which contribute to system separation. Locating UPFC in one of the lines belonging to the critical cutset strengthens the system and improves transient stability. By the assumption of coherent areas, the system kinetic and potential energies are given by (the derivation is given in [10])

$$W_1 = \frac{1}{2} M_{eq} \left(\frac{\mathrm{d}\delta_k}{\mathrm{d}t}\right)^2 \tag{10}$$

$$W_2 = T_k \int_{\delta_{ko}}^{\delta_k} (P_k - P_{ks})\,\mathrm{d}\delta_k \tag{11}$$

where M_{eq} is the equivalent inertia constant, δ_k is the angle across any line k in the critical cutset, T_k is a constant, P_k is power flow in line k, δ_{ko} is the initial value of δ_k and P_{ks} is the steady state value of P_k.

$$M_{eq} = \frac{M_I M_{II}}{M_I + M_{II}}, \quad M_I = \sum_{i \in area I} M_i, \quad M_{II} = \sum_{i \in area II} M_i$$

The energy function given by (10) and (11) is also applicable for the detailed (1.1) model of the generators; a lossless network is assumed in the derivation of the energy function.

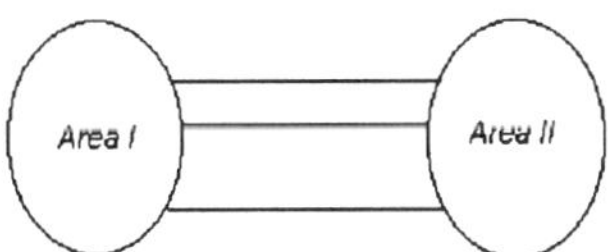

Fig. 2 Coherent areas

The energy margin is given by

$$W_{em} = T_k \int_{\delta_{kcl}}^{\delta_{ku}}(P_k - P_{ks})\,\mathrm{d}\delta_k - \frac{1}{2}M_{eq}\left(\frac{\mathrm{d}\delta_{kcl}}{\mathrm{d}t}\right)^2 \tag{12}$$

where δ_{ku} is the angle across the line at which $P_k = P_{ks}$. The second term on the RHS of (12) is independent of control. The angle δ_{ku} depends on control. The expression for the energy margin is similar to that for the SMIB system given by (8). The control strategy derived for the SMIB system is extended to multimachine systems and the power flow P_k on a critical line is maximized to improve transient stability.

If a UPFC is placed in one of the lines belonging to the critical cutset, maximizing energy margin is equivalent to programming the power angle relationship in the transmission line in which UPFC is situated. The system external to the line in which UPFC is situated, is represented by a Thevenin equivalent network on either side of the line. Fig. 1 is also the equivalent circuit for the multimachine case, with the only difference being that the power flow P_k considered is the power at the input port of UPFC.

The power flow P_k in the line in which UPFC is situated, is given by

$$P_k = AV^2 + BI^2 + CV\cos\phi + DV\sin\phi + FI\cos\psi + GI\sin\psi$$
$$+ HVI\cos(\psi - \phi) + KVI\sin(\psi - \phi) + L \qquad (13)$$

The expressions for the coefficients A to L are given in the appendix. P_k is maximized subject to the constraints (5), (6) and (7).

CASE STUDIES

SMIB System

Case studies are conducted for the SMIB system to study the variation of control variables w.r.t. δ and the effect of UPFC location on the performance. The following values are assumed for the SMIB system: $E_1=E_2=1$, $R_1=R_2=0$, $X_1=X_2=0.5$, $V_{max}=0.5$, $I_{max}=0.5$, where $R_1+jX_1=Z_1\angle\theta_1$ and $R_2+jX_2=Z_2\angle\theta_2$.

If $R_1=R_2=0$, then $A=B=H=K=m=n=s=0$. For this case, it can be shown that the equations giving the necessary conditions for maximum power can be simplified to quartic equations in a single variable. Therefore the control variables for global maximum of power can be obtained. The control variables ϕ and ψ for global maximum of power are plotted in figs. 3 and 4. The voltage and current magnitudes for global maximum are the limits V_{max} and I_{max} respectively for all values of δ. The effective series impedance $R_{se}+jX_{se}$ and shunt admittance $G_{sh}+jB_{sh}$ of the UPFC are plotted in figs. 5 and 6.

The power-angle curves with and without UPFC are shown in fig. 7. The effect of location of UPFC on transient stability can be quantified by computing the area below the power-angle curve $\int_0^\pi P_e \mathrm{d}\delta$ for different locations. Fig. 8 gives a plot of area below the power-angle curve w.r.t. X_1 where $X_1+X_2=1$. It can be seen that there is no significant variation in the area below the power-angle w.r.t. the location of UPFC in a line. For the values of the system parameters chosen, the optimal location of the UPFC is at the midpoint of the line. The area below the power-angle curve without UPFC is 2.

The power through the DC link (power transferred from the series branch to the shunt branch) for maximum value of power P_e is 0.125 for all values of δ. If the power through the DC link is constrained to be zero, the series and the shunt branches inject reactive voltage and reactive current respectively. The plot of the injected reactive voltage and reactive current are shown in figs. 9 and 10 respectively. The variation of current is continuous; but the voltage magnitude jumps from 0.5 to 0.191 at $\delta=155°$. Fig. 11 shows the plot of critical energy w.r.t. the steady state power for different cases. The last case is for DC link power at zero and V and I constrained to be at their limits. It can be seen that not constraining V and I to be at their limits will be helpful at lower values of steady state power.

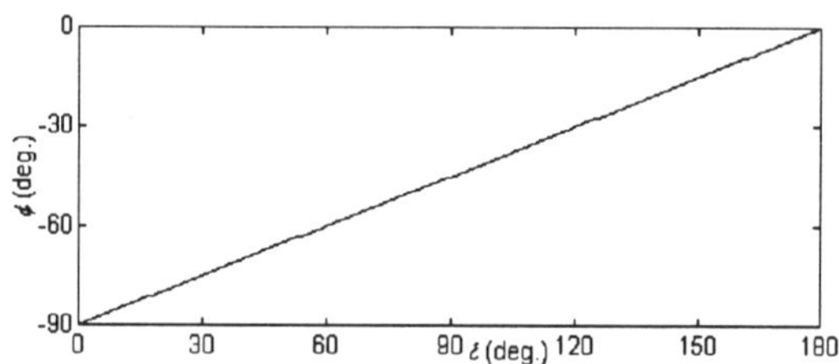

Fig. 3 Phase angle of series voltage injected by UPFC

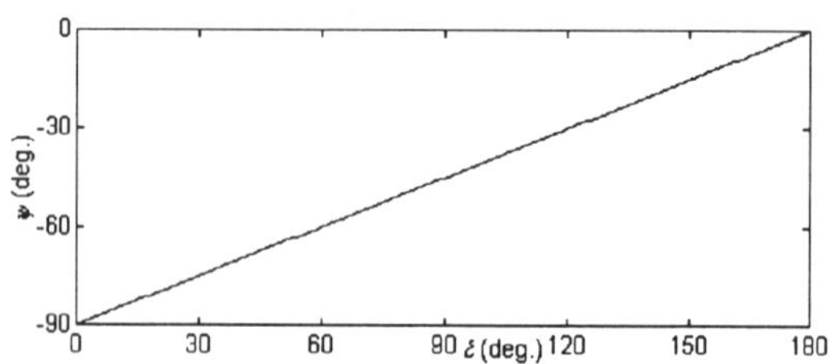

Fig. 4 Phase angle of shunt current injected by UPFC

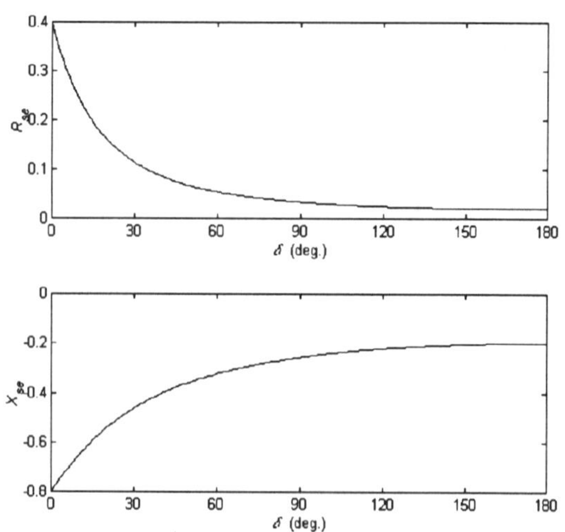

Fig. 5 Effective series resistance and reactance of UPFC

Multimachine System

The New England system (the system data are given in [2]) is considered for the multimachine study. The generators are represented by detailed (1.1) model with excitation system. Loads are modelled as constant impedances. Network losses are ignored.

Fig. 12 shows the swing curves for a fault at #14 cleared by opening the line 14-34 at 0.346 s. The system is critically unstable; generator 2 separates from the rest of the system. It can be seen from fig. 13 that the angle across the lines 11-12 and 18-19 become unbounded. A UPFC is located in the line 11-12 at bus #11. The power flow in the line 11-12 is maximized in the post-fault period using UPFC. A rating of 0.5 pu is used for the series voltage and shunt current of the UPFC. With UPFC, the system is stable for the same fault as shown by the swing curves in fig. 14 and the critical clearing time increases from 0.345-0.346 s to 0.356-0.357 s. Fig. 15 gives the plot of the power transferred from the series branch to the shunt branch of the UPFC; the discontinuity in the plot is due to the discontinuities in the magnitude and angle of the shunt

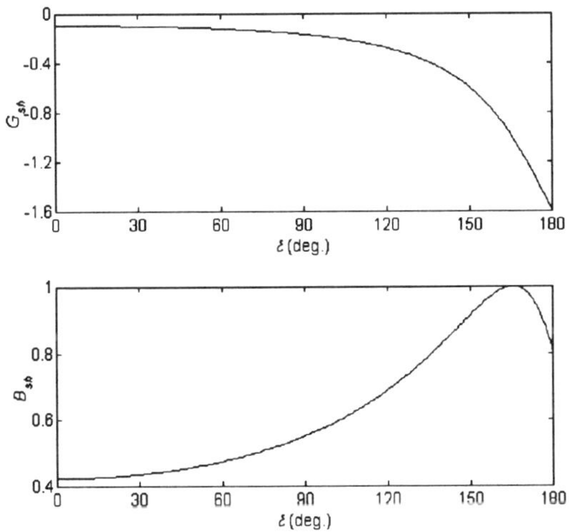

Fig. 6 Effective shunt conductance and susceptance of UPFC

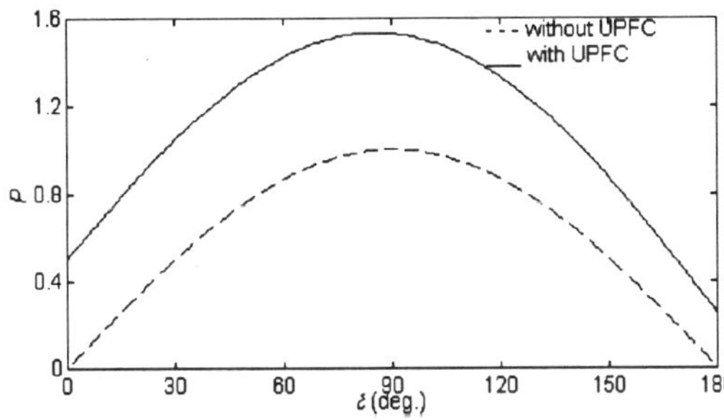

Fig. 7 Power-angle curve

current injected by the UPFC. The power transfer on the line with and without UPFC are plotted in fig. 16.

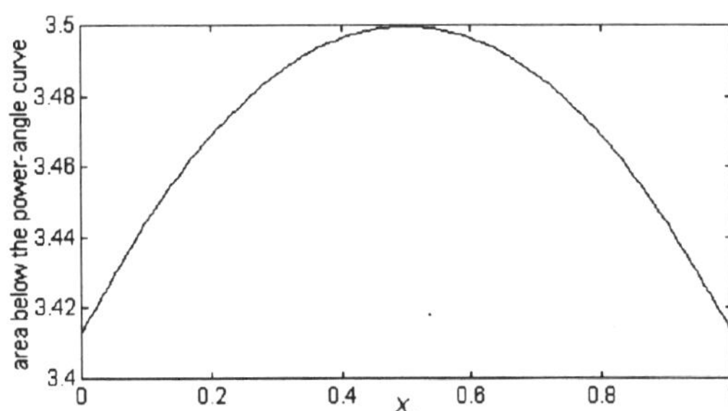

Fig. 8 Variation of area below the power-angle curve w.r.t. location of UPFC

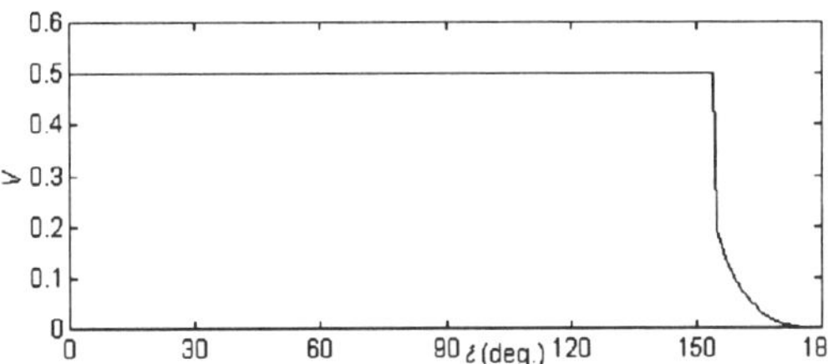

Fig. 9 Magnitude of the series voltage injected by UPFC (DC power=0)

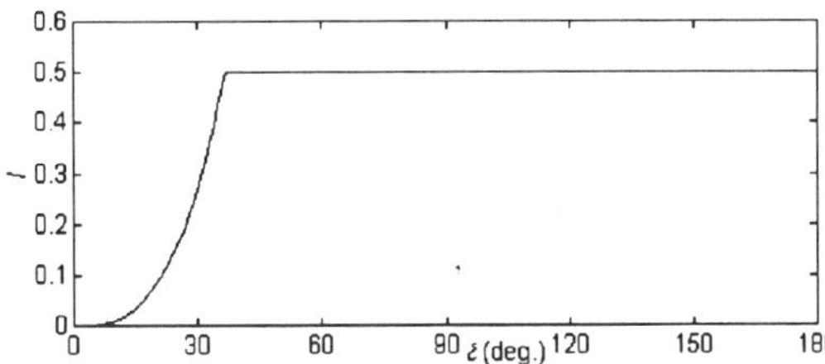

Fig. 10 Magnitude of the shunt current injected by UPFC (DC power=0)

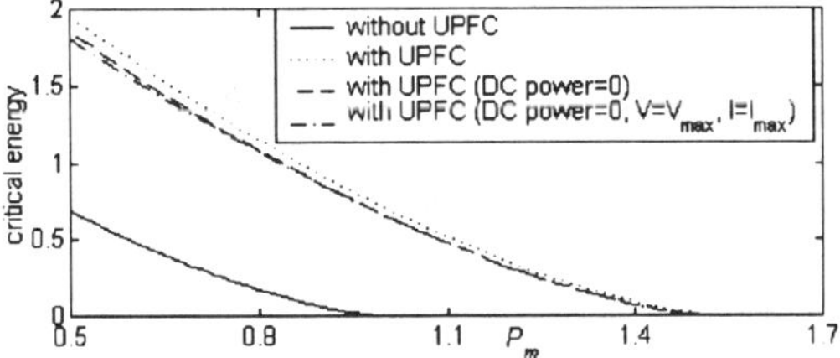

Fig. 11 Variation of critical energy w.r.t. steady state power for different cases

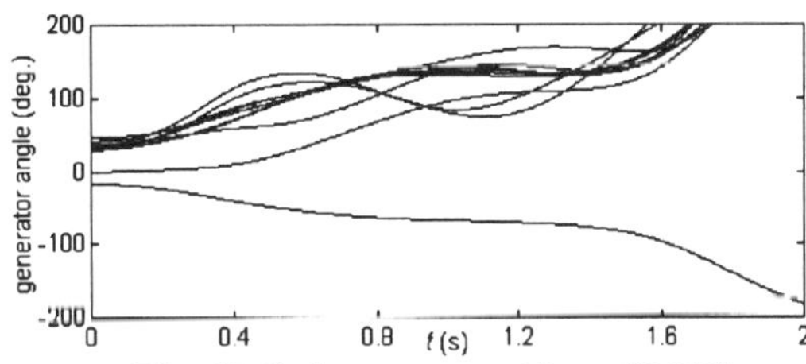

Fig. 12 Swing curves without UPFC

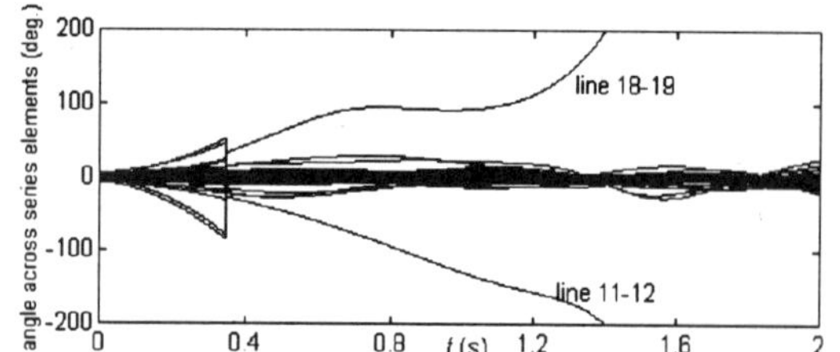

Fig. 13 Angle across series elements for unstable case

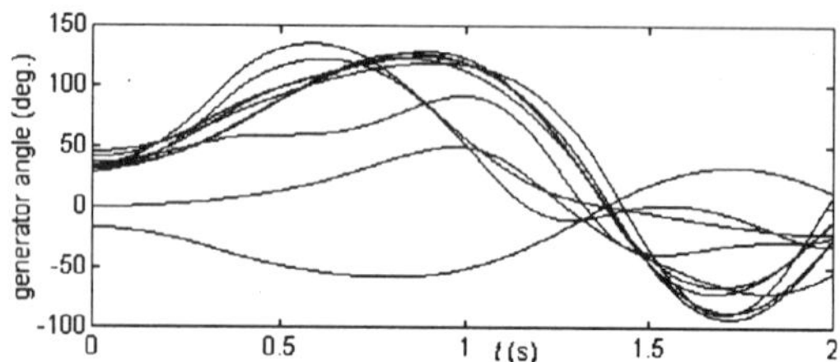

Fig. 14 Swing curves with UPFC

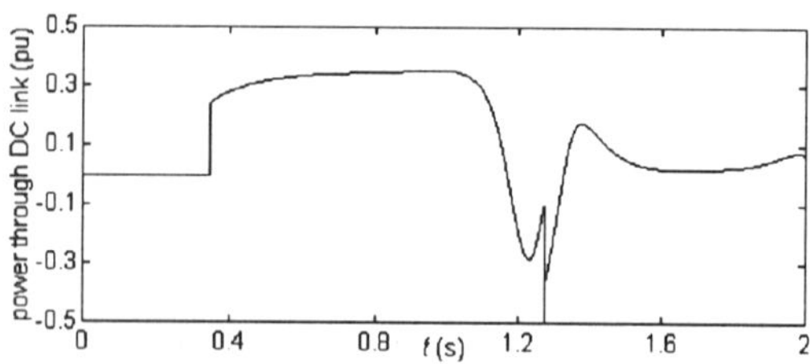

Fig. 15 Power through the DC link

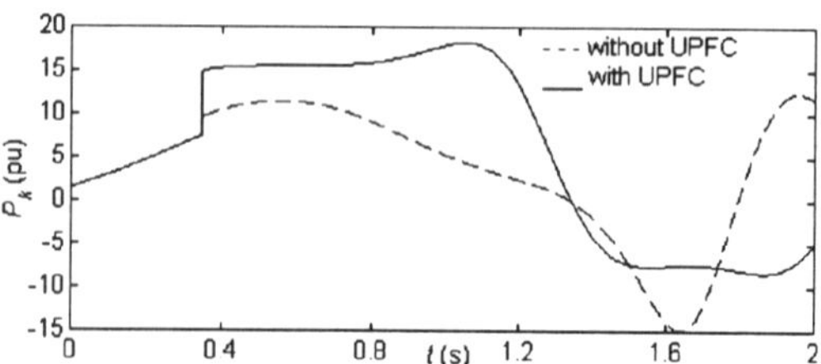

Fig. 16 Power transfer on the line

DISCUSSION

The control variables are chosen as the magnitude and angle of the injected series voltage and shunt current. The constraints on the ratings of the series and shunt converters can be easily expressed due to the selection of these control variables. This is in contrast to using the effective shunt susceptance B_{sh} as a control variable for the shunt branch in [7]. Moreover unlike in [7], B_{sh} need not be constant (at the limit) at all values of δ as shown in fig. 6.

When the power through the DC link is constrained to be zero, the optimal magnitudes of the injected voltage and current are not at the limits for large and small values of δ respectively. This illustrates that for small values of δ a SSSC will be helpful in increasing the power transfer, whereas for large values of δ a STATCOM will be more effective.

The control variables resulting in global maximum of power can be easily obtained for a lossless ($R_1=R_2=0$) circuit. For the multimachine system, though the network is lossless, R_1 and R_2 are not zero due to the real power loads. The solution obtained for the lossless circuit is used as starting values for the iterative solution for the circuit with losses.

It was found from different case studies on the New England system that for many contingencies, generator 2 separates from the rest of the system because of its large inertia constant. For this mode of instability, the angle across the lines 11-12 and 18-19 become unbounded. Locating UPFC in one of these lines strengthens the system and aids in improving transient stability.

CONCLUSION

A control strategy is derived for maximal improvement in transient stability by maximizing the energy margin. The control strategy is to maximize the power transfer from one area to the other, and is applicable to any controller which can affect the power flow in a line. The control scheme is applied to a UPFC considering the constraints on the ratings of the converters. Suitable location of the UPFC for transient stability improvement can be one of the lines across which the angle becomes unbounded in case of instability; these lines can be identified by simulation studies.

APPENDIX

$$c = -\frac{E_1}{Z}\cos(\delta+\theta)$$

$$d = -\frac{E_1}{Z}\sin(\delta+\theta)$$

$$f = -\frac{E_1 Z_2}{Z}\cos(\delta-\theta_2+\theta)$$

$$g = -\frac{E_1 Z_2}{Z}\sin(\delta-\theta_2+\theta)$$

$$l = \frac{E_1^2}{Z}\cos\theta - \frac{E_1 E_2}{Z}\cos(\delta+\theta)$$

$$m = \frac{1}{Z_2}\cos\theta_2 - \frac{Z_1}{ZZ_2}\cos(\theta+\theta_2-\theta_1)$$

$$n = \frac{Z_1 Z_2}{Z}\cos(\theta_1+\theta_2-\theta)$$

$$o = -\frac{E_1}{Z_2}\cos(\theta_2-\delta) + \frac{E_1 Z_1}{ZZ_2}\cos(\theta+\theta_2-\delta-\theta_1)$$
$$- \frac{E_2 Z_1}{ZZ_2}\cos(\theta-\theta_1+\theta_2) + \frac{E_2}{Z_2}\cos\theta_2$$

$$p = \frac{E_1}{Z_2}\sin(\theta_2 - \delta) - \frac{E_1 Z_1}{Z Z_2}\sin(\theta + \theta_2 - \delta - \theta_1)$$
$$+ \frac{E_2 Z_1}{Z Z_2}\sin(\theta - \theta_1 + \theta_2) - \frac{E_2}{Z_2}\sin\theta_2$$

$$q = E_1\cos\delta - \frac{E_1 Z_1}{Z}\cos(\delta + \theta_1 - \theta) + \frac{E_2 Z_1}{Z}\cos(\theta_1 - \theta)$$

$$r = E_1\sin\delta - \frac{E_1 Z_1}{Z}\sin(\delta + \theta_1 - \theta) + \frac{E_2 Z_1}{Z}\sin(\theta_1 - \theta)$$

$$s = \frac{2Z_1}{Z}\sin(\theta_1 - \theta)$$

$$A = -\frac{Z_1}{Z^2}\cos\theta_1$$

$$B = -\frac{Z_1 Z_2^2}{Z^2}\cos\theta_1$$

$$C = -\frac{E_1}{Z}\cos(\delta + \theta) + \frac{E_1 Z_1}{Z^2}\cos(\delta + \theta_1) + \frac{E_1 Z_1}{Z^2}\cos(\theta_1 - \delta)$$
$$- \frac{2E_2 Z_1}{Z^2}\cos\theta_1$$

$$D = -\frac{E_1}{Z}\sin(\delta + \theta) + \frac{E_1 Z_1}{Z^2}\sin(\delta + \theta_1) - \frac{E_1 Z_1}{Z^2}\sin(\theta_1 - \delta)$$

$$F = -\frac{E_1 Z_2}{Z}\cos(\delta - \theta_2 + \theta) + \frac{E_1 Z_1 Z_2}{Z^2}\cos(\delta + \theta_1 - \theta_2)$$
$$- \frac{E_2 Z_1 Z_2}{Z^2}\cos(\theta_1 - \theta_2) + \frac{E_1 Z_1 Z_2}{Z^2}\cos(\theta_1 + \theta_2 - \delta)$$
$$- \frac{E_2 Z_1 Z_2}{Z^2}\cos(\theta_1 + \theta_2)$$

$$G = -\frac{E_1 Z_2}{Z}\sin(\delta - \theta_2 + \theta) + \frac{E_1 Z_1 Z_2}{Z^2}\sin(\delta + \theta_1 - \theta_2)$$
$$- \frac{E_2 Z_1 Z_2}{Z^2}\sin(\theta_1 - \theta_2) - \frac{E_1 Z_1 Z_2}{Z^2}\sin(\theta_1 + \theta_2 - \delta)$$
$$+ \frac{E_2 Z_1 Z_2}{Z^2}\sin(\theta_1 + \theta_2)$$

$$H = -\frac{Z_1 Z_2}{Z^2}\cos(\theta_1 - \theta_2) - \frac{Z_1 Z_2}{Z^2}\cos(\theta_1 + \theta_2)$$

$$K = -\frac{Z_1 Z_2}{Z^2}\sin(\theta_1 - \theta_2) + \frac{Z_1 Z_2}{Z^2}\sin(\theta_1 + \theta_2)$$

$$L = \frac{E_1^2}{Z}\cos\theta - \frac{E_1 E_2}{Z}\cos(\delta + \theta) - \frac{E_1^2 Z_1}{Z^2}\cos\theta_1$$
$$+ \frac{E_1 E_2 Z_1}{Z^2}\cos(\delta + \theta_1) + \frac{E_1 E_2 Z_1}{Z^2}\cos(\theta_1 - \delta)$$
$$- \frac{E_2^2 Z_1}{Z^2}\cos\theta_1$$

where $Z\angle\theta = Z_1\angle\theta_1 + Z_2\angle\theta_2$.

ACKNOWLEDGEMENT

The financial support received from the Dept. of Science and Technology, Govt. of India under the project titled "Dynamic security assessment and control of power grids" is gratefully acknowledged.

REFERENCES

1. Task Force on Discrete Supplementary Controls of the Dynamic System Performance Working Group of the IEEE Power System Engineering Committee, 1978, "A Description of Discrete Supplementary Controls for Stability", IEEE Tran. Power Apparatus and Systems, PAS-97, 149-165.

2. K.R. Padiyar, 1996, "Power System Dynamics: Stability and Control", John Wiley and Interline.

3. CIGRE Task Force 38.02.17, 2000, "Advanced Angle Stability Controls", Technical Brochure, Ref. no. 155.

4. Narain G. Hingorani and Laszlo Gyugyi, 2000, "Understanding FACTS: Concepts and Technology of Flexible AC Transmission Systems", IEEE Press, New York.

5. K.R. Padiyar and A.M. Kulkarni, 1998, "Control Design and Simulation of Unified Power Flow Controller", IEEE Tran. Power Delivery, 13, 1348-1354.

6. K.R. Padiyar and K. Uma Rao, 1999, "Modeling and Control of Unified Power Flow Controller for Transient Stability", Int. J. Electrical Power and Energy Systems, 21, 1-11.

7. R. Mihalic, P. Zunko and D. Povh, 1996, "Improvement of Transient Stability Using Unified Power Flow Controller", IEEE Tran. Power Delivery, 11, 485-492.

8. J. Bian, D.G. Ramey, R.J. Nelson and A. Edris, 1997, "A Study of Equipment Sizes and Constraints for a Unified Power Flow Controller", IEEE Tran. Power Delivery, 12, 1385-1391.

9. K.R. Padiyar and K. Uma Rao, 1997, "Discrete Control of Series Compensation for Stability Improvement in Power Systems", Int. J. Electrical Power and Energy Systems, 19, 311-319.

10. K.R. Padiyar and S. Krishna, 2001, "On-Line Detection of Loss of Synchronism Using Locally Measurable Quantities", Preprint.

11. Ian McCausland, 1969, "Introduction to Optimal Control", John Wiley & Sons.

OPTIMAL POWER FLOW CONTROL BY CONVERTER BASED FACTS CONTROLLERS

X-P Zhang
University of Warwick, UK

E J Handschin
University of Dortmund, Germany

INTRODUCTION

In recent years, energy, environment, deregulation of power utilities have delayed the construction of both generation facilities and new transmission lines. These problems have necessitated a change in the traditional concepts and practices of power systems. There are emerging technologies available, which can help electric companies to deal with above problems. One of such technologies is FACTS (Flexible AC Transmission System), which is a family of power electronic products including Static Synchronous Compensator (STATCOM) [1], Static Synchronous Series Compensator (SSSC) [2], and Unified Power Flow Controller (UPFC) [3], etc. These are called converter based FACTS controllers.

Better utilisation of existing power system capacities by installing new power electronic controllers such as Flexible AC Transmission System (FACTS) has become imperative. FACTS devices are able to change, in a fast and effective way, the network parameters in order to achieve a better system performance. FACTS devices, such as phase shifter, shunt or series compensation and the most recent developed converter-based power electronic controllers, make it possible to control circuit impedance, voltage angle and power flow for optimal operation of power systems, facilitate the development of competitive electric energy markets, and stipulate the unbundling the power generation from transmission and mandate open access to transmission services, etc.

However, in contrast to the practical applications of the STATCOM, SSSC and UPFC in power systems, very few publications have been focused on the mathematical modelling of these converter based FACTS controllers in optimal power flow analysis. In [17], a UPFC model has been proposed, and the model has been implemented in a Successive QP. In [18], mathematical models for TCSC, IPC and UPFC have been established, and the OPF problem with these FACTS controllers has been solved by Newton s method. Modelling of STATCOM and SSSC in OPF has not been reported yet.

In this paper, general mathematical models for the converter based FACTS controllers such as STATCOM, SSSC, and UPFC suitable for optimal power flow study are established. There are several solution methods for OPF available [6-16], which include interior point methods. In this paper the optimal power flow problem with these converter based FACTS

controllers is solved by the newly developed Nonlinear Interior Point Methods [15, 16] since they have been the most successful OPF methods.

MATHEMATICAL MODELS OF CONVERTER BASED FACTS CONTROLLERS

Static Synchronous Compensators (STATCOM)

The STATCOM like its conventional counterpart, the SVC (Static Var Compensator), is used to control transmission voltage by reactive power shunt compensation.

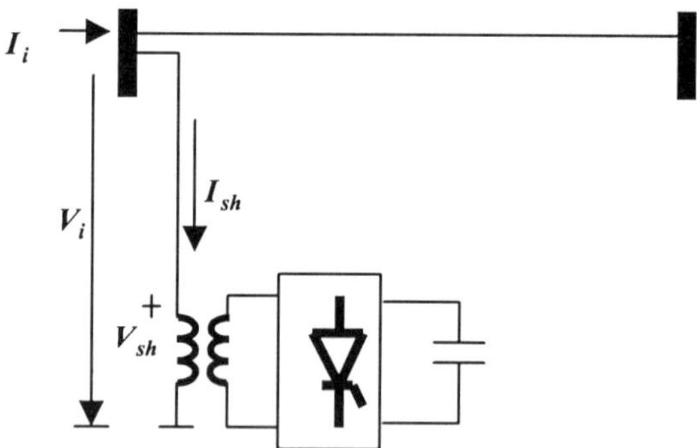

Fig.1 STATCOM operation principle

Typically, a STATCOM consists of a coupling transformer, an inverter and a DC capacitor, which is shown in Figure 1. For such an arrangement, in ideal steady state analysis, it can be assumed that active power exchange between AC system and the STATCOM can be neglected, and only reactive power can be exchanged between them.

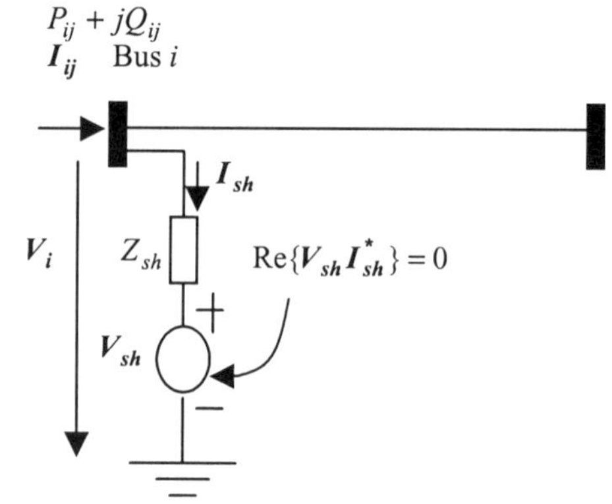

Fig.2 STATCOM equivalent circuit

AC-DC Power Transmission, 28-30 November 2001
Conference Publication No. 485 © IEE 2001

However, if a storage device is connected with the STATCOM, it can then exchange active power with the AC system.

Based on the operating principle shown in Fig. 1, the equivalent circuit of the STATCOM can be derived, which is given in Figure 2. In principle, the STATCOM output voltage can be regulated such that the reactive power of the STATCOM can be changed.

According to the equivalent circuit of the STATCOM shown in Fig.2, suppose $V_{sh} = V_{sh} \angle \theta_{sh}$, $V_i = V_i \angle \theta_i$, then the power flow constraints of the STATCOM are:

$$P_{ij} = V_i^2 g_{sh} - V_i V_{sh} (g_{sh} \cos(\theta_i - \theta_{sh}) + b_{sh} \sin(\theta_i - \theta_{sh})) \tag{1}$$

$$Q_{ij} = -V_i^2 b_{sh} - V_i V_{sh} (g_{sh} \sin(\theta_i - \theta_{sh}) - b_{sh} \cos(\theta_i - \theta_{sh})) \tag{2}$$

where $g_{sh} + j b_{sh} = 1/Z_{sh}$

Operating constraint of the STATCOM (active power exchange via the DC link) is:

$$PE = \mathrm{Re}(V_{sh} I_{sh}^*) = 0 \quad \text{or}$$

$$V_{sh}^2 g_{sh} - V_i V_{sh} (g_{sh} \cos(\theta_i - \theta_{sh}) - b_{sh} \sin(\theta_i - \theta_{sh})) = 0 \tag{3}$$

The bus voltage control, and active and reactive power flow control constraints are as follows,

$$V_i - V_i^{Specified} = 0 \tag{4}$$

where $V_i^{Specified}$ is the specified bus voltage.

The equivalent voltage injection $V_{sh} \angle \theta_{sh}$ bound constraints:

$$V_{sh}^{\min} \le V_{sh} \le V_{sh}^{\max} \tag{5}$$

$$\theta_{sh}^{\min} \le \theta_{sh} \le \theta_{sh}^{\max} \tag{6}$$

Static Synchronous Series Compensators (SSSC)

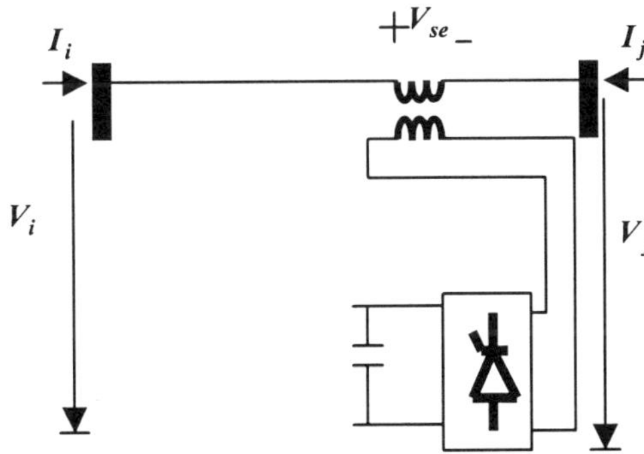

Fig.3 SSSC operation principle

Similarly, a SSSC usually consists of a coupling transformer, an inverter and a capacitor. However, the

SSSC is series connected with a transmission line through the coupling transformer. Operation principle of a SSSC is shown in Figure 3. In principle, the inserted series voltage V_{se} can be regulated to change the impedance (more precisely reactance) of the transmission line. Therefore the power flow of transmission line can be controlled.

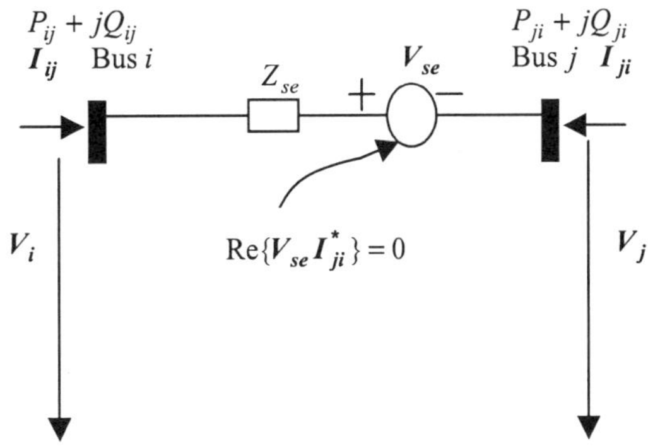

Fig.4 SSSC equivalent circuit

An equivalent circuit of the SSSC as shown in Fig.4 can be derived based on the operation principle of the SSSC. According to the equivalent circuit, suppose $V_{se} = V_{se} \angle \theta_{se}$, $V_i = V_i \angle \theta_i$ $V_j = V_j \angle \theta_j$, then the power flow constraints of the SSSC are:

$$P_{ij} = V_i^2 g_{ii} - V_i V_j (g_{ij} \cos \theta_{ij} + b_{ij} \sin \theta_{ij}) - V_i V_{se} (g_{ij} \cos(\theta_i - \theta_{se}) + b_{ij} \sin(\theta_i - \theta_{se})) \tag{7}$$

$$Q_{ij} = -V_i^2 b_{ii} - V_i V_j (g_{ij} \sin \theta_{ij} - b_{ij} \cos \theta_{ij}) - V_i V_{se} (g_{ij} \sin(\theta_i - \theta_{se}) - b_{ij} \cos(\theta_i - \theta_{se})) \tag{8}$$

$$P_{ji} = V_j^2 g_{jj} - V_i V_j (g_{ij} \cos \theta_{ji} + b_{ij} \sin \theta_{ji}) + V_j V_{se} (g_{ij} \cos(\theta_j - \theta_{se}) + b_{ij} \sin(\theta_j - \theta_{se})) \tag{9}$$

$$Q_{ji} = -V_j^2 b_{jj} - V_i V_j (g_{ij} \sin \theta_{ji} - b_{ij} \cos \theta_{ji}) + V_j V_{se} (g_{ij} \sin(\theta_j - \theta_{se}) - b_{ij} \cos(\theta_j - \theta_{se})) \tag{10}$$

where, $g_{ij} + j b_{ij} = 1/Z_{se}$, $g_{ii} = g_{ij}$, $b_{ii} = b_{ij}$, $g_{jj} = g_{ij}$, $b_{jj} = b_{ij}$

Operating constraint of the SSSC (active power exchange via the DC link) is:

$$PE = \mathrm{Re}(V_{se} I_{ji}^*) = 0 \quad \text{or}$$

$$-V_i V_{se} (g_{ij} \cos(\theta_i - \theta_{se}) - b_{ij} \sin(\theta_i - \theta_{se})) + V_j V_{se} (g_{ij} \cos(\theta_j - \theta_{se}) - b_{ij} \sin(\theta_j - \theta_{se})) = 0 \tag{11}$$

The active power flow control constraint is as follows,

$$P_{ji} - P_{ji}^{Specified} = 0 \tag{12}$$

where $P_{ji}^{Specified}$ is specified active power flow.

The equivalent voltage injection $V_{sh} \angle \theta_{sh}$ bound constraints:

$$V_{se}^{\min} \le V_{se} \le V_{se}^{\max} \tag{13}$$

$$\theta_{se}^{\min} \le \theta_{se} \le \theta_{se}^{\max} \qquad (14)$$

Unified Power Flow Controllers (UPFC)

The basic operation principle diagram is shown in Fig.5, which was described in [3,4,5]. The UPFC consists of two switching converters based on gate-turn-off (GTO) valves. The two inverters are connected by a common DC link. The series inverter is coupled to the transmission line via a series transformer. The shunt inverter is coupled to the transmission line via a shunt-connected transformer. The shunt inverter can generate or absorb controllable reactive power, and it can provide active power exchange to the series inverter to satisfy operating control requirements. In the steady state operation, the main objective of UPFC is to control voltage and power flow.

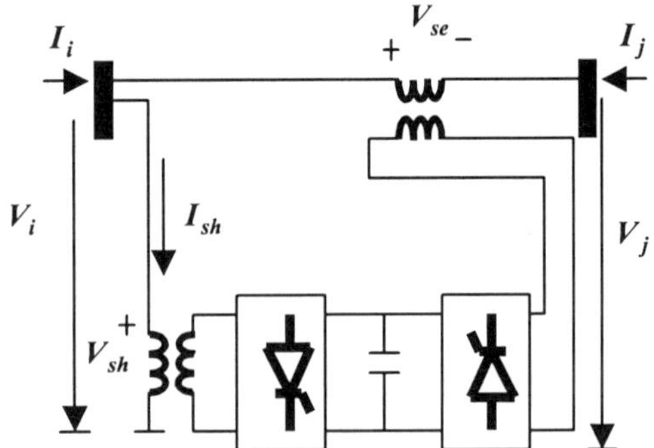

Fig. 5 UPFC operation principle

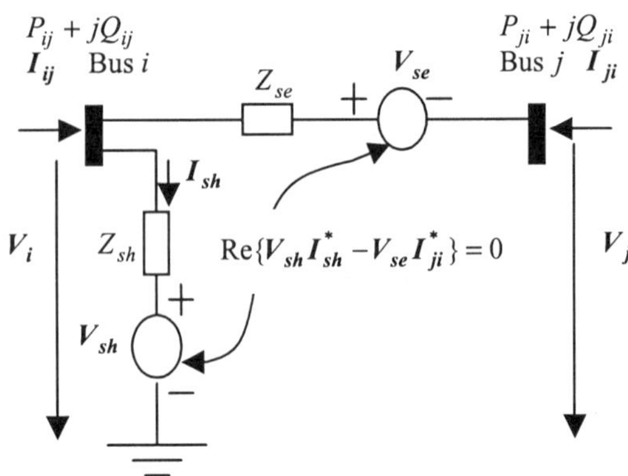

Fig.6 UPFC equivalent circuit

The equivalent circuit of UPFC is shown in Fig. 6. According to the equivalent circuit, suppose $V_{se} = V_{se}\angle\theta_{se}$, $V_{sh} = V_{sh}\angle\theta_{sh}$, $V_i = V_i\angle\theta_i$ $V_j = V_j\angle\theta_j$, then the power flow constraints of the UPFC are:

$$
\begin{aligned}
P_{ij} = {}& V_i^2 g_{ii} - V_i V_j (g_{ij}\cos\theta_{ij} + b_{ij}\sin\theta_{ij}) \\
& - V_i V_{se}(g_{ij}\cos(\theta_i - \theta_{se}) + b_{ij}\sin(\theta_i - \theta_{se})) \\
& - V_i V_{sh}(g_{sh}\cos(\theta_i - \theta_{sh}) + b_{sh}\sin(\theta_i - \theta_{sh}))
\end{aligned} \qquad (15)
$$

$$
\begin{aligned}
Q_{ij} = {}& -V_i^2 b_{ii} - V_i V_j (g_{ij}\sin\theta_{ij} - b_{ij}\cos\theta_{ij}) \\
& - V_i V_{se}(g_{ij}\sin(\theta_i - \theta_{se}) - b_{ij}\cos(\theta_i - \theta_{se})) \\
& - V_i V_{sh}(g_{sh}\sin(\theta_i - \theta_{sh}) - b_{sh}\cos(\theta_i - \theta_{sh}))
\end{aligned} \qquad (16)
$$

$$
\begin{aligned}
P_{ji} = {}& V_j^2 g_{jj} - V_i V_j (g_{ij}\cos\theta_{ji} + b_{ij}\sin\theta_{ji}) \\
& + V_j V_{se}(g_{ij}\cos(\theta_j - \theta_{se}) + b_{ij}\sin(\theta_j - \theta_{se}))
\end{aligned} \qquad (17)
$$

$$
\begin{aligned}
Q_{ji} = {}& -V_j^2 b_{jj} - V_i V_j (g_{ij}\sin\theta_{ji} - b_{ij}\cos\theta_{ji}) \\
& + V_j V_{se}(g_{ij}\sin(\theta_j - \theta_{se}) - b_{ij}\cos(\theta_j - \theta_{se}))
\end{aligned} \qquad (18)
$$

where

$$g_{sh} + jb_{sh} = 1/Z_{sh}, \; g_{ij} + jb_{ij} = 1/Z_{se}, \; g_{ii} = g_{ij} + g_{sh},$$
$$b_{ii} = b_{ij} + b_{sh}, \quad g_{jj} = g_{ij}, \quad b_{jj} = b_{ij}$$

Operating constraint of the UPFC (active power exchange between two inverters via the DC link) is:

$$PE = \mathrm{Re}(V_{sh}I_{sh}^{*} - V_{se}I_{ji}^{*}) = 0 \quad \text{or}$$

$$
\begin{aligned}
PE = {}& V_{sh}^2 g_{sh} - V_i V_{sh}(g_{sh}\cos(\theta_i - \theta_{sh}) \\
& - b_{sh}\sin(\theta_i - \theta_{sh})) + V_{se}^2 g_{ij} \\
& - V_i V_{se}(g_{ij}\cos(\theta_i - \theta_{se}) - b_{ij}\sin(\theta_i - \theta_{se})) \\
& + V_j V_{se}(g_{ij}\cos(\theta_j - \theta_{se}) - b_{ij}\sin(\theta_j - \theta_{se})) \\
= {}& 0
\end{aligned} \qquad (19)
$$

The bus voltage control, and active and reactive power flow control constraints are as follows,

$$V_i - V_i^{Specified} = 0 \qquad (20)$$

$$P_{ji} - P_{ji}^{Specified} = 0 \qquad (21)$$

$$Q_{ji} - Q_{ji}^{Specified} = 0 \qquad (22)$$

where $V_i^{Specified}$ is the specified bus voltage.

$P_{ji}^{Specified}$, $Q_{ji}^{Specified}$ are specified active and reactive power flow, respectively.

The basic operating control mode is given by (20)-(22). However, implementation of a comprehensive UPFC model with various advanced control modes can be found in [19].

The two equivalent voltage injection $V_{sh}\angle\theta_{sh}$, $V_{se}\angle\theta_{se}$ bound constraints:

$$V_{sh}^{\min} \le V_{sh} \le V_{sh}^{\max} \qquad (23)$$

$$\theta_{sh}^{\min} \le \theta_{sh} \le \theta_{sh}^{\max} \qquad (24)$$

$$V_{se}^{\min} \le V_{se} \le V_{se}^{\max} \qquad (25)$$

$$\theta_{se}^{\min} \le \theta_{se} \le \theta_{se}^{\max} \qquad (26)$$

NONLINEAR INTERIOR POINT OPF WITH CONVERTER BASED FACTS CONTROLLERS

Mathematical Formulation of the OPF

Mathematically, as an example, an OPF for minimization of the total operating cost can be formulated as follows,

Objective: $Min \ f(x) = \sum_i^{Ng}(\alpha_i * Pg_i^2 + \beta_i * Pg_i + \gamma_i)$ (27)

Subject to the following constraints:

$$g(x) = 0 \qquad (28)$$

$$h^{min} \le h(x) \le h^{max} \qquad (29)$$

where $x = [\theta sh, Vsh, \theta se, Vse, V, \theta, T, Pg, Qg]^T$

α_i, β_i, γ_i - coefficients of quadratic production cost functions of generator.

$g(x)$ - equality constraints including bus power flow equations, operating and control constraints of FACTS controllers given by (1)-(4), (7)-(12), ((15)-(22).

$h(x)$ - inequality-constraints including line flow constraints, simple inequality constraints of variables such as voltage-magnitudes, generator active power, generator reactive power, transformer tap ratio and bound constraints of the FACTS variables in (5), (6), (13), (14), (23)-(26).

θsh - angle of shunt voltage source of the FACTS

Vsh - magnitude of shunt voltage source of the FACTS

θse - angle of series voltage source of the FACTS

Vse - magnitude of series voltage source of the FACTS

θ - bus angle

V - bus voltage magnitude

T - tap ratio vector of transformer

Pg - bus active generation

Qg - bus reactive generation

The nonlinear OPF problem given in (27)-(29) can be solved by the Nonlinear Interior Point Methods [15,16], which include three important achievements in optimisation: Fiacco & McCormick s barrier method for optimisation with inequalities, Lagrange s method for optimisation with equalities and Newton s method for solving nonlinear equations.

By applying Fiacco & McCormick s barrier method, we transform the OPF problem (27)-(29) into the following equivalent OPF problem,

Objective: $Min \left\{ f(x) - \mu \sum_i^{Nh} \ln(sl_i) - \mu \sum_i^{Nh} \ln(su_i) \right\}$ (30)

Subject to the following constraints:

$$g(x) = 0 \qquad (31)$$

$$h(x) - sl - h^{min} = 0 \qquad (32)$$

$$h(x) + su - h^{max} = 0 \qquad (33)$$

where $\mu > 0$.

The lagrangian function for equalities optimisation for problem (30) is

$$L = f(x) - \mu \sum \ln(sl) - \mu \sum \ln(su) - \lambda^T g(x)$$
$$- \pi l^T (h(x) - sl - h^{min}) - \pi u^T (h(x) + su - h^{max}) \qquad (34)$$

where $\lambda, \pi l, \pi u$ are lagrangian multipliers for constraints (31), (32), (33), respectively.

The Karush-Kuhn-Tucker (KKT) first order conditions for the Lagrangian function of (34) are,

$$\nabla_x L_\mu = \nabla f(x) - \nabla g(x)^T \lambda - \nabla h(x)^T \pi l - \nabla h^T \pi u = 0 \ (35)$$

$$\nabla_\lambda L_\mu = \quad -g(x) \quad = 0 \qquad (36)$$

$$\nabla_{\pi l} L_\mu = \quad -(h(x) - sl - h^{min}) \quad = 0 \qquad (37)$$

$$\nabla_{\pi u} L_\mu = \quad -(h(x) + su - h^{max}) \quad = 0 \qquad (38)$$

$$\nabla_{sl} L_\mu = \quad \mu e + Sl * \Pi l \quad = 0 \qquad (39)$$

$$\nabla_{su} L_\mu = \quad \mu e - Su * \Pi u \quad = 0 \qquad (40)$$

where $Sl = diag(sl_j)$, $Su = diag(su_j)$, $\Pi l = diag(\pi l_j)$, $\Pi u = diag(\pi u_j)$.

The Newton equation for the nonlinear interior point OPF algorithm derived above may be expressed as the following compact form,

$$\begin{bmatrix} -\Pi l^{-1} Sl & 0 & -\nabla h & 0 \\ 0 & \Pi u^{-1} Su & -\nabla h & 0 \\ -\nabla h^T & -\nabla h^T & H & -J^T \\ 0 & 0 & -J & 0 \end{bmatrix} \begin{bmatrix} \Delta \pi l \\ \Delta \pi u \\ \Delta x \\ \Delta \lambda \end{bmatrix}$$
$$= \begin{bmatrix} -\nabla_{\pi l} L_\mu + \Pi l^{-1} \nabla_{Sl} L_\mu \\ -\nabla_{\pi u} L_\mu + \Pi u^{-1} \nabla_{Su} L_\mu \\ -\nabla_x L_\mu \\ g(x) \end{bmatrix} \qquad (41)$$

$$\Delta sl = \Pi l^{-1}(-\nabla_{sl} L_\mu - Sl \Delta \pi l) \qquad (42)$$

$$\Delta su = \Pi u^{-1}(\nabla_{su} L_\mu - Su \Delta \pi u) \qquad (43)$$

where $H(x, \lambda, \pi l, \pi u)$

$$= \nabla^2 f(x) - \lambda \nabla^2 g(x) - (\pi l + \pi u) \nabla^2 h(x) \ , J(x) = \frac{\partial g(x)}{\partial x}$$

By solving the Newton equation of (41)-(43), $\Delta \pi l$, $\Delta \pi u$, Δx, $\Delta \lambda$, Δsl, Δsu can be obtained. Then the Newton solution can be updated as follows,

$$sl = sl + \sigma \alpha_p \Delta sl \qquad (44)$$

$$su = su + \sigma \alpha_p \Delta su \qquad (45)$$

$$x = x + \sigma \alpha_p \Delta x \qquad (46)$$

$$\pi l = \pi l + \sigma \alpha_d \Delta \pi l \qquad (47)$$

$$\pi u = \pi u + \sigma \alpha_d \Delta \pi u \qquad (48)$$

$$\lambda = \lambda + \sigma \alpha_d \Delta \lambda \qquad (49)$$

where $\sigma = 0.995 \sim 0.99995$. α_p, α_d are primal and dual step length respectively. They can be determined by

$$\alpha_p = \min\{\min(\frac{sl}{-\Delta sl}), \ \min(\frac{su}{-\Delta su}), \ 1.0\} \qquad (50)$$

$$\alpha_d = \min\{\min(\frac{-\pi l}{\Delta \pi l}), \ \min(\frac{\pi u}{-\Delta \pi u}), \ 1.0\} \qquad (51)$$

for those $\Delta sl < 0$, $\Delta su < 0$, $\Delta \pi l > 0$, $\Delta \pi u < 0$

The complementary gap for the nonlinear interior point OPF is,

$$Cgap = su^T \pi u - sl^T \pi l \tag{52}$$

The barrier parameter can be determined by,

$$\mu = \frac{\beta * Cgap}{2 * n} \tag{53}$$

where β =0.01 ~ 0.2, n is the number of inequality constraints.

Solution Procedure of the OPF

The solution procedure for the nonlinear interior point OPF with FACTS is summarized as follows:

Step 0: Set iteration count $K = 0$, $\mu = \mu 0$, and initialise the OPF solution.

Step 1: If KKT conditions are satisfied & complementary gap is less than a tolerance, output results. Otherwise go to step 2.

Step 2: Form and solve Newton equation in (41), then (42) and (43).

Step 3: Update Newton solution by equation (44)-(49).

Step 4: Compute complementary gap by (52).

Step 5: Determine barrier parameter by (53).

Step 6: $K=K+1$, go to step 1.

NUMERICAL EXAMPLES

Test Systems

Test cases in this paper are carried out on IEEE 30 bus system and IEEE118 bus system. The IEEE 30 bus system has 6 generators, 4 OLTC transformers and 37 transmission lines. The IEEE 118 bus system has 18 controllable active power generation, 54 reactive power generation, 9 OLTC transformers, 177 transmission lines Thermal constraints of transmission lines are included. For all cases in this paper, the convergence tolerances are 5.0e-4 for complementary gap and 1.0e-4 (0.01MW/MVar) for maximal absolute bus power mismatch, respectively.

Optimal Power Flow Control by Converter based FACTS Controllers

In order to show the power flow control capability of the nonlinear interior point OPF algorithm proposed, four cases based on the IEEE 30 bus system are carried out.

Case 1: This is a base case without FACTS.

Case 2: This is similar to the case 1 except that there are three STATCOMs installed for control of voltages of bus 3, 12, and 25. The control settings of these bus voltages are 1.0 p.u.

Case 3: This is similar to the case 1 except that there are two STATCOMs installed for control of voltage of bus 3 and 12, and there is a SSSC between bus 25 and 27 for control active power flow of that transmission line.

Case 4: This is similar to the case 1 except that there is a STATCOM installed at bus 12 for control of voltage, there is a SSSC installed on transmission line 25-27 for control of active power flow, and there is a UPFC installed for control of voltage of bus 3 and active and reactive power flow of line 3-4.

The test results of the above cases are summarized in Table 1.

Table 1 Test results of the IEEE 30 bus system

	Case 1	Case 2	Case 3	Case 4
The Number of FACTS	-	3	3	3
The Total Number of Control Objectives by FACTS	-	3	3	5
The Total Number of Controllable Active and Reactive Power Flow by FACTS	-	- -	1 P	2 P 1 Q
The Total Number of Controllable Bus Voltage by FACTS	-	3	2	2
The Number of Iterations	12	10	14	13

Note: P, Q represent active & reactive power flow

Case studies are further carried out on the IEEE 118 bus system. Three cases are presented as follows,

Case 5: This is a base case IEEE 118 bus system without FACTS.

Case 6: This is similar to the case 1 except that there are three STATCOMs installed for control of voltages of bus 21, 45, and 94. The control settings of these bus voltages are 1.0 p.u.

Case 7: This is similar to the case 1 except that there is a STATCOM installed at bus 45 for control of voltage, there is a SSSC installed on transmission line 21-20 for control of active power flow, and there is a UPFC installed for control of voltage of bus 94 and active and reactive power flow of line 94-95

Test results based on the cases 5, 6, 7 are summarized in Table 2. In all cases 1-7, active power flow settings are over 120% of their corresponding base case active power flows.

Table 2 Test results of the IEEE 118 bus system

	Case 5	Case 6	Case 7
The Number of FACTS	-	3	3
The Number of Iterations	13	13	16

Discussion of the Results

From these results on the IEEE 30 and 118 bus systems, it can be seen: (a) numerical results demonstrate the feasibility as well as the effectiveness of the STATCOM, SSSC and UPFC models established and the OPF method proposed; (b) by using FACTS controllers, the transfer capability of transmission lines can be increased significantly; (c) In practical application of FACTS controllers, STATCOM can be used to control voltage, SSSC can be used to control power flow, while UPFC can be used to control both voltage and power flow in the same time.

GENERAL CONCLUSIONS

Mathematical models of converter based FACTS controllers such as STATCOM, SSSC and UPFC have been established in this paper. Furthermore, the above models have been implemented in a nonlinear interior point OPF algorithm.

Numerical results demonstrate the feasibility as well as the effectiveness of the OPF method proposed and the FACTS models established. OPF with modelling of FACTS controllers will be a very useful tool to operate and manage the electric power systems in efficient ways.

REFERENCES

1. Schauder C D, et al, 1995, "Development of a ±100MVar Static Condenser for Voltage Control of Transmission Systems", <u>IEEE Transactions on Power Delivery</u>, PWRD-10, (3), pp.1486-1493.
2. Gyugyi L, Schauder CD, and Sen K K, 1997, "Static synchronous series compensator: a solid-state approach to the series compensation of transmission lines", <u>IEEE Trans. on Power Delivery</u>, PWRD-12, (1), pp.406-417.
3. Gyugyi L, Schauder CD, Williams S L, Rietman TR, Torgerson D R, and Edris A, 1995, "The unified power flow controller: a new approach to power transmission control", <u>IEEE Trans. on Power Delivery</u>, PWRD-10, (2), pp.1085-1097.
4. Sen K K, and Stacey E J, 1998, "UPFC – Unified power flow controller: theory, modelling and applications", <u>IEEE Trans. on Power Delivery</u>, PWRD-13, (4), pp.1453-1460.
5. Nabavi-Niaki A and Iravani M R, 1996, "Steady-state and dynamic models of unified power flow controller (UPFC) for power system studies", <u>IEEE Trans. on Power Systems</u>, PWRS-11, (4), pp.1937-1943.
6. Dommel H W and Tinney W F, 1968, "Optimal power flow solutions", <u>IEEE Trans. on PAS</u>, PAS-87, (10), pp.1866-1876.
7. Sun D I, Ashley B, Brewer B, Hughes A and Tinney W F, 1984, "Optimal power flow by Newton approach", <u>IEEE Trans. on PAS</u>, PAS-103, (10), pp.2864-2880.
8. Stott B and Marinho J L, 1979, "Linear programming for power system network security applications", <u>IEEE Trans. on PAS</u>, PAS-98, (3), pp.837-848.
9. Burchett R C, Happ H H and Veirath D R, 1984, "Quadratically convergent optimal power flow", <u>IEEE Trans. on PAS</u>, PAS-103, (11), pp.3267-3275.
10. Glavitsch H and Spoerry M, 1983, "Quadratic loss formula for reactive dispatch", <u>IEEE Trans. on PAS</u>, PAS-102, (12), pp.3850-3858.
11. Vargas L S, Quintana V H and Vannelli A, 1993, "A tutorial description of an interior point method and its applications to security-constrained economic dispatch", <u>IEEE Trans. on Power Systems</u>, PWRS-8, (3), pp.1315-1323.
12. Zhang X-P and Chen Z, 1997, "Security-constrained economic dispatch through interior point methods", <u>Automation of Electric Power Systems</u>, 21,(6), pp.27-29.
13. Zhang X-P, and Chen Z, 1997, "An efficient interior point algorithm for security-constrained economic dispatch", The International Conference on Electrical Engineering (ICEE 97), July 29-August 1, 1997, Matsue, Japan, pp.480-483.
14. Momoh J A, Guo S X, Ogbuobiri E C and Adapa R, 1994, "The quadratic interior point method solving power system optimization problems", <u>IEEE Trans. on Power Systems</u>, PWRS-9, (3), pp.1327-1336.
15. Granville S, 1994, "Optimal reactive power dispatch through interior point methods", <u>IEEE Transactions on Power Systems</u>, PWRS-9, (1), pp.136-146.
16. Wu Y-C, Debs A and Marsten R E, 1994, "A direct nonlinear predictor-corrector primal-dual interior point algorithm for optimal power flows", <u>IEEE Trans. on Power Systems</u>, PWRS-9, (2), pp.876-883.
17. Handschin E and Lehmkoester C, 1999, "Optimal power flow for deregulated systems with FACTS-Devices", 13th PSCC, June 28 – July 2nd, 1999, Trondheim, Norway, pp.1270-1276.
18. Acha E and Ambriz-Perez H, 1999, "FACTS devices modelling in optimal power flow using Newton s method", 13th PSCC, June 28 – July 2nd, 1999, Trondheim, Norway, pp.1277-1284.
19. Zhang X-P and Handschin E, "Advanced implementation of UPFC in a nonlinear interior point OPF", to appear in <u>IEE Proceedings–Generation, Transmission & Distribution</u>.

NON-INTRUSIVE CONTROL SYSTEM ARCHITECTURE FOR AC POWER TRANSMISSION

D. Westermann; Ch. Rehtanz

ABB High Voltage Technologies Ltd., Switzerland; ABB Corporate Research, Switzerland

Abstract: Any network expansion demands intensive planning studies and redesign of control and protection systems in order to maintain a reliable system behaviour. Applying a control architecture, which enables the operation of a controlled transmission path without affecting the rest of the system, can eliminate these problems. This non-intrusiveness is the key issue of the proposed Non Intrusive System Control (NISC[TM]) architecture. In this paper the basic requirements and structure of this new control architecture are described first. A second focus is given to the problem of controller interactions in abnormal operation situations where the NISC[TM] architecture helps to avoid malfunctioning or adverse reactions. The overall efficiency and robustness is shown in connection with a case study.

Index Terms: Non-intrusive system control and system integration, real time control, power flow control, damping control, power transmission, FACTS.

I. INTRODUCTION

While designing power system control the employed methodologies are focusing on the impact of single devices in the systemic behavior. In particular in the area of rapid network controllers (i.e. FACTS-devices) the corresponding design techniques are differ from steady state operation to system dynamic considerations. However, most of these techniques are limited to device properties rather than designing the entire system on functional basis. The design of a solution for a transmission problem by starting from a functional specification offers more degrees of freedom. Herein, impedance control, voltage and current injection are considered as single functions. This design process demands a portfolio of modularized components comprising switched elements as well as power electronic subsystems (see Figure 1). Both are suggested to the integration into well-known and field proven devices.

Most of the known system design approaches are based upon information of the entire system, i.e. detailed knowledge of the structure and parameters of all other network components is mandatory for the design process. This is not only related to a huge effort during the design phase but also more and more limited due to the deregulation.

Furthermore, the design methods may yield a complete set of new parameters for all controllers of the entire system. Both, new controlled and uncontrolled AC transmission paths will always affect the dynamics and behavior of the rest of system. In conclusion, it is mandatory to provide a system behavior that is not inadvertently affecting the entire system. Exceptions are related to the provision of certain control functions as ancillary services.

The proposed control architecture, called Non-Intrusive System Control (NISC[TM]), enables a most effective system expansion and more effective network utilization by considering the needed transmission functions first. In a second step the hardware modules are assembled accordingly (Figure 1).

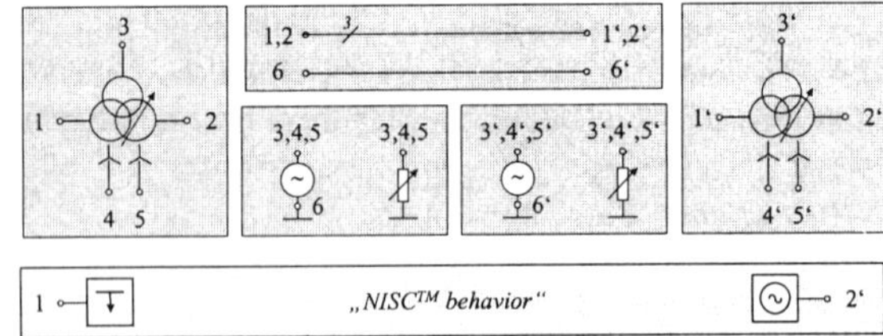

Figure 1:   Model of a controllable transmission line with the NISC[TM] approach and underlying building block philosophy

For the operation of a new transmission path the NISC[TM] architecture avoids adverse control interactions within the entire system while minimizing the redesign effort for already implemented controllers (e.g. automatic voltage regulators, power system stabilizers etc.). Additionally, the proposed architecture allows for a proper reaction on critical events and avoids insufficient and hence wrong operation after the power system state changes. This due to an umbrella approach where both, normal and abnormal operation situations are considered at the same time. This is to overcome adverse control interactions as result out of system design processes incorporating global parameterization for a fixed topology [1], [2].

Against this background the NISC[TM] architecture as control philosophy demands a certain amount of controllability. This can be achieved by integrating controlled impedances or voltage sources and transformers [3] or having a four-conductor transmission line with symmetry

AC-DC Power Transmission, 28-30 November 2001
Conference Publication No. 485 © IEE 2001

compensation [4]. A special design of transmission lines yields certain surge impedance in order to avoid bulk series compensation equipment [5]. Furthermore, controlled series resonance circuits can be added for decoupling the sending and receiving end in terms for short circuit current contributions [6]. As a result the transmission path can be designed according to a building block concept and hence a huge variety of controllers can be created without a need for standardized network control devices.

II. CONTROL SYSTEM STRUCTURE

A. Specification

The problem is that conventional controller design of controllable transmission paths demands to incorporate the entire system. In most of the cases this results in a redesign of other network controllers. Easy scalability over different control ranges and flexibility to add ancillary services is needed. However, today the number of controlled paths is limited since the control systems cannot cope with potential adverse interaction between the controllers. Overall network controllers that would desire a complete new high-speed network control system can overcome this problem. But even in this case the adverse interaction cannot definitely be avoided.

A second approach is to design a controller exclusively working on local input variables. In this case it behaves "neutral" to the static and dynamic behavior of the transmission and/or distribution grid. This reflects the basic problem solution capability of NISC^{TM} architecture. For the realization of such a controller design the following specifications are defined:

- New controller design does not require a redesign of already installed network controllers
- Several network controllers work together with the same control approach
- Robustness according to requirements of power system operation (change of operational points during time periods of days and years)
- Modular controller design for system control and ancillary services; scalable for different control ranges
- No need for long distance communication links
- No misbehavior in contingency situations

B. Architecture

Generally, one has to distinguish between predefined robust controllers for regular operation and contingency situations. In the following the controller for regular operation is referred to the function $\mathfrak{S}_1(\underline{u}_1)$. This function comprises several control algorithms for controlling the

transmission path, e.g. active power flow control, reactive power flow control, voltage control, etc. The contingency case is covered by function $\mathfrak{S}_2(\underline{x},\underline{u}_2)$. This function affects the regular device control in order to adapt its behavior according to changing network conditions, in particular during contingencies. The overall structure of a NISC^{TM} controller is shown in Figure 2.

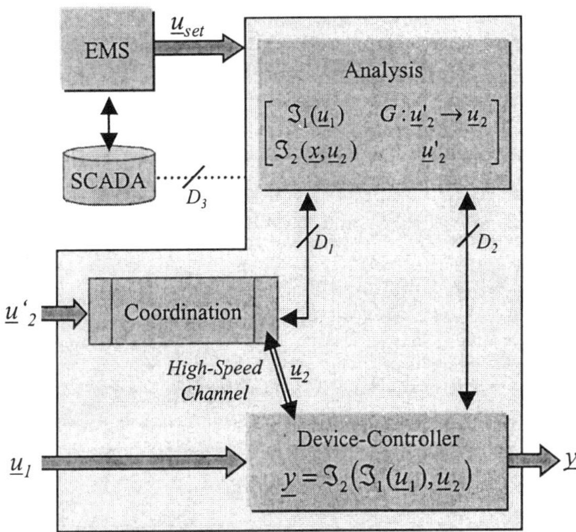

Figure 2: NISC^{TM} Architecture

In the simplest case the contingency controller does not affect the regular control function. For the initial design of the controller the function of the regular controller can be separated:

$$\mathfrak{S}_2(\mathfrak{S}_1(\underline{u}_1),\underline{u}_2) \equiv \mathfrak{S}_1(\underline{u}_1) \tag{1}$$

The design of the regular control function is traditionally based on a thorough network analysis where conventional robust controller design methodologies are applied e.g. H_∞ [7], [8], [9]. For practical applications it is hard to get the dynamic system model to design the controller. The effort for this procedure is one reason for the limited use of network controllers in practice. Therefore the controller should be designed more or less independently from detailed system studies for each application. But at first the stability for such designs independent from their special desired control characteristics must be ensured. First approaches are discussed in chapter III.

If the controller has a certain desired characteristic for all operational points, the design can be done once without applying neither structural nor parameter changes during online operation. If not, the controller performance has to be checked in regular intervals and control parameters have to be updated accordingly. Therefore, D_2 (see Figure 2) serves as a data channel used for downloading the updated control parameters. The approach for this control characteristic is discussed in chapter III. However, the overall objective of this controller design methodology is to get rid of the connection between controller and SCADA-EMS-System D_3. The information exchange shall be reduced to the set points $\underline{u}_{set}$ for the network controllers.

The contingency controller supervises the regular controller to prevent it from malfunctioning. This means coordination between the considered controlled transmission path and the entire system. One possible realization is a coordination instance that derives (from measurement values $\underline{u}_2$') the contingency case e.g. short circuit, line tripping, outages, overloading, under-voltage, etc. The result is an additional input $\underline{u}_2$ for the device controller $\underline{u}$, on which the regular control system structure is adapted to the contingency situation.

The coordination is time variant and depends on the actual network parameters and topology. Therefore the proposed NISC^{TM} architecture is despite its functional similarity not directly belonging to the class of adaptive controllers. The major difference lies in the mapping $G : \underline{u}'_2 \rightarrow \underline{u}_2$ that defines what kinds of measurement quantities are mapped on which additional input quantity for the device controller. In particular in comparison to centralized real time network control systems, within this approach the amount of high speed data transmission is drastically reduced. No additional broadband SCADA system is needed for the realization [10], [11]. Future optimization potential of the NISC^{TM} architecture lies in totally reducing the high speed data channel by substituting the coordination instance with a special signal processing unit on the device level. The major task of this signal-processing unit is to establish a mapping

$$H : \underline{u}_1 \rightarrow \underline{u}_2 \qquad (2)$$

and thereby deriving the contingency case out of locally available measurements.

In conclusion the ideal NISC^{TM} architecture concentrates all high-speed data processing, measurement and reaction schemes at the device level. Slow processes and methodologies are placed on the system level. Real-time broadband communication between these two major components is not needed.

III. NISC^{TM}-APPROACH FOR REGULAR OPERATION

The non-intrusiveness will be explained in the following according to Figure 3. The NISC^{TM}-approach ensures that there are no new instability regions due to adding a new component.

The ideal goal of the NISC^{TM} control design is to avoid the frequent update of the controller while ensuring certain robustness. There are several approaches possible to realize such a controller for the standard function of controlling the power flow or the voltage with the additional network element.

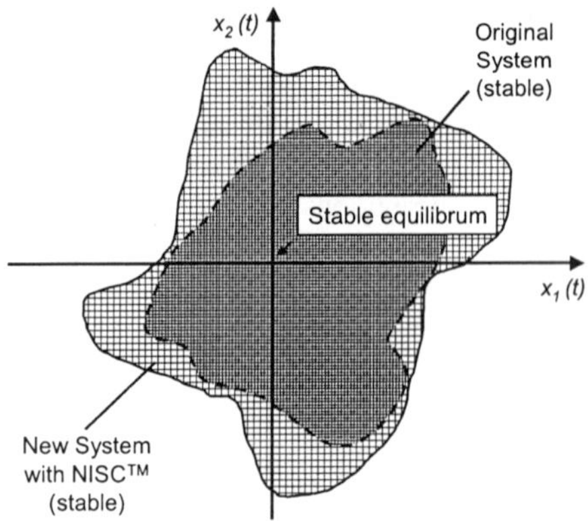

Figure 3:   Areas of stable operation points enlarged by adding new controllers with NISC^{TM}-approach

The first approach is coming from the theory of passivity. If a stable power system without the new controllable device is assumed, the system is passive if an energy function $V(T)$ exists for time points $T \geq 0$ [12].

$$V(T) \leq V(0) + \int_0^T y(t)u(t)dt \qquad \forall \ u(.), T \geq 0 \qquad (3)$$

If the additional network controller fulfills the same requirement and is also passive, then both systems in parallel or in a feedback loop are also passive and therefore stable. This means, that the additional component does not affect the stability itself if there is no energy input from this system. For the normal operation of fixing operational points this is sufficient, but this approach does not tell anything about the damping of the resulting system. Also for additional components with storage characteristic this is not applicable.

Another nearly similar approach is the Controlled Lyapunov Function (CLF) for a system with the structure:

$$\dot{x} = f(x,u) = f_0(x) + \sum_{i=1}^{m} u_i f_i(x) \qquad (4)$$

If the power system without control input is stable, it can be shown that there exists a positive energy function $V_{PS}(x)$ with $\dot{V}_{PS} \leq 0$. The system with the network controller is stable if, when V_{PS} is combined with the energy function of the controllable element V_{CO}, the resulting function V is a Lyapunov function for the new system. This holds if:

$$\dot{V} = \dot{V}_{PS} + \dot{V}_{CO} \leq \dot{V}_{CO} \leq 0 \qquad (5)$$

In [13] this is shown with the example of a controllable series device. It is shown that adding the new component enlarges the stability area of the resulting system. To get an improved damping characteristic is a question of the controller design. The resulting controller must be checked to fulfill the above requirements for CLF. The results so far are adaptable for the basic control function. The robustness of the controller depends on the model of the device and is independent from the system's model so far the system can be assumed to be stable. Therefore a robust control design is desired.

To design a robust controller for specific characteristics it is desired to make the design based on a typical structural environment and not with a detailed system study. An approach for such a design is shown in [14] where the structure of the system is known, but not the exact parameter values.

With these approaches within the NISC™ architecture a redesign of the controller can be avoided and the stable operation together with other controllers can be guaranteed. The stability is guaranteed and the robustness depends only on the device model. As a result the area of stable operation points remains the same after integrating a new controlled transmission path that adds stable operation points.

IV. NISC™-APPROACH FOR CONTINGENCIES

A. General Functionality

The major difficulty for the application of network controllers is that it cannot be assured that they behave correctly during abnormal operation situations or contingency cases. In particular this is true for rapid series compensating or controlling devices (class of FACTS devices). Many application studies have shown that the technical advantages of e.g. power flow controllers can only be profitably utilized in connection with a purposeful extension of the control and protection system. The critical factor is the dynamic behavior of the power system. This gets worsened and furthermore an overall endangering of the steady state and dynamical system security is expected if the operation of network controllers like FACTS devices is not coordinated.

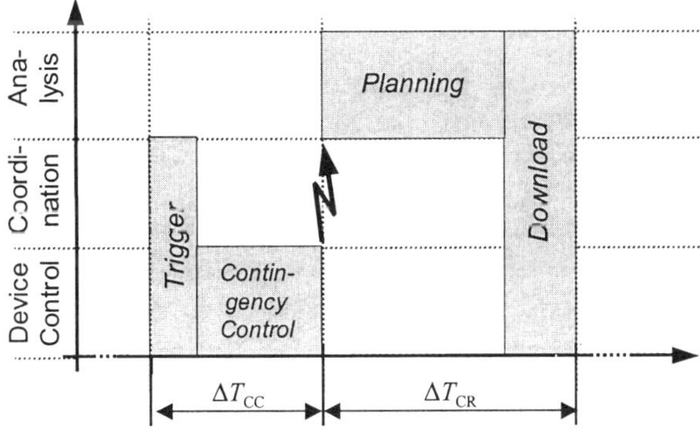

Figure 4: Typical contingency control cycle within the NISC™ architecture

The coordination has to be done according to changing operating situations or critical events in the power system. The NISC™ architecture solved this problem due to its preventive coordination mechanism. This control is activated by a trigger signal reflecting a contingency event in the entire system. This broadcast activates the corresponding local contingency control method within the device controllers (see Figure 4). After the contingency has been cleared the device controllers request a new planning and download cycle since the network topology or operation condition could have changed.

The analysis of the contingency cycle time and the regular cycle time shows that an online coordination of several network controllers cannot be achieved.

$$\Delta T_{CC} << \Delta T_{CR} \tag{6}$$

To realize a dynamic system analysis online is not possible due to the centralized databases and analysis time effort. Therefore the underlying concept of coordination is referred to as preventive coordination since the coordination is done before execution starts.

B. Preventive Coordination - Realization Example

Typical contingency events that require coordinating control measures for network controllers are:

- Short circuits in transmission elements,
- Overloading of electrical devices,
- Outages of electrical devices,
- Fast changes of power flows, e.g. related to the disturbance of power plants.

To realize these coordinated control tasks it is necessary to detect the abnormal situation in the power system. One realization possibility is a hybrid heuristic / analytic event-reaction control system (see Figure 5). The knowledge about the operation of controllable devices in critical situations can be formulated in generic rules. Two rules of this type for the case of a rapid power flow controller (PFC) are the following:

1. **IF** "short circuit on PFC path" or "short circuit on a parallel path ",
    THEN "slow down the set point control of the PFC"

This coordinating measure prevents excessive power oscillations after a short circuit followed by automatic reclosing. The reason for this is that the power flows heavily change during the short circuit. Because of the short response time of the device the PFCs immediately respond and try to meet the preset set point values. Hence, the manipulated variables of the PFC are strongly increased within a short period of time and reach their limits even before the error is clarified and the automatic reclosing is started.

After fault clearing the enlarged state variables of the PFCs controllers may lead to power oscillations. A countermeasure is to slow down the power flow controllers. In addition, a PFC can commutate the power flow on parallel paths. Consequently, changing the setpoint values for active- and reactive-power flows on the control path can influence apparent power flows on parallel paths. In this way device overloads on parallel paths can be avoided by changing the set point values.

2. **IF** "device overloading on a parallel path of PFC" **THEN** "modify the P-set point of PFCs controller"

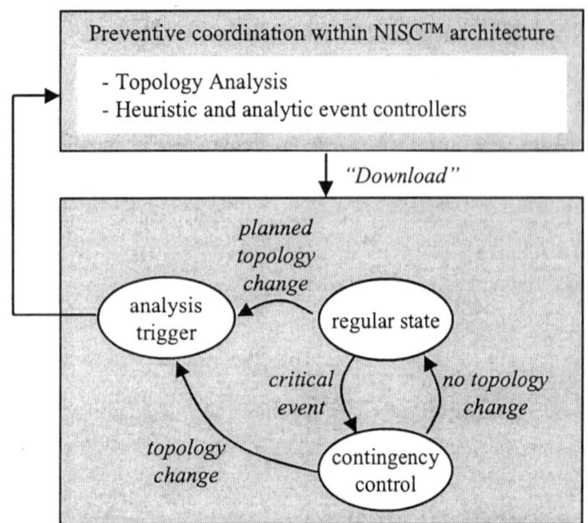

Figure 5: Process for preventive coordination within NISC™ architecture

For the automatic transformation of these generic rules into concrete control actions an autonomous system structure is proposed in 10. At first the network topology has to be analyzed automatically. The result is a mapping of all parallel paths to PFC path. This information is used to generate rule bases for the device controllers on the device control level for each device.

The rulebases are part of the downloaded information. Specific rule bases that represent coordinating measures for contingency cases are topology depended. They only can handle the "next" events. After a steady-state situation is reached after the contingency and those rule bases must be updated according to the new operation situation. Even for planned operation events (like switching) the rule bases must be updated accordingly (see Figure 5).

V. EXAMPLES

Two examples illustrate the effectiveness of the coordinating measures within the NISC™ architecture. Therefor a typical network situation according to Figure 6 is considered. The situation reflects the interconnected operation of three different systems (A, B and C). System C is connected to system B by means of a PFC in line $BC2$ for interconnection control. Within system B a PFC $B1B2$ serves for intra-area control in order to avoid congestions. System A acts as a backbone to system B. For equalizing the exchanged energy system A employs a PFC in line $AB1$.

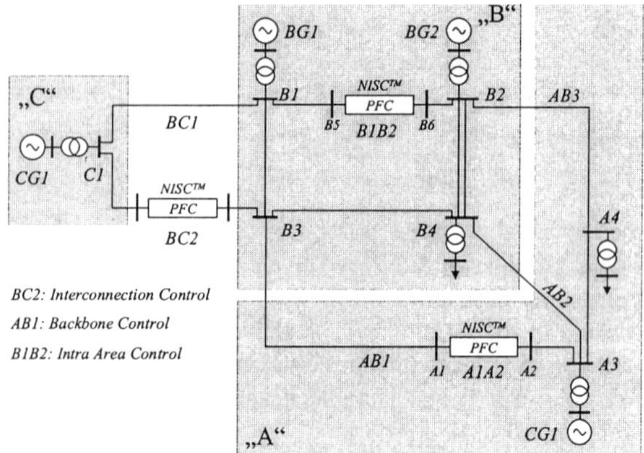

Figure 6: Typical network situation within an interconnected power system for illustrating the properties of preventive coordination within NISC™ architecture

For the numerical simulations the rule bases and global information for the PFCs integral controllers were generated with the described method. By means of a simulation environment in MATLAB/SIMULINK the two scenarios

- three-phase short-circuit on interconnecting line $AB2$ close to station $A3$ at $t = 0.1$ for 100ms with auto-reclosure for 220 ms,
- line overloading caused by a rapid increase of a load in node $B3$

are used to show the properties of the preventive coordination of PFC within an interconnected system.

By the topology analysis Line $AB2$ was identified as a part of a parallel path of PFC $A1A2$ and $B1B2$. The modules for generic rule 1 directly react upon the short circuit and slow down their set point controllers. The effect is visible in Figure 7:

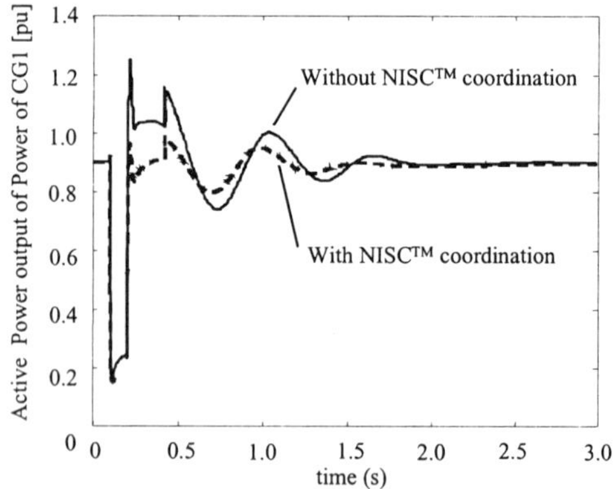

Figure 7: Active power of power unit $CG1$

The controllers of PFC $A1A2$ and $B1B2$ only cause a small increase of the variables during the short circuit and the following auto reclosing. The oscillations are better damped than with NISC™. No special damping controller is active in the system.

In the case of a stepwise load increase the load frequency controllers act in order to cover the supplementary power requirement. The three PFC freeze the power flows on the controlled paths. Consequently, they cannot be used for transmitting the surplus of generated power. The capacity of Line *AB2* is approximately utilized with 94 % before load increase. With a sudden load increase of 0.24 pu at t = 1 s, the load frequency control contribution of *CG1* has to be transmitted over line *AB2*. Without preventive coordination a certain overload occurs (see Figure 8).

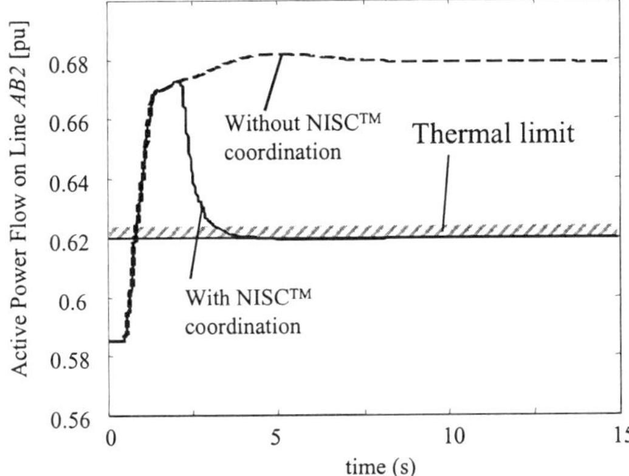

Figure 8: Active power flow on line *AB2*

Since this line is recognized as an element of a parallel path to *A1A2*, its controller counteracts to the overloading by increasing the setpoint value for active power flow. Subsequently, a power flow commutation occurs and AB2 is relieved. At t = 15 s load is restored to its original value. The control system resets itself if its coordinating control actions are not longer needed.

VI. CONCLUSIONS

This paper describes a control system architecture that enables the control of a transmission path virtually without affecting the rest of the system. Hence it is introduced as non-intrusive system control architecture (NISC™-architecture). The idea is to design the control of a transmission path independent of the special system situation as far as possible. Therefore a high degree of robustness and an effective design procedure can be achieved. Two case scenarios have been analyzed in order to show the effectiveness of the preventive coordination methodology within the NISC™ architecture.

VII. REFERENCES

1. Povh, D; Haubrich, H.; et.al.: "Global settings of FACTS controllers in power systems", CIGRE Session Paper 14-305, 1996
2. Larsen, E.V.; Sunchez-Gasca,J.J., et.al.: "Concepts for design of FACTS controllers to damp power swings", IEEE Transactions on Power Systems, Vol. 10, No. 2, May 1995
3. Beer, A; Rahmani, M.; Stemmler, H.; Westermann, D.: "Hybrid FACTS device applications for tailor-made solutions", 13th Power Systems Computations Conference, Trondheim, Norway, 1999
4. Glavitsch, H.; Rahmani, M.: "Increased transmission capacity by forced symmetrization", IEEE Transactions on Power Systems, Vol.: 13 Iss.: 1 , Feb. 1998
5. Esmeraldo, P.C.V.; Gabaglia, C.P.R.; Aleksandrov, G.N.; Gerasimov, I.A.; Evdokunin, G.N.: "A proposed design for the new Furnas 500 kV transmission lines-the High Surge Impedance Loading Line", IEEE Transactions on Power Delivery, Vol.: 14 Iss.: 1 , pp. 278-286, Jan. 1999
6. Brochu, J.: "Interphase Power Controllers", Polytechnic International Press, Montreal, January 1999
7. Taranto, G.N.; Shiau, J.-K.; Chow, J.H.; Othman, H.A.: "Robust decentralised design for multiple FACTS damping controllers", IEE Proceedings Generation, Transmission and Distribution, Vol.: 144, Iss,: 1, pp. 61-67, Jan. 1997
8. Li Wang; Ming-Hsin Tsai,: "Design of a H00 static VAr controller for the damping of generator oscillations", International Conference on Power System Technology, 1998. Proceedings. POWERCON '98., Vol.: 2, pp. 785-789, 1998
9. Ngamroo, I.; Mitani, Y.; Tsuji, K.: "Robust load frequency control by solid-state phase shifter based on H00 control design", Power Engineering Society 1999 Winter Meeting, IEEE , Vol.: 1, pp 725 -730, 1999
10. Becker, C.; Leder, C.; Rehtanz, C.: "Autonomous systems for preventive co-ordinating control of FACTS devices", IEEE International Workshop on Liberalization and Modernization of Power Systems: Operation and Control Problems, SEI, Irkutsk, Russia, August 2000
11. Handschin, E.; Kuhlmann, D.; Westermann, D.: "Advanced energy management systems using open market architekture", International Symposium on Modern Electric Power Systems, MEPS´96, Wroclaw, Polen, September 1996
12. Ortega, R.; Loria, A.; Nicklasson, P.J.; Sira-Ramirez, H.: "Passivity-based Control of Euler-Lagrange Systems", Springer, Netherlands, 1998
13. Andersson, G.; Ghandari, M.; Hiskens, I.A.: "Control Lyapunov Functions for controlled series devices", Invited paper to VII SEPOPE, Curitiba, Brazil, 23 - 28, May 2000
14. Bulliger, E.; Allgöwer, F.: "Adaptive λ-tracking for nonlinear systems with higher relative degree"; Proceedings of the Conference on Decision and Control 2000, Sydney Australia

CONTROL STRATEGY FOR UNSYMMETRICAL OPERATION OF HVDC-VSC BASED ON THE IMPROVED INSTANTANEOUS REACTIVE POWER THEORY

Guibin Zhang[1], Zheng Xu [1], Guangzhu Wang[2]

1. Zhejiang University, P.R. China; 2. Shandong University, P.R. China

INTRODUCTION

During an unsymmetrical fault period the ac voltage will become unsymmetrical, for a temporally fault this period may last more then thirty cycles, which is much larger than the response time of the HVDC-VSC. In order to improve the stability and reliability of the power system, it is desired that the HVDC-VSC system remain operation in the fault period. Besides, even in normal operation conditions, the voltage of a distribution system may be unsymmetrical. It is well known that the VSC (Voltage Source Converter) is quite sensitive to the negative sequence component of the ac voltage. When there is a negative sequence component in the ac voltage, a second harmonic dc voltage will be produced, and the reactive power output of the VSC will ripple at the frequency of 100Hz. For HVDC-VSC system, a ripple of dc voltage will lead to a ripple of the transmitted power, and thus cause other VSCs in the system operate unsymmetrically, even those VSCs are connected to a symmetrical ac system. So unsymmetrical operation can spread through the dc line. What most serious is that if the amplitude of the negative sequence component is high enough, safety of the valves and stability of the HVDC-VSC controller will be endangered. So it is necessary to study control strategy for unsymmetrical operation of HVDC-VSC system. In this paper, a feedback linearization control strategy for symmetrical system is developed first, and then the unsymmetrical condition is considered, and a feedback linearzation control strategy for unsymmetrical system is proposed. The validity of the proposed control strategy has been verified by digital simulation using NETOMAC [1].

BASIC PRINCIPLE FOR OPERATION AND CONTROL OF HVDC-VSC[2]

Assume a two-level, six-pulse voltage source converter (VSC) as shown in Fig. 1 is considered, and the fundamental component of the ac bus voltage is $\dot{U}_s$. The fundamentle componet of the VSC's output voltage is $\dot{U}_c$, which lag behind $\dot{U}_s$ an angle of δ, and the converter reactance is X. If the harmonics is ignored, then

$$P = \frac{U_s U_c}{X} \sin\delta ,\qquad (1)$$

$$Q = \frac{U_s (U_s - U_c \cos\delta)}{X} .\qquad (2)$$

Here P and Q are the active and reactive power absorbed by the VSC. From (1) and (2), it can be concluded that P and Q can be controlled independently by δ and U_c. While pulse width modulation (PWM) is adopted, δ is just the phase angle of the fundamental component of PWM, and U_c is proportional to the modulation index M of PWM. So it is possible to control P and Q independently by δ and M.

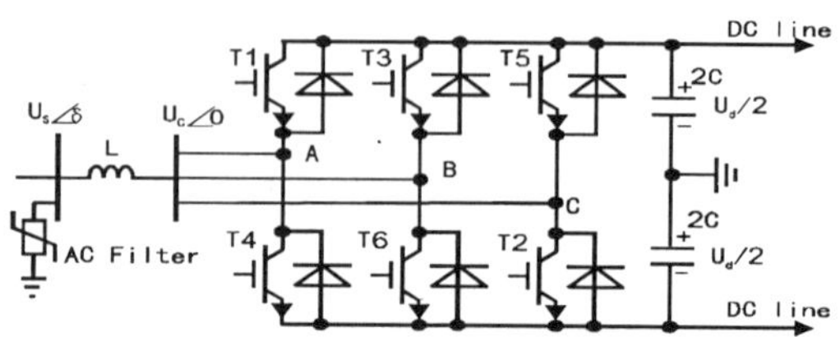

Fig.1 Voltage source converter

For HVDC-VSC, there are three basic control modes for the VSC. a) Constant dc voltage control, b) Constant dc current control, c) Constant ac voltage control. In mode a) and b), the reactive power sent to the ac system by the VSC is also controlled. Mode c) is usually used when the HVDC-VSC supplies a passive ac network. For HVDC-VSC, it is necessary for one station to adopt the constant dc voltage control. This paper will study a HVDC-VSC system with its two side both connected to active ac network.

THE STEADY-STATE MODEL OF HVDC-VSC WHILE THE AC VOLTAGE IS SYMMETRICAL

According to the basic principle described above, the active and reactive power outputs of the VSC can be controlled by controlling the amplitude and phase angle of its output voltage. But, if so, the VSC will be a nonlinear and coupled control object. Here the space vector of the voltage across the equivalent impedance of the VSC is chosen as the control variable, and the indirect current control strategy is adopted, thus a steady state linear and decoupled control strategy is realized by feedback linearzation.

The space vector is defined as follows

$$\vec{V} = \sqrt{\frac{2}{3}} (u_a + u_b e^{j120°} + u_c e^{j240°}) = U e^{j\omega t} ,\qquad (3)$$

$$\vec{I} = \sqrt{\frac{2}{3}} (i_a + i_b e^{j120°} + i_c e^{j240°}) = I e^{j\omega t} .\qquad (4)$$

According to the instantaneous reactive power theory of three-phase circuit, we have

$$P + jQ = \vec{V} * \vec{I}^* .\qquad (5)$$

No special explanation, all the three-phase voltage and current referenced below means the corresponding space vector.

AC-DC Power Transmission, 28-30 November 2001
Conference Publication No. 485 © IEE 2001

In order to develop the steady-state model of HVDC-VSC, a simplified physical model is presented as shown in Fig. 2, where the VSCs are simplified as proportional amplifiers, and the loss of the two VSCs is represented by resistors R_1 and R_2 respectively. $R_{\lim 1}$ and $R_{\lim 2}$ are start current limiting resistors, which should be cut off after startup.

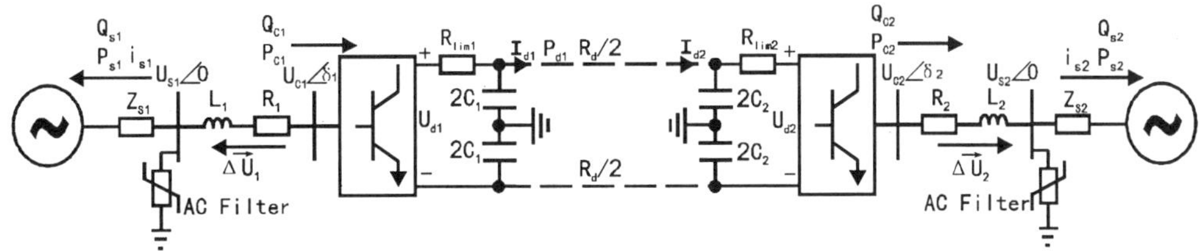

Fig.2 Steady-state physical model for HVDC-VSC

In Fig. 2, it is assumed that the constant dc voltage control mode is adopted for the left side (named dc voltage controlling station), and constant dc current control mode is adopted for the right side (named dc current controlling station). Then P_{s1} and Q_{s1} are the active and reactive power sent to the ac system by the dc voltage controlling station. P_{s2} and Q_{s2} are the active and reactive power sent to the ac system by the dc current controlling station. For the convenience of study, we let $X_1 = \omega L_1$, $X_2 = \omega L_2$, $\phi_1 = \arctan \dfrac{X_1}{R_1}$,

$\phi_2 = \arctan \dfrac{X_2}{R_2}$ $Z_1 = \sqrt{R_1^2 + X_1^2}$, $Z_2 = \sqrt{R_2^2 + X_2^2}$.

For the dc voltage controlling station, we assume the voltage across R_1 and X_1 is $\Delta \vec{U}_1$, and its reference value is $\Delta \vec{U}_{1ref}$. $\Delta \vec{U}_{1ref}$ can be decomposed into two orthogonal voltage space vectors $\Delta \vec{U}_{1pref}$ and $\Delta \vec{U}_{1qref}$ which determine the active and reactive current injected into the ac system respectively as shown in Fig. 3, where $\hat{\phi}_1$ is the estimated value of ϕ_1.

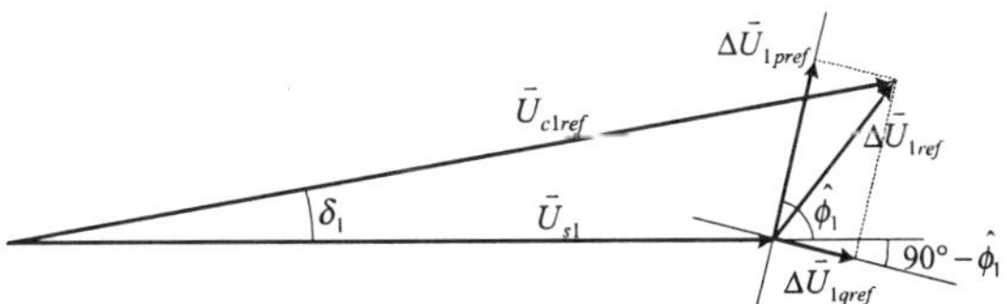

Fig. 3 Space vector diagram for dc voltage controlling station

We let

$$\Delta \vec{U}_{1pref} = K_{p1} \frac{U_{d1base}}{U_{d1}} U_{s1} e^{j(\omega t + \hat{\phi})}, \tag{6}$$

$$\Delta \vec{U}_{1qref} = K_{q1} \frac{U_{d1base}}{U_{d1}} U_{s1} e^{j(\omega t + \hat{\phi} - 90°)}, \tag{7}$$

$$\vec{U}_{c1ref} = \Delta \vec{U}_{1ref} + \vec{U}_{s1}. \tag{8}$$

In (6), (7) and (8) U_{d1base} is the dc voltage value used in calculating the modulation index, and $\vec{U}_{c1ref}$ is reference value of $\vec{U}_{c1}$. Assume the dc voltage utilization ratio of the adopted PWM method is 1 (such as voltage space vector PWM), and the modulation index is M_1 ($0 \le M_1 \le 1$), then we have

$$M_1 = \frac{\sqrt{2} U_{c1ref}}{U_{d1base}}. \tag{9}$$

So, the real value of $\vec{U}_{c1}$ is

$$\vec{U}_{c1} = \frac{M_1}{\sqrt{2}} U_{d1} \frac{\vec{U}_{c1ref}}{U_{c1ref}}. \tag{10}$$

Substituting (6), (7), (8), and (9) into (10), after some simplification we get

$$\vec{U}_{c1} = \frac{U_{d1}}{U_{d1base}} \vec{U}_{s1} + K_{p1} U_{s1} e^{j(\omega t + \hat{\phi})} + K_{q1} U_{s1} e^{j(\omega t + \hat{\phi} - 90°)} \tag{11}$$

According to (5) we have

$$P_{s1} + jQ_{s1} = \vec{U}_{s1} \left(\frac{\vec{U}_{c1} - \vec{U}_{s1}}{R_1 + jX_1} \right)^*. \tag{12}$$

Substituting (11) into (12), we get

$$P_{s1} = \frac{U_{s1}^2}{Z_1} \left[\frac{U_{d1} - U_{d1base}}{U_{d1base}} \cos\phi_1 + K_{q1} \sin(\hat{\phi}_1 - \phi_1) + K_{p1} \cos(\phi_1 - \hat{\phi}_1) \right], \tag{13}$$

$$Q_{s1} = \frac{U_{s1}^2}{Z_1} \left[\frac{U_{d1} - U_{d1base}}{U_{d1base}} \sin\phi_1 + K_{q1} \cos(\hat{\phi}_1 - \phi_1) + K_{p1} \sin(\phi_1 - \hat{\phi}_1) \right]. \tag{14}$$

And we also can get

$$P_{s1} = -P_{d1} - P_{loss1}, \tag{15}$$

$$P_{d1} = U_{d1} I_{d1}. \tag{16}$$

Here, P_{loss1} is the VSC's power loss i.e. the difference of the VSC's input and output active power. Eq. (13), (14), (15) and (16) are the basic relationship of the dc voltage controlling station.

Similarly we can get the basic equations of the dc current controlling station as follows

$$P_{s2} = \frac{U_{s2}^2}{Z_2} \left[\frac{U_{d2} - U_{d2base}}{U_{d2base}} \cos\phi_2 + K_{q2} \sin(\hat{\phi}_2 - \phi_2) + K_{p2} \cos(\phi_2 - \hat{\phi})_2 \right], \tag{17}$$

$$Q_{s2} = \frac{U_{s2}^2}{Z_2} \left[\frac{U_{d2} - U_{d2base}}{U_{d2base}} \sin\phi_2 + K_{q2} \cos(\hat{\phi}_2 - \phi_2) + K_{p2} \sin(\phi_2 - \hat{\phi}_2) \right], \tag{18}$$

$$P_{s2} = P_{d2} - P_{loss2}, \tag{19}$$

$$P_{d2} = U_{d2} I_{d2}. \tag{20}$$

FEEDBACK LINEARZATION CONTROL STRATEGY FOR SYMMETRICAL SYSTEM

For the dc voltage controlling station, according to (15) and (16) P_{s1} can be real time detected. So, to get linear relationships between the two controlling and the two controlled variables, we let

$$K_{p10} = \frac{P_{s1} Z_1}{U_{s1}^2}, \tag{21}$$

$$K_{p1} = K_{p10} - \Delta K_{p1}. \tag{22}$$

Substituting (21) and (22) into (13), and taking into account of $\hat{\phi}_1 - \phi_1 \approx 0$ (in rads), we have

$$\frac{U_{d1} - U_{d1base}}{U_{d1base}} \cos\phi_1 + K_{q1} \sin(\hat{\phi}_1 - \phi_1) \\ - \Delta K_{p1} \cos(\phi_1 - \hat{\phi}_1) \approx 0 \tag{23}$$

Generally, there is $U_{d1base} = U_{d1}$ or $U_{d1base} = U_{d1ref}$. If we let $U_{d1base} = U_{d1}$, U_{d1} will be eliminated from (23), which means U_{d1} will be uncontrollable. So, it seems $U_{d1base} = U_{d1ref}$ should be adopted, thus we get

$$\frac{U_{d1} - U_{d1ref}}{U_{d1ref}} \cos\phi_1 + K_{q1} \sin(\hat{\phi}_1 - \phi_1) \\ - \Delta K_{p1} \cos(\phi_1 - \hat{\phi}_1) \approx 0 \tag{24}$$

Eq. (24) shows that with $U_{d1base} = U_{d1}$ the system itself already has an inherent proportion control loop, but the proportion coefficient can't be adjusted, and desirable stability can't be attained. So we introduce an additional coefficient K_1, and let

$$U_{d1base} = \frac{U_{d1} U_{d1ref}}{K_1 U_{d1} + (1 - K_1) U_{d1ref}}, \tag{25}$$

Substituting (25) into (23), we have

$$K_1 \frac{U_{d1} - U_{d1ref}}{U_{d1ref}} \cos\phi_1 + K_{q1} \sin(\hat{\phi}_1 - \phi_1) \\ - \Delta K_{p1} \cos(\phi_1 - \hat{\phi}_1) \approx 0 \tag{26}$$

Thus, the proportion coefficient of the internal proportion control loop can be adjusted by K_1, and desirable stability may be attained. However, it is impossible to eliminate the steady-state error with proportion control strategy only, so additional control strategy is necessary. According to (26), we have

$$\frac{\partial U_{d1}}{\partial \Delta K_{p1}} = \frac{U_{d1ref} \cos(\phi_1 - \hat{\phi}_1)}{K_1 \cos\phi_1} > 0, \tag{27}$$

$$\frac{\partial U_{d1}}{\partial K_{q1}} = \frac{U_{d1ref} \sin(\phi_1 - \hat{\phi}_1)}{K_1 \cos\phi_1} \approx 0. \tag{28}$$

According to (14) we have

$$\frac{\partial Q_{s1}}{\partial K_{p1}} = \frac{U_{s1}^2 \sin(\phi_1 - \hat{\phi}_1)}{Z_1} \approx 0, \tag{29}$$

$$\frac{\partial Q_{s1}}{\partial K_{q1}} = \frac{U_{s1}^2 \cos(\hat{\phi}_1 - \phi_1)}{Z_1} > 0. \tag{30}$$

Thus, by introducing the feedback quantity K_{p10}, linear relationship is realized between ΔK_{p1} and U_{d1}. And (27), (28), (29) and (30) show that U_{d1} should be controlled by ΔK_{p1}, and Q_{s1} should be controlled by K_{q1}, and PI controllers can meet the control requirements. Further more, if ignore the estimation error of $\hat{\phi}$ (i.e. $\phi = \hat{\phi}$) the control of U_{d1} and Q_{s1} is linear and decoupled.

According to the discussion above, a control strategy combining PI controllers and proper feedback linearzation measure for the dc voltage controlling station has been developed, as shown in Fig. 4, where feedforward strategy is also adopted for the control of Q_{s1} in order to raise its responding speed. The feedforward block is designed according to (14) and the facts that $U_{d1} - U_{d1base} \approx 0$ and $\hat{\phi}_1 - \phi_1 \approx 0$ (in rads).

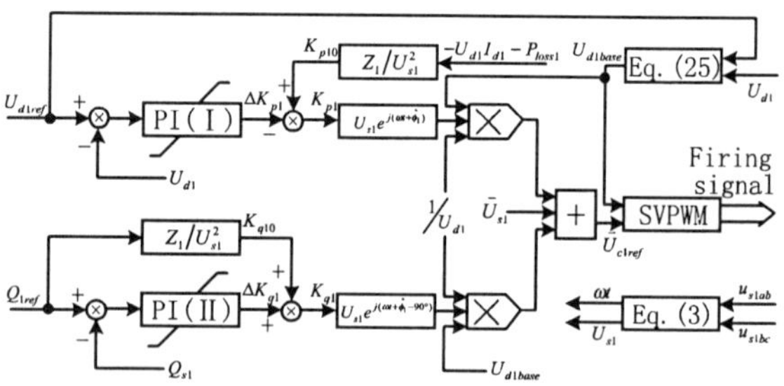

Fig. 4 Control diagram for dc voltage controlling station for symmetrical system

Similarly, for dc current controlling station, we let

$$U_{d2base} = U_{d2} \tag{31}$$

$$K_{p20} = -\frac{P_{loss2} Z_2}{U_{s2}^2}, \tag{32}$$

$$K_{p2} = K_{p20} + \Delta K_{p2}. \tag{33}$$

Substituting (19), (20), (31), (32) and (33) into (17), we get

$$I_{d2} = \frac{U_{s2}^2}{U_{d2} Z_2} \Big[K_{q2} \sin(\hat{\phi}_2 - \phi_2) \\ + \Delta K_{p2} \cos(\phi_2 - \hat{\phi})_2 \Big] \tag{34}$$

Substituting (31) into (18), we get

$$Q_{s2} = \frac{U_{s2}^2}{Z_2} \Big[K_{q2} \cos(\hat{\phi}_2 - \phi_2) \\ + K_{p2} \sin(\phi_2 - \hat{\phi}_2) \Big] \tag{35}$$

According to (34) and (35) we have

$$\frac{\partial I_{d2}}{\partial \Delta K_{p2}} = \frac{U_{s2}^2 \cos(\phi_2 - \hat{\phi}_2)}{U_{d2} Z_2} > 0, \tag{36}$$

$$\frac{\partial I_{d2}}{\partial K_{q2}} = \frac{U_{s2}^2 \sin(\hat{\phi}_2 - \phi_2)}{U_{d2} Z_2} \approx 0, \tag{37}$$

$$\frac{\partial Q_{s2}}{\partial K_{p2}} = \frac{U_{s2}^2 \sin(\phi_2 - \hat{\phi}_2)}{Z_2} \approx 0 , \qquad (38)$$

$$\frac{\partial Q_{s2}}{\partial K_{q2}} = \frac{U_{s2}^2 \cos(\hat{\phi}_2 - \phi_2)}{Z_2} > 0 . \qquad (39)$$

Thus, by introducing the feedback quantity K_{p20}, linear relationship is realized between ΔK_{p2} and I_{d2}. And (36), (37), (38) and (39) show that I_{d2} should be controlled by ΔK_{p2}, and Q_{s2} should be controlled by K_{q2}, and PI controllers can meet the control requirements. Further more, if ignore the estimation error of $\hat{\phi}_2$ (i.e. $\phi_2 = \hat{\phi}_2$) the control of I_{d2} and Q_{s2} is linear and decoupled.

According to the discussion above, a control strategy combining PI controllers and feedback linearzation measure for the dc current controlling station has been developed, as shown in Fig. 5, where feedforward strategy is also adopted for the control of both I_{d2} and Q_{s2} in order to raise their response speed. The feedforward blocks are designed according to (34) and (35) as well as the fact that $\hat{\phi}_2 - \phi_2 \approx 0$ (in rads). And a one-order inertial element is adopted in the feedorward block of dc current control system to avoid too high overshoot.

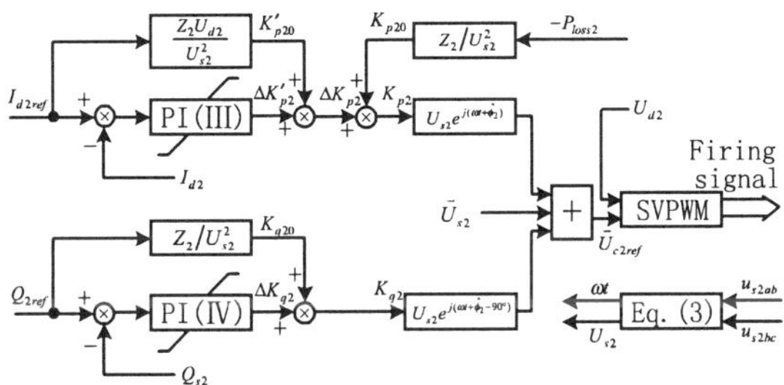

Fig. 5 Control diagram for dc current controlling station for symmetrical system

ROBUSTNESS OF THE PROPOSED CONTRL STATEGY FOR SYMMETRICAL SYSTEM

The proposed strategy above needs the circuit parameters $Z_1, Z_2, \hat{\phi}_1$ and $\hat{\phi}_2$. Because usually the values of Z_1 and Z_2 are mainly decided by the values of L_1 and L_2, which can be measured precisely, thus the estimated errors of Z_1 and Z_2 will be very small and their effects on the control performance are negligible.

However, the values of ϕ_1 and ϕ_2 are greatly affected by R_1 and R_2, which usually varied with the load of VSC. According to (27-30) and (36-39), the estimated errors of ϕ_1 and ϕ_2 have obvious effects on the coupling degree of the control system, so it's necessary to analyzing the robustness of the proposed control system while there is apparent estimation errors for R_1

and R_2.

For the system simulated in this paper, while the VSC worked at rated condition, $\phi_1 = \phi_2 = 1.321 rad$. Suppose the errors of the estimated values of R_1 and R_2 are $\pm 50\%$, the estimating errors of ϕ_1 and ϕ_2 will be about $0.1231 rad$ or $-0.1152 rad$. And it can be easily calculate that $|\cos 0.1231| \gg |\sin 0.1231|$, from (27)~(30) and (36)~(39), we can concluded that even the estimated error of R_1 and R_2 are as much as $\pm 50\%$, the proposed decopled control strategy is still feasible. So the proposed control strategy has good robustness, this will be further confirmed in the following simulation study.

FEEDBACK LINEARZATION CONTROL STRATEGY FOR UNSYMMETRICAL SYSTEM

To develop an unsymmetrical control strategy to overcome the negative effect caused by the unsymmetrical voltage of the ac system, the power characteristic of this case is analyzed as follows. Suppose the fundamental voltage of the ac system is

$$\bar{V} = \bar{V}_1^+ + \bar{V}_1^- = A_1^+ e^{j(\omega_1 t + \phi_1^+)} + A_1^- e^{j(-\omega_1 t - \phi_1^-)} , \qquad (40)$$

and the fundamental current injected into the VSC is

$$\bar{I} = \bar{I}_1^+ + \bar{I}_1^- = B_1^+ e^{j(\omega_1 t + \phi_2^+)} + B_1^- e^{j(-\omega_1 t - \phi_2^-)} \qquad (41)$$

according to Eq. (5), we have

$$p + jq = \bar{V} * \bar{I}^* = \bar{p} + \tilde{p} + j(\bar{q} + \tilde{q}) , \qquad (42)$$

here

$$\bar{p} = A_1^+ B_1^+ \cos(\phi_1^+ - \phi_2^+) + A_1^- B_1^- \cos(\phi_2^- - \phi_1^-) , \qquad (43)$$

$$\begin{aligned}\tilde{p} = {}& A_1^+ B_1^- \cos(2\omega_1 t + \phi_1^+ + \phi_2^-) \\ & + A_1^- B_1^+ \cos(2\omega_1 t + \phi_1^- + \phi_2^+) \end{aligned} , \qquad (44)$$

$$\bar{q} = A_1^+ B_1^+ \sin(\phi_1^+ - \phi_2^+) + A_1^- B_1^- \sin(\phi_2^- - \phi_1^-) , \qquad (45)$$

$$\begin{aligned}\tilde{q} = {}& A_1^+ B_1^- \sin(2\omega_1 t + \phi_1^+ + \phi_2^-) \\ & - A_1^- B_1^+ \sin(2\omega_1 t + \phi_1^- + \phi_2^+) \end{aligned} . \qquad (46)$$

So

$$\tilde{p} \equiv 0 \Leftrightarrow \begin{cases} A_1^+ B_1^- = -A_1^- B_1^+ \\ \phi_1^+ + \phi_2^- = \phi_1^- + \phi_2^+ + 2k\pi \end{cases}$$
$$\Leftrightarrow \begin{cases} A_1^+ B_1^- = -A_1^- B_1^+ \\ \phi_1^+ - \phi_2^+ = \phi_1^- - \phi_2^- + 2k\pi \end{cases} \qquad (47)$$

According to (46), to eliminate the reactive power oscillation, the condition as follow must be satisfied.

$$\tilde{q} \equiv 0 \Leftrightarrow \begin{cases} A_1^+ B_1^- = A_1^- B_1^+ \\ \phi_1^+ + \phi_2^- = \phi_1^- + \phi_2^+ + 2k\pi \end{cases}$$
$$\Leftrightarrow \begin{cases} A_1^+ B_1^- = A_1^- B_1^+ \\ \phi_1^+ - \phi_2^+ = \phi_1^- - \phi_2^- + 2k\pi \end{cases} \qquad (48)$$

Obviously, Eq (47) and (48) conflict with each other, so it seems the oscillation of active and reactive power can't be eliminated simultaneously. However, in fact, the conventional reactive power theory and the instantaneous reactive power theory are complete different concepts, the conventional reactive power theory describes the active power exchanging between the load and source, but the instantaneous reactive power theory describes the active power exchanging between phases. For power system the reactive power we usually care about is the conventional one. Because for a three-phase, symmetrical, sinusoidal and positive sequence system, the reactive power defined by the conventional reactive power theory equal to that defined by the instantaneous reactive power theory, eq. (5) may be used to detect the reactive power defined by both kind of reactive power theory. But if the system is unsymmetrical, it is not suitable to use (5) to detect the reactive power defined by conventional reactive power theory, for example, the sign of the calculated reactive power with (5) is relate to not only the phase angle but also the selection of phase sequence. So for the convenience of control, we redefine the reactive power as follows

$$p_{new} + jq_{new} = \bar{V}_1^+ * \bar{I}^* + \left(\bar{V}_1^-\right)^* * \bar{I}$$
$$= \bar{p}_{new} + \tilde{p}_{new} + j(q_{new} + \tilde{q}_{new}) \tag{49}$$

Here

$$\bar{p}_{new} = \bar{p} = A_1^+ B_1^+ \cos(\phi_1^+ - \phi_2^+)$$
$$+ A_1^- B_1^- \cos(\phi_1^- - \phi_2^-) \tag{50}$$

$$\tilde{p}_{new} = \tilde{p} = A_1^+ B_1^- \cos(2\omega_1 t + \phi_1^+ + \phi_2^-)$$
$$+ A_1^- B_1^+ \cos(2\omega_1 t + \phi_1^- + \phi_2^+) \tag{51}$$

$$\bar{q}_{new} = A_1^+ B_1^+ \sin(\phi_1^+ - \phi_2^+) + A_1^- B_1^- \sin(\phi_1^- - \phi_2^-) \neq \bar{q} \tag{52}$$

$$\tilde{q}_{new} = A_1^+ B_1^- \sin(2\omega_1 t + \phi_1^+ + \phi_2^-)$$
$$+ A_1^- B_1^+ \sin(2\omega_1 t + \phi_1^- + \phi_2^+) \neq \tilde{q} \tag{53}$$

While the system is symmetrical, the reactive defined by the new definition still equal to that defined by conventional reactive power theory. The notable advantage of the new definition is that the sign of the reactive power defined by (49) always have the same physical meaning as the conventional definition.

Accord to (51), to eliminate the active power oscillation the following condition must be satisfied.

$$\tilde{p}_{new} \equiv 0 \Leftrightarrow \begin{cases} A_1^+ B_1^- = -A_1^- B_1^+ \\ \phi_1^+ + \phi_2^- = \phi_1^- + \phi_2^+ + 2k\pi \end{cases}$$
$$\Leftrightarrow \begin{cases} A_1^+ B_1^- = -A_1^- B_1^+ \\ \phi_1^+ - \phi_2^+ = \phi_1^- - \phi_2^- + 2k\pi \end{cases} \tag{54}$$

According to (53), to eliminate the reactive power oscillation, the condition as follow must be satisfied.

$$\tilde{q}_{new} \equiv 0 \Leftrightarrow \begin{cases} A_1^+ B_1^- = -A_1^- B_1^+ \\ \phi_1^+ + \phi_2^- = \phi_1^- + \phi_2^+ + 2k\pi \end{cases}$$
$$\Leftrightarrow \begin{cases} A_1^+ B_1^- = -A_1^- B_1^+ \\ \phi_1^+ - \phi_2^+ = \phi_1^- - \phi_2^- + 2k\pi \end{cases} \tag{55}$$

The condition of (55) is the same as (54), so the active and reactive power oscillation can be eliminated simultaneously according to the new definition. According to (49), to use the new definition, the fundamental positive and negative components must be real time detected. An effective detecting method for this purpose has been presented in reference [3].

According to (54) or (55), as well as the proposed feedback linearization control strategy for symmetrical system, we can get the linearization control strategy for unsymmetrical system as shown in Fig 6 and Fig. 7.

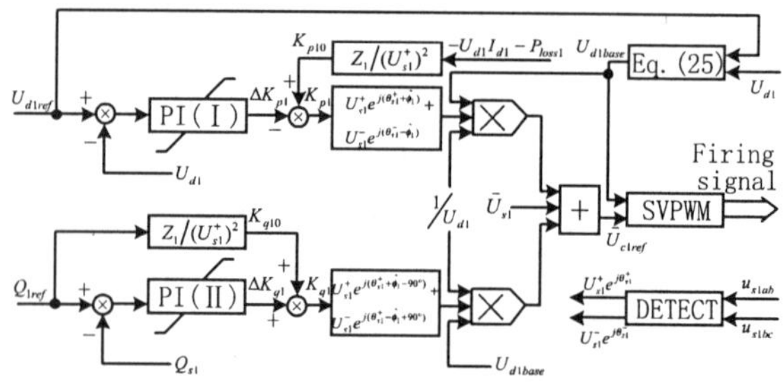

Fig. 6 Control diagram for dc voltage controlling station for unsymmetrical system

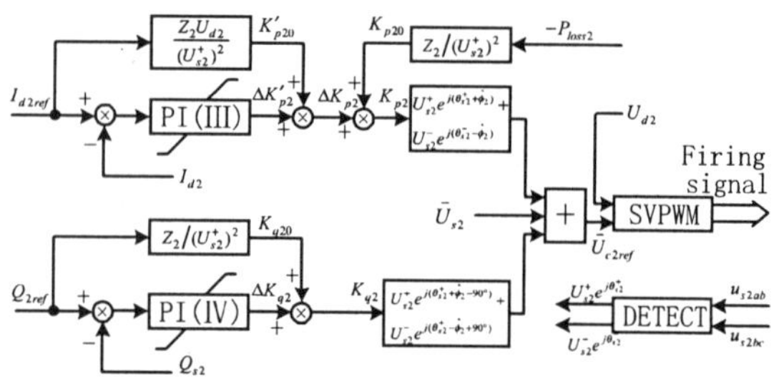

Fig. 7 Control diagram for dc current controlling station for unsymmetrical system

In Fig. 6 and Fig. 7, $U_{s1}^+ e^{j\theta_{s1}^+}$, $U_{s1}^- e^{j\theta_{s1}^-}$, $U_{s2}^+ e^{j\theta_{s2}^+}$ and $U_{s2}^- e^{j\theta_{s2}^-}$ are the detected fundamental positive and negative sequence components. The detecting block is realized according to reference [3]. And it can be easily testified that the control strategy shown in Fig.6 and Fig.7 satisfy the condition of (54) and (55). What should be stressed is that in Fig.6 and Fig.7 the reactive power is calculated according to (49) but not (5).

SIMULATION STUDY

To validate the established steady-state model and the proposed control strategy, Simulations of the system shown in Fig. 2 have been done with NETOMAC. The parameters of the system are as follows. $U_{s1} = U_{s2} = 10kV$, $R_1 = R_2 = 0.8\Omega$, $L_1 = L_2 = 10mH$, $C_1 = C_2 = 100\mu F$, $L_0 = 1.036mH/km$, $R_0 = 0.27\Omega/km$, the transmission distance is $10km$, the rated dc voltage is $\pm 10kV$, and the rated capacity of each VSC is $3MVA$. A switching frequency of $1.8kHz$ is adopted in the simulations. When the proposed controllers shown in Fig. 4 and Fig.5 are adopted, the simulation results with step changes of setting values are shown in Fig. 8(a). When the proposed controllers shown in Fig. 6 and Fig.7 are adopted, the simulation results with step changes of setting values are shown in Fig. 8(b) and Fig 8(c). For Fig. 8(a) 5% of fundamental negative sequence component is added to ac side 2 at the 6th cycle of the simulation procedure. For Fig. 8(b) 10% of fundamental negative sequence component is added to both ac side 1 and 2 at the 6th cycle of the simulation procedure. For Fig 8 (c), it is supposed that there are -50% of estimation errors with R_1 and R_2.

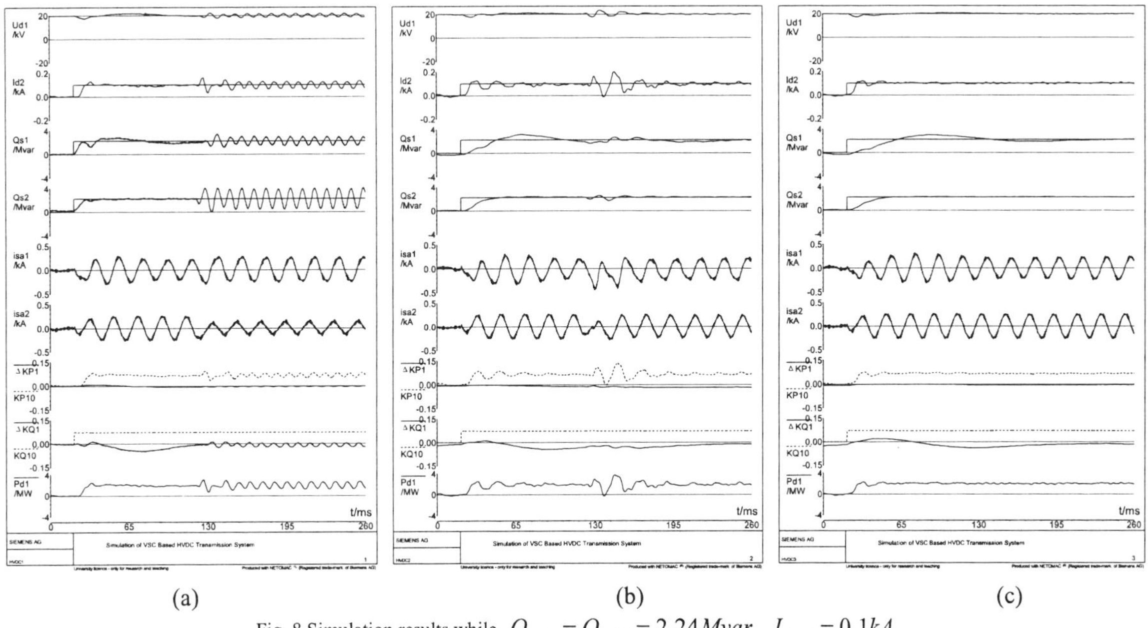

(a)   (b)   (c)

Fig. 8 Simulation results while $Q_{ref1} = Q_{ref2} = 2.24Mvar \quad I_{dref} = 0.1kA$

In Fig. 8, the subscript 1 means variables in the dc voltage controlling side, such as U_{d1}, Q_{s1}, i_{sa1} and so on, the subscript 2 means variables in the dc current controlling side, such as U_{d2}, Q_{s2}, i_{sa2} and so on. The real line curves in graphic area 7 and 8 (sequence from top down) are ΔK_{p1} and ΔK_{q1} respectively (almost superposed with t axis), which are the outputs of the two PI controllers of the dc voltage controlling station. And the dashed curves in graphic area 7 and 8 are K_{p01} and K_{q01} respectively, which are the outputs of the feedback block and feedforward block of the dc voltage controlling station.

Following conclusions can be drawn from Fig. 8: a) Even all the setting values are changed in step mode at the same time, the proposed controller strategy still have high response speed and desirable stability. b) The proposed control strategy for unsymmetrical operation of HVDC-VSC system can overcome the affect caused by the fundamental negative sequence component of ac voltage effectively. c) Even the estimated errors of R_1 and R_2 are as much as -50%, the proposed control strategy still works perfectly, which means the proposed control strategy has good robustness. d) In steady state ΔK_{p1} and ΔK_{q1}, are close to zero, which prove the correctness of the established steady-state model.

CONCLUSION

With the selected control variable, the steady-state model of HVDC-VSC system has been developed in this paper. On this basis, feedback linearzation control strategies for symmetrical and unsymmetrical system are developed. Feedforward blocks are also adopted to raise the response speed. Simulation results show the validity of the proposed control strategies, the correctness of the established steady-state model and the robustness of the control system. Because the steady-state model is independent of the structure of the VSC, the proposed control strategy is universally applicable.

ACKNOWLEDGMENTS

Project 59707005 supported by National Science Foundation of China.
Project No. G1998020312 supported by National Key Basic Research Special Fund of China.

REFERENCE

[1] Asplund G, Eriksson K et al. Dc Transmission based on voltage source converters[C]. CIGRE, 1998.
[2] Lei X, Lerch E, Povh D, Ruhle O. A large integrated power system software package – NETOMAC[C]. Proceedings of POWERCON'98, Beijing, China:17-22.
[3] Guibin Zhang, Zheng Xu. A New Real-time Negative and Positive Sequence Components Detecting Method Based on Space Vector. Proceedings of IEEE PES Winter meeting 2001,Columbus, Ohio USA. Jan.28- Feb.1.

SEQUENTIAL DECENTRALISED CONTROL OF FACTS DEVICES IN LARGE POWER SYSTEMS

M M Farsangi, Y H Song
Brunel University, UK

ABSTRACT

A procedure for sequential design of decentralised FACTS controllers for a multimachine power system is presented. It is shown how to include a simple estimate of the effect of closing subsequent loops into the design problem for the loop, which is to be closed. The focus of this paper is on the design performance where frequency domain methods prove to be very useful. A comprehensive and systematic way of designing decentralised controllers is presented and the results are illustrated using a 16-machine 5-area system.

Key words: FACTS devices, decentralised control, power system damping controller.

INTRODUCTION

Power system stability is a complex subject that has challenged power system engineers for many years. Poorly damped low-frequency (0.1 Hz – 2 Hz) inter-area oscillations are inherent in interconnected power systems. The phenomenon is very complex as it involves several electromechanical oscillatory subsystems often comprising several groups of machines distributed over neighbouring utilities. These modes are poorly damped and hence, pose a threat to the economic and reliable means of power exchange through the lines connecting neighbouring utilities. The nature of the interaction makes the damping control design task challenging. The design methodology, therefore, should aim at improving the damping performance of one mode while ensuring the least interaction with other modes. In addition, the controller should guarantee stable operation over the full range of system operating conditions. This type of robust control action has been sought by H_∞ loop shaping method. The H_∞ loop shaping method is used to arrive at a decentralised controller structure, which provides robust stability and performance over a range of uncertainties of interest. Some practical approaches to the design of decentralised controllers have evolved as: simultaneous design using parameter optimisation for a fixed controller structure, independent design and sequential design. The sequential design, which involves closing and tuning one loop at a time, has been used to design the controller for UPFC, SPFC and SVC (refer to Song and Johns (1) for details about FACTS), which is probably the most common design procedure in real world applications. In sequential design a limited degree of failure tolerance is guaranteed. If stability has been achieved after the design of each loop, then the system

will remain stable if loops fail or are taken out of service in the reverse order of how they were designed. However, the problems with sequential design are as follows (Hovd and Skogoslad (4)):

- The final controller design and thus the control quality achieved may depend on the order in which the controller in the individual loops is designed.

- Only one output is usually considered at a time and the closing of subsequent loops may alter the response of previously designed loops and thus make iteration necessary.

- The transfer function between input u_i and y_i may contain right half-plane (RHP) zeros that do not correspond to RHP transmission zeros of G(s).

Thus, the usefulness of a sequential design procedure will depend on how successfully it addresses the above issues. While the efforts of Djukanovic et al (5) have been devoted to the design of PSS using sequential design by applying μ-approach without considering the above mentioned problems, in order to take care of the interaction, this paper aims to use the conventional rule (introduced by Hovd and Skogoslad (4)) to deal with the above problems and presents a design procedure based on obtaining a simple priori estimates of the final loop shape. Since the objective of the supplementary controller is to damp out system oscillations in a particular frequency band, it is necessary to reformulate the controller design problem, which is explained in next section.

STANDARD H∞ MIX-SENSITIVITY AND DAMPING CONTROL PROBLEM

The design objective in a standard H∞ mix-sensitivity robust control problem (Figure 1) is to find a controller, k, so that the infinity norm of a mixed and weighted closed loop transfer function is minimised:

$$\min_{K(s)} \left\| \begin{matrix} W_1(s)S(s) \\ W_3(s)T(s) \end{matrix} \right\|_\infty$$

The controller synthesised using this method has two important properties: it will guarantee the closed-loop stability in the presence of uncertainty as long as the uncertainty is bounded by W_3 and it will make the tracking error signal e as small as prescribed by W_1. In other words, it will make the output y track the command signal r properly (Skogoslad (2) and Zhou et al (3)).

In power system, most of the tracking problems are taken care of by primary controllers. Since the objective of the supplementary controller is to damp out system oscillations in a particular frequency band,

AC-DC Power Transmission, 28-30 November 2001
Conference Publication No. 485 © IEE 2001

the design emphasis is different from that of tracking a reference input signal. Therefore, it is necessary to reformulate the controller design problem based on the H∞ optimisation technique (Zhao and Jiang (6 and7)). Then, the selection of weighting functions for damping controller design is different from that of tracking controllers. In this new formulation, the feedback loop with appropriate weighting functions will take the form as illustrated in Figure 2. The objective of the controller is to provide additional damping to the system, or equivalently, to reduce the resonance peak of the close-loop transfer function V(s), which is defined as:

$$V(s) = \frac{y(s)}{r(s)} = \frac{G(s)}{1 + k(s)G(s)}$$

where $G(s) = G_o(s)(1 + \Delta_m(s))$

and the transfer function from r(s) to u(s) is

$$\frac{u(s)}{r(s)} = \frac{k(s)G(s)}{1 + k(s)G(s)} = K(s)V(s)$$

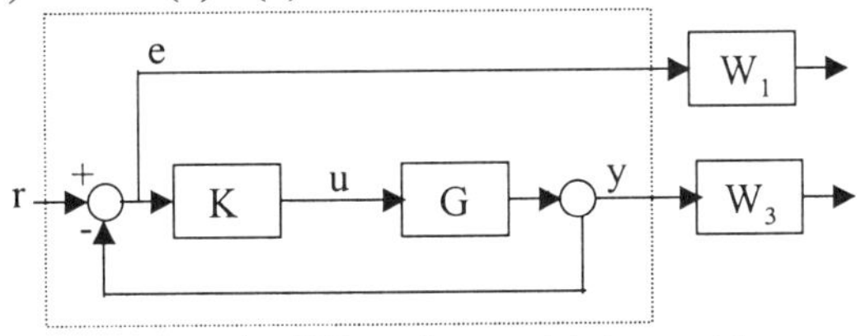

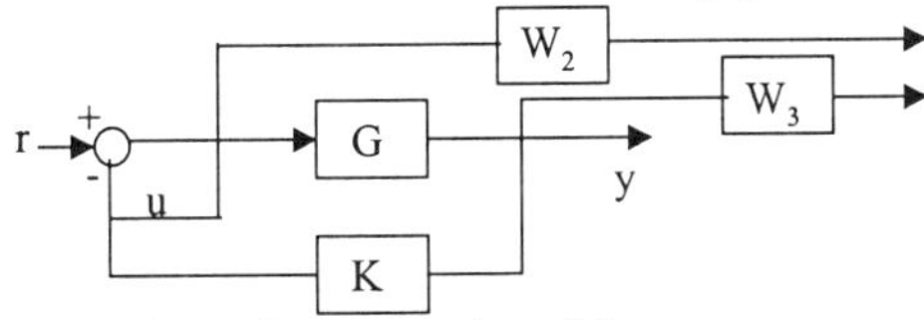

FIGURE 1- Standard H∞ mix-sensitivity problem

FIGURE 2- Damping control problem

The weighting function $W_3(s)$ is the weight or the penalty that is placed on the closed-loop transfer function to achieve the desired damping performance. The weighting function $W_2(s)$ is the weight on u(s), the output of the controller, which influences the robustness of the closed-loop system against model variations. The damping and robustness requirements can be achieved, such that the transfer function of the closed-loop system satisfies the following norm inequality:

$$\left\| \begin{matrix} W_2(s)V(s)k(s) \\ W_3(s)V(s) \end{matrix} \right\|_\infty \prec 1 \qquad (1)$$

by proper choice of the weighting functions $W_2(s)$ and $W_3(s)$, the designed controller will be able to meet both the damping and robustness requirements.

SOLUTION TO THE SEQUENTIAL METHOD PROBLEMS

In this section some simple facts to derive a priori estimates of the individual loop shapes (in terms $g_{ii}k_i$), are used and shown how this may be used to estimate the interactions caused by the undesigned loops. Figure 3 shows the decentralised diagonal control of a plant, which the matrix G(s) denotes a square plant of dimension $n \times n$. The decentralised controller is

assumed to be diagonal with diagonal elements k_i $i = 1,...n$. The matrix consisting of the diagonal elements of G is denoted $\tilde{G} = diag(G_{ii})$. The sensitivity function is $S = (I + GK)^{-1}$ and the complementary sensitivity is $T = I - S = GK(I + GK)^{-1}$. Loop i is the SISO feedback system consisting of g_{ii} and k_i. The sensitivity functions and complementary sensitivity functions for the individual loops are collected in the diagonal matrices: $S = diag(s_i) = (I - \tilde{G}K)^{-1}$, $T = diag(t_i) = \tilde{G}K(I - \tilde{G}K)^{-1}$.

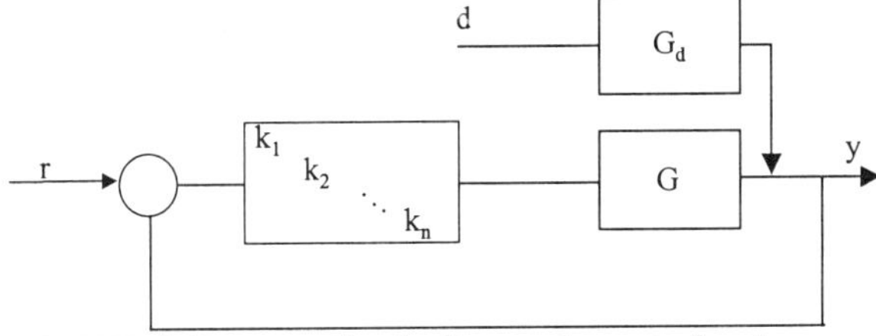

FIGURE 3-Decentralised diagonal control of a n×n plant

By considering the feedback system in Figure 3 (Hovd and Skogoslad (4) and Skogoslad (2)):

$$y = G(s)u + G_d(s)d = GK(r - y) + G_d d \text{ or}$$

$$y = (I + GK)^{-1}GKr + (I + GK)^{-1}G_d d$$

$$e = y - r = -Sr + SG_d d$$

To satisfy the performance objectives, for any single set point $|r_j| \prec 1$ at least required: $\left| [S_{ij}] \right| \prec 1$ (2)

and for single disturbances $|d_k| \prec 1$ at least required

$$\left| [SG_d]_{ik} \right| \prec 1 \qquad (3)$$

The aim is that to express these performance requirements in terms of the individual design loops. The matrix $(G(s) - \tilde{G}(s))\tilde{G}^{-1}(s)$ plays a central role in interaction analysis. For simplicity of notation, the matrix $E(s) = (G(s) - \tilde{G}(s))\tilde{G}^{-1}(s)$ is defined, which represents the relative interactions. A very important relationship for decentralised control is given by the following factorisation of the return different operator:

$$\underbrace{(I + GK)}_{Overall} = \underbrace{(I + E\tilde{T})}_{interactions} \underbrace{(I + \tilde{G}K)}_{individual \; loops}$$

or equivalently in terms of the sensitivity function

$$S = \tilde{S}(I + E\tilde{T})^{-1} \quad \text{where}$$

$$\tilde{S} \overset{\Delta}{=} (I + \tilde{G}K)^{-1} = diag\left\{ \frac{1}{1 + G_{ii}k_i} \right\} \quad \text{and} \quad \tilde{T} = I - \tilde{S}$$

contain the sensitivity and complementary sensitivity functions for the individual loops.

At frequencies where feedback is effective we have

$$\tilde{T} \approx I$$

$$(I + E\tilde{T})^{-1} \approx \tilde{G}(s)\tilde{G}^{-1}(s) = \Gamma$$

$$E = \Gamma^{-1} - I$$

$$S = (I + \tilde{S}(\Gamma - I))^{-1}\tilde{S}\Gamma \qquad S \approx \tilde{S}\Gamma$$

where $\Gamma = \{\gamma_{ij}\}$ is the Performance Relative Gain Array (PRGA) and the control error becomes:

$$e = y - r \approx -\widetilde{S}\Gamma r + \widetilde{S}\Gamma G_d \quad \omega \prec \omega_B \qquad (4) \text{ or}$$

$$e_i \approx -(1 + G_{ii}k_i)^{-1}\gamma_{ij}r_j + (1 + G_{ii}k_i)^{-1}\delta_{ik}d_k \quad \omega \prec \omega_B$$

which shows that Γ is important when evaluating performance with decentralised control. For a setpoint change r_i in loop i that the performance relative gain, γ_{ii} gives the approximate change in offset caused by closing all the loops.

The performance requirements in Equations (2) and (3) may then be rewritten in terms of bounds on the individual loop gains, $g_{ii}k_i$:

Setpoint: $\left|\dfrac{\gamma_{ij}}{g_{ii}k_i}\right| \prec 1 \Leftrightarrow |g_{ii}k_i| \succ |\gamma_{ij}| \qquad \omega \prec \omega_B$

disturbances: $\left|\dfrac{\delta_{ij}}{g_{ii}k_i}\right| \prec 1 \Leftrightarrow |g_{ii}k_i| \succ |\delta_{ij}| \qquad \omega \prec \omega_B$

For a perfect control, it is required $e = y - r = 0$; that is desired $e \approx 0r + 0d$. Thus to have a good control it is required to have $\widetilde{S} \approx 0$ and $(I + E\widetilde{T})^{-1} \approx 0$. Therefore, in this sequential design, it is attempted to reduce the severity of second problem by using simple estimate of how the undesigned loops will affect the output of the loop to be designed. Thus, instead of considering G, the $(I + ET)^{-1}$ is added to G as additive term to express the interaction from the undesigned loops when evaluating performance (Figure 4). To evaluate this matrix, an estimation of t_i for the undesigned loops are required. The severity of first problem is reduced by closing the fast loops first. This is done by PRGA that PRGA elements larger than one imply interactions. Since RHP-zeros imply poor control performance, problem 3 may affect the order of loop closing. If G has no multivariable RHP zeros, but presence of RHP zeros in the individual element make it necessary to design the fastest loop last.

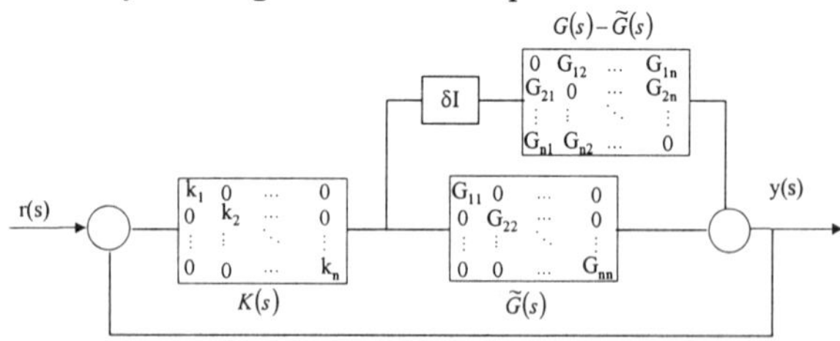

FIGURE 4-Decentralised diagonal control of a n×n plant

DESIGN PROCEDURE

The objective of the controller design is to design SISO controllers, k_i, that minimise some performance objective and consider a norm (Equation 1) of the weighted sensitivity function of the overall system where $W_2 = diag\{w_{2i}\}$ and $W_3 = diag\{w_{3i}\}$.

A step-by-step procedure to sequential design of SVC, SPFC and UPFC is outlined as follows:

Step 1: Find the set of critical oscillation modes that need to be damped.

Step 2: Once the FACTS are located and input signals are chosen, define the order of loop design by using PRGA.

Step 3: The initial estimates for the complementary sensitivity functions for the individual loops are chosen to be second order of the form $t_i(s) = \dfrac{1}{\left(\dfrac{s}{w_i} + 1\right)^2}$.

Step 4: Design of controller k_i related to loop i by considering only output y_i.

Step 5: Design of controller k_{i+1} related to loop $i+1$ by considering output y_i to y_{i+1}.

Step 6: Design of the last controller is carried out by considering the overall system.

STUDY SYSTEM

The system shown in Figure 5, consisting of 16 machines and 68 buses is used. This is a reduced order model of the New England New York interconnected system. The first nine machines are the simple representation of the New England systems generation. Machines 10 to 13 represent the New York power system. The last three machines are the dynamic equivalent of three large neighbouring areas interconnected to New York power system. Eigenvalue analysis of the system shows that the system has the following inter-area modes shown in Table 1. All the inter-area modes are not observable in a particular signal nor they all controllable from a single bus location. It is a general practice in power systems to damp out one mode by one device. In the case of multiple inter-area modes, more than one device is employed. It is found that the following locations are effective for one particular mode:

Mode 1 is very dominant at the input-output characteristic of the system when UPFC located at bus 40,40-48. The SPFC location at line 9-36 and SVC at bus 42 are very good indicator to influence modes 2 and 3 respectively. These can be seen in Figures 6,7 and 8.

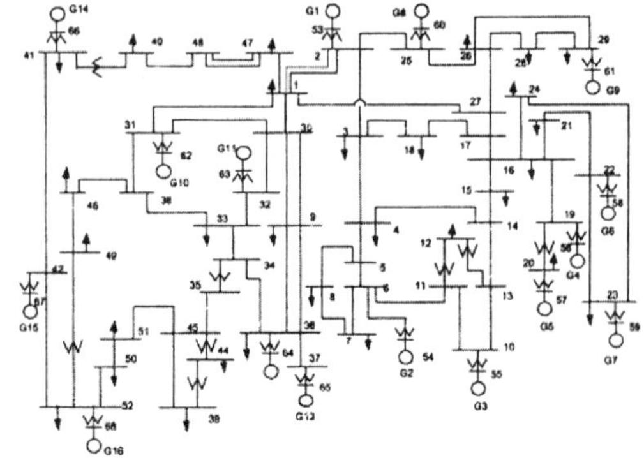

FIGURE 5-Structure of the power system model

TABLE 1- Effects of FACTS devices on inter-area modes

M o d e	Without FACTS devices		With FACTS devices	
	Eigenvalue	ζ	eigenvalue	ζ
1	$-0.1154 \pm 2.325I$	0.049	$-0.1813 \pm 2.288i$	0.078
2	$-0.1306 \pm 3.426I$	0.038	$-0.1468 \pm 3.416i$	0.042
3	$-0.2493 \pm 4.972I$	0.05	$-0.6097 \pm 4.661i$	0.129

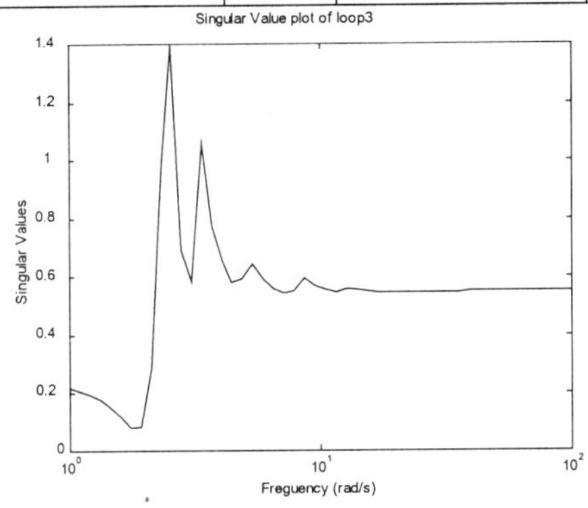

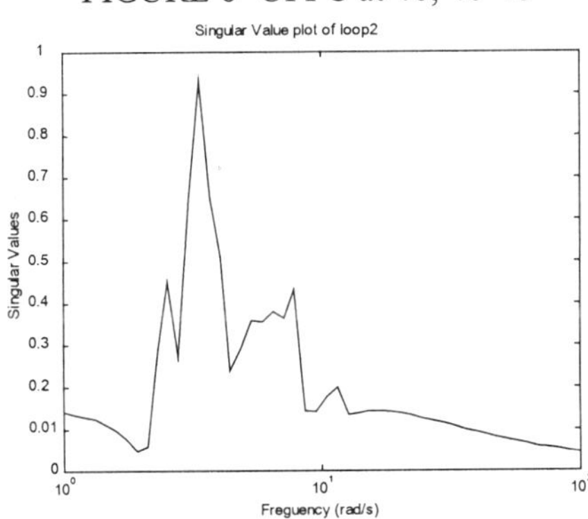

FIGURE 6- UPFC at 40, 40-48

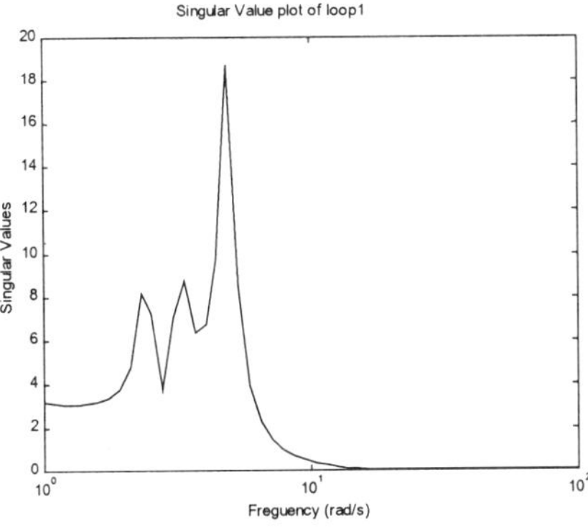

FIGURE 7- SPFC at 9-36

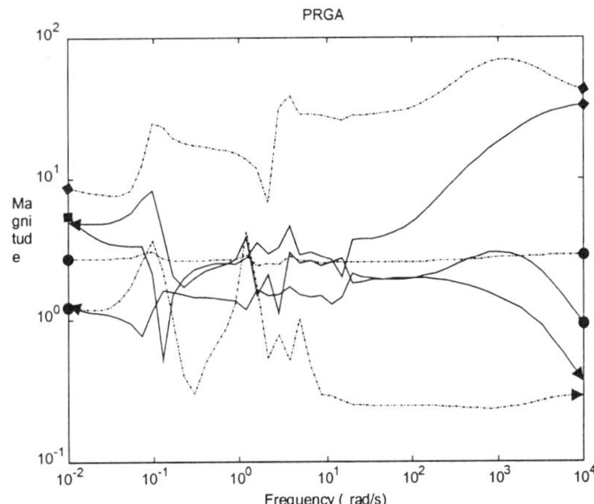

FIGURE 8- SVC at bus 42

NUMERICAL STUDIES

The procedure, explained before, is used to design a controller for SVC (loop 1), SPFC (loop 2) and UPFC (loop 3). Firstly, the PRGA for the study system is plotted (Figure 9). Since, the current study system has no multivariable RHP zeros and no zeros in the individual element, from Figure 9, it can be concluded that the order of loop closing should be: loop 3, loop 2 and loop 1. PRGA elements larger than 1 imply interactions, and this figure shows that there is a severe interaction from loops 1 and 2 into loop 3.

Controller design for loop 3

In order to have a simplified estimate of the effect of closing of the other loops, an estimate of the complementary sensitivity function for the individual loops is needed. Therefore, the initial estimate for the complementary sensitivity function for loop 3 is chosen as:

FIGURE 9-Elements of PRGA: Lines ended by square; dashed line y31, Solid line y32; Lines ended by circle: dashed line y21; Solid line y23Lines ended by arrow: dashed line y12; Solid line y13

$t_3(s) = \dfrac{1}{(0.1s+1)^2}$. By calculating, $(I+ET)^{-1}$ and adding this matrix as an additive term, that expresses the interaction from the undesigned loop when evaluating performance. Figure 10 shows how the interaction from the undesigned loops affect on the loop 3. As shown in this figure, by considering the interaction from the undesigned loop, the dominant mode will be different. Therefore, care will be given to damp out two modes by loop 3. The choice of weights is based on the suggestions made by Hu et al (8) and Beaven et al (9). Here the weighting functions are synthesised such that they have a reasonably wide band to cover the possible shifts in the resonance frequency of the system under different operating conditions and interactions. The following weighting function is chosen.

$$W2 = \frac{s^4 + 3.5s^3 + 22.5s^2 + 27s + 98}{s^4 + 11s^3 + 31s^2 + 84s + 98}$$

$$W_3 = \frac{s^2 + 80s + 1600}{3.5s^2 + 420s + 12600} \text{ (for all three loops)}$$

The controller obtained is of high order and is reduced to a lower order without greatly affecting the damping performance. A 3 states controller for UPFC is derived as follows:

$$K_{s\_UPFC} = \frac{0.02s^2 + 0.3s + 2.8}{0.0114s^3 + 0.0936s^2 + 1.21s + 1}$$

The corresponding loop shape is shown Figure 11.

Controller design for loop 2

Design controller for loop 2 starts by considering of K_3 in the place and by replacing the estimate of t_3 by the actual design for loop 3. The initial estimate for the complementary sensitivity function for loop 2 is chosen as $t_2(s) = \dfrac{1}{(0.25s + 1)^2}$. Figure 12 shows the effect of interaction on the loop 2. Once again, as shown in this figure, by considering the interaction from the undesigned loop the following weight is chosen:

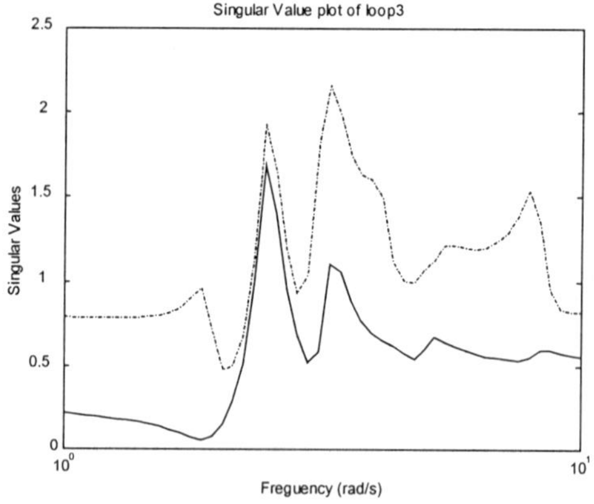

FIGURE 10-Solid line: without considering interaction; Dashed line: with considering interaction

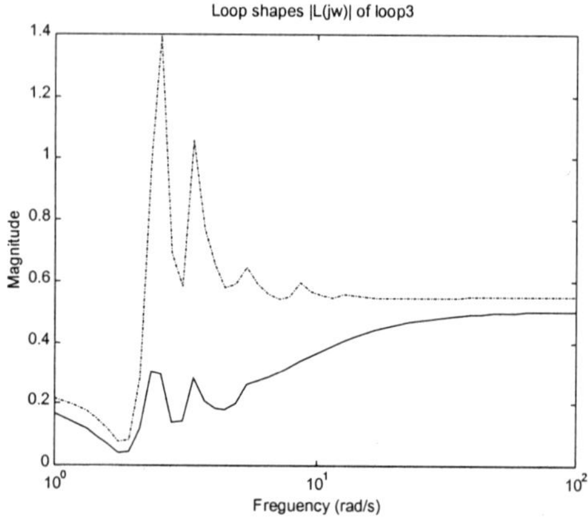

FIGURE 11-Solid line: shaped plant; Dash-Dotted line: initial plant

$$W2 = \frac{s^4 + 4s^3 + 72s^2 + 97s + 155}{s^4 + 20s^3 + 169s^2 + 500s + 155}$$

The order of K_s should be reduced to derive reduced controllers. A 4 states controller SPFC is derived as follows:

$$K_{s\_SPFC} = \frac{0.01s^3 + 0.04s^2 + 0.45s + 0.333}{0.001s^4 + 0.01374s^3 + 0.093s^2 + 1.3317s + 1}$$

The corresponding loop shape is illustrated in Figure 13.

Controller design for loop 1

At this stage, all loops are included by replacing the estimate of t_2 by the actual design for loop 2. The initial estimate for the complementary sensitivity function for loop 1 is chosen as $t_1(s) = \dfrac{1}{(0.125s + 1)^2}$ and also the following weighting function is selected.

$$W2 = \frac{s^2 + s + 25}{s^2 + 10s + 14}$$

The order of K_s should be reduced to derive reduced controllers. A 6 states controller SVC is derived as follows:

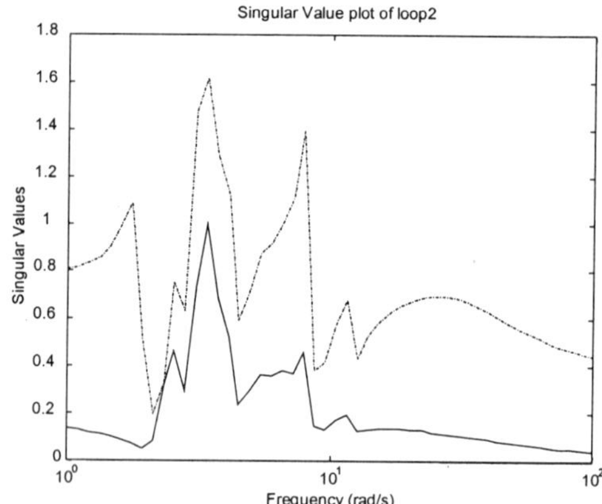

FIGURE 12-Solid line: without considering interaction; Dashed line: with considering interaction

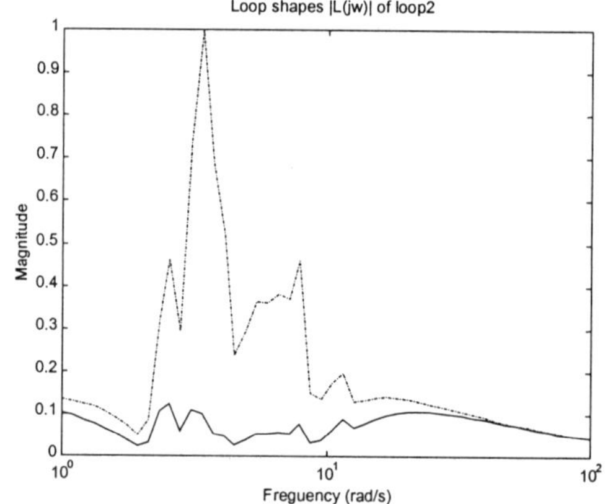

FIGURE 13-Solid line: shaped plant; Dash-Dotted line: initial plant

$$K_{s\_SVC} = \frac{0.001s^5 + 0.012s^4 + 0.061s^3 + 0.31s^2 + 0.28s + 1.26}{0.0001s^6 + 0.002s^5 + 0.026s^4 + 0.09s^3 + 0.64s^2 + 0.66s + 3.21}$$

The corresponding loop shape is shown in Figure 14. In order to verify the performance of the controller in the face of the system nonlinearity, a nonlinear simulation was performed. A 3-phase fault at bus 1 is assumed in one of the tie lines between bus 1 and bus 2. The fault is cleared by the removal of the faulted circuit. The simulation is performed for 20 s. The variations of machine angles with reference to machine 15 are computed and shown in Figure 15.

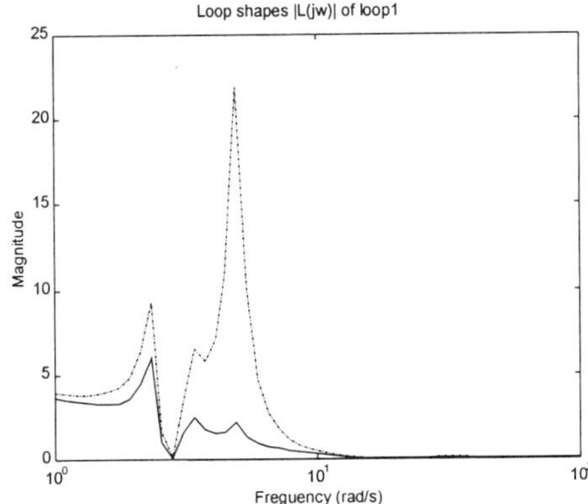

FIGURE 14-Solid line: shaped plant; Dash-Dotted line: initial plant

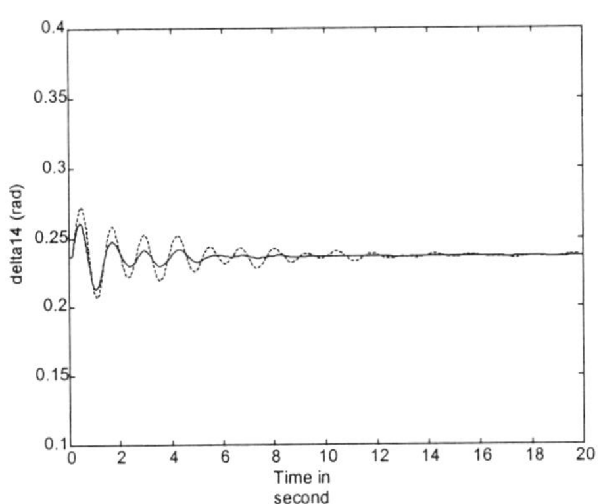

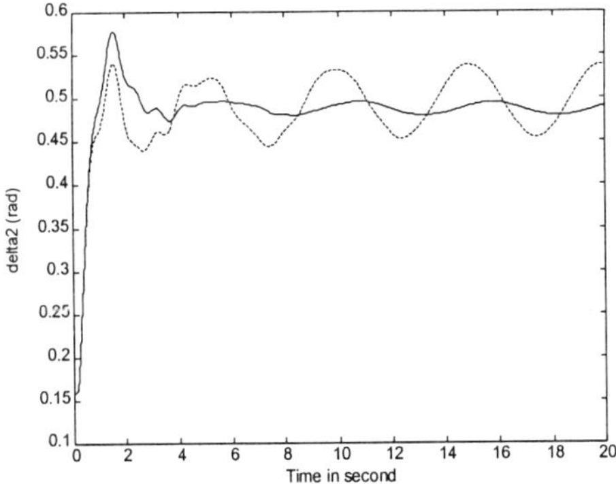

FIGURE 15-Dynamic response of the system following a 3-phase fault at bus 1. Solid line: with controllers and dashed line: without controllers

CONCLUSIONS

In this paper, a robust sequential design is applied to a large power system that suffers from very lightly damped inter-area oscillations. In order to deal with the sequential design problems, the performance requirements, Equations (1) and (2), are expressed in terms of the individual designs. The Equation (4) shows that Γ is important when evaluating performance with decentralised control. $(I + ET)^{-1}$ is added to G as additive term to express the interaction from the undesigned loops when evaluating performance. The performance evaluations in the frequency and time domain, show that coordinated actions of three FACTS devices.

The work reported in this paper leads to important conclusion as follows: pure H∞ synthesis only enforces closed-loop stability and does not allow for direct placement of the closed-loop poles in more specific regions of the left-half plane (see Table 1). Since the pole location is related to the time response and transient behaviour of the feedback system, it is often desirable to impose additional damping and clustering constraints on the closed loop dynamics. For a system with multiple inter-area modes, pole-placement must be used with caution. It is expected that damping controller for a particular device at a particular location would improve the damping of a particular mode. The damping of other modes would not be improved with a single device. This is another limitation of sequential design to incorporate additional pole-placement constraint simultaneously. How to deal with this problem is the current research of the authors.

REFERENCES

1. Song, YH., and Johns, AT., 1999, "Flexible AC Transmission Systems (FACTS)", IEE Power and Energy Series 30

2. Skogoslad, S., 1996, "Multivariable feedback control: analysis and design", John Wiley

3. Zhou, K., Dole, JC., and Glover, K., 1996 "Robust and optimal control", Prentice-Hall, Inc

4. Hovd M and Skogoslad S, 1994, "Sequential design of decentralised controller", Automatica, 30, 1601- 1607

5. Djukanovic M, Khammash M and Vittal V, 1999, "Sequential synthesis of structured singular value based decentralized controllers in power systems", IEEE Trans. Power System, 14(2),635-641

6. Zhao Q and Jiang J, 1995: "Robust SVC controller design for improving power system damping" IEEE Trans. Power System, 10(4), 1927-1932

7. Zhao Q and Jiang J, 1998: "A TCSC damping controller design using robust control theory" Electrical Power & Energy System, 20(1), 25-33

8. Hu J, Bohn C and Wu HR, 2000, "Systematic H∞ weighting function selection and its application to the real-time control of a vertical take-off aircraft". Control Engineering Practice, 8, 241-252

9. Beaven RW, Wright MT and SAEWARD DR, 1994, "weighting function selection in the H∞ design process", Automatica,, 30, 1307-1317

THE SPECIFICATION, DESIGN AND PROTECTION OF PHASE SHIFTING TRANSFORMERS FOR THE ENHANCED INTERCONNECTION BETWEEN NORTHERN IRELAND AND THE REPUBLIC OF IRELAND

R Sweeney[1], G Stewart[1], P O'Donoghue[2] and P Smith[2]

NIE Powerteam, Northern Ireland[1] and ESB International, Ireland[2]

INTRODUCTION

NIE is the electricity transmission and distribution company for Northern Ireland and the ESB is the electricity generating, transmission and distribution company for the Republic of Ireland. The interconnection of both electricity networks brings benefits to each company in the form of sharing the generation reserves, facilitating the liberalising of the electricity market in the island of Ireland and generally improving the quality and continuity of electricity to their respective customers.

At present there is a 275kV AC double circuit interconnection and two standby 110kV AC interconnections between the NIE and ESB networks as shown in Figure 1. The 110kV AC interconnections are run normally open and only closed in the event of a system emergency on either network. The 110kV AC interconnections need to run normally open because it its currently not possible to control the power flows through them and under certain network conditions it is possible that these interconnections could be overloaded. The installation of phase shifting transformers on these lines will control power flows and will enable the 110kV interconnections to run normally closed and thus enable both NIE and ESB to accrue the benefits of full interconnection.

PHASE SHIFTING TRANSFORMER FUNDAMENTALS

Interconnected Power Systems

Power system theory shows that when power flows between two networks there is a voltage drop and a phase angle shift between the sending end and the receiving end voltages. If the systems are connected together in two or more parallel paths so that a loop exists, any difference in the impedance's will cause unbalanced line loading. For two interconnected networks it can be shown that:

MW flow is proportional to voltage phase angle difference.
MVAr flow is proportional to scalar voltage difference.

If a PST is inserted into one of the lines it is possible to control MW and MVAr flow in the line through the use of a phase shift tap changer and voltage tap changer.

Types of PST

PST's can be either non-symmetric type or symmetric type. Non-symmetric devices add a quadrature voltage to the input voltage and the output voltage being the vector sum of these two perpendicular voltages (therefore the output voltage is boosted by a small amount and this "boost" reaches a maximum on extreme taps).

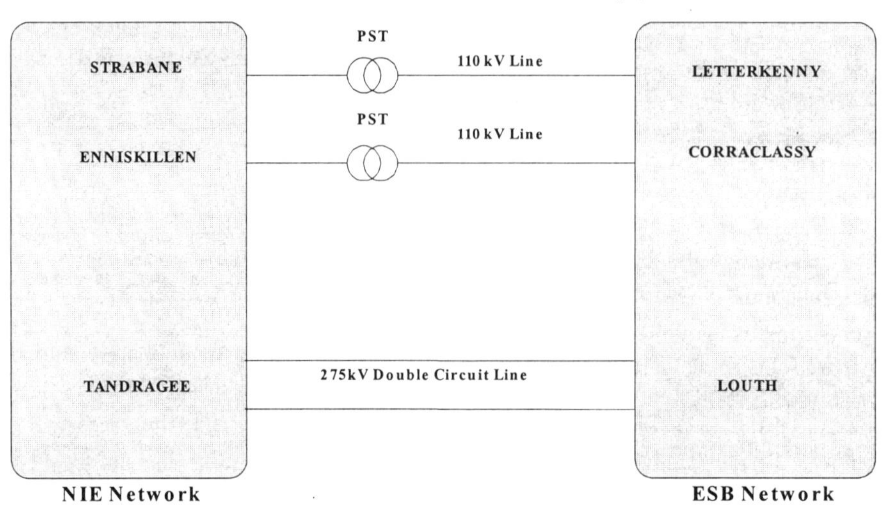

Figure 1 – AC Interconnection Between NIE and ESB

AC-DC Power Transmission, 28-30 November 2001
Conference Publication No. 485 © IEE 2001

Symmetric PST's are different in that they have constant or nearly constant input and output voltages across the phase shift range (the input and output voltage vectors would describe circular arcs through the phase shifting range). This is achieved by adding a symmetrical quadrature voltage to the input and output phase angle, as shown in Figure 2.

Depending on the direction of the power flow two different operation conditions are defined: Advance Operation – being the phase angle α that results when the load (L) leads the source (S) voltage. Retard Operation – being the phase angle α that results when the load lags the source voltage.

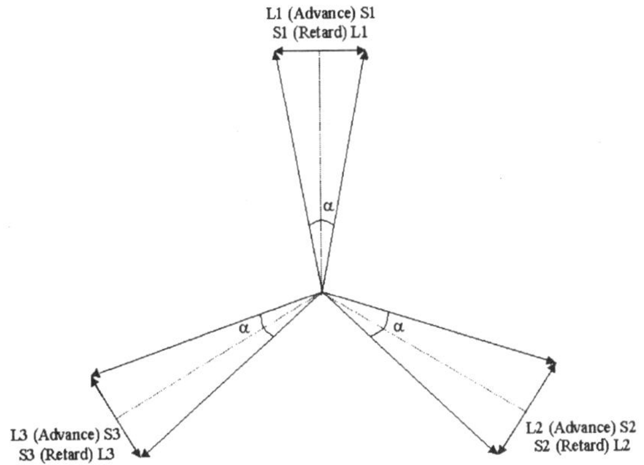

Figure 2 – Symmetrical PST

PST Under Load

A PST will in practice, have an internal impedance and this will modify the characteristics of the device. Take for example the equivalent circuit for a PST in Figure3

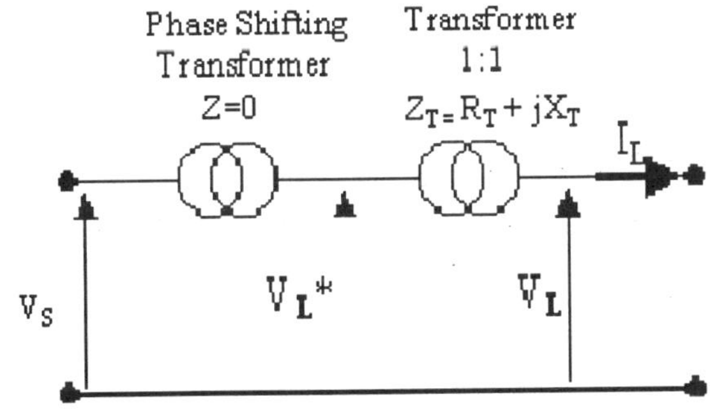

Figure 3 – PST Equivalent Circuit

$V_L^* =$ Load Voltage (No load)
$V_L =$ Load Voltage (Loaded)
$\cos \varphi_L =$ Load power factor
$Z_T =$ Transformer Impedance
$I_L =$ Load current
$V_{S(a)} =$ Source Voltage (advanced)
$V_{S(r)} =$ Source Voltage (retarded)

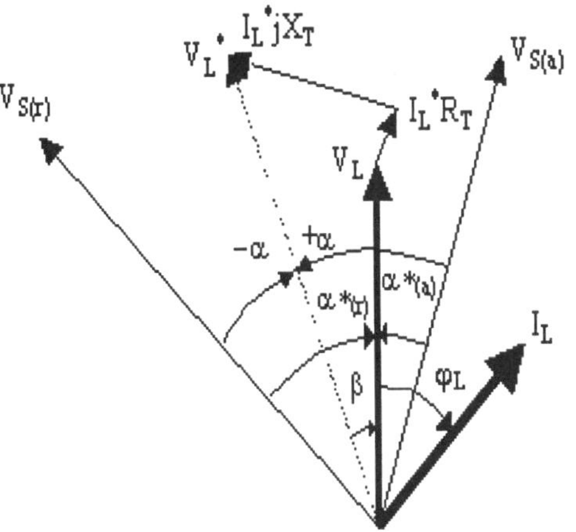

Figure 4 – PST Under Load

$\beta =$ Transformer Load Angle
$\alpha =$ Phase Shift Angle (No load)
  + Advanced (Leading)
  - Retard (Lagging)
$\alpha^* =$ Phase Shift Angle (loaded)

From Figure 4 it can be shown that:

Advanced phase shift angle (loaded) $\alpha^*_{(a)} = \alpha - \beta$ ..(1)
Retard phase shift angle (loaded) $\alpha^*_{(r)} = -(\alpha - \beta)$ ....(2)

At unity power factor load it can be proven that the transformer load angle β can be described in terms of percentage impedance:
$$\beta = \tan^{-1}\left\{\frac{z\%}{100}\right\} \quad\dots\dots\dots\dots\dots\dots\dots\dots\dots\dots\dots\dots\dots(3)$$

These three formulae are important for the design of a PST because under loaded conditions the impedance of the device will reduce the advanced phase angle range of operation.

PST DESIGN AND SPECIFICATION

The choice of impedance of the PST is dependent on a number of issues:

% Impedance

- Transient stability requirements –The impedance of the PSTs have an influence on the ability of NIE and ESB to transfer power between their networks. This could be significant during network emergencies, to ensure that networks remain within transient stability limits.

  Network studies showed that the networks would be stable for PST's with impedances below 20%. It was recommended to minimise the impedance to provide an additional margin of stability.

- Losses – It is envisaged that for normal system operation power flows between NIE and ESB would be via the main 275kV interconnection and the normal function of the PST's would be to inhibit MW and MVAr flow that would normally occur on an interconnected network if the PSTs were absent. The enquiry document for the PSTs therefore included a value for reactive losses.

- Short Circuit – Studies indicated that fault levels would be maintained within equipment ratings even for a device with zero impedance.

- Sensitivity of PSTs to impedance – A number of load flow studies were undertaken to ensure that the chosen impedance would not result in large changes in power flows under circuit outage conditions. The results showed that for a circuit outage, the power flows were relatively insensitive to PST impedance. Reducing the impedance from 10% to 3% increased the power flow through the PST in the limiting case by only 3MW.

Further system analysis showed that with the PSTs initially operating at a OMW transfer set-point they would be capable of restoring the flow to within their maximum continuous rating for any single simultaneous circuit outage on the combined NIE and ESB networks.

On the basis of the above it was considered that the impedance of the PSTs should be specified as less than 10% on all tap combinations.

No-Load phase shift advance and retard angles

Load flow studies were performed for operating conditions associated with the extreme ranges of power transfer between NIE and ESB. To determine the no-load phase shift angles. In these studies the PSTs were blocking power flow in the 110kV interconnectors. This resulted in a no-load advance angle of +45° and a no-load retard angle of –38°. Manufacturers of PST's require this information to design the exciter winding and to assist with the tap-changer selection.

Load Phase shift advance and retard angles

Similarly, load flow studies were undertaken for those operating conditions associated with the extreme ranges of power transfer between NIE and ESB to determine the load phase shift angles. With a PST impedance of 10% the studies required a load advance angle of +27° and a load retard angle of –17°, with reference to controlling the power on the line side of the PST. Manufacturers of PSTs require this information to ensure that under full load conditions the voltage drop across the PST's will still permit the specified load advance and load retard angle range.

Figure 5 compares the specified advance and retard angles with the actual angles of the designed PST. The actual PST's will have symmetrical no-load advance and retard angles of 45°. This symmetry was due to the symmetrical design of the tap-changer. The actual load advance and retard angles are asymmetrical. The internal impedance of the PST has the effect of reducing

the advance angle with increasing load and increasing the retard angle with increasing load.

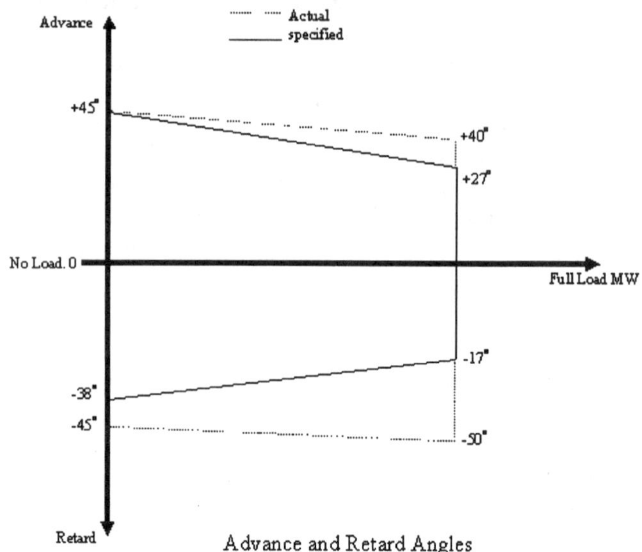

Figure 5 – Load and No-Load Phase Angles

Power flow step size

A quadrature tap-changer has a discrete number of steps to control the flow of power over the no-load advance and retard angle range. It is necessary for the client to advise the PST manufacturer of the number of steps that are required. A balance needs to be determined between a practicable number of steps and the magnitude of the MW change per step, as this has an operational impact on the two networks.

Load flow studies based on a peak demand of 490MW export were performed for varying power transfers through the PST and showed that, on average a 20MW change in power flow would require a 5° change in phase angle across the PST. To provide the power system operators with an appropriate level of control i.e. 5MW, a 1.25% step tap changer was selected.

With a no-load range of 90° a 72 step tap changer was needed. It is not practicable to purchase a single tap-changer with this number of steps. The number of connections, leads and contacts would impact on the reliability of the PST. There would also be the issues of cost and potential problems with transport due to the weight of the PST.

Taking these factors into consideration, a 35 position tap-changer with an advance retard switch (ARS) was selected. The ARS switch reverses the polarity of the quadrature voltage in the PST and almost doubles the tap-changer range. The combination of tap changer and ARS switch resulted in a 65 step device that would provide approximately 5.6 MW/step power flow control. This option also helped to minimise the number of connections, leads and tap changer contacts.

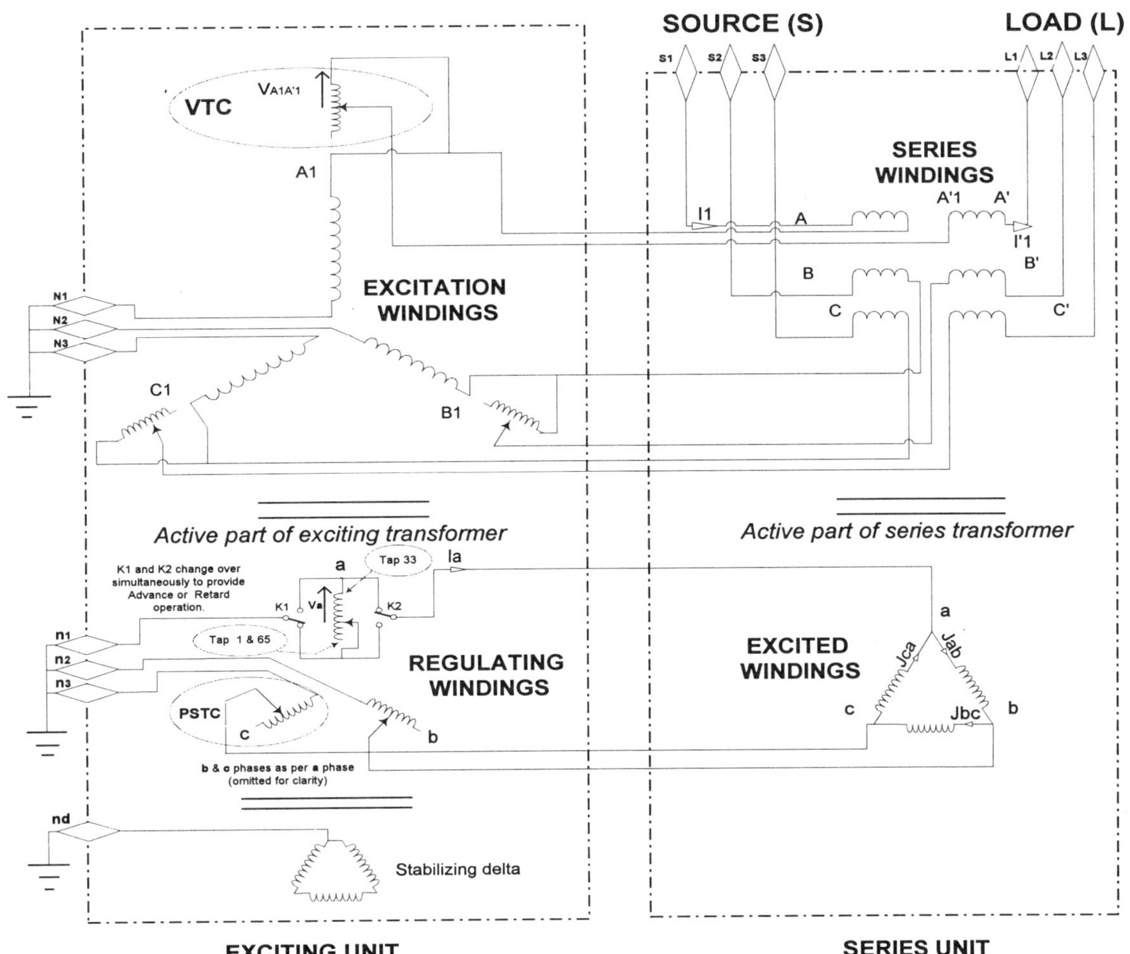

Figure 6 Winding Connections for Phase Shifting Transformer

The final PST design was of shell type design having two cores, in two separate tanks; the series and the exciting unit, as shown in Figure 6.

Voltage step size (MVAr)

As the voltages on each network may vary by $\pm10\%$ and with the need to synchronise the two networks, a tapping range of $\pm20\%$ is required. With the tap-changers that are commercially available and by selecting from a regular step fraction a 35 step tap changer was specified, to give a step size of $1.33°$ and a range of $\pm22.67\%$.

With the PST in service the voltage tap-changer will be used to control MVAr flow between the two networks. NIE and ESB require the ability to control reactive power flows between the respective in the range $\pm$ 30 MVAr. System studies were performed to investigate what the expected MVAr flow would be per step. The analysis produced an average figure of 6.0 MVAr/step which is similar in magnitude to the MW/step for the phase shifting tap changer.

MVA Rating

The Letterkenny–Strabane and Enniskillen–Corraclassey 110kV overhead lines have winter ratings of 125 MVA. The ratings of the PSTs were thus matched to the overhead lines.

Voltage Rating

The 110kV voltage on the NIE and ESB systems may vary by $\pm$ 10% and the maximum voltage could rise to 121kV. The PSTs were therefore rated to the closest IEC standard of 123kV.

Rated Power	125MVA
Cooling	ONAN/OWAF
Rated Voltage	123kV
Nominal Winding Voltage	110kV/110kV
Voltage Tapping Variations	
Range	$\pm$ 22.67%
Step Voltage	1.33%/step (6MVAr/step)
Phase Tapping Variations	
No-Load Range	$\pm45°$
Full-Load Range	$+40°$ to $-50°$ (5.6MW/step)
Impedance Voltage	3.5% to 9.5%
Standards	IEC

Table 1 - Summary of PST Specification

TAPCHANGER INTERACTIONS

PSTs which control both MW and MVAr must be given special consideration as the two tap changers are not totally independent of each other.

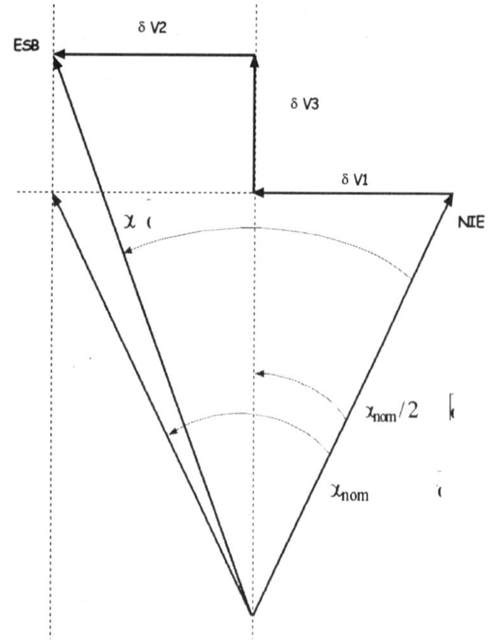

Figure 7 – Tapchanger Interactions

The phasor diagram in Figure 7 illustrates that increasing the voltage tap position will have the effect of slightly reducing the phase angle α

No Load Voltages

$V_{ESB} = V_{NIE} + \delta V_1 + \delta V_2 + \delta V_3.$

$\delta V_1 = \delta V_2 =$ Voltage shift due to PSTC tapping.

$\delta V_3 =$ Voltage shift due to VTC tapping.

$V_{ESB} = w\ q[V_{NIE}.Cos(\alpha_{nom}/2) + \delta V_3]^2 + [V_{NIE}.Sin(\alpha_{nom}/2)]^2 r$

$\delta V_3 = V_{NIE}.Cos(\alpha_{nom}/2). K_{nom}$, K_{nom} being the voltage nominal tapping adjustment, in this case, of the range $+/- 0.2267$ in 35 taps.

$V_{ESB} = V_{NIE}.Cos(\alpha_{nom}/2).w\ q[1 + K_{nom}]^2 + [Tan(\alpha_{nom}/2)]^2 r$

The actual no-load voltage ratio is therefore:

$\text{Ratio}_{(act)} = Cos(\alpha_{nom}/2).wq[1 + K_{nom}]^2 + [Tan(\alpha_{nom}/2)]^2 r$

$\alpha_{(act)} = \alpha_{nom}/2 + Tan^{-1} q\ Tan(\alpha_{nom}/2)/[1 + K_{nom}] r$

An example of the effect of voltage control on phase angle under no-load conditions is given in Figure 8, where the nominal phase shift $\alpha = 45°$ at voltage tap 18 (principal tap), to $\alpha = 50.7°$ at voltage tap 1 an to $\alpha = 41.2°$ at voltage tap 35.

Under loaded conditions the transformer provided an additional shift in phase due to it's own internal impedance which alters the effective phase shift further. The range of variation from the nominal $45°$ extends from to $+53.9°$ to $-34.1°$ depending on the voltage tap position and loading of the transformer.

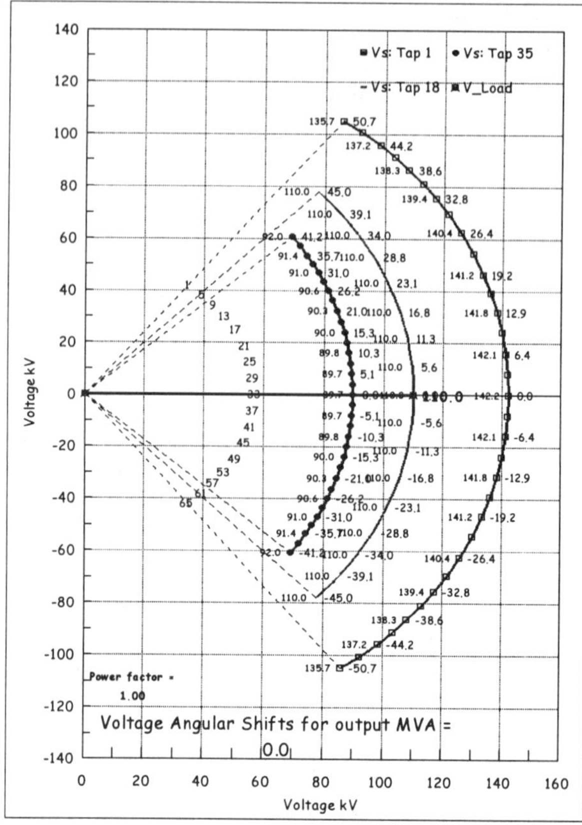

Figure 8 - No Load Condition

PST PROTECTION

Transformer Failures

Most transformer failures in the past have occurred near to the tap changer and were most likely to be associated with the tap changer leads. The tap changing and winding switching arrangements for a phase shifting transformer can be particularly complex, due to the high number of taps provided. Therefore, special attention should be paid to the possibility of a fault The most likely form of short circuit is an earth fault. High-speed protection, with adequate sensitivity must be provided to respond to an internal earth fault at likely locations within the PST units.

In considering the PST design, where the connections between the exciter transformer and series transformers may go through phase segregated bus ducts, a phase-phase fault is considered to be of low probability. However, there may be a heightened risk of a phase fault occurring if the bus ducts are not phase segregated, or amongst winding tail connections to one of the tap changers assemblies, or at the ARS switch.

Some types of failure within a transformer, such as an inter-turn fault or deteriorating insulation, can be classed as an incipient fault. These may not lead immediately to a major electrical fault. Buchholz gas alarm devices were specified to detect these incipient faults.

The transformer tank and tap changer compartments were to be protected with pressure relief devices and buchholz surge devices.

Thermal Protection

Transformer loading is hourly, seasonally and weather dependent, as is a transformer's thermal withstand limit. The winding temperature indicators for each winding were specified to the forced cooling equipment and provide winding temperature alarms and trips.

The thermal protection of the PSTs requires particular attention, as there are numerous windings located in two separate transformer cooling systems.

A standard temperature indicator can only measure the top oil temperature by a suitable transducer, ie thermocouple or mercury filled capillary. This oil temperature indicator can be modified by a CT driven heater to simulate the winding temperature. Whilst this is indicative of the transformer thermal state, the most critical temperature is that deep within the core and windings (Core Temperature), but the transducer would be difficult and expensive to place as it must be laid into the winding and may adversely effect the reliability of the PST.

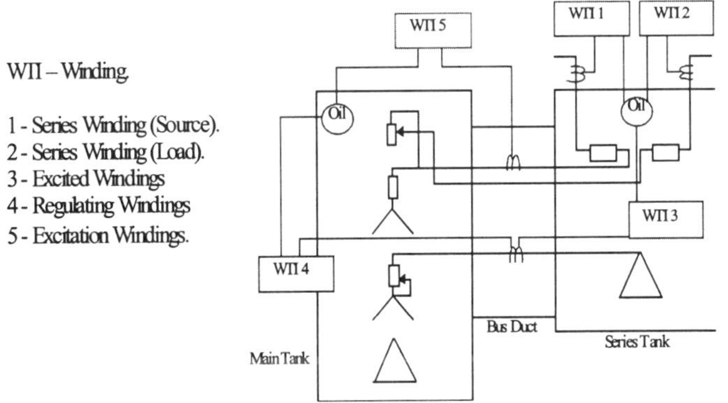

Figure 9 - Winding Temperature Protection

For a PST of this configuration about the zero phase shift taps there could be a small proportion of the regulating winding conducting the full load current of the delta windings, for this design, approximately 677 amps. This could cause an extremely unequal temperature distribution across the winding with poor heat dissipation from the portion of the windings under load. With a small area of winding transferring heat to a large volume of oil, the top oil temperature would not be indicative of the winding loading, which may lead to the WTI scheme described above being inaccurate, allowing the regulating winding to exceed thermal limits. In this situation the expense and complexity of a core mounted temperature probe to measure the actual temperature of this winding may be justified.

The PST manufacturer, offered a shell type transformer and demonstrated the windings about the star point were at the extreme ends of the pancake winding. Each pancake by virtue of shell type design has a separate ducted cooling system with vertical oil flow. The fine winding of the phase angle tap changer regulating winding comprised of two pancake windings with the neutral ends at the extreme outsides thus benefiting from maximum cooling. In addition, the tappings about zero phase shift are in the top oil temperature and therefore the conventional WTI thermal model should remain acceptably accurate. After this analysis the proposed scheme for the thermal protection of the PSTs was to use conventional WTIs as shown in Figure 9.

Overvoltage

The risk of overfluxing of the PST was considered to be low and no specific protection was recommended. However, a risk of severe overvoltage was identified in the event of tap changer runaway. Overvoltage protection was recommended for both sides of the PST's to address this risk.

Overcurrent

High-speed differential protection (87P) for the primary windings of the series and excitation transformers was recommended, as illustrated in Figure 10.

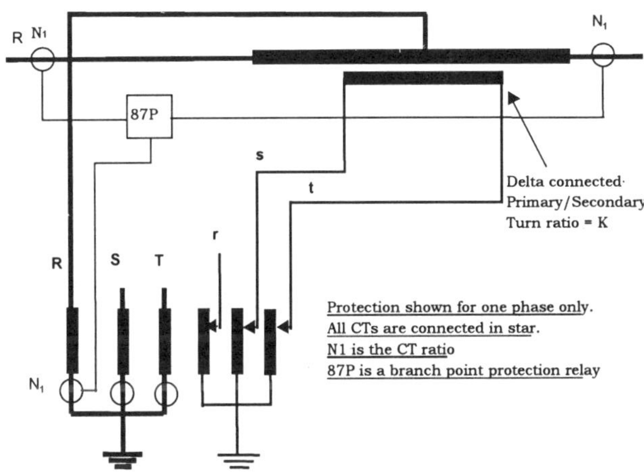

Figure 10 - Primary Winding Differential Protection

This is a circulating current scheme, similar to that often employed for auto-transformers. No bias is required for magnetic inrush currents as they are viewed as through currents by the protection. Similarly ampere-turn balance is maintained during voltage and phase shifting tapchanger operations.

A biased of differential protection (87S) for the series transformer unit, as illustrated in Figure 11 was also proposed. This protection summates the input source current and the output line current of the series unit primary windings. The summated current is then compared (after phase and ratio correction) with the series unit secondary current. By locating CT's in the neutral tails of the exciting transformer secondary windings, some phase fault protection coverage can be provided for these windings.

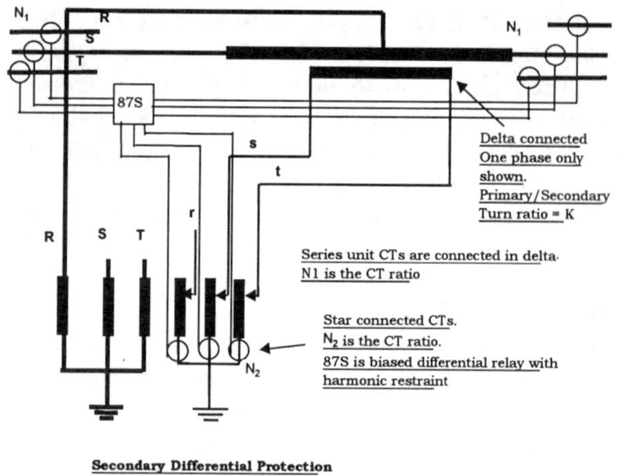

Figure 11 – Secondary Winding Differential Protection

The protective relay must be biased so that it will remain stable under through fault conditions and for magnetic inrush currents. Again, no bias is required for the voltage and phase shifting tapchanger operations as ampere-turn balance is maintained.

Figure 11 indicates that the two sets of CTP CT's should be connected in delta and then directly paralleled before connection to a bias input of the differential relay. The direct delta connection of the

CTP CT's reflects a US tradition. The European tradition has commonly been to use interposing CT's, for both phase and ratio correction, rather than attaining phase correction by the secondary connections of main CT's.

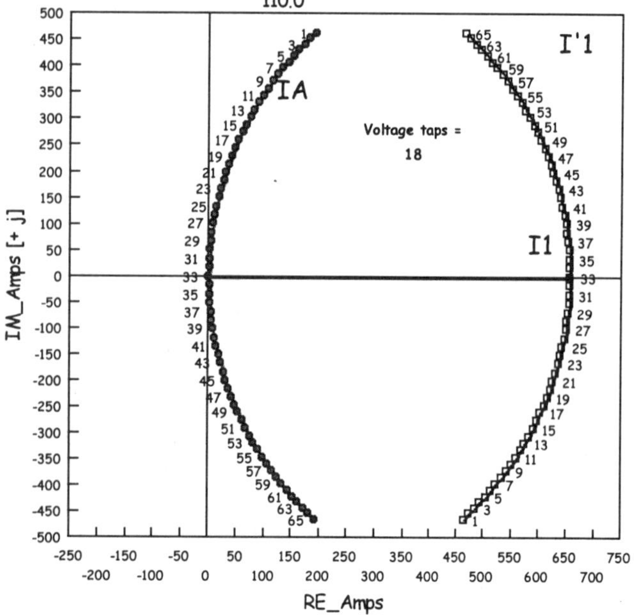

Figure 12 - PST Winding Current Flows for Change in Tapchanger Phase Angle

With the very low input impedance's of modern relays, greater through fault security will be offered with the two sets of CTP CT's connected to separate sets of biased current input circuits, rather than commoning the CT's to one set of relay inputs.

PST Winding Current Flows

The single phase current flows in the PST windings are shown in Figure 12.

An example of the relationship between the input current I1, the output current I'1 and the neutral current IA, is shown graphically in Figure 13, for the case when the voltage tapchanger remains on the principal tap position and the phase shifting tapchanger is varied.

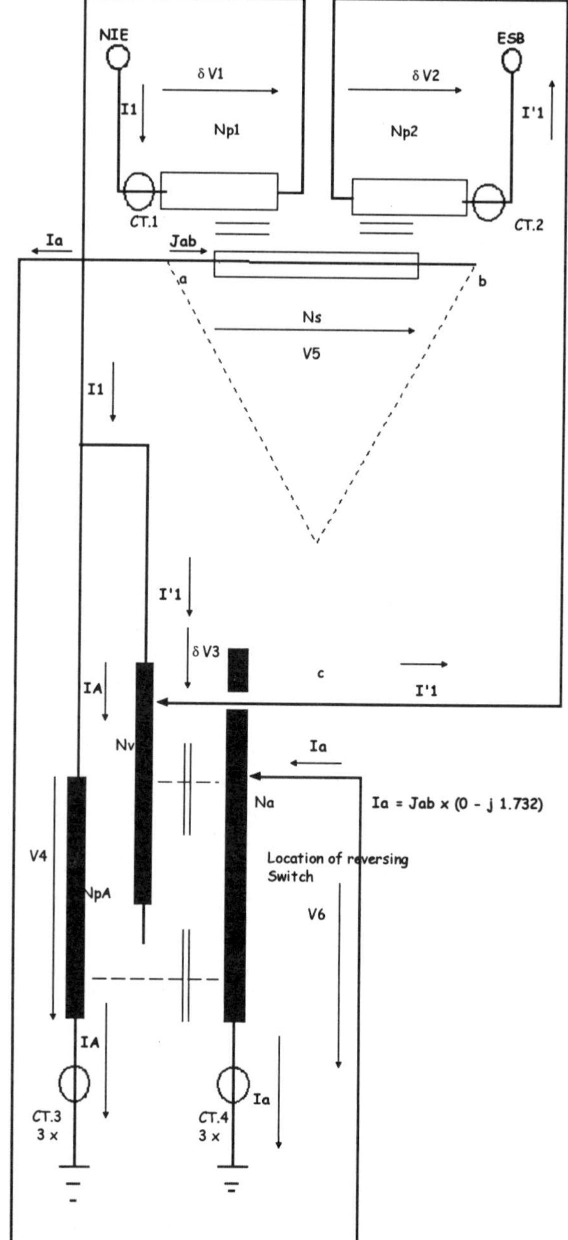

Figure 13 - PST Winding Current Flows

Back-up Overcurrent and Earth Fault

The PST transformer core is of shell-type construction. A shell-type core is essentially a 5-limb magnetic circuit and the zero sequence shunt magnetising reactance of the PST would typically be of a similar value to the positive sequence magnetising reactance. This could be too high for the effective earthing of the excitation transformer primary windings. The provision of a stabilising delta winding means that the PST will have a low zero sequence impedance and will therefore be a source of zero sequence current for 110kV system earth faults.

Primary winding stand-by earth fault protection was provided and will co-ordinate with the 110kV system earth fault protection. It is made insensitive to 2nd harmonics, by a time delayed characteristic. It is set to offer greater sensitivity to earth faults near the transformer neutral than might be possible with the differential protection.

Earth fault protection was also recommended for the earth point of the secondary exciter windings. This relay has a short time delay as it does not need to co-ordinate with other protection schemes.

Back-up over-current and earth fault relays were also recommended to prevent the PSTs from being subjected to sustained through fault conditions in excess of their short-time withstand limits.

CONCLUSIONS

1. PST's are generally designed to be site specific and many system studies need to be performed to adequately specify these devices. In particular, the load and no-load phase angle ranges need to be determined. The % impedance will also influence the performance of the PST

2. Utilities need to ensure that there is an adequate number of steps per tap changer to allow tolerable network MW and MVAr flows per step change.

3. Special attention needs to be given to adequately protect the PST for phase and earth faults. Care needs to be taken to ensure that the design of the PST can adequately accommodate the CT's necessary for the protection schemes.

4. Special attention also needs to be given to ensure the PST has adequate thermal protection.

ACKNOWLEDGEMENTS

The authors wish to thank the following for their assistance: PB Power for assistance with system studies and technical specifications, Jeumont Schneider for PST design insights.

REFERENCES

1. Brown, F.B. et al.- "The First 525kV Phase Shifting Transformer Conception to Service - Salt River Project". Doble Engineering Company 1997.
2. ANSI/IEEE PC 57.13S – "Guide for the Application, Specification and Testing of Phase Shifting Transformers".
3. ABB – "Protective Relaying Theory and Applications." 1994, Chapter 10.
4. Fyvie J.D. – "Design and Testing of Phase-Angle Regulating Transformers, April 1992, Canadian Electrical Association.
5. Seitlinger, W. – Phase shifting Transformers Discussion of specific Characteristics, 1998, CIGRE Session 12-306.
6. CIGRE Working Group WG12.05 report of 1983.
7. Hindle,P. "Review of PST Protection Requirements". June 2000.
8. Stedall, B. "NIE-ESB Joint Studies Aimed at Improved use and Development of Interconnection - Final Report". February 2000.

AUTHORS ADDRESS

The principal author can be contacted at:

Robert Sweeney
Northern Ireland Electricity,
120 Malone Road,
Belfast,
Northern Ireland,
BT9 5HT.

Authors can be contacted by e-mail at:
Email : bob.sweeney@nie.co.uk
Email : gerry.stewart@nie.co.uk
Email : Patrick.ODonoghue@esbi.ie
Email : paul.smith@ngrid.ie

NEW SWITCHING TECHNOLOGIES IN MEDIUM-VOLTAGE SYSTEMS - APPLICATION AND REQUIREMENTS

C. Heinrich, H. Schmitt, I. Hoever

Siemens AG, Germany

ABSTRACT

Deregulation is setting new standards also in the energy market. It is creating new opportunities for power supply corporations and industrial suppliers. The direct competition in power distribution exerts severe cost pressure on the system operation with the result that systems are expanded only if absolutely necessary. Consequently, the quality of power supply is stagnating or even deteriorating.

To be able to act more flexible and cost-effective and to meet the requirements of the changed market the operators of public distribution systems and industrial networks ask for new switching solutions. These solutions include fast acting solid state circuit-breakers and also DC transmission systems based on electronic power modules and superconducting fault current limiters, and can be used, for example, to link different system sections flexibly without impairing stability or short-circuit behavior.

INTRODUCTION

The deregulation of the energy market leads to changes of the operation of industrial networks and public distribution systems (Fig. 1):
Direct competition in energy supply will exert a strong cost pressure on the system operation.
In addition to this, industrial plants tend to increase the process automation. Fluctuations in supply voltages and short-term system interruptions may cause problems in the production processes, extending to failures and even damage to the manufacturing plants.
Furthermore, independent power producers like windparks or combined cycle stations will feed their energy into public as well as industrial networks, increasing their short-circuit rating and in some cases exceeding the short circuit withstand capability of installed equipment.
These changes require the application of new technologies providing new possibilities to prevent interruptions of power supply, to improve the power quality and to reduce short-circuit currents.

The paper describes the system requirements and the application of new switching technologies, including fast acting circuit-breakers, superconducting fault current limiters and flexible direct current links for electrical medium-voltage systems.

AC-DC Power Transmission, 28-30 November 2001
Conference Publication No. 485 © IEE 2001

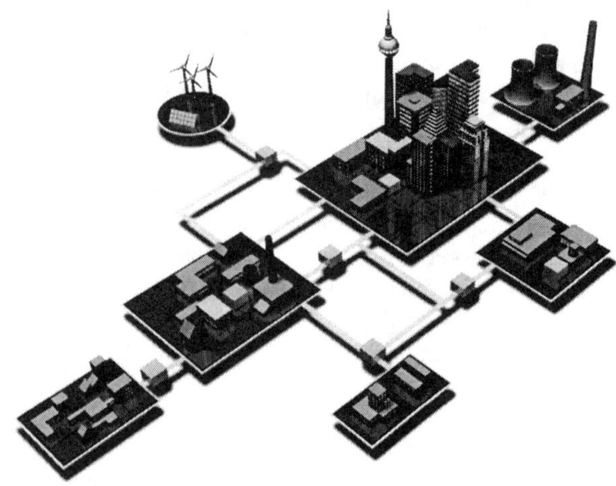

Fig. 1: Distribution network design

SYSTEM REQUIREMENTS

Public networks

Changed conditions for network operation in the public area, mainly caused by the deregulated energy market, lead to an increased cost pressure. The reaction of the operators is an increase of maintenance intervals and a higher utilisation of transformers, cables and other equipment, leading to more stress and resulting in a decreased life time expectance.
Furthermore network operators tent to outsource complete divisions like consulting or maintenance.

This situation is combined with a cut of investment in new equipment (i.e. for the replacement of old equipment and for the extension of the existing network). It may be expected that the quality of power supply decreases with impact on the reliability and therefore the continuity of supplied processes [1].
In countries with a historically high level of power supply the reliability and power quality is usually much higher than required by standards [2]. Especially there the network operators will refer to this fact using it as an excuse for the reduced level.
Private customers may feel only a little loss of comfort, caused by an increased average number of interruptions or minutes lost. Trade and industrial customers may suffer from short and long term outages, leading to interruption or loss of production processes. Follow-up costs of these

interruptions may be compensated by appropriate insurances, but to avoid these interruptions the customer may install additional equipment to compensate the decreased reliability.

Another change with impact on network design and network operation is the increasing number of independent power producers (IPP), feeding their energy into MV networks. Their contribution to the short-circuit rating on top of the existing network short-circuit rating may exceed the carrying or breaking capability of existing equipment like cables or switchgear. New solutions are needed to limit the increased short-circuit currents in case of failure to a tolerable value.

Beside of the increased short-circuit currents decentralized generation leads to a change in load flow within the systems. In the past the energy flows from large power stations straight to the customers. Nowadays the variety of power stations and their variable energy production (depending on wind, sunshine, or energy demand in general) contributes to a load flow with new demands for operation control and system protection.

Industrial networks

Industrial plants tend to increase the process automation; mainly the number of variable speed drives or processes, controlled by computers or automatic controls, increase significantly. These production lines react much more sensitive to disturbances in voltage supply and may be interrupted even at short term sags or outages (interruptions longer than a halfwave). This may lead to a loss of production and high follow-up costs.

The sensitivity of variable speed drives against short term disturbances is enhanced by the tendency to reduce the built-in capacity of the drives. This leads to increased requirements for a high power quality of the feeding network. Furthermore, industrial systems are characterized by a high energy density because of the high energy consumption within a relative small area.

APPLICATION OF NEW TECHNOLOGY

The paper describes the principle applications of new switching technologies in electrical systems:

- superconducting fault current limiters
- direct current links
- fast acting solid state circuit-breakers,

with main focus on the application of fast acting switching.

Superconducting Fault Current Limiter

The superconducting fault current limiter is practically non-resistive, that is under normal operating conditions as long as the temperature remains below a critical value. If a short-circuit occurs, the current and therefore the temperature in the superconducting material rises and the resistance increases to a value significant over zero. This process lasts for just a millisecond and limits the current to about three times the rated current, which the downstream circuit-breaker or even load-break switchgear is easily able to interrupt. Subsequently, the current limiter regenerates automatically in less than two seconds and is superconducting again.

The superconducting fault current limiter may be used to protect existing or new systems from inadmissible high short-circuit currents. At system extension or integration of additional generation the short-circuit power of medium-voltage systems may increase to such an extent, that the breaking capacity of existing switchgear and the short-circuit strength of the electrical system are exceeded. The installation of current limiters has the advantage that no switchgear or other equipment need to be replaced.

Medium-voltage direct current (MVDC) link

The application of a MVDC link is the only possibility to couple asynchronous systems of different frequency or phase angle as well as systems with different levels of stability. While the coupling of systems with different frequencies may be applied only to a limited number of applications, phase shift between neighboring subsystems can be found at various points: at the interconnection of short and long line distribution systems or systems of different utilities. Because of its ability to control the flow of active and reactive power respectively the MVDC link is used for controlled exchange and trading.

Coupling of these networks is possible at any phase angel without impact on the stability of each system. The same feature may be used for connecting IPP to existing networks.

HVDC links have already been successfully used for considerable time on the high-voltage level. Based on the progress in power electronics and drives, such systems are now economically attractive for medium voltage applications. State-of-the-art power electronics provide numerous opportunities for optimizing distribution systems. This is of significant importance, especially for a changing energy market where the required quality of supply can vary from place to place.

With SIPLINK (Siemens Multifunctional Power Link), Siemens provides a MVDC system for distribution networks. SIPLINK is used to control the load flow and optimizes the voltage stability by providing active or reactive power [3].

In addition, SIPLINK also improves the quality of supply in terms of reliability and voltage quality on the distribution level. This feature may be enforced by integrating an energy storage into the DC link providing an uninterruptible power supply.

The first SIPLINK has been installed at the switchgear factory in Frankfurt/M (Germany). The system comprises two units: The first one allows the coupling of two existing power supplies to exchange power, resulting in reduced reserve power for the complete power demand. The second one builds up an independent 60-Hz-system for the test laboratory, and therefore opens the possibility to conduct various tests in a 60 Hz environment (Fig. 2).

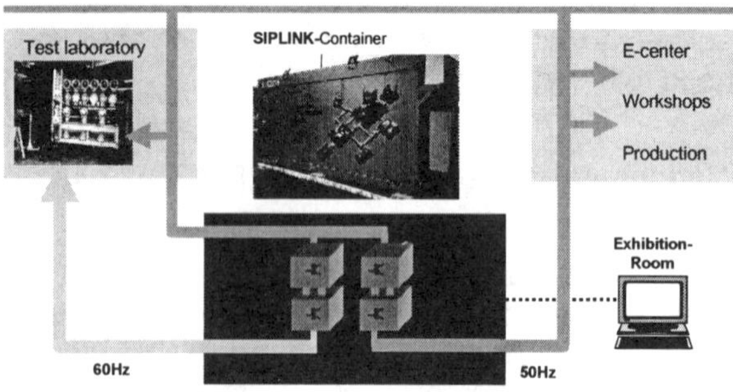

Fig. 2: Application of SIPLINK at the switchgear factory in Frankfurt/M

Another application of a 2 MVA SIPLINK is used to improve the power (voltage) quality in a 20 kV distribution system. Furthermore, it is used to control the power flow between adjacent sub-networks depending on the demand and economical aspects [5]

Fast Acting Solid state circuit-breaker

The most important module of the new switching technologies is the fast acting solid state circuit-breaker, a switching technology based on power-electronic components. Solid state circuit-breakers are able to interrupt currents much faster than conventional switching devices. Clearing a fault in a very short time means, it will have less or practically no effect on other loads or processes in adjacent feeders. Even sensitive consumers will withstand such a short term interruption of supply without negative impact. Particularly in industrial

networks, this will prevent high follow-up costs caused by production losses.

Another important application for fast acting solid state circuit-breakers is the current limiting switching. Due to the ability to interrupt currents extremely fast and outside of the current zero, they may interrupt a short-circuit current before its peak value. This way, it is possible to eliminate short-circuit currents or prevent additional contributions to subsystem.

The Electronic Siemens switch (**ESI**-Switch) is designed for the reduction of failure durations or the limitation of short-circuit currents depending on the used solid state devices. They can also be used for synchronous making and breaking operation. This suppresses impulse or inrush currents during switching on. Furthermore the switching of direct currents is possible, giving the chance to improve the operation of DC systems like railway networks or on-board supply for motor vessels. The different design of the **ESI**-Switch depending on the application is described in the following chapters.

Fast switching

Fast switching requires the interruption of currents at the first zero crossing after failure ignition. For this purpose phase-commutated light triggered thyristors are used in the **ESI**-Switch. They are highly reliable devices with long term experiences in different applications like converters, switches and HVDC systems. They are available for high blocking voltages and large rated currents. High-power thyristors exhibit relative low on-state voltage for reduced power losses during operation. There is no control unit or other electronic device necessary on high-voltage potential (Fig. 3).

Fig. 3: Thyristor-modules of **ESI**-Switch

The most common application of thyristor based solid state circuit-breakers is the transfer switch (SSTS) (Fig. 4). It allows a sensitive load to be transferred from the standard supply to an alternate supply in less than a quarter cycle. During normal operation the load is supplied by the standard source 1. If source 1 fails (interruption of supply or under-voltage below a definable limit, i. e. sag) switch 1 will open and switch 2 will close almost simultaneously. The change of the current flow causes switch 1 to open and disconnect the load from the faulty source 1.

A SSTS in combination with two independent feeders provides the complete functionality of an UPS-system.

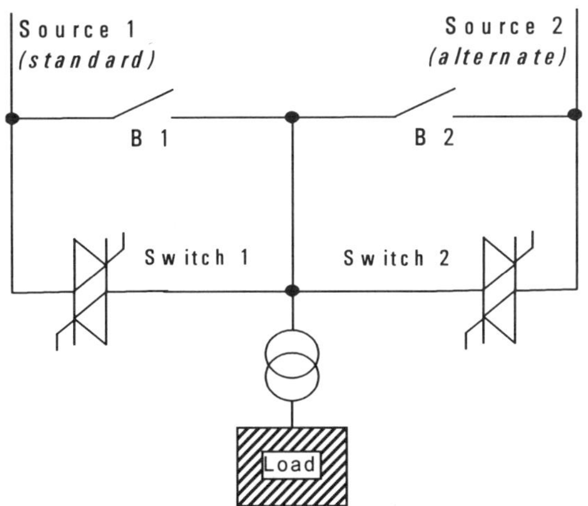

Fig. 4: Application of **ESI**-Switch as transfer switch

For industrial applications in areas of public networks with lower reliability the **ESI**-Switch allows a very fast disconnection of a healthy plant system incorporating in-plant generation from the faulty public system (Fig. 5).

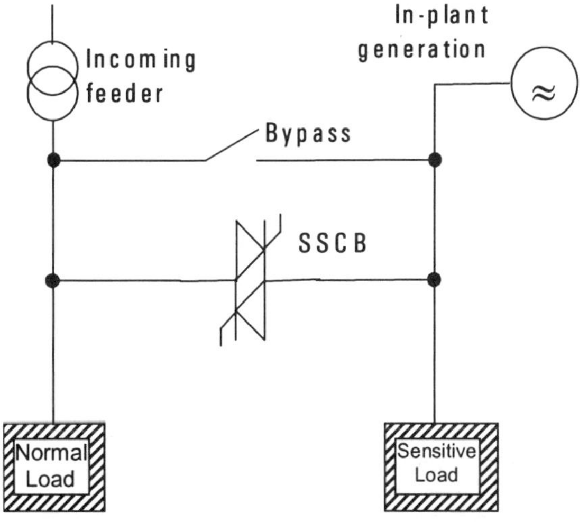

Fig. 5: **ESI**-Switch for disconnection of in-plant generation from incoming feeder

A fast sag-detector recognises a voltage drop below a definable limit, allowing typical time for detection and disconnection less than half-cycle after fault ignition. Depending on the installed generation either the whole plant load or part of them (the most sensitive part) may be supplied. This allows in fact an uninterrupted power supply for all sensitive processes, using the existing in-plant generation instead of an energy storage.

An **ESI**-Switch has been already installed in the switchgear factory in Frankfurt/M (Germany). It operates as circuit-breaker in the incoming feeder and is able to interrupt failures in less than 20 ms including arc extinguishing. The supply of the factory in that case is secured by the second incoming feeder.

In addition to the mentioned advantages the **ESI**-Switch may be used for synchronous switching operation for switching on. This reduces the stress for other equipment by avoiding large inrush currents of transformers or impulse currents of capacitors. For directly connected motors a soft-starter can be realised, reducing the transient current during start-up.

Fast and current limiting switching

In order to prevent short-circuit currents or to switch direct currents, self-commutated solid state devices are used as the active part of the **ESI**-Switch. They are able to interrupt a rising short-circuit current before it reaches the peak short-circuit current I_P or at any instant of the direct current.

Among a variety of self-commutated solid state devices the 'hard' driven GTO or Gate Commutated Thyristor (GCT) are the best suited devices for MV application (Table 1). They possess very short operating times of only a few microseconds. GCT are available as integrated devices (IGCT) with complete gate unit close to the GCT.

The switching capability of these self-commutated devices for interrupting short-circuit currents has been tested in laboratory. For this purposes a square wave generator was used. It represents a distribution network with its high inductance. It is able to generate currents up to 6 kA with a rise time of approximately 200 µs. These values are comparable to those of distribution networks [4].

This test circuit is used to investigate the switching capability of solid state devices with respect to the specific conditions in distribution systems. Especially the switching-overvoltage behavior in accordance with the current chopping effect of current limiting switching has to be investigated carefully.

TABLE 1: Comparison of different self-commutated semiconductor devices

	GCT / IGCT	IGBT	Thyristor with commutation circuit
Power losses	1 p.u.	1.8 p.u.	0.8 p.u.
For switching off	Gate unit (LV capacitors)	Gate	HV capacitors
Rated operating current	Up to 2000 A	Up to 800 A	Up to 4000 A
Rated switching current	4 – 6 kA	1.5 kA	Up to 8 kA (depending on capacitors)

The results of a 3 kA breaking test are presented in Fig. 7. This figure shows the voltage and current characteristics of one IGCT disc when switching off a current of 3 kA. The energy consumption of the semiconductor for this breaking duty is also shown.

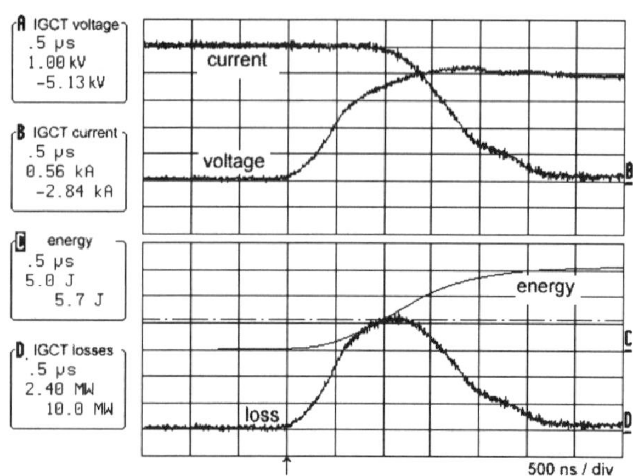

Fig. 7: Voltage and current of a 3 kA breaking test

The laboratory tests have shown, that IGCTs are well suited to interrupt short-circuit currents while rising. This gives the opportunity to increase the utilisation of distribution systems under normal operation conditions without exceeding the short-circuit handling capability of the installed switchgear.

INDUSTRIAL APPLICATION

The **ESI**-Switch has been used in an industrial environment at a manufacturing plant. To proof the reliability of devices for high voltage applications a large number of specimen have been tested to ensure the statistical lower level of failure. During these tests at the instant of failure of specimen a short circuit current started, endangering the feeding transformer and other equipment. Instead of using a fuse and replacing it after each test a one-phase **ESI**-Switch was installed (Fig. 8).

Fig. 8: Modul of the **ESI**-Switch during application at manufacturer

It has been triggered shortly after ignition of the short circuit current and interrupting it within less than 1 ms (Fig. 9).

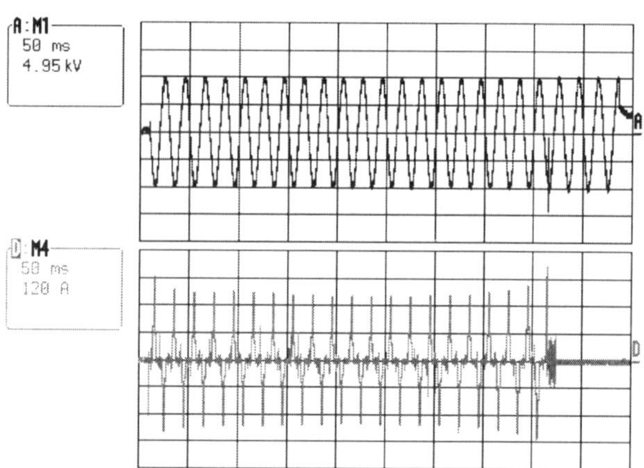

Fig. 9: Example of a test cycle with short circuit interruption after specimen failure

SUMMARY

A high level of power quality will become more and more important in the future - especially for industrial networks.

Due to the deregulation of the energy markets utilities and network operators will commercialize the quality of power supply as a product. The costs of this product depend on the demands of the customer and the processes to be supplied and will certainly increase with the requirements.

New switching technologies will improve the power quality. This will be used as a benefit from the utilities and network operators as well as directly from the industrial customers to ensure the power

quality requested for important processes by their own.
The superconducting fault current limiter, SIPLINK and **ESI**-Switch are such kind of new technologies.

REFERENCES

1. C. Heinrich, H. Schmitt, "Netzbetrieb auf der Basis neuer Schalttechnologien für Mittelspannungsnetze", <u>Elektritzitätswirtschaft Jg.99 (2000), Heft 17-18</u>

2. EN 50160, "Voltage characteristics of electricity supply by public distribution systems"

3. J. Flottemesch, U. Hildmann, M. Weinhold, „Kontrolle der Leistungsflüsse in Verteilungsnetzen mit Hilfe von Leistungselektronik", ETG-Fachtagung 2000, Hannover

4. C. Heinrich, H. Schmitt, "Integration of new switching technologies in medium-voltage systems", CIRED 2001

5. B. M. Buchholz, „Netzoptimierung am Beispiel EDIson", conference „Life Needs Power" at the Hannover Fair 2001

Design of a Combined Converter-Switch for Converting Existing AC Transmission into HVDC (and back)

E. Imal,
Fatih University
Turkey

M. Bagriyanik,
Istanbul Tech. Un.,
Turkey

P. Ali-Zada,
Fatih University,
Turkey

K. Rajdablı,
Kema Consulting,
USA

R. Alı-Zada,
Azerbaijan Poly Un.,
Azerbijan

Key words: star-delta rectifier, inverter, VSC Transm., DG.

Abstract: The paper brief is one out of the first approaches in Turkey to popularize VSC Transmission. It deals with:
1. Lab design and PSCAD modeling of a special bridge controlled AC – DC three load (star-delta) rectifier. It uses 9 group of valves (instead of 6 group valves of common bridge), three lines of former AC transmission, the earth and three DC loads between the lines and earth (or a star-delta inverter).
2. Lab design and PSCAD modeling of a special controlled bridge DC – AC star-delta inverter between three DC source and AC network. It also uses 9 group of valves (instead of common 6 group valves), three input DC lines (and the earth) and three outputs AC transmission lines.

Comparison of the bridges construction and modeling results of traditional VSC Transmission (6 valves rectifier, traditional DC line and 6 valves inverter) and designed AC-DC-AC transmission (9 valves rectifier, AC line and 9 valves inverter).

The proposed star-delta rectifier and inverter can be implemented in common electrical applications such as HVDC transmission, supply of isolated loads, asynchronous grid connection, infeed of small-scale distributed generators (DG), infeed to city centers, for DC grids, converting the AC lines into VSC Transmission, etc.

I. INTRODUCTION

As it is known, using HVDC to interconnect two points in a power grid, in many cases is the best economic alternative and furthermore it has excellent environmental benefits. With a HVDC system, the power flow can be controlled rapidly and accurately as to both the power level and the direction. From the other site, in areas where permits for building new AC transmission lines are hard or impossible to get permission, the transmission capacity of existing distribution systems can be increased in a very effective way by converting the AC lines into VSC Transmission.

The other and very important issues of the day are DC grids creation and the infeed of small-scale DG (usually running asynchronously) to main AC grid. New FACTS techniques (AC-DC and DC-AC converters etc) afford an opportunity to realize the above mentioned problems into practice.

AC-DC Power Transmission, 28-30 November 2001
Conference Publication No. 485 © IEE 2001

Thus, the main advantages [1] of DC transmission compare to AC one are the following:
- Transmission power higher for the same cross-section of conductor.
- Asynchronous interconnection of two AC system.
- Smaller environmental impact.
- No limits in transmitting distance.
- Statistically higher availability and reliability rate.
- Economically more efficient etc.

From AC-DC-AC technology point of view there are three ways of achieving the conversation [1]:

1. Natural Commutated Converters (operate with 50-60Hz) are widely used systems today in the HVDC. The main component here is thyristors. It is possible to change the DC voltage of the bridge by means of a control angle and, consequently, to control the transmitting power rapidly and efficiently.
2. Capacitor Commutated Converters (CCC) are characterized by the use of commutation capacitors inserted in series between the converter transformers and the thyristors.
3. Forced Commutated Converters valves operate with high frequency and use semiconductors with the ability not only to turn-on but also to turn–off their currents at any time. They are known as Voltage Source Converters (VSC). Two types semiconductors are used in VSC: Gate Turn-Off Thyristor (GTO) or Isolated Gate Bipolar Thyristor (IGBT). The operation of VSC is achieved by Pulse Width Modulation (PWM), which offers possibility to control both active and reactive power independently and almost instantaneously.

II. BODY

Along this paragraph firs of all several traditional rectifying schemes will be described, so in the given application one can be aware of the comparison between the various circuits. Fig. 1 presents three autonomous single-phase AC/DC bridges (rectangle scheme). To make things clear here and later are omitted all AC and

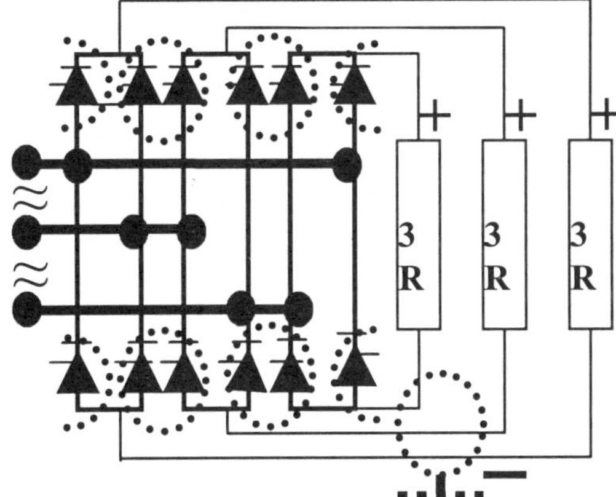

Fig. 1. Three autonomous single-phase DC bridge
Rectifiers (rectangle design).

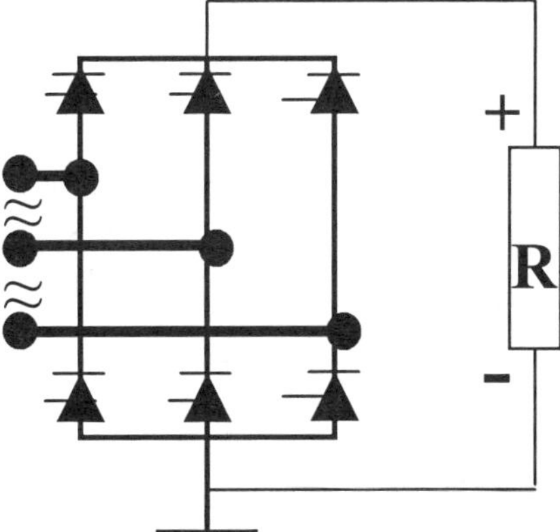

Fig. 2. Three phase DC bridge rectifier
(rectangle design).

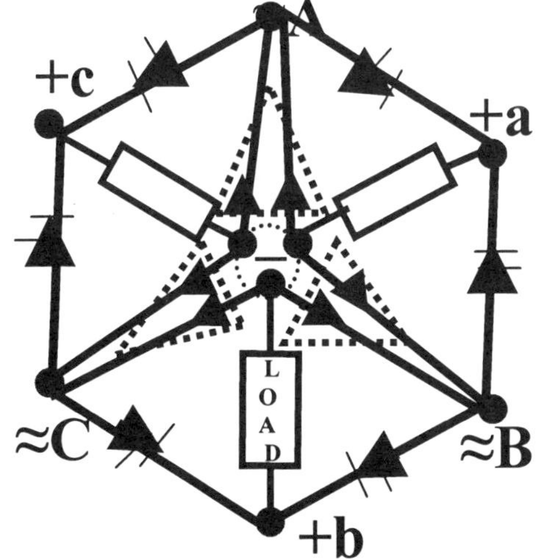

Fig. 3. Transformation of three one
phase bridges into a star-delta
bridge (hexahedron design).

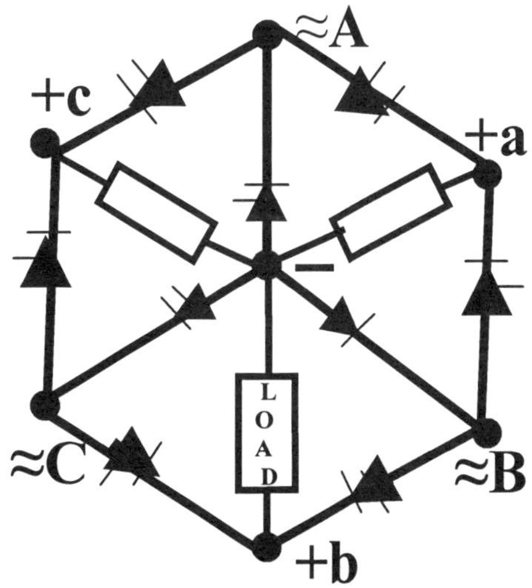

Fig. 4. A star-delta rectifier bridge
(hexahedron design).

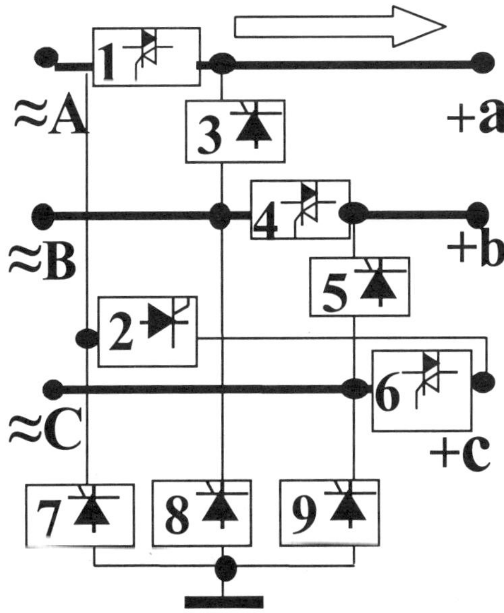

Fig. 5. Star-delta controlled bridge RECTIFIER-switch
(normal design).

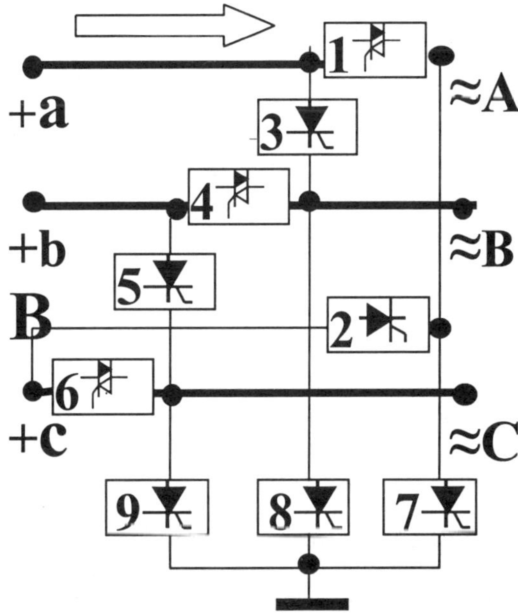

Fig.6. Star-delta controlled bridge INVERTER-switch
(normal design).

DC filters, anti-parallel diodes, snubber circuits, smoothing reactors, shunt capacitors, control systems and the other special for the converters attributes etc.

One may say that the well-known fully controlled three-phase AC/DC bridge rectifier (Fig. 2) can be designed by grouping in Fig.1 (as it is shown by doted lines) the elements of the parallel shoulders of three autonomous single-phase AC/DC bridges. If only the bottom elements of these AC/DC bridges of Fig. 1 are grouped and joined, it can be designed [2,3] the star-delta bridge rectifier (Fig. 3 and 4), which is under consideration in this paper. The same star-delta bridge converter can be designed from well-known controlled three-phase AC/DC bridge (Fig. 2) by splitting each three upper thyristors (divide each into two but cheaper half current thyristors) in terms to supply DC power for three separated loads 3R each. Both thyristors of each new pear control by the same gate signal as their parent thyristor.

The same approach was applied to similar to Fig.1 and 2 but inverter bridges and similar to Fig.3 and 4 the star-delta inverter bridge was designed. Finally it is presented on the Fig. 5 and 6 more pictorial and convenient schemes of the both converters: the rectifier and the inverter for AC-DC-AC power VSC Transmission. Turn on/off of each single thyristor is operated from the same traditional angle control equipment as the traditional 6 valves rectifier or inverter bridges.

The main advantage of the designed converters is the following. As stated above (Fig. 5 and 6), the designed AC-DC and DC-AC converters have got three extra bi-directional thyristors per converter (in each phase or line of the converters). During AC-DC-AC (VSC Transmission) application they help fully utilize three lines of existing AC as DC transmission (increasing the transmission capacity during asynchronous grid connection). But when a synchronous connection between the systems is possible they help to make quick twist switching to direct AC-AC synchronous transmission.

So, in the rectifier (inverter) there are three common and three bi-directional types thyristors, rated on the nominal current, and three common type thyristors, rated on the half-nominal current. If all thyristors of the converters are bi-directional, the AC-DC-AC transmission will be bi-directional, too.

Lab low voltage (220/380V) and current (around 1-5A) probe tests of these star-delta converters separately and for AC-DC-AC one directional transmission connection on L-R load have shown satisfactory results.

With the aim to get more wide information about transient and steady state regimes of the converters it was made multi variants analysis by the help of modeling software PSCAD V3.0.2 [Educational Edition].

As the basic AC-DC-AC transmission it was taken the well-known low voltage HVDC educational example: "VSC transmission based on 6-pulse 110kV STATCOM example; it provides up to 75 MW power transmission to the AC receiving end". It was added the same type three IGBTs valves into rectifier and inverter bridges, as well as, the transmission cable was substituted by overhead line with the same resistance (Fig. 7 and 8).

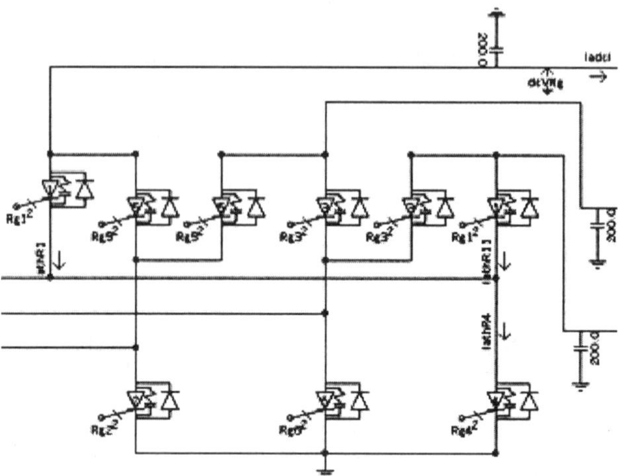

Fig,7 PSCAD model of star-delta rectifier.

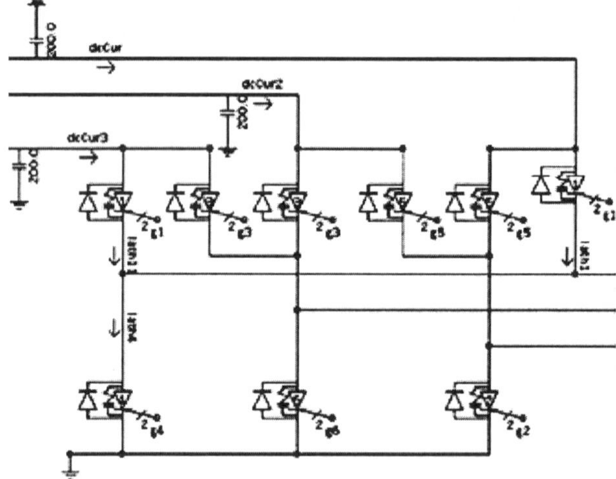

Fig. 8 PSCAD model of star-delta inverter.

Apparent, P-active and Q-reactive powers of the AC-DC-AC transmission during switching-on transient process (crossover to steady state) are presented on Fig. 9.

On Fig. 10 is presented inverter-input line-earth DC voltage of the AC-DC-AC transmission system during switching-on transient process (crossover to steady state).

The comparisons have shown that they are just similar to the same powers and voltages of basic VSC Transmission educational example. Multi variant analyses by the help of the PSCAD model have verified this AC-DC-AC system working availability.

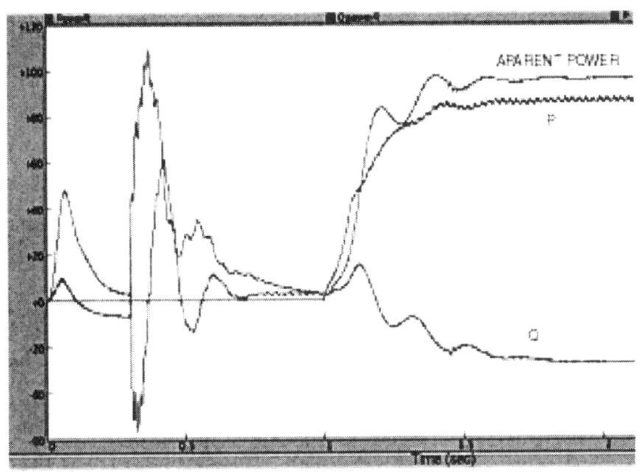

Fig. 9. PSCAD model data: P, Q and apparent power.

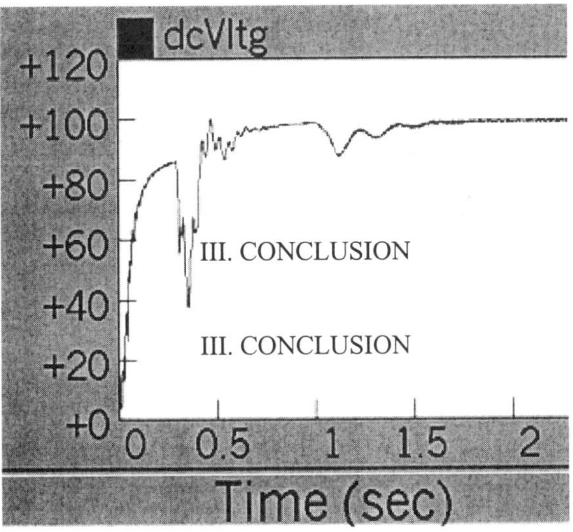

Fig. 10. Transient line-earth DC voltage.

III. CONCLUSION

The proposed star-delta rectifier and inverter can be implemented in common electrical applications such as VSC Transmission, supply of isolated loads, asynchronous grid connection, infeed of small-scale generation, infeed to city centers, for DC grids, converting the AC lines intoVSC Transmission, etc. [4].

IV. REFERENCES

[1] R. Ruderval, J. Carpentier, R. Sharma, "High Voltage Direct Voltage (HVDC) Transmission Systems", Technology Review Paper, Presented at Energy week 2000, Washington , USA, March 7, 2000.

[2] K.H. Kulizade, P.G. Ali-Zade, A.C. Kuliev, Cascade converter electric drive, USSR PATENT, SU 543118, 1974/1976.

[3]. P.G. ALI-ZADE, H.M. KULIEV, Thyristor Converter, PATENT, Azerbaijan Republic İ 2000 0207 01.11.2000 (Priority from 27 Jan. 1998).

[4] M. Bahrman, A-A. Edris, R. Haley, "Asynchronous back-to-back HVDC link with voltage source converter", Presented at Minnesota Power System Conference, Nov 1999,USA.

POWER SYSTEM APPLICATIONS FOR THE CHAIN-CELL CONVERTER

D W Sandells[1], T C Green[1], M Osborne[2], A Power[2].

[1]Imperial College of Science, Technology and Medicine. [2]National Grid Company plc.

ABSTRACT

The Static VAr Compensator (SVC) recently commissioned by National Grid Company plc incorporates a three-phase chain-cell STATCOM. The installation demonstrates the advantages of the chain-cell converter in providing dynamically variable exchange of reactive power with a transmission system to facilitate voltage control. However, the chain-cell converter in this form is not able to exchange significant real power with the transmission system. Such ability would be of great benefit in operations ranging from damping transients through to load levelling.

This paper will demonstrate that the chain-cell converter, when modified for the task of both real and reactive power exchange, has the potential to compensate for many transient and dynamic effects in a transmission system. The required cell ratings are discussed in terms of the usable stored energy of the converter and compared to other converters used for interfacing storage elements. A state-space model is also presented for incorporation into system studies.

I.   INTRODUCTION

The purpose of a transmission system is to transmit electrical power efficiently and reliably from sites of generation to sites of demand. This is a complex and dynamic process because of the following issues:

- Generation output will vary based on market prices and maintenance programmes.
- Demand profiles require varying quantities of power at different times.
- System faults must be catered for and contingency provided.
- Non-linear loads inject unwanted current and voltage harmonics into the transmission system.
- The transmission system is a high-order and non-linear dynamic system, modelling of such a system is computationally intensive.

To ensure integrity of supply, transmission systems have to be robust to outages and abnormal conditions. This is achieved through the installation of multiple circuits, the application of power flow control techniques and the employment of reactive compensation (when voltage levels require support). Flexible AC Transmission System (FACTS) controllers are compensators that utilise power electronics to enhance transmission system performance, Hingorani and Gyugi (1).

Central to a transmission system are the power flow characteristics of an inductive line. Voltage-sag in a transmission line is caused by excessive reactive power flow. Excessive real power flows cause circuits to reach thermal limits and can also cause stability problems and frequency variations, Weedy (2). Compensators, both traditional and FACTS, have been developed that address transmission system problems through injection or control of reactive power, real power and sometimes both. As well as the distinction between real and reactive power problems distinctions can also be made on the basis of the time duration of the problem. The range of duration is indicated in table 1.

Problem	Problem Duration	Solution Required
Distortion due to non-linear loads	Sub-cycle (< 20 ms)	Harmonic Compensation
Fault/ Fault-recovery	Small number cycles (1ms to 100 ms)	Transient Stabilisation
Inter-area power oscillations, loss of power lines	Medium number cycles (20 ms to 20s)	Dynamic Stabilisation
Change in generation level or load demand	Large number cycles (20 minutes to 12 hours)	VAr compensation & load levelling

Table 1- Transmission System Problems and their Time Durations

Problems that occur within the dynamic timeframe (for a small to medium number of cycles) can be tackled by the implementation of fast acting FACTS controllers. Many conventional reactive compensators are limited because their design does not facilitate real power processing, Ye and Kazerani (3). A FACTS controller designed to compensate for reactive power flows, such as a STATCOM, Sen (4), will have limited impact where the problem requires the compensator to apply dynamic stabilisation which is best achieved by the injection of real power, Zhang et al (5).

This paper presents three FACTS controllers based around the chain-cell converter, two of which are unique developments. The real power handling capability of each chain-cell converter is examined and suggestions on how to improve their capabilities are presented. Emphasis is placed on making the topologies suitable for handling short duration problems. Simulation models of the chain-cell converters are presented.

AC-DC Power Transmission, 28-30 November 2001
Conference Publication No. 485 © IEE 2001

II. THE STANDARD CHAIN-CELL TOPOLOGY

The standard chain-cell converter, Figure 1, is a series arrangement of individual cells, each cell consisting of four power electronic switches and a DC capacitor that is regulated to a specific DC voltage, Marchesoni and Mazzucchelli (6). The switches in the converter are commutated so that each element produces a quasi-square voltage waveform at fundamental frequency, Figure 2. The sum of several waveforms, each with a different duty-cycle, creates a sinewave approximation at system frequency (50Hz), Figure 3.

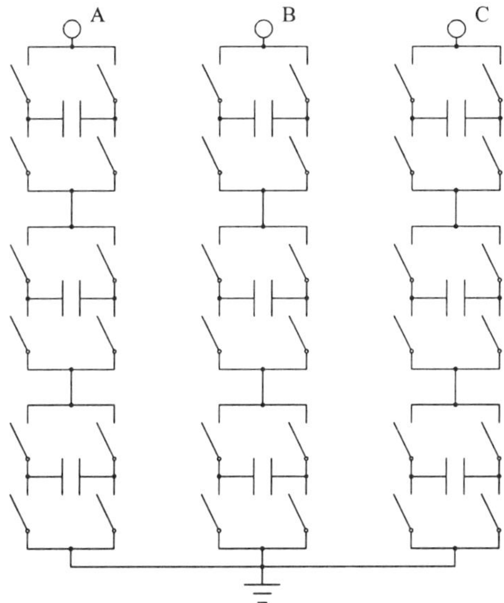

Figure 1- A Three-Cell, Three-Phase, Capacitive Chain-Cell Converter

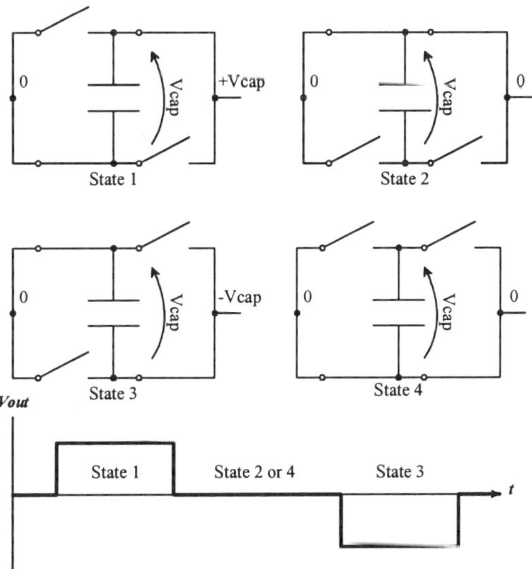

Figure 2 - Switching Commutations for Chain Cell Element

The chain cell converter topology has a number of advantages over other multi-level (and multi-pulse) converters.

- The converter may be constructed of individual cell modules that are identical. Modular design promotes ease of construction and simplicity of part replacement.
- The chain-cell converter can be designed with redundancy. If a cell fails the remaining cells can absorb the increased voltage stress.
- The chain-cell converter can produce waveforms with low harmonic content, without the need for the expensive and complex transformers that are required for multi-pulse converter designs. An increase in the number of voltage levels improves the harmonic performance.
- Each chain-cell element switches at fundamental frequency, thereby keeping switching losses low.
- Component count rises linearly with number of voltage levels. The component count for other multi-level converters has a square-law relation to the number of levels.

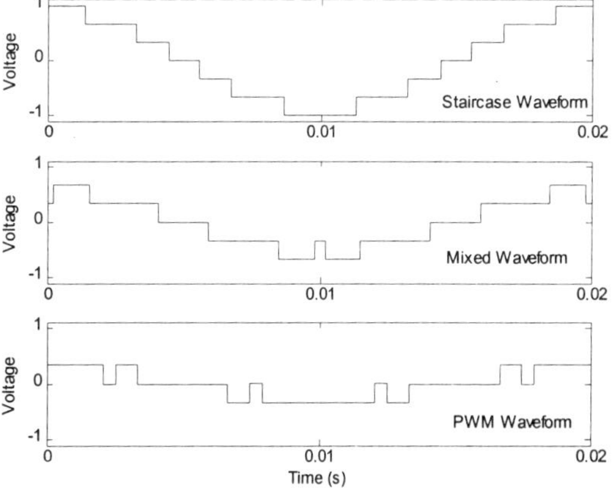

Figure 3 - Example Waveforms from Three-Cell Converter

The capacitive chain-cell converter has already been implemented within a FACTS controller design, to provide a STATCOM installed in the National Grid Transmission system, Hanson (7). The converter is not designed to provide real power support and is controlled to have zero energy exchange over a cycle. The reasons behind these limitations are discussed in detail in section IV.

The chain-cell converter topology is not limited to using capacitive elements on its DC-side. The fundamental concept is the connection, disconnection and reversal of energy storage elements, such that a complex waveform can be built up from three-level elements, thereby allowing energy flow to be controlled. Any DC storage element is potentially suitable for integrating into a cell topology. In the next section two new chain cell topologies are presented.

III. THE NEW CHAIN-CELL CONVERTER TOPOLOGIES

The two new chain-cell converter topologies consider alternative types of energy storage. The first examines a

294

Battery Energy Storage System (BESS), Leung and Sutanto (8), and the second a Super-conducting Magnetic Energy Storage (SMES), Luongo (9). BESS and SMES devices are usually connected to distribution/transmission systems through a Power Conditioning System (PCS), Casadei et al (10), but the modified chain-cell converters are an interesting alternative. A comparison of the standard PCS to the chain-cell is given in Section V.

BESS Chain Cell Converter

The BESS topology is similar to that shown in Figure 1, except that batteries or fuel cells replace the capacitors. A battery element is composed of multiple Voltage Regulated Lead-Acid (VRLA) battery cells. VRLA cells have a nominal voltage of only 2 V but are normally manufactured into series blocks of six cells. These series blocks can be further connected in series to create voltage sources in the kVA range, Miller et al (11).

VRLA batteries do not release all of their stored energy if they are discharged rapidly. The MVA-h rating of the proposed controller is therefore higher if the batteries are discharged over a period of hours rather than minutes. Consequently, the BESS chain-cell converter is more suitable for steady state response such as load-levelling/peak-lopping compensation; but when required can also be used to compensate for short duration problems. To make best use of a BESS chain cell converter it should be placed in a transmission system at a point where its load levelling and short duration compensation capabilities can both be utilised.

SMES Chain-Cell Converter

The SMES chain-cell topology shown in Figure 4 is significantly different from the other chain-cell designs; because the cells are parallel-connected current sources rather than series-connected voltage sources. The elements when switched in circuit now supply current in either a positive or negative sense; the individual cell currents are summed to create quasi-sinusoidal current waveforms. These waveforms can be injected into the transmission system at a node.

SMES coils are high efficiency inductors in which ohmic losses are extremely low but cooling effort is required. The most effective and reliable SMES coils are the low temperature variety, which need cooling below liquid nitrogen levels with cooling plant. There are significant power losses in the vapour-cooled leads that cross the temperature boundary to interface the coils to the rest of the circuit, Hassan et al (12). In the SMES chain-cell converter a set of leads is needed for each cell and therefore losses are increased when compared to a more traditional SMES design. However advantages of the chain-cell topology are retained (see II).

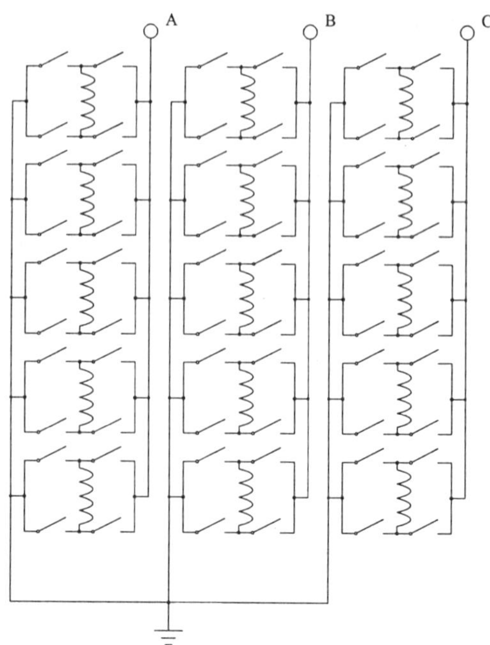

Figure 4 - A Five-Cell, Three-Phase, SMES Chain Cell Converter

IV. TRANSFER OF ENERGY WITH STORAGE ELEMENTS

For all chain-cell converters, when real power is transferred between the converter and the transmission system there is a change in the stored energy within the converter. For the capacitive and SMES converters a change in stored energy causes a change in cell voltage and a change in cell current respectively. The change occurs in any cell that is switched into circuit. Taking the capacitive case as an example, the change in voltage across one cell is given by.

$$\Delta V_{capacitor} = \frac{1}{C} \int i(t)_{switched}\, dt \tag{1}$$

Where $i(t)_{switched}$ is the current waveform that flows through the cell element.

The converter phase current is given by:

$$i(t)_{converter} = \sqrt{2} I_{r.m.s} \sin(\omega t + \theta) \tag{2}$$

Depending upon the state the cell switches are in, Figure 2, $i(t)_{switched}$ is equal to $i(t)_{converter}$ (states 1 and 3) or 0 (states 2 and 4).

If the current through a cell is in quadrature to the voltage produced by that cell, then the net change in capacitor voltage over a cycle is zero, Figure 5. In this case only reactive power is being exchanged between cell and transmission system.

If current and voltage waveforms are not in quadrature then real power is exchanged with each cell and a net change in cell voltage occurs. The worst-case situation,

Figure 6, is when a cell is switched at maximum duty-cycle directly in phase with the system voltage waveform. The change in capacitor voltage over a cycle is given by equation (3).

$$\Delta V_{max} = \frac{4.\sqrt{2}I_{r.m.s.}}{\omega C} \qquad (3)$$

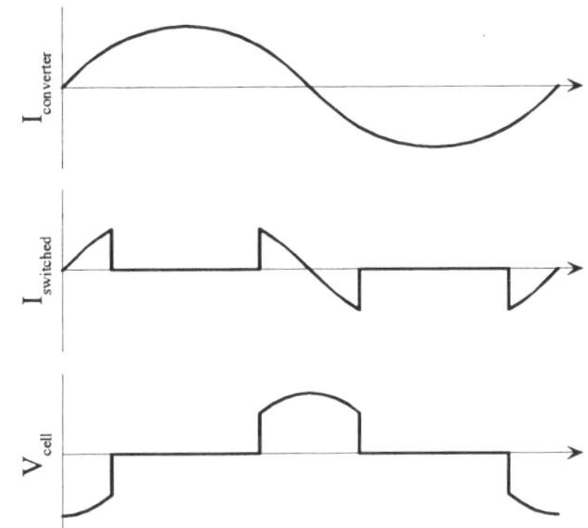

Figure 5 - Cell Waveforms for Reactive Power Exchange

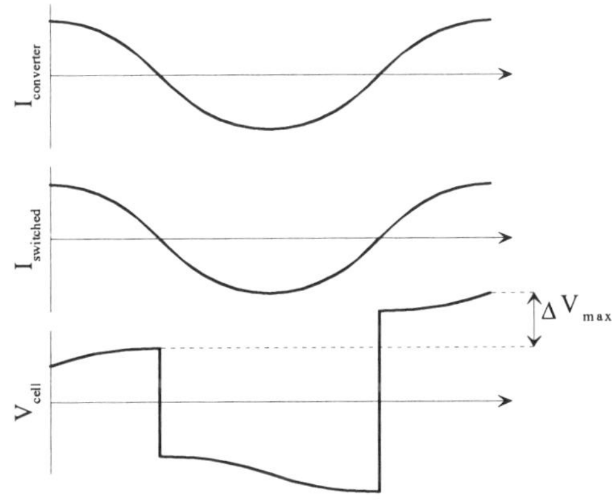

Figure 6 - Worst Case Voltage Change per Cycle

Similar equations can be developed for the SMES converter, resulting in a worse case change in SMES current over a cycle of:

$$\Delta I_{SMES} = \frac{4.\sqrt{2}V_{r.m.s.}}{\omega L} \qquad (4)$$

Where all r.m.s. quantities are phase values.

For all chain-cell converter topologies, the cells that are switched in for the longest duty-cycle will suffer the largest changes in stored energy and consequently the largest voltage/current variations. However, assuming that switch-balancing techniques are used (which keep the average change in stored energy per cell equal, Joos et al (13)), the net change in stored energy per cell is given by.

$$\Delta E_{cell} = \frac{I_{r.m.s.}V_{r.m.s.}\cos\phi}{k}t \qquad (5)$$

Where ϕ is the power factor angle, t is the time duration and k is the number of cells in the converter phase.

A chain-cell FACTS controller must be designed so that changes in cell voltage or current do not have a detrimental effect on its operation. This can be done in the following ways.

1. The converter can be operated so that there is no net energy exchange over a cycle, i.e., ΔE_{cell} is zero. Setting ϕ equal to 90° achieves this. In practice, ϕ may need to vary slightly from 90° so that losses in the converter are compensated.
2. The energy storage elements in the chain-cell converter can be made large so that the voltage or current does not change significantly. If net energy exchange is required then the store will need to be significantly over-sized so that removal of rated energy will produce only a small change in the voltage or current
3. The converter may be designed with the expectation that the cell voltage or current will vary. The converter can then be controlled such that the energy output is not affected by the change in voltage or current.

Each option for handling cell voltage/current variations has advantages and disadvantages. Option 1 has been applied to the National Grid design of chain cell converter. This design is restrictive as it only allows the converter to exchange reactive power with the transmission system such that the FACTS controller acts as a STATCOM. An advantage with this design is that component values for the energy storage elements are modestly sized.

An advantage with option 2 is that the converter is simple to control, but a disadvantage is that each energy storage element will need to be extremely large. The required storage element sizes are so large that the converter design is unachievable, inefficient or uneconomic with currently available components.

Option 3 is discussed in detail in V. It has advantages over both the other options, as it facilitates real power exchange and yet keeps the sizes of the energy storage elements within reasonable limits.

V. CONVERTER WITH VARYING ENERGY STORAGE

In this section the SMES chain-cell converter will be used for all examples but the arguments are equally applicable to the capacitive case.

If it is possible to exchange a significant amount of energy with a chain cell converter then that converter has the ability to transiently handle real power. This provides greater operational flexibility for a FACTS compensator as the compensator can then successfully deal with real (as well as reactive) power problems.

For the chain cell converter there is a choice to be made over the fraction of the energy to be recovered and the ratio of the maximum and minimum currents.

$$E_{re\,cov\,ered} = \tfrac{1}{2} L \left(I_{max}{}^2 - I_{min}{}^2 \right) \qquad (6)$$

To make best use of the energy storage in each SMES cell, the current should be allowed to change over a large range. The maximum current will dictate the rating of the semiconductors and the superconductors. The minimum current is set by the magnitude of AC current injection required. Assuming that over-modulation is not used when choosing the firing angles of the cells, the current in each cell must exceed its share of the phase current, equation (7).

$$I_{cell} \geq \frac{\sqrt{2} I_{r.m.s.}}{k} \qquad (7)$$

If the maximum cell current (and therefore the converter rating) is set to twice the minimum current then by using Equation (6), it can be shown that 75% of the stored energy can be transferred between cells and the transmission system, before I_{cell} becomes too small to support nominal current.

The cell currents differ depending upon the level of stored energy in each cell, therefore the firing angles for each cell must change depending upon the stored energy levels, to keep the output waveform at the required magnitude. Detailed analysis of firing angle optimisation is given in Sandells and Green (14). It is sufficient here to show example waveforms for the SMES chain cell converter, Figure 7.

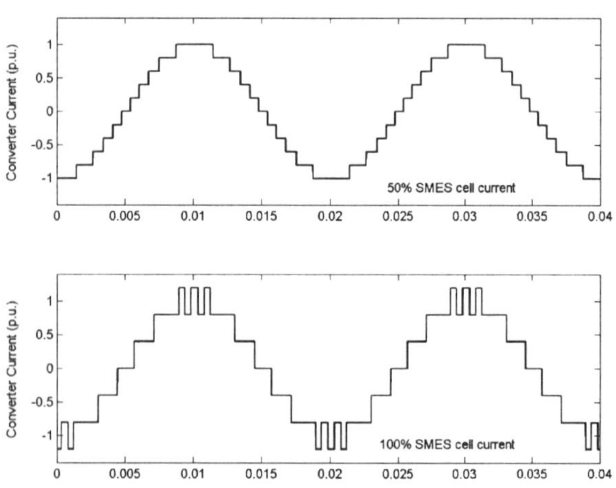

Figure 7 - Five-Cell SMES Converter Operating at Cell Current Limits

The example waveforms show that the FACTS controller can output rated current provided that each cell current remains within the defined limits. When the cell current is low the waveform is of staircase type. When the cell current is high the waveform changes to that of a mixed waveform. Both waveforms have the same fundamental magnitude but with different harmonic content.

The trade-off for adding energy storage to each cell is that the power electronics need to be rated higher than the nominal cell power, with a ratio of 2 for 75% energy use. The ratio of 2 is a reasonable choice as it makes the chain cell converter rating equivalent to the rating of a conventional power conditioning system. In a conventional PCS system one fully rated DC/DC converter is used to provide a regulated link and a second fully rated DC/AC converter is used to interface to the system voltages, Casadei et al (10), Figure 8. The total PCS rating is therefore twice the nominal power.

In a standard PCS system the power is 'processed' twice, once through the DC/DC converter and once through the DC/AC converter. In a chain-cell topology the power is processed once and more effective use is made of the power electronics.

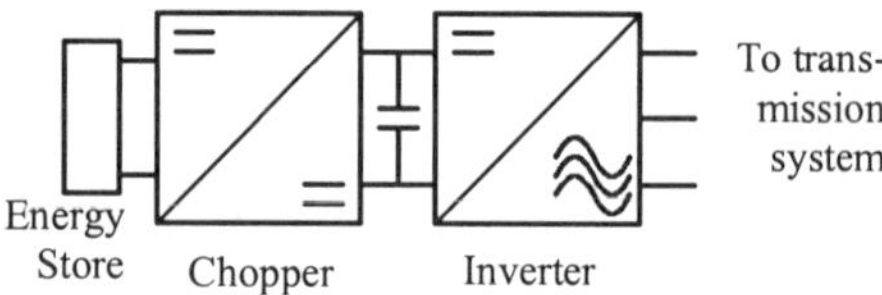

Figure 8 - Standard PCS Design

In principle, the PCS system allows the SMES to be run down to zero current and all stored energy can be recovered, in practice 90% recovery is normal, Hassan et al (12). For the chain cell converter the percentage of available stored energy can be increased from 75%, but this requires the power electronics within each chain cell element to be further over-rated. More energy can also be made available if over-modulation techniques are used when controlling the firing angles of the chain cell converter, but this leads to an increase in the harmonic content of the output current.

An advantage of the double-rated chain-cell converter over the conventional PCS is that while it is fully charged it has the capability of supplying up to twice the nominal current. This can provide extra transient real power support or damping. This would help towards making the economic case for the introduction of this type of power electronic solution. The conventional power conditioning system can only ever sink/source current up to the nominal value, even though the total power converter rating is double that.

The ability of the chain cell converter to operate with varying stored energy makes it suitable for handling real power. This allows the FACTS controller to be used to compensate for short duration problems that require real power flow.

VI. MODELLING OF CHAIN-CELL CONVERTER TOPOLOGIES

The chain cell converter can be modelled with the use of state-space techniques. Analysis for a capacitive chain will be presented here. The formulation of state-space models for the BESS and SMES converters follows a similar pattern.

The system can be modelled in state-space form using the energy storage variables. Equation (8) is the state-space equation for the circuit shown in Figure 9.

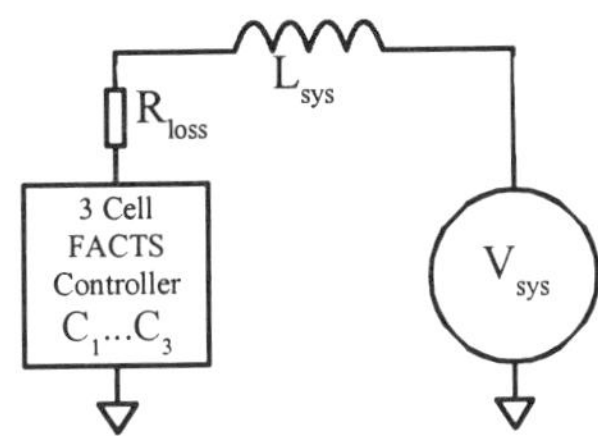

Figure 9 - Three-Cell Converter Model

$$\begin{bmatrix} \frac{dV_{C1}}{dt} \\ \frac{dV_{C2}}{dt} \\ \frac{dV_{C3}}{dt} \\ \frac{dI_{sys}}{dt} \end{bmatrix} = \begin{bmatrix} 0 & 0 & 0 & \frac{b_1}{C_1} \\ 0 & 0 & 0 & \frac{b_2}{C_1} \\ 0 & 0 & 0 & \frac{b_3}{C_1} \\ \frac{-b_1}{L_{sys}} & \frac{-b_2}{L_{sys}} & \frac{-b_3}{L_{sys}} & \frac{-R_{loss}}{L_{sys}} \end{bmatrix} \begin{bmatrix} V_{C1} \\ V_{C2} \\ V_{C3} \\ I_{sys} \end{bmatrix} + \begin{bmatrix} 0 \\ 0 \\ 0 \\ 1 \end{bmatrix} \begin{bmatrix} \frac{V_{sys}}{L_{sys}} \end{bmatrix}$$

$$\begin{bmatrix} V_{chain} \end{bmatrix} = \begin{bmatrix} 1 & 1 & 1 & 0 \\ 0 & 0 & 0 & 0 \\ 0 & 0 & 0 & 0 \\ 0 & 0 & 0 & 0 \end{bmatrix} \begin{bmatrix} V_{C1} \\ V_{C2} \\ V_{C3} \\ I_{sys} \end{bmatrix} + \begin{bmatrix} 0 \\ 0 \\ 0 \\ 0 \end{bmatrix} \begin{bmatrix} \frac{V_{sys}}{L_{sys}} \end{bmatrix}$$

$$(8)$$

The voltages across each cell capacitor and the current in the interface inductor are chosen as the state variables. Equation (8) shows that most non-zero elements within the state matrix depend upon a variable b_x, $b_x \in \{1,0,-1\}$. The value of b_x determines whether a cell output is a positive, negative or zero voltage. The value of b_x is determined by the internal control system of the chain cell converter and is time varying. This means that Equation (8) defines a time varying system, precluding the application of time invariant controller design methods. However Equation (8) may be successfully implemented in computer simulations of the chain-cell converter when modelling the FACTS controller. The results shown in Figure 7 were derived from the five-cell, SMES converter model, which was implemented using the MATLAB-SIMULINK programming language. The state-space model described allows for accurate computer simulation of the chain-cell converter topology.

VII. CONCLUSIONS

This paper has discussed the use of chain cell converters within transmission systems. The advantages and disadvantages of different energy storage systems within a chain-cell topology have been discussed. Two new variations of the chain-cell topology have been presented. The capability of chain-cell converters to deal with short duration problems has been analysed and the ratings required for a given energy capability discussed. Finally, state-space models of the chain-cell topology have been determined, these allow for computer simulation and modelling of the chain-cell topology.

BIBLIOGRAPHY

1. Hingorani N and Gyugyi L, 1999, "Understanding FACTS: Concepts and Technology of Flexible AC Transmission Systems", IEEE Press, New York, USA

2. Weedy B, 1995, "Electric Power Systems", Wiley Press, Chichester, UK

3. Ye Y and Kazerani M, 2000, "Operating Constraints of FACTS devices", PES Summer Meeting, 3, 1579-1584

4. Sen K, 1999, "STATCOM – STATic synchronous COMpensator: Theory, Modelling and Applications", PES Winter Meeting, 2, 1177-1183

5. Zhang L, Shen C, Yang Z, Crow M, Arsoy A, Liu Y, Atcitty S, 2001, "A Comparison of the Dynamic Performance of FACTS with Energy Storage to a Unified Power Flow Controller", PES Winter Meeting, 2, 611-616

6. Marchesoni M and Mazzucchelli M, 1993, "Multilevel converters for high power AC drives: a review", Int. Symposium on Industrial Electronics, 38-43

7. Hanson D, 1998, "A Transmission SVC for National Grid PLC incorporating a ±55MVAr STATCOM", IEE Colloquium: Flexible AC Transmission Systems – The FACTS, 5/1-5/8

8. Leung K and Sutanto D, 1997, "Improving Power System Operation and Control utilizing Energy Storage", Proc. 4th International Conference on Advances in Power System Control Operation and Management, 626-631

9. Luongo C, 1996, "Superconducting Storage Systems: An Overview", IEEE Trans. on Magnetics, 32, 2214-2223

10. Casadei D, Grandi G, Reggiani U and Serra G, 1998, "Analysis of a Power Conditioning System for Superconducting Magnetic Energy Storage (SMES)", Proc. Int. Symposium on Industrial Electronics, 2, 546-551

11. Miller N, Zrebiec R, Delmerico R, Hunt G and Achenbach H, 1996, "A VRLA Battery Energy Storage System for Metlakatla, Alaska", 11th Annual Battery Conference on Applications and Advances, 241-248

12. Hassan I, Bucci R and Swe K, 1993, "400 MW SMES Power Conditioning System Development and Simulation", IEEE Trans. on Power Electronics, 8, 237-249

13. Joos G, Huang X and Ooi B, 1998, "Direct-coupled multilevel cascaded series VAr Compensators", IEEE Trans. on Industry Applications, 34, 1156 -1163

14. Sandells D and Green T, 2001, "Optimisation of Firing Angles for the Chain Cell Converter", Power Electronics Specialist Conference 2001

TOPOLOGIES FOR VSC TRANSMISSION

B R Andersen, L Xu, K T G Wong*

ALSTOM T&D Ltd – Power Electronic Systems, * now with ALSTOM Research and Technology Centre

ABSTRACT

The increasing rating and improved performance of self-commutated semiconductor devices have made dc power transmission based on Voltage-Sourced Convertors (VSCs) possible. This technology is called VSC Transmission. The main components in a dc scheme are depicted and their functions explained. The features of three main categories of convertor topology suitable for dc transmission are described. Three specific convertors viz. two-level, three-level diode-clamped and four-level floating-capacitor convertors for a 300MW scheme are compared in terms of costs, dc capacitor volume, commutation inductance and footprint. The floating capacitor convertor is shown to yield the lowest system cost.

INTRODUCTION

High-Voltage Direct Current (HVDC) is often the economic means for delivering electric power over long distances and/or for interconnecting two unsynchronised ac networks, which may be at different frequencies. HVDC schemes totalling 60GW with individual scheme ratings between 50MW and 6300MW have been installed world-wide, and many more installations are being considered.

Until recently all commercial HVDC schemes employed the well-established line-commutated current-sourced convertors, with the thyristor being the present day switching device. However, self-commutated Voltage-Sourced Convertors (VSCs) [1, 2, 3] with ratings suitable for power transmission are now possible because of the emergence of semiconductor devices, such as Insulated Gate Bipolar Transistor (IGBT), with high voltage and current ratings. HVDC applications of this technology are referred to as VSC Transmission schemes.

VSC Transmission technology is still evolving. The main advantage of VSC Transmission is its ability to control the reactive power output at the two terminals independently of each other and at the same time independently of the real power flow. This makes it suitable for connection to weak ac networks or even networks without other voltage sources. The disadvantages, however, include high power losses and high capital costs compared with conventional HVDC.

To enable VSC Transmission to compete with conventional HVDC outside a niche area, a number of technical advancements in the VSC technology are required. This paper discusses a breakthrough - the use of a floating capacitor, multilevel convertor topology. This technology reduces the convertor power losses and alleviates the technical difficulties commonly associated with the design of very high voltage self-commutated, semiconductor switches and other equipment.

FUNDAMENTALS OF VSC TRANSMISSION

The fundamentals of VSC transmission operation may be explained in simple terms by considering each terminal as a voltage source connected to the ac mains via a three-phase reactor and interconnected by a dc link as schematically shown in Fig. 1. Both the amplitude and the phase angle of the output fundamental voltage V_2 are controlled with respect to the source voltage V_1. In essence, the voltage drop ΔV across the reactor X can be controlled to determine the active and reactive power flows.

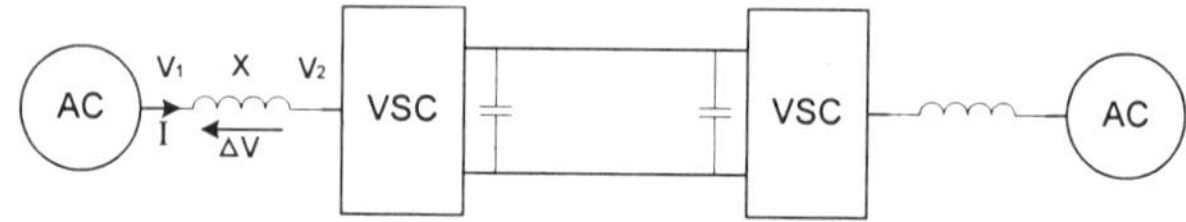

Fig 1: Basic VSC Transmission

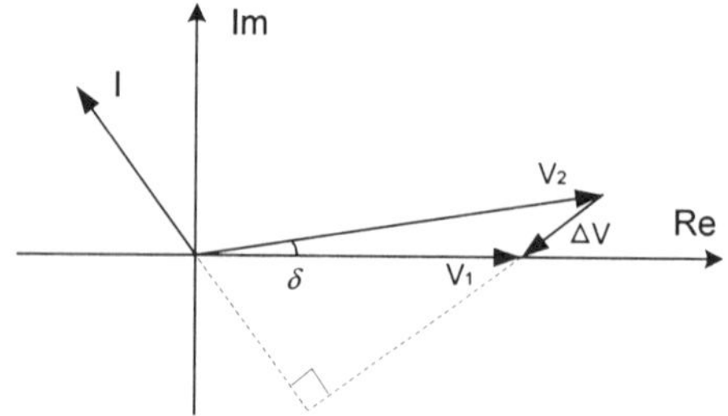

Fig 2: Phasor diagram

Fig. 2 shows the fundamental frequency phasor representation for the inverter. According to Figs. 1 and 2, the active and reactive power exchange (P&Q) as seen from the AC system terminals can be expressed respectively as:

$$P = \frac{V_2 \sin\delta}{X} \cdot V_1$$

$$Q = \frac{V_2 \cos\delta - V_1}{X} \cdot V_1 \tag{1}$$

where δ is the phase angle between V_1 and V_2.

AC-DC Power Transmission, 28-30 November 2001
Conference Publication No. 485 © IEE 2001

From Eq. (1), it can be seen that the active and reactive power are controlled independently through $V_2 \sin\delta$ and $V_2 \cos\delta$, respectively. Therefore, by adjusting $V_2 \sin\delta$ and $V_2 \cos\delta$ properly, a voltage-sourced convertor is able to operate at any power factor. The PQ diagram of VSC operation is shown in Fig. 3. A convertor of a given MVA rating can operate at any point within the circle shown.

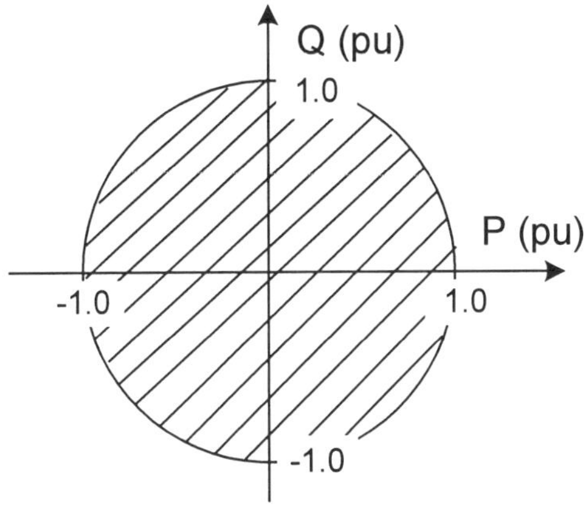

Fig3: PQ diagram of VSC transmission

In contrast to conventional HVDC, a VSC with self-commutated semiconductor devices does not need an ac source to commutate. Therefore, VSC Transmission can transmit power to weak systems as well as dead loads where there is no rotating machine available.

The operation of a VSC generates harmonics that depend on the convertor topology and the switching strategy used. Using multilevel convertor topology and PWM control, the harmonics generated by VSC are at high frequencies and, therefore, the harmonic filter can be smaller compared to conventional HVDC scheme. Consequently, the footprint for VSC transmission is smaller than for a conventional HVDC convertor. The ability to operate at a satisfactory power factor and output waveshape without the need for large ac harmonic filters is an additional advantage when operating in a weak system.

The filter may also include a series element, designed in conjunction with the shunt elements to limit the dv/dt and magnitude of the repetitive pulses. Without this feature the convertor transformer may have to be of a special design, in order to ensure that the repetitive pulses do not have a detrimental impact on the life of the transformer. The series reactance forms part of X in Eq (1).

VSC TRANSMISSION

Fig. 4 shows a more detailed circuit of one half of a VSC Transmission scheme, comprising the dc line, the reservoir dc capacitors, the VSC, a high-frequency blocking filter, a convertor transformer and a shunt harmonic filter. The other half of the scheme is by-and-

large the mirror image about the mid-point of the dc line. Switchgear and surge arresters are not shown.

The dc link, which may be a cable or overhead line, has two conductors, one of positive polarity and the other negative polarity. These carry currents of the same magnitude but opposite sense. Consideration is given to the configuration of the conductors to minimise the magnetic field along the line. The dc voltage is generally determined by the distance of power transmission, the power level, line losses and sometimes the ac system voltage if a scheme without transformers is employed.

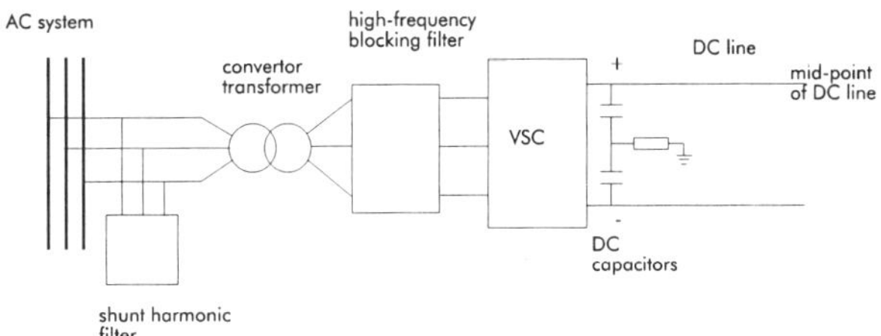

Fig.4: Half of a VSC Transmission scheme

A large dc capacitor is connected to the terminals of the VSC. These dc capacitors store the energy necessary for the dynamics of the system and govern the voltage ripple on the dc line. The middle point of the bank is earthed, normally via a high impedance which limits the fault current on the dc side.

The VSC is required to generate near-sinusoidal 3-phase voltages on its ac terminals. This can be achieved either by interconnecting several convertors by interface transformers, by using multilevel convertors, by using PWM switching or by a combination of these methods.

High-voltage semiconductor switches comprising a large number of self-commutated devices in series have been developed. By suitable gating and the use of snubbers, all the devices of each switch are turned on (and off) simultaneously with the minimum stress on the individual devices.

Where PWM is employed, converter ac connections carry significant harmonics at frequencies close to the switching frequency and multiples thereof. A high-frequency filter is therefore normally installed to prevent these harmonics from causing undesirable effects in the ac system.

The convertor transformer serves several purposes. Firstly, where the preferred convertor ac voltage is different from that of the ac system, it provides the necessary voltage transformation. Secondly, its leakage reactance forms part of X in Eq.(1), taking part in the dynamics of the convertor. Thirdly, by having appropriate winding arrangements, the convertor transformer blocks or traps some of the harmonics from the convertor. It can therefore ease the filtering requirements on the ac system side.

In some applications it may be desirable to use a shunt harmonic filter at the interface of the ac system and the VSC Transmission scheme. Appearing as a capacitor at fundamental frequency, it generates reactive power to offset that absorbed by the convertor transformer and high-frequency blocking filter. This reactive power support would otherwise have been provided by the VSC and so convertor VA rating is correspondingly reduced. The filter, if used, will of course also contribute to achieving the scheme's harmonic specification.

SELF-COMMUTATED POWER ELECTRONIC DEVICES

Until recently, self-commutated convertors were associated mainly with industrial motor drives. The main switching elements are Gate-Turn-Off (GTO) thyristor, Insulated Gate Bipolar Transistor (IGBT) and Metal-oxide Semiconductor Field Effect Transistor (MOSFET). As the ratings of these devices were initially small, it was technically difficult and uneconomical to form the switches suitable for convertors in power transmission applications.

A power transmission system has features, which are different from those of an industrial drive system. Generally speaking, equipment for a transmission system must continue in service whilst the ac system experiences faults, disturbances and transients. High availability and high efficiency are other requirements, which convertors of a power transmission system must meet. These requirements are in turn imposed on the semiconductor devices leading to the desire for:

- High blocking voltage
- High turn-off current
- Low conduction and switching losses
- Short turn-on and turn-off times
- Suitability for series connection
- Good dv/dt and di/dt capability
- Good thermal characteristics
- Low failure rate

Hitherto, no device can meet all these requirements because optimising one area often renders another area suboptimal. At best, a device can satisfy only some of these requirements and, in this respect, GTO thyristors and IGBT have been the potential candidates. Because of the relative ease of control and suitability for high frequency switching, IGBT is emerging as the chosen device for most VSC applications.

By the mid-1990s, GTO thyristors of a 4.5kV rated voltage were available and IGBTs of similar voltage rating were under development. Towards the late 1990s, self-commutated convertors were recognised as an alternative to line-commutated ones for dc power transmission for low power applications.

CONVERTOR TOPOLOGIES

For most convertors used in power systems, a sinusoidal ac output voltage is required. The basic unrefined ac output waveform of a convertor is determined by the topology of the convertor and three main categories of topology [4] suitable for dc power transmission exist. These are:

- Two-level topology
- Multilevel diode-clamped topology
- Multilevel floating-capacitor topology.

The two-level topology has been widely used in many applications at a wide range of power levels. Convertors of over 50MVA have been built for compensating reactive power. Fig. 5 (a) shows one phase of a two-level convertor.

In order to improve the quality of the output, PWM can be used to produce an output waveform with a dominant fundamental component, at the expense of significant high-order harmonics. A typical PWM switched waveform, using a carrier based control method with a frequency of 1000Hz is shown in Fig. 5 (b). For the purpose of this illustration the dc capacitor has been assumed to have an infinite capacitance (i.e. no voltage ripple). A harmonic analysis of the waveform in a 50Hz application shown in Fig. 5 (b) is shown in Fig. 5 (c).

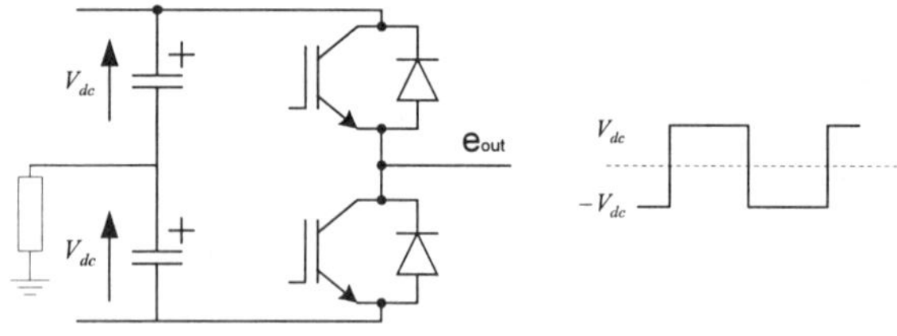

(a) One phase of a two-level convertor

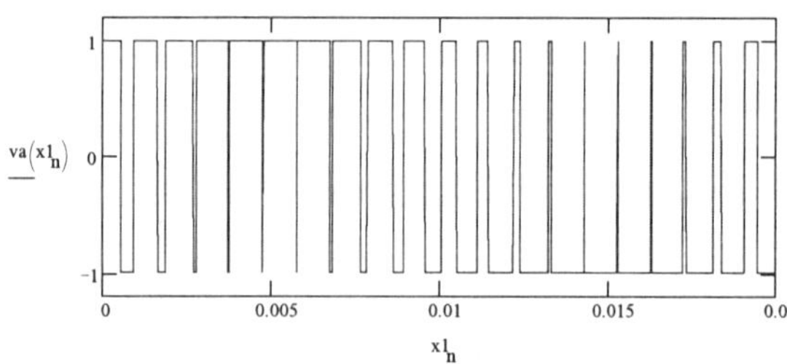

(b) Two-level output waveform

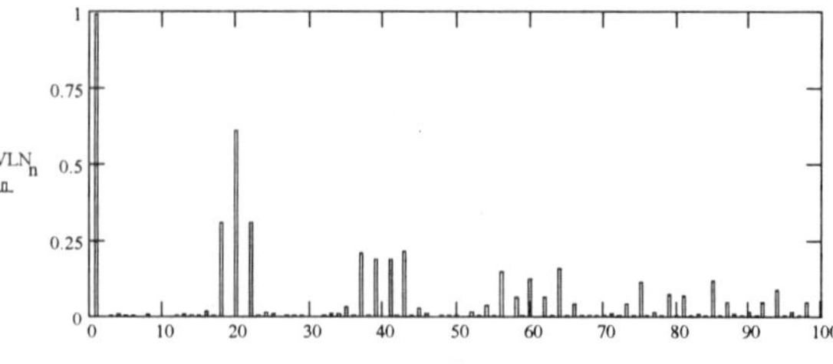

(c) Harmonic spectrum of two-level output waveform

Fig.5: Two-level topology

The advantages of the two-level topology are:

- Simple circuitry
- Small dc capacitors
- Small footprint
- Semiconductor switches have the same duty.

The disadvantages of the two-level topology are:

- Large blocking voltage of semiconductor switches
- Crude basic ac waveforms.

By using a number of dc capacitors in series and additional diodes, a multilevel diode-clamped convertor can be formed [4]. Fig. 6 (a) shows one phase of a three-level circuit. For a three-phase unit, the dc capacitors are usually shared by the phases.

Again, PWM can be used to improve the output waveform quality. A typical PWM switched waveform, using a carrier based control method with a frequency of 1000Hz is shown in Fig. 6 (b). The dc capacitor has been assumed to have an infinite capacitance (i.e. no voltage ripple). A harmonic analysis of the waveform in Fig. 6 (b) is shown in Fig. 6 (c).

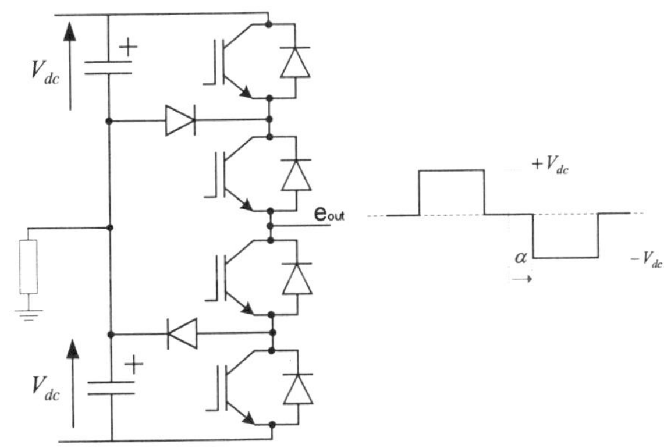

(a)   One phase of a three-level diode-clamped convertor

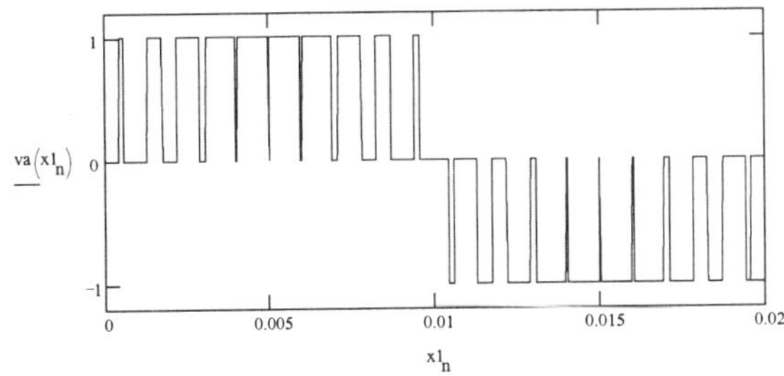

(b)   Three-level output waveform

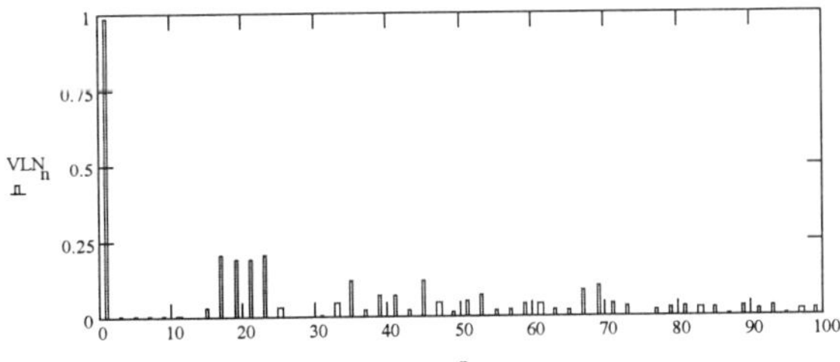

(c) Harmonic spectrum of three-level output waveform

Fig.6: Three-level diode-clamped topology

The advantages of the diode-clamped topology are:

- Reasonably small dc capacitors
- Small footprint
- Good basic ac waveform.

However, the disadvantages are:

- Inherent difficulty in keeping dc capacitor voltages constant
- Complex circuitry for large number of levels; the number of added diodes increases rapidly with the number of levels
- Semiconductor switches have different duties.

The multilevel floating-capacitor topology produces the same ac waveform as the multilevel diode-clamped topology. This topology has no additional diodes but has additional dc capacitors known as floating capacitors [5]. One phase of a 3-level floating-capacitor convertor is shown in Fig. 7. For a three-phase unit, the dc main capacitors are shared by the phases but the floating capacitors, marked f, are not. The PWM switched waveform and the Fourier analysis is identical to the 3-level diode clamped circuit (Fig. 6 (b) and (c)). The advantages of the multilevel floating-capacitor topology are:

- Relatively simple circuitry
- Semiconductor switches have the same duty
- Good basic ac waveform.

With the volume of capacitors largely proportional to the square of their nominal voltages, the disadvantage of this topology is the large footprint incurred by the floating capacitors.

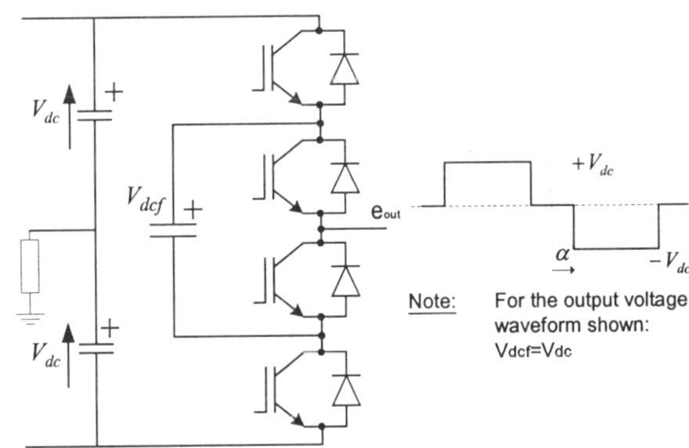

Fig.7: Three-level floating capacitor topology

COMPARISON OF TOPOLOGIES

As part of its development work, ALSTOM has compared in detail several different convertor topologies. This work has concluded that for practical and economic reasons multi-level topologies with more than 4 levels are unlikely to be attractive unless the power level is substantially higher than 300MW. This section gives the results of a comparison of three systems with different topologies, i.e., 2-level, 3-level diode-clamped and 4-level floating-capacitor convertors. Other studies concluded that the 3-level floating capacitor topology was marginally superior to the 3-level diode clamped topology. Furthermore, since a 4-level topology can be readily implemented using the floating capacitor topology, it was decided to limit the comparison in this paper to these three topologies. The topologies are compared in terms of the capital costs, capitialised losses, dc capacitor volume, commutation inductances, and the footprint of the convertor station. The system rating considered is 300MW, ±150kVdc.

An equivalent switching frequency of 1050Hz has been assumed for all three convertor topologies. Fig. 8 compares the normalised capital cost, capitalised losses and the total cost, where the capital cost of the 2-level convertor is defined as 1 pu. The different duty on the semiconductor devices in the different topologies has been taken into account when evaluating the total capital cost. In particular, for the 2-level and 3-level topologies the higher switching losses, associated with the higher switching frequency for each device, reduce the current capability of the semiconductor necessitating the use of higher rated semiconductor devices. Not withstanding this, the 2-level convertor results in lower capital cost than the 3-level diode-clamped and 4-level floating-capacitor convertors.

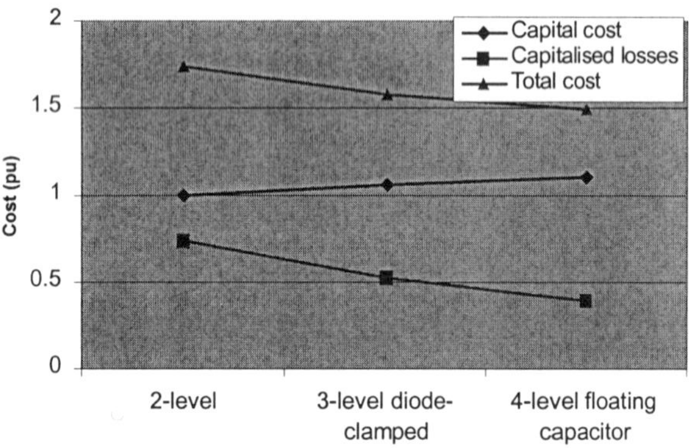

Fig. 8: Comparison of system cost

The high switching power loss, which is capitalised at the rate of 3000Euro/kW, makes the 2-level convertor most costly overall. In contrast, in spite of having the highest capital cost, the low power losses for the 4-level floating-capacitor makes this topology the most economically attractive. At a loss capitalisation of 1000Euro/kW the 3 topologies yield almost identical total cost.

The comparison of dc capacitor volume is shown in Fig. 9. Again the value for the 2-level convertor is defined as 1 pu. The dc capacitor volume has been calculated to limit the voltage ripple to less than 5%. Fig. 9 shows that the 4-level floating-capacitor convertor requires much higher dc capacitor volume than the 2-level and 3-level diode-clamped convertors. This is partly because of the additional floating capacitors needed but

primarily because of the relatively lower PWM carrier frequency associated with each capacitor.

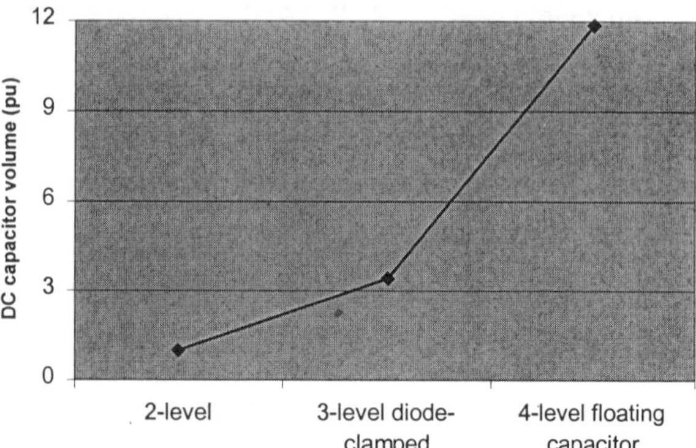

Fig. 9: Comparison of dc capacitor volume

One concern relative to the multi-level topologies was whether the commutating loop stray inductance would increase dramatically, as the number of levels increased. A large commutating loop inductance would necessitate a significant increase in snubber capacitance, which would increase the switching loss, thereby negating some of the benefits of the multi-level topology. The investigation, which included the creation of 3D drawings of the complete convertor circuits, concluded that this concern was unfounded. Fig. 10 compares the commutation inductances for the three topologies where the commutation inductance of the 2-level convertor is again defined as 1 pu. It can be seen that the multi-level circuits do not shown a significant increase in commutating loop inductance. This is largely due to the fact that the large dc capacitor banks used in these circuits consist of many parallel connected capacitor units, which can be arranged to provide low stray inductance between their terminals. As shown in Fig. 10, the 3-level diode-clamped convertor results in lower commutation loop inductance than 2-level and 4-level floating-capacitor convertors. This is because for the 3-level diode-clamped convertor, only half of the stray inductance of the dc capacitor appears to each valve during commutation.

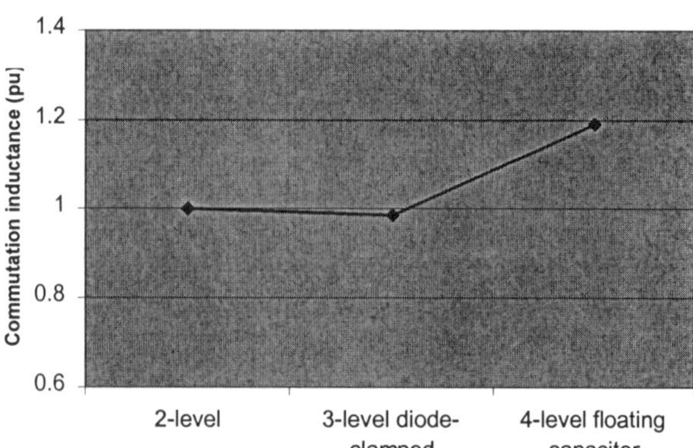

Fig. 10: Comparison of commutation inductance

The normalised footprints of the convertor stations are compared in Fig. 11 where again the footprint of the 2-level convertor is defined as 1 pu. It can be seen that the 4-level floating-capacitor topology has a much bigger footprint than the 2-level and 3-level diode-clamped convertors. The large footprint is mainly due to the volume of the dc capacitors required.

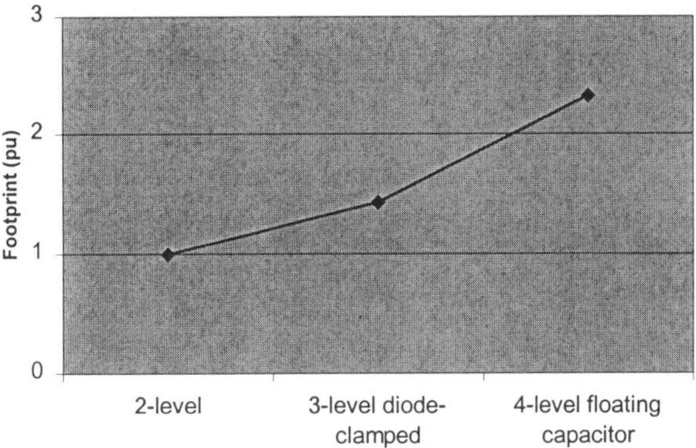

Fig. 11: Comparison of footprint

Nevertheless, during each switching, for a dc voltage of 300kV, the voltage steps for the 2-level, 3-level diode-clamped and 4-level floating-capacitor are 300kV, 150kV, and 100kV, respectively. Therefore, the high frequency blocking filter connected between the VSC and the convertor transformer can be smaller for the 4-level floating-capacitor topology than for the 2-level and 3-level diode-clamped topologies. For the purpose of this comparison the same dv/dt at the transformer terminal, thus giving identical transformer stresses, is maintained.

Thus, the inductance of the series blocking reactor for the 4-level floating-capacitor can be approximately two thirds of that for the 3-level diode-clamped and one third of that for the 2-level convertor.

300MW EXAMPLE

The example shown here is a 300MW scheme with a dc voltage of ±150kV. Fig. 12 shows the single-line circuit diagram of one terminal of the VSC Transmission system. The system consists of a dc link, a 4-level floating capacitor convertor, a series reactor, two high frequency filters and a 2-winding convertor transformer.

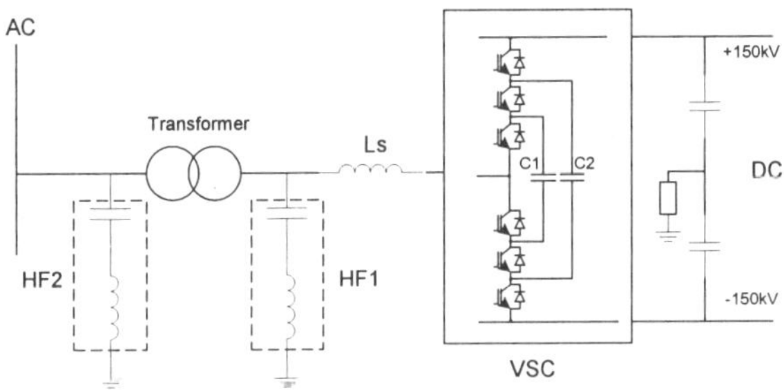

Fig. 12: Single-line circuit diagram of 300MW VSC transmission system

- For the 4-level floating-capacitor convertor, each valve is rated at one-third of the total dc link voltage, i.e., 100kV, and it consists of a number of

series connected IGBTs. The two floating capacitors C1 and C2 are rated at 100kV and 200kV, respectively.

- The series reactor Ls and the high frequency filter HF1 are used to protect the transformer against high repetitive dv/dt. Since the 4-level convertor gives a smaller voltage step than the 2-level convertor, Ls can be smaller and HF1 can be tuned to a higher frequency.
- The transformer provides isolation and voltage matching between the VSC output and the ac system. It also provides the major part of the reactance, which defines the power flow between the ac network and the VSC.
- Harmonic filter HF2, together with the transformer provides filtering of the switching harmonics to meet the harmonic requirements of the ac network. HF2 can be tuned to the dominant harmonic frequency of the VSC. Compared with a conventional HVDC system, the harmonic filters are relatively small. Typically, the total var rating of the two filters added together can be less than 20% of the rated power.

The 3-D view of one terminal of a 300MW VSC transmission system is shown in Fig. 13. The size of the valve hall is approximately 22m by 18.5m with an interior clear height of 7.5m.

ADVANTAGES AND DISADVANTAGES OF VSC TRANSMISSION

Since a VSC generates its own ac output voltage, a VSC transmission link permits the flow of power between two ac systems, one of which need not have an internal source voltage. It is therefore suitable for feeding power to ac systems with no or few synchronous generators. Similarly, it is suitable for use with induction generators, as may be used in windfarms. Linking several ac systems together also becomes straight forward.

The power flow between a VSC and an ac system is determined mainly by δ in Eq.(1). At the same time, the reactive power flow can be varied by changing V_2. These two variables can be adjusted independently and quickly to provide the ac system with the necessary frequency and voltage control.

By using topologies with greater numbers of switching levels, the basic ac output waveforms of VSCs resembles a sinusoid more closely. Where PWM is used, the harmonics of the waveform are mostly of high orders and can be absorbed relatively easily by its filters whose ratings are typically less than 20% of the rating of the VSC.

Power reversal is achieved by reversing the direction of the current in the convertor. Without the need for reverse the polarity of the dc voltage multi-terminal schemes become easier to implement.

VSC stations lend themselves to a modular design, with the modules fully assembled and tested at the manufacturer's works before being delivered to site. This means shorter overall project times.

VSC Transmission relies on self-commutated semiconductor devices. Despite recent significant improvements, these devices are more lossy than line-commutated thyristors. This potentially limits the extent of applying PWM to achieve better waveforms and results in larger cooling plant.

As is evident from the waveforms, PWM leads to large and rapid changes of voltage within the convertor and at its terminals. Electromagnetic compatibility must be addressed and precautions taken to guarantee normal operation of nearby telephone lines and electronic/electrical equipment.

CONCLUSION

The emergence of VSC Transmissions has opened new markets for HVDC that were previously uneconomic and commercial VSC Transmission schemes are now in operation. The biggest obstacle to their wider application arises from the relatively high power losses of the switching devices used. Nevertheless, new devices are being developed and some of these are promising low conduction loss and reasonable switching loss. These loss reductions can be further enhanced by the use of floating capacitor convertors, thereby making the application of VSC Transmission far more economically attractive.

REFERENCES

1. Mohan, N., Undeland, T.M., and Robbins, W.P., 1995, 'Power Electronics: Convertors, Applications and Design', John Wiley & Sons Inc.
2. Andersen, B., Barker, C., 'A New Era in HVDC?', March 2000, IEE Review, pp.33-39.
3. Gyugyi, L., 'Control of shunt compensation with reference to new design concept', Nov.1981, IEE Proceedings, Vol.128, Pt.C, No.6, pp.374-381.
4. Newton, C., Sumner, M., 'Multi-level convertors: a real solution to medium/high-voltage drives?', Feb.1998, IEE Power Engineering Journal, pp.21-26.
5. Meynard, T.A., and Foch, H. "Multi-level Conversion: High Voltage Choppers and Voltage – Source Inverter", Proc. Of IEEE PESC'92, 1992, pp. 397-403.

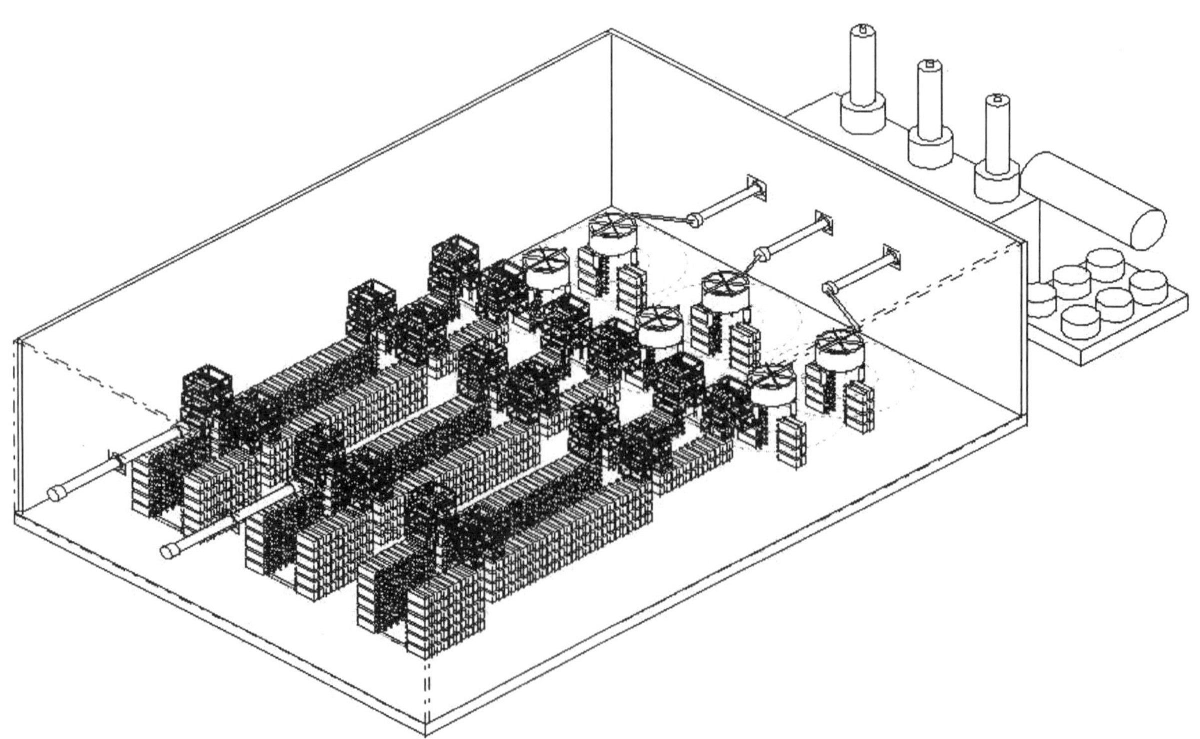

Fig. 13: Isometric view of one terminal of a 300MW VSC transmission system

MODELLING OF THE DYNAMIC CHARACTERISTICS OF THE DC LINE FOR VSC TRANSMISSION SCHEME

S Poullain, F Héliodore, A Henni, JL Thomas and E Courbon

Power Electronics Research Team
ALSTOM technology, France

ABSTRACT

In order to face increasing of electric power exchanges, interconnection of networks appears like an adequate solution. Associated technical problems (as deregulated electrical networks, ...) must be taken into account in the conception of any system of electric power transmission. In particular, physical behaviour of transmission line must be viewed in order to include it in any system of control. In this paper, specific modelling of transmission line is developed and an example of such a control system is presented.

KEY WORDS

DC line, telegraphists equation, state-space model, transfer function, VSC-HVDC system.

I.   INTRODUCTION

Nowadays the requirement in electric power increases continuously. The interconnection of the networks can be a viable solution to respond to this need. However the interconnection in alternative voltage cannot be always feasible especially to deal problems of the deregulated electrical network, IGBT based VSC transmission schemes (e.g. VSC-HVDC system) may be such interesting solution, especially for power network < 500 MVA.

This kind of transmission is profitable especially for long distances or in the case of the underwater connections where the use of electric cables is required. In these cases, the dynamic characteristics of the DC line used must be established to analyse its impacts on the VSC transmission.

Numerous methods of transmission line transient simulation have been approached in time domain ([2], [3] [4], [5]). They are mainly models of knowledge.

Our objective is to set up a model devoted to control the VSC transmission. This paper deals with dynamic model of transmission line for the control in term of transfer function. Two methods of dynamic modelling are developed for analysing the frequency behaviour of the transmission line.

The paper is organised as follows. At first, the VSC – HVDC transmission is presented. The necessity of modelling the DC transmission is highlighted. After that, the physical model is presented in term of transfer function. Its complex form for the control is pointed out. An equivalent model using spatial discretization is then presented. As an example, the proposed model is simulated with a simple RC scheme (resistor and capacitor in parallel) as load. To highlight the resonance frequency problem of the transmission line behaviour, the simulation of the VSC – HVDC system with a long transmission line is presented. Finally, conclusions are drawn about the necessary control of the VSC transmission and, in consequence, on the needs of specific modelling of transmission lines.

II.   PROBLEM FORMULATION

The VSC transmission consists of two stations of conversion connected by transmission line. A standard VSC transmission scheme is shown on the figure 1.

Combining such transmission schemes with the availability of high switching frequencies (e.g. 1 kHz), the delivered power quality could be significantly enhanced for both transient (high speed control) and steady-state (reduction of harmonics content) operations. However, with respect to the harmonic content (i.e. high bandwidth DC bus voltage controllers (U_{C1} and U_{C2} control) and high switching frequency), it becomes necessary to describe the physical behaviour of the VSC transmission scheme in a large frequency range.

The optimisation of transmission VSC – HVDC performances appears, according to Figure 1, very dependent on the dynamic characteristics of the link between stations, and thus conditions the structure of the VSC – HVDC control. It is thus particularly significant to obtain an accurate dynamic model of cable. In this paper, only a linear model is developed.

AC-DC Power Transmission, 28-30 November 2001
Conference Publication No. 485 © IEE 2001

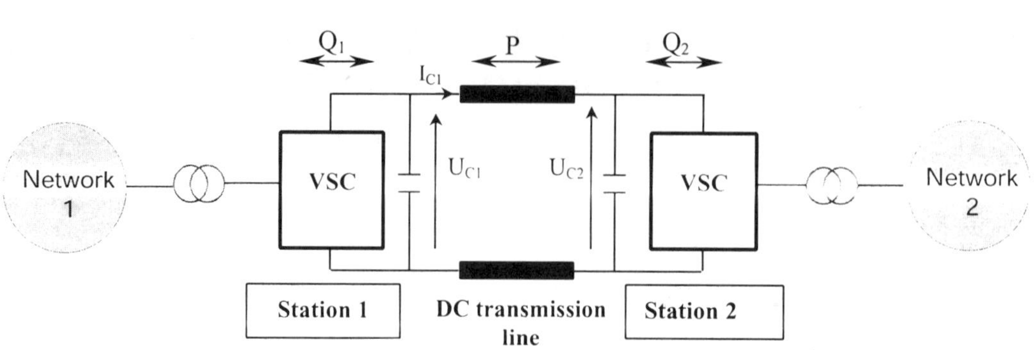

Figure 1 : VSC – HVDC Transmission Structure

III. PHYSICAL MODEL

Let us consider a transmission line represented by its length l (z, the space variable) and R, L, C are respectively resistance, self inductance and the capacity per unit of length. At one line end applies an electromotive force $E_0(t)$, the other end is closed on an impedance Z, like it shows in Figure 2.

The transmission line physical equations are given as follows :

$$\begin{cases} \dfrac{\partial V}{\partial z} = -L\dfrac{\partial i}{\partial t} - Ri \\[2mm] \dfrac{\partial i}{\partial z} = C\dfrac{\partial V}{\partial t} \end{cases} \qquad (1)$$

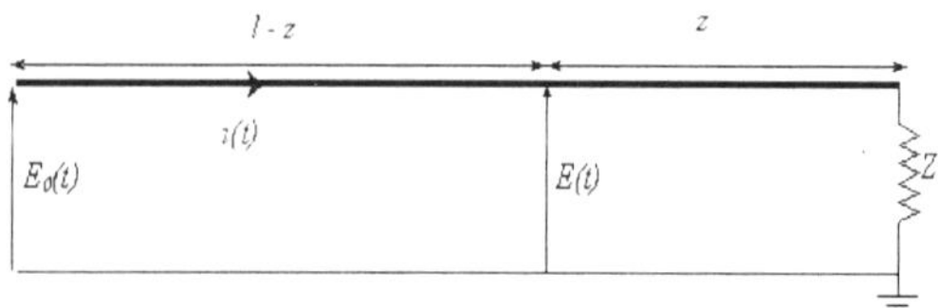

Fig.2. Transmission line physical model

They are partial differential equations, their combination give us a well-known equation called *"telegraphists equation"* :

$$\frac{\partial^2 V}{\partial z^2} = LC\frac{\partial^2 V}{\partial t^2} + RC\frac{\partial V}{\partial t} \qquad (2)$$

Thus, the transfer function which represents the ratio between the input and the output voltages of the transmission line is given by :

$$H(p) = \frac{\gamma(p).ch(\lambda(p).x) + sh(\lambda(p).x)}{\gamma(p).ch(\lambda(p).l) + sh(\lambda(p).l)} \qquad (3)$$

where

$$\gamma(p) = Z(p)\sqrt{\frac{C.p}{R+L.p}} = \frac{Z(p)}{Z_c(p)} \qquad (4)$$

$$\lambda(p) = \sqrt{C.p.(R+L.p)} \ , \qquad (5)$$

$Z_C(p)$ is the line characteristic impedance and $\lambda(p)$ is the constant of propagation.

As an example, the bode diagram of the transmission line transfer function with a resistive load is presented in Figure 3 to analyse the behaviour of the transmission line in frequency domain :

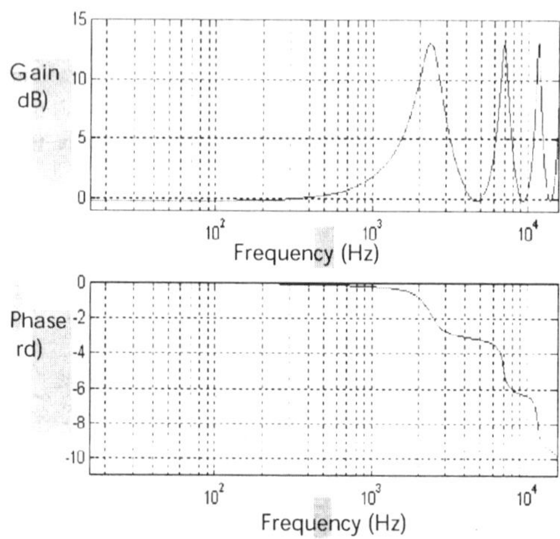

Fig.3 : Dynamic behaviour of the transmission line model with resistive (R = 200 Ω)

We note that the transmission line presents resonance phenomena generated by the space dependence of the DC line (wave interferences occur). The first resonance frequency appears around 2300Hz. Other resonance frequencies are into half wavelength.

Notice that the RC parallel load behaves like a low-pass filter where the frequency resonance are shifted towards the low frequency and the gain is attenuated at the high frequency (Figure 4).

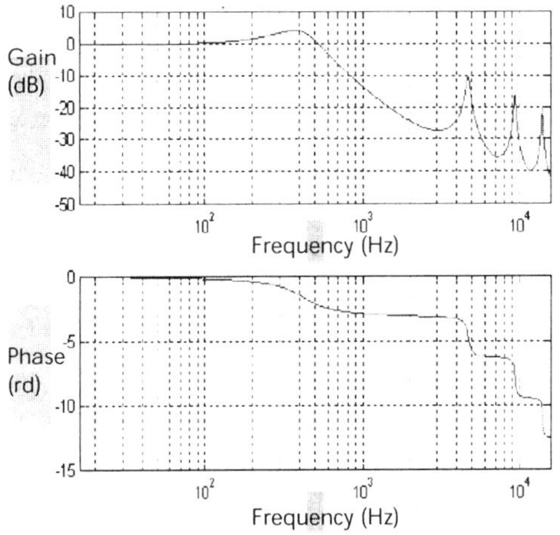

Fig.4 : Dynamic behaviour of the transmission line model with parallel RC load (R = 200 Ω, C = 50 μF)

We obtained the dynamic characteristics of the transmission line and the main conclusion of this analysis is the existence of waves interferences due to the length of transmission line.

However, the transfer function has a complex form and solving partial differential equations cannot be done easily in a control procedure. To sort out this problem, an equivalent model dedicated to control must be developed (ordinary differential equations).

IV. EQUIVALENT MODEL BASED ON LINEAR SPACE DISCRETISATION

The physical behaviour of the cable can be described by a succession of chained infinitesimal elementary cells. This technique is useful to a setting in a state space form of the transmission line model.

IV.1 Elementary cell

For highly accurate modelling, symmetrical structures of cells must be taken, but they have an interest when the effect of a mutual inductance is taken into account between lines.

Thus, we consider a non symmetrical (R,L,C) structure which is appropriate for our study. Indeed, we are interested in relations between the input and the output voltages of the transmission line.

Figure 5 shows the structure of the elementary cell which interpret the behaviour of the transmission line for an elementary length.

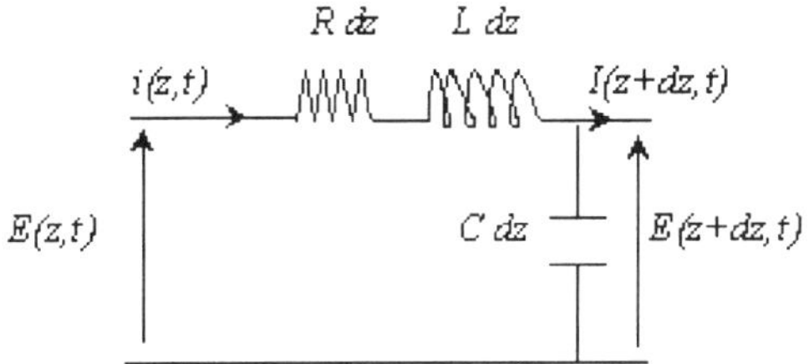

Figure 5 : Elementary cell structure

The state space system of equations is as follows:

$$\frac{d}{dt}\begin{bmatrix} i_e(k) \\ u_s(k) \end{bmatrix} = \begin{bmatrix} -\dfrac{R'}{L'} & -\dfrac{1}{L'} \\ \dfrac{1}{C'} & 0 \end{bmatrix}\begin{bmatrix} i_e(k) \\ u_s(k) \end{bmatrix} + \begin{bmatrix} 0 & \dfrac{1}{L'} \\ -\dfrac{1}{C'} & 0 \end{bmatrix}\begin{bmatrix} i_s(k) \\ u_e(k) \end{bmatrix}$$

(6)

Notice that the elementary cell gives a linear system.

IV.2 State-space model

To obtain the equivalent model of the transmission line, the cells are chained like it is showed in Figure 6. From the state space model, we can deduce the transfer function Us(n) / Ue(1)

$$\dot{X} = A.X + B.U \qquad FT = u_s(n)/u_e(1)$$

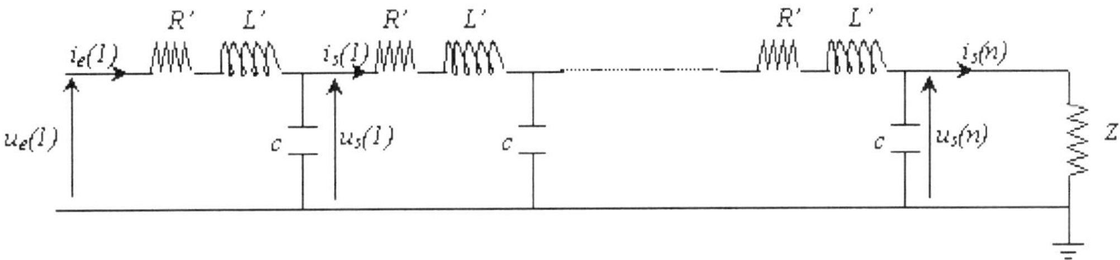

Fig. 6 : Equivalent model of transmission line based on linear spatial discretisation

Different configurations of cells have been considered and good agreements (7 % incertitude) can be obtained for 10 cells (Figure 7). It's worthwhile to precise that no optimisation has been done with respect to the nature of the load.

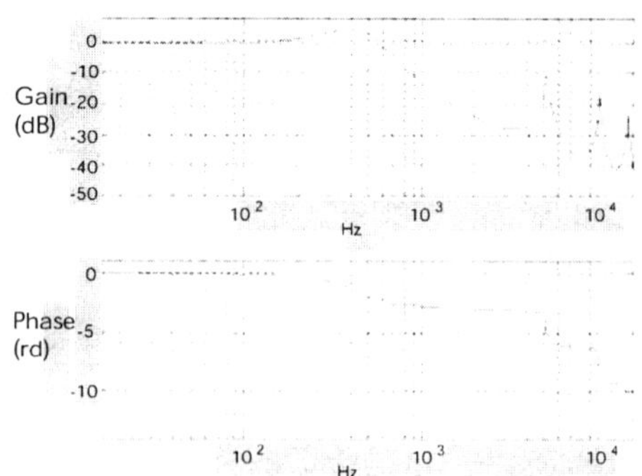

Figure 7 : Dynamic behaviour of transmission line with parallel RC load (state space model, 10 cells)

Next step is now to consider the influence of such a transmission line in a VSC – HVDC system.

V. INFLUENCE OF THE DC TRANSMISSION LINE ON THE VSC-HVDC SYSTEM BEHAVIOUR

In order to highlight the influence of the DC transmission line on the behaviour of the VSC-HVDC system, the developed equivalent state space model of the cable is introduced in a Matlab/Simulink simulation scheme of a VSC-HVDC system. This simulation scheme is based on an equivalent continuous-time model of a standard VSC-HVDC transmission scheme, presented in [6]. In such a model, switching effects resulting from PWM "station 1" and "station 2" are not taken into account. As a consequence, the simulation results presented here illustrate mainly the influence of the lowest resonance frequencies of the DC line on the "fundamental" dynamics of the system.

The VSC-HVDC system control structure is based on nonlinear state feedback laws performing full linearized and decoupled closed loop control of the line current components of each "station", as well as closed loop control of "station 1" DC-Bus voltage. In this structure, "station 1" is dedicated to the control of the reactive power Q1 and the DC-Bus voltage u_{c1}, as "station 2" is used for the control of the reactive power Q2 and the active power P. The details of this control structure can be found in [6].

Then, our goal is to highlight the behaviour of the VSC-HVDC system state variables ("stations" currents, DC line current, DC-Bus voltages) versus the length (i.e. resonance frequencies) of the DC transmission line while keeping a constant controllers

tuning. More particularly, the behaviour of "station 2" DC-Bus voltage u_{c2} which is "open-loop regulated" (there is no voltage loop on "station 2" side) is examined. All simulation results are obtained for a 200 MW step disturbance of active power P, by action on the direct component of "station 2" current, at the instant t = 0.1s. "Station 1" DC-Bus voltage reference set point is 300 kV and "stations 1 and 2" reactive power reference set point is zero, i.e. unity power factor operation.

The three DC line lengths considered are: 10, 100 and 500 kms. The number of cells of the DC line state space model is set to 10 for each length. As we are concerned by the effects of the lowest resonance frequencies through the equivalent continuous-time model of the VSC-HVDC system, this number of cells is adequate to describe the frequency behaviour of the DC line, even for a 500 kms length cable.

Figure 8 shows the resulting tracking of the "station 1" current direct component i_{d1} for the three cable lengths. As the length increases, it can be observed both higher current overshoot and lower current damping.

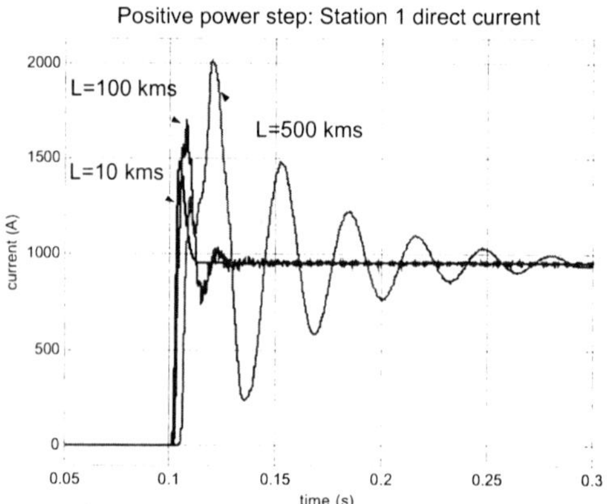

Figure 8: Tracking of "station 1" current direct component for 10, 100 and 500 kms cables

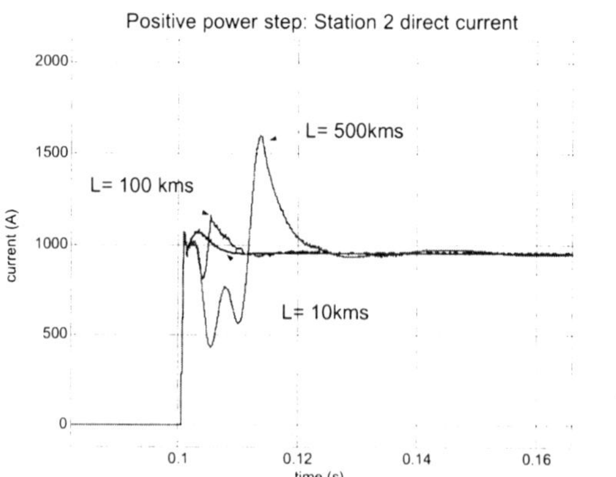

Figure 9: Tracking of "station 2" current direct component for 10, 100 and 500 kms cables

The responses of the current direct component, i_{d2}, of "station 2" are depicted in Figure 9. Increasing the cable length leads also to higher current overshoot.

The regulation behaviour of the "station 1" DC-Bus voltage u_{c1} is illustrated by Figure 10. Note that the current i_{d1} corresponds to the output of the "station 1" DC-Bus voltage controller. Then, for a good behaviour of u_{c1} regulation, the i_{d1} response is directly related to the evolution of the DC line current i_{c1} shown in Figure 11. When the cable length increases, due to DC line resonance frequencies the DC line current exhibits high current overshoot and low current damping, then resulting in the i_{d1} evolution of Figure 8 which can be not acceptable for the converter safety.

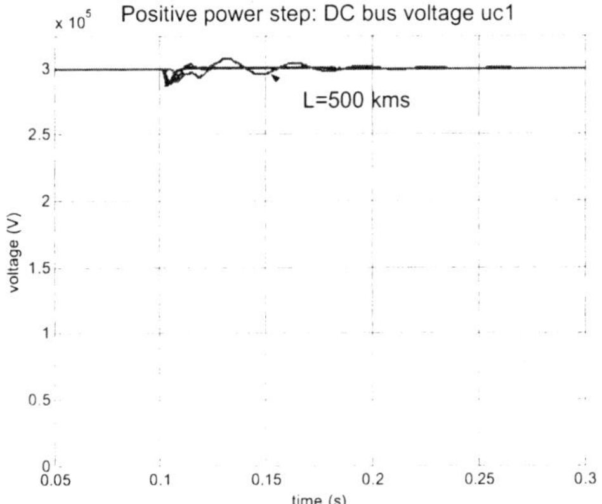

Figure 10: Regulation of "station 1" DC-Bus voltage for 10, 100 and 500 kms cables

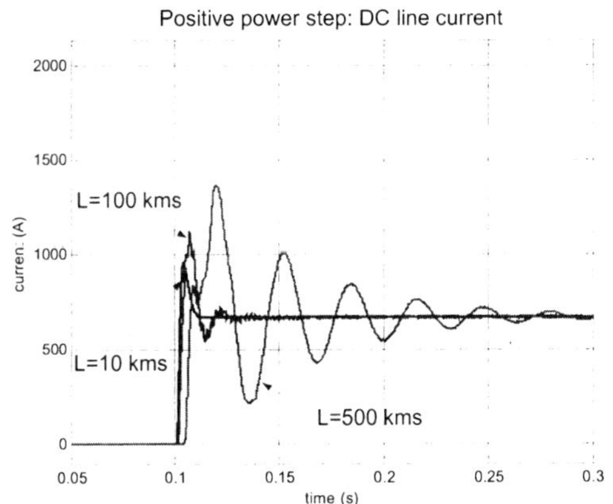

Figure 11: Evolution of DC line current for 10, 100 and 500 kms cables

The "open loop" regulation of the "station 2" DC-Bus voltage u_{c2} is illustrated in Figure 12. It can be observed higher voltage overshoots with respect to u_{c1} voltage. These overshoots can result in saturation of the "station 2" control variables as it is the case for the 500 kms length cable when the u_{c2} lower voltage is limited around 280 kV (see Figure 13). In such a case, it results a temporary "open loop" control of "station 2" current components as illustrated in Figure 9 and in a lost of

the decoupling property of the control structure as shown in Figure 14.

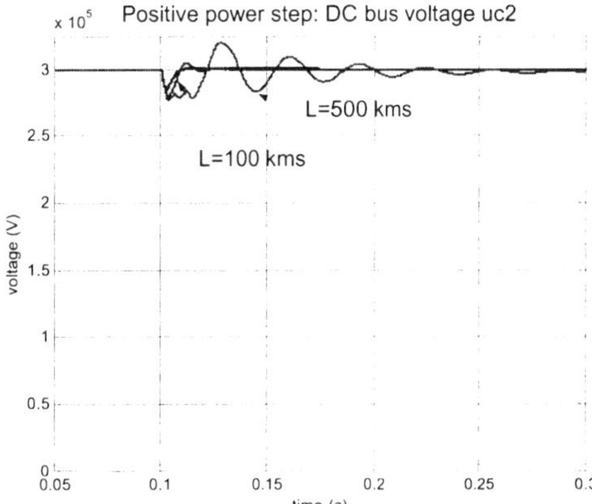

Figure 12: Evolution of "station 2" DC-Bus voltage for 10, 100 and 500 kms cables

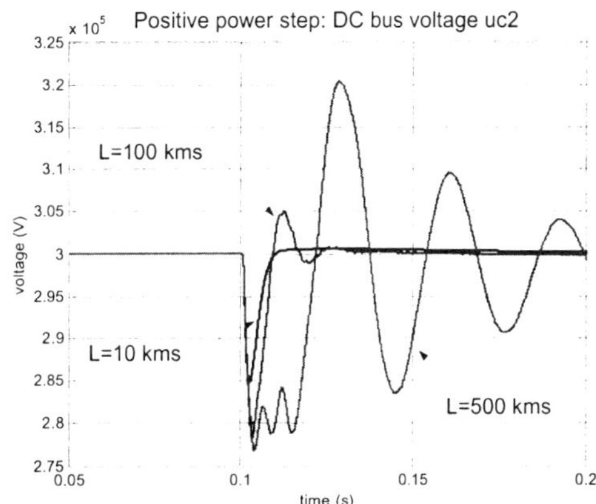

Figure 13: Evolution of "station 2" DC-Bus voltage for 10, 100 and 500 kms cables: Zoom of Figure 12

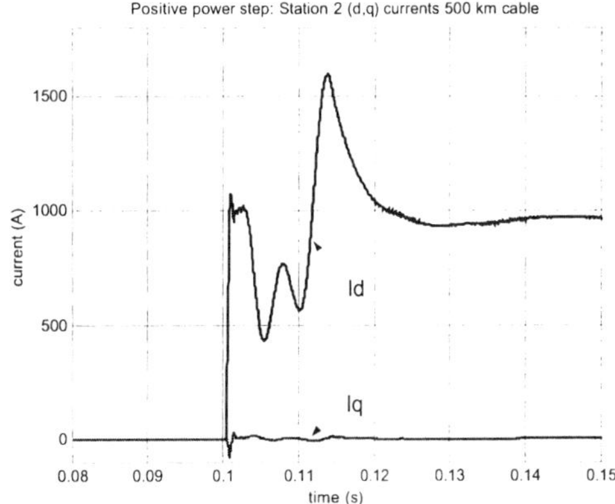

Figure 14: Tracking of "station 2" current direct and quadrature components for a 500 kms cable

Finally, Figure 15 presents the DC line current evolution where "station 1 and 2" currents and "station 1" DC-Bus voltage "closed loop" dynamics are divided by 2. As DC line resonance frequencies are less excited, the current overshoot is reduced as well as the current damping is increased.

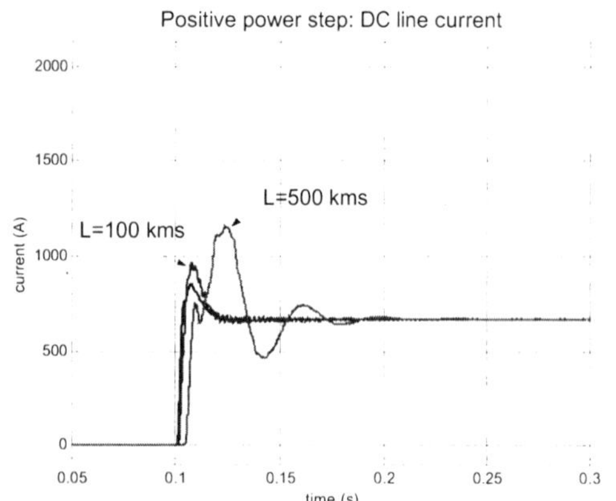

Figure 15: Evolution of DC line current for 10, 100 and 500 kms cables – Currents and DC-Bus voltage "closed loop" dynamics divided by 2

VI. CONCLUSIONS

An equivalent state space model of a DC transmission line, based on linear space discretisation, has been proposed and validated in regards of a physical model behaviour (Telegraphists equation).

Such a model can be used for simulation purposes, as well for continuous-time as for discrete-time simulation schemes. Furthermore, this model can be used for more in depth analysis of the full VSC-HVDC system and then for the synthesis of high performances controllers, involving if requested an observer in order to take into account the variation of "station 2" DC-Bus voltage.

The influence of the DC transmission line length (i.e. the DC line frequency response) on the behaviour of the VSC-HVDC system have been highlighted through simulation results. These results have been obtained considering "ideal" actuators, i.e. the "stations 1 and 2" PWM switching effects have not been taken into account. However, the proposed DC line state space model can be used, without any modification, for the study of PWM switching effects, i.e. the influence of higher DC line resonance frequencies, in simulation schemes.

VII. ACKNOWLEDGEMENTS

The authors would like to express their gratitude to B. Andersen of ALSTOM T&D, for initiating and supporting this work.

VIII. REFERENCES

1. Angot A, "Compléments de mathématiques à l'usage des ingénieurs de l'électrotechnique et des télécommunications". 2nd edition,1982, MASSON, Paris, France

2. Dommel H M, "Digital computer solution of electromagnetic transients in single and multiple networks", IEEE transaction on power apparatus and systems, vol. pas-88, n°4, April 1969

3. Dufour C, Le-Huy H, " Highly accurate modelling of frequency – dependent balanced transmission lines", IEEE transaction on power delivery, vol. 15, n°2, April 2000

4. El Hakimi A, Le-Huy H, Viarouge P, "Modélisation de lignes de transmission d'énergie électrique: solution semi-analytique pour simulation temps réel", Proc. Canadian Conference on Electrical and Computer Engineering- CCECE- Halifax, Nova Scotia, Sept. 1994, pp.1-4.

5. Ferri G, "Transmission line modelling by ladder networks", 22nd International Confrence on Microelectronics, vol. 2, pp. 755-757, Nis, Serbia, 14-17 May, 2000

6. J-L. Thomas, S. Poullain, and A. Benchaib, "Analysis of a robust DC-Bus voltage control system for a VSC transmission scheme", 7th International Conference on AC-DC Power Transmission, IEE, London, 28-30 November 2001.

A CASE STUDY ON THE ALTERNATIVE SOLUTIONS FOR THE LOAD FLOW PROBLEM IN A 50 KV CABLE GRID

Y H Fu*, J F Groeman*, H van der Geest** and H Zee**

* KEMA T&D Power, The Netherlands **Noord West Net, The Netherlands

ABSTRACT

This paper describes a case study on the alternative solutions for a load flow problem in a Nuon 50 kV cable grid in North Holland. This part of the cable grid is typically parallel fed with short distances. The grid is designed to be operated safely under the condition when one of the cable links is out of service, the so-called N-1 condition. Under a certain circumstance, this grid will have an overload problem due to unbalanced load flow, meaning some cables are overloaded while others are under loaded. The transmission capacity of this grid will be insufficient from the year 2015 anyway, under N-1 condition. The ultimate solution is to lay extra cables in this grid, but this is a very costly option. The purpose of this paper is to present the alternative solutions studied to solve the load flow problems so that the investment of laying extra cables can be postponed after year 2010.

Three alternative solutions have been studied: inserting a variable impedance, inserting a phase shifting transformer and inserting a UPFC. The effects of the three alternative solutions, both positive and negative, have been compared. The best solution among them has resulted, based on the technical aspects, operational aspects and economical aspects.

1   INTRODUCTION TO THE LOAD FLOW PROBLEM

The 50 kV Purmerend cable grid

The 50 kV Purmerend underground cable grid in the Dutch province of North Holland is shown in figure 1. The power is mainly fed from the 150 kV substation at Wijdewormer (WW). The power is further distributed through three 50 kV substations at Purmerend Schaepmanstraat (PS), Purmerend Kwadijkerkoogweg (PK) en Edam (ED). There is one combined heat and power generator (WKC) connected at the PK 50 kV substation. The cable lengths in this grid are relatively short (between 3 to 14 km).

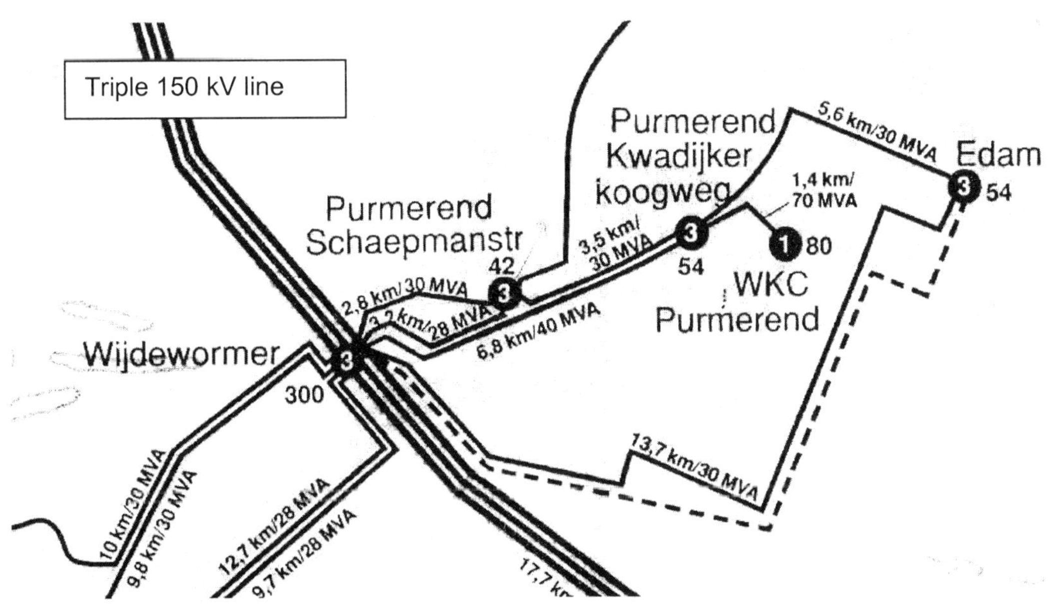

Figure 1   The 50 kV Purmerend cable grid in North Holland.

AC-DC Power Transmission, 28-30 November 2001
Conference Publication No. 485 © IEE 2001

The total load in this grid amounts 100 MVA with a power factor of larger than 0.9. The grid is designed to be operated safely under the condition when one of the cable link is out of service, the so-called N-1 condition, not taking into account the combined heat and power generator. In that case, 110% overloading is allowed on the remaining connections.

Table 1 Load prognosis (MVA)

Year Substation	01 02	02 03	04 05	06 07	08 09	10 11	12 13	14 15	16 17
PS	27,4	28,2	29,8	31,4	33,0	34,5	36,1	37,7	39,3
PK	33,0	33,6	34,8	35,9	37,1	38,2	39,4	40,6	41,7
ED	36,9	37,8	39,6	41,3	43,1	44,9	46,7	48,4	50,2

The load prognosis

The area Purmerend is a steady growing region for both residencial and industrial loads. The load prognoses of up to year 2017 are given in table 1.

The load flow problem in the 50 kV Purmerend grid

This grid will have already problem under one particular condition. This condition is: if the combined heat and power generator is out of service for whatever reason, and one of the two cable links between PS-WW is not available, the other cable link will be more than 120% overloaded. Around the year 2005, this particular overload situation will be more than 140%. This implies that this grid can not meet the N-1 design criterion any more.

Around the year 2010, the unavailability of any cable in this grid will at least 120% overload in another cable, while other cables in the grid still have the capacity to transmit power. This is due to the unbalanced load flow. Some examples are shown in Table 2.

The problem will become worse with the increase of the load. Around the year 2015, the load will be so high that the cable transmission capacity in this grid will be exceeded. The only solution is then to lay extra cables.

Table 2 Examples of load flow problem in 2010

Disturbed link	Overloaded link	Overload
WW-PS 1	WW-PS 2	150%
WW-PS 2	WW-PS 1	158%
WW-PK	WW-PS 1	123%
	WW-PS 2	123%
	PS-PK	128%

2 POSSIBLE SOLUTIONS

The 50 kV Purmerend cable grid can be characterised as a meshed network with parallel cable links. It is a strong and mainly resistive grid. The solution for the load flow problem is to unbalance the load flow among the remaining cable links when one cable link is unavailable. Generally there are two ways to control the load flow: by changing the impedance of a cable link or by changing the voltage drop between two substations (amplitude and phase angle). Of course these changes should always be within the operational tolerance for the voltage. Thanks to the small impedance in this network, the latter criterion is not a limiting factor.

The conventional solutions for this load flow problem are laying extra cables or re-routing the cables, for example by connecting the cable between WW-PK, which presently by-passes PS, to substation PS. Laying extra cables is a final solution but costly. Such a solution is usually postponed until no other alternative available. Re-routing the cable will only help until year 2005.

Therefore the following three alternative solutions have been studied:

a)   Inserting a series impedance
b)   Inserting a phase shifting transformer
c)   Inserting an UPFC

Series impedance solution

First the most suitable location where to insert the series impedance should be determined. The most critical cable link in this grid is the link between WW-PS with the shortest length and the lowest capacity. Inserting a series impedance in this cable link will force the load flow through other parallel links with more capacity and therefore balance the load flow. By taken into account the available space in the substation, it was found that the suitable location is in the 50 kV substation of PS between the cable link WW-PS.

Then the value of such series impedance is determined by taking into account all possible scenarios of cable unavailability, the voltage tolerance and the load growth. The impedance should be a reactor with an optimum value that varies with the load growth. This solution works until the load situation year 2010 with all cables loaded below 110% under N-1 condition. For example for the load condition of year 2010, a 1.2 Ω reactor will be an optimum value.

The realisation of such series reactor can have several options, for example:

- A reactor with different taps and by-pass circuit breaker. The reactor is normally by-passed. When one of the cables is unavailable, the reactor with a pre-selected tap position is then inserted until the disturbed cable is restored. The reactor value and the number of taps should be determined by detailed system studies. Replacing the tap selector plus circuit-breaker by thyristor-controlled tap selection would increase the response speed.

- A GTO-thyristor controlled series reactor with a continuously variable inductive value. The circuit is as in figure 2. Normally the total series impedance is zero as the voltage drops

across the inductor and the capacitor cancel out each other. When one of the cables is unavailable, the total series impedance will be regulated to an optimum value within half a cycle (by controlling the turn-off phase angle of the GTOs, the net reactance can be controlled continuously between zero and the value of the inductance).

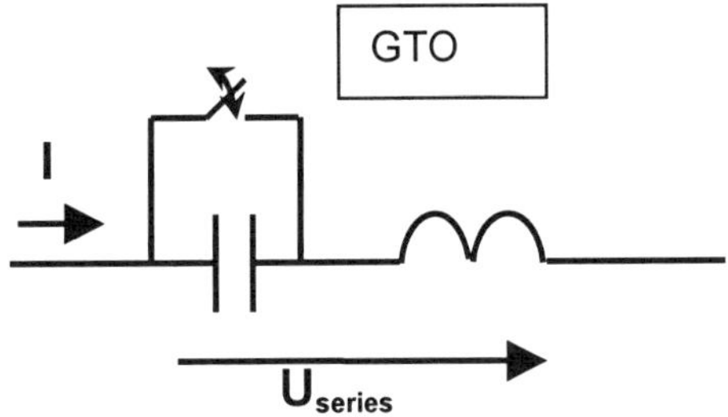

Figure 2   Circuit of GTO-thyristor controlled series reactor.

Phase shifting transformer solution

A phase shifting transformer can shift the voltage in perpendicular direction by inserting a certain series voltage, 90° phase shifted to the bus voltage. The most suitable location for the phase shifting transformer is in the 50 kV substation of PS between the cable link PS-PK. Through load flow studies it was found that a phase shifting transformer alone could not balance the load flow sufficiently in this grid when one of the cables is unavailable. This is due to the effect that both the P and Q loadflow are affected by the phase shifting. The influence on P and Q are not independently controllable, and the optimum phase shifting angle leads to an improvement of the overload condition, but the improvement is insufficient to comply with the criteria.

Therefore a shunt capacitor bank is needed to completely balance the load. This solution works until the load situation year 2010 with all cables loaded below 110% under N-1 condition. The optimum value is with a phase angle shift of 1.13° and a shunt capacitor bank of 17 Mvar at bus PS for the load situation of year 2010.

The phase shifting transformer can be realised by conventional transformer technology where the phase angle can be stepwise changed as well as with power electronic tap changers where the phase angle can be continuously changed.

UPFC solution

A UPFC (Unified Power Flow Controller) can inject a certain series voltage in any phase angle. In another words, an UPFC can change the voltage amplitude and phase angle independently. The choice of the location is the same, i.e. in the 50 kV substation of PS between the cable link PS-PK to install the UPFC. Load flow studies resulted that a small UPFC, rated less than 5% the cable rating, will be sufficient to balance the load. This solution works until the load situation year 2010 with all cables loaded below 110% under N-1 condition. The optimum value is a series injection voltage of 0.4 kV with a phase shift of 0.55° for the load situation of year 2010.

The UPFC can be realised by a back-to-back power electronic inverter and converter. The inverter draws the power from the grid. The converter inserts the required voltage back to the grid in series through a DC link. The inserted voltage can be changed continuously for amplitude and phase angle separately.

3 COMPARISON OF SOLUTIONS

Detailed calculations have shown that all three alternative solutions can solve the load flow problem in the 50 kV Purmerend cable grid for the load situation up to 2010. There are advantages and disadvantages for each alternative solution. These three alternative solutions can be compared with each other in the table 3.

Table 3   Comparison of the three alternative solutions for the load flow problem in the 50 kV Purmerend cable grid.

	Series reactor in the link WW-PS1/2	Phase shifting transformer in the link PS-PK	UPFC in the link PS-PK
Advantages	+ Able to balance the load below 110% until 2010 under N-1 condition + Simple solution, easy to realise + Possible current limiting function	+ Limited ability to balance the load	+ Able to balance the load below 110% until 2010 under N-1 condition + Sophisticated solution with fast control + Can also be used for fast voltage regulating when it is needed
Disadvantages	– Possible reduction of reliability due to series device	– Possible reduction of reliability due to series device – Need a shunt capacitor bank, can be costly – Difficult for control	– Possible reduction of reliability due to series device – Complex equipment

4 CONCLUSIONS

From this case study of the three alternative solutions for the load flow problem in the 50 kV Purmerend cable grid, it can be concluded that all three alternative solutions can be used to solve the problem until year 2010. This means that the laying extra cables can be postponed until then. Comparing these three alternative solutions, the series reactor solution is the most direct and simple solution and the UPFC is a more sophisticated solution that requires new technology. The phase shifting transformer solution will not be recommended because it requires an additional shunt capacitor bank. Of course one should make economic comparisons between laying extra cables and install a series reactor or a UPFC before a decision can be made. Most probably the installation of a series reactor with taps will be the most economic solution.

ANALYSIS AND IDENTIFICATION OF HVDC SYSTEM FAULTS USING WAVELET MODULUS MAXIMA

L. Shang, G. Herold; J. Jaeger, R. Krebs, A. Kumar

Friedrich-Alexander University of Erlangen–Nuremberg; Siemens AG, Germany

ABSTRACT

In this paper, different HVDC system faults are analysed and the rationing criteria based on wavelet modulus maxima for the identification of the HVDC system faults are proposed. The simulation results are discussed. The results show that the application of wavelet technique leads to a proper and more reliable solution for fault identification. The results also provide a good basis for the new high-speed protection of HVDC lines.

KEYWORD

HVDC fault, HVDC line protection, Identification, wavelet, wavelet modulus maxima

1. INTRODUCTION

The detection and fast clearance of faults in HVDC lines are important for a safe operation of power systems. The protection principle based on travelling wave theory provides the fastest protection. Long HVDC lines cannot be sufficiently modelled with concentrated parameters as assumed in traditional protection systems. Therefore, long HVDC lines have to be represented as distributed elements and a protection for long HVDC lines should be developed based on travelling wave theory.

According to travelling wave theory, voltage and current travelling waves appear on the line when fault occurs. The fault generated travelling waves contain sufficient fault information that can be used for high-speed fault identification and line protection. In AC transmission lines, the amplitude of fault generated travelling waves changes with the voltage angles. There is a problem for the travelling wave protection when faults occur near voltage zerocrossing. However, there is no such problem for DC transmission lines so that travelling wave protection is ideally suited for HVDC lines.

In HVDC systems, commutation failures in the converter station and single-phase short circuit faults at the AC side are similar to HVDC line faults. It is an important requirement of HVDC line protection that different fault types be identified and the correct decision be made as fast as possible.

However, a fast and reliable fault identification is still a big challenge. It is not easy to identify HVDC faults by using pure frequency domain based methods or pure time domain based methods. The pure frequency domain based methods are not suitable for the time-varying transients and the pure time domain based methods are very easily influenced by noise.

The wavelet transform provides a new approach for analysing time-varying transients. It has the capability of analysing signals simultaneously in time and frequency domain. Moreover, it can adjust analysis windows automatically according to frequency, namely, shorter windows for higher frequency and vice visa. Hence it is suitable for characteristic identification and travelling wave protection [1-3]. However, wavelet based fast identification and protection in HVDC systems is a relatively new field.

In this paper, the behaviour of different faults in HVDC power systems will be analysed through wavelet transform and the identification criteria based on wavelet techniques will be proposed. Based on the above investigations, the high-speed HVDC line protection will be developed. The simulations are carried out with MATLAB$^{©}$. The results show that the wavelet techniques lead to a new way for the fault identification and the protection in HVDC systems.

2. FAULTS IN HVDC SYSTEMS

For the analysis and the identification of HVDC system faults, different cases are studied. A standard model of 12-pulse HVDC system under the MATLAB$^{®}$ environment is used for the simulation. Figure 1 shows the simulation model in which a 1000 MW (500 kV, 2kA) DC line is used to transmit power from a 500 kV, 5000 MVA, 60 Hz network to a 345 kV, 10 000 MVA, 50 Hz network. The DC line is 300 km long and the speed of the travelling wave is 296112 km/s.

Figure 2 shows the voltages and currents of DC line at the rectifier terminal when (a) DC line short circuit, (b) commutation failure at the inverter station, (c) single-phase short circuit on the AC side of inverter station, and (d) normal operation condition as a reference case.

From Figure 2, we can see that different types of faults lead to similar transient processes. It is not easy to identify the faults and to make correct protection decision fast within 3-5ms by using traditional methods. It is even more difficult if there is noise.

AC-DC Power Transmission, 28-30 November 2001
Conference Publication No. 485 © IEE 2001

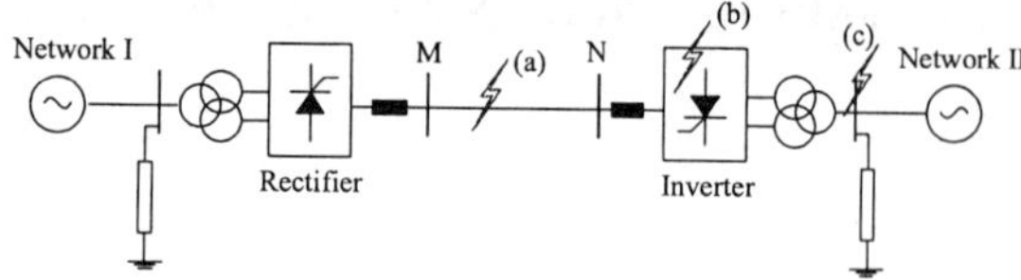

Figure 1. Simulation model

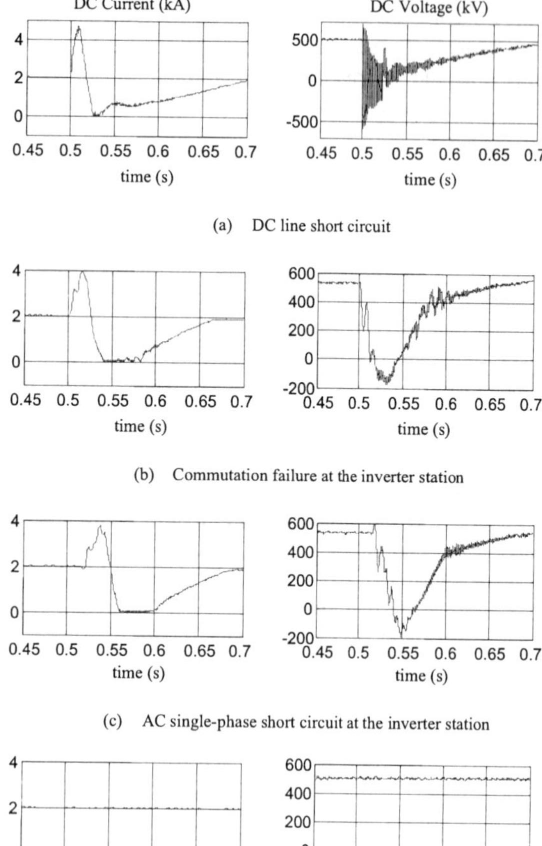

(a)   DC line short circuit

(b)   Commutation failure at the inverter station

(c)   AC single-phase short circuit at the inverter station

(d)   Normal operation condition

Figure 2. Faults in HVDC systems

3.   WAVELET TRANSFORM

The Wavelet transform transfers a time varying signal into a time-scale plane and thus can represent the original signal with time as well as frequency information. Each scale in wavelet transform corresponds to a certain frequency band and the time window widths are changed with scale or frequency automatically. Such multiresolution property is particularly suitable for analysing transient signals. Another important reason why wavelet transform is attractive for engineers is because there are fast calculation algorithms based on filter bank structure.

There are different algorithm structures for the wavelet transform. Considering a better time location and a better information keeping, we use the a trous structure but without down-sampling blocks following the high-pass filters. Figure 3 shows our filter bank where H_0 and H_1 are low-pass filters and high-pass filters respectively. The outputs of high-pass filters are the wavelet transform of the original signal, called as wavelet coefficients.

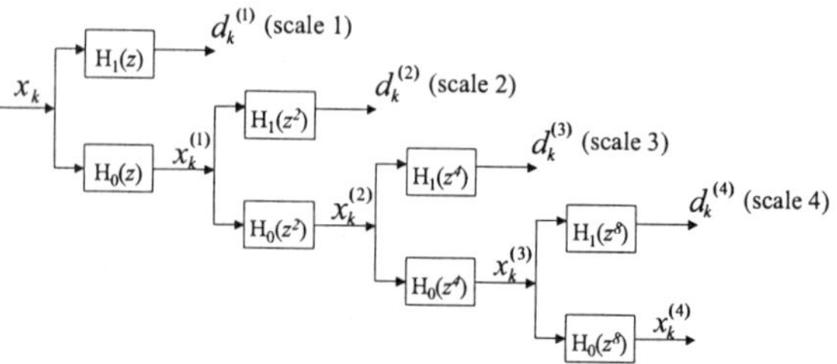

Figure 3. Wavelet filter bank

The absolute local maximum values of wavelet coefficients are called wavelet modulus maxima. If the mother wavelet is the first derivative of a smooth function, the edge of a signal can be represented well by its wavelet modulus maxima. Under above condition, the wavelet modulus maxima occur at an edge point, the polarity of the maxima shows the change direction of the edge, and the amplitude represents the changing intensity of the edge. It is proved that the wavelet modulus maxima satisfy the following relation.

$$\left| W_{max} x(t) \right| \leq As^{\alpha} \tag{1}$$

Where, $W_{max}\, x(t)$ is the wavelet modulus maxima of signal $x(t)$, A is a constant, s is scale and α is Lipschitz exponent.

This relation means that the wavelet modulus maxima of an edge ($\alpha = 0$ or $\alpha > 0$) remain unchanged or increase in value while the wavelet modulus maxima caused by white noise ($\alpha < 0$) decrease in value when scale increases. Additionally, the number of wavelet modulus maxima caused by white noise decrease sharply when scale increase [4]. This makes a strong denoising function possible. The simulation results show that the wavelet modulus maxima represent the edges well even when signals are mixed with 15% noise [5].

3   WAVELET ANALYSIS OF HVDC SYSTEM FAULTS

The transients of the study cases above will be analysed through wavelet transform. Each case is sampled at 80 kHz and 512 samples are taken for wavelet transform in 4 scales. The Mallat wavelet is used as the mother wavelet. HVDC line currents, HVDC line voltages and corresponding reverse voltage travelling waves are used as the input signals of the wavelet filter bank. Reverse voltage travelling waves are calculated with Equation (2).

$$u_r = (u_{DC} - Z_c i_{DC})/2 \qquad (2)$$

Where u_r is reverse voltage travelling wave, u_{DC} and i_{DC} are the DC voltage and DC current respectively, and Z_c is the surge impedance of the HVDC line.

3.1 NORMAL OPERATIONS

The DC current and DC voltage are steady with small changes around the rated values during the normal operation conditions. Figure 4 shows the DC current at the terminal M and its wavelet modulus maxima in four scales. Figure 5 and Figure 6 display the DC voltage and the reverse voltage travelling wave at the terminal M and their wavelet modulus maxima in four scales. The wavelet modulus maxima occur regularly and with small values: less than 0.04 for DC current, less than 40 for DC voltage and less than 20 for reverse voltage travelling wave.

3.2 HVDC LINE FAULTS

HVDC line faults at different locations with different fault resistances are simulated. One of the HVDC line faults occurs at 100 km from terminal M and with zero fault resistance. Figure 7 shows the DC current at the terminal M and its wavelet modulus maxima in four scales. Figure 8 and Figure 9 display the DC voltage and the reverse voltage travelling wave at the terminal M and their wavelet modulus maxima in four scales.

It can be seen that:
- The wavelet modulus maxima occur at every arriving and reflection instant of the travelling waves. The polarities appear regularly: positive and negative in turns. The values of the wavelet modulus maxima are much larger than ones during the normal operation conditions.
- The negative wavelet modulus maxima of DC current are relative small. They are too small to measure the time delay for the fault location, although the positive ones are large enough for the fault detection.
- The wavelet modulus maxima of the DC voltage provide a secure fault detection and time location.
- The wavelet modulus maxima of the reverse voltage travelling wave provide a similar effect as the DC voltage. Notice that the reverse voltage travelling wave is nearly equal to zero during normal operation conditions, and it can be more easily processed than the DC voltage.
- The first four values of the wavelet modulus maxima of the reverse voltage travelling wave are about 1000. The fault can be securely detected if the pick-up value is set to 100.

3.3 COMMUTATION FAILURES

Figure 10 shows the wavelet modulus maxima of the reverse voltage travelling wave at the terminal M during a commutation failure at the inverter station.

It can be seen that:
- All wavelet modulus maxima are less than 40.
- The polarities of the wavelet modulus maxima remain the same 3ms after the disturbance arrives at terminal M. During this time, the values of the wavelet modulus maxima are less than 15.
- The commutation failure can be identified clearly from the HVDC line fault and the normal operation with the setting (e.g. 100) value and the polarity change.

3.4 AC SINGLE-PHASE FAULTS

Figure 11 shows the wavelet modulus maxima of the reverse voltage travelling wave at the terminal M during an AC single-phase fault at the inverter station.

Similar to the commutation failure, all wavelet modulus maxima in AC single-phase fault are less than 40, the polarities of modulus maxima remain the same for 3ms. Therefore, the AC fault can also be identified surely from the HVDC line fault and the normal operation with the setting (e.g. 100) value and the polarity change.

Unlike the commutation failure, the wavelet modulus maxima in AC single-phase fault occur with more density and relative larger value specially 3ms after the disturbance arrives at terminal M, and the polarities become positive and negative in turns after the unified polarity changes. Considering a fast identification in 3ms, the difference of the wavelet modulus maxima can be better used. Figure 12 shows the energy during 3ms on each scale. It can be seen that the difference between a commutation failure and an AC fault can be discriminated with the energy.

4 WAVELET IDENTIFICATION OF HVDC SYSTEM FAULTS

With the help of the above analysis of HVDC faults through wavelet transform, the criteria for the identification can be obtained.
- For HVDC line fault identification, the amplitude of the first wavelet modulus maxima of reverse voltage travelling wave, denoted as $|W_{max}U_r|$, should be larger than the setting value $K_{setting}$. The setting can be taken as 100 for 500 kV HVDC lines.

$$|W_{max}U_r| > K_{setting} \qquad (3)$$

- For identifying commutation failures and AC faults from the normal operation condition, the pattern of polarity change is used. During the normal operations, the polarities regularly change positive and negative in turns. It will be a commutation failure or an AC fault if the polarities remain the same for 3ms.

- For differentiation between a commutation failure and an AC fault, the energy of the wavelet modulus maxima on the scale 4 during 3ms are used. The energy, denoted as $E(W_{max}U_r)$, should be larger than the setting value $E_{setting}$. The setting can be taken as 40 for 500 kV HVDC lines.

$$E(W_{max}U_r) > E_{setting} \qquad (4)$$

Based on the identification of HVDC faults, a new high-speed HVDC line protection can be developed. In this protection, the fault location can be also obtained at the same time as the fault identification. The fault location is calculated according to Equation 5. The key here is the measurement of the time delay Δt. Considering that the reflection from the fault location (reflection factor is negative) is different than the reflection from the line terminal (reflection factor is positive), we measure the time delay between the first two opposite polarity wavelet modulus maxima.. Namely, if the first is negative, the next one should be taken with positive polarity.

$$L = \frac{v \times \Delta t}{2} \qquad (5)$$

Where L is fault distance in km from the measuring point, Δt is the time delay in s and v is travelling wave speed in km/s.

For example, a DC line fault with 60 ohm fault resistance occurs at 200 km from terminal M. The first modulus maxima is 723, larger than the setting (100) so that the DC line fault is detected. In this case, the reflected travelling wave from the terminal N arrives at the terminal M earlier than the reflection from fault location. With the help of the polarity, it is easy to identify them. The time delay between the first two opposite polarity wavelet modulus maxima (here corresponding to the first and the third modulus maxima) is 0.0013 s, thus the fault is located 199.87 km away from terminal M, as shown in Figure 13. The fault is detected and located correctly.

5 CONCLUSION

For high-speed HVDC line protection based on travelling waves, methods for transient signal analysis are necessary. Particularly, the identification with similar HVDC transients caused by faults is decisive. The wavelet transform provides a new possibility for this. In this paper, the different HVDC system faults are analysed and the criteria for the identification and the line protection are proposed. The simulation results show that the proposed approach based on the wavelet modulus maxima can make a definite identification of HVDC line faults, commutation failures and AC single-

phase faults. The application of wavelet techniques leads to a faster, easier and more reliable solution for the identification of HVDC system faults and the development of new high-speed protections of HVDC lines.

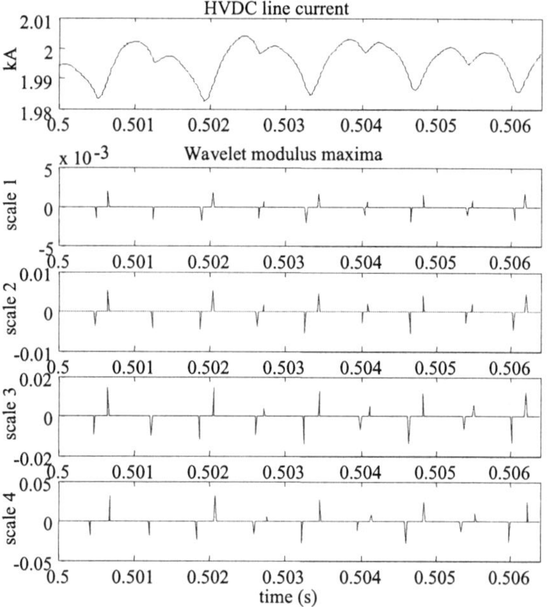

Figure 4. DC current during normal operation

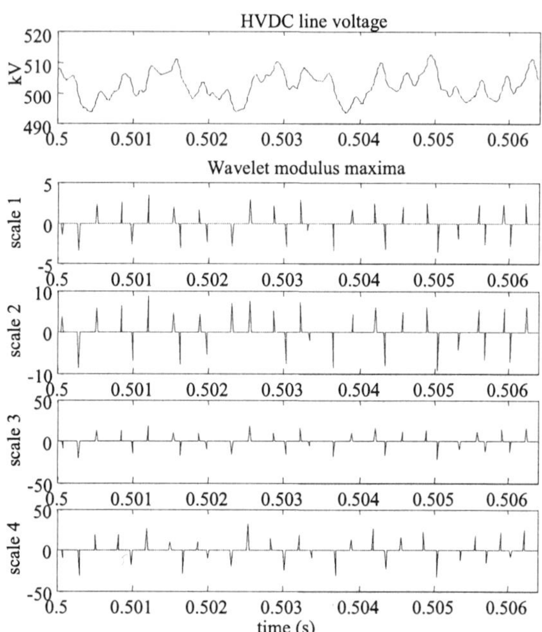

Figure 5. DC voltage during normal operation

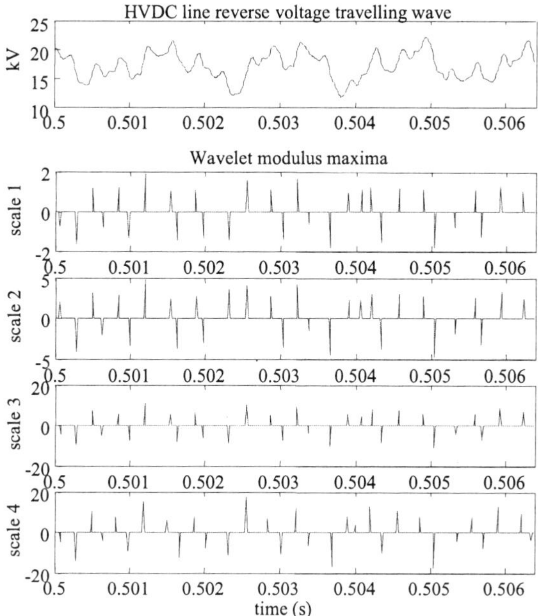

Figure 6. Reverse voltage travelling wave during normal operation

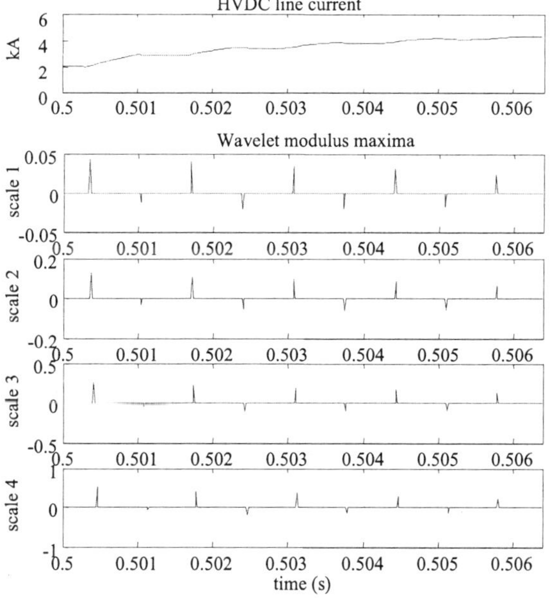

Figure 7. DC current during HVDC line fault

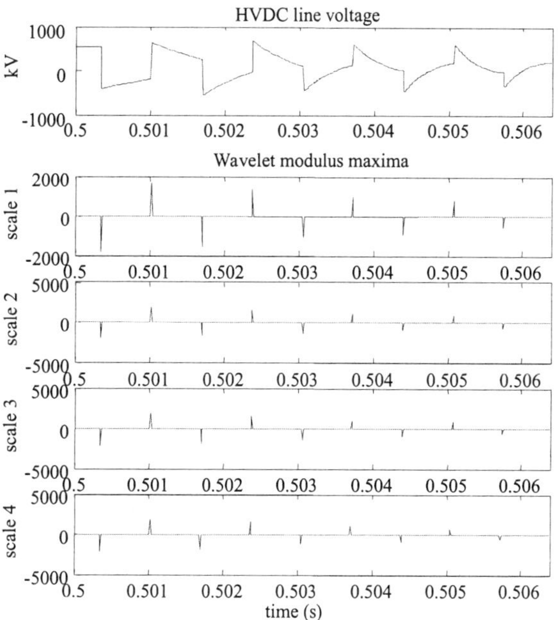

Figure 8. DC voltage during HVDC line fault

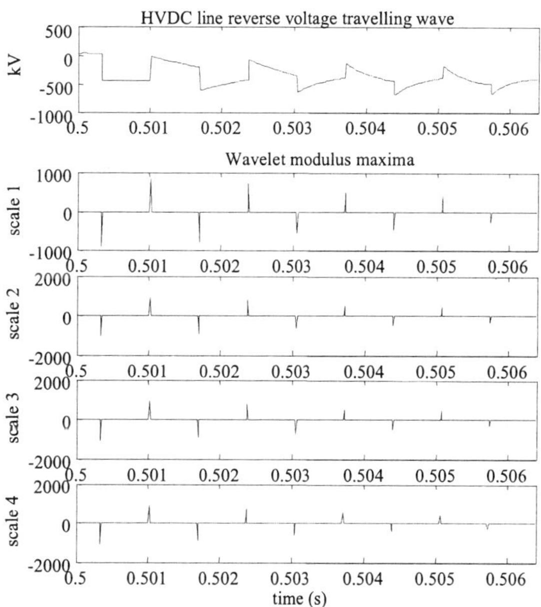

Figure 9. Reverse voltage traveling wave during HVDC line fault

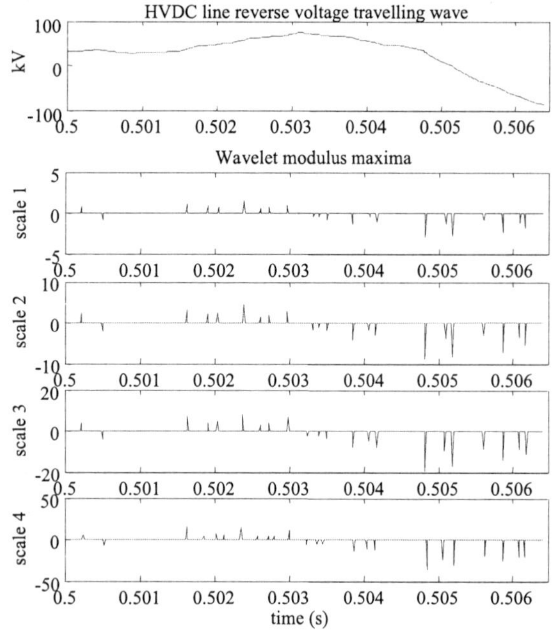

Figure 10. Reverse voltage travelling wave during commutation failure

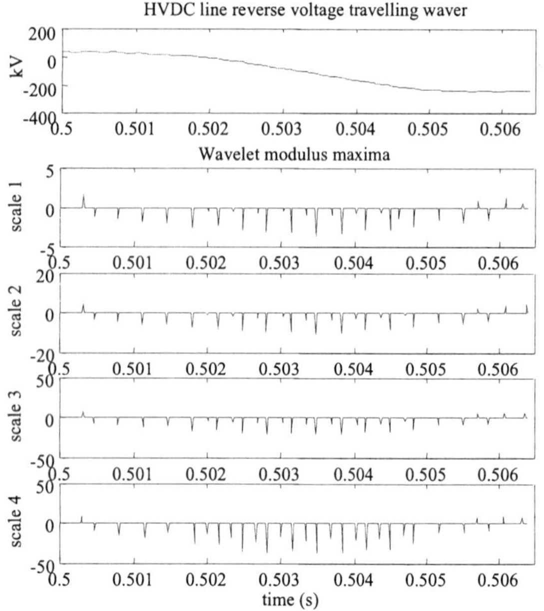

Figure 11. Reverse voltage travelling wave during AC single-phase fault

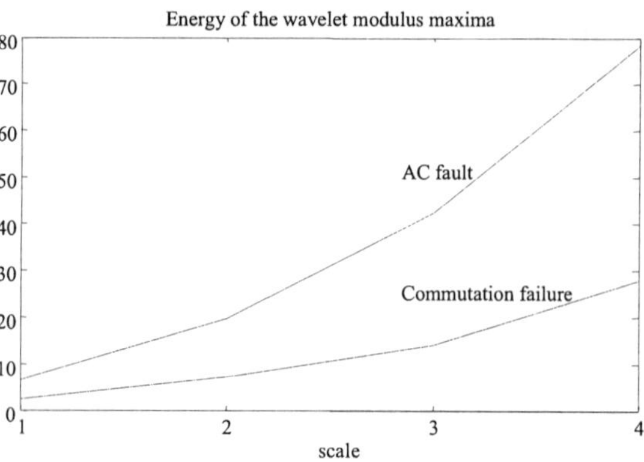

Figure 12. Energy of the wavelet modulus maxima

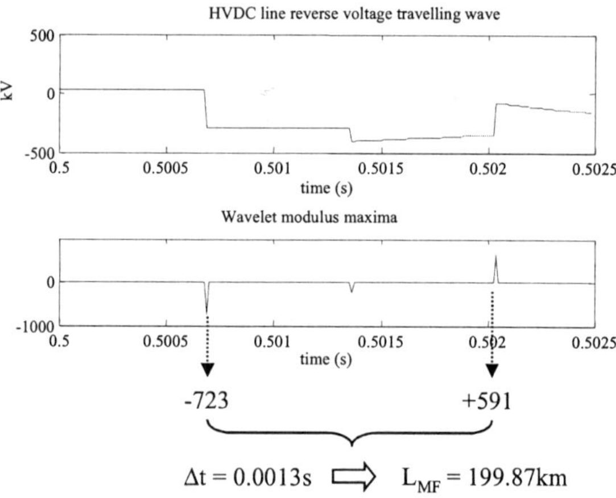

Figure 13. A DC line fault: 200km from terminal M

REFERENCES

[1] Magnago, F.H. and Abur, A., 1998, "Fault location using wavelets", IEEE Trans. on Power Delivery, 13-4, 1475-1480.

[2] Dong, X. Zh., Ge, Y. Zh. and Xu, B. y., 2000, "Fault position relay based on current travelling waves and wavelets", IEEE PES 2000 Winter Meeting, 23-27.

[3] Shang, L., Herold, G. and Jaeger, J., 2000, "A new approach to high-speed protection for transmission line using wavelet technique", PSP 2000, 85-90.

[4] Mallat, S. and Hwang, W. L., 1992, "Singularity detection and processing", IEEE Trans. on Information Theory, 38-2, 617-643.

[5] Shang, L., Herold, G. and Jaeger, J., 2001, "High-speed protection for transmission line based on transient signal analysis using wavelets", IEE DPSP 2001, 173-176.

FAST RELIABLE UNIFIED POWER FLOW CONTROLLER (UPFC) ALGORITHM

E. M. Saied
M. A. EL-Shibini
Faculty of Engineering, Egypt.

ABSTRACT

Growth of electric power demand leads to growth in electric power systems. So ac transmission systems must be controlled fast enough and practical to handle the variation of power system conditions. Many traditional controllers were used before with the design of transmission systems but greater operating flexibility and better utilization of these transmission systems could not achieved with these traditional controllers. Unified power flow controller (UPFC) can provide fast and real-time control of power systems. In this paper an algorithm of a unified power flow controller is suggested. Through this suggested algorithm, the UPFC is able to control both real and reactive power flow of transmission line independently. This is achieved through deriving of analytical equations leading to direct relationship between the power flow to be controlled in a certain transmission line and the controller parameters that presented in the equivalent circuit describing the steady state model of the UPFC.

INTRODUCTION

Ac transmission systems could not be controlled practically and fast when traditional controllers are used. Traditional controllers that used before, like series and shunt compensators, voltage regulators, phase shifting and tap changing transformers E. Handschin et al (1-5) could not realize the required level of control of these transmission systems. Recently, the unified power flow controller (UPFC) is widely used for real-time control of ac transmission systems. Basic operating principles of the (UPFC) was described before L. Gyugyi et al (6-8). The unified power flow controller (UPFC) is a power electronics system, which can provide control of the active and reactive power flow of the ac transmission systems Papic et al (9). Generally it consists of a converter represented by two voltage-sourced branches in parallel and series. The main function of the parallel branch is to control the active power required by the series branch. It also control the reactive power as a reactive compensation for the line (9), Yasuo Morioka et al (10). The series branch connected provides the main function of the UPFC by injecting an ac controlled voltage magnitude and phase angle. Due to the flow of transmission line current through the series converter leads to an exchange in active and reactive power with the ac system (10). The UPFC load flow model is presented before as a controlling tool incorporated into the Newton-Raphson load flow algorithm (8), M. Noroozian et al (11-13). Many studies were made before in order to achieve the suitable and optimal representation of the UPFC model with the Newton-Raphson load-flow algorithm. The drawback of these represented models is mainly for its difficulty and heavy computation burden.

In this paper, control of both real and reactive power flow of transmission line is achieved through a suggested UPFC load flow algorithm. This suggested algorithm and through the incorporation with the existing Gauss-Seidel load flow algorithm, the control of the transmitted power is made using the series converter of the UPFC while keeping the shunt converter as a normal power system element. The suggested technique is implemented on a 5-bus, 2-generator power system and numerical results obtained are satisfactory.

THEORTICAL FORMULATIONS

The Unified Power Flow Controller Model and Algorithm

Fig.1 shows the equivalent circuit of the UPFC, which used to derive its steady state model. This equivalent circuit consists of series and parallel branches. The series branch composed of voltage source converter coupled to the ac system by a series transformer (13). The elements of this series branch are represented by an ideal current source $(V_{cs} y_{cs})$ where (V_{cs}) is the fundamental component of the voltage waveform at the ac converter terminal, and (y_{cs}) is the positive sequence leakage admittance of the UPFC transformers.

AC-DC Power Transmission, 28-30 November 2001
Conference Publication No. 485 © IEE 2001

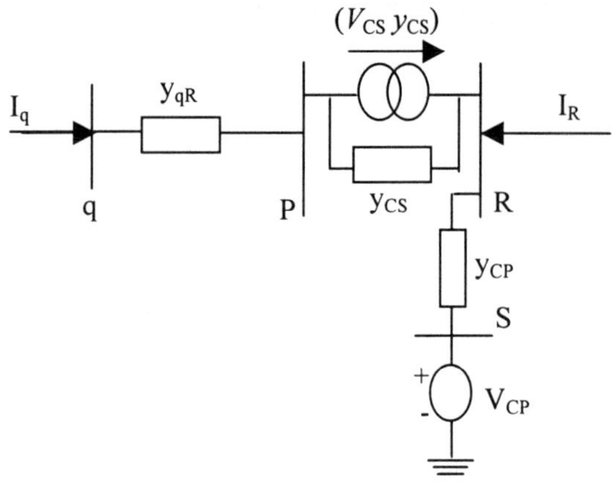

Fig.1 UPFC equivalent circuit

The parallel branch composed of a voltage source converter (V_{cp}) coupled to the ac system with a shunt transformer which represented by (y_{cp}). The two voltage source are

$$V_{cs} = V_{cs\,real} + jV_{cs\,imag} \tag{1}$$

$$V_{s} = V_{cp\,real} + jV_{cp\,imag} \tag{2}$$

Where, $V_{cs\,real}$ and $V_{cs\,imag}$ are the real and imaginary controllable components of the ideal series voltage source which are the in-phase, and the quadrature components respectively with respect to the voltage at bus R. ($V_s = V_{cp}$) is the voltage at bus S, considering bus S as a normal bus in the system and dealing with the shunt converter as a normal power system element. The in-phase component of the series injected voltage ($V_{cs\,real}$) determines the amount of reactive power flow to be controlled, while the quadrature component of the series injected voltage ($V_{cs\,imag}$) determines the amount of active power flow to be controlled.

The transfer admittance matrix for the UPFC when applying Kirchhoff's laws to the electric circuit in Fig.1 is given by:

$$\begin{bmatrix} I_R + y_{cs}V_{cs} \\ I_P - y_{cs}V_{cs} \end{bmatrix} = \begin{bmatrix} y_{cs} + y_{cp} & -y_{cs} & -y_{cp} \\ -y_{cs} & y_{cs} & 0 \end{bmatrix} \begin{bmatrix} V_R \\ V_P \\ V_S \end{bmatrix} \tag{3}$$

Before using the UPFC,
The apparent power flow through the line connected between buses R and q (SRq) is;

$$s_{Rq} = (V_R - V_q)y_{Rq}V_R^* \tag{4}$$

After using the UPFC,

If $S_{Rq}^{\backslash}$ is the required controlled power, then

$$S_{Rq}^{\backslash} = \left[(V_R^{\backslash} + V_{CS}) - V_q^{\backslash} \right] y_{Rq} V_R^{\backslash *}$$

$$= \left[(V_R + \Delta V_R + V_{CS} - V_q - \Delta V_q)y_{Rq}(V_R^* + \Delta V_R^*) \right]$$

$$= \left[(V_R - V_q) + (\Delta V_R - \Delta V_q) + V_{CS} \right] y_{Rq}(V_R^* + \Delta V_R^*)$$

$$= (V_R - V_q)y_{Rq}V_R^* + (\Delta V_R - \Delta V_q)y_{Rq}V_R^*$$

$$+ V_{CS}y_{Rq}V_R^* + (V_R - V_q)y_{Rq}\Delta V_R^*$$

$$+ (\Delta V_R - \Delta V_q)y_{Rq}\Delta V_R^* + V_{CS}y_{Rq}\Delta V_R^* \tag{5}$$

Therefore the amount of change in power flow,

$$\Delta s_{Rq} = S_{Rq}^{\backslash} - S_{Rq}$$

$$= \left[(V_R + \Delta V_R) - (V_q + \Delta V_q) \right] y_{Rq}V_R^*$$

$$+ (\Delta V_R - \Delta V_q)y_{Rq}V_R^* + (V_R^* + \Delta V_R^*)y_{Rq}V_{CS} \tag{6}$$

Where the change in active power flow ΔP_{Rq} is the real part of equation (6) and the change in reactive power flow ΔQ_{Rq} is the negative of the imaginary part of equation (6). Fig.2 shows a simplified Phasor diagram illustrate the direction of V_{CS} and its components V_{CSreal}; V_{Csimag} with respect to buses R and q.

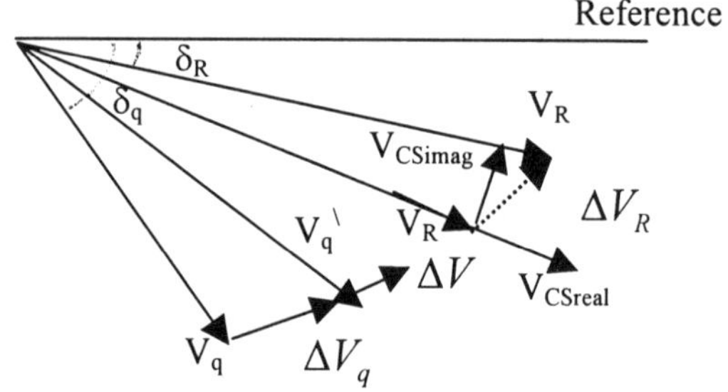

Fig.2 Phasor Diagram shows V_{CS} direction w.r.t. buses q and R

If $V_{Cs\,real}$ and V_{CSimag} are the in-phase and quadrature components with respect to V_R, then a direct linear relationship obtained between the reactive power flow between bus R and q, Q_{Rq} and the in-phase component of V_{CS} (V_{CSreal}). Also a direct linear relation between the active power flow, P_{Rq} and the quadrature component of V_{CS} (V_{CSimag}). This linear relation can be represented by a group of straight lines. Every line represent the relation between the power flow and V_{CS} components at certain value of the other component (Fig.3). These straight lines have the same slope and so closed to each other, which can be assumed as one straight line. Therefore according to these resultant direct relations and through the modified Gauss-Seidel load flow program which including the UPFC equations, the following relations are obtained,

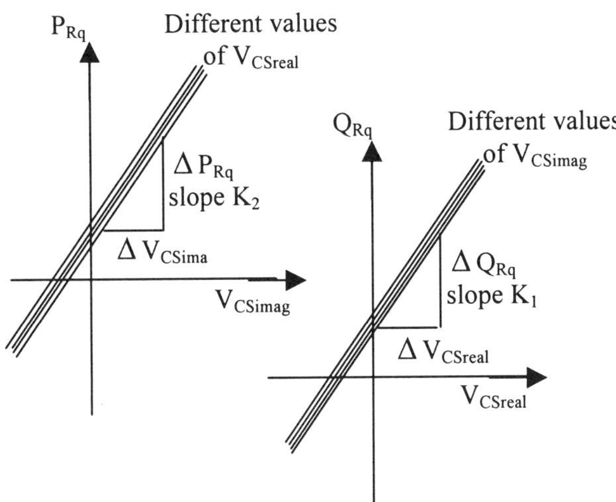

Fig.3 The linear relationships between the powers flow and V_{CS} components

From Fig.3,

$$K_1 = \frac{\Delta Q_{Rq}}{\Delta V_{CSreal}} \qquad (7)$$

$$K_2 = \frac{\Delta P_{Rq}}{\Delta V_{CSimag}} \qquad (8)$$

Therefore initially, from the normal load flow program, power flow in line R-q are obtained (P_{Rq0}, Q_{Rq0}), the corresponding $V_{Csreal\,0}$ and $V_{Csimag\,0}$ are obtained from Fig.3.
Let $P^\backslash_{Rq}$ and $Q^\backslash_{Rq}$ are the desired power flow in line p-q, therefore,

$$\Delta p^\backslash_{Rq} = P^\backslash_{Rq} - P_{Rq0} \qquad (9)$$

$$\Delta Q^\backslash_{Rq} = Q^\backslash_{Rq} - Q_{Rq0} \qquad (10)$$

Consequently $\Delta V^\backslash_{CSreal}$ $\Delta V^\backslash_{CSimag}$ will be,

$$\Delta V^\backslash_{CSreal} = \frac{1}{K_1} \Delta Q^\backslash_{Rq} \qquad (11)$$

$$\Delta V^\backslash_{CSimag} = \frac{1}{K_2} \Delta Q^\backslash_{Rq} \qquad (12)$$

Therefore,

$$V^\backslash_{CSreal} = V_{CSreal\,0} + \Delta V^\backslash_{CSreal} \qquad (13)$$

$$V^\backslash_{CSimag} = V_{CSimag\,0} + \Delta V^\backslash_{CSimag} \qquad (14)$$

Insert $V^\backslash_{Csreal}$ and $V^\backslash_{Csimag}$ as an input data to the modified Gauss-Seidel load flow program, the

resultant power flow in line R-q must equal the desired power $P^\backslash_{Rq}$ and $Q^\backslash_{Rq}$. Also this result must satisfy result obtained from equations (5) and (6).

APPLICATIONS AND RESULTS

A simple power system composed of 2-generators, 5-buses and 7 lines is used to demonstrate the idea, Fig.4. Buses 3, 4 and 5 are load buses. The UPFC with its equivalent circuit represented in section 2, is represented between buses 2 and 5 in order to control the power flow from bus 2 to bus 5 (line 5).

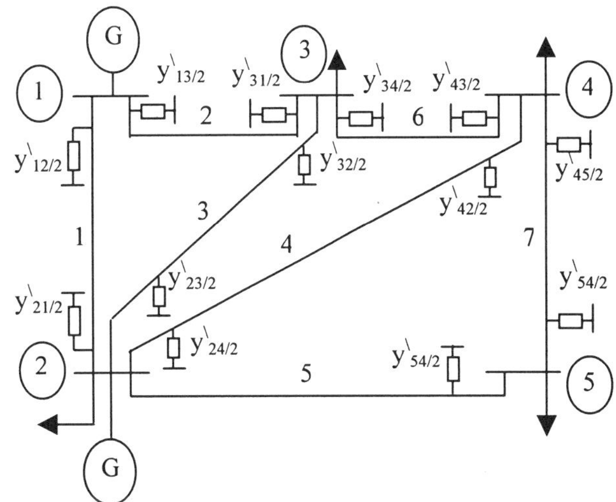

Fig.4 A Simple Power System Model

Renaming the buses q and R in Fig.1 by buses 2 and 5 and buses P and S by 6 and 7 in Fig 4, 5. The bus-currents and voltages equations of the power system with the UPFC connected between buses 2 and 5 using the Guass-Seidel load flow (LF) solution with bus 1 slack bus are represented in appendix A.

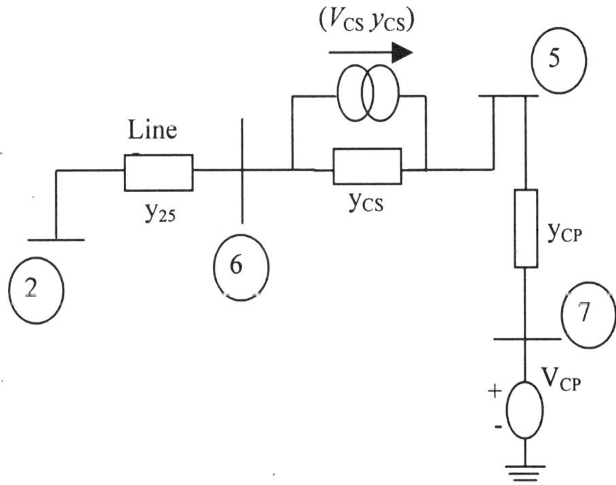

Fig.5 The equivalent circuit between bus 2 and 5

RESULTS

As a result of applying normal ac load flow on the 5-bus system, the active and reactive power flow from bus 2 and 5 are found $P_{25\,0}$ = 0.5417 p.u. and $Q_{25\,0}$ = 0.1264 p.u. Applying the suggested algorithm on the 5-bus system to control the power flow from bus 2 and 5, taking the values of series and parallel converters impedances are equal Z_{CS}= Z_{CP}= 0+j0.05 p.u. A group of parallel lines that describing the relationships between the reactive power flow Q_{25} and V_{Csreal}, for different values of $V_{Csimag.}$ (Table 1, Fig.6). Table 2 and Fig.7 illustrate the linear relationships between the active power flow P_{25} and $V_{Csimag.}$ with different values of V_{Csreal} .

TABLE 1: Variation of Q_{25} with V_{Csreal}

$V_{Cs\,imag.}$ (p.u)	$V_{Cs\,real}$ (p.u)	Q_{25} (p.u)
0	0.15	0.3031
0	0.25	0.4360
0	0.35	0.5668
0.35	0.15	0.2452
0.35	0.25	0.3773
0.35	0.35	0.5074
0.70	0.15	0.1871
0.70	0.25	0.3181
0.70	0.35	0.4480

TABLE 2: Variation of P_{25} with $V_{Csimag.}$

$V_{Cs\,real}$ (p.u)	$V_{Cs\,imag.}$ (p.u)	P_{25} (p.u)
0	0.3	0.8860
0	0.4	0.9981
0	0.5	1.1100
0.3	0.3	0.9330
0.3	0.4	1.0450
0.3	0.5	1.1570
0.4	0.3	0.9832
0.4	0.4	1.0952
0.4	0.5	1.2072

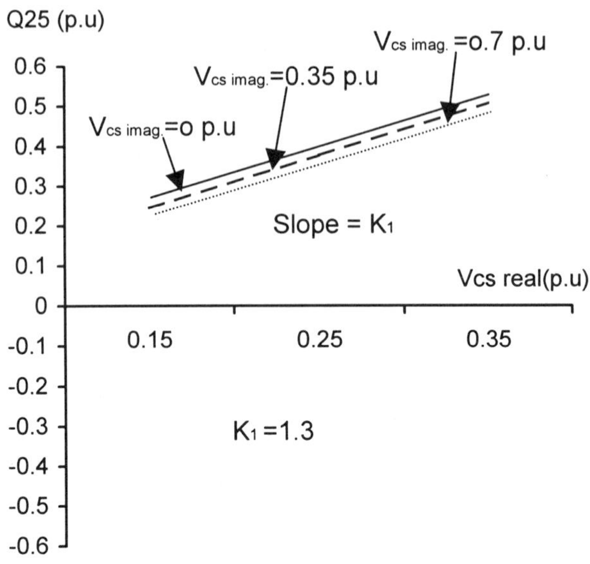

Fig. 6 Relationship between Q25 and the in-phase component of Vcs (Vcs real)

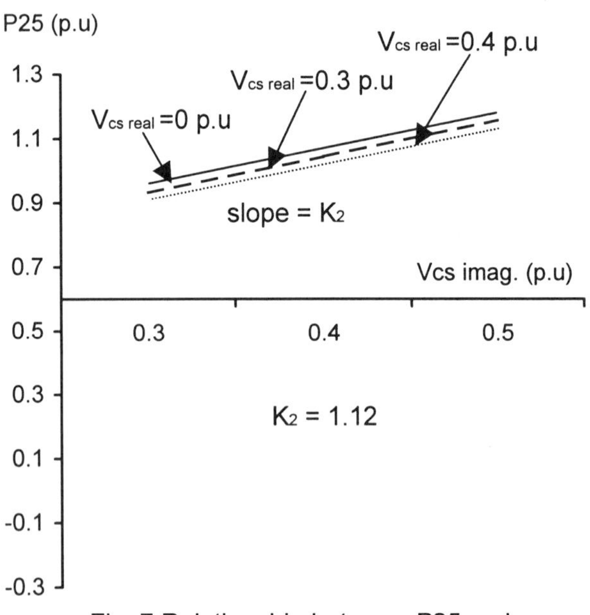

Fig. 7 Relationship between P25 and the quadrature component of Vcs (Vcs imag.)

Now, if it is required to increase the active power flow in line 5 by 50% of its normal value, that's to say $P^{\backslash}_{25}=0.8125$ p.u., and increase the reactive power flow in line 5 to 300% of its normal value, that's to say $Q^{\backslash}_{25}=0.3792$ p.u., therefore,

$\Delta P_{25} = P^{\backslash}_{25} - P_{25\,0} = 0.8125 - 0.5417 = 0.2708$ *p.u.*
and $\Delta Q_{25} = Q^{\backslash}_{25} - Q_{25\,0} = 0.3792 - 0.1264 = 0.2528$ *p.u.*

Therefore,

$$\Delta V^{\backslash}_{CSreal} = \frac{1}{K_1} \cdot \Delta Q_{25} = \frac{0.2528}{1.30} = 0.1944\, p.u.$$

$$\Delta V^{\backslash}_{CSimag.} = \frac{1}{K_2} \cdot \Delta P_{25} = \frac{0.2708}{1.12} = 0.2417\, p.u.$$

$$V^{\backslash}_{CSreal} = V_{CSreal\,0} + \Delta V^{\backslash}_{CSreal} = 0.1944\, p.u.$$

$$V^{\backslash}_{CSimag.} = V_{CSimag.\,0} + \Delta V^{\backslash}_{CSimag.} = 0.2417\, p.u.$$

Therefore according to the suggested algorithm program using $V^{\backslash}_{Csreal}$ and $V^{\backslash}_{Csimag.}$ as an input data, the resultant controlled active and reactive power flow output from this suggested program are, $P_{25}=0.8539$ p.u., and $Q_{25}=0.3295$ p.u. Comparing between the required power flow and that obtained from UPFC load flow suggested program, the error is found,
error in $P_{25}=0.8435 - 0.8125 = 0.031$ *p.u.*
error in $Q_{25}=0.3792 - 0.3295 = 0.0497$ *p.u.*

This results are good and reasonable as compared with that calculated through equations (5) and (6) where the value of $V^{\backslash}_2$ and $V^{\backslash}_5$ are calculated from the load flow solution when using the UPFC algorithm.

CONCLUSIONS

In this paper a UPFC power flow algorithm has been suggested and presented. This algorithm is incorporation with the Gauss-Seidel load flow algorithm. The suggested algorithm is capable of solving and well controlling of any size power network. A set of analytical equation have been derived through this suggested algorithm to provide good relations between the power system and UPFC parameters. Through this algorithm, direct and linear relations between the power flow to be controlled and the UPFC parameters are obtained. This good results make the use of this suggested UPFC algorithm easy, reliable and efficient. The proposed algorithm is applied on a simple power system with good, satisfactory, and high accurate results are obtained.

APPENDIX A

The bus current equations:

$$\begin{bmatrix} I_1 \\ I_2 \\ I_3 \\ I_4 \\ I_5 + y_{CS}V_{CS} \\ I_6 - y_{CS}V_{CS} \\ I_7 \end{bmatrix} = \begin{bmatrix} Y_{11} & Y_{12} & Y_{13} & Y_{14} & Y_{15} & Y_{16} & Y_{17} \\ Y_{21} & Y_{22} & Y_{23} & Y_{24} & Y_{25} & Y_{26} & Y_{27} \\ Y_{31} & Y_{32} & Y_{33} & Y_{34} & Y_{35} & Y_{36} & Y_{37} \\ Y_{41} & Y_{42} & Y_{43} & Y_{44} & Y_{45} & Y_{46} & Y_{47} \\ Y_{51} & Y_{52} & Y_{53} & Y_{54} & Y_{55} & Y_{56} & Y_{57} \\ Y_{61} & Y_{62} & Y_{63} & Y_{64} & Y_{65} & Y_{66} & Y_{67} \\ Y_{71} & Y_{72} & Y_{73} & Y_{74} & Y_{75} & Y_{76} & Y_{77} \end{bmatrix} \begin{bmatrix} V_1 \\ V_2 \\ V_3 \\ V_4 \\ V_5 \\ V_6 \\ V_7 \end{bmatrix}$$

$$(1A)$$

where the diagonal elements of the bus admittance matrix are:

$$Y_{11} = y_{12} + \frac{y^{\backslash}_{12}}{2} + y_{13} + \frac{y^{\backslash}_{13}}{2}\ ,$$

$$Y_{22} = y_{21} + \frac{y^{\backslash}_{21}}{2} + y_{23} + \frac{y^{\backslash}_{23}}{2} + y_{24} + \frac{y^{\backslash}_{24}}{2} + y_{25} + \frac{y^{\backslash}_{25}}{2},$$

$$Y_{33} = y_{31} + \frac{y^{\backslash}_{31}}{2} + y_{43} + \frac{y^{\backslash}_{43}}{2} + y_{45} + \frac{y^{\backslash}_{45}}{2},$$

$$Y_{44} = y_{42} + \frac{y^{\backslash}_{42}}{2} + y_{43} + \frac{y^{\backslash}_{43}}{2} + y_{45} + \frac{y^{\backslash}_{45}}{2},$$

$$Y_{55} = y_{54} + \frac{y^{\backslash}_{54}}{2} + y_{CS} + y_{CP},$$

$$Y_{66} = y_{25} + y_{CS} + \frac{y^{/}_{52}}{2} \text{ and } Y_{77} = y_{CP},$$

And the off-diagonal elements of buss admittance matrix are:

$$Y_{12} = -y_{12}, Y_{13} = -y_{13}, Y_{21} = -y_{21}, Y_{23} = -y_{23},$$

$$Y_{24} = -y_{24}, Y_{26} = -y_{25}, Y_{31} = -y_{31}, Y_{32} = -y_{32},$$

$$Y_{34} = -y_{34}, Y_{42} = -y_{42}, Y_{43} = -y_{43}, Y_{45} = -y_{45},$$

$$Y_{54} = -y_{54}, Y_{56} = -y_{CS}, Y_{57} = -y_{CP}, Y_{62} = -y_{25},$$

$$Y_{65} = -y_{CS}, Y_{75} = -y_{CP},$$

All the remainder elements of the bus admittance matrix are zeroes.

Where, ($y^{/}/2$) is the line charging admittance.
Therefore, the Gauss-Seidel, iterative bus voltage equations will be as normal bus voltage equations except for buses 5 and 6, where,

$$V^{k+1}_5 = V^{k+1}_{5(normal\ LF)} + \frac{y_{CS}}{Y_{55}} V_{CS} \qquad (2A)$$

$$\text{and } V^{k+1}_6 = V^{k+1}_{6(normal\ LF)} - \frac{y_{CS}}{Y_{66}} V_{CS} \qquad (3A)$$

REFERENCES

1- E. Handschin, E. Kliskys, 1995, "Transformer Tap Position Estimation and Bad Data Detection Using Dynamic Signal Modelling", IEEE Transactions on Power Systems, 10, No. 2, 810-817.

2- K. Imbof, ABB Netcom AG, F. Oesch, EGL, and I. Nordanlycke, 1996, "Modelling of tap-changer Transformers in an Energy Management System", IEEE Transactions on Power Systems, 11, No. 1, 428-434.

3- R. M. Mathur and R. S. Basati, 1981, "A Thyritor Controlled Static Phase-Shifter for AC Power Transmission", IEEE Trans. PAS-100, No. 5, 2650-2655.

4- B. K. Patel, H. S. Smith, T. S. Hewes Jr., W. J. Marsh, 1986, "Application of Phase Shifting Transformers for Daniel-Mcknight 500-Kv Interconnections", IEEE Transactions on Power Delivery, PWRD-1, No. 3, 167-173.

5- Boon Teck Ooi, Shu Zu Dai, Francisco D. Galiana, 1993, "A Solid-State PWM Phase-Shifter", IEEE Transactions on Power Delivery, No. 2, 573-579.

6- L. Gyugyi, 1992, "Unified Power-Flow Control Concept for Flexible AC Transmission Systems", IEE Proc.-C, 139, 323-333.

7- L. Gyugyi, C. D. Schauder, S. L. Williams, T. R. Rietman. D. R. Torgerson, and A. Edris, 1995, "The Unified Power Flow Controller A New Approach to Power Transmission Control", IEEE Transaction on Power Delivery, No. 2, 1085-1097.

8- A. Nabavi – Niaki and M. R. Iravani, 1996, "Steady-State and Dynamic Models of Unified Power Flow Controller (UPFC) for Power System Studies", IEEE/ PES Winter Meeting 96 WM 257-6PWRS.

9- Papic, P. Zunko, D. Povh, and M. Weinhold, 1997, "Basic Control of Unified Power Flow Controller", IEEE Transactions on Power Systems, No. 4, 1734-1739.

10-Yasuo Morioka, Masanao Kato, Yasuhiro Mishima, and Yoshiki Nakachi, 1999, "Implementation of Unified Power Flow Controller and Verification for Transmission Capability Improvement", IEEE Transactions on Power Systems, No. 2, 575-581.

11-M. Noroozian, L. Angquist, M. Ghandhari, and G. Anderson, 1995, "Use of UPFC for Optimal Power Flow Control", IEEE / KTH Stockholm Power Tech. Conference, Stockholm, Sweden, 501-511.

12-C. R. Fuerte – Esquivel, E. Acha, 1996, "Newton-Raphson Algorithm for The Reliable Solution of Large Power Networks with Embedded FACTS Devices", IEE Proc.-Gener. Transm. Distrib. No. 5, 447-454.

13-C. R. Fuerte – Esquivel, E. Acha, and H. Ambriz – Perez, 2000 "A Comprehensive Newton-Raphson UPFC Model for the Quadratic Power Flow Solution of Practical Power Networks", IEEE Trans. On Power Systems, No. 1, 102-109.

ENHANCED CONTROL SYSTEM FOR A STATIC SYNCHRONOUS SERIES COMPENSATOR WITH ENERGY STORAGE

I Papič[*], A M Gole[**]

[*] University of Ljubljana, Slovenia
[**] University of Manitoba, Canada

ABSTRACT

A strategy is presented for the decoupled control of real and reactive power flows in an energy storage based static series synchronous compensator (SSSC). State-variable based design methods are employed in the basic design, which uses only proportional control. Increased robustness to ac parameter changes is implemented by enhancing this basic design with a control structure that internally uses a state-variable model of the plant. The design is corroborated via electromagnetic transient simulation.

KEYWORDS

Voltage Sourced Converter, Static Series Synchronous Compensator, Robust Control, Decoupled Control

INTRODUCTION

The static synchronous series compensator (SSSC) is a FACTS (Flexible AC Transmission System) device that injects a controllable fundamental frequency voltage in series with a transmission line. This allows for rapid control of the transmitted real and reactive powers over the line, which can help in stabilization of power and voltage swings and also increase the power transmission capability [1,2]. Adding an energy storage device such as a battery can be shown to further increase the ability to damp oscillations [3,4]. The SSSC device is based on Voltage Sourced Converter (VSC) technology.

An important requirement for good control performance is the decoupling between the real and reactive power control loops. Another requirement is to have a rapid oscillation-free response. Earlier publications, i.e. [3], have presented control strategies which are decoupled only in the steady state so that there is a still a transient in the reactive power following a sudden change in the real power order. This paper discusses the implementation of a novel control method based on state-space techniques that addresses both the above requirements and also takes into account limits such as maximum ceiling on the voltage output by the VSC. In addition to the above-mentioned transient decoupling other control scenarios that are easily realizable with the state-space approach include increasing the damping factor and/or eliminating the oscillatory mode.

AC-DC Power Transmission, 28-30 November 2001
Conference Publication No. 485 © IEE 2001

Although the paper describes the application of this design method to the energy-storage SSSC, it could readily be extended to other FACTS devices such as the STATCOM and the Unified Power Flow Controller (UPFC).

CONTROL SYSTEM DESIGN

In order to design the controller, a state-variable form of the SSSC and the ac system is required. This model can then be used to select the control system parameters of the preliminary basic design. It is also used a component in the enhanced robust control method described later.

Mathematical Model of the SSSC and Ac Network

The converter, which is capable of active and reactive power exchange with the ac system, is represented by sinusoidal voltage sources. It is assumed that the dc-circuit consists of energy storage with a constant dc-voltage. The balanced three-phase system can be transformed into a synchronously rotating orthogonal system (d-q coordinates). with the d-axis coincident with the instantaneous sending-end voltage vector ($v_{id}=|v_i|$, $v_{iq}=0$). The d-axis current component contributes to the instantaneous active power $p(t)$ and the q-axis current component represents the instantaneous reactive power $q(t)$, both at the sending-end of the transmission line.

$$p(t) = \frac{3}{2} v_d i_d \; ; q(t) = \frac{3}{2} v_d i_q \qquad (1)$$

Fig. 1 shows a generalized model of the SSSC with energy storage and series inductance and resistance ($R/X=0.1$) in three lines representing the impedance of the transmission line and coupling transformer. The following per-unit system (Eqn. 2) has been adopted, in which the primed variables denote pu quantities and ω_B is the nominal angular frequency.

$$i_x' = \frac{i_x}{i_B} ; u_{ix}' = \frac{u_{ix}}{u_B} ; u_{ox}' = \frac{u_{ox}}{u_B} ; u_{sx}' = \frac{u_{sx}}{u_B}$$

$$z_B = \frac{u_B}{i_B} ; L' = \frac{\omega_B L}{z_B} ; R' = \frac{R}{z_B} \qquad (2)$$

$$x = a, b, c$$

In terms of instantaneous variables shown in Fig. 1, the ac-side circuit equation can be written (Eqn. 3), where $p.=\mathrm{d}/\mathrm{d}t$. When re-formulated in d-q variables as Eqn. 4, the cross-coupling of the d and q axis currents is evident (off-diagonal element ω). The zero-sequence term is not considered on account of the presumed three-wire connection.

$$p.\begin{bmatrix} i'_a \\ i'_b \\ i'_c \end{bmatrix} = \begin{bmatrix} \dfrac{-R'\omega_B}{L'} & 0 & 0 \\ 0 & \dfrac{-R'\omega_B}{L'} & 0 \\ 0 & 0 & \dfrac{-R'\omega_B}{L'} \end{bmatrix} \begin{bmatrix} i'_a \\ i'_b \\ i'_c \end{bmatrix} + \dfrac{\omega_B}{L'} \begin{bmatrix} v'_{ia} - v'_{oa} + v'_{sa} \\ v'_{ib} - v'_{ob} + v'_{sb} \\ v'_{ic} - v'_{oc} + v'_{sc} \end{bmatrix} \quad (3)$$

$$p.\begin{bmatrix} i'_d \\ i'_q \end{bmatrix} = \begin{bmatrix} \dfrac{-R'\omega_B}{L'} & \omega \\ -\omega & \dfrac{-R'\omega_B}{L'} \end{bmatrix} \begin{bmatrix} i'_d \\ i'_q \end{bmatrix} + \dfrac{\omega_B}{L'} \begin{bmatrix} v'_{id} - v'_{od} + v'_{sd} \\ v'_{iq} - v'_{oq} + v'_{sq} \end{bmatrix} \quad (4)$$

Basic State-Space Control System

The inputs to the control system are reference values for the real and reactive (d and q axis) currents – the vector **w** in Fig 2 and defined later in Eqn. 7. The series VSC voltage (v'_{sd}, v'_{sq}) is rapidly controllable and is the instrument through which the control is achieved (vector **u** in Fig. 2). It is thus the output from the control system.

Note that the d and q variables are coupled and if a simplistic control method is used, any attempt at changing i_d causes a change in i_q and vice-versa. It is relatively easy to decouple the d and q currents in the steady state [3]; however it is also desirable to have no coupling during the transient state.

The following section shows how a control system can be designed which is decoupled in the steady-state as well as in the transient state [5,6]. Furthermore, the controller is simple as only a proportional gain is used. However, the design relies on an accurate knowledge of the network parameters (R, L, v_i, v_o etc.), which may not always be the case. The following section will introduce refinements to overcome this shortcoming.

Eqn. 4 can be rewritten in the classical form as Eqn. 5, with state vector **x** and input vector **u**.

$$\dot{\mathbf{x}} = \mathbf{Ax} + \mathbf{Bu}$$

$$\mathbf{x} = \begin{bmatrix} i'_d \\ i'_q \end{bmatrix} \quad \mathbf{u} = \begin{bmatrix} v'_{id} - v'_{od} + v'_{sd} \\ v'_{iq} - v'_{oq} + v'_{sq} \end{bmatrix} \quad (5)$$

$$\mathbf{A} = \begin{bmatrix} \dfrac{-R'\omega_B}{L'} & \omega \\ -\omega & \dfrac{-R'\omega_B}{L'} \end{bmatrix} \quad \mathbf{B} = \begin{bmatrix} \dfrac{\omega_B}{L'} & 0 \\ 0 & \dfrac{\omega_B}{L'} \end{bmatrix}$$

The eigenvalues of the matrix **A**, calculated below in Eqn. 6, provide valuable information as to the nature of the system's response. The real part ($-R'\omega_B/L'$)

represents the damping and the imaginary part (ω) is the angular frequency of oscillation.

$$\lambda_{1,2} = -\frac{R'\omega_B}{L'} \pm j\omega \quad (6)$$

The basic scheme of the state-space control system is given in Fig. 2. The vector **w** (Eqn. 7) is the controller's input, i.e. the desired d and q current orders. The negative feedback loop is closed via the control matrix **F**. The matrix **V** is an adjustable matrix that under steady state conditions assures **x**=**w**.

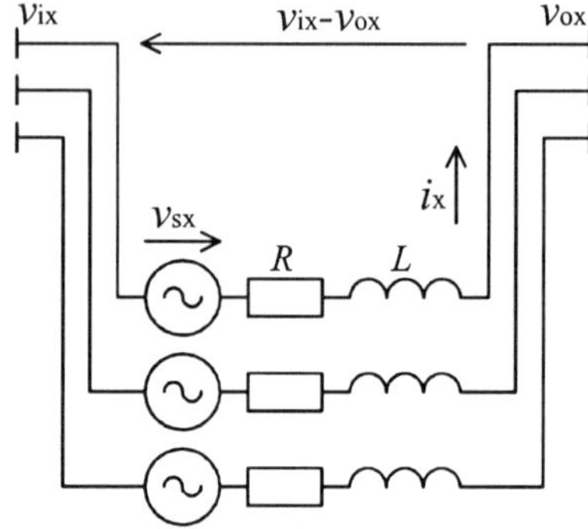

Fig. 1. Equivalent Circuit Diagram of the SSSC with energy storage

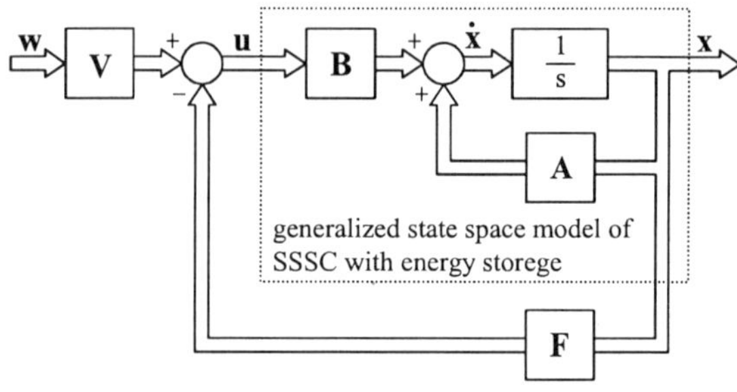

Fig. 2. Basic scheme of state-space controller together with generalized model of SSSC with energy storage

$$\mathbf{w} = \begin{bmatrix} i'^*_d & i'^*_q \end{bmatrix}^T \quad (7)$$

The objective of the control system design is to select appropriate matrices **F** and **V**. The control system's output vector **u** (which contains the necessary d and q axis components for the injected VSC voltage) is given by Eqn. 8. The new resultant state equation is given in Eqn. 9.

$$\mathbf{u} = \mathbf{Vw} - \mathbf{Fx} \quad (8)$$

$$\dot{\mathbf{x}} = (\mathbf{A} - \mathbf{BF})\mathbf{x} + \mathbf{BVw} \quad (9)$$

The state matrices for the system with controller are now given by $\widetilde{\mathbf{A}}$ and $\widetilde{\mathbf{B}}$ defined in Eqn. 10.

$$\widetilde{\mathbf{A}} = (\mathbf{A} - \mathbf{BF})$$
$$\widetilde{\mathbf{B}} = \mathbf{BV} \qquad (10)$$

The selection of $\widetilde{\mathbf{A}}$ is determined by the objective at hand. In the most typical scenario, it may be desired to have an oscillation free response with high damping. In this case, we would select purely real eigenvalues as shown in Eqn. 11. The factor K represents the multiple by which the damping of the original system is increased. The network planning or performance engineer could select the actual value of K from detailed transient stability studies. The other requirement from $\widetilde{\mathbf{A}}$ is that of decoupling the d and q quantities. Hence the matrix $\widetilde{\mathbf{A}}$ shown in Eqn. 12 is selected, which has zero diagonals (no cross-coupling) as well as the desired eigenvalues. Once $\widetilde{\mathbf{A}}$ is chosen, the control matrices $\mathbf{F}$ and $\mathbf{V}$ are readily calculated. The additional information required for calculating $\mathbf{V}$ is the requirement of zero steady state error. The ensuing calculation is summarized in Eqn 14.

$$\lambda_{1,2} = -K \frac{R' \omega_B}{L'}; \; K > 1 \qquad (11)$$

$$\widetilde{\mathbf{A}} = \begin{bmatrix} -K \frac{R' \omega_B}{L'} & 0 \\ 0 & -K \frac{R' \omega_B}{L'} \end{bmatrix} \qquad (12)$$

$$\mathbf{F} = \mathbf{B}^{-1}(\mathbf{A} - \widetilde{\mathbf{A}}) \qquad (13)$$

$$0 = (\mathbf{A} - \mathbf{BF})\mathbf{x} + \mathbf{BVw} \rightarrow \mathbf{x} = \mathbf{w}$$
$$\mathbf{V} = \mathbf{F} - \mathbf{B}^{-1}\mathbf{A} = -\mathbf{B}^{-1}\widetilde{\mathbf{A}} \qquad (14)$$

After performing these calculations, the $\mathbf{F}$ and $\mathbf{V}$ matrices emerge as shown in Eqn. 15.

$$\mathbf{F} = \begin{bmatrix} (K-1)R' & L' \\ -L' & (K-1)R' \end{bmatrix} \quad \mathbf{V} = \begin{bmatrix} KR' & 0 \\ 0 & KR' \end{bmatrix} \quad (15)$$

Alternatively, if only the system damping is to be increased by a factor K without changing the resonant frequency, $\widetilde{\mathbf{A}}$ is as shown in Eqn. 16 and the matrices $\mathbf{F}$ and $\mathbf{V}$ are determined by Eqn.17.

$$\widetilde{\mathbf{A}} = \begin{bmatrix} -K \frac{R' \omega_B}{L'} & \omega \\ -\omega & -K \frac{R' \omega_B}{L'} \end{bmatrix} \qquad (16)$$

$$\mathbf{F} = \begin{bmatrix} (K-1)R' & 0 \\ 0 & (K-1)R' \end{bmatrix} \quad \mathbf{V} = \begin{bmatrix} KR' & -L' \\ L' & KR' \end{bmatrix} \quad (17)$$

Note that the different designs for $\mathbf{F}$ and $\mathbf{V}$ correspond to different desired objectives. It should be realized that the particular selected choice affects the energy exchanged with the energy storage unit (battery) and the resultant peak injected voltage transient.

Consideration of Converter Voltage Ceiling

The design procedure described above does not consider nonlinearities and hence ignores limits such as the maximum voltage, which may be injected by the VSC. In order to prevent the control system from ordering too large an injected voltage, the raw order $\mathbf{u}$ is transformed to the final order $\mathbf{u}_L$ as shown in Fig. 3. The components of the voltage vector are transformed into polar co-ordinates, the magnitude of the vector is limited and the phase angle remains unchanged. At the output the vector is transformed back to d-q components. The elements of matrices $\mathbf{F}$ and $\mathbf{V}$ do not change. In practice the voltage vector is only occasionally limited during transient conditions. When the limitation is effective the system is not linear anymore and it is expected that the decoupling between both state variables will not be completely achieved.

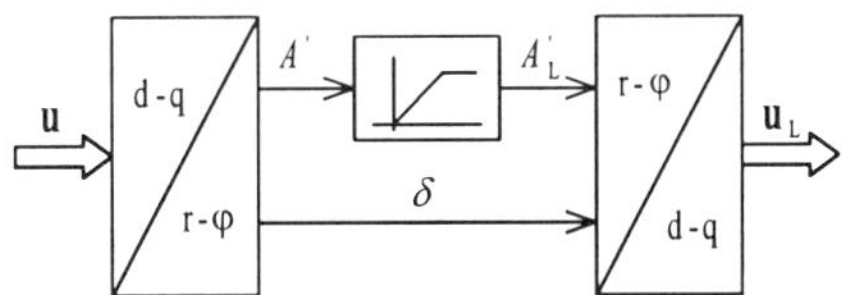

Fig. 3. Implementation of limitation of converter output voltage

Validation of Control Method

The derived state-space control algorithm was tested by digital simulation using the PSCAD/EMTDC transient simulation program. The ac system was as in Fig. 1 and had the following parameters: $(\omega_B = 100\pi, \; L' = 0.1, \; R' = 0.01, \; \delta = 20°)$. For this particular study, a simplified model for the VSC was used which consisted of a controllable voltage source. The control system was designed to provide an oscillation-free response with increased damping ($K=3$). The d and q axis current orders were independently changed in a stepwise fashion and the resulting response is plotted in Fig. 4.

It can be clearly seen that the response is well damped and there is no coupling between the d and q currents (i.e., change in i_d does not affect i_q and vice-versa). The original time constant of the system describing the operation of the SSSC with energy storage without a controller equals 31.8 ms. The factor K of the state-space controller is inversely proportional to the time constant, so that at the chosen value $K=3$ the time constant of the system with a controller is reduced, as expected, to $T=10.6$ ms.

A higher value for the factor K could potentially provide a quicker response. However, this could lead to an encroachment of the ceiling voltage limit and result in degraded performance.

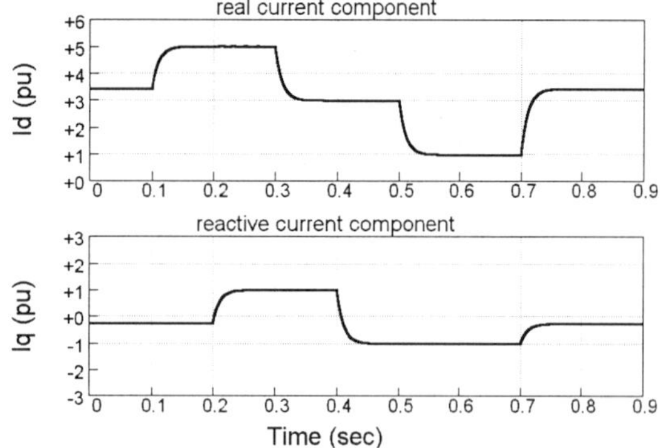

Fig. 4. State-space controller performance in the case of non-oscillatory settings (*K*=3)

Because of the use of proportional gains only, the controller performance requires knowledge of system parameters, i.e. impedance of the transmission line and the voltages on the other side of the transmission line. As an example the system response is shown in Fig. 5, where the line impedance settings in the controller (matrix A) are 5% higher than the impedance of the network. This could happen, for example, when the impedance data available is incorrect.

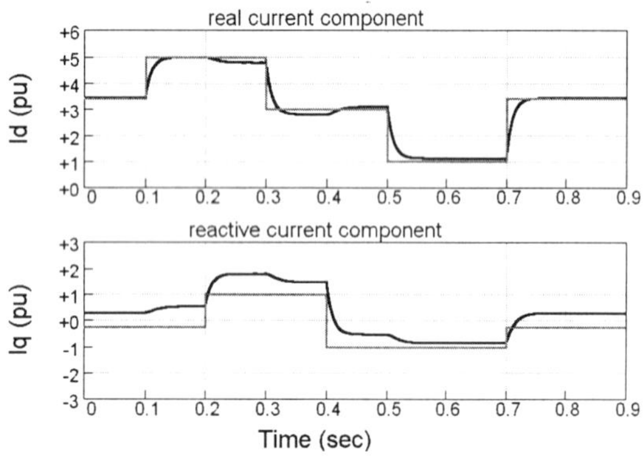

Fig. 5. State-space controller performance in the case of 5% error in system impedance (*K*=3)

Note that there the performance is now substantially poorer- not only is there some coupling, but there is also a steady-state error (evident particularly in the i_q plot). The next section describes enhancements to rectify this problem.

ENHANCED CONTROL SYSTEM

The following sections discus the design an validation of a robust control system that overcomes the deficiencies of steady state error and poor dynamic response of the basic control system described previously.

Robust Control System

The new control scheme shown in Fig. 6 includes an additional control path in the form of a state variable 'model' of the ac network. The parameters in the 'model' are the best estimates of the true ac network parameters. The output current vector **x'** from the model is compared with the response of the actual system **x** and the error **e** is used to modify the current order vector **w** to **w'**. Note that the additional control elements include the matrices **A** and **B** of the model and the two integrators required for integrating the state variables in the model.

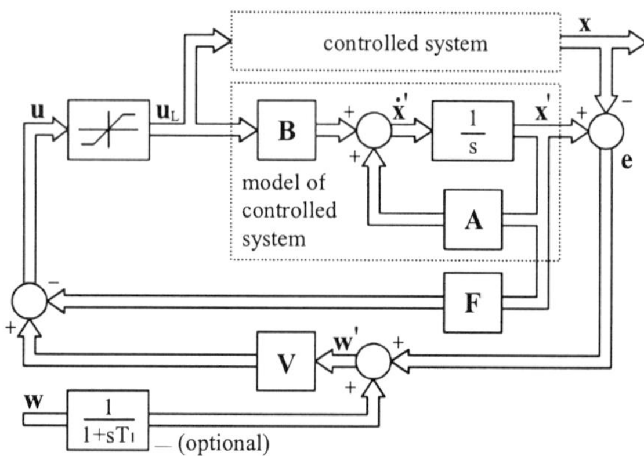

Fig. 6. Scheme of enhanced state-space controller

The additional control loop enables compensation of all uncertainties in the actual system and the effects of the errors in the system parameters. From Fig. 6 it can be readily shown that the error is eliminated under steady state conditions. Because **V** and **F** are designed using the 'model' parameters, the model's steady state response **x'** equals the input **w'**, and Eqn. 18 applies.

$$\mathbf{x'} = \mathbf{w'}$$
$$\mathbf{w'} = \mathbf{w} + \mathbf{e} = \mathbf{w} + \mathbf{x'} - \mathbf{x}$$
$$\text{hence}$$
$$\mathbf{x} = \mathbf{w}$$

(18)

The above approach does not completely address the problem that the incorrect parameters may cause the dynamic response to be different from the designed objective. Further improvements can be made to the dynamic response by limiting the rate of change of the control order (by using a first order lag-labeled 'optional' in Fig. 6) and by making *K* very large. The large *K* results in an extremely fast rise time, and so that variations in this rise time caused by parameter changes are no longer important. The time constant of the overall response is thus essentially determined by the externally connected first order lag.

Although the control system was designed for an energy storage type SSSC, the same ideas could be extended to other types of Voltage Sourced Converter based devices. The difference would arise from the fact the assumed constant voltage dc source would then be replaced by some other dynamic model.

Detailed SSSC model and simulation

In the next stage the device was modeled in detail using a quasi 48-pulse voltage source converter with the ability of voltage phase position and magnitude control. The maximum series injected voltage equals 0.3 pu. The dc-side circuit consists of a constant dc-voltage source. The voltage magnitude control is achieved by using three-level voltage source converters. The ac network is as described in the previous section. The SSSC model is located at the sending end of the transmission line. The traces of current d- and q-component flowing through the transmission line are shown again. The device follows the same changes of the reference current vector as in the case presented in Fig. 4. Simulations are performed for different cases regarding the controller settings and considering the use of the first-order block for smoothing the current reference vector. In Fig. 7 the system response in the case of known system parameters and without the use of smoothing of input reference current is shown. The controller factor setting is $K=3$, meaning that the system has a time constant $T=10.6$ ms.

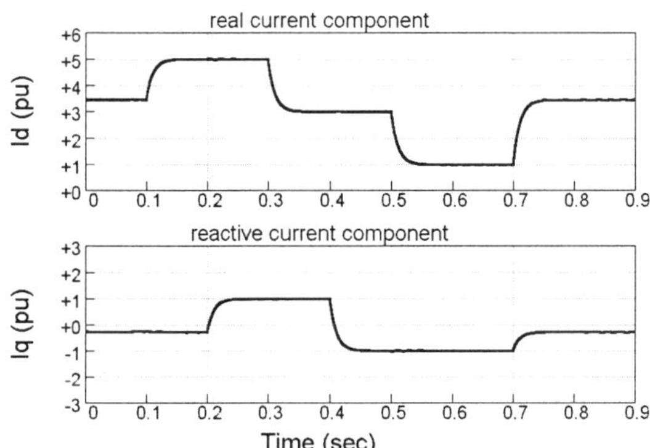

Fig. 7. State-space controller performance using detailed SSSC model; system parameters are known ($K=3$)

The response in the case with increased damping $K=50$ is represented in Fig. 8a. As can be observed response is virtually instantaneous. Also the decoupling is good as the duration for which the controller is saturated at the ceiling is negligibly small. Fig 8b shows the zoomed-in plots of the magnitude order and the actual injected series voltage over two cycles. The plot is centred at 0.2 s, the instant at which i_q^* is changed (see Fig. 8a). Notice the attainment of the ceiling voltage over approximately 2 ms. The ripple in the injected ac voltage waveform is due to the 48-pulse nature of the VSC converter.

In the following case, the influence of incorrect impedance values in the controller settings is investigated. In the case for which results are shown in Fig. 9, the impedance used in the controller is 50% lower than the impedance of the transmission line. The response is qualitatively the same, however, the time

constant changes to $T=21.3$ ms. Notice that the steady state error is now eliminated. As mentioned, the controller is also not sensitive to the incorrect value for the voltage vector on the other side of the transmission line.

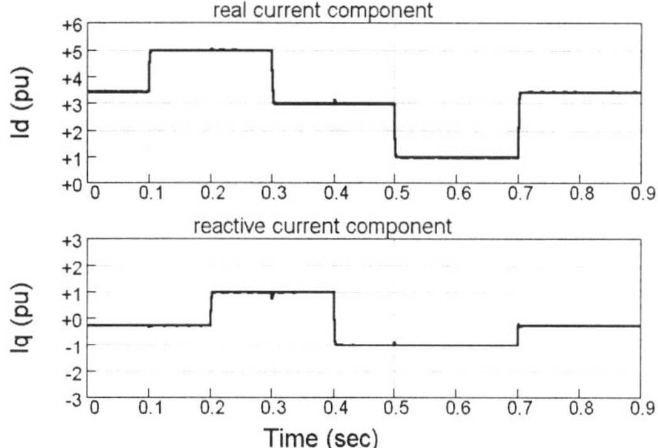

Fig. 8a. State-space controller performance using detailed SSSC model; system parameters are known ($K=50$)

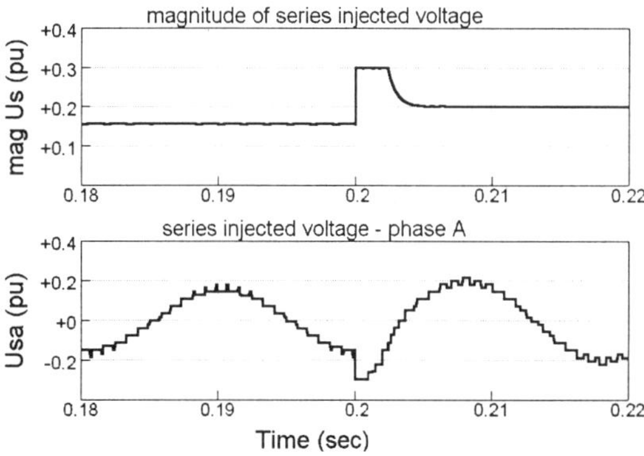

Fig. 8b. Series injected voltage during transition into new operating point

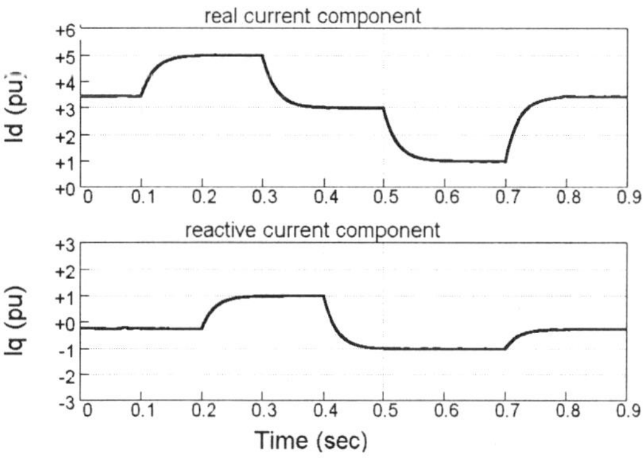

Fig. 9. State-space controller performance using detailed SSSC model; line impedance setting: 50 % lower ($K=3$)

To overcome the problem with the changed dynamic response when the accurate system impedance is not known, the use of the first-order block for limiting the rate of rise of the reference current is introduced as described in the previous section. Damping is increased to a value of 50 meaning that the system response predominately depends on the time constant selected in the first-order block. Fig. 10 shows simulation results for this case, with incorrect parameter settings identical to those used to generate the plots in Fig. 9. The time constant of the first-order block is set to $T_1=10$ ms, close to that in the case with correct parameters (Fig. 4). As expected, the observed time constant of the entire system is essentially the same (10 ms) and is not affected even with significant errors in the estimation of system parameters.

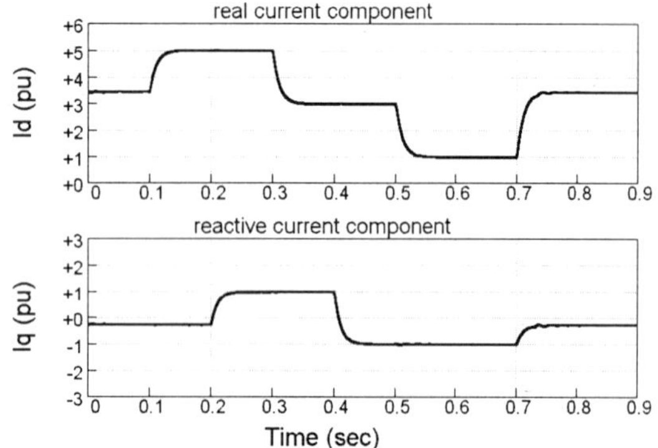

Fig. 10. State-space controller performance using detailed SSSC model; line impedance setting: 50 % lower, transmission angle setting: 25° (network transmission angle: 20°, K=50, smoothing of reference current vector)

POTENTIAL APPLICATIONS

Earlier researchers have shown that an SSSC with energy storage can be used to control the real and reactive power flow in a transmission line, as well as to damp electromechanical oscillations. The energy storage aspect allows a device of considerably smaller rating than one without energy storage. The new control method introduces transient as well as steady state decoupling and also allows a precise selection of performance parameters. It is expected that this will result in an improved performance in such devices.

CONCLUSIONS

A proportional control system was designed for a Static Synchronous Series Compensator with energy storage. The controller can be designed to provide the transient response (damping/oscillation frequency) of choice, but relies on an accurate knowledge of system parameters.

Making its performance less sensitive to the network parameters made the control system more robust. This was achieved with an additional corrective loop that modifies the ordered inputs as dictated by the error between the responses of the actual system and its 'model'. Further modifications to the damping and the externally introduced time constant allow even the dynamic response to be independent of the system parameters.

The performance of the control system was successfully verified using the tool of electromagnetic transient simulation. Although the design was primarily intended for an energy storage type SSSC, the same ideas could be extended to other types of Voltage Sourced Converter based devices.

REFERENCES

1. Gyugyi L., Schauder C.D., Sen K.K., 1997, "Static Synchronous Series Compensator: A Solide State Approach to the Series Compensation of Transmission Lines", IEEE Transactions on Power Delivery, 12/1, 406-417,

2. Mihalič R., Papič I., 1998, "Static Synchronous Series Compensator - A Mean for Dynamic Power Flow Control in Electric Power Systems", Electric Power Systems Research, 45, 65-72,

3. Zhang L., Crow M.L., Yang Z., Chen S., 2001, "The Steady State Characteristics of an SSSC Integrated with Energy Storage", Proceedings on IEEE PES Winter Meeting, 3, 1310 -1315,

4. Arsoy A., Liu Y., Chen S., Yang Z., Crow M.L., Ribeiro P.F., 2001, "Dynamic Performance of a Static Synchronous Compensator with Energy Storage", Proceedings on IEEE PES Winter Meeting, 2, 605 -610,

5. Unbehauen H., 1987, Regelungstechnik II, Friedr. Vieweg & Sohn Verlagsgesellschaft, Braunschweig, 62-97,

6. Papič I., 1998, Unified device for power flow control in electrical power networks, doctoral thesis, Faculty of Electrical Engineering, University of Ljubljana, 103-108.

ADAPTIVE FUZZY SLIDING MODE CONTROL OF SVC AND TCSC FOR IMPROVING THE DYNAMIC PERFORMANCE OF POWER SYSTEMS

R. Ghazi, A. Azemi, K. Pour Badakhshan

Ferdowsi University of Mashhad, Iran

Abstract

The FACTS devices have recently received great attention in the damping of a power system. In this paper, the simultaneous application of TCSC and SVC is proposed to improve the power system dynamics . In order to obtain a better performance, different control methodologies can be used to control the FACTS devices. In this paper the adaptive fuzzy sliding mode control (AFSMC) system is employed for the control of SVC and TCSC to improve the dynamic stability. The results show that the proposed scheme with AFSMC system can greatly improve the dynamic performance of the sample power system.

Keywords

Power System Stabilisation, Adaptive Fuzzy Sliding Mode Control (AFSMC), Static Var Compensator (SVC), Thyristor Controlled Series Capacitor (TCSC), and FACTS devices.

1. INTRODUCTION

It is well known that shunt and series compensation can be used to improve the power system stability. With the improvement in current and voltage handling capabilities of power electronic devices that have allowed for the development of FACTS, the possibility has been arisen of using different types of controllers for efficient shunt and series compensation. Thus thyristor based FACTS such as SVC's and TCSC's have been used by several utilities to compensate their systems [1]. More recently various types of controllers for series and shunt compensator have been proposed and developed. For the control of SVC and TCSC different control systems have been used [2]. In many cases linear control of devices is efficient in solving the dynamic problem caused by small disturbances. In the case of larger disturbances the traditional control systems based on linearised system model are of limited value also do not guarantee the insensitivity of plant parameter variations, which are very important in improving the dynamic performance of power systems. To solve the problems the variable structure control (VSC) strategy using sliding mode has been used before. This control theory has been focus of many researchers for the control of non-linear systems for many years [3,4]. The sliding mode controller has been used to improve the power system stability due to its robust response characteristic [5,6]. As the selection of switching surface is the most important part, in the design of the control system, in this paper a new switching surface of the sliding mode based on lyapunov approach is proposed. In general sliding mode control, the upper bound of uncertainties, which include parameter variations and external disturbances, must be available, which is difficult to obtain in advance. Therefore , fuzzy sliding mode control is used in which the fuzzy inference mechanism can estimate the uncertainty, so the chattering and steady state error can be reduced. The fuzzy sliding mode control can be further improved by adapting the centers of the membership functions. Therefore the proposed AFSMC system combines the advantages of the sliding mode control, the fuzzy inference mechanism and the adaptive algorithm. This control system has been used to control the nonlinear systems [7] and to control the motor drives [8]. In this paper the AFSMC system is employed for the control of SVC and TCSC to improve the stability of the single machine infinite bus (SMIB) system. In order to demonstrate the effectiveness of the proposed controller and also the effectiveness of the simultaneous application of SVC and TCSC, computer simulations are carried out under the application of large disturbances. To show the effectiveness and its superiority over the conventional controllers, such as PID, comparison has been made using the simulation results. The results show that the proposed AFSMC system can greatly improve the dynamic stability .

2. SYSTEM DESCRIPTION AND MODELLING

The system considered is a synchronous generator connected to an infinite bus through a double circuit transmission line as shown in Fig. 1. To improve the dynamic performance of this system, SVC and TCSC have been used as shunt and series compensation. The dynamic behavior of the studied system can be described by mathematical model. The synchronous generator is described by a 4th order nonlinear equation [9].

AC-DC Power Transmission, 28-30 November 2001
Conference Publication No. 485 © IEE 2001

2-1. Excitation System

The equation of the excitation system is obtained as follows

$$\dot{E}_{Fd} = [-E_{Fd} + K_a(V_{ref} - V_t)]/T_a \qquad (1)$$

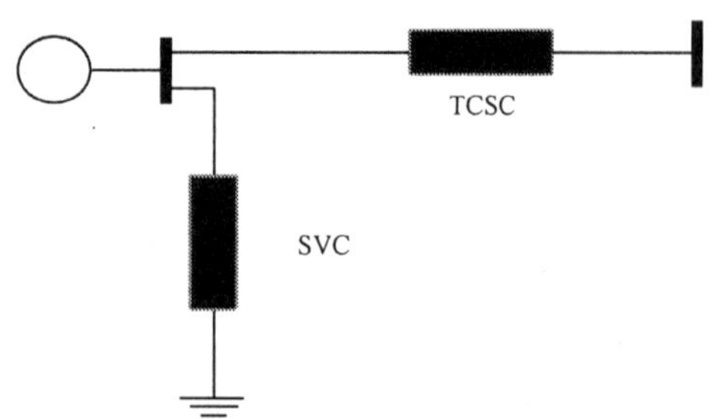

Fig. 1. Studied system with FACTS devices.

2-2. Static Var Compensator(SVC)

The applied SVC is shown in Fig. 2 which is a FC/TCR type and connected to the generator terminal. The magnitude of the SVC admittance is obtained by :

$$B_s = B_c - B_L(\alpha) \qquad (2)$$

$$B_L(a) = \frac{2\pi - 2\alpha + \sin 2\alpha}{2\pi X_L} \qquad (3)$$

According to the variations in terminal voltage and angular speed, the suceptance of the inductor B_L and then B_s can be regulated as shown in Fig. 3

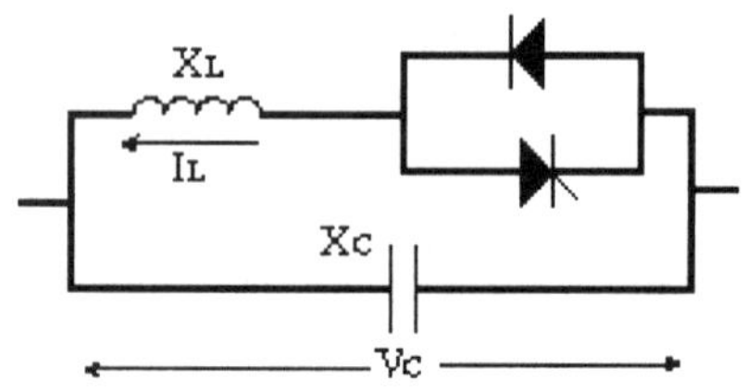

Fig. 2 FC/TCR type of SVC and TCSC

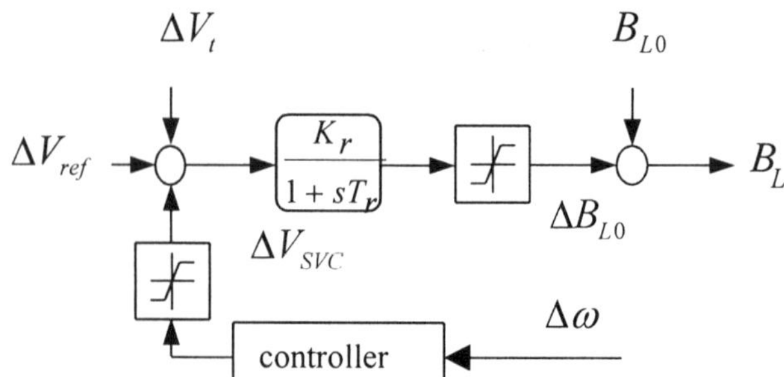

Fig. 3 Block diagram of the SVC controller

The equations of SVC are obtained as follows:

$$\dot{B}_L = [k_r(\Delta V_{ref} - \Delta V_t + \Delta V_{svc}) - B_L]/T_r \qquad (4)$$

$$B_L = B_{L0} + \Delta B_L \qquad (5)$$

2-3. Thyristor Controlled Series Compensator (TCSC)

The TCSC has the arrangement of FC/TCR which is connected in series with transmission line. According to the change of angular speed by adjusting the firing angle of the thyristor, the equivalent reactance of the TCSC is controlled between two limits [10]. The block diagram of the TCSC model is shown in Fig.4.

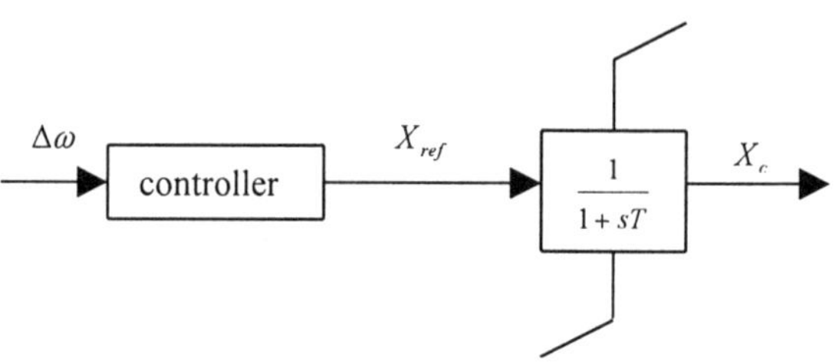

Fig. 4 Block diagram of the TCSC

3. CONTROL SYSTEM DESIGN

In this study the two aforementioned types of FACTS devices have been applied to the same system of reference [11]. The parameters of the system are found in this reference. The design procedure of the different controllers is described in the following sections.

3-1. Sliding Mode Control (SMC)

The variable structure control is a branch of nonlinear control theory which has been applied to the stabilisation of power systems due to its robust response characteristics. The nonlinear state equation of power system of Fig. 1 is given as :

$$\dot{X} = A(X) + B(X)U \qquad (6)$$

The control signal U should be determined so that the state trajectory from any point in state space plane drives to switching surface S=CX=0. For convenience an exponential law is adopted for the switching surface as:

$$\dot{S} = -Q_s \, \mathrm{sgn}(S) - K_s S \qquad (7)$$

where sgn is the sign function and Q_s, K_s are positive constants. On the other hand

335

$$\dot{S} = C\dot{X} = C[A(X) + B(X)U] \tag{8}$$

from the above equations the following control law is obtained,

$$U = -[CB(X)]^{-1}[CA(X) + Q_s \, \mathrm{sgn}(S) + K_s S] \tag{9}$$

constants K_s and Q_s can be determined by trial and error. In this paper a new switching surface of the sliding mode based on lyapunov approach is proposed [12]. Consider the system given in equation (6). Assuming that (A,B) is controllable, there exists a stabilising feedback gain K such that A+BK=As is asymptotically stable. It follows that there exists a positive define matrix P that solves the lyapunov equation.

$$PAs + As^{T}P = -Q \tag{10}$$

now we choose the switching surface

$$S(X) = DB^{T}PX = 0 \tag{11}$$

where D is a nonsingular matrix. A special choice of $D=(B^{T}PB)^{-1}$ will diagonalize the control coefficient matrix to the dynamics for S:

$$\dot{S} = DB^{T}PAX + U \tag{12}$$

so we have the following control law;

$$U = -(DB^{T}PA + K_s DB^{T}P)X - Q_s \, \mathrm{sgn}(S) \tag{13}$$

In this case in obtaining the desired response the parameter Q_s can be determined independent of other parameters.

3-2. Fuzzy Sliding Mode Control (FSMC)

In this paper a fuzzy sliding mode controller is proposed in which a fuzzy inference mechanism is used to estimate the value of Q_s in equation (13). The control block diagram of fuzzy sliding mode controller is shown in Fig. 5. The membership functions for the fuzzy sets corresponding to switching surface S , $\dot{S}$ and Qs are defined in Fig. 6
The fuzzy inference rules are as follows :

IF {(S is P AND $\dot{S}$ is P) OR (S is N AND $\dot{S}$ is N)} THEN Q_s is PH

IF {(S is P AND $\dot{S}$ is Z) OR (S is N AND $\dot{S}$ is Z)} THEN Q_s is PB

IF {(S is P AND $\dot{S}$ is N) OR (S is N AND $\dot{S}$ is P)} THEN Q_s is PM

IF {(S is Z AND $\dot{S}$ is P) OR (S is Z AND $\dot{S}$ is N)} THEN Q_s is PS

IF (S is Z AND $\dot{S}$ is Z) THEN Q_s is ZE

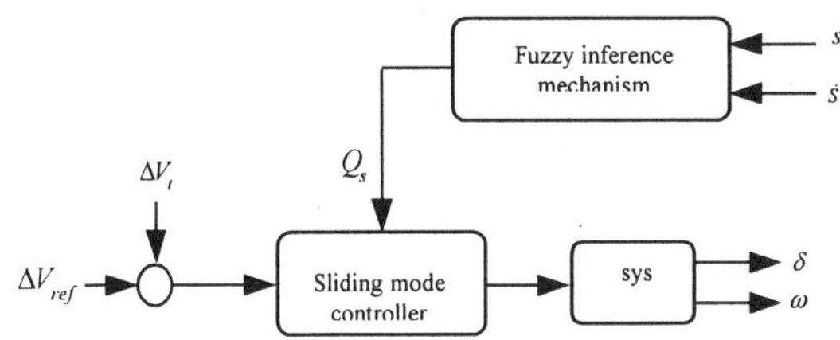

Fig. 5 Block diagram of FSMC

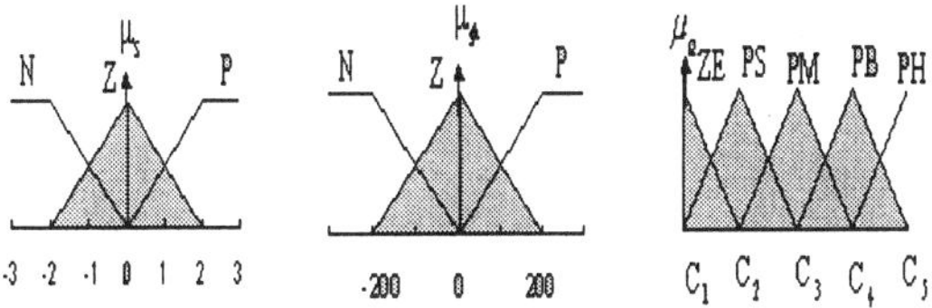

Fig . 6 Membership functions of fuzzy sets

Fuzzy output Qs can be calculated by the center of area defuzzification as :

$$Q_s = \frac{\sum_{i=1}^{5} W_i V_i}{\sum_{i=1}^{5} W_i} = \frac{[V_1 .. V_5]\begin{bmatrix} W_1 \\ \vdots \\ W_5 \end{bmatrix}}{\sum_{i=1}^{5} W_i} \tag{14}$$

where $V=[V_1,...,V_5]$ is a adjustable parameter vector, V_1 through V_5 are the center of the membership functions of Q_s. It should be noted that the appropriate determination of these parameters is an important issue to obtain the correct value for Q_s. These parameters can be adjusted according to the different operating conditions (that is states being close to or far from the switching surface) to get the optimum value for Q_s, as described in the following section.

3-3. Adaptive Fuzzy Sliding Mode Control (AFSMC)

In this paper to determine the optimum value of Q_s, an adaptive algorithm is proposed. According to the operating conditions and based on the proposed algorithm the centers of the membership functions of the fuzzy inference mechanism are adapted by the following adaptive law [8].

$$\dot{V} = \alpha \, V B |S| W \tag{15}$$

in which α is a positive constant. It is obvious that when parameter variations or disturbances occur, the fuzzy inference mechanism and the adaptive law will be excited to find the new value of Q_s.

4. SIMULATION RESULTS

In order to illustrate the effectiveness of the proposed AFSMC system and also the effectiveness of the proposed AFSMC system and also the effectiveness of the simultaneous application of SVC and TCSC, time domain simulations are performed based on a nonlinear system model, using Matlab software. Dynamic responses of the generator to a 4 cycle three phase fault which occurs at the infinite bus, under different controllers for SVC and TCSC are considered. To show the effectiveness of the controller and its superiority over the conventional controller, comparison has been made between PID and AFSM controller. Generator angle and speed deviations obtained using PID controller and AFSM controller are shown in Fig. 7 and 8 respectively. The results show that the proposed control system con greatly improve the dynamic performance of the power system. To demonstrate the superiority of the AFSM controller, simulation studies are also performed for LQG, VSC, FSMC as well as PID and AFSMC under the same disturbance. The results relating to the deviation of rotor angle are shown in Table 1.

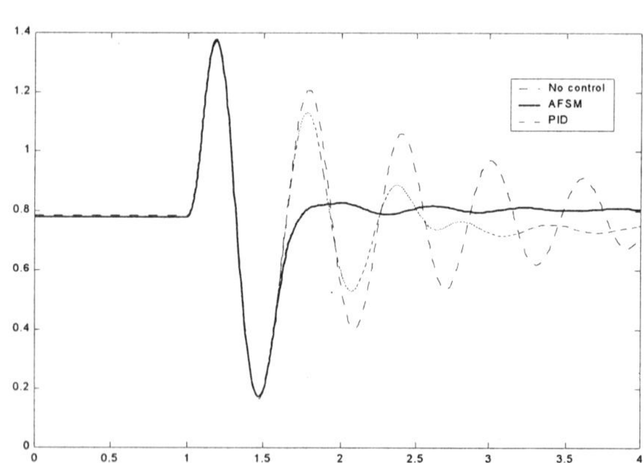

Fig 7 Rotor angle variations

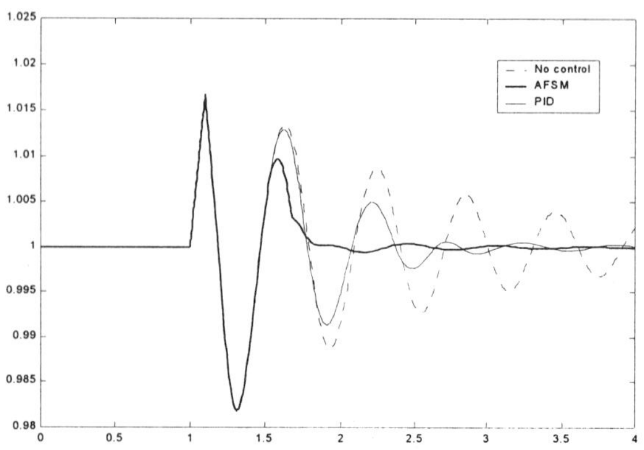

Fig 8 Speed variations

Controller	Settling time (second)	Overshoot (Percent)
No controller	6	75.66
PID	1.5	75.60
LQG	1	57.5
VSC	1.5	76
FSMC	0.7	76
AFSMC	0.6	59

Table. 1 Comparison of controllers (For δ)

5. CONCLUSION

Damping of generator oscillations is investigated using two types of FACTS devices using adaptive fuzzy sliding control system. The results show that these FACTS based type stabilizers can offer good damping characteristic under 3-phase fault condition. The results show the role and effectiveness of the proposed controller in damping power system when the power system experiences a large disturbance, such as a 3 phase fault.

6. REFERENCES

1- Anagquist.L, Lundin.B and Samuelsson.J, 1993 "Power oscillation damping using controlled reactive power compensation" IEEE Trans. On Power System , vol.8, no.2, pp 687-700

2- Kawng.M. S., Jong.K. P, 2000 "On The Robust LQG control of TCSC for damping power system oscillation" IEEE Trans. On Power System , vol.15, no.4.

3- Welbing Gaoa James C. Hung 1993, "Variable structure Control of nonlinear System : A new Approach" IEEE Trans on Ind. Electronic Vol. 40, No.1.

4- Raymond A. Decarlo Stanslaw H. Zak , Gregory P. Mathewes, 1988 "Variable Structure Control of nonlinear multivariable System: A tutorial", Proc. IEEE vol. 76, no 3, pp 212- 233.

5- P.K. Dash, N.C. Sahoo, S. Elangovan and A.C. Liew, 1996 " Sliding mode control of a static var compensator for synchronous generator stabilization", Electrical power and energy system, vol. 18, No.1, pp 55-64.

6- Wang. Y, Mohler R.R, Spee R. and mihelstadt W. feb. 1992 "Variable structure facts controllers for power system transient stability" IEEE Trans. On Power System vol. 7, No.1, pp 307-313.

7- Yoo B. and Ham W.,1998, "Adaptive fuzzy sliding mode control of nonlinear system" IEEE Trans. On Fuzzy System, vol. 6 No.2.

8- Lin F. J. and Chiu S.L.,1998, " Adaptive fuzzy sliding mode control for PM synchronous servo motor drive" IEE Proc. Control theory, vol. 145, no.1, pp 63-72.

9- Anderson P.M. and Fouad A. 1966 "Power system control and stability" Iowa State University Press, Ames, IOWA.

10- Larsen E.V., Clarke, Miske S.A, Urbane T. 1994 "Characteristics and rating Considerations of thyristor Controlled Series Compensator" IEEE PD vol.4 pp 992-1000.

11- Chang C. H. and Hsu Y.Y May 1992 "Damping of generator oscillations using an adaptive static Var compensator" IEEE Trans. On Power System, vol.7, No.2, pp 718-725.

12- Chung Su Wu, Sergey V. Drakunove and Umit Ozguner 1996 "Constructing discontinuity surfaces for variable structure system: A Lyapunov approach" Automatica, vol. 32, No.6, pp 925-928.

A CO-ORDINATED LARGE-SIGNAL MODULATION STRATEGY FOR MULTI-INFEED HVDC SYSTEMS

W D Yang, Z Xu, Z X Han

Zhejiang University, P.R. China

ABATRACT

A coordinated large-signal modulation strategy for multi-infeed HVDC systems is presented in this paper. With the HVDC systems being treated as a variable admittance connected at the commutation busbars, the control law of HVDC system's transfer power can be derived using optimal control theory. A two level hierarchical control structure is assumed for the rectifier and inverter side ac system respectively. The control of constituent HVDC subsystems can be better coordinated by the consideration of the ac commutation voltage included in the optimized performance index, and both the performance of rectifier and inverter side ac systems are considered.

INTRODUCTION

The emergence of multi-infeed HVDC systems has motivated extension of HVDC modulation control widely used in single-infeed HVDC system to such configurations. With more HVDC links located in the same ac system, the degree of controllability of the hybrid AC/DC power system will be improved greatly, and new possibilities will be provided by using HVDC link for emergency power support to the connected ac systems. However, interactions between AC/DC and DC/DC systems may be very complicated at closely connected busbars due to very fast control actions, sometime with a possible negative overall effect as shown by Szechtman et al (1). In order to obtain better system performance, it is necessary to coordinate the separate dc control actions, especially when the electrical connection between separate converters is very strong.

Previous researchers have seldom studied the use of several HVDC links for emergency power support to the connected ac system when a large disturbance occurs, and only a few of them have addressed the issue of coordination between different HVDC controls. It is given by Pilotto et al(2), as multiple HVDC systems are added to the power modulation activities, it becomes necessary for their control actions to be integrated and centrally coordinated, thus unforeseen interactions between different HVDC subsystems can be avoided. Furthermore, when the problem of augmenting power system perfomance is approached from a system-wide viewpoint, both the inverter and rectifier side system responses should be considered. One reason for the above two points is that, in some cases, a simultaneous fast power increase of several HVDC systems can lead

to a rapid and high requirement for reactive power, that will in turn cause a depressed ac bus voltage and commutation failure at the inverter side. Another reason is that, if the power abstraction by the HVDC systems increases so fast, the system performance of the rectifier side ac system may be degraded. Therefore, it is necessary for the rectifier side ac system to have the ability to provide such a load increase while maintaining its frequency at an acceptable range.

The main feature of the proposed method in this paper is that HVDC systems can be treated as a variable admittance connected at the inverter or rectifier ac bus. After deriving the analytical relationship between the variable admittance and output power of each generator, the conventional generator dynamic equations can be expressed with a variable admittance of HVDC systems as an additional state as indicated by Hammons et al(3). In turn, the traditional optimal control theory can be used to derive the modulated dc power. A two level hierarchical control structure is assumed with the local feedback controllers associated with each subsystem constituting the lower level and the transient coordinating controller placed at the upper level. Additionally, more emphasis was paid to the inverter commutation bus voltage in the performance index to maintain them at a constant level and both the performance of rectifier and inverter side ac systems are considered. A tradeoff is made between the fast response speed of the HVDC modulation and the frequency deviation of different ac systems, so that the frequency deviation of them is within a permitted range. To illustrate the feasibility of the proposed method, simulation studies on a typical double infeed HVDC power systems were performed.

THE SYSTEM STUDIED

For the testing of the proposed method, a simplified double converter multi-infeed HVDC system model as shown in Fig.1 is used in this paper. The two HVDC subsystems have the same parameters (together with their control parameters) as that given by Szechtman et al (4) and Wess and Ring (5). However, its inverter side ac system is replaced by the ac system as shown in Fig.1, with the two commutating buses connected by admittance. The rectifier side ac system of dc link 2 is also substituted by a 3-machine ac system.

To simplify the comparison, the conventional control used in this paper is as follows: the rectifier has a

AC-DC Power Transmission, 28-30 November 2001
Conference Publication No. 485 © IEE 2001

constant current control, and the inverter is subjected to constant extinction angle control, and both of them are PI controllers. Additionally, we assume the two DC links have the same controller parameters and DC power transfer capacities.

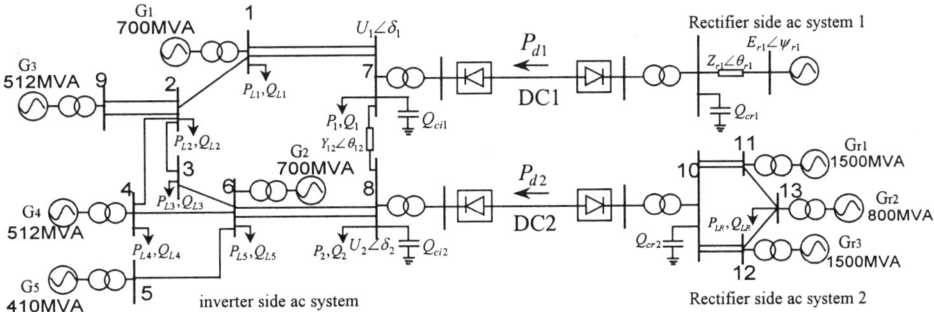

Fig. 1 Simplified double-infeed HVDC systems model

ANALYTICAL RELATIONSHIP BETWEEN THE GENERATOR OUTPUT POWER AND THE EQUIVALENT ADMITTANCE OF HVDC NODE

For a hybrid AC/DC system as illustrated in Fig. 1, its inverter side ac system can be expressed as a general network that includes n generator nodes, m load nodes, p HVDC nodes and some physical nodes. For theoretical purposes, generators are modeled by constant voltage electromotive forces behind transient reactance, and static load model as given by IEEE Committee(6) is employed in this paper. Only the generator internal nodes and HVDC nodes are retained and all other nodes are eliminated through network reduction as shown by Pai (7). For simplicity, we first suppose only one HVDC node (connected at node p) is included in the hybrid AC/DC system in the following derivation.

For the reduced network (see more detail in (3)), we have

$$\begin{bmatrix} I_G \\ I_p \end{bmatrix} = \begin{bmatrix} Y_{GG} & Y_{Gp} \\ Y_{pG} & Y_{pp} \end{bmatrix} \begin{bmatrix} E_G \\ U_P \end{bmatrix} \tag{1}$$

The electrical output power of the ith generator ($i = 1, 2, \cdots, n$) with the effect of HVDC node p taken into account, is given in (3) as follows:

$$P_{ei} = E_i^2 G_{ii} + \frac{g_{DC}}{1 + g_{DC}^2} \beta_{ip}^2 B_{pp} +$$

$$+ \sum_{i \neq j}^{n} (E_i E_j B_{ij} - \frac{1}{1 + g_{DC}^2} \beta_{ip} \beta_{pj} B_{pp}) \sin \delta_{ij} + \tag{2}$$

$$+ \sum_{i \neq j}^{n} \frac{g_{DC}}{1 + g_{DC}^2} \beta_{ip} \beta_{pj} B_{pp} \cos \delta_{ij}$$

Where, $B_{ij}, B_{ip}, B_{pj}, B_{pp}$ are elements of the admittance matrix in equation(1), and

$$\beta_{ip} = \frac{E_i B_{ip}}{B_{pp}}, g_{DC} = \frac{G_{DC}}{B_{pp}} \tag{3}$$

G_{DC} is the equivalent admittance of HVDC system at node p and can be expressed as

$$G_{DC} = P_{DC} / U_p^2 \tag{4}$$

P_{DC} is the injected active power by HVDC system;

U_p is the ac voltage at HVDC converter bus;

Usually, $g_{DC} \leq 0.3$, utilizing the Taylor series, equation (2) can be simplified to

$$P_{ei} = E_i^2 G_{ii} + g_{DC}(\beta_{ip}^2 B_{pp} + \sum_{i \neq j}^{n} \beta_{ip} \beta_{pj} B_{pp} \cos \delta_{ij}) +$$

$$+ \sum_{i \neq j}^{n} (E_i E_j B_{ij} - \beta_{ip} \beta_{pj} B_{pp}) \sin \delta_{ij} \tag{5}$$

In the above derivation, only one HVDC node is considered. When there are m HVDC nodes existing in the inverter side ac system, the dimension of the admittance matrix in (1) will reach m+n, and the analytical expression for the output power P_{ei} of generator i will be very complicated. However, it does exist and can be derived, and will have a form as follows:

$$P_{ei} = E_i^2 G_{ii} + \sum_{j=1}^{m} g_{DCj} G_{ij}(\delta_1, \cdots, \delta_n) + F_i(\delta_1, \cdots, \delta_n) \tag{6}$$

where $G_{ij}(\delta_1, \cdots, \delta_n)$, $F_i(\delta_1, \cdots, \delta_n)$ are functions of δ_i, $i = 1, \cdots, n$, $j = 1, \cdots, m$.

ASSUMPTIONS

Due to the location and severity of a fault, normal functioning of a dc line may be disrupted for the duration of the fault. In this paper, we only consider the case when dc system can function smoothly.

It was given by Kimbark (8), with the dynamic processes of a dc link neglected, a dc system can be treated as a one order inertial element, and the state equation of dc power can be written:

$$\dot{P}_{dc} = \frac{1}{T_d}(-P_{dc} + P_{dcref} + u_{dc}) \tag{7}$$

Where, P_{dcref} is the given value of the dc power, T_d is the equivalent time constant of dc system, u_{dc} is the control variable of dc system.

From (3) and (4), we have $g_{DC} = \frac{G_{DC}}{B_{pp}} = \frac{P_{DC}}{B_{pp} U_p^2}$, if assume U_p is constant, then we get

$$\dot{g}_{DC} = \frac{1}{T_d}(-g_{DC} + g_{DCref}) + \frac{1}{T_d U_p^2 B_{pp}} u_{dc} \qquad (8)$$

where, $g_{DCref} = \dfrac{P_{dcref}}{U_p^2 B_{pp}}$.

SYSTEM MODEL

In an n-machine system, the conventional dynamical model of the ith machine with steam valving control may be described by the state equations in the form of

$$\dot{P}_{mi} = -\frac{1}{T_{si}}P_{mi} + \frac{1}{T_{si}}P_{mi0} + \frac{1}{T_{si}}u_i \qquad (9)$$

$$\dot{\omega}_i = -\frac{D_i}{2H_i}\omega_i + \frac{\omega_0}{2H_i}(P_{mi} - P_{ei}) \qquad (10)$$

$$\dot{\delta}_i = \omega_i \qquad (11)$$

Where, u_i is the valve opening in p.u..
Substituting equation (6) into (10), we have

$$\dot{\omega}_i = -\frac{D_i}{2H_i}\omega_i + \frac{\omega_0}{2H_i}(P_{mi} - E_i^2 G_{ii} - \sum_{j=1}^{m} g_{DCj}G_{ij}(\delta_1,\cdots,\delta_n)$$
$$-F_i(\delta_1,\cdots,\delta_n)) \qquad (12)$$

For the sake of convenience and simplicity, the test system as shown in Fig.1 will be considered in this paper. We also assume generator 1 and 2 has a constant input power, and will have a HVDC link modulation control instead of the steam valve control, the other generators will have their conventional steam valve control. At the inverter side ac system, we will design controllers for generator 1 and 2 only. With all the above assumptions, and assuming that each synchronous generator (together with its connected HVDC link) constitutes a subsystem, equation (8) (12) and (11) will be changed to the form as

$$\dot{X}_i = A_i X_i + B_i U_i + f_i(X) \qquad (13)$$

where,

$$X_i = \begin{bmatrix} g_{DCi}, \omega_i, \delta_i \end{bmatrix}^T , \quad U_i = u_{dci}, \quad i = 1,2 \qquad (14)$$

$$A_i = \begin{bmatrix} -\dfrac{1}{T_{di}} & 0 & 0 \\ 0 & -\dfrac{D_i}{2H_i} & 0 \\ 0 & 1 & 0 \end{bmatrix}, \quad B_i = \begin{bmatrix} \dfrac{1}{T_{di}U_{pi}^2 B_{ppi}} \\ 0 \\ 0 \end{bmatrix} \qquad (15)$$

$$f_i(x) = \begin{bmatrix} \dfrac{1}{T_{di}}g_{DCiref} \\ \dfrac{\omega_0}{2H_i}(P_{mi} - E_i^2 G_{ii} - \sum_{j=1}^{m} g_{DCij}G_{ij}(\delta_1,\cdots,\delta_3) - K_i(\delta_1,\cdots,\delta_3)) \\ 0 \end{bmatrix}$$

$$(16)$$

where $G_j(\delta_1,\cdots,\delta_n)$, $K_j(\delta_1,\cdots,\delta_n)$, $j=1,2$ can be easily obtained by expanding the dimension of the admittance matrix in (1) from m+1 to m+2 , and with the effect of HVDC node 4 and 5 (as shown in Fig.1) incorporated into the electrical output power of the ith generator, $i = 1,2,3$ in this case, and they are not shown here due to the space limitation.

Note that equation (13) consists of two parts, the first part incorporated in matrix A_i and B_i is a linear function of the variables from subsystem i only, therefore, a local part. The second part represented as f_i vector is a nonlinear function of the variables of the subsystem i and the neighboring subsystem j, representing the nonlinear interactions among the subsystems.

For the rectifier side ac system, its system state equation can be expressed as (9) to (11). Suppose at the rectifier side ac system, the equivalent admittance of HVDC 2 is treated as a constant, and the steam valve opening is selected as the control variable, thus we have $X_i = \begin{bmatrix} P_{mi}, \omega_i, \delta_i \end{bmatrix}^T$, $U_i = u_i$, $i = 1,2,3$, the matrix A_i, B_i and the vector f_i can be easily obtained from (9) to (11).

TWO LEVEL CONTROL STRATEGY

Over the past years, a considerable research effort has been devoted to design hierarchical controllers for power systems as given by Domijan and Song (9), Ngan et al (10), and Rubaai (11). They are based on the idea that an large interconnected power system can be divided into many subsystems, and the control and monitoring structures in modern power system are basically designed in a distributed way. In this paper, the idea of using hierarchical control to enhance the transient stability of power system reported in (11) has been extended to multi-infeed HVDC systems.

It is shown by (9), the objective of hierarchical control is to find control series U_i, $i = 1,\cdots,n$, such that the following quadratic objects function is minimized:

$$J = \sum_{i=1}^{n} \int_{t_0}^{t_f} \frac{1}{2} t[\|x_i^2(t)\|Q_i(t) + \|u_i^2(t)\|R_i(t)]dt \qquad (17)$$

According to the method proposed in (11), a transient coordinating controller should generates a coordination vector which contains the initial state trajectories as well as the Lagrange multipliers with values at the mth iteration as

$$X_i = X_i^{0m}$$

$$\Pi_i = \Pi_i^{0m} \tag{18}$$

At every iteration, the item $f_i(X^{om})$ in (13) can be taken as a constant, and the optimized local control strategy for each subsystem can be decided by using the transmitted data from the upper level. By doing this, the optimization problem (17) reduces to that of solving a number of lower level de-coupled local sub-problem as given by (11), each of which can be expressed as

$$MinJ_i^1 = \frac{1}{2}\int_0^{t_f}[X_i^T Q_i X_i + U_i^T R_i U_i +$$
$$+(X_i - X_i^{0m})^T W_i(X_i - X_i^{0m})]dt \tag{19}$$

subject to the constraints

$$\dot{X}_i = A_i X_i + B_i U_i + f_i(X^{0m}) \tag{20}$$

$$X_i = X_i^{0m} \tag{21}$$

By using the maximum principle, local control law of the ith sub-problem can be obtained by:

$$U_i = -R_i^{-1}B_i^T P_i = -R_i^{-1}B_i^T(g_i + k_i X_i) \tag{22}$$

where P_i is the co-state vector, k_i is the solution of the following Riccati equation:

$$\dot{k}_i = -k_i A_i - A_i^T k_i + k_i B_i R_i^{-1} B_i^T k_i - Q_i - W_i$$
$$k_i(t_f) = 0 \tag{23}$$

g_i is the solution of the following adjoint vector equation:

$$\dot{g}_i = -[A_i^T - k_i B_i R_i^{-1} B_i^T]g_i - \Pi_i^{01} - k_i f_i(X^{0m}) + W_i X^{om}$$
$$g_i(t_f) = 0 \tag{24}$$

As shown in (11), updating at the upper level for the initial state trajectories and Lagrange multipliers from iteration m to m+1 is performed as:

$$X_i^{0(m+1)} = X_i^m \tag{25}$$

$$\Pi_i^{0(m+1)} = W_i(X_i^m - X_i^{0m}) + \frac{\partial}{\partial X_i} f_i^T\Big|_{X_i = X_i^{om}} P_i^m \tag{26}$$

Remark: The advantage of the feedback control law (22) is clearly indicated in (11), it is valid for all operating and fault conditions, i.e., the control strategy (22) is robust. From equation (3) (4), we also know that the ac commutation voltage is reflected by g_{DCi} (i=1,2). Therefore, with the state variable g_{DCi} included in the performance index (19), the dynamic responses of the two HVDC nodes' ac commutation voltage can be optimized also, which is beneficial to the recovery of the two HVDC subsystems. So it can be concluded that control law (22) is also a coordinated control law from the viewpoint of fast recovery of HVDC systems after an disturbance.

The structure in Fig.2 shows the algorithm realization of the proposed hierarchical control strategy. The control scheme can be realized through successive exchanges of information between the two levels. In the lower level, using the transmitted data Π_i^{om} and X^{om} from the upper level and the local information X_i^m, a decentralized optimal control can be realized including the effects of coordination of the upper level. Next, the updated subsystem state variables and the costate vector P_i will be transmitted to the upper level to calculate the updated value of $X_i^{o(m+1)}$ and $\Pi_i^{o(m+1)}$ according to (25)(26). In this way, a continuous real time control is achieved.

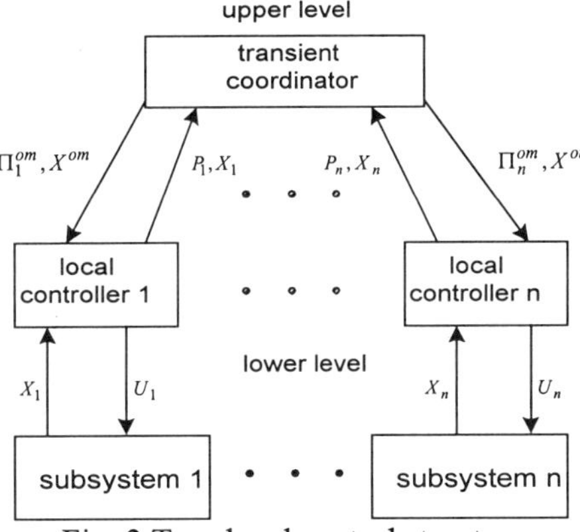

Fig. 2 Two-level control structure

The detailed two-level control structure for the whole AC/DC power system is shown in Fig.3. The rectifier and inverter side ac system has no direct couplings, except that they are mutually affected by the dc link 2. Thus, we have designed a corresponding transient coordinator for the inverter side and rectifier side ac system 2 respectively, and they are responsible for the control coordination of the two ac systems, the effect of dc systems on ac system is represented by their equivalent admittance.

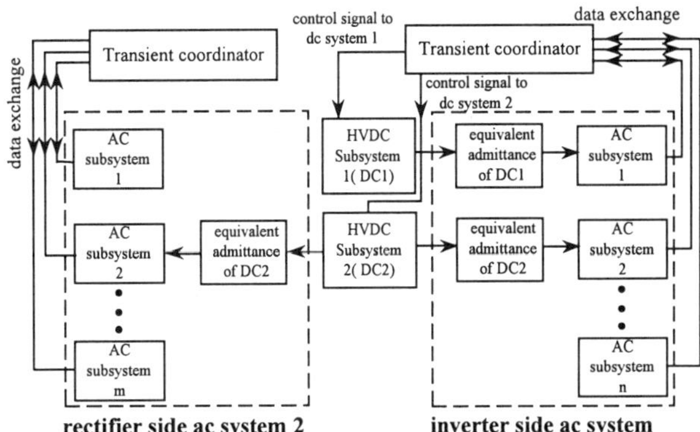

Fig. 3 Detailed two-level control structure

After the modulated signal, i.e. $u_{dc1}(t)$ and $u_{dc2}(t)$ (divided by dc voltage) is derived, they will be added to the reference current of the rectifier current controller. In this way, the reference quantity of rectifier is biased, the power transmitted through dc system is rapidly increased or decreased to prevent instability in ac system during a severe disturbance.

CASE STUDY

In this section, the system model as shown in Fig. 1 will be used for the testing of the proposed control strategy. Table 1 and 2 gives the line and load parameters of the inverter and rectifier side ac system. The generator parameters were given by Anderson and Fouad (12). Standard excitation and governor models are used.

TABLE1-Line parameters

From bus	To bus	R(p.u.)	X(p.u.)
1	2	0.0242	0.2463
2	3	0.0397	0.4082
3	6	0.0242	0.2463
2	4	0.0484	0.4925
4	6	0.0148	0.1505
9	2	0.047	0□4789
1	7	0.0268	0□2737
7	8	0.0201	0□2053
6	8	0.0134	0□1368
5	6	0.0403	0□4105
1	8	0.0161	0.164
10	11	0.0119	0.1236
11	13	0.0397	0.4082
13	12	0.0159	0.1644
12	10	0.0144	0.1484

TABLE2-Load parameters

Bus	Active power P_L(MW)	Reactive power Q_L(Mvar)
1	535	200
2	525	330
3	425	190
4	605	190
6	615	185
7	680	156.7
8	560	160
13	715	185

A step increase of the load at bus 1 by 300 MW and 160 Mvar, which corresponds nearly to 10 per cent of the total real load of inverter side ac system, is introduced. The following three cases are recorded for the above disturbance:

1)The two dc links are operated with only constant current and constant extinction angle control, all the generators use conventional governors without fast valving control;

The simulation results presented in Fig.4 show the transient performance of inverter side and rectifier side ac system. Only the frequency deviation of the inverter side ac system, power angle of generator 2 to 5 in the inverter side ac system is demonstrated. It can be seen from Fig.4, oscillation of the above variables after the occurring of such a disturbance is serious. The inverter side ac system frequency is decelerated substantially because it is controlled only by the generator's control system, which is comparatively slow to the fast dc control, which in this case, does not contribute to the balance of power. The oscillation of the rectifier side ac system 2 is comparatively small which is not shown due to the space constraints.

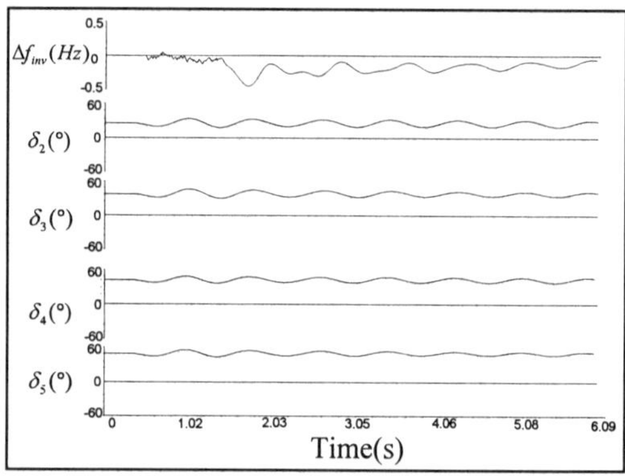

Fig.4 Inverter side ac system responses when the two DC links are operated with constant current and constant extinction angle control

2)All the conditions are same as in 1), except that the two dc links are operated with the proposed coordinated hierarchical control strategy.

Except the variables shown in Fig.4, the frequency deviation and the power angle of generator G_{2r}, G_{3r} of the rectifier side ac system are also demonstrated in Fig.5. With the proposed control strategy adopted, the power imbalance at the inverter side ac system can be quickly compensated by the power support from the two dc links, and oscillations of the inverter side ac system variables are minimized to a greater extent. However, the variable oscillation of the rectifier side ac system is obvious. Because the dc link 2 has draw so much additional power from the rectifier side ac system, whose regulating facilities (generators) in this case, are slow to increase their output power.

3)Both the rectifier and inverter side ac system are operated with the proposed coordinated hierarchical control strategy, and coordination between them is considered.

In order to get a better system performance, more emphasis is paid to the inverter commutation bus voltage in the performance index to maintain them at a constant level. This is realized by carefully selecting the weighting matrices, so that more emphasis can be given to the state variable g_{DCi}, which has a relation with the HVDC ac commutation voltage in the cost function. Additionally, both the inverter and rectifier side system responses is considered, and a tradeoff is made between the fast response speed of the HVDC modulation and the frequency deviation of different ac sides, so that the frequency deviation of them is within a permitted range. This is realized by implementing the hierarchical control strategy at the rectifier and inverter side ac system simultaneously. Besides that, we also let the power modulation quantity of dc link 2 changed slowly from a lower level (the derived modulation quantity multiplied by a coefficient less than 1) to the desired modulation quantity in a time period. That time period can be obtained by utilizing the optimization function embedded in the NETOMAC program as shown by Lei (13).

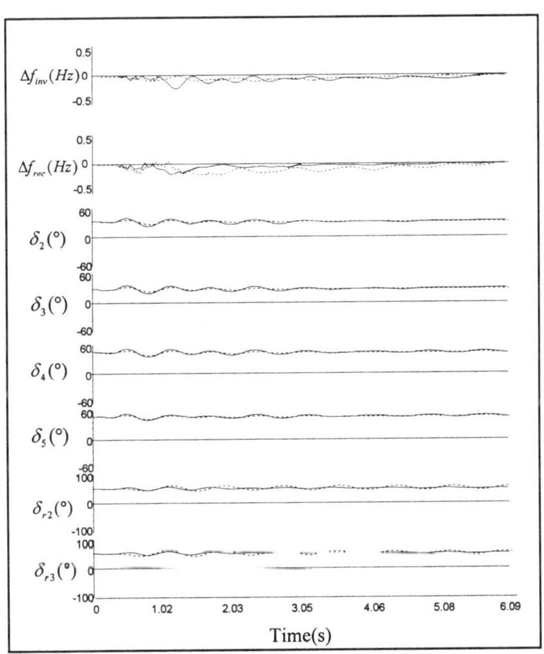

The two dc links are operated with the proposed
------ coordinated hierarchical control strategy, all the generators in the rectifier side ac system 2 use conventional governors

Both the rectified and inverter side ac system are operated
——— with the proposed coordinated hierarchical control strategy, and coordination between them is considered

Fig.5 Rectifier and inverter side ac system responses

Fig.5 clearly shows that, when both the rectifier and inverter side ac system are operated with the proposed hierarchical control strategy, the system performance of the rectifier side ac system has been improved a lot, while the inverter side system performance degraded to some extent in this case. However, because a tradeoff has been made between the fast response speed of the HVDC modulation and the frequency deviation of different ac sides, the system performance of both side ac system is within a permitted range.

CONCLUSIONS
In this paper, a two level coordinated hierarchical control strategy has been proposed to improve the performance of hybrid AC/DC power system during a large disturbance. Due to the inherent inclusion of the ac commutation voltage in the performance index, the strategy has the advantage that both the control of the constituent HVDC subsystem control can be coordinated. Additionally, the system performances of the rectifier and inverter side ac systems are both considered. Simulation results on a double-infeed test system also show the validity of the proposed control scheme.

ACKNOWLEDGMENTS
Project No. G1998020312 supported by National Key Basic Research Special Fund of China.

REFERENCES
1. Szechtman M., Pilotto L. A. S., Ping W. W., et al., 1992, "The behavior of several HVDC links terminating in the same load area", Paper 14-201,CIGRE General Session 1992, Group 14, Paris, France.
2. Pilotto L. A. S., Szechtman M., Wey A., et al., 1995, "Synchronizing and Damping Torque Modulation Controllers for Multi-infeed HVDC Systems", IEEE Trans. on Power Delivery, 10, 1505-1513
3. Hammons T. J., Yeo R. L., Gwee C. L., et al., 2000, "Enhancement of power system transient response by control of HVDC converter power", Electric Machines and Power Systems, 28, 219-241
4. Szechtman M., et al., 1991, "First benchmark model for HVDC control studies, Electra, 135, 54-67
5. Wess T. and Ring H., 1988, "FHG controls for the HVDC Benchmark model study", FGH report presented to Cigre WG 14.02
6. IEEE Committee, 1995, "Standard load models for power flow and dynamic performance simulation", IEEE Trans. on Power Sys., 10, 1302-1313
7. Pai M. A., 1979, "Computer techniques in power system analysis", Tata McGraw-Hill Publishing Company Limited , NewDelhi, India
8. Kimbark W., 1970, "Direct current transmission", Volume 1. John Wiley, New York, USA
9. Domijan A. and Song Z., 2000, "Incorporation of hierarchical control with FACTS technologies in power system", Int.J. Power and Energy Systems, 20, 20-25
10. Ngan H.W., David A.K. and Lo K.L, 1992, "Decentralized hierarchical optimal control of dynamic instability in AC/DC power systems", Electr. Power Energy Syst., 14, 358-363
11. Rubaai A., 1991, "Transient stability control: a multi-level hierarchical approach", IEEE Trans on Power Sys., 6, 262-268
12. Anderson P.M. and Fouad A.A., 1977, "Power system control and stability (Volume I)", The Iowa State University Press, USA
13. Lei X. Z., Lerch E. and Povh D., et al., 1997, "Optimization-a new tool in a simulation program system", IEEE Trans on Po wer Sys., 12, 598-604

DEVELOPMENT OF A DYNAMIC POWER SYSTEM LOAD MODEL

F T Dai J V Milanović N Jenkins

V Roberts

UMIST   UK

EA Technology Ltd. UK

Abstract: The paper addresses the issue of measurement based power system load model development. The majority of power system loads respond dynamically to voltage disturbances and such contribute to overall system dynamics. The induction motors represent major portion of system loads that exhibit dynamic behaviour following the disturbance. In this paper dynamic behaviours of an induction motor and a combination of induction motor and static load were investigated under different disturbances and operating conditions in the laboratory. A first order generic dynamic load model is developed based on the test results. The model proposed is in a transfer function form and it is suitable for direct inclusion in the existing power system stability software. The robustness of the proposed model is also assessed.

Key words: load model, system identification, induction motor, power system stability.

INTRODUCTION

The dynamic response of power system loads has long been recognised as an important factor contributing to the overall power system dynamic stability. Reliable power system load models are therefore, essential for a good understanding of power system dynamic behaviour[1,2].

Power system load model can be considered as a set of mathematical equations that describe the relationship between the real and reactive power at a given bus bar in the system and the voltage and frequency at the same bus bar. If this relationship is described using algebraic equations the model is said to be static if however, the differential equations are used the model is dynamic. Developed load models (static or dynamic) are incorporated in the software packages, which enable the system planners and operators to evaluate power system response to a wide range of disturbances under different operating conditions and to operate the system in the most secure manner.

A large portion of power system loads consists of induction motors (IM) which respond dynamically to system disturbances. The tendency therefore, is to develop dynamic load models as they describe more closely actual load behaviour. Several dynamic load models have been proposed and used in the past in small-disturbance, transient stability and voltage stability studies[3,4]. The influence of the power system operating condition and the measurement method on

load model parameters was also analysed for different load models [5,6].

For development of power system load model two approaches are generally used. The component-based approach and the measurement based approach.

The component-based approach builds up the model based on the prior knowledge of the composition of loads and the corresponding specific models of its main components. The difficulty of this approach for a large utility is the collection of statistics information of load components.

In the measurement-based approach, the dynamic process of aggregate load is recorded during a disturbance by the disturbance-recording device installed at the load bus, and load model is further determined either on-line or off-line[7,8,9]. In this case the prior knowledge of actual load composition is not needed. The main difficulty associated with measurement based approach is that it is generally confined to small disturbances that are allowed to be performed in actual power system. The models developed from the results of measurements of the load responses to those small disturbances are strictly valid for small disturbance studies. Alternatively, load models can be extracted from the results of permanent monitoring of system performance following various naturally occurring disturbances. In this case however, more advanced system identification and signal processing techniques are needed.

In this study measurement based approach was used for load model development. The aim of the study was to investigate the dynamic behaviour of the IM load and a combination of static load and IM under different voltage disturbances and to propose a generic low-order dynamic load model that can be used for power system stability studies. A generic model for dynamic load is developed using numerous tests performed in the laboratory. The model is validated using Matlab / Simulink environment and the robustness of the developed model is discussed.

LABORATORY TEST SYSTEM

The block diagram of the test rig used for measurement of load responses to various disturbances is shown in Fig. 1. Three phase voltages and three line currents were taken from the voltage transformer (VT) and the current transformer (CT) respectively, and input to the PC-based measurement system. The block diagram of the measurement system used is shown in Fig. 2. Voltages

AC-DC Power Transmission, 28-30 November 2001
Conference Publication No. 485 © IEE 2001

and currents were sampled continuously at a sampling frequency of 1KHz. When a voltage step occurs, the system triggers automatically to record the three-phase voltages and currents for a certain period of time. The recorded data are then processed off-line using Matlab to calculate the RMS value of voltage (V), real power (P) and reactive power (Q). These values are then used for model development in Matlab / Simulink environment and for further off-line analysis.

The induction motor used in tests was a standard three phase, squirrel-cage machine of the following ratings: 5.5kW, 415V, 50Hz, and 1450rpm.

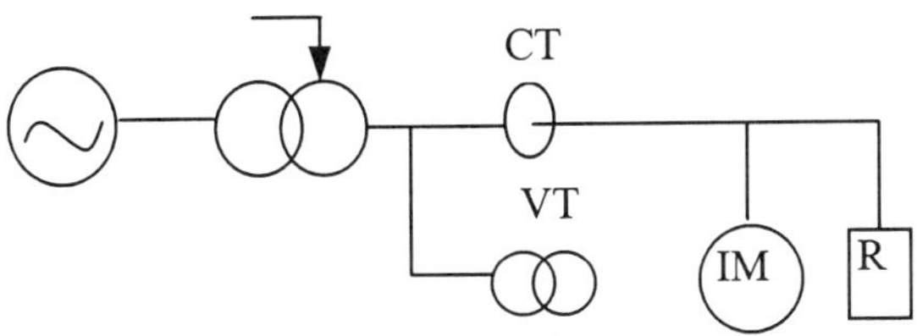

Fig. 1 Single line diagram of the laboratory test rig set up

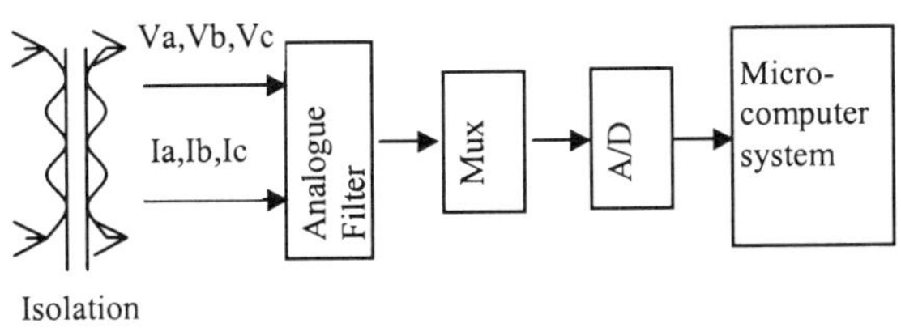

Fig. 2 Diagram of the PC-based measurement system for load model development

CASE STUDIES CONSIDERED AND TEST RESULTS

The responses of the IM and a combination of IM and static load to voltage step were studied under different operating conditions and using different voltage steps. The following case studies were considered: 1) Three different ways of producing voltage step at load terminals were used. 2) The dynamic behaviour of IM load is compared with the behaviour of a combination of IM and a static load. 3) The influence of the source impedance on the load response was also studied by inserting source resistors between the voltage source and the load. 4) The dynamic recovery processes of the induction motors with different inertia were compared. 5) The dynamic behaviour of differently loaded induction motor was studied. 6) Finally, the responses of the dynamic load were analysed for the voltage steps of different magnitude.

IM Responses for Differently Produced Voltage Steps

Fig.3 shows the responses of voltage and the real and reactive power of the IM for different ways of producing voltage step at load terminals. The methods used for producing voltage step were: 1) Tripping out one of the two transformers running in parallel. 2) Inserting the resistance (R) between transformer and mains. 3) Tapping down transformer that supplies the load.

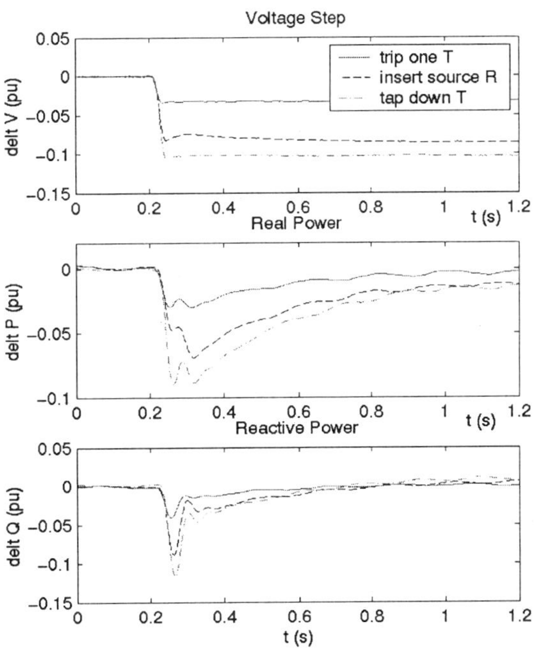

Fig. 3 Load responses obtained by using different methods for producing voltage steps

From Fig. 3 it can be seen that tripping out one of the transformers and tapping down the transformer makes qualitatively no big difference to the voltage and power responses. Those voltage steps were of approximately the same magnitude. The insertion of the resistance however, results in smaller voltage step due to the smaller voltage drop on the resistance. The immediate drop of the real and reactive power (transient responses) is also reduced though, the steady state values subsequently reached are almost the same. The power responses following the insertion of R show more oscillatory behaviour during the initial 200 ms.

Responses of Different Load

The comparisons of the responses of pure IM load and a combination of IM/static load are presented in Fig.4 for the case of 10% voltage step produced by taping down single transformer. From Fig. 4 it can be seen that the addition of static load component has no influences on the dynamic behaviour of the whole load. The static component of the load (in this case set of resistors) contributes to different steady state characteristic of the real power. Transient characteristics and the recovery process of both, P and Q are not influenced by load static component.

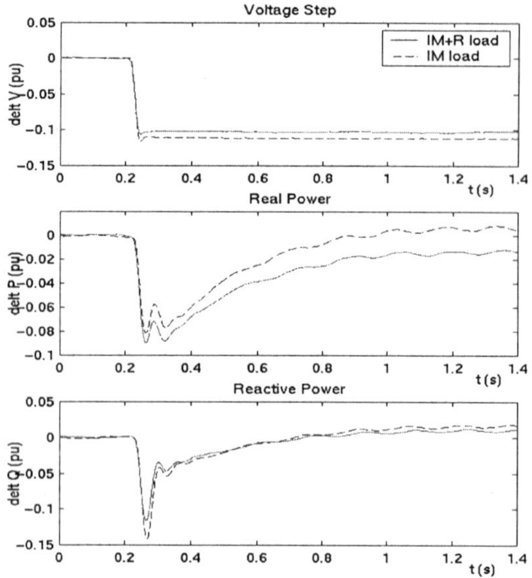

Fig. 4 Load responses of pure IM load and a combination of IM/static load.

Load Responses with Different Source Resistance

The simulation results obtained previously using PSCAD/EMTDC have shown that the network resistance influences significantly the load response. The IM responses to voltage step with and without resistance inserted in front of the transformer are therefore also compared. The results of these measurements are shown in Fig. 5.

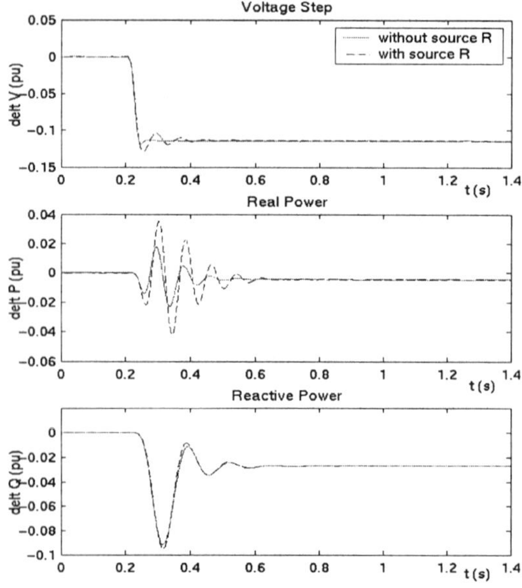

Fig. 5 IM responses with and without source resistance

The induction motor used for this test was half loaded. It can be seen that the existence of the source resistance affects the real power recovery by reducing magnitude of oscillations. The reactive power response however, is not influenced at all by the presence of additional resistance in the network.

Responses of Induction Motor with Different Inertia

In order to investigate the influence of the inertia of the IM (and such indirectly the size of the motor) on the dynamic recovery processes two different cases were considered. In the first case the induction motor operated in the open circuit condition. In the second case the IM was coupled with the generator (operating in open circuit, non-excited condition). In this case the equivalent inertia was significantly increased. The measured responses of the IM to voltage step for these two cases are shown in Fig. 6. It can be seen that the dynamic recovery process is much more oscillatory if the inertia is small whereas, it becomes of the 1st order form (exponential recovery) when the inertia is bigger. The influence of the inertia is much more evident in the case of the real power recovery.

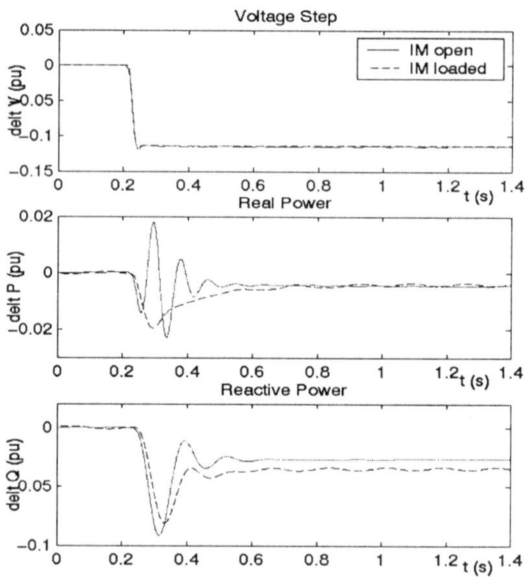

Fig. 6 Responses of the IM with different inertia

Responses of Differently Loaded IM

To assess the influence of the IM loading on the recovery process the IM was coupled with the synchronous generator (SG) and the following tests were carried out. Case 1: Open circuited and not excited SG - the measured IM current was 9% of the rated value. Case 2: Open-circuited and excited SG - the measured IM current was 12% of the rated value. Case 3: Half loaded SG - the measured IM current was 44% of the rated value. The measured load responses to voltage step are shown in Fig. 7.

It can be seen from this figure that the complete dynamic recovery process of the real power is significantly influenced by the loading condition. In the case of the reactive power the immediate drop (transient response) is not significantly affected by the loading of the motor however, the recovery process after the immediate drop shows the same, large dependency on loading as in the case of the real power.

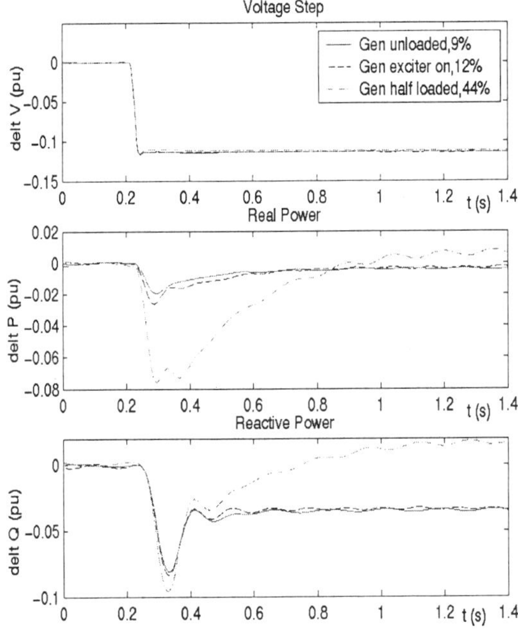

Fig. 7 IM responses for different loading

Load Responses to Voltage Steps of Different Magnitude

The dynamic behaviour of the IM was also investigated following voltage steps of different magnitude. The test results of this experiment are shown in Fig. 8. It can be seen that the magnitude of the voltage steps affects not only the new steady state but also the transient recovery process. The steady state values reached by the real power following the voltage steps of different magnitude are very similar except for voltage step of 30%. The transient characteristics and the duration of recovery however, increase with the increase in voltage step magnitude. In the case of reactive power, the transient behaviour is the same as in the case of the real power. In this case however, the new steady state values are also significantly influenced by the voltage step magnitude. Qualitatively however, the responses of both real and reactive power are not generally significantly influenced by the voltage step magnitude. This means that possibly the same form of the model (with different parameters though) could be used for modelling those responses.

GENERIC LOAD MODEL DEVELOPMENT

Following the numerous tests performed and documented in the previous section a load model was developed with the aim to be used in power system stability studies. Due to the space limitation full load model development procedure is omitted here and only a brief overview is given. The full model development procedure can be found in [10].

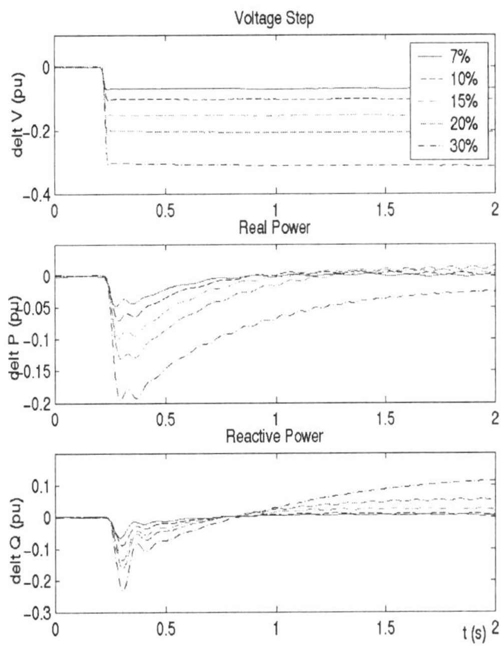

Fig. 8 Test results for different magnitudes of voltage steps

The first order input-output transfer function form of the load model is adopted in this study. The mathematical description of the model is given by Eqn. (1).

$$\Delta P = \frac{b_1 s + b_0}{s + a_0} \Delta V$$
$$\Delta Q = \frac{b_1 s + b_0}{s + a_0} \Delta V \tag{1}$$

Once the model structure is set, the next step is to estimate the unknown parameters of the model using system identification method. Model parameter estimation is accomplished in discrete time domain. In discrete time-domain, the system is described using the difference equations rather than differential equations. The basic model has the following form:

$$y(t) + a_1 y(t\text{-}1) + \ldots + a_{na} y(t\text{-}na) = b_1 u(t\text{-}nk) + \ldots + b_{nb} u(t\text{-}nk\text{-}nb+1) \tag{2}$$

where y(t) is the current value and y(t-k) and u(t-k) are past output and input respectively.

The parameters of the model were estimated using the least square method and then transformed to the continuous time domain to get the familiar transfer function form of the model given by Eqn. (1).

Model Parameters and Model Validation

The voltage, real and reactive power are used as input and output for load model development in Matlab. The responses obtained with different voltage step sizes (Fig .8) were used for parameter identification. From the responses of real power and reactive power it can be seen that the recovery process can be approximated by 1st order form. The estimated parameters of the load models for real and reactive power are given in Table 1 for different magnitudes of voltage step.

TABLE 1 - Parameters for 1st order form of model

		b_1	b_0	A_0
P	Case 1(7%)	0.816	-0.09	4.7
	Case 2(10%)	0.824	-0.193	4.084
	Case 3(15%)	0.774	-0.21	3.266
	Case 4(20%)	0.758	-0.183	2.547
	Case 5(30%)	0.7	0.1274	2.087
Q	Case 1(7%)	0.68	-0.318	4.704
	Case 2(10%)	0.678	-0.35	4.09
	Case 3(15%)	0.676	-0.536	3.25
	Case 4(20%)	0.616	-0.634	2.39
	Case 5(30%)	0.565	-0.65	1.532

The measured and simulated results of the 1st order model for two different voltage step magnitudes are compared in Fig. 9 for the real power. It can be seen that the fit is quite good.

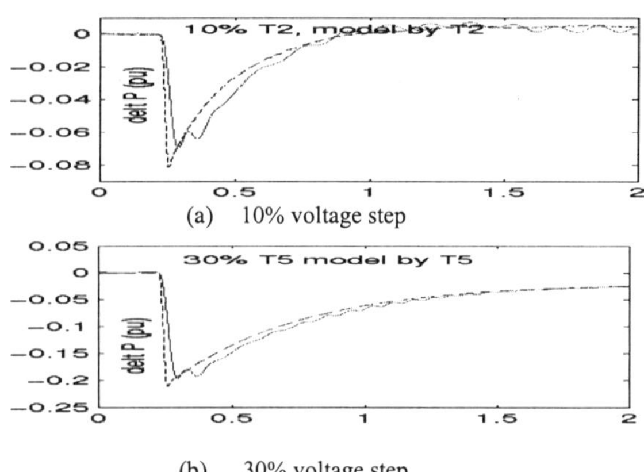

(a)   10% voltage step

(b)   30% voltage step

Fig. 9 Comparison of measured and simulated results using 1st order load model

It can be seen from Table 1 that the model parameters change with the changes of voltage steps. These changes are shown for both real and reactive power in Fig. 10. From Fig. 11(b) it can be seen that the parameter b_0 even changes sign from "-" to "+" when the voltage step is above certain level.

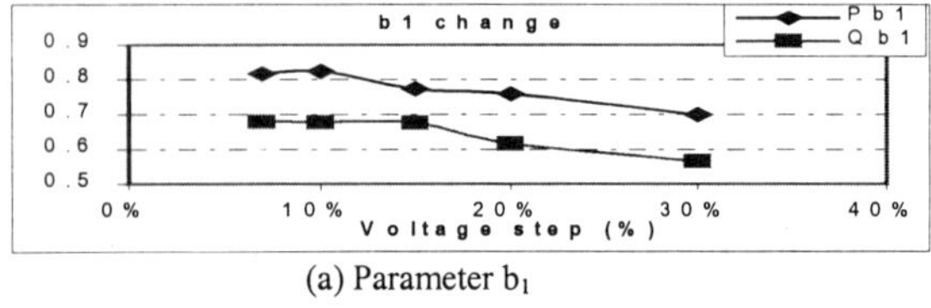

(a) Parameter b_1

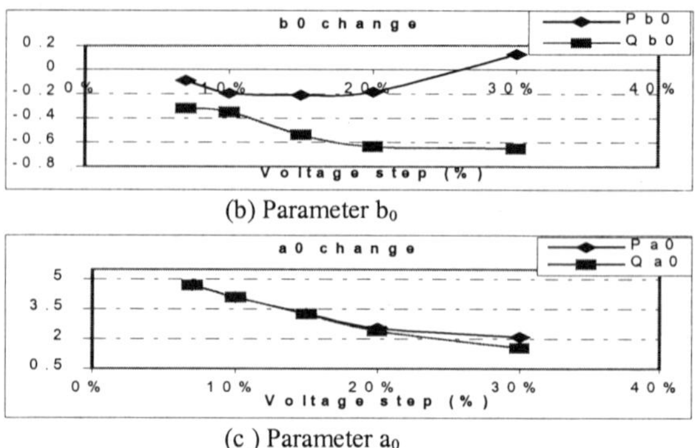

(b) Parameter b_0

(c) Parameter a_0

Fig. 10 Parameter variation with the voltage step

Steady state characteristics. Characteristics of the steady state and the transient state of the load for voltage steps of different magnitudes were analysed using the load model. The steady state characteristics of the load can be calculated using the following simplified expression based on the load model.

$$\frac{\Delta P(\infty)}{\Delta U(\infty)} = \underset{s \to 0}{Lim}\, G(s) = \underset{s \to 0}{Lim}\, \frac{b_1 s + b_0}{s + a_0} = b_0 / a_0 \quad (3)$$

The steady state characteristics for the real power are shown in Fig. 11.

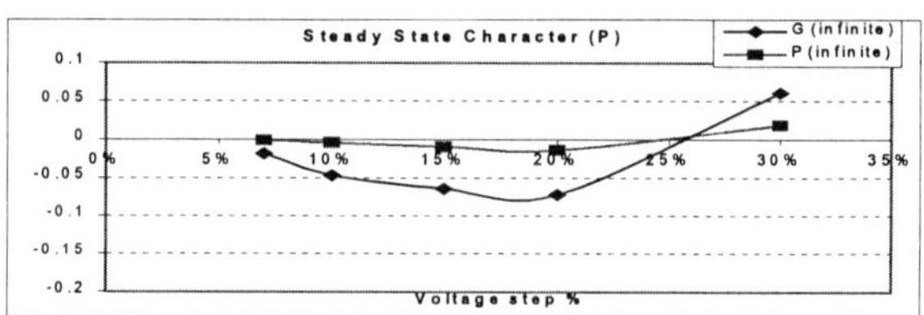

Fig. 11 Steady state characteristics for real power

It can be seen from the figure that though $G(\infty)$ is non-linear function of voltage step, the change in real power ΔP as the function of voltage step can be approximated with linear function. This approximation is valid for voltage steps up to 20%. If however, voltage step increases the assumption of linearity becomes non valid.

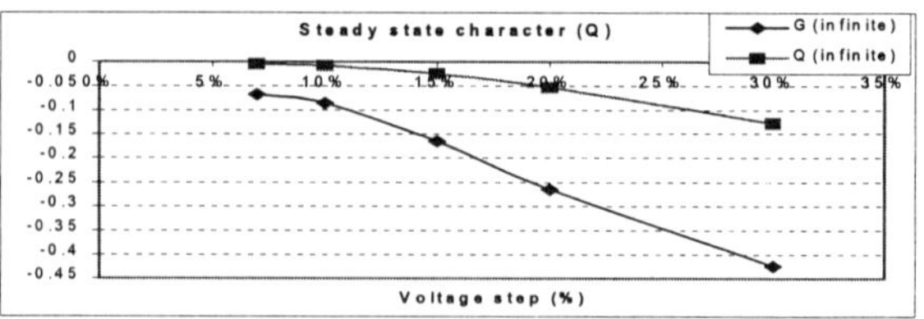

Fig. 12 Steady state characteristics for reactive power

Fig. 12 shows the steady state characteristics of the reactive power. The reactive power transfer function $G(\infty)$ is also non-linear function of voltage step for the steady state behaviour. The change in reactive power ΔQ however still can be approximated with a linear function for small voltage steps.

Transient characteristics. The transient characteristics of the load can be calculated using the following simplified expression based on the load model :

$$\frac{\Delta P(0_+)}{\Delta U(0_+)} = \underset{s \to \infty}{Lim}\, G(s) = \underset{s \to \infty}{Lim}\, \frac{b_1 s + b_0}{s + a_0} = b_1 \qquad (4)$$

Transient characteristics for real power are depicted in Fig. 13 and for reactive power in Fig. 14.

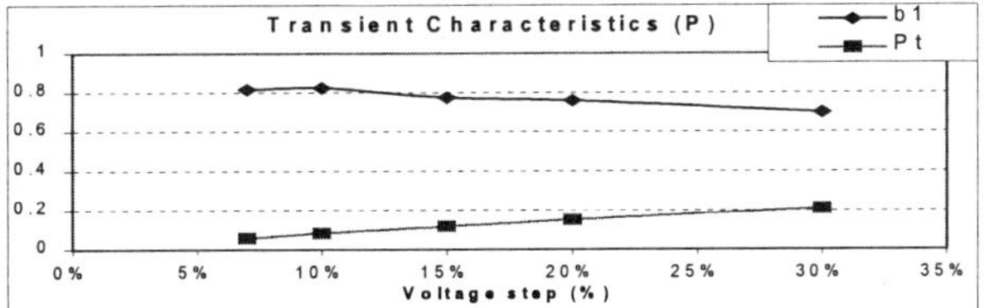

Fig. 13 Transient characteristics for real power

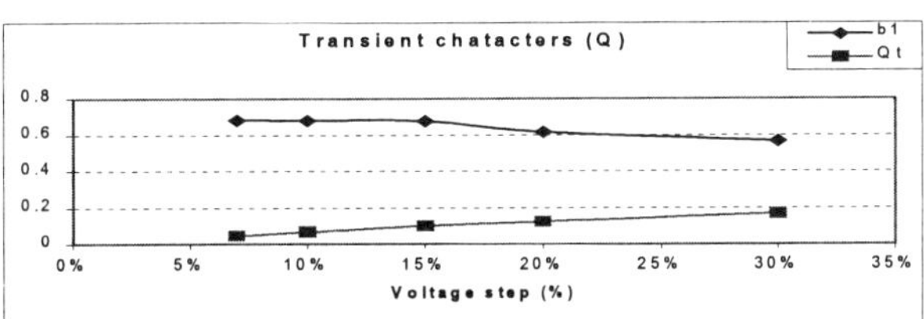

Fig. 14 Transient characteristics for reactive power

It can be seen from these figures that the transient characteristics for both, P and Q are linear functions of voltage step.

The performed analysis shows that for voltage steps of up to 20% the linear dependence between power responses and model parameters can be assumed and such load modelling simplified in terms of reduced requirement for parameter identification for different voltage steps.

CONCLUSIONS

The paper presented results of extensive laboratory testing of induction motor load and a combination of induction motor and static load. Based on the experimental results a generic dynamic load model in the form of the first order transfer function is proposed. The model proposed can be directly incorporated in existing software for power system stability studies. The sensitivity of load model parameters to size of voltage step is also analysed. Based on the analysis performed the following general conclusions are reached:

- Static load component contributes only to the steady state characteristics of the load.
- Source impedance has big effects on the dynamic recovery process of real power of induction motor if it is light loaded.
- Inertia of induction motor affects the dynamic recovery process significantly. The recovery process of the real power can be approximated using the 1st order model if the inertia is large. Otherwise, higher order models are needed.

- Dynamic recovery process is affected by IM loading.
- Because of the non-linearity of steady state and transient characteristics of dynamic loads model developed from a small disturbance tests can only be used for voltage step sizes bellow 20%.
- The 1st order dynamic model can be used for modelling sufficiently well both, real and reactive power recovery for reasonably wide range of operating conditions and voltage step sizes.

ACKNOWLEDGEMENT

The work on this project is funded by the EPSRC grant GR/M38179 and EA Technology Ltd..

REFERENCES

[1] IEEE Task Force on Load Representation for Dynamic Performance, 1993, " Load Representation for Dynamic Performance Analysis", IEEE Transactions on Power Systems, Vol. 8, No. 2, pp.472-482

[2] IEEE Task Force on Load Representation for Dynamic Performance, 1994, "Standard Load Models for Power Flow and Dynamic Performance Simulation" *94 SM 579-3 PWRS*

[3] J.V.Milanovic and I.A.Hiskens, 1995, "Effects of Load Dynamics on Power System Damping", IEEE Transactions on Power Systems, Vol. 10, No. 2, pp. 1022-1028.

[4] D.JHill, 1993, "Nonlinear dynamic load models with recovery for voltage stability studies", IEEE Trans. Power Systems, Vol. 8 No 1, pp 166-176.

[5] W.Xu and Y.Mansour, 1994, ``Voltage stability analysis using generic dynamic load models", IEEE Trans. Power Systems, Vol. 9, No. 1, pp 479-493.

[6] F.T.Dai, J.V.Milanovic, N.Jenkins and V.Roberts, 2001, "The Influence of Voltage Variations on Estimated Load Model Parameters" 16th International Conference & Exhibition on Electricity Distribution, CIRED2001, Amsterdam, Netherlands, Paper 2.33.

[7] F.T. Dai, J.V. Milanovic and N.Jenkins, 2000, "The Influence Of Measurement Delays on Estimated Load Model Parameters", International Conference on Harmonics and Quality of Power, ICHQP 2000, Orlando, Florida, USA, pp. 324-328.

[8] Task Force 02.05 of Study committee 38(CIGRE), "Load Modelling and Dynamics" ELECTRA No. 130. pp. 124 – 141

[9] Chia-Jen Lin, Yung-Tien Chen, etc. 1993, "Dynamic Load Models in Power Systems Using the Measurement Approach" IEEE Transactions on Power Systems, Vol. 8, No. 1, pp. 309

[10] F.T.Dai, J.V.Milanovic, N.Jenkins and V.Roberts, 2001, "Load Model Development Based on Measured Data" to be presented at UPEC 2001, 12-14 September 2001, Swansea, UK

INFLUENCE OF FACTS ON POWER SYSTEM VOLTAGE STABILITY

Mukhedkar R.A., Davies T.S., Nouri H.

University of the West of England, Bristol, United Kingdom

ABSTRACT

One of the major causes of voltage collapse is the reactive power limit of the system. Improving the system's reactive power handling capacity via Flexible Alternating Current Transmission System (FACTS) devices is a remedy for prevention of voltage instability and hence voltage collapse. This paper discusses the influence of a multilevel STATCOM employed in an EMTDC simulation network and its associated mathematical model. The paper then discusses the effects of STATCOM on some of the network's online voltage collapse indices.

INTRODUCTION

Electricity is a vital commodity considering the role it plays in the day to day functioning of modern life. Electrical energy cannot be stored in large quantities and thus continuity of supply has a much higher value than the energy consumed. An increase in system interconnections due to demand without expansion of the existing transmission and generation infrastructure to some degree is matched with developments and advances in power electronics technology in an attempt for the system to operate close to the steady state margin at a moderate cost. Koessler (1) indicates a lack of reactive power as one of the causes of voltage instability and eventual collapse. Vargas et al (2), have pointed out the need for timely and local reactive power compensation to avert the situation of voltage collapse and Flexible Alternating Current Transmission Systems (FACTS) devices can be used to achieve this.

The usage of modern power electronics and advanced control systems to manipulate all fundamental characteristics of the power system such as: power flow, series and shunt impedance of transmission line, network topology and voltage profile, termed as FACTS was first introduced by EPRI in 1986.

This paper discusses the incorporation of a multilevel STATCOM, initiated by Peng et al (3) in EMTDC and its mathematical model. The STATCOM is controlled to provide voltage support at the point of connection. Reactive compensation of the transmission line increases the quantity of transmitted power and

consequently reduces the margin separating the normal operating point from the point of collapse. This paper also investigates the effect of introducing STATCOM into the system on some online voltage collapse indices.

POWER SYSTEM MODEL

A typical four bus interconnected system as shown in the figure 1 is considered for this study. The system is symmetrical in terms of transmission, generation and load. The simulated system consists of Thevenin sources with internal impedance of 0.1014 pu and no PV control at bus 1 and 3 and variable impedance load of 0.5 power factor at bus 2 and 4. The initial resistance is 2.5 pu, reactance is 1.443 pu and the reduction step is 0.225 pu every 2.5 second. Effects due to the introduction of FACTS in power systems follow Mukhedkar et al (4) previously published work on the online voltage collapse indices.

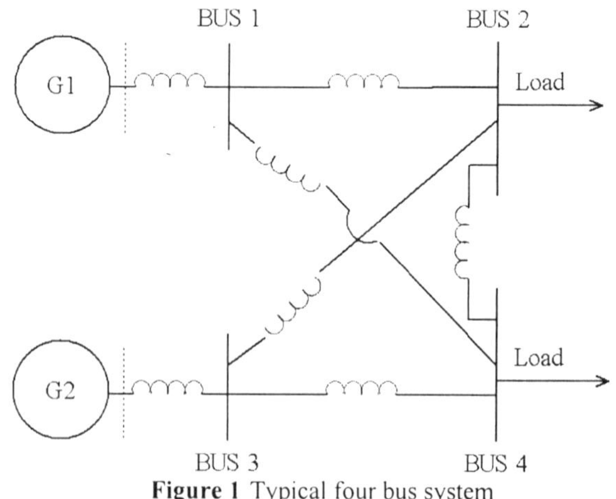

Figure 1 Typical four bus system

The indices relative to which the performance of the power system is studied are as follows:
i. The voltage existence index suggested by Kessel et al (5)

$$L_j = \left| 1 + \frac{U_{Oj}}{U_j} \right| \qquad \text{--- (1)}$$

This index is modified to accommodate the assumption of the source voltage being constant at 1 pu.

AC-DC Power Transmission, 28-30 November 2001
Conference Publication No. 485 © IEE 2001

ii. The line apparent power loadability index derived in (4) based on the line real and reactive power loadability index suggested by Moghavvemi et al (6)

$$L_S = \frac{(2ZSin(2\theta)S_R)}{\left[U_S^2 \sqrt{Sin^2(\theta)Cos(2(\theta + \delta_R - \delta_S)) + Sin^4(\theta + \delta_R - \delta_S)}\right]}$$

--- (2)

iii. Transmission line loss sensitivity index

$$\frac{\partial Q_L}{\partial U_R} = -2Sin(\theta)\left|\frac{S_R^*}{U_R^*}\right|Cos(\theta - \phi)$$ --- (3)

$$Q_L = (Q_S - Q_R)$$

This is valid for lines with zero susceptance.

where

U_S, P_S, Q_S - Sending end voltage, real and reactive power

U_R, P_R, Q_R - Receiving end voltage, real and reactive power

S_R - Receiving end apparent power

$\underline{U}_{Oj}$ - Bus equivalent voltage

$\underline{U}_j$ - Voltage at load bus j

R, X - Transmission line impedance

θ - Transmission line impedance angle

δ - Angle between the sending and the receiving end voltage

ϕ - Load power factor angle

STATCOM MODEL

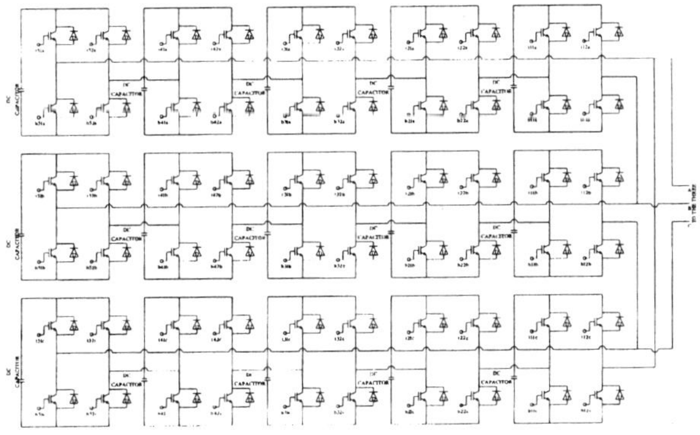

Figure 2 Five level three-phase delta connected STATCOM

The STATCOM consists of a voltage source converter (VSC) as the building block. Five VSC modules are connected in series to form a single phase unit, and three such units are connected in delta to form a three phase delta connected STATCOM as shown in Figure 2. The delta configuration is selected since the capacitive reactance is 3 times that required in star configuration. Also in the event of unbalanced

operation the circulating currents will not enter the power system.

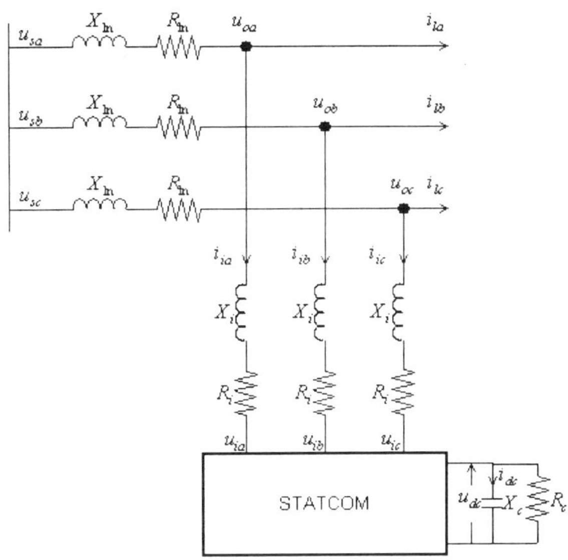

Figure 3 a. STATCOM equivalent circuit

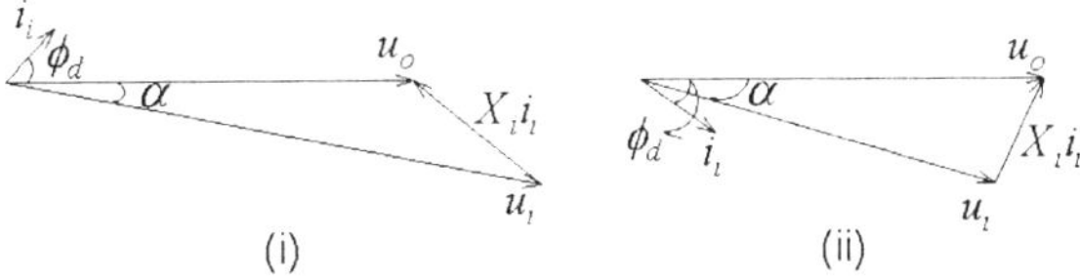

Figure 3 b. Phasor diagrams

The exchange of real and reactive power between the power system and the STATCOM depends on the magnitude and the angle of the STATCOM output voltage with respect to the power system bus voltage as clear from the phasor diagram in figure 3b.

The value of the STATCOM dc link capacitor depends on the reactive power support intended and the allowable fluctuations in the dc link voltage, and is given by

$$C_j = \frac{\Delta Q_j}{\Delta U_{dc}} = \frac{\int_{\theta_j(t)}^{T/4} \sqrt{2}I_c Cos\omega t\, dt}{2\varepsilon U_{dc}} = \frac{\sqrt{2}I_c(1 - Sin\theta_j)}{2\omega\varepsilon U_{dc}}$$ --- (4)

where

C_j - Capacitance on level j

ΔQ_j - Charge on the capacitor in a quarter cycle

ΔU_{dc} - Change in the dc link voltage

I_c - Rated VSC current

θ_j - Switching angle

ε - Allowed variation in dc link voltage

ω - $2\pi f$ where f is the system frequency

The converters are switched to obtain a staircase waveform very close to the sinusoidal waveform as shown in the figure 4. Considering the quarter wave symmetry of the waveform all the even harmonics are excluded, and the triplen harmonics are eliminated by the virtue of the system being a balanced three-phase system. Thus elimination technique is used to eliminate all odd harmonics upto 13. This is limited only by the number of VSC connected in series. Generalised harmonic elimination technique suggested by Patel et al (8) is used to obtain switching sequence for different modulation indices. Thus the technique suitable for programmed PWM switching is successfully extended

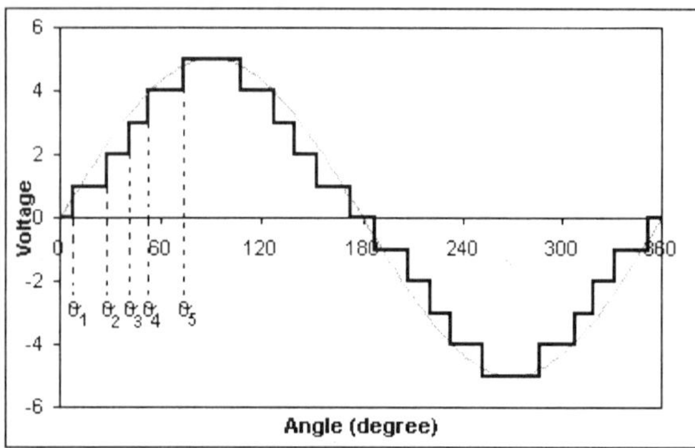

Figure 4 Stepped waveform

for staircase wave formation. Newton's algorithm is used to numerically solve the non-linear equations to obtain the triggering angles as follows

$$F(\theta) = T \qquad \qquad ---(5)$$

$$F^j + \left[\frac{\partial F}{\partial \theta}\right]^j d\theta^j = T$$

$$d\theta^j = inv\left[\frac{\partial F}{\partial \theta}\right]^j \left[T - F^j\right] \qquad ---(6)$$

where

$$F^j = \begin{bmatrix} Cos\theta_1 + Cos\theta_2 + Cos\theta_3 + Cos\theta_4 + Cos\theta_5 \\ Cos(5\theta_1) + Cos(5\theta_2) + Cos(5\theta_3) + Cos(5\theta_4) + Cos(5\theta_5) \\ Cos(7\theta_1) + Cos(7\theta_2) + Cos(7\theta_3) + Cos(7\theta_4) + Cos(7\theta_5) \\ Cos(11\theta_1) + Cos(11\theta_2) + Cos(11\theta_3) + Cos(11\theta_4) + Cos(11\theta_5) \\ Cos(13\theta_1) + Cos(13\theta_2) + Cos(13\theta_3) + Cos(13\theta_4) + Cos(13\theta_5) \end{bmatrix}$$

$$\left[\frac{\partial F}{\partial \theta}\right]^j = -\begin{bmatrix} Sin\theta_1 & Sin\theta_2 & Sin\theta_3 & Sin\theta_4 & Sin\theta_5 \\ 5Sin(5\theta_1) & 5Sin(5\theta_2) & 5Sin(5\theta_3) & 5Sin(5\theta_4) & 5Sin(5\theta_5) \\ 7Sin(7\theta_1) & 7Sin(7\theta_2) & 7Sin(7\theta_3) & 7Sin(7\theta_4) & 7Sin(7\theta_5) \\ 11Sin(11\theta_1) & 11Sin(11\theta_2) & 11Sin(11\theta_3) & 11Sin(11\theta_4) & 11Sin(11\theta_5) \\ 13Sin(13\theta_1) & 13Sin(13\theta_2) & 13Sin(13\theta_3) & 13Sin(13\theta_4) & 13Sin(13\theta_5) \end{bmatrix}$$

$$T = \begin{bmatrix} m\dfrac{5\pi}{4} & 0 & 0 & 0 & 0 \end{bmatrix}^T$$

m - Modulation index $= \dfrac{u_i}{u_{dc}}$

The selected angles for simulation are 0.134, 0.481, 0.712, 0.917, 1.275 radians.

CONTROL SYSTEM MODEL

The well established state equation for STATCOM in the rotating dq reference frame is as follows, Papic (9) and Schauder et al (10)

$$p\begin{bmatrix} i_{id} \\ i_{iq} \\ u_{dc} \end{bmatrix} = A\begin{bmatrix} i_{id} \\ i_{iq} \\ u_{dc} \end{bmatrix} - \frac{\omega_b}{X_i}\begin{bmatrix} u_{od} \\ u_{oq} \end{bmatrix} \qquad ---(7)$$

where

$$A = \begin{bmatrix} -\dfrac{R_i\omega_b}{X_i} & \omega & \dfrac{m_d\omega_b}{X_i} \\ -\omega & -\dfrac{R_i\omega_b}{X_i} & \dfrac{m_q\omega_b}{X_i} \\ -\dfrac{3}{2}\omega_b X_c m_d & -\dfrac{3}{2}\omega_b X_c m_q & -\dfrac{\omega_b X_c}{R_c} \end{bmatrix}$$

$m_d = mCos\alpha$

$m_q = mSin\alpha$

α - phase angle between u_i and u_o (rad)

u_i, u_o - STATCOM internal and external terminal voltage (pu)

R_i, X_i - Interface resistance and reactance (pu)

R_c, X_c - Parallel resistance and capacitive reactance (pu)

i_i - STATCOM current on AC side (pu)

u_{dc} - DC link voltage

ω_b, ω - Base and actual angular speed (rad/sec)

d, q   - Subscript for direct and quadrature axis component

There are two adjustable parameters and three state variables, hence only two state variables can be independently controlled. The mode of operation adopted in this paper uses phase shift "α" as the control parameter and the modulation index "m" is kept fixed. Thus controlling the dc link voltage indirectly controls the output ac voltage and the reactive power. Thus linearising the equation with "α" as an input variable.

$$p\begin{bmatrix} \Delta i_{id} \\ \Delta i_{iq} \\ \Delta u_{dc} \end{bmatrix} = A_\Delta\begin{bmatrix} \Delta i_{id} \\ \Delta i_{iq} \\ \Delta u_{dc} \end{bmatrix} - B_\Delta\begin{bmatrix} \Delta u_{od} \\ \Delta u_{oq} \\ \Delta\alpha \end{bmatrix}$$

where

$A_\Delta = A$

$$B_\Delta = \begin{bmatrix} -\dfrac{\omega_b}{X_i} & 0 & -\dfrac{m\omega_b u_{dco}Sin(\alpha_o)}{X_i} \\ 0 & -\dfrac{\omega_b}{X_i} & \dfrac{m\omega_b u_{dco}Cos(\alpha_o)}{X_i} \\ 0 & 0 & \dfrac{3}{2}mX_c\omega_b\big(i_{ido}Sin(\alpha_o) - i_{iqo}Cos(\alpha_o)\big) \end{bmatrix}$$

The transfer function between the input parameter "α" and the reactive component of the output current can be determined from the above equation. The control depicted in figure 5 is incorporated in the simulated system.

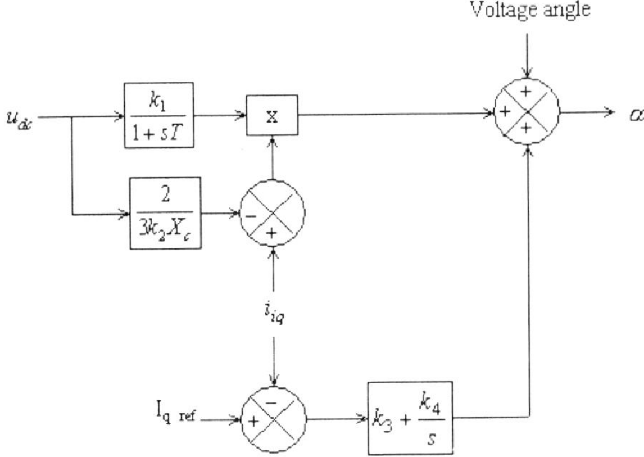

Figure 5 Control diagram

RESULTS AND DISCUSSION

The voltage at the load bus 2 and 4 are as shown in figure 6. STATCOM capacity limit is reached at time of 12.5 second, and the effect is observed as the voltage at the load bus in the compensated configuration goes below the uncompensated configuration.

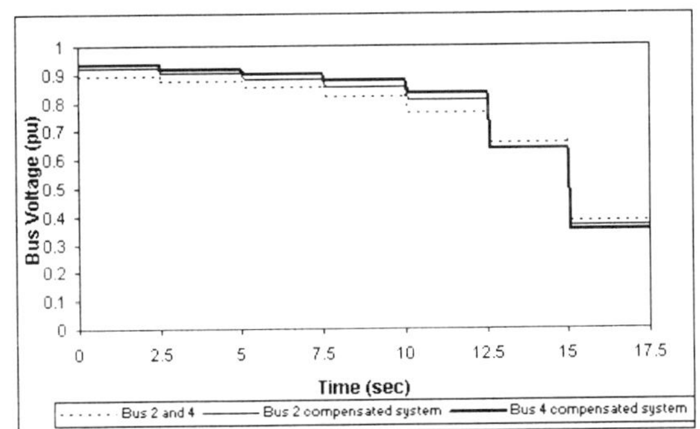

Figure 6. Load bus voltage profile

The bus load ability index L in figure 7 also reflects a similar effect. In addition it can be conclude that in the compensated system the compensated bus 4 is more voltage stable than the bus 2 until time 12.5 second

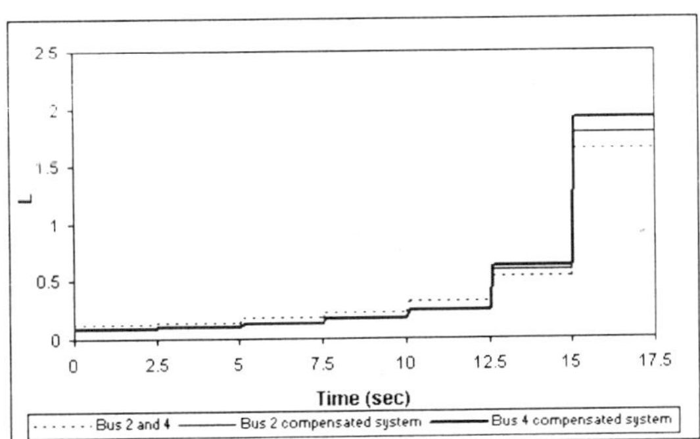

Figure 7. Load bus load ability index

after which, the STATCOM limit is reached and bus 2 is more stable than bus 4. The STATCOM now appears as an inductive load (figure 9).

The line load ability index (figure 8) when used to relate to the load bus condition, indicates that bus 2 and 4 in the uncompensated system and the bus 2 in the compensated system have same stability. The stability of the compensated bus 4 is considerably improved in the capacitive regime of operation of STATCOM, and deteriorates after the STATCOM limit is reached.

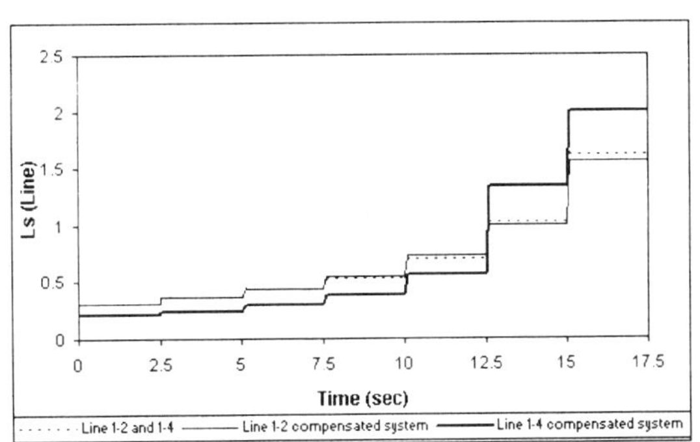

Figure 8. Transmission line loading index

The line reactive power loss sensitivity to the receiving end voltage shows similar behavior.

From the above observations, the capacity margin of the STATCOM can be considered as an indication of voltage instability or collapse. It can also be concluded, that though the change from the stable to the unstable domain for the system with compensation is more abrupt as observed from figures 7 and 8, the indices faithfully reflect the overall condition of the system

The deterioration of Bus 4 stability is attributed to the relaxed limits of the STATCOM controller (figure 5). The controller continues to follow the demand of the system as the system loading is increased. The change in the mode of operation of the STATCOM and hence the occurrence of the limit on its output is observed in the form of reduced voltage stability on Bus 4, which is clearly reflected by the indices considered (figures 7 and 8). In practice, due to limits on the STATCOM controller output, this effect will not be observed. In such a system the value of the indices for the

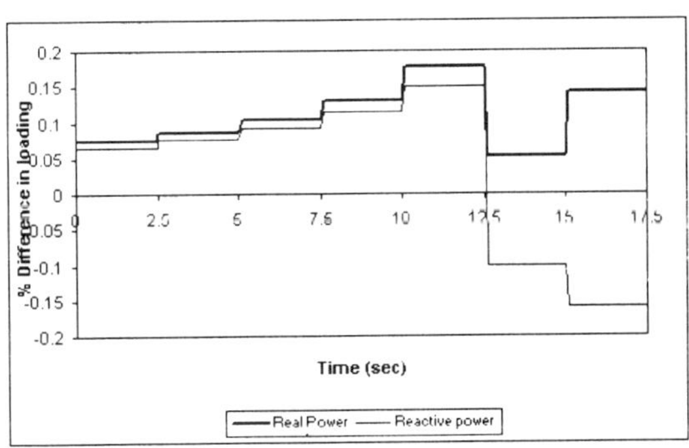

Figure 9. Percentage difference in the real and reactive power loading of the system

compensated bus will continue to remain lower than the uncompensated bus.

The percentage difference in the real and reactive power loading of the system with respect to the loading of the uncompensated system is shown in figure 9. It is observed that both the real and reactive power handling capacity of the system is improved in the compensated system in the active region.

CONCLUSION

The limit of STATCOM capacity determines the level of voltage at the load bus.

The bus load ability index suggests that the compensated bus is more voltage stable compared to adjacent buses. This effect is reversed when the STATCOM limit is reached. The line load ability index indicates that the stability of the compensated bus is considerably improved in the capacitive regime of the operation of STATCOM, and deteriorates when the STATCOM limit is reached. The index based on line reactive power loss sensitivity to the receiving end voltage follows the preceding results.

The percentage difference in the real and reactive power loading of the system shows that STATCOM improves both the real and reactive power handling capacity of the system.

Analysis of the study suggests that the capacity margin of the STATCOM can be considered as an indication of voltage instability or collapse.

The de-stabilising effect on Bus 4 of the system is related to relaxed limits of the STATCOM controller, where as in practice limits are fixed.

REFERENCES

1. Koessler, R.J, 1997, " Voltage Instability/Collapse - An Overview", IEE Colloquium on Voltage Collapse - Dig. No. 1997/101, 1/1-1/6
2. Vargas, L.S., Cañizares, C.A., 2000, "Time dependance of controls to avoid voltage collapse", IEEE Transactions on Power Systems, 15-04, 1367-1375.
3. Peng, F.Z., Lai, J.S., McKeever, J, VanCoevering, J., 1995, "A multilevel voltage-sourced inverter with separate DC sources for Static Var generation", Proceedings of PESC IEEE 26th Annual Conference, 2541-2548.
4. Mukhedkar, R.A., Nouri, H., Davies, T.S., 2001, "A comparative study and performance analysis of online static voltage collapse indices in interconnected systems", 36th Universities Power Engineering Conference.
5. Kessel, P., Glavitsch, H., 1986, "Estimating voltage stability of a power system", IEEE transactions on Power Delivery, PWDR-1-3, 346-353.
6. Moghavvemi, M., Omar, F.M., 1998, "Technique for contingency monitoring and voltage collapse prediction", IEE Proceedings - Generation, Transmission, Distribution, 145-6, 634-640.
7. Moghavvemi, M., Omar, F.M., 2001, " Technique for assessment of voltage stability in ill-conditioned radial distribution network", IEEE Power Engineering Review, 58-60.
8. Patel, H.S., Hoft, R.G., 1973, "Generalised techniques of harmonic elimination and voltage control in thyristor inverters: Part I- Harmonic elimination", IEEE Transactions on Industry Applications, IA-9-3, 310-317.
9. Papic, I., 2000, "Mathematical analysis of FACTS devices based on a voltage source converter Part 1: mathematical model", Electric Power Systems Research, 56, 139-148.
10. Schauder, C., Mehta, H., 1991, "Vector analysis of advanced static var compensators", 5th IEE AC/DC Transmission and Distribution Conference, 266-272.

INTERACTION OF THE STATCOM AND ITS ASSOCIATED TRANSFORMER NON-LINEARITY: TIME DOMAIN MODELLING AND ANALYSIS

N. Garcia   M. Madrigal   E. Acha

Department of Electronics & Electrical Engineering
University of Glasgow, Scotland, U.K.

Abstract: A comprehensive time domain model of a synchronous, three-phase, STATic COMpensator (STATCOM) is presented in this paper. The STATCOM model comprises the Voltage Source Converter (VSC), which incorporates Pulse Width Modulation (PWM) control techniques, and the coupling transformer, which includes the non-linear, magnetizing branch. The switching pattern used in this model is a unipolar PWM scheme. The coupling transformer is modelled as an equivalent circuit connected in a delta-star configuration. Besides saturation, core losses are also included in the transformer model. The model is applied to study the interactions that take place between the different components of the STATCOM, namely the VSC, the coupling transformer, the DC-link capacitor and the control system. The model is developed from first principles and the ensuing formulation is coded in C++ using a UNIX platform.

Keywords: STATCOM, VSC, PWM, transformers, adaptive control and time domain.

I. INTRODUCTION

Recent break-throughs in power electronics technology have enabled the development of a variety of sophisticated controllers used to solve long-standing technical and economic problems found in electrical power systems at both the transmission and the distribution levels. These emerging controllers are grouped under the headings of FACTS and Custom Power technology, respectively. At the transmission level, the issues are increased power transfers with no net loss of stability margins, effective power flow control and enhanced reactive power support, whereas at the distribution level the issues are total reliability and quality of power supply [1]. However, the incorporation of such novel controllers in the power network is generating a number of questions which require research work. Among these is how transformers with the ability to saturate affect the performance of such power electronic controllers. In general, the interaction of these controllers, many of which operate at high frequencies, with conventional plant equipment such as transformers, banks of capacitors and transmission lines and cables is not yet well known.

Thus far, most of the work reported on STATCOM modelling and analysis has concentrated on the associated VSC, with the coupling transformer only represented by a linear equivalent circuit. No concern has been shown for possible interactions between harmonics generated by saturated transformers and harmonics generated by VSCs. The rational behind this engineering simplification is that the transformers are designed to operate below their saturating regions. However, energization tests conducted in a full size UPFC prototype generated large inrush currents in the shunt-connected transformer after fault clearing [2]. These tests are closely related to STATCOM energization. Aiming at advancing the understanding of interactions between STATCOMs and their associated transformers, this paper presents a comprehensive STATCOM model where the magnetic non-linearity of the coupling transformer and STATCOM control are represented in detail. The control system uses a novel adaptive control scheme, which shows to be very effective in controlling the voltage in the DC-link capacitor and at the Point of Common Coupling (PCC).

II. STATCOM MODEL

The main parts of the STATCOM model shown in Fig. 1 are the three-phase delta-star transformer, VSC and the control system.

A. Transformer model

The delta-star transformer shown by Fig. 1 is modelled as a three-phase transformer bank built with three single-phase transformers. Applying Kirchhoff voltage and current laws to the circuit of Fig. 2 the following three basic equations can be obtained,

$$v = r_p i_p + l_p \frac{di_p}{dt} + v_2 \tag{1}$$

$$v_2 = a^2 r_s i_s + a^2 l_s \frac{di_s}{dt} + e_1 \tag{2}$$

$$i_p = i_c + i_m + i_s \tag{3}$$

where

$$a = \frac{N_1}{N_2} \tag{4}$$

$$e_1 = a e_2 \tag{5}$$

and the core losses can be expressed as

$$v_2 = r_c i_c \tag{6}$$

AC-DC Power Transmission, 28-30 November 2001
Conference Publication No. 485 © IEE 2001

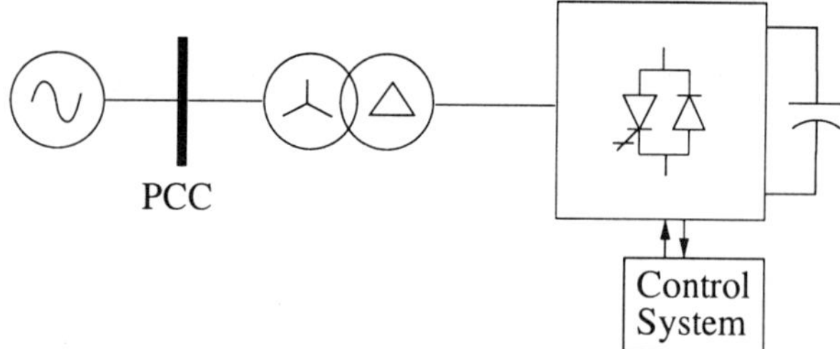

Fig. 1. Schematic diagram for the STATCOM.

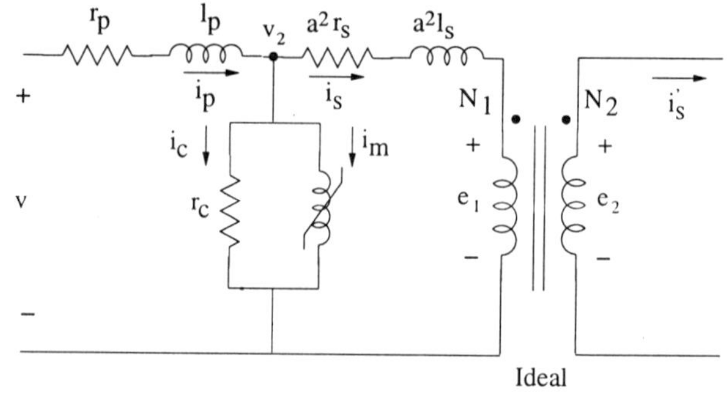

Fig. 2. Equivalent circuit for a single-phase transformer.

Solving equations (1) and (2) for i_p and i_s respectively,

$$\frac{di_p}{dt} = \frac{v - (r_p + r_c)\,i_p + r_c\,(i_s + i_m)}{l_p} \qquad (7)$$

$$\frac{di_s}{dt} = \frac{-ae_2 - (a^2 r_s + r_c)\,i_s + r_c\,(i_p - i_m)}{a^2 l_s} \qquad (8)$$

Besides, the flux linkage associated with the magnetic non-linear characteristic of the core can be expressed as

$$v_2 = \frac{d\lambda_m}{dt} \qquad (9)$$

Substituting eqs. (3) and (6) in (9), the following is obtained,

$$\frac{d\lambda_m}{dt} = r_c\,(i_p - i_m - i_s) \qquad (10)$$

Therefore, the three-phase model for the delta-star bank connected is given by

$$\frac{d}{dt}
\begin{bmatrix}
\lambda_{ap} \\ \lambda_{bp} \\ \lambda_{cp} \\ \lambda_{as} \\ \lambda_{bs} \\ \lambda_{cs} \\ \lambda_{ma} \\ \lambda_{mb} \\ \lambda_{mc}
\end{bmatrix}
= A
\begin{bmatrix}
\lambda_{ap} \\ \lambda_{bp} \\ \lambda_{cp} \\ \lambda_{as} \\ \lambda_{bs} \\ \lambda_{cs} \\ f(\lambda_{ma}) \\ f(\lambda_{mb}) \\ f(\lambda_{mc})
\end{bmatrix}
+
\begin{bmatrix}
v_a \\ v_b \\ v_c \\ -ae_{2a} \\ -ae_{2b} \\ -ae_{2c} \\ 0 \\ 0 \\ 0
\end{bmatrix}
\qquad (11)$$

where the matrix A is defined as,

and e_{2a}, e_{2b} and e_{2c} are the line voltages on the delta side of the transformer.

The single-phase transformer saturation characteristic is modelled by means of a polynomial characteristic. Fig. 3 shows the magnetization curve which is used in the test cases presented in this paper. The saturation curve is defined as

$$i = 0.7576\lambda + 0.69\lambda^7 \quad p.u. \qquad (13)$$

B. VSC model

Neglecting losses in the semiconductor switches, the VSC model may be represented by the following voltage and current relationship [3] [4],

$$\begin{bmatrix} e_a \\ e_b \\ e_c \end{bmatrix} = \begin{bmatrix} s_a \\ s_b \\ s_c \end{bmatrix} \times v_{dc} \qquad (14)$$

and

$$i_{dc} = \begin{bmatrix} i_a & i_b & i_c \end{bmatrix} \times \begin{bmatrix} s_a \\ s_b \\ s_c \end{bmatrix} \qquad (15)$$

where s_a, s_b and s_c are the switching functions that govern the VSC.

The ordinary differential equation describing the DC-link is given by

$$A =
\begin{bmatrix}
\frac{-(r_p+r_c)}{l_p} & 0 & 0 & \frac{r_c}{a^2 l_s} & 0 & 0 & r_c & 0 & 0 \\
0 & \frac{-(r_p+r_c)}{l_p} & 0 & 0 & \frac{r_c}{a^2 l_s} & 0 & 0 & r_c & 0 \\
0 & 0 & \frac{-(r_p+r_c)}{l_p} & 0 & 0 & \frac{r_c}{a^2 l_s} & 0 & 0 & r_c \\
\frac{r_c}{l_p} & 0 & 0 & \frac{-(a^2 r_s + r_c)}{a^2 l_s} & 0 & 0 & -r_c & 0 & 0 \\
0 & \frac{r_c}{l_p} & 0 & 0 & \frac{-(a^2 r_s + r_c)}{a^2 l_s} & 0 & 0 & -r_c & 0 \\
0 & 0 & \frac{r_c}{l_p} & 0 & 0 & \frac{-(a^2 r_s + r_c)}{a^2 l_s} & 0 & 0 & -r_c \\
\frac{r_c}{l_p} & 0 & 0 & \frac{-r_c}{a^2 l_s} & 0 & 0 & -r_c & 0 & 0 \\
0 & \frac{r_c}{l_p} & 0 & 0 & \frac{-r_c}{a^2 l_s} & 0 & 0 & -r_c & 0 \\
0 & 0 & \frac{r_c}{l_p} & 0 & 0 & \frac{-r_c}{a^2 l_s} & 0 & 0 & -r_c
\end{bmatrix}
\qquad (12)$$

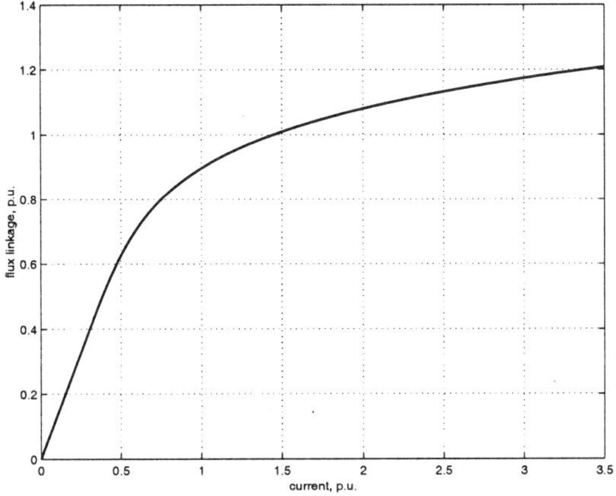

Fig. 3. Magnetization curve for the single-phase transformer.

$$\frac{dv_{dc}}{dt} = \frac{1}{C} i_{dc} \tag{16}$$

Regarding the PWM scheme, the switching signals can be obtained by comparing a sawtooth wave of frequency f_s and three sine waves of main frequency f_1 and, therefore, a frequency-modulation ratio m_f where,

$$m_f = \frac{f_s}{f_1} \tag{17}$$

and when the amplitude of the three sine waves is varied ($V_{control}$) with respect to the amplitude of the sawtooth signal ($V_{sawtooth}$), the amplitude modulation ratio is defined as

$$m_a = \frac{V_{control}}{V_{sawtooth}} \tag{18}$$

C. Control system

The main function of the control system is to operate the VSC and to produce a synchronous output voltage waveform which not only forces the exchange of reactive power but also the active power required to maintain a constant DC-link voltage. A basic controlled approach for the reactive output current can be implemented indirectly via varying the DC capacitor voltage, or as another approach, directly by the internal voltage control mechanics in which case the voltage in the capacitor is kept constant (PWM converter) [1]. In [5] various feedback control strategies are presented to control the reactive power interchange by indirectly controlling the voltage in the capacitor. The design of a controller based on mathematical analysis and simulation for a STATCOM which varies the voltage in the DC capacitor is presented in [4]. A decoupled control of the real and reactive power exchanged between the power converter and the electric system in a STATCOM using PWM is presented in [6]. The control system presented in this paper and shown by Fig. 4 proposes an alternative solution. The main functions of this control are to maintain constant the voltage at the PCC of the STATCOM and also to keep constant the voltage in the DC capacitor. These two tasks are done

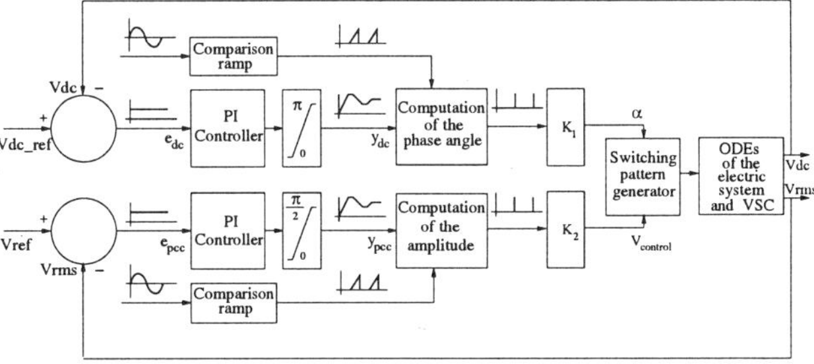

Fig. 4. Structure of the control system.

by means of two PI controllers working independently. Since the output signals of these controllers will be used for the switching pattern generator of the VSC, these output signals must be the phase angle and amplitude of the sine waves used by the PWM scheme.

From Fig. 4, the error signals are

$$e_{dc} = V_{dc-ref} - V_{dc} \tag{19}$$

$$e_{pcc} = V_{ref} - V_{rms} \tag{20}$$

Then, the ODEs for the two controllers can be expressed as

$$\frac{dy_{dc}}{dt} = \frac{K x_{dc}(T) - y_{dc}}{T_p} \tag{21}$$

$$\frac{dy_{pcc}}{dt} = \frac{K x_{pcc}(T) - y_{pcc}}{T_p} \tag{22}$$

and

$$x_{dc}(T) = x_{dc}(T - T_0) + e_{dc} \tag{23}$$

$$x_{pcc}(T) = x_{pcc}(T - T_0) + e_{pcc} \tag{24}$$

where
K constant of the PI controller
T_p time constant of the PI controller
Since the rms values for the voltages are updated every T seconds, T being the period of the system, $x_{dc}(T)$ and $x_{pcc}(T)$ are also updated each period T_0.

The signals y_{dc} and y_{pcc} are compared with ramps and multiplied by constants K_1 and K_2, respectively, in order to obtain the angle α and the amplitude $V_{control}$ within adequate limits for the PWM scheme.

The complete model of the STATCOM shown in Fig. 1 is described by a set of twelve ODEs, see Appendix A.

III. CASE STUDIES

Fig. 5 shows a three-phase system. The system feeds a linear load through a transmission line equivalent, a fixed capacitor bank for reactive power compensation is used, and a STATCOM for voltage support. The system was used for two cases of study: (1) Neglecting the transformer's saturation characteristic, (2) Including the transformer's saturation characteristic. For both cases, the STATCOM operates after three cycles of the simulation to maintain the voltage at the PCC at 1.0 p.u.

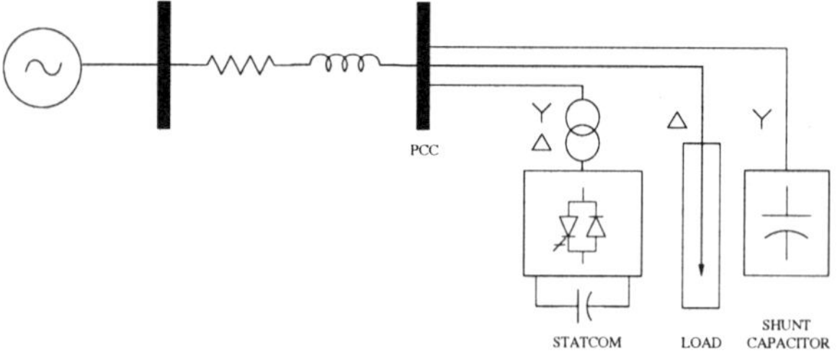

Fig. 5. Electric test system.

A. Analysis of the transient response

Figs. 6 and 7 show the currents in the STATCOM for both cases using a modulation ratio of $m_f = 9$. These results show that including the transformer's saturation the currents present a high magnitude during the transient period. This effect can be related to the inrush current due to the transformer non-linearity and not necessarily to the interaction between the saturation and the VSC. However, Fig. 8 shows the currents when $m_f = 27$, although the same saturation characteristic was used, this result shows higher transient current magnitudes than in Fig. 7. This effect can be related to the interaction between the saturation and the high harmonic frequencies generated by the VSC. It is important to mention that for all the simulations, the transformer was operating in its non-linear region which is not practical in steady state, but gives a general idea of the effects that the transformer's non-linearity can produce when interacting with the harmonic frequencies generated by the VSC.

B. Harmonic analysis

Keeping in mind that the same transformer's non-linearity interacts in a different fashion with the VSC as a function of the modulation ratio m_f, Fig. 9 shows the THD current in the STATCOM as a function of m_f. In the same figure, the current magnitude at the fundamental frequency is also shown as a way of comparison, which remains almost constant for both cases. The results show that the THD decreases as m_f increases when saturation is not included, this is an expected result since the harmonic frequencies are controlled by the VSC. For the case when the saturation is included, after the modulation ratio $m_f = 27$ the THD remains almost constant, this effect can be associated to the fact that the transformer remains saturated and also to the interaction with the VSC, even though a high modulation ratio is used. This effect is less appreciated in the THD voltage in the PCC shown in Fig. 10 where the THD decreases as m_f increases. This effect can be related to the fact that the STATCOM is used to maintain the voltage in the PCC at a constant value.

Figs. 11 and 12 show the harmonic content of the current in the STATCOM and the voltage in the PCC, respectively. It can be observed from these figures that the 3rd harmonic produced by the saturation is always present. Besides, typical harmonics associated with the

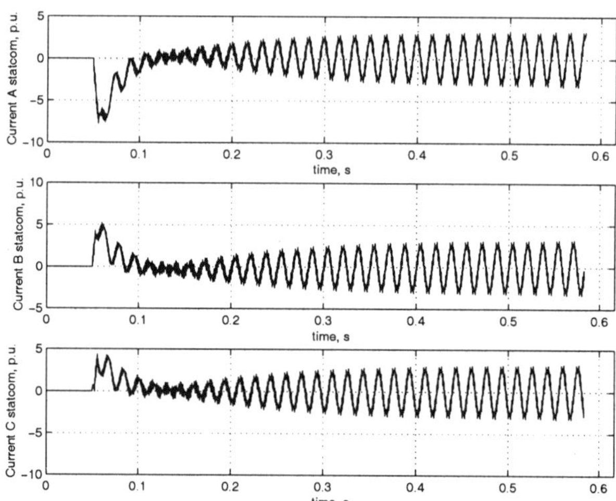

Fig. 6. Currents in the STATCOM, without transformer saturation and mf=9.

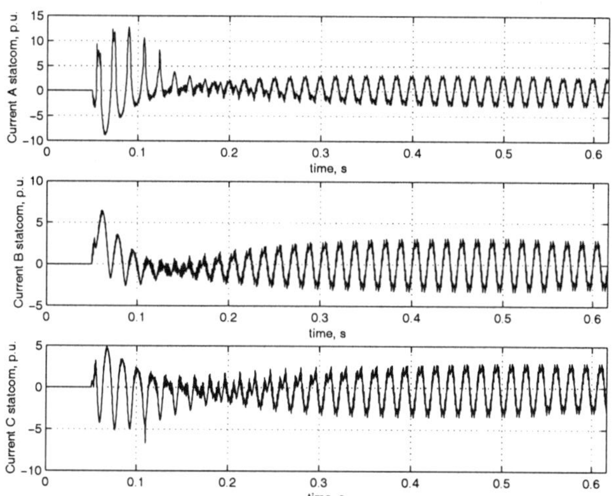

Fig. 7. Currents in the STATCOM, with transformer saturation and mf=9.

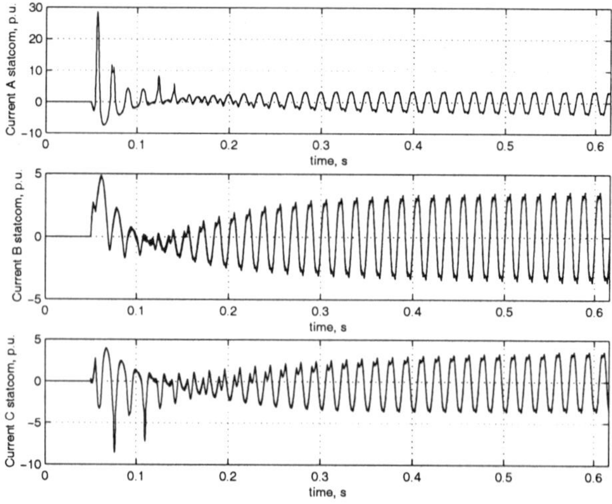

Fig. 8. Currents in the STATCOM, with transformer saturation and mf=27.

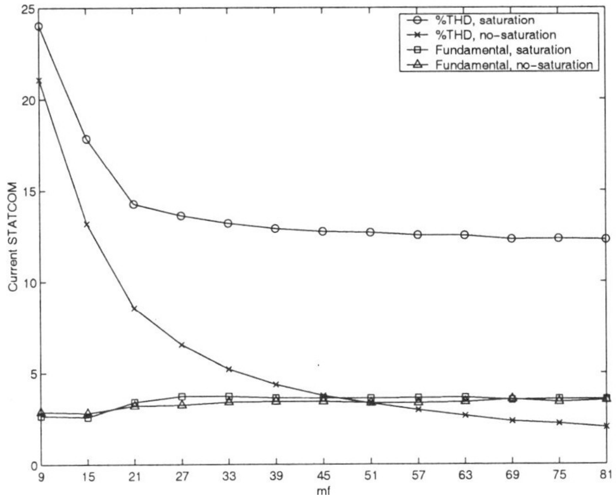

Fig. 9. THD current in the STATCOM.

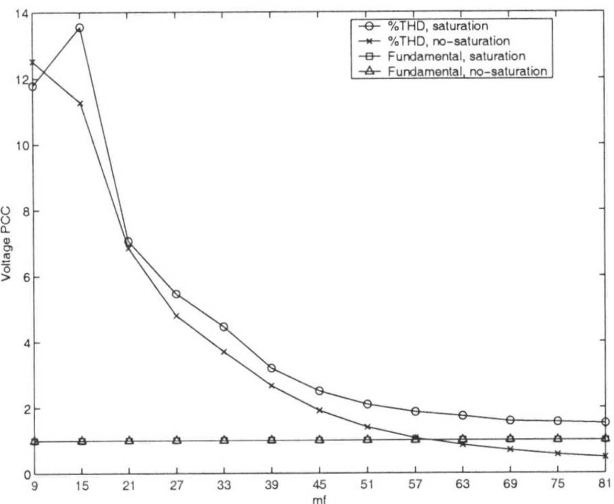

Fig. 10. THD voltage in the PCC.

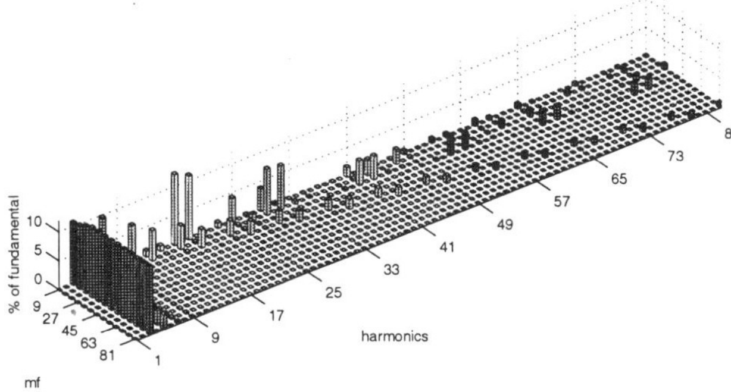

Fig. 11. Harmonic current spectrum in the STATCOM.

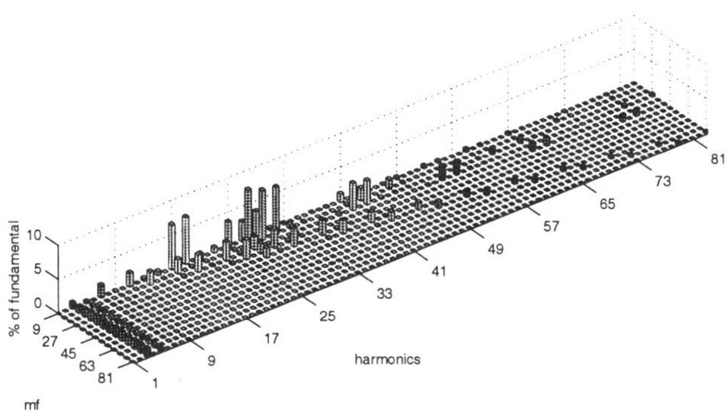

Fig. 12. Harmonic voltage spectrum in the PCC.

a driving point impedance as seen from the STATCOM was carried out, where the resonance peak was detected at 4.74 Hz, so the magnification of the $27th$ harmonic magnitude was not due to a resonance problem with the rest of the system.

IV. CONCLUSIONS

A complete time domain model for the STATCOM has been presented in this paper. The model includes the controllers and its associated transformer. The control scheme was applied to control the DC voltage in the capacitor and the voltage in the PCC. The transformer model was obtained from a three-phase transformer bank considering the core losses and the non-linear characteristic. The STATCOM was used to analyse the interaction between the VSC and the non-linear characteristic of the transformer.

The results show that the inrush currents of the transformer and the harmonics generated by the VSC may interact producing a very high current in the STATCOM during the transient period, even though a high modulation ratio is used by the VSC.

During the STATCOM steady state operation, it was found that some harmonics generated by the transformer may interact with those harmonics generated by the VSC producing a kind of resonance at high harmonic frequencies, such as the $27th$, as found in this paper.

conmutation frequency of the VSC are also detected. A particular case is the harmonic 27 for $m_f < 33$, which can be associated with the interaction between the transformer and the VSC.

The most appreciated harmonics in the voltage at the PCC, Fig. 12, are those associated with the switching frequency, which are significantly reduced when modulation ratio m_f is increased (i.e. the VSC generates odd harmonic sidebands centred around m_f [7]). However, for the harmonic currents in the STATCOM, Fig. 11, the $3rd$ harmonic and the others associated with the saturation appear at any m_f with an almost constant magnitude, but those harmonics associated to the VSC are reduced as m_f increases.

It can also be observed, e.g. $m_f = 9$, that in both harmonic spectrums the $27th$, $3rd$, $21st$, $15th$, $13th$ and $5th$ harmonics (in the same order in magnitude) do not correspond to those generated by the VSC. These harmonics correspond to the transformer's non-linearity, with the particularity that the magnitudes do not follow the expected order ($3rd$, $5th$, $7th$, ...), this effect is due to the interaction with the VSC. To support this conclusion,

REFERENCES

[1] Hingorani N. G. and Gyugyi L., 2000, *Understanding FACTS: Concepts and Technology of Flexible AC Trabsmission Systems*, IEEE Press.

[2] Schauder C., Gyugyi L., Lund M., Hamai D., Rietman T., Torgerson D. and Edris A., 1998, "Operation of the Unified Power Flow Controller (UPFC) Under Practical Constraints", IEEE Trans. on Power Delivery, 13, 630-639.

[3] Acha E. and Madrigal M., 2001, *Power Systems Harmonics*, John Wiley & Sons.

[4] Han B., Karady G., Park J. and Moon S., 1998, "Interaction Analysis Model for Transmission Static Compensator with EMTP", IEEE Trans. on Power Delivery, 13, 1297-1302.

[5] Rao P., Crow M.L. and Yang Z., 2000, "STATCOM Control for Power System Voltage Control Applications", IEEE Trans. on Power Delivery, 15, 1311-1317.

[6] Garcia P. and Garcia A., 2000, "Control System for a PWM-Based STATCOM", IEEE Trans. on Power Delivery, 15, 1252-1257.

[7] Mohan N., Undeland T.M. and Robbins W.P., 1995, *Power Electronics: Converters, Applications, and design*, John Wiley & Sons.

Appendix A

$$\frac{d}{dt}\begin{bmatrix} \lambda_{ap} \\ \lambda_{bp} \\ \lambda_{cp} \\ \lambda_{as} \\ \lambda_{bs} \\ \lambda_{cs} \\ \lambda_{ma} \\ \lambda_{mb} \\ \lambda_{mc} \\ v_{dc} \\ y_{dc} \\ y_{pcc} \end{bmatrix} = A \begin{bmatrix} \lambda_{ap} \\ \lambda_{bp} \\ \lambda_{cp} \\ \lambda_{as} \\ \lambda_{bs} \\ \lambda_{cs} \\ f(\lambda_{ma}) \\ f(\lambda_{mb}) \\ f(\lambda_{mc}) \\ v_{dc} \\ y_{dc} \\ y_{pcc} \end{bmatrix} + \begin{bmatrix} v_a \\ v_b \\ v_c \\ 0 \\ 0 \\ 0 \\ 0 \\ 0 \\ 0 \\ \frac{Kx_{dc}}{T_p} \\ \frac{Kx_{pcc}}{T_p} \end{bmatrix} \qquad (25)$$

where

$$A = \begin{bmatrix}
-\frac{(r_p+r_c)}{l_p} & 0 & 0 & \frac{r_c}{a^2 l_s} & 0 & 0 & r_c & 0 & 0 & 0 & 0 & 0 \\
0 & \frac{-(r_p+r_c)}{l_p} & 0 & 0 & \frac{r_c}{a^2 l_s} & 0 & 0 & r_c & 0 & 0 & 0 & 0 \\
0 & 0 & \frac{-(r_p+r_c)}{l_p} & 0 & 0 & \frac{r_c}{a^2 l_s} & 0 & 0 & r_c & 0 & 0 & 0 \\
\frac{r_c}{l_p} & 0 & 0 & \frac{-(a^2 r_s+r_c)}{a^2 l_s} & 0 & 0 & -r_c & 0 & 0 & -as_a & 0 & 0 \\
0 & \frac{r_c}{l_p} & 0 & 0 & \frac{-(a^2 r_s+r_c)}{a^2 l_s} & 0 & 0 & -r_c & 0 & -as_b & 0 & 0 \\
0 & 0 & \frac{r_c}{l_p} & 0 & 0 & \frac{-(a^2 r_s+r_c)}{a^2 l_s} & 0 & 0 & -r_c & -as_c & 0 & 0 \\
\frac{r_c}{l_p} & 0 & 0 & \frac{-r_c}{a^2 l_s} & 0 & 0 & -r_c & 0 & 0 & 0 & 0 & 0 \\
0 & \frac{r_c}{l_p} & 0 & 0 & \frac{-r_c}{a^2 l_s} & 0 & 0 & -r_c & 0 & 0 & 0 & 0 \\
0 & 0 & \frac{r_c}{l_p} & 0 & 0 & \frac{-r_c}{a^2 l_s} & 0 & 0 & -r_c & 0 & 0 & 0 \\
0 & 0 & 0 & \frac{s_a}{a^2 l_s C} & \frac{s_b}{a^2 l_s C} & \frac{s_c}{a^2 l_s C} & 0 & 0 & 0 & 0 & 0 & 0 \\
0 & 0 & 0 & 0 & 0 & 0 & 0 & 0 & 0 & 0 & \frac{-1}{T_p} & 0 \\
0 & 0 & 0 & 0 & 0 & 0 & 0 & 0 & 0 & 0 & 0 & \frac{-1}{T_p}
\end{bmatrix}$$

$$(26)$$

BIOGRAPHIES

Norberto Garcia received his first degree in electrical engineering from the Universidad Michoacana de San Nicolás de Hidalgo in 1993. He obtained his Msc degree from the Universidad Michoacana in February 1999. He has been working since 1994 as an Associate Research Engineer at the Facultad de Ingeniería Eléctrica, UM-SNH, Morelia, México. At present he carries out PhD studies in power systems at the University of Glasgow, Scotland, U.K., norberto@elec.gla.ac.uk

Manuel Madrigal was born in Purépero Mich., México. Obtained his BSc (Hons) in electrical engineering at Instituto Tecnológico de Morelia, México in 1993 and MSc at Universidad Autónoma de Nuevo León, México in 1996. Since 1996, he is conducting research in harmonics and power quality and teaching at Instituto Tecnológico de Morelia. At present he carries out PhD studies in power system harmonics analysis at the University of Glasgow, Scotland, U.K. e-mail: madrigal@elec.gla.ac.uk

Enrique Acha was born in México. He graduated from Universidad Michoacana in 1979 and obtained his PhD degree from the University of Canterbury, Christchurch, New Zealand in 1988. He was a posdoctoral fellow at the University of Toronto, Canada. He holds a permanent appointment at the University of Glasgow. He is a chairman of the inter-university Glasgow-Strathclyde Centre for Economic Renewable Power Delivery and head of the FACTS Research Laboratory. e-mail: e.acha@elec.gla.ac.uk